D0850563

Proteins, Enzymes, Genes

Proteins, Enzymes, Genes

THE INTERPLAY OF CHEMISTRY AND BIOLOGY

Joseph S. Fruton

Yale University Press *New Haven and London*

Published with assistance from the Louis Stern Memorial Fund.

Designed by Gregg Chase. Set in Minion type by à la page, New Haven, Connecticut. Printed in the United States of America by BookCrafters, Inc., Chelsea, Michigan.

Library of Congress Cataloging-in-Publication Data

Fruton, Joseph S. (Joseph Stewart), 1912–
Proteins, enzymes, genes : the interplay of chemistry and biology / Joseph S. Fruton.
p. cm.
Includes bibliographical references and index.
ISBN 0-300-07608-8 (cloth: alk. paper)
1. Biochemistry—History. 2. Molecular biology—History. 3. Proteins—Research—History. 4. Enzymes—Research—History. 5. Genes—Research—History. I. Title.
QP511.F783 1999
572′.09—dc21 98-38892

A catalogue record for this book is available from the British Library.

The paper in this book meets the guidelines for permanence and durability of the Committee on Production Guidelines for Book Longevity of the Council on Library Resources.

10 9 8 7 6 5 4 3 2 1

To the memory of Gerty Theresa Cori
1896–1957

Dorothy Mary Crowfoot Hodgkin
1910–1994

Mary Ellen Jones
1922–1996

Dorothy Moyle Needham
1896–1987

CONTENTS

PREFACE

In this book I present, as accurately as the documentary evidence available to me permits, a historical account of the most significant features of the development of reliable knowledge about the chemistry of life. I hope that my account will be useful to those who are or will be engaged in biological or chemical research, and who seek understanding of the historical processes in the emergence of the conceptual structure and practical methodology of their scientific specialties. I also hope that this book will be of interest to professional historians of the chemical and biological sciences, as well as to philosophers and sociologists of science.

This book is an outgrowth of one published in 1972 under the title *Molecules and Life: Historical Essays on the Interplay of Chemistry and Biology.* Extensively revised portions of that book (which went out of print in 1980) are included in the present volume, and much new material has been added in order to take account of two important developments during the past twenty-five years: (1) the growth of interest among many professional historians in the development of the biochemical sciences and in the emergence of molecular biology as an important research discipline, and (2) the striking transformation of the conceptual structure and methodological practice in the scientific specialties which deal with the chemistry of living organisms.

I have relied chiefly on the published description and discussion of scientific work shortly after it was done and have quoted, sometimes at length, from contemporary scientific writings. The quotations are meant to be read as part of the running text, and are intended to provide a more accurate statement of the authors' considered views than one based on hindsight. All the quotations are in English; except where otherwise stated, the translations from French, German, and Russian are my own, and often differ in detail from others that may be available.

I am especially indebted to Frederic L. Holmes for his patient guidance and generous hospitality, to Sofia Simmonds for her invaluable corrections and criticisms, and to Joanna Gorman for her open-hearted help in preparing the manuscript for publication. I also thank John Wiley and Sons, the publishers of *Molecules and Life,* for transferring to me the copyright for that volume.

Proteins, Enzymes, Genes

INTRODUCTION

In this book, I attempt to document, within a reasonable compass, the widely acknowledged but often disparaged interdependence of the development of thought and practice in the study of living organisms and of the chemical substances that constitute their anatomical structure and are involved in their physiological activities. Many of the challenges to chemists have come from the observations and experiments of biologists and physicians, and the chemical answers often have been fruitful in further biological and medical studies. In recent times, questions asked by chemists have frequently been answered by the use of concepts and methods derived from biological investigations. In organic-chemical work, an essential objective has been the isolation of a substance in a "pure" state, but, as Konrad Bloch has noted, the discovery of an impurity may lead to new biological findings.[1] Moreover, although chemistry is now classified as a physical science, organic chemistry as a research discipline is more akin to art than to physics. Its analytical methods may now include mass spectrometry and nuclear magnetic resonance spectroscopy, chemical structures are now determined by means of X-ray crystallography, and chromatography now constitutes the principal method for the separation of chemical compounds, but the art of chemical synthesis of complex compounds from simpler molecules, inherited from the nineteenth century and further developed in the twentieth, still remains the most distinctive feature of present-day organic chemistry.[2] The comparable and more challenging "synthesis" (or "reconstitution")—from the myriad discrete chemical substances identified as components of biological systems to the vastly more complex organized assemblies found in recent times to be involved in fundamental intra- and intercellular processes—has only begun during the past few decades.[3]

It is perhaps fair to say that a book dealing with the historical development of an area of biological or chemical investigation is likely to be evaluated differently by a working member of that specialty than by a historian engaged in the study of its development. If the scientist wrote the book, the historian might deplore the excess of technical detail, the tendency to glorify or denigrate the work of some individuals, and the celebration of a seemingly unbroken progress toward the high ground of the present. If the historian was the author, the scientist might complain that the book was riddled with technical and historical inaccuracies, questionable interpretations, and glib generalizations. Sociologically minded historians have considered historical writings by scientists in terms of a presumed intent to defend or promote the social status of a professional specialty, a view that can also be applied to the historians or sociologists who seek to defend or promote their particular approaches. I hope that this book may contribute to narrowing the intellectual gap that still separates the historically minded scientific worker in the biochemical sciences from the historians engaged in the study of the development of these sciences.

My activity as a biochemical worker and teacher ended during the 1980s, but I am still listed as a member of the Biochemistry Section of the National Academy of Sciences USA (NAS) and cannot deny my interest in defending and promoting the social status of the academic discipline with which I am officially identified. My subsequent effort has mainly been devoted to historical studies. I have been a member of the History of Science Society for nearly fifty years and, together with like-minded colleagues, I sought (unsuccessfully) to promote the establishment in the NAS of a Section of the History and Philosophy of Science. From that experience I learned that, while medical scientists and many physicists and chemists are ready to support the efforts of professional historians of science, the attitude of most of my scientific colleagues, especially those working in the so-called biomedical sciences, tends to be negative in that regard. Also, from my association with professional historians of the natural sciences, I have come to appreciate the ambiguity of their academic status vis-à-vis the general historians in American university faculties. Although, as a scholarly profession, the history of science has come a long way since 1912, when George Sarton founded the journal *Isis*,[4] today an aspiring young historian of science may still be under considerable pressure to emphasize the intellectual, social, or political features of a scientific development, at the expense of a thorough treatment of its technical aspects. I offer these impressions as a partial explanation of the gap referred to above.

For my own part, I cannot identify myself as a member of the present-day company of historians of the biochemical sciences, for I deplore the inadequacy of much of the recent literature in this field. I appreciate the necessity of adhering to the canons of the historical profession and have attempted to be accurate in the matter of historical events, but I have eschewed generalizations such as those relating to "paradigms," "Gestaltswitches," or "creativity." I do not find it necessary to invoke the concept of *material culture* in referring to the instruments, chemicals, or organisms used in particular researches, nor do I use the term *moral economy* in writing about the social relations among scientific investigators.

For this reason, the first chapter of this book opens with a discussion of recent writings on the origins of molecular biology, with particular emphasis on the distinction that needs to be made between research disciplines and academic disciplines. The chapter then deals with the place of sociology in the study of the history of science, continues with an examination of the recently fashionable concept of the social construction of scientific knowledge, and ends with a brief discussion of the biographical approach, with its attendant neglect of scientists who did not gain public renown. The second long chapter offers what I believe to be a necessary corrective to Robert Kohler's oft-cited account of the institutional development of academic biochemistry,[5] and in the third chapter I discuss some of the philosophical aspects of chemical and biological thought that are relevant to the theme of this book.

These three chapters are intended to provide a background for the succeeding discussion of the "internal" historical development of ideas and methods in the interplay of chemistry and biology. The material in these later chapters is not organized according to a "logical" conceptual structure such as those found in textbooks of biochemistry (now generously illustrated in full color). These texts begin with the current knowledge of the physical properties, organic-chemical structure, and biological function of proteins and nucleic acids, and then present the details of the chemical aspects of diverse biological processes. The material in this book is arranged instead according to my perception of the historical sequence in the emergence of concepts whose formulation and elaboration proved to be fruitful in the discovery of previously unknown biochemical substances and processes, and in the invention of new methods for their chemical characterization and for the study of their biological function.

I begin with the concept of ferments because it represents, in my view, the first attempt to explain biological change in chemical terms. The next two chapters deal with the parallel emergence, during the nineteenth century, of the role assigned to the "albuminoid" constituents (now termed proteins) of "protoplasm" and of the concept of *vital force,* now embodied in the term *bioenergetics.* These developments led, later in the nineteenth century, to the study of the "metabolism" *(Stoffwechsel)* of the chemical constituents of animals, plants, and microorganisms, and the outcome of these efforts provides the theme of the seventh chapter. It was not until the first decade of the twentieth century that there emerged the two aspects of the interplay of chemistry and biology that now occupy the center of the stage—the roles of nucleic acids in biological processes and of chemical "messengers" in the regulation of such processes—and therefore form the subjects of the final two chapters. In attempting to bring the reader some sense of the present state of knowledge about those biological problems for which chemical explanations have become available, I have, of necessity, sketched the recent advances only briefly, but I have included references to books and journal articles published during the 1990s. Many of the sections of each of these chapters begin with an account of a contribution that I deem to have been of seminal importance in clarifying a particular problem whose earlier study is then described. Also, there is a considerable chronological overlap in the discussion of the development of thought and experiment on the various problems under investigation during a particular time period. I must add that this book does not constitute a comprehensive history of the biochemical sciences. I have knowingly omitted a discussion of some currently active areas of biochemical research—for example, the efforts to define the chemical basis of embryonic differentiation.

I recognize that in separating the discussion of the institutional, biographical, and philosophical aspects of the development of biochemical thought and practice from the account of the discoveries and inventions that characterized that development, I may be criticized for implying that the growth of scientific knowledge can be interpreted solely in terms of theory and experiment, without regard to the social circumstances and personal qualities of the men and women engaged in the scientific enterprise. Mirko Grmek has rightly included such an "internalist" view among the "methodological myths" in the historiography of the sciences, but he lists among the other myths the view that historical understanding can come only from the study of the "external" factors in scientific research.[6] Other noted historians of science have debated this matter inconclusively.[7] It seems to me that what is at issue are the limitations of all the historical approaches, and that none of them can fully capture the complexity of scientific research as a disciplinary, collaborative, competitive, and individual activity whose character has constantly been subjected to social and political pressures. Because this book deals with the scientific work and thought of very many men and women, living under the most diverse "external" pressures at different times and in different places, facile generalizations such as those offered by some sociologically minded historians of science are likely to be wrong, and therefore best omitted. Certainly, when the historian deals with the career of an individual scientist, or of a closely linked small group of investigators, their social circumstances and personal qualities are likely to give additional insight into the sources of their productivity, but this approach is certain to lead to error when applied to the study of scientific populations of the kind considered in this book. Moreover, from the point of view of a biochemist turned historian, what is remarkable is that over the centuries, from Francis Bacon to Francis Crick, the efforts of these many people led to the acquisition of a considerable body of reliable, systematic, and widely accepted knowledge about the chemistry of life. The fact that much of that knowledge has been used for the destruction and degradation of human life, as well as for more worthy purposes, poses even more important historical questions, which have been addressed more forcefully by scientists than by present-day historians of the chemical and biological sciences.

I also recognize that my "internalist" approach in Chapters 4–9 may be criticized by some professional historians of science as being tainted with "presentism," "Whiggism," or "positivism." I find the rhetorical use of these terms, in their derogatory sense, to be a form of "snobbism." I share with most professional historians of science the view that the evaluation of a particular research contribution should be based on its contemporary context, but I also follow their example in calling attention to the consequences of such contributions. Denigrations of these discredited stances are hypocritical: too many present-day historians of science have chosen to study the careers of particular individuals because their work and thought are *now* judged to have been important in the development of lines of research that are *now* being actively pursued. In my view, Herbert Butterfield's stricture about "Whiggism" (which he did not follow) may properly be violated.[8]

For me, one of the most fascinating aspects of the interplay of chemical and biological thought and action during the past two centuries has been the fact that so much has been learned about the workings of living things by following theories that turned out to be wrong. I am therefore inclined to the view of the philosophical maverick Paul Feyerabend, who wrote that "scientists do not solve problems because they have a magic wand—methodology, or a theory of rationality—but because they have studied a problem for a long time."[9] So far as the post-1945 interplay of chemistry and biology is concerned, at least in the area of investigation now termed the "biomedical" sciences, to that quip may be added, "and by putting more men and women on the job." A consequence has been more intense competition and, as Efraim Racker put it: "Rejoice when other scientists do not believe what you know to be true. It will give you extra time to work on it in peace. When they start claiming that they have discovered it before you, look for a new project."[10]

Although Feyerabend's dictum that "anything goes" in scientific speculation is too extreme, all biochemical speculations, no matter how fruitful or unproductive, or whether they gained for a time the status of "doctrines" or "dogmas," merit consideration because they frequently reveal something about the tacit preconceptions a scientist has brought to his or her work. One of my reasons for emphasizing the interplay of biological and chemical thought is that throughout that interplay professional biologists and chemists have frequently studied the same problem at the same time, but with different preconceptions, and the differences in their explanations have raised tensions among members of the same or separate academic disciplines. By the same token, when a scientist-historian and a professional historian examine the same historical problem, they are likely to bring to the task different preconceptions of what is important. Perhaps, as in the case of biochemistry itself, the tensions generated by such differences in attitude and approach to the study of the history of the biochemical sciences may be sources of vitality in that enterprise.[11]

As regards the methodology of historical research, Stephen Brush has generously suggested that "those scientists who are willing to learn historical methods and study original sources have a continuing essential role to play" in the study of the history of science.[12] I assume that he included published scientific reports among the "original sources," but some well-trained historians of science appear to have taken seriously Peter Medawar's provocative argument that the scientific paper is a fraud because "it misrepresents the processes of thought that accompanied or gave rise to the work that is described in the paper."[13] Medawar's statement, in a radio broadcast to the general public, omitted the qualification that a scientific reader familiar with the background of the subject of a research report can discern much about the "processes of thought" that underlie the work. Indeed, this has been recognized by many professional historians of physics and chemistry but does not appear to have been accepted by some of the present-day historians of the biochemical sciences. Instead, most of their documentation is derived from unpublished material (personal correspondence, research notes, confidential reports, and so on) deposited in archives. I do not question the immense importance of such "original sources" in historical studies, and some of the fruits of my own examination of archival material may be found in this book, but I have also found in some recent historical writings egregious errors in the use of such material to document statements about people and events. Moreover, a scientist's recollections of the past, whether in memoirs or oral interviews, tend to reflect an adjustment in thought to later developments, and if not checked through informed examination of the scientist's research publications, and those of his contemporaries, are likely to lead the historian astray. To this I would add that too often items of idle gossip, whether transmitted by word of mouth or in personal correspondence, have been mistakenly taken to correspond to the facts of the matter.

There can be no doubt that honest and thorough study of archival material is essential for biographical research and, in recent times, has cast shadows on the halos assigned to some "great men of science." In this book I single out many scientists for special admiration, although my adulation of some of them is less than complete. My reservations are based on the examination of their published writings and confirmed by my own archival research as well as that reported by others. In this respect I dissent from the opinions of those of my scientific colleagues whose biographical writings incline toward hagiography, and side with those historians who, in admiring the scientific achievements of famous men, such as Claude Bernard or Louis Pasteur, have provided insights into the frailties which often attend scientific genius.[14] Clearly, if one expects a professional historian to present an account of the development of a scientific specialty as accurately as possible, even though the historian's interpretation may challenge inherited views, by the same token scientists have an equal obligation to strive for accuracy in their historical writings. Some of the prominent participants in the recent efflorescence of the biochemical sciences have been deficient in that regard, and I mention in later pages several examples. Such errors may be considered trivial, but because they have crept into the writings of professional historians of the biochemical sciences, they require correction.

As will be evident to readers of this book, it is not a popularization (I find the French term *vulgarisation* more appropriate) that, as Ludwig Wittgenstein is reported to have said, is "intended to make you believe that you understand a thing which actually you do not understand, and to gratify what I believe to be one of the lowest desires of modern people, namely the superficial curiosity about the latest discoveries in science."[15] In bringing into the 1990s this account of the historical development of the interplay of chemistry and biology, I do not wish to simplify the complexity of the present reliable knowledge of the chemistry of life, nor to suggest that the outstanding achievements in recent decades represent the culmination of an effort that began two centuries ago, but to convey a sense of the continuity of that effort. I hope that historically minded readers unfamiliar with the details of the recent achievements will find in the citation of key experimental papers and review articles a useful guide to further study. I also hope that they will recognize that, as before, new stimuli to chemical research have come from biological discoveries, just as new interpretations of long-known biological phenomena have come from new chemical ideas and from the use of new physical and chemical techniques and tools. The complexity of what we now accept as reliable knowledge of the chemical structure and workings of living organisms offers challenges not only to the present and future generations of scientific investigators but also to scholars interested in the historical development of that knowledge.

CHAPTER 1

On the History of Scientific Disciplines

The phrase "the molecular vision of life" has been used to denote the distinctive quality of the new discipline of molecular biology,[1] and a French molecular biologist has stated that

> molecular biology is not the description of the living in terms of molecules. If that were the case, it would be necessary to include not only biochemistry but also all the work done during the nineteenth century, in chemistry as well as physiology, that has permitted the characterization of the molecules of living [organisms]. With such an extensive definition, Pasteur himself would have been a molecular biologist! It is the totality of the techniques and discoveries that have permitted the molecular analysis of the most intimate processes of life, those which assure its perennial character [*pérennité*] and reproduction. In terms of disciplines, molecular biology is the fruit of the encounter of two branches of biology that developed at the beginning of the twentieth century, genetics and biochemistry. The two disciplines each had defined the privileged object in their researches, the gene for genetics, the proteins and the enzymes for biochemistry.[2]

The same author has also written that the prime attribute of the "new vision" is the recognition of the transmission of *information* by means of the transmission of a *code* to produce *sequences,* all words that "were unknown to biologists of the 1940s."[3]

This approach to the history of molecular biology, as seen from Paris, raises many questions that merit critical attention. I had previously written about the "origins of molecular biology," and Michel Morange stated in a footnote that "we often agree, in this book, with the opinions offered by the biochemist Joseph Fruton. His remarks are as well 'trails' for historians."[4] In what follows I repeat some of these opinions, as they relate to the language used to describe the "new molecular vision of life" and to the emergence of molecular biology as a prominent scientific discipline.

I ask two questions in considering the changes in the language to which Morange referred: (1) How were these changes related to the development of the conceptual structure and practical methodology of particular areas of the interplay of chemistry and biology? (2) How was the choice of new words influenced by the social interests and professional backgrounds of the protagonists of the new discipline graced with a "new vision of life"? The two questions are closely related but call for somewhat different approaches.

In writing about "disciplines" in the chemical and biological sciences, I deem it necessary to distinguish between research disciplines and institutional or academic disciplines.[5] The first classification denotes the individual and collective activity whose main objective is the investigation of scientific problems which are of common interest to the people who participate in that activity. The term *institutional*, on the other hand, relates to the forms of the social organization, support, and control of that activity in universities and in schools of medicine, pharmacy, agriculture, or technology, in scientific academies, and in private and governmental research institutes, hospitals, and industrial establishments. The term also relates to the societies, meetings, and publications set up to promote particular research activities, and to the social obligations and customs adopted by the participants, or imposed upon them by their employers or patrons. The term *academic* refers to institutions whose primary social function has been the formal instruction of candidates for various degrees, such as those in medicine, pharmacy, agriculture, engineering, or the natural sciences, as well as social studies or the liberal or fine arts.

The principal difference between research disciplines and institutional or academic disciplines lies in the fact that in the first kind decisions arise from the interactions among scientists, usually of various national or institutional affiliation, whereas in the second, decisions are made by such people as university administrators, government officials, or private patrons in local or national settings. To be sure, the decisions of bureaucrats are frequently influenced by the entrepreneurship of scientists, and many scientists have themselves become bureaucratic administrators, but in social organizations such as universities or research institutes these decisions usually involve considerations other than those which bring together scientists who share common research interests. Such considerations include budgetary constraints, current fashions, institutional prestige, national interest, or public demands and prejudices. Whether it was an adviser to a ruler of a nineteenth-century state or an official in a twentieth-century granting agency, his or her decision may have helped to initiate, accelerate, inhibit, or abort the performance of particular research efforts, but so far as the course of the development of reliable knowledge in an area of scientific inquiry is concerned, what counts most is the manner in which the results of such efforts have affected the thought and work of other scientific investigators.

At any given time, the biologists, chemists, and physicists who participated in the interplay that preceded the emergence of the new "molecular vision of life" may have been associated with different institutions, or with different parts (divisions, departments, sections) of the same institution. Nevertheless, these people were united in their search for explanations of particular biological phenomena in the language of the chemistry and physics of their time through the discovery of previously unknown objects or processes (or of previously unknown properties of known objects and processes), the formulation and testing of new hypotheses, and the invention of new or improved methods and instruments. The ideas, empirical results, and techniques developed in various localities and institutional settings were communicated in various ways (publications, conferences, personal letters and conversations) and led to the establishment of new journals and of new professional societies. The choice of particular scientific problems for investigation was usually, though not always, influenced by the expectations of private patrons and public officials that such research would contribute to the solution of contemporary problems in medicine, agriculture, industry, and the military. The manner in which the interplay of chemical and biological thought and action was conducted, however, depended on the state of the development of theory and practice in each of the specialties involved, and on the insight and skill of the individual participants.

Since about 1800, and particularly after 1945, the growth of reliable and socially useful knowledge that has emerged from the interplay of biological and chemical thought and action has been both the result and cause of a rapid increase in the number and size of scientific research groups. Separate areas of investigation have proliferated and new research disciplines have emerged within what we understand to be the general fields of chemistry and biology. Thus, in chemistry, to the branches known during most of the nineteenth century as general, inorganic, or organic chemistry were added such specialties as physical, electro-, thermo-, photo-, colloid, natural product, or coordination chemistry, with the later addition of physical-organic and polymer chemistry. The development of most of these chemical specialties was closely linked to the rapid growth of large industrial enterprises in many countries, at first in Great Britain, Germany, France, and Switzerland, and later in the United States and Japan. In its relation to medicine, chemistry appeared during the nineteenth century in the form of medical chemistry (later clinical chemistry or laboratory medicine), nutrition, or pharmacology. Similarly, in the fields of zoology and botany (once known as parts of natural history), with their close ties to both medicine and agriculture, the studies of animals (especially humans) and of plants were each divided into the academic specialties of anatomy, physiology, and pathology, with the subsequent emergence of such disciplines as evolutionary biology, embryology (later developmental biology), histology, cytology (later cell biology), bacteriology (later microbiology and virology), genetics, immunology, ecology, endocrinology, and neurobiology. As in the case of the chemical specialties, the development of some of the biological ones—for example, microbiology and genetics—was greatly promoted by the links to the fermentation industry and to the agricultural practices of animal husbandry and plant hybridization. As will be evident from the account offered in this book, throughout the development of the various chemical and biological specialties, there were continuous interactions among them, so that the professional separation and competition that attended these interactions were repeatedly confounded by changes in theory and practice within each of these specialties.

I emphasize in the body of this book the development of the interplay of chemical and biological theory and practice rather than the social organization of scientific research not because I consider the latter to be less important for an understanding of the historical processes whereby "science" has assumed an increasingly significant role in human life. Rather, my emphasis reflects a belief that such understanding requires an appreciation of the fact that, during the course of human history, the search for explanations of natural phenomena, for whatever practical or intellectual reasons, has continued through centuries of profound social reorganization. I have chosen for special attention the period after about 1800 because by that time the knowledge inherited from such ancient practical arts as nutrition or fermentation, as well as from medieval alchemy and sixteenth-century "iatrochemistry," had begun to be transformed, during the eighteenth century, into a more comprehensive and more coherent set of chemical concepts, with a more distinctive methodology of experimental practice. Moreover, the first half of the nineteenth century was marked by the emergence of experimental physiology, as well as by the increased use of such instruments as the microscope and of staining techniques to study the internal structure of cells.

At the beginning of the nineteenth century Jöns Jacob Berzelius called the part of chemistry dealing with the substances obtained from the tissues and fluids of animals and plants (and the processes in which these substances are involved) organic chemistry, and there was a succession of books about animal chemistry.[6] This area of investigation became a prominent feature of research for medical physiologists, notably Claude Bernard, and the parallel efforts of chemists, especially Justus

Liebig, led to the emergence in Germany of *physiologische Chemie,* with Felix Hoppe-Seyler as its chief protagonist after about 1870. The English equivalent of this term was adopted in the United States, while British physiologists preferred the term *chemical physiology,* and in France it came to be *chimie biologique.* Later the institutional separation of this discipline from medical physiology was marked by a preference for designations such as *biological chemistry* or *biochemistry,* as defined by Frederick Gowland Hopkins. Some "biochemists" tended to approach the major problems in their field as organic chemists, with an emphasis on the structure and chemical properties of substances isolated from biological systems, whereas others tended to emphasize the study of the physiological function of such substances.

With the rise to prominence of physical chemistry (and its derivative colloid chemistry), many biochemists preferred to apply the concepts and methods of that specialty to the study of biological problems rather than follow in the organic-chemical tradition developed during the nineteenth century. The research of such biochemists led to the designation of their specialty as biophysical chemistry (later changed to molecular biophysics). Moreover, after 1945, some organic chemists with an interest in the mechanisms of enzyme action considered their research to be in the field of bio-organic chemistry.

Of special importance in relation to the "origins of molecular biology" were the contributions of a group of nineteenth-century physiologists, among them Hermann Helmholtz, Adolf Fick, and Emil du Bois-Reymond, whose use of physical theory and the invention of physical instruments for the study of biological processes provided the background for what came to be called biophysics. Later, Archibald Vivian Hill and Selig Hecht, among others, carried that tradition forward, although some biophysicists preferred the term *general physiology.*[7] After 1945, however, some ambiguity about the meaning of the term *biophysics* arose from the entry into biology of several former theoretical physicists who sought to leap from what they knew into the biological unknown without touching foot in chemistry; some of them later identified themselves as molecular biologists. The interactions among these various kinds of nineteenth- and twentieth-century biologists, chemists, and physicists will be examined more closely in subsequent chapters. At this point I only note that during the course of this interplay, members of several biological disciplines were repeatedly attracted to new developments in chemistry and physics, and brought into their specialties words used in the latter disciplines. So far as I know, the advocates of the older fashions did not consider such usage a "new vision," but the similarity to the rhetoric of some historians of molecular biology is unmistakable. Thus, after the recognition that colloidal "albuminoid" substances are components of the "protoplasm" of living organisms, during the 1860s Thomas Henry Huxley proclaimed such substances to be the "physical basis of life." Later in the nineteenth century, much attention was given to the idea of "living proteins," and after the organic chemist Eduard Buchner prepared from yeast a material he called zymase, there emerged what some historians have called the enzyme theory of life. Until about 1930, colloid chemistry, based on the principles of physical chemistry, was a prominent feature of biological thought and experimental investigation.

The now-immense literature on the "origins of molecular biology" amply attests to the historical importance of the emergence of this discipline. The definition of its scope has changed, however, with the development of the concepts and methods in the study of the chemical structure and biological function of nucleic acids, proteins, and enzymes. The detailed features of that development will be examined in later chapters of this book, and here I only consider some possible reasons for the replacement of the word *chemical* (as, for example, in *chemical biology*) by *molecular.* Certainly, one of the consequences of the emergence of molecular biology as an important scientific discipline has been considerable tension as regards its institutional relationship to what had come to be called biochemistry. I believe, however, that the roots of that tension are embedded in the interplay of nineteenth-century biological thought and the conceptions of the "molecule" as formulated by leading chemists and physicists of the time. I repeat here the suggestion I offered in 1976 that

> after about 1860, the molecular explanations that biologists offered for physiological phenomena tended to distinguish between molecules as physical or chemical entities. One style of speculation was based on the definition of the molecule as the smallest unit of a substance that moves as a whole, and specific biological function was seen as the expression of differences in a continuously variable motion within a specific arrangement of molecular units. [For example, in 1876 Adolf Fick wrote: "In the muscle and nerve fibers the protoplasmic molecules are regularly arranged, so that the propagation of the characteristic process proceeds regularly in one direction for long distances."] The other defined a molecule as the smallest portion of a substance that retains its properties in chemical reactions, and biological phenomena were considered to be a consequence of the specific properties of different kinds of interacting molecules. This approach required, however, some knowledge of the chemical structure of these molecules. The two definitions of molecules as physical or chemical entities were not mutually exclusive, and occasionally attempts were made to combine them. Never-

theless, the dichotomy between the two styles of speculation is a striking feature of biological thought during the 19th century and carried forward into the 20th century. This dichotomy has a more ancient lineage, as in the relation of iatrochemistry to iatrophysics in the 17th and 18th centuries.

After about 1860, the undoubted success of men like Helmholtz and du Bois-Reymond in demonstrating the explanatory power of the new biophysics, and the apparent inability of organic chemistry to elucidate the constitution of albuminoid protoplasm, clearly made the physicalist mode of biological thought the more attractive one. Even for physiological processes that did not appear to lend themselves experimentally to the biophysical approach, the molecular explanations of protoplasmic activity that were offered during the latter half of the 19th century reflected a predilection for this style of speculation. For example, the embryologist Wilhelm His considered that in the fertilization of the egg by sperm there was a transmitted motion rather than the transfer of specific material, and the cytologist Eduard Strasburger used similar language in describing the influence of the cell nucleus on the cytoplasm.[8]

To these examples may be added the fact that the physician Henry Bence-Jones referred to "chemical or molecular disease" and that the physiologist Michael Foster wrote: "The physiological function of any substance must depend ultimately on its molecular (including its chemical) nature."[9] However oversimplified this interpretation of the scientific attitudes of nineteenth-century biologists may be, I find in recent accounts of the emergence of a new "molecular vision of life" much that recapitulates some of the story of the preceding interaction of physical and biological thought. What is different, owing to the efforts of various kinds of nineteenth- and twentieth-century chemists, is that one no longer speaks of an "albuminoid protoplasm" whose chemical dissection could only yield artifacts, but rather of discrete substances like proteins, enzymes, and nucleic acids (as well as polysaccharides, peptides, nucleotides, steroids, and so on) whose chemical structure has been determined and whose biological function, intracellular organization, and metabolic conversions can be studied further experimentally.

Why then the replacement of the word *chemical* by *molecular?* Part of the answer lies, I believe, in the historical development of physical theory about heat as a mode of molecular motion. During the last quarter of the nineteenth century, much attention was given to the difficulty of providing a satisfactory kinetic theory of gases in terms of classical mechanics that was also consistent with the second law of thermodynamics as formulated by Rudolf Clausius. The difficulty had been recognized by Clerk Maxwell and led Ludwig Boltzmann to develop a statistical interpretation of the second law.[10] The subsequent development of the quantum mechanics introduced by Max Planck, its application by Niels Bohr to Ernest Rutherford's model of the atom, and its elaboration during the 1920s by Werner Heisenberg, Paul Dirac, Wolfgang Pauli, Louis de Broglie, and Erwin Schrödinger (plus others), had to take account, however, of the discovery of subatomic "particles" (or "wavicles"), and such terms were used in the language of theoretical physics in addition to the "molecules" (or "corpuscles") of earlier physicists. Later, in addition to the valence-bond theory of the chemical bond, there was also the molecular-orbital theory associated with the names of Friedrich Hund, Robert Mulliken, and Erich Hückel.[11]

Two of the noted theoretical physicists named above—Bohr and Schrödinger—have received much attention in historical accounts of the emergence of molecular biology, the first because of his influence on the thought of Max Delbrück, the second because of the importance attached to Delbrück's 1935 "quantum mechanical model" of the gene in Schrödinger's 1944 book *What is Life?*[12] One of the many other theoretical physicists who participated before World War II in the development of quantum mechanics was Léon Brillouin, whose book *Science and Information Theory* (1956; 2d ed. 1961) attracted wide attention among biologists. This book was in the tradition of Boltzmann's statistical interpretation of the second law, reflected the previous contributions of Claude Shannon and John von Neumann to information theory, and led to some controversy about the relation between entropy and information.[13]

It would seem, therefore, that part of the answer to the question of why the words *chemical* and *chemistry* had fallen into disfavor among biologists lies in the public prestige of such theoretical physicists as Bohr and Schrödinger who wrote about the "nature of life." It is not the whole answer, however, for another tradition, within experimental physics, played a far more significant role. That tradition began with the discovery of X-rays by Wilhelm Conrad Röntgen and the finding by Max von Laue, Walter Friedrich, and Paul Knipping of the diffraction of X-rays by crystals. The subsequent development of X-ray spectroscopy by William Henry Bragg, his son William Lawrence Bragg, and their associates into a powerful technique for the study of the arrangement of atoms in molecules provided the solution of problems in the determination of the structure of many molecules of biological interest.[14] With successive improvements in the analysis of X-ray diffraction spectra, it became possible to study the three-dimensional structure of proteins and nucleic acids. The results of these studies will be discussed later in this book; for the present purpose, what is important is that among the first notable successes were those of a small group attached to the Cavendish Laboratory of physics at Cambridge, and that during the 1950s this research group became the nucleus of the renowned Medical Research Council Laboratory of Molecular Biology.[15] When Max Perutz, John Kendrew, Francis Crick, and James Watson were awarded Nobel Prizes in 1962 for work done at the MRC laboratory, the public recognition accorded molecular biology was vastly increased. In subsequent years, molecular biology was defined as embracing both the "informational" and "structural" schools,[16] but more recently, many of the former physicists and physical chemists in the latter group have declared themselves to be structural biologists.

In the writings of some historians of science, research centers other than the one in Cambridge, such as those at the California Institute of Technology (Linus Pauling, George Beadle, Max Delbrück) and the Pasteur Institute in Paris (André Lwoff, Jacques Monod, François Jacob) have been cited as birthplaces of molecular biology. Also, in addition to the investigators mentioned above, others have been credited with using the term before its widespread acceptance. Special attention has been given to Warren Weaver of the Rockefeller Foundation, who noted in 1970 that he had used it in a 1938 confidential report: "Among the studies to which the Foundation is giving support is a series in a relatively new field, which may be called molecular biology, in which delicate modern techniques are being used to investigate ever more minute details of certain life processes."[17] Another scientist who is credited with having anticipated the widespread acceptance of the term was William Astbury, who wrote, with characteristic exuberance, "We are at the dawn of a new era, the era of 'molecular biology' as I like to call it, and there is an urgency about the need for more intensive application, and especially structural analysis, that is not sufficiently appreciated."[18]

My admiration of the scientific achievements of these men—and, in the case of Weaver, of his program for the support of what he called publicly during the 1930s quantitative biology (hence the name of the Cold Spring Harbor symposia)—is unbounded. I find, however, in most of the historical accounts special pleading for the roles of particular individuals or groups in the generation of the new "molecular vision of life." Moreover, before about 1960, that vision was not only "molecular" but also "chemical." George Beadle and Edward Tatum, for example, considered their work to be a contribution to "chemical genetics," and Erwin Chargaff described the work in his laboratory as "genetic chemistry." In the series of important biological symposia at the McCollum-Pratt Institute at Johns Hopkins, the one held in 1956 was entitled "The Chemical Basis of Heredity."[19] Among the numerous participants were Beadle, Benzer, Jacob, Chargaff, Crick, Watson, and Delbrück. The Watson-Crick model of DNA was discussed, but the important experimental reports of Matthew Meselson and Franklin Stahl and of the biochemist Arthur Kornberg in support of the model had not yet appeared. Nor had the pioneering experimental work of the biochemists Marshall Nirenberg and Severo Ochoa, or of the organic chemist Gobind Khorana, leading to the decipherment of the "genetic code," been published. Although the biochemist Frederick Sanger had already shown how one can determine the amino acid "sequence" of a protein, he had not yet devised a method for determining the nucleotide sequence of DNA. These achievements were in the tradition of the interplay of biological and chemical thought and action. Prominent theoretical physicists (through ignorance or prejudice) disdained biochemical complexity, spoke of "bags of enzymes," sought "new biological laws," or found new metaphors in information theory and statistical quantum mechanics. The language of the interplay may have changed, but the course of the development of knowledge about the "chemical basis of heredity" still depended on the empirical discovery of previously unknown chemical constituents of biological systems, as well as on the invention of new methods and instruments or improvement of older ones for the study of chemical structure and chemical interaction. Nor was anything changed, so far as the development of scientific knowledge was concerned, by the fact that many biologists preferred words used by theoretical physicists in writing about the "nature of life" to the more specific (and unfamiliar) ones used by biochemists or organic chemists, and that these biologists, like the general public, were charmed by the symmetry of the "double helix" or the simplicity of the "Central Dogma." In this respect, as I suggested above, the attitudes of such biologists mirrored those of their nineteenth-century ancestors. My view of writings such as those of Lily Kay and Michel Morange about molecular biology as a "new vision of life" is that these writings are deficient as historical efforts in simply taking for granted the reliable scientific knowledge gained through much trial and much error in the centuries before 1950 and in employing the time-honored professional strategy of denigrating the value of the contributions of their predecessors. To me, the words *molecular* and *biology* mean something more than parts of the designation of a new institutional discipline. Indeed, if the term *molecular biology* were taken to denote an area of research in which biological thought and experiment, not only in genetics but also in other branches of biology, has interacted with chemical thought and experiment, not only about DNA, RNA, proteins, and enzymes but also about the other known material constituents of biological systems, and the processes in which these chemical substances have been found to be involved, I, for one, would readily consider the chemist and microbiologist Pasteur to have been a molecular biologist.[20]

That there were tensions in academic institutions between molecular biologists and biochemists after 1950 is undeniable, but it should be noted that by the 1980s the ensuing development of the interplay of chemistry and biology had led many adversaries to recognize the continuity of that historical process.[21] One sign was the change, in 1987, of the name of the American Society of Biological Chemists to the American Society for Biochemistry and Molecular Biology. I leave that matter to the more dispassionate judgment of future historians, but deny the validity of the assertions that "competition with molecular biologists was a distinctive feature of the history of biochemistry after 1950. Biologists and chemists regarded biochemists as narrow-minded specialists, who were neither proper chemists nor biologists and who were interested only in the petty details of metabolic pathways. A. V. Hill wrote: 'The trouble with so many biochemists or physiological chemists, or whatever one calls them, is that they either know no chemistry or no physiology or no biology.'"[22]

The Sociological Perspective

I began this chapter with a discussion of the emergence of molecular biology as a scientific discipline in order to indicate some of the challenges faced by anyone who seeks to promote understanding of the historical development of reliable knowledge about the natural world, in particular through the interplay of biological and chemical thought and action. One of these challenges has come, in the recent past, from historians of the natural sciences who have differed in their response to new fashions in sociological thought about the political factors in the institutional development of scientific disciplines.[23] The controversy was highlighted in 1980 by the report that, in his Sarton Lecture, the noted historian of science Charles Gillispie had described the social history of science as devoid of "hard" science, with a "lust" for the personal and the anecdotal, and that students of science history have tended to be political scientists whose lack of scientific knowledges produces a "depiction of scientists as hucksters of weapons and research." The report elicited a discussion about "internal" versus "external" history of science.[24] One feature of the debate was the disparagement by some social historians of the writings of scientists about the historical development of their specialties, and books by chemists received particular attention.[25] Moreover, in advocating the primacy of social factors in the development of scientific disciplines, this approach was contrasted with that of "intellectual history," although since 1945 prominent historians of science had emphasized in their writings the intimate connection between scientific thought and technical practice.[26] Frederic Holmes has put the issue well:

> The meaning of a tennis match or a football game is in part determined by the fans, the patrons, the officials, the construction of the field, the stadium, and the television broadcasters. To study the socio-economic backgrounds of the players, the audience, and the intermediaries, how the facilities are maintained, players hired and traded, is certainly to extend one's understanding of the sport. All this remains hollow, however, unless one is first interested in the game; unless one takes pleasure in following the play-by-play, appreciates a fine shot, and cares about the difference between a forced and unforced error. In the case of the history of science, while attentive to the ground rules and institutions without which the game could not be played, I would like to direct my own central efforts to more penetrating portrayals and interpretations of the play-by-play on the field.[27]

Holmes's approach has not been adopted by many aspiring historians of science, probably because few of them are willing to engage in the time-consuming and often difficult study of the technical details in the research under examination. Instead, since about 1960 there has been a succession of fashions which absolved historians of science from undertaking such demanding tasks, and it was repeatedly stated that the historian of science did not need to acquire an intimate knowledge of the empirical results and practical methods in that area of research. It was also fashionable to label such "internalist" studies as parochial, progressivist, or positivist.

As will be evident from the account in the next chapter, the interplay of chemical and biological thought and action since 1800 gave rise to new research disciplines under a variety of social circumstances in different countries, and in a diversity of institutional settings. Even when considered only in the context of the academic profession (as distinct, say, from scientists in industrial or military establishments), the question arises whether the ideals, aims, values, and norms in the social behavior of university chemists and biologists have been much the same over the years, regardless of their institutional or national affiliation.[28] That question has been examined most incisively since the 1930s by Robert Merton, whose writings, and those of his students and like-minded colleagues, established the sociology of science as a respected discipline in the United States.[29]

During the 1930s, the attitudes of many scientists and scholars to the social factors in the development of scientific knowledge were profoundly affected by the worldwide economic depression and the rise of fascism in Europe. In Britain and the United States, that impact found expression in statements about the social responsibility of scientists, the need for greater state support of scientific research, and the association of the ideals of the democratic state with those of scientific inquiry. For example, John Desmond Bernal wrote:

> The events of the past few years have led to a critical examination of the function of science in society. It used to be believed that the results of scientific investigation would lead to continuous progressive improvements in the conditions of life; but first the War and then the economic crisis have shown that science can be used as easily for destruction and wasteful purposes, and voices have been raised demanding the cessation of scientific research as the only means of preserving a tolerable civilization. Scientists themselves, faced with these criticisms, have been forced to consider, effectively for the first time, how the work they are doing is connected with the social and economic developments which are occurring around them.[30]

Bernal, a physical chemist with wide-ranging scientific interests, was one of the pioneers in the use of X-ray crystallography to determine the three-dimensional structure of biologically important molecules. In his approach to the "social function of science," he was the chief spokesman for a group that included the biologists C. H. Waddington and Lancelot Hogben, the biochemical geneticist J. B. S. Haldane, the biochemical embryologist Joseph Needham, the plant biochemist N. W. Pirie, the nuclear physicist P. M. S. Blackett, and the mathematician Hyman Levy. These men were attracted, in different measure, to the Marxist interpretation of the historical development of the natural sciences.[31] They were opposed by an equally distinguished group (led by the physical chemist Michael Polanyi) who founded the Society for the Freedom of Science, devoted to the defense of "pure science."

In an essay published in 1942, and later entitled "Science and Democratic Social Structure," Merton echoed Bernal's dismay at the upsurge of "incipient and actual attacks upon the integrity of science," and wrote:

> The institutional goal of science is the extension of certified knowledge. The technical methods employed toward this end provide the relevant definition of knowledge: empirically confirmed and logically consistent predictions. The institutional imperatives (norms) derive from the goal and the methods. The entire structure of technical and moral norms implements the final objective. The technical norm of empirical evidence, adequate, valid, and reliable, is a prerequisite for sustained true prediction; the technical norm of logical consistency, a prerequisite for systematic and valid prediction. The mores of science possess a methodological rationale but they are binding, not only because they are procedurally efficient, but because they are believed right and good. They are moral as well as technical prerequisites. Four

> sets of institutional imperatives—universalism, communism, disinterestedness, organized scepticism—comprise the ethos of modern science.[32]

Briefly, (1) universalism requires that the question of the validity of a scientific report be independent of the personality, race, creed, or nationality of the persons who produced it; (2) communism obliges scientists to share freely their theories and empirical findings, which then become common property; (3) disinterestedness requires that scientists not seek personal advantage (professional recognition, public prestige, financial reward) from their research; (4) organized skepticism implies that all published scientific work must be subject to critical scrutiny and that each scientist is solely responsible for the validity of his or her claim. To these norms, "intellectual honesty," "integrity," and "impersonality" were added later. But as many chemists and biologists know from their own experience, and from the study of the historical development of their specialties, "deviation" from these norms had long been a feature of the real life in research laboratories. It is not surprising, therefore, that Merton's idealized view of the ethos of science was later redefined.[33]

With the recent growth of the scientific population in the "biomedical sciences" in the United States, and increased competition for recognition and reward, "scientific integrity" has become a subject of congressional inquiry and regulatory action by the National Institutes of Health and the National Science Foundation. Universities that receive grants from these agencies must set up mechanisms for the investigation of accusations of such "misconduct" as plagiarism and the fabrication or doctoring of experimental data.[34] Although not included in the official listing of types of scientific misconduct, improper citation of previous work has long been a source of grievance among chemists, biochemists, and biologists. Such departures from the Mertonian ideal of honest and accurate allocation of credit have usually consisted in the omission or, at best, perfunctory or misleading citation of the published reports of other scientists in order to buttress a claim for priority.

In declarations by leading American scientists on this subject, it is rightly stated that the prime responsibility for educating the junior members of an academic research group in the norms of scientific conduct rests with its leader, who is expected to serve as a "role model." I find a modicum of tartufery in this counsel now that numerous noted university molecular biologists have followed into the marketplace the many noted academic organic chemists who entered it many years ago. The norm of openness and generous sharing of knowledge often gives way to the processing of patents or to the retention of trade secrets, and press conferences often take precedence over the publication of a critically evaluated article in a reputable scientific journal. When such practices are not only accepted but actively encouraged by university administrators, the research students and associates of a faculty member who has entered the commercial world may find it difficult to reconcile the message in lectures on scientific ethics with the behavior of his or her "role model." I know that many academic research groups are led by outstanding biochemists and molecular biologists who need no exhortations about "scientific integrity," who take seriously their responsibility as educators of future scientists, and who are alert to the possibility that a junior member of the group may do sloppy work or even commit fraud in order to provide data in support of the wishful thinking of the group leader.[35]

During the 1970s and 1980s, a rather different approach to the Mertonian norms was taken by a group of European sociologists who sought to establish a "new" sociology of scientific knowledge (SSK), based on the premise that the content of knowledge about the natural world had been derived from its social context and therefore did not deserve the privileged status previously accorded it among the various forms of human knowledge or belief. Although this approach was rejected by many historians of science, it influenced the discourse about the sociological aspects of the historical development of scientific disciplines and therefore merits serious attention.

The SSK movement first emerged at Edinburgh,[36] where its most prolific members were Barry Barnes, David Bloor, Steven Shapin, and Andrew Pickering. The many other active participants included Harry Collins and Trevor Pinch (Bath), Michael Mulkay and Nigel Gilbert (York), and Steve Woolgar (Brunel) in Britain, Bruno Latour and Michel Callon (Paris), and Karin Knorr-Cetina (most recently in Bielefeld). In attempting to advance their cause, members of the SSK group offered philosophical arguments and empirical evidence. As to the first, they found support in the writings of Thomas Kuhn, who had written in his famous book that "many of my generalizations are about the sociology or social psychology of scientists." According to Barnes, "Kuhn has made one of the few fundamental contributions to the sociology of knowledge. It fell to him to provide, at the time it was most needed in the 1960s, a clear indication of how our own forms of natural knowledge could be understood sociologically."[37]

As is well known, Kuhn had proposed that the historical development of scientific knowledge has been marked by periods of "normal science," when the principal research activity in a discipline is one of "puzzle solving" within the framework of a "paradigm" (an accepted mode or pattern), but when "anomalies" arise, a "revolution" may ensue, with the emergence of a new paradigm with theories that are "incommensurable" with those of the superseded paradigm. Some years later, he replaced the term *paradigm* with the phrase *disciplinary matrix* and still later with *lexicon* ("the [mental] module in which members of a speech community store the community's kind terms").[38]

In the opinion of some members of the SSK group, Kuhn had provided a radical alternative to Merton's formulation of the problems in the sociology of science. Apart from the criticisms of the Mertonian norms, there were statements such as that of Roger Krohn that "the Mertonian sociology of science is limited and naive because it does not practice a systematic sociology of knowledge, but only an anecdotal one. Because it carries no sense that scientific knowledge is problematic, science is removed as a category of exception before the analysis has begun."[39] I consider the different approaches of Merton and of Kuhn to have been complementary.[40]

In an attempt to establish the philosophical basis of the SSK enterprise, David Bloor offered what he called "the strong programme in the sociology of knowledge" and argued for an explanation of a scientist's beliefs in terms of his personal and social interest. To support his argument, Bloor invoked selected and, in the opinion of his critics, distorted versions of the writings of Emile Durkheim and Ludwig Wittgenstein, and his general claims have been disputed by philosophers and by other sociologists.[41]

In seeking empirical evidence in support of the idea of the "social construction of scientific facts," some members of the SSK group have used the methods developed by anthropologists to study technologically primitive societies. As applied to the examination of the practices of scientists, such "ethnomethodology" involved the recording of conversations among members of particular laboratory groups, observation of the work done by these people, oral interviews, and the "deconstruction" of successive drafts of research articles prepared for publication. Among the books that emerged from these studies, those by Latour and Woolgar *(Laboratory Life)* and by Knorr-Cetina *(The Manufacture of Knowledge)* were reported to have been the most frequently cited books in the SSK field.[42]

Laboratory Life was based on the observation, between October 1975 and August 1977, by the anthropologist Latour of the group headed by the neuroendocrinologist Roger Guillemin at the Salk Institute for Biological Studies at La Jolla, California. The scientific research discussed most extensively in *Laboratory Life* dealt with the isolation and determination of the chemical structure of the thyrotropin releasing hormone (TRH) secreted by the hypothalamus. Thyrotropin is generated by the anterior pituitary and stimulates the secretion of thyroxin by the thyroid gland. The work on TRH had been done by Guillemin's group during the 1960s at Baylor

University in Houston, and a parallel effort on the same problem was then under way in the laboratory of Andrew Schally (a former postdoctoral associate of Guillemin) at the Veterans Administration Hospital affiliated with Tulane University in New Orleans. The immediate aims of these two efforts were achieved by 1970, and Guillemin and Schally were corecipients, with Rosalyn Yalow, of the 1977 Nobel Prize in physiology or medicine.

Latour, who did not actually "observe" the work on TRH, provided in *Laboratory Life* a "reconstructed" form, mostly based on the published literature.[43] I will not retell the story but must emphasize that the research undertaken by the two groups was of major scientific significance, and difficult problems were overcome with originality and experimental skill. The isolation of TRH, present in hypothalami in minuscule amounts, was a formidable task, and the answer to the question of its chemical nature came from a traditional method of organic chemistry, namely the analysis of the chemical composition of the isolated product and the artificial synthesis of a substance which exhibits the physical, chemical, and biological properties of that product. Given their preconceptions, perhaps it is unfair to suggest that Latour and Woolgar might have learned much from Vincent du Vigneaud's account of the work of his group during the 1940s and 1950s on the chemical structure of the peptide hormones oxytocin and vasopressin of the posterior pituitary.[44] The success of that work did much to encourage the efforts of investigators such as Guillemin and Schally, but there appears to have been a notable lack of interest on the part of the authors of *Laboratory Life,* as of other members of the SSK movement, in the historical record of the development of theory and practice in the area of science under ethnographic examination. As in the case of oxytocin and vasopressin, the definitive proof of the chemical structure of TRH was obtained through the work of organic chemists familiar with the best methods then available for the synthesis of peptides. Whereas du Vigneaud was himself experienced in that field and had able young organic chemists in his laboratory, Guillemin depended on a collaboration with peptide chemists (notably Rolf Studer) in Basle, while Schally enlisted the help of the noted organic chemist Karl Folkers in Austin. One consequence of the cavalier treatment accorded in *Laboratory Life* to the contributions of these organic chemists appeared in the otherwise admirable article by the philosopher Ian Hacking, who wrote, "Instead of discovering the chemical structure of TRH by analyzing it, the two groups discovered it by synthesizing the substance. . . . But how could a couple of authors [Latour and Woolgar] in their right minds ever imagine that these laboratories did not discover the structure of TRH? The labs chemically constructed a synthetic version of TRH, yes, but what has this to do with social construction? The brute fact is that TRH is . . . pyroGlu-His-Pro-NH_2."[45] According to Latour and Woolgar,

> There are, so far as we know, no a priori reasons for supposing that scientists' practice is any more rational than that of outsiders. . . . Our discussion is informed by the conviction that a body of practices widely regarded by outsiders as well organized, logical and coherent, in fact consists of a disordered array of observations with which scientists struggle to produce order. . . . Without a bioassay . . . a substance could not be said to exist. The bioassay is not merely a means of obtaining some independently given entity; the bioassay constitutes the construction of the substance. . . . By remaining obstinate, our anthropological observer resisted the temptation to be convinced by the facts. Instead, he was able to portray laboratory activity as the organisation of persuasion through literary inscription.[46]

In his next book, dealing with Pasteur, Latour restated (somewhat obscurely) the constructivist thesis: "Did the microbe exist before Pasteur? From a practical point of view—I insist practical and not 'theoretical'—it must be acknowledged that it did not," and also suggested that Pasteur enlisted "the microbe" as an "ally" in his campaign to gain fame. In the same book, there is also a display of nostalgia and moral outrage: "It is because we have other interests and follow other roads that we find the myth of reason and of science to be unacceptable, intolerable, even immoral. We are no longer, alas, at the end of the 19th century, the most beautiful of the centuries, but at the end of the 20th, and the main source of pathology and mortality is reason itself, its pomps, its works, and its armaments."[47] Latour's later book *Science in Action* (1987), a provocative exercise in mystification and clowning, expands upon these themes.

Knorr-Cetina's *The Manufacture of Knowledge* deals with her study during October 1976–October 1977 of activities at a laboratory in Berkeley engaged in research that led to the publication of a paper entitled "Potato protein concentrates: The influence of various methods of recovery upon yield, compositional and functional characteristics." The names of the institution and the authors were not revealed, but the published paper is in the public domain and, with the help of *Chemical Abstracts*, was found in *The Journal of Food Processing and Preservation* 1 (1977), 235–247. The work was done at the Western Regional Research Center of the Agricultural Research Service, U.S. Department of Agriculture, Berkeley, and the authors of the papers were listed as D. Knorr, G. O. Kohler, and A. A. Betschart. The latter two scientists were then regular members of the staff (George O. Kohler was head of the laboratory and Antoinette A. Betschart was one of his associates), and the published paper reported that D. Knorr had been "visiting from the Department of Food Technology, University of Agriculture, Vienna, Austria." There was a later publication on the same subject, with Dietrich Knorr as the sole author. I also learned from *Wer ist Wer* (1992, p. 726) that Karin Knorr-Cetina had studied anthropology and sociology in Vienna, and that in 1969 she married Prof. Dr. Dietrich Knorr.

As an item of the human diet, the potato is prominent in the long history of poverty and famine.[48] Its protein content is low (1–5 percent dry weight), but its abundance made it an object of study by agricultural chemists who were seeking relatively inexpensive ways to meet the protein requirements in the diet of poverty-stricken peoples, especially in central Africa. The paper by D. Knorr and colleagues may therefore be considered to have been a contribution to that worthy effort. In the context of research in the field of protein chemistry during the mid-1970s, however, it added little, if anything, to knowledge in that field.

I will not dwell on Knorr-Cetina's preconceptions and interpretations, since I agree with Woolgar's statement in his review of *The Manufacture of Knowledge* that "once having taken account of difference in style and order of presentation, it becomes clear that the main themes of the argument are astonishingly similar to those of *Laboratory Life*."[49] In my opinion, despite the difference in the quality and significance of the work done in the two laboratories, both sociological studies produced little that was new and much that was false. The empirical evidence was too limited in institutional scope, period of observation, and attention to the historical record to warrant the acceptance of the philosophical views and sociological preconceptions which underlay the interpretation of the empirical findings.

During the 1980s there appeared other reports of studies that resembled in some respects those of Latour and Knorr-Cetina. The areas of investigation ranged from radio astronomy to learning in planarian worms, and included such paranormal phenomena as "spoonbending."[50] As might have been expected, the social constructivist theme in the SSK movement was subjected to searching critique and lost some its adherents.[51] Instead, renewed attention has been given to the issue of social conduct, or, as some now prefer to call it, moral economy.[52]

There can be little doubt that deviation from the ideal norms of personal and group behavior has long been a feature of the historical development of the interplay of chemistry and biology and has been accentuated when members of a research discipline enter the commercial marketplace. This was evident in the rise of organic chemistry and the pharmaceutical industry in Wilhelmine Germany, and in the United States after World War I. On the other hand, before about 1950, researchers in the biological specialties, though conscious of their ties to medicine and agriculture, had largely eschewed aspirations of wealth. The scientific lives of the remarkable group of American geneticists of the pre–World War II era, with that of Barbara McClintock as a shining example, bear witness to the possibility of maintaining high standards of personal and group conduct while serving to improve agricultural practice.[53] Today the temptations and opportunities for gaining wealth from the patenting of new products and methods are vastly greater than ever before, and so has become the frequency of reported instances of scientific misconduct. I hope that I am correct in my belief that most investigators in the chemical and biological disciplines have retained a commitment to the idea of a "vocation," and that they will continue to attract able young men and women who seek to develop innovative lines of research and who, in adapting to the social pressures of their time, will maintain high standards of personal scientific conduct and their independence of thought and action.

The Biographical Approach

In my opinion, one of the lines of historical research on the interplay of biology and chemistry that merits closer attention than it has hitherto received is the examination of the opportunities for young people from the "lower" social classes to rise from the status of an indentured apprentice or hired technician to that of scientific investigator. With some exceptions, such technical workers had long been "invisible" to historians bred in the tradition of "intellectual history," suggesting an attitude that exalts the thinking of the master above the labor of the servant.[54] That attitude has changed somewhat in recent years, with efforts to learn more about the people who produced the results that conferred upon their employers the status of "great scientists," and about the ways in which leaders of research groups interacted with their technical assistants and student apprentices.[55]

The proliferation during the past two centuries of relatively large research groups, first in the chemical sciences and more recently in the biological sciences, has been accompanied by marked changes in their social stratification. Before about 1800, the "elite" was drawn largely from the propertied classes, such as the landed gentry in Robert Boyle's entourage, and the "proletariat" came principally from families of various kinds of artisans. After the French Revolution, the broadening of educational opportunity in Europe was spurred by the emergence of large-scale industrial enterprises and the expansion of university teaching staffs, and many more men of relatively humble social origin could choose to aspire to scientific careers. In the German states, Liebig's program at Giessen led some of the sons of pharmacists to become research chemists. In Victorian England, the programs at the Royal College of Chemistry in London and at Owens College in Manchester encouraged many students to enter the chemical profession. In the United States, the establishment of state universities, like the ones in Michigan, Wisconsin, and Minnesota, made it possible for men and women from farming families to prepare for careers in scientific research, and in New York City during the first decades of the twentieth century, many daughters and sons of lower middle-class Jewish immigrants found at Hunter College and City College the incentive to become scientists. More recently, there has been an influx of able young people from many countries, especially from some of the East Asian nations. Regrettably, although for a time the opportunities appeared to be unlimited, descendants of the slaves brought to the Western Hemisphere from Africa remain underrepresented in the scientific community. Further, the hope that this disparity might soon be corrected has been dimmed by the realization that American universities are now producing more Ph.D. recipients in the natural sciences than are likely to find academic opportunities for independent research.[56]

I have called attention to the "little people" in chemical and biological research because I deplore the tendency to treat them solely as items in statistical data and because, as will be evident from the historical account in later chapters, many important contributions were made by men and women whose insight and skill were chiefly responsible for the achievements credited to their employers or mentors. What is needed, in my opinion, is a greater effort by historians to enlarge the scope of what they call prosopography to include the scientific lives of these hitherto neglected individuals. Although the program of such studies has been outlined, and numerous bio-bibliographic sources are available, more needs to be done. The value of historical investigations of this kind is evident from the many recent writings about women scientists.[57]

Apart from the fact that, for many of these "lesser" scientists, the finding of documents such as research notebooks and personal correspondence is likely to be difficult, there is also the question raised above regarding the past attitude of historians to the relation of thought to practice in laboratory research. In its modern manifestation, this attitude is expressed in the interpretation of scientific creativity in terms of such mental traits as imagination, intuition, or inspiration. These qualities of thought have long been considered to play a large role in scientific innovation and have been invoked by philosophers, psychologists, and scientist-historians, as well as by professional historians of science. It has also been noted, however, that imitation often precedes innovation, that accidental discovery is not infrequent, and that personal attributes like independence of mind and nonconformity with an established consensus are also significant in the creative process.[58] The search for the sources of "great thoughts" by "great men"[59] may be worthwhile, but scientific innovation includes the invention of new instruments and techniques as well as finding suitable living organisms and development of good methods for their cultivation, all tasks that may require much hard work. The craftsmanship, skill, and precision in such preparatory work, and in the actual conduct of experiments, are no less important for the growth of reliable scientific knowledge than the mental attributes cited above. Many famous chemists and biologists were outstanding in both respects—one thinks of such men as Robert Bunsen and Carl Ludwig—and in more recent times it has often been the case that the validity of a brilliant insight was established by "invisible technicians." I distort the meaning of the famous aphorism about "dwarfs perched on the shoulders of giants"[60] in suggesting that, in such cases, "a giant is perched on the shoulders of dwarfs."

As regards the value of the biographical approach in the study of the historical development of research disciplines, historians have been ambivalent in their estimate, for too many of the accounts of the life and work of famous scientists have been exercises in hagiography, and of doubtful accuracy.[61] All too often, disciples of a noted scientist approach his life in too worshipful a spirit, and one is reminded of Oscar Wilde's quip: "What is true in a man's life is not what he does but the legend which grows up around him. . . . You must never destroy legends." That is why some of the most valuable recent biographies of famous nineteenth-century scientists have demythologized them to a considerable extent.[62] Such interpretations, based on scrupulous examination of all relevant documents, also require familiarity with the state of knowledge and the meaning of scientific terms in a particular specialty at the time of the subject's research, as well as with the details of the techniques that were used in that research.

Writers on the methodology of historical investigation have stressed the fact that what matters is the interpretation that a particular historian places on the available documentary material, and that such interpretation is usually based on a previously formulated hypothesis.[63] Thus if a historian believes that a scientific discipline is primarily a political entity and bases the evaluation of a scientist's career in terms of that person's effectiveness as a political "discipline builder," the selection of documentary material and its interpretation is likely to reflect that preconception. Moreover, while much progress has been made in the collection and cataloging of papers of noted twentieth-century chemists and biologists, and in recording interviews,[64] the reliability of such source materials is so variable, and the difficulty of checking the accuracy of their use so considerable, that some skepticism about interpretations offered in support of a particular historical hypothesis is essential.

Other questions relating to the biographical approach arise when the objective is to "understand the whole person" by examining the private life of a scientist from a particular sociological, psychological, or medical point of view. It has been customary to treat the more intimate aspects of a famous scientist's life with greater circumspection than is now common practice among biographers of stage or screen stars, literary figures, or political leaders. Such delicacy may perhaps be justified on the ground that there does not appear to be any correlation between moral virtue and scientific achievement. To these questions must be added the matter of the social consequences of the knowledge a scientist has provided. It is taken for granted that if a chemist has synthesized a new vitamin, or a microbiologist found in a soil organism a new antibiotic agent, such contributions merit praise. The issue becomes murky, however, when a chemist has developed a compound that a botanist finds to be an effective defoliant, or that a pharmacologist finds to be a nerve poison, and such chemicals are then used in warfare or in "racial purification." This aspect of the "social responsibility" of scientists has received much attention from biographers of such scientists as Francis Galton in relation to the eugenics movement or Fritz Haber in relation to the use of poison gas during World War I.[65]

Historians of the chemical and biological sciences can also refer to autobiographies, the literary equivalent of a painter's self-portrait. These writings are so variable in reliability, however, that the only features they may be said to have in common are testimony to the vanity of the authors and an indication of how they sought to portray themselves to posterity. As has been noted by scholarly critics of the art of autobiography, it bears a close if ambiguous relationship to the art of the novel.[66] Scientific autobiographies intended for the general public have frequently oversimplified and overdramatized the contributions of the author. Occasionally, such accounts become bestsellers and have been used by philosophers or sociologists in their discourse about the nature of scientific investigation. The list of autobiographies by chemists and biologists is extensive (I have added my contribution), and in the hands of critical historians these accounts can serve as useful starting points for further study.[67] I would add that in a sense, scientists create a part of their own biographies in their published research reports. To be sure, such papers distort the actual course of a scientific investigation, but the style of presentation often tells much about the author's attitude toward the norms and values of his or her time and place.

Finally, in connection with this plurality of biographical approaches to problems in the history of science, I welcome the recent appearance of Alistair Crombie's remarkable three-volume treatise on the historical development of commitments and styles of European scientific thinking.[68] There he traces the emergence of six styles of scientific reasoning—mathematical, experimental, analogical modeling, taxonomic, probabilistic and statistical, and historical:

> These six styles and their objects are all different, sometimes incommensurable, assuming fundamentally different physical worlds, but frequently they are combined in any particular research. By identifying the regularities that become its object of inquiry, and by defining its questions and acceptable evidence and answers, a style both creates its own subject-matter and is created by it. A change of style introduces not only new subject-matter, but also new questions about the same subject-matter. . . . Different styles introduce new questions about the existence of their theoretical objects; are these real or products of methods of measurement or sampling, or even of language?[69]

CHAPTER 2

The Institutional Settings

The interplay of chemistry and biology has developed in many nations and in a variety of institutions—in university schools of medicine, agriculture, pharmacy, and technology, in private or government research institutes, and in industrial establishments. It must be emphasized, however, that the individual scientists and research groups in each of the various chemical and biological specialties form parts of an international effort to study particular scientific problems. This linkage should not be obscured by the fact that in the following account the institutions are grouped according to their geographical locations. Moreover, it should be recognized that what has made a scientific institution noteworthy is the quality and significance of the work of its members as investigators and teachers and that, in themselves, institutions are important in the development of scientific knowledge only to the extent that they may provide a favorable setting for such work by able scientists. As John Ziman has stated: "Any research organization requires generous measures of the following: social *space* for personal initiative and creativity; *time* for ideas to grow to maturity; *openness* to debate and criticism; hospitality towards *novelty;* and respect for specialized *expertise.*"[1]

Chemistry in France and Germany

The institutional settings in which chemical and biological education and research interacted during the nineteenth century were, in many respects, markedly different from those in previous decades. The contrast in the teaching of chemistry was especially striking. In leading eighteenth-century European universities, chemistry had been part of the medical curriculum, the other subjects usually being the theory and practice of medicine, "institutes of medicine" (later called physiology), anatomy, and botany. Herman Boerhaave, the leading university teacher of chemistry during the first decades of the century, also taught medicine and botany at

Leiden. Later the position of leadership shifted to Edinburgh; in 1755 William Cullen became professor of chemistry and physic there, and he was succeeded eleven years later by Joseph Black, his most celebrated pupil.[2]

Eighteenth-century physicians did not need to do practical chemistry, however; this was the task of pharmacists, who transmitted the art of chemical manipulation to their apprentices, and to affluent amateurs for whom chemistry was a diversion. A major center of such teaching was Paris, where pharmaceutical chemists such as Guillaume François Rouelle, his younger brother Hilaire Martin Rouelle, and Pierre Joseph Macquer taught chemistry to the men (among them Antoine Lavoisier) who fashioned what came to be called the Chemical Revolution. The subservience of chemistry to pharmacy, and its relationship to botany, is evident from the fact that these teachers worked at the illustrious Jardin du Roi; in 1790 it was renamed the Jardin des Plantes, and three years later it became the Musée d'Histoire Naturelle. This institution continued to be an important center of chemical research well into the nineteenth century, especially through the work of Michel Eugène Chevreul.[3] As for the medical faculty, before the French Revolution Jean Bucquet taught pharmacy and chemistry there; his students included Antoine François Fourcroy, who succeeded Macquer at the royal garden in 1784. Among Fourcroy's laboratory assistants was Louis Nicolas Vauquelin, his successor as professor of chemistry in the medical faculty and the museum. Some of the others who preserved the tradition of practical chemistry during the Revolution, the Directory, and the First Empire, and who were principal participants in the interaction of chemistry and biology, were French pharmacists such as Henri Braconnot, Jean-Pierre Robiquet, Joseph Pelletier, and Joseph Caventou.[4] Some of these men also set up factories for the manufacture of chemicals. After 1803, chemistry was also taught at independent schools of pharmacy in Paris, Montpellier, and Strasbourg.[5]

In part because of the Napoleonic 1808 "reform" that placed higher education under the domination of Paris, that city (with its Sorbonne, Collège de France, Ecole Polytechnique, Société d'Arcueil, Ecole Normale Supérieure, Musée d'Histoire Naturelle, and Conservatoire des Arts et Métiers) became the leading center of chemical education and research in France. During the first three decades of the nineteenth century, practical chemistry was still taught in private laboratories, such as those headed by Joseph Louis Gay-Lussac, Louis Jacques Thenard, Vauquelin, and Chevreul, but by the 1840s such instruction was increasingly given in public institutions. The leading figure was Jean-Baptiste Dumas; among the others were Jean-Baptiste Boussingault and Théophile Jules Pelouze.[6] Two of the most original French chemists of the 1840s and 1850s, however—Auguste Laurent and Charles Frédéric

Gerhardt—were not named to professorships in Paris after their service in Bordeaux and Montpellier, respectively.[7]

During the second half of the nineteenth century, the most prominent French chemists were Marcelin Berthelot (at the Collège de France), Charles Adolphe Wurtz (at the Sorbonne, then dean of the medical faculty), Charles Friedel (at the Sorbonne) and, of course, Louis Pasteur (at the Ecole Normale Supérieure and the Sorbonne).[8] Between 1900 and about 1950, the French contributions to the development of organic chemistry were less prominent than before, compared with those of German, British, Swiss, and American chemists. In particular, the emergence of many excellent German centers of organic-chemical research and instruction elicited a strongly nationalistic response, and what had begun during the 1840s as a contest between Dumas and Liebig was exacerbated by the outcome of the Franco-Prussian War.[9] There were, however, several notable academic investigators, among them Edouard Grimaux (Ecole Polytechnique), Henri Sainte-Claire Deville and Robert Lespieau (Ecole Normale Supérieure), Henry Le Châtelier and Georges Urbain (Sorbonne) in Paris, Albin Haller in Nancy (later Paris), François Marie Raoult in Grenoble, Victor Grignard in Lyon, and Paul Sabatier in Toulouse; also, Ernest Fourneau worked at the Pasteur Institute. Jean Perrin (Sorbonne) was more inclined toward physics than toward chemistry.[10] After World War II, with the establishment of the Centre National de la Recherche Scientifique (CNRS), organic chemistry once again became an important component of scientific research and education in France and, in addition to active laboratories in Paris and its suburbs, other universities (notably the one in Strasbourg) gained international prominence in this field.[11]

In the eighteenth-century German states, as in France, chemical instruction was offered by pharmacists, such as Andreas Sigismund Marggraf, Johann Christian Wiegleb, and Martin Heinrich Klaproth, with emphasis on the uses of chemistry in pharmacy, metallurgy, and mining, and in the tradition of the seventeenth-century adept Johann Rudolph Glauber.[12] After 1800, although the Germans and Austrians followed the French example in promoting the institutional status of chemistry, and some of their leading chemists (including Eilhard Mitscherlich and Justus Liebig) had studied in Paris, they set up many important university centers of chemical training. At first there were relatively modest efforts in Göttingen under Friedrich Stromeyer, in Heidelberg under Leopold Gmelin, and in Berlin under Mitscherlich.[13] A decisive change came, however, after 1825, when Liebig established his teaching and research laboratory in Giessen. An outstanding organic chemist, Liebig was an energetic (and frequently combative) entrepreneur whose writings on the application of chemistry to agriculture and physiology made him a world-renowned scientific figure.[14] Of at least equal importance were the laboratories of Friedrich Wöhler in Göttingen and of Robert Wilhelm Bunsen in Marburg and Heidelberg.[15] What happened in Germany was reflected on a smaller scale in Sweden, where pharmacy, metallurgy, and mining had provided much of the stimulus for the eighteenth-century work of Carl Wilhelm Scheele and of Torbern Olof Bergman. During most of the first half of the nineteenth century the acknowledged leader of European chemistry was Jöns Jacob Berzelius, who taught at Stockholm; his most famous pupil was Friedrich Wöhler, and the others included the Dutch chemist Gerrit Jan Mulder.[16]

The rise after 1840 of the German states and, after 1871, of a united Germany, to preeminence in chemical research and education was a consequence of at least three mutually reinforcing factors: the worldwide recognition of the importance of the contributions made by German chemists to the development of the conceptual structure and practical methodology of organic chemistry, the rapid growth after 1850 of large-scale chemical industries, and, after the Franco-Prussian War, the collaboration of government officials, university professors, and industrial leaders in developing a national policy for the promotion of scientific research and education.

Liebig played a considerable role in relation to the first two factors, because of his own scientific contributions during the 1820s and 1830s, the fame of his Giessen laboratory, and his encouragement of such men as Heinrich Emanuel Merck and Ernst Sell to establish chemical factories.[17] I think it fair to say, however, that during the 1840s Liebig's reputation was somewhat higher in England and the United States than in Germany. His writings on the application of chemistry to agriculture and physiology received much unfavorable criticism, and his decision in 1852 to move from Giessen to Munich, where he enjoyed a more comfortable but scientifically less productive life, marked his withdrawal from the development of organic chemical thought and practice. That decision may have been influenced by the acceptance of the new ideas, so different from his own, offered by Laurent and Gerhardt in France, by Alexander Williamson, William Odling, Edward Frankland, and Archibald Scott Couper in England, and by his former student August Kekulé (at Ghent, only later in Bonn). Moreover, Liebig's chief protégé, August Wilhelm Hofmann, had gone in 1845 to London, where he headed the private Royal College of Chemistry, later made part of the government-funded School of Mines. It was in Hofmann's London laboratory that a young student, William Henry Perkin, accidentally made the first of the aniline dyes ("mauve").[18]

Although Hermann Kolbe (first in Marburg, then in Leipzig), who had made pioneering contributions to synthetic organic chemistry, derided the new theories of chemical valence and structure, other German chemists, notably Adolf Baeyer, eagerly embraced Kekulé's theories as the basis for their experimental work. In his successive posts in Berlin (the Gewerbeinstitut), Strassburg, and especially in Munich, Baeyer's research groups included many future luminaries of organic chemistry, the most illustrious of whom was Emil Fischer (Erlangen, Würzburg, Berlin).[19] Other former junior associates of Baeyer who won distinction as university professors in Germany, Austria-Hungary, or Switzerland were Carl Graebe (Königsberg, Geneva), Carl Liebermann (Charlottenburg), Victor Meyer (Zurich, Göttingen, Heidelberg), Paul Friedländer (Darmstadt), Theodor Curtius (Kiel, Bonn, Heidelberg), Eugen Bamberger (Zurich), Guido Goldschmiedt (Prague, Vienna), Hans von Pechmann (Tübingen), Johannes Thiele (Strassburg), Richard Willstätter (Zurich, Berlin, Munich), and Heinrich Wieland (Freiburg, Munich). These men, in turn, led sizable research groups that included many future professors of organic chemistry at German universities; among Fischer's students, for example, were Ludwig Knorr (Jena), Carl Harries (Kiel), Otto Diels (Kiel), and Karl Freudenberg (Heidelberg).[20]

If Baeyer's chemical progeny formed a large proportion of the German professoriate in organic chemistry, former students of other leading nineteenth-century chemists (notably Wöhler, Kolbe, and Hofmann) also led productive research groups—for example, Adolf Claus (Freiburg), Rudolph Fittig (Tübingen, Strassburg), Otto Wallach (Göttingen), Karl Friedrich von Auwers (Greifswald, Marburg), Adolf Windaus (Göttingen), and Arthur Hantzsch.[21]

During the years before 1914, many aspiring young organic chemists came to the German laboratories from abroad, especially from Great Britain, the United States, Russia, and Japan, and the flow resumed after the end of World War I.[22] On becoming university professors in their home countries, some of these students imitated the quasi-military features of the German system of research and instruction, in particular the frequently dictatorial rule of an *ordentlicher* (full) *Professor* solely responsible for the teaching in the discipline represented by his *Lehrstuhl* (professorship) and, as *Institutsdirektor,* also in charge of the research of the staff. In the larger institutes, such as those in Berlin and Munich, there were sections headed by faculty members holding the rank of *ausserordentlicher* (associate) *Professor* or *Privatdozent.* With some exceptions, the research work for which the director gained fame was done by *Assistenten* and by candidates for doctoral degrees, and, depending on his liberality, the junior associates were given the opportunity to conduct their own research programs with the help of such students. Thus Baeyer's style throughout his academic career was to encourage the independent efforts of his junior colleagues, whereas that of Emil Fischer, while in Berlin, was less liberal in that regard. The position of the Privatdozenten was especially precarious, because they received no salary for their service as teachers and were expected to receive only student fees. Consequently, unless a Privatdozent had private financial means, he was obliged to accept appointment as an Assistent of the institute director.[23]

During 1850–1914, most of the directors of the German chemical institutes were elaborating methods for the determination of the structure and for the synthesis of organic compounds. This emphasis on chemical practice reflected a special relation with the burgeoning industry increasingly engaged in the manufacture of synthetic dyes and drugs. It also affected adversely the institutional development of such chemical specialties as theoretical chemistry *(allgemeine Chemie)*, of the kind of experimental chemistry based on the use of physical instruments, and of the specialty of inorganic chemistry. Two notable exceptions were the university laboratories of Wöhler in Göttingen and of Bunsen in Heidelberg. Both men had done important organic-chemical research early in their scientific careers, but Wöhler returned to his interest in inorganic chemistry, and Bunsen chose to develop new methods for gas analysis, electrochemistry, and (with the physicist Kirchhoff) for spectrum analysis. The decision of these two gifted chemists not to participate further in the development of organic chemistry suggests a negative response to the excitement generated by the new ideas of chemical valence and structure, and perhaps also to the increasing commercialization of their discipline.

In the large chemical institutes, inorganic chemistry was usually taught by a separate corps of investigators. Thus in Baeyer's Munich institute there was an inorganic section successively headed by Clemens Zimmermann, Gerhard Krüss, Friedrich Wilhelm Muthmann, Karl Andreas Hofmann, and Wilhelm Prandtl.[24] Also, the university in Munich had a separate Lehrstuhl for "applied" chemistry, successively occupied by Albert Hilger and Theodor Paul, both with ties to pharmacy. Moreover, by 1900, the institutional status of technical schools (such as the Berlin Gewerbeinstitut, where Baeyer had taught during the 1860s) for students of pharmacy, food chemistry, fermentation chemistry, agricultural chemistry, and chemical engineering had been upgraded, and many outstanding chemists established highly productive research laboratories there.

The most successful challenge to the academic dominance of the German organic chemists came during the 1890s from a group of chemists and physicists who were interested in problems relating to thermodynamics, theory of solutions, chemical affinity, electrochemistry, and chemical kinetics. The movement was led by the Baltic German Wilhelm Ostwald, who was named in 1887 to the Lehrstuhl of physical chemistry at Leipzig. The other two were the noted Dutch chemist Jacobus Henricus van't Hoff and the Swedish chemist Svante August Arrhenius.[25]

They received support from leading German physicists, especially Friedrich Kohlrausch and Ludwig Boltzmann.

The use of the term *physical chemistry* to denote the application of physical theory and methods in chemical studies dates from the eighteenth century. During the first decades of the nineteenth century, the electrochemical experiments of Humphry Davy and Michael Faraday influenced the development of chemical theory. By 1880 thermodynamic theory, as developed by Hermann Helmholtz, Rudolf Clausius, and others, had begun to be applied to chemical problems by August Horstmann, Josiah Willard Gibbs, and van't Hoff, and the work of the pharmacist Ludwig Wilhelmy marked the beginning of the study of chemical kinetics. If to these theoretical and experimental achievements one adds the calorimetric studies of Julius Thomsen and Marcelin Berthelot, the contributions of Bunsen and Kirchhoff to spectroscopy, and the work of Hermann Kopp on the physical properties of organic compounds, it is evident that the success of Ostwald's triumvirate represented the formal recognition as an academic discipline of a research effort that had long been under way. Thus what had been established in 1882 at Berlin as the "second chemical institute" (the "first" was the one headed by A. W. Hofmann) was renamed in 1905, when Walther Nernst took over its directorship, as the Physikalisch-chemisches Institut. It is also noteworthy that Nernst wrote: "As the chief purpose of instruction we strive to encourage students to carry out independent physical-chemical research in accordance with the directions and aims given to them. It will therefore continue to be the custom in the institute to emphasize the personal responsibility of the students for their work and also in the designation of authorship of publications."[26] This attitude was rather different from that in Emil Fischer's Berlin institute. It should also be noted that during the period of the Weimar Republic, Berlin was a world center in physical chemistry and attracted many postdoctoral students from abroad.[27]

I turn next to the reciprocal interaction, during 1850–1914, of German university chemists and those in the chemical industry. As mentioned above, Liebig and some of his pupils played a significant role in the early stages of the creation of large-scale chemical plants. For example, Karl Clemm and his brother Gustav successively enlarged the scope of their enterprise from artificial fertilizers to the manufacture of soda and sulfuric acid and then to synthetic dyes; in 1865 their company became the Badische Anilin und Sodafabrik (BASF) in Ludwigshafen. Three years later the dye chemist Heinrich Caro, a friend of Baeyer's, joined the firm, and the synthesis of alizarin, by Baeyer's associates Carl Graebe and Carl Liebermann, provided the impetus for the transformation of a struggling enterprise into a world leader in the manufacture of synthetic dyes. At about the same time, the seeds were planted for other future giants in the manufacture not only of dyes but also of synthetic drugs. The most successful of these companies were the Friedrich Bayer factories in Elberfeld and Leverkusen, the firm of Meister, Lucius & Brüning in Höchst (a suburb of Frankfurt am Main), and Cassella & Co. in that city; the last-named firm was later taken over by the Höchst company. After World War I, these four companies joined to form the mammoth I. G. Farben (Interessengemeinschaft Farbenindustrie Aktiengesellschaft), and its director became Carl Duisberg (the head of Bayer) who had worked with Pechmann in Munich and had been one of Emil Fischer's close friends. Another former student in Baeyer's institute, Arthur von Weinberg, became an executive of the Cassella firm and later of I. G. Farben. The collaboration of one of Emil Fischer's favorite students, Ludwig Knorr, with the Höchst firm marked its entry into the manufacture of synthetic drugs. Among the other prominent German producers of pharmaceutical agents were Merck in Darmstadt, Boehringer in Mannheim, and Schering in Berlin.[28]

The fruitful interaction of university chemists with manufacturers of dyes and drugs was not the only factor in the emergence of Wilhelmine Germany to commercial dominance. The occupation of Alsace-Lorraine after 1871 gave German and Swiss companies control of the thriving dye industry in the Mulhouse region and had as a side effect the rise of Basle as an important center of chemical manufacture, with three companies—Geigy, CIBA (later CIBA-Geigy), and Sandoz—as the leading enterprises. As in Germany, these companies established close relations with university chemists. The principal Swiss center of chemical education and research was Zurich, with its university and the Eidgenössische Technische Hochschule (ETH). At the university, the chemical institute had a succession of outstanding professors, including Alfred Werner and Paul Karrer, while at the ETH the organic chemical institute was successively headed by Eugen Bamberger, Hermann Staudinger, Leopold Ruzicka, and Vladimir Prelog. Ruzicka brought to the ETH Tadeusz Reichstein, who later became the head of the organic chemical institute at Basle. The researches of Ruzicka, Prelog, and Reichstein on natural products played a large role in the development of the above Basle companies (plus Hoffmann–La Roche) as leading manufacturers of pharmaceutical agents. In his work on essential oils, Ruzicka was for a time associated with the Naef perfumery company in Geneva.[29]

In addition to the scientific achievements of German chemists before World War I and the commercially successful use of the knowledge they provided in the manufacture of dyes and drugs, there was a third and perhaps most important "external" factor in the emergence of Wilhelmine Germany as the nation to which young scientists came from many countries for doctoral and postdoctoral education and brought home the German model of the "research university." That factor was the role of government officials in promoting, guiding, and controlling the development of the science faculties in German universities. Because the ordentliche Professoren were employees of the state, the opinions of such officials about the relative merits of candidates for professorships were frequently decisive.[30] The most famous of these officials was Friedrich Althoff, whose activities between 1882 and 1907 as head of the personnel division for universities of the Prussian Ministry of Ecclesiastical, Educational, and Medical Affairs (often referred to as the Kultusministerium) made him the ruler of higher education in Prussia.[31] In 1871, after serving as a medical orderly in the Franco-Prussian War, the thirty-four-year-old Althoff became part of the bureaucracy charged with administering the German occupation of Alsace and, as deputy to Franz von Roggenbach, the *Kommissar* for the university in Strasbourg (renamed Strassburg), was responsible for implementing his chief's plan to transform it into a showcase research university by attracting outstanding younger scholars and scientists to its professorial faculty.[32] Among them were the organic chemist Baeyer, the physiological chemist Felix Hoppe-Seyler, and the pharmacologist Oswald Schmiedeberg. Althoff had studied law before the war, but lacked the usual credentials (doctorate, *Habilitation* [lectureship], experience as Privatdozent) for a professorship. His performance in Strassburg, however, led to his appointment in Berlin, to his decisive role in strengthening the faculties of the Prussian universities and in the advancement of the academic careers of German professors. Althoff had great respect for Emil Fischer, and brought van't Hoff to Berlin. He concerned himself with the problems arising from the emergence of new biological disciplines, such as the developmental biology promoted by Wilhelm Roux. He was especially interested in medical research and education and furthered the development of important medical centers such as the Charité hospital in Berlin. The establishment in 1888 of the Pasteur Institute in Paris was followed three years later by the opening in Berlin of the Robert Koch Institute for Infectious Diseases.[33] To finance his enterprises, Althoff was ready to accept the help of bank directors and industrial magnates.

A glimpse of the manner in which Althoff conducted his business is provided by the following account by Johannes Reinke, professor of botany at Kiel, of the circumstances of the appointment of Theodor Curtius there. Upon the departure of Albert Ladenburg to Breslau in 1889,

> the ministry rejected the list, prepared under the influence of Ladenburg, of three full professors from elsewhere, and called upon the faculty for new

> proposals. Ladenburg had already left, and the faculty elected me to be chairman of the committee to prepare the new proposals. In the meantime, the ministry had asked us to bear in mind three older associate professors at Prussian universities because they had long merited promotion to full professorships. I sought the opinion of Baeyer in Munich, Emil Fischer in Würzburg, and Lothar Meyer. . . . All agreed that although the three men recommended by the ministry were able people, objections could be made to all of them, and that far more preferable candidates were available. . . . We therefore turned down the three desired men, and proposed three Bavarian Privatdozenten, von Pechmann and Claisen in Munich, Theodor Curtius in Erlangen. I prepared the report in rather concise sentences, and it was sent off to Berlin. Shortly afterward, I received a telegram from Althoff, asking me to come to his office on a certain day and to arrange to travel immediately thereafter to southern Germany. Now I knew what was afoot [*wie der Hase lief*]. In Berlin, Althoff met me in his office with the report of our faculty in his hand. He asked: "Herr Professor, this report is yours?" To my reply that it expressed the opinion of the entire faculty, he said: "The report is yours, I know your style! I admit that we have examined the matter again and have come to the same conclusion as you; we wish however to reverse the order. The Bavarian Junker with the monocle is not suitable for Kiel, and Claisen is a too timid little man; we think that Curtius is the right man for you. We are doubtful, however, about his capacity as a teacher, as you also indicated in your report. Since I have convinced myself that you understand chemistry better than I do, I ask you to travel to Erlangen, to attend Curtius's lecture, and if he pleases you personally and as a Dozent, you will have him."[34]

Reinke did as he was ordered, found Curtius to be an excellent lecturer and teacher in the laboratory, and returned to Berlin to report this to Althoff.

An autocrat in an autocratic state, Althoff won the favor of the Kaiser and thus was able to withstand criticism from many quarters. In particular, not only did university professors chafe at Althoff's dictatorial methods, but many of them also despaired at the weakening of the "idea of the university" as formulated by Wilhelm von Humboldt in 1810 for the new university in Berlin.[35] In that "idea," born out of the German romanticism of the Napoleonic era, the linking of teaching and scholarship was to be free of outside interference *(Lehrfreiheit und Lernfreiheit)* and of concern with the practical problems of contemporary social life. In fact, of course, the uprisings in 1848, the industrialization of the German states, and the development of the natural, medical, and engineering sciences, all contributed to making illusory the yearning for what Humboldt and the idealist German philosophers of his time considered to be the "Greek" model of an institution of higher learning. Many of Althoff's critics looked longingly at Oxford, where the "proper" balance had been maintained between the liberal arts and the natural sciences.

The Emergence of Research Institutes

In addition to his effort to enhance the international prestige of the Prussian universities in the natural sciences, medicine, and technology, thus incurring the wrath of many of the "mandarins" in the German professoriate, Althoff played a large role in the creation of research institutions supported financially by bankers and industrialists. He responded favorably to the proposal of the electrical manufacturer Werner Siemens for the establishment of what in 1888 came to be the Physikalisch-technische Reichsanstalt, wholly devoted to research and independent of university control. Althoff took a warm interest in Paul Ehrlich and helped him in 1899 to set up an institute for experimental therapy in Frankfurt am Main; the funds came largely from the Speyer family and the Höchst firm. Althoff also looked favorably on the appeals of leading chemical manufacturers, such as Carl Duisberg, who offered plans for a Chemische Reichsanstalt, and those of some biologists for a research institution in their field. These efforts were unsuccessful, in part owing to the downturn in the German economy during the first decade of the next century. The hopes of the chemists and biologists remained undimmed, however, and they pointed with alarm to the establishment in the United States of the Carnegie Institution of Washington and of the Rockefeller Institute for Medical Research, with what were at that time large endowments. From Althoff's conversations and correspondence with Emil Fischer, Walther Nernst, and others, there emerged an ambitious plan for the construction of a set of research institutes in Dahlem, a suburb of Berlin, to be financed by private donations and to be administered by the Kaiser-Wilhelm-Gesellschaft zur Förderung der Wissenschaften.[36]

After Althoff retired in 1907 (he died a year later), Friedrich Schmidt-Ott became the Kultusminister and, with the Kaiser's encouragement, carried the plan forward. The theologian Adolf von Harnack and Emil Fischer were Schmidt-Ott's principal advisers, the fund-raising drive was more successful than had been expected, and the society was formally launched in 1911. Fischer must have been gratified, and his attitude toward his American competitors is suggested by his statement that "I consider it likely that because of their greater wealth the Americans will beat us in several fields, and I have expressed this opinion at every opportunity. However, we can withstand this competition for a time because of our greater inventiveness and more distinguished individual achievements."[37] Fischer's anti-American prejudice appears to have been directed at biochemists such as Thomas Burr Osborne, who had dared to encroach on his field of research; on the other hand, Fischer conducted a warm correspondence with Theodore William Richards. It must also be noted that, in contrast with the anti-Semitism of most German university professors, Fischer was liberal in that regard, as is indicated in the appointments in the first Kaiser Wilhelm institutes. Thus in the one for chemistry (Ernst Beckmann, director), the section of organic chemistry was headed by Richard Willstätter, and Otto Hahn and Lise Meitner were named to the section of radiochemistry. The institute for physical chemistry, funded by the Jewish banker and industrialist Leopold Koppel, was headed by Fritz Haber. In the one for experimental therapy (August Wassermann, director), a section of biochemistry was set up for Carl Neuberg, and a separate laboratory was given to Otto Warburg, who had received his D.phil. for work in Fischer's laboratory. Other institutes established before 1918 included one for coal research in Mühlheim/Ruhr, headed by Franz Fischer, one for biology headed by Carl Correns, and Max Rubner's laboratory in Berlin, which was named the Kaiser Wilhelm Institute for the Physiology of Work. An institute of physiology was approved in 1914, with Fischer's former associate Emil Abderhalden as the director, but that decision was not implemented. Among the new postwar institutes was the one in Dresden for leather research, directed by Max Bergmann, who had been Fischer's chief associate until the latter's death in 1919. Another, supported by the textile industry, was the institute for fiber *(Faserstoff)* chemistry, headed by Reginald Herzog. Others were the institute for medical research in Heidelberg, with Johannes von Kries as director, and one for cell biology in Dahlem. The latter was built for Otto Warburg with a large grant from the Rockefeller Foundation, which later also financed the construction of a new institute for physics, headed by Peter Debye. Shortly afterward, a part of that institute was taken over by the war ministry, Debye was dismissed, and the scientific part was jointly headed by Otto Hahn and Werner Heisenberg; in 1942 the latter was named sole director.

During 1930–1937 and 1945–1946, Max Planck was the president of the Kaiser Wilhelm Society, and after his death in 1947, the society was renamed the Max-Planck Gesellschaft. By the 1970s, there were about fifty institutes and special laboratories under its jurisdiction, scattered among many German cities. After 1918 Schmidt-Ott was replaced as Kultusminister by Carl Heinrich Becker, under whose administration there was established the Notgemeinschaft der deutschen Wissenschaften (later renamed the Deutsche Forschungsgemeinschaft), an autonomous body that supported most of the research in German universities. In contrast to the policy adopted for the Kaiser Wilhelm institutes, grants were made by committees composed of university scientists and scholars, and the private patrons were excluded from participation in these decisions. For this reason, such industrialists as Carl Duisberg, who had been influential in the funding of the Kaiser Wilhelm Society, declined to support the Notgemeinschaft, on the ground that the grants would go for research in areas of little practical interest to German industry. Except for the years 1933–1939, when the Nazis dismissed hundreds of scientists and scholars from their posts at German-speaking universities and research institutes, and the period of World War II, these two agencies played a major role in promoting high-caliber scientific research.[38] It should be noted, however, that during the period of Hitler's rule, even when the worst crimes were being committed in the cause of "racial purity," outstanding biological and chemical research was being conducted at German universities and research institutes by groups led by such men as Heinrich Wieland, Adolf Butenandt, Otto Warburg, and Fritz von Wettstein.[39] These men, each in his own way, made accommodations with and, on occasion, resisted the intrusion of Nazi barbarism into their scientific lives.

Although the stimulus for the creation of the Kaiser Wilhelm Society may have come in part from the Carnegie and Rockefeller endowments in the United States, important research institutes independent of universities had existed in several European countries during the nineteenth century. Among them was the Royal Institution of Great Britain in London, founded in 1799 by public subscription on the initiative of Benjamin Thompson, Count Rumford. Gifted investigators, notably Humphry Davy and Michael Faraday, worked there during the first half of the century, and were succeeded by John Tyndall, James Dewar, and William Henry Bragg. The principal educational function of the Royal Institution was to provide public enlightenment by "diffusing the knowledge and facilitating the general introduction of useful mechanical inventions and improvements, and for teaching by courses of philosophical lectures and experiments the application of science to the common purposes of life." In 1896 the chemist and industrialist Ludwig Mond, a German Jew who made his fortune in England, gave a large donation for the establishment of the Davy-Faraday Research Laboratory at the Royal Institution.[40]

The amateur tradition of the Royal Society of London, and the emphasis in Oxford and Cambridge on liberal rather than professional education, were factors in the temporary decline, during the first decades of the nineteenth century, of British distinction in chemistry, compared with developments in France and Germany. During the 1820s the leading British chemists included, in addition to Davy,

William Hyde Wollaston and William Prout, both of whom were physicians who had private laboratories. By 1830 Davy and Wollaston had died, and the principal luminary, apart from Faraday, was John Dalton, who flourished in Manchester, one of the centers of the Industrial Revolution, where middle-class virtues were more conducive to the development of professional science allied to technology. Dalton was associated with the Manchester Literary and Philosophical Society and had a laboratory there. One of his pupils, and later president of the society, was James Prescott Joule, who had a private laboratory. What later came to be called the Manchester school of chemistry was created by Henry Enfield Roscoe, a student of Bunsen's, at Owens College (later the University of Manchester). With his associate Carl Schorlemmer, Roscoe developed a program of instruction which attracted many students and, through his involvement with local industry in civic affairs, gained great public recognition. The prestige of the Manchester chair in chemistry was further enhanced by Baeyer's student William Henry Perkin Jr. and his successor Arthur Lapworth, one of the founders of physical-organic chemistry. During the twentieth century, among the holders of that chair were Robert Robinson and Alexander Todd, who became professors of chemistry at Oxford and Cambridge, respectively.[41] At Oxford a university chemical laboratory was built during the 1860s and the Dyson Perrins Laboratory of Organic Chemistry was opened in 1915 with the financial aid of the Lea and Perrins firm, maker of the famous Worcestershire sauce. In addition, beginning in 1848 there were also chemical laboratories attached to several undergraduate colleges (Magdalen, Balliol, Trinity, Christ Church, Queen's, and Jesus). At Cambridge a university chemical laboratory was not provided until 1887, and college laboratories were set up, beginning in the 1850s, at St. John's, Gonville and Caius, Girton, Newnham, and Downing. As the university laboratories became well established, by the 1940s most of the college laboratories were closed.[42]

I return to the institutional status of chemical education and research in London during the first decades of the nineteenth century. The first university professors of chemistry were Edward Turner (appointed 1828 at University College) and John Frederic Danielli (appointed 1831 at King's College). During the 1840s and 1850s the influence of the French and German schools of chemistry became evident not only in the appointment of A. W. Hofmann as head of the Royal College of Chemistry, and in the contributions of Williamson, Couper, Odling, and Frankland and their association with Kekulé in the emergence of the concepts of valence and structure, but also in the increased professionalization of British chemists. Thomas Graham, for example, who had come from Glasgow in 1837 to succeed Turner at University College, became the first president of the Chemical Society, founded in 1841.

Moreover, the medical schools of several London hospitals (notably St. Bartholomew's, St. George's, and Guy's) engaged the services of such men as Odling and Frankland and offered them research facilities.[43]

After Hofmann returned to Germany in 1865, he was replaced by Frankland as head of the Royal College of Chemistry, which was moved to South Kensington and, after several changes in name, became in 1911 part of the Imperial College of Science and Technology. Among its professors of chemistry was Henry Edward Armstrong, one of the most colorful figures in British chemistry. A student of Frankland's and Kolbe's, Armstrong was an outspoken critic of prominent physical chemists and a stimulating teacher. During the 1890s, several of the leading twentieth-century British chemists were students in his laboratory, among them Arthur Lapworth and Thomas Martin Lowry. After World War I, there was a succession of outstanding organic chemists at London universities, the most notable of whom was Christopher Kelk Ingold (at University College).[44]

The Royal Institution was only one of the major nonuniversity centers for chemical research in nineteenth-century London. Among those set up before 1900 was the British Institute for Preventive Medicine, established in 1891 at the suggestion of Henry Roscoe and funded by public subscription; Ludwig Mond was one of the principal donors. The stimulus clearly came from the founding of the Pasteur Institute, of which more later. After the death in 1912 of Joseph Lister, one of the cofounders, it was renamed the Lister Institute. The main functions of the institute were related to infectious diseases, but during the years of its existence (it closed in 1975), much significant work was done there on enzymes, proteins, and vitamins. Among the early investigators were Allen MacFadyen, Arthur Harden, Henry Dakin, Charles Martin, and Harriette Chick; later chemists included John Gulland, Alexander Todd, Richard Synge, and James Baddiley.[45]

After World War I, what had been set up in 1914 as the National Institute of Medical Research (NIMR) in Hampstead (then a London suburb) became a world-renowned center under the successive directorship of Henry Dale, Charles Harington, Peter Medawar, and Harold Himsworth. The institute was the chief research facility administered by the Medical Research Council (MRC), whose first two executive secretaries were Walter Fletcher and Edward Mellanby. All these men had a warm appreciation of the importance of chemistry in medical research. Dale, a noted pharmacologist, had been a research director at the Wellcome pharmaceutical firm, Harington had won fame for his work on the chemistry of thyroxine, and both Fletcher and Mellanby had been associated with Gowland Hopkins.[46] In 1950 the institute moved to Mill Hill in north London. Among the outstanding organic chemists on the staff during the 1930s and 1940s were Otto Rosenheim, Harold King, and John Cornforth.

Before the opening of the Pasteur and Lister institutes, there appeared in 1875 a chemical and physiological laboratory at Jacob Christian Jacobsen's Carlsberg brewery in Copenhagen. In the following year, he established the Carlsberg Foundation, which took over the administration of the laboratory and whose board of trustees was chosen by the Danish Academy of Sciences and Letters. The successive heads of the chemical division included Johann Kjeldahl, Søren P. L. Sørensen, and Kaj Linderstrøm-Lang, all prominent in the twentieth-century development of protein and enzyme chemistry, while in the physiological division there were, over the years, such notables as the fermentation chemist Emil Christian Hansen, the geneticists Øjvind Winge and Mogens Westergaard, and the cell chemist Heinz Holter. After World War II, the Carlsberg Laboratory attracted many postdoctoral students, especially from the United States.[47]

Because of the prominence given (especially by French historians of science) to the role of the Pasteur Institute in the emergence of molecular biology, the earlier stages in the development of that institution merit special interest. The institute bearing Pasteur's name grew out of his success in 1885 in saving the life of a boy who had been bitten by a rabid dog. A committee was appointed by the Académie des Sciences to organize the funding, and voluntary contributions flowed from many sources, including the Russian Czar. The institute was inaugurated in 1888, Pasteur served as its director until his death in 1895, and among its first members were Emile Roux, Elie Mechnikov, and Emile Duclaux. In addition to their research and the preparation of vaccines, instruction in biological chemistry (by Duclaux) was transferred from the Sorbonne to the institute. During the period 1900–1940, the staff included Gabriel Bertrand, Félix d'Hérelle, Albert Calmette, Jules Bordet, and Ernest Fourneau, with an emphasis on research in microbiology, immunology, and chemotherapy. After World War II the Pasteur Institute gained great fame owing to the work of André Lwoff, Jacques Monod, and François Jacob. After the award in 1965 of Nobel Prizes to these three men, there was a revolt against the outworn administrative practices of the past, and in 1971 Monod became the director of the institute.[48]

Another medical research institute established during the 1890s was one in St. Petersburg for the Polish biochemist Marceli Nencki. After a stay in Baeyer's laboratory, Nencki had been professor of medical chemistry at Berne (1872–1891), where he attracted many foreign students, among them John Jacob Abel, Martin Hahn, and Allan MacFadyen. During the second half of the nineteenth century, chemistry flourished in Czarist Russia. As evidence for the importance of individual scientists in determining the distinction of institutions, the case of the provincial city of Kazan is especially noteworthy. Although there were outstanding chemists in Moscow and St. Petersburg, it was in Kazan that there emerged one of the distinguished European schools of organic chemistry, under the successive leadership of Nikolai Zinin, Aleksandr Butlerov, Vladimir Markovnikov, and Aleksandr Zaitsev. That school, beginning with Zinin, embraced and, in the case of Butlerov, contributed to the development of the new ideas about valency and chemical structure offered at mid-century.[49]

If one must, in studying the historical development of the interplay of chemistry and biology, take account of the role of various institutions devoted principally to medical research and education, or to the manufacture of drugs or products of alcoholic fermentation, equal attention needs to be paid to the influence of outstanding work by chemists and biologists at agricultural colleges and experiment stations. With the development during the first decades of the nineteenth century of analytical methods for the determination of the elementary composition of plant and animal foodstuffs and of soils, such analyses formed the principal activity at institutions devoted to agricultural chemistry. One of the early practitioners was Heinrich Einhof, at Albrecht Daniel Thaer's institute in Möglin (Mittenwald), established in 1806. In subsequent years, numerous other agricultural experiment stations were set up in Germany; at the one in Möckern (near Leipzig), established in 1861, Heinrich Ritthausen did pioneering work on the isolation and characterization of plant proteins. In France the leading agricultural chemist was Jean-Baptiste Boussingault, whose analyses of food materials during the 1830s and 1840s formed the empirical basis for the theories of nutrition advanced by Dumas and Liebig. Liebig rejected some of the prevailing views about the fertilization of soils and proposed artificial manures that, for a time, attracted the attention of agriculturists in Great Britain and the United States, but his claims were discounted after the work of Joseph Henry Gilbert at the Rothamsted Experimental Station set up by John Bennet Lawes in Harpenden, England. That institution has continued to the present; the famous research on plant viruses by Charles Frederick Bawden and Norman Wingate Pirie was done there during the 1930s. During the second half of the nineteenth century, numerous state agricultural stations were established in the United States; among the more prominent ones were those in Connecticut (1875) and Wisconsin (1883). The first was headed by Samuel William Johnson, a professor of agricultural chemistry at Yale, and his son-in-law Thomas Burr Osborne did outstanding work there on the amino acid composition of plant proteins. The Wisconsin group gained prominence during the first half of the twentieth century through the work of Elmer Verner

McCollum and Harry Steenbock on vitamins. At the Rutgers experiment station, the work of Selman Waksman's group led to the discovery of new antibacterial agents in soil microorganisms. American state experiment stations also figure largely in the development and application of Mendelian genetics; some of the notable investigators, whose work will be discussed in a later chapter, were Lewis Stadler, Leslie Dunn, Donald Jones, Marcus Rhoades, and Barbara McClintock.[50]

Physiological Chemistry and Chemical Physiology

For our inquiry into the interplay of chemistry and biology it is significant that the professionalization of chemistry during the first decades of the nineteenth century was paralleled by comparable changes in the university status of several biological specialties, especially animal physiology, and in their relationship to the practical arts of medicine and agriculture, as well as to the time-honored field of natural history. These changes were initially most evident in the German-speaking states and in France. The first steps were taken by Carl Schultze in Freiburg and by Jan Evangelista Purkyně in Breslau, who were followed by Johannes Müller in Berlin. Although Müller's own outlook was largely that of a naturalist, he encouraged his students to use chemical, physical, and microscopic methods in physiological research. The remarkable group (Theodor Schwann, Hermann Helmholtz, Ernst Brücke, Rudolf Virchow, Emil du Bois-Reymond) educated in his institute included many of the leaders of nineteenth-century science. The contrast in the style of research in Müller's laboratory, compared with that of his chemical contemporary Liebig in Giessen, is striking. Müller's students were exposed to his wide-ranging biological interests and to a variety of experimental methods, whereas the tasks assigned by Liebig were more narrowly defined, and the practical experience was largely limited to elementary analysis.[51]

During the second half of the nineteenth century, the most famous German institute of physiology was that of Carl Ludwig in Leipzig. His postdoctoral education had begun in Marburg, where he learned Bunsen's new methods of gas analysis, which Ludwig later modified for his studies on blood gases. In 1847 he visited the Müller institute in Berlin, and found in Helmholtz, Brücke, and especially du Bois-Reymond kindred spirits in their attitude to the place of physics and chemistry in the study of physiological problems. In Ludwig's own professorial career, first at Zurich (1849–1855), then at Vienna (1855–1865), and finally at Leipzig (1865–1895), he exhibited remarkable acumen in attacking a variety of such problems, and in the use or invention of methods and instruments, as well as skill in surgical and histological techniques. As a teacher and author of a famous textbook of human physiology

(1852), Ludwig began his exposition, as he put it in a letter to du Bois-Reymond, "with the complex chemical atoms of which the human body is composed, at the same time disclosing everything referring to their chemical constitution, conversion products, and physical characteristics. This prepares the way for the explanation of many physiological processes." In the Leipzig institute, there were three departments: physiology, chemistry, and anatomy/histology. Many students came from abroad to study with Ludwig and upon their return to their home countries played a large role in the institutional development of the physiological sciences there. Among these people were the Russians Ivan Mikhailovich Sechenov and Ivan Petrovich Pavlov, the Americans Franklin P. Mall, Henry P. Bowditch, Charles S. Minot, and William H. Welch, and the Englishman Walter H. Gaskell.[52]

In addition to Ludwig's institute in Leipzig there were, during the second half of the nineteenth century, other notable physiological institutes in the German-speaking countries: in Berlin (du Bois-Reymond), Bonn (Eduard Pflüger), Munich (Carl Voit), Heidelberg (Willy Kühne), Strassburg (Friedrich Leopold Goltz), Vienna (Brücke), and Prague (Ewald Hering). The research interests of these men ranged from electrophysiology to physiological chemistry and microscopic anatomy. Also, by that time, the German physiologists had largely freed themselves of the domination of anatomists on the medical faculty and, in many cases where physiology and anatomy were still combined, the professorships had been taken over by physiologists.[53]

Among the other physiological institutes at German-speaking universities the one headed by du Bois-Reymond in Berlin is noteworthy for the breadth and quality of the research conducted there. For his own work in electrophysiology du Bois-Reymond devised improved instruments, and he also had an aquarium for electrical fishes. During the period of his professorship (1858–1896), the junior associates at the Berlin institute included Franz Boll (the discoverer of "visual purple"), Eugen Baumann (noted for his work on homogentisic acid and the iodine-containing component of the thyroid), Albrecht Kossel (the pioneer in the chemical study of Miescher's "nuclein"), Ludwig Brieger (who worked on "ptomaines"), and Hans Thierfelder (who did outstanding work on phosphatides and cerebrosides). There were many guest workers from abroad, including the American physician James Bryan Herrick.[54]

Of special interest in relation to the development of the interplay of chemistry and biology is the fact that some of the physiological institutes mentioned above were headed by men who figure largely in historical accounts of the emergence of physiological chemistry as an independent academic discipline. Among them were

Kühne, Voit, and Brücke. Moreover, during the first part of the twentieth century, the research programs in the physiological institutes of Kossel (Heidelberg, 1901–1923) and Emil Abderhalden (Halle, 1911–1945) were largely chemical in nature.[55]

In addition to the physiological institutes in the medical faculties of German-speaking universities, those for pathology, pharmacology, and hygiene also emerged during the latter half of the nineteenth century as centers for the fruitful interplay of chemical and biological thought and experiment. Especially noteworthy in this regard was the Berlin institute for pathology, headed by Rudolph Virchow (1847–1849, 1856–1902). As a physician, he was devoted both to the clinic and the laboratory, and insisted that his institute be connected to the Charité hospital in order to permit clinical observation with the aid of microscopy and of chemical and physical methods. In this sense, Virchow represents a pioneer in the development of the hospital clinic as an important institution for scientific investigation. In his most famous book, *Die Cellularpathologie* (1858), Virchow advocated the consideration of the cell as the fundamental unit in health and disease, and the interpretation of changes in cellular morphology in chemical and physical terms. It was in Virchow's institute that Felix Hoppe (later Hoppe-Seyler) worked on the chemistry of hemoglobin and was succeeded in the chemical laboratory by Willy Kühne, Oskar Liebreich, Ernst Leopold Salkowski, and Peter Rona.[56] A parallel development was the emergence of clinical chemistry as an adjunct to the diagnosis of human disease. Among the pioneers were Johann Franz Simon in Berlin, and especially Johann Florian Heller in Vienna, where the renowned pathologist Carl von Rokitansky emphasized the importance of chemical analysis in medical practice.[57]

The development of experimental physiology led, during the second half of the nineteenth century, to the transformation into experimental pharmacology of what had been known in the medical curriculum of German-speaking universities as *materia medica,* with its ties to pharmacy, botany, and chemistry. The impetus came in part from the work of François Magendie and Claude Bernard on the physiological action of toxic agents. The leading spirit in establishing pharmacology as an academic discipline was Oswald Schmiedeberg, whose mentor Rudolf Buchheim had in 1847 set up such an institute in Dorpat (now Tartu). As the ordenticher Professor for this subject at Strassburg (1872–1918), Schmiedeberg was an active participant in research on many problems of contemporary biochemical interest. His laboratory attracted many guests from abroad; for the development of pharmacology in the United States, his most important student was John Jacob Abel, himself an outstanding research biochemist and, as professor of pharmacology at Johns Hopkins (1895–1932), a leading figure in the institutional development of American biochemistry. Another student, Arthur Cushny, became professor of pharmacology at University College London (1905–1918) and Edinburgh (1918–1926). Schmiedeberg was also largely responsible for the appointment in 1896 of Franz Hofmeister as Hoppe-Seyler's successor in Strassburg.[58]

The field denoted as "hygiene" in the medical faculties of German universities, and later elsewhere, was established as an academic discipline in response to the recognition of the importance, in the promotion of public health, of sanitation, nutrition, and the control of infectious diseases. The view during the second half of the nineteenth century that this field had a relation to chemistry is indicated by the fact that, upon Hoppe-Seyler's appointment at Strassburg, his institute was responsible for instruction and research in hygiene, as well as in physiological chemistry. That the problems of human nutrition were considered to lie within the domains of physiology, chemistry, and hygiene is suggested by the change in Max Rubner's professorial title from hygiene (1891–1909) to physiology (1909–1922) and by the fact that Max Pettenkofer, Liebig's colleague in Munich, was professor of hygiene there (1865–1894). Institutes of hygiene were also assigned responsibility for research and instruction in the emerging fields of bacteriology and parasitology, as is apparent from the appointment to professorships of hygiene of Robert Koch in Berlin (1885–1891), Karl Flügge in Breslau (1887–1909), and Friedrich Löffler in Greifswald (1888–1913). In Munich, Pettenkofer was succeeded by the bacteriologist Hans Buchner (1894–1902). If, as I have suggested before and as will be evident from my account in subsequent pages, much of the groundwork for the later fruitful interplay of chemistry and biology was laid in laboratories of nineteenth-century German universities, why were there so few independent institutes for what was then called "physiological chemistry" before 1914? Like other questions of this sort, a variety of answers can be produced. In the case of the slow emergence of independent institutes of biochemistry in German universities, it was not only a matter of academic politics but also a reflection of the state of the development of biological and chemical theory and practice during the second half of the nineteenth century.

Before 1914 there were only two German universities in which physiological chemistry was considered to be worthy of independent status—Tübingen and Strassburg—although several physiological chemists were granted the rank of honorary ordentlicher Professor in a few other institutes, usually physiology. More often, independent research was conducted by men holding the rank of ausserordentlicher Professor (in many cases a salaried position), as in Ludwig's institute in Leipzig, or that of du Bois-Reymond in Berlin. The humble beginnings in Tübingen, under Ludwig Sigwart and Julius Schlossberger, were largely reflections of the influence of

Berzelius and Liebig, and it was not until 1861, when Felix Hoppe came there from Berlin to be professor of applied chemistry in the medical faculty, that a more productive program of research and teaching was instituted. At that time, there was much discussion at the university regarding the proposal, which was approved in 1863, to establish a separate faculty of science. Hoppe accepted an invitation to join it, and, as a consequence, he became closely associated with Adolf Strecker, the professor of general chemistry, one of the few outstanding Liebig pupils of the 1840s. Among Hoppe-Seyler's students was Friedrich Miescher, later to be recognized for his isolation of "nuclein." As has been noted, Hoppe-Seyler was brought in 1872 by Althoff to the medical faculty at Strassburg. From that laboratory there emerged many important scientific contributions, as well as several outstanding investigators, notably Eugen Baumann and Albrecht Kossel. That tradition was continued by Franz Hofmeister, whose scientific progeny included such men as Franz Knoop, Gustav Embden, and Otto Loewi.[59] In 1918, when Strassburg became Strasbourg once again, the German professors (as well as the Austrian Hofmeister) left. At Tübingen, Hoppe-Seyler's successors were Gustav Hüfner (1872–1908; he worked on hemoglobin), Hans Thierfelder (1908–1928, cerebrosides), Franz Knoop (1928–1945, intermediary metabolism), and Adolf Butenandt (1945–1956, steroid hormones). Ernst Klenk (neurochemistry) also worked in Tübingen (1923–1936).[60]

It should be noted that in some German-speaking universities outside Germany independent institutes for physiological (or medical) chemistry had been established before 1914. Among them were those of Carl Schmidt in Dorpat (1852–1894), Karl Hugo Huppert in Prague (1872–1903), Marceli Nencki in Berne (1877–1891), Gustav von Bunge in Basle (1886–1920), and Fritz Pregl in Graz (1913–1930).[61]

The opposition before 1914 to appeals for additional independent institutes of physiological chemistry (or biochemistry, as it began to be called) came from several sources, most loudly from some professors of physiology. When in 1877 Hoppe-Seyler stated in the first issue of his *Zeitschrift für physiologische Chemie* that "biochemistry . . . has grown to a science that has not only placed itself on a par with biophysics, but in activity and success competes with it for rank," he was clearly making a claim for increased recognition of the independence of his academic discipline. That statement was countered by Pflüger, editor of an established journal that had published many papers by physiological chemists, who wrote that "the division of physiology into physiological physics and chemistry is philosophically inadmissible and impossible in practice" and appealed to his colleagues "not to underestimate the danger . . . but to oppose strongly the divisive forces which now threaten the one great and glorious science of physiology." Pflüger's views were echoed by du Bois-

Reymond and his former pupil Ludimar Hermann (then professor of physiology at Zurich, later in Königsberg), as well as by Max Verworn, the professor of physiology in Göttingen.[62]

Among the many unsuccessful attempts to set up independent institutes for physiological chemistry, the case of Göttingen is of special interest because of the role of Friedrich Wöhler, who became professor of chemistry there in 1836 and chose to be a member of the medical faculty (his doctorate had been in medicine). Six years later, the newly appointed professor of physiology, Rudolph Wagner, set up a laboratory for physiological and pathological chemistry in his institute, and instruction in that field was given by Wagner's associates Julius Vogel and Friedrich Theodor Frerichs. In 1852, after both men had left Göttingen, the laboratory was turned over to Georg Staedeler, an ausserordentlicher Professor in Wöhler's institute. In the following year, however, Staedeler became professor of chemistry in Zurich and was succeeded by another Wöhler student, Carl Boedeker, while Wöhler was asked by the university administration to add physiological chemistry to his professorial responsibilities. His letter declining the request merits quotation:

> Physiological chemistry, a specialty that is not represented in Göttingen at present, is a discipline that has emerged only recently. . . . It is the chemical part of physiology, the science of the composition and formation of the substances which constitute the blood and organs of the animal body, the science of metabolism or the chemical processes whereby nutrients are converted into the constituents of the various organs, muscle, nerves, brain, etc., and whereby these are decomposed into excretory products. . . . The student learns these general relations, which belong, or should belong, to the basic knowledge of the physician, partly in lectures on general chemistry and partly in those on general physiology. However, the field has become so large and so difficult that it requires of the teacher such special chemical and physiological knowledge that neither the chemist, whose science is already branched in many directions, nor the physiologist who has barely mastered it, can do more than handle it in only fragmentary fashion. Whoever wishes to teach this subject thoroughly and successfully and to contribute to its advances through his own researches—it is now clear—must make it his sole lifework.[63]

It would seem that Wöhler, though a friend of Liebig's and his scientific collaborator during the 1820s and 1830s, did not share Liebig's penchant for entrepreneurship, for despite the above statement the cause of physiological chemistry did not fare well in Göttingen. Wagner died in 1864 and was succeeded as professor of physiology by

Georg Meissner (1864–1901), who was followed by Max Verworn (1901–1910), both of whom shared Pflüger's views about the proper academic status of physiological chemistry. Göttingen did not have an ordentlicher Professor for that subject until 1946, in the person of Hans Joachim Deuticke, a pupil of Gustav Embden.

In Leipzig, during the tenure of the professorship of physiology by Carl Ludwig (1865–1895) and Ewald Hering (1895–1916), there was a succession of able heads of the section for physiological chemistry—Gustav Hüfner (1865–1872), Edmund Drechsel (1872–1892), and Max Siegfried (1892–1916). The separation of the section from the physiological institute did not occur, however, until 1916. After Siegfried's death in 1920 (he had been made a full professor in 1919), Karl Thomas became the head of the new institute.[64] Similar recognition of the independent status of physiological chemistry was accorded during the 1920s at five other German universities—Freiburg, Münster, Jena, Erlangen, and (in 1928) Berlin. In 1931 Robert Feulgen was made professor of physiological chemistry at Giessen. During the Nazi regime, two more (Kiel, Cologne) were added, and after 1945 the remaining fourteen followed suit. The delay in the case of Berlin is especially striking because between the two wars that city was a leading world center of biochemical research at the Kaiser Wilhelm Institute laboratories of Otto Warburg, Otto Meyerhof, and Carl Neuberg, laboratories in which such future notables as Hans Krebs, Karl Lohmann, Fritz Lipmann, and Hugo Theorell had worked. It should be added, however, that Berlin was the place where the outstanding biochemist Leonor Michaelis failed to obtain a permanent academic post.[65] During the 1920s German biochemists welcomed a statement by Gowland Hopkins, in his introduction to a lecture at the 1926 International Physiological Congress in Stockholm. After noting that the response of Pflüger to Hoppe-Seyler "still found sympathy," Hopkins said:

> The specialized school at Strassburg under the guidance of such men as Hoppe-Seyler and Franz Hofmeister supplied the world with fundamental biochemical knowledge of a kind that when the School began its work was quite generally thought by the orthodox physiologists of the day to be beyond the reach of experimental science. Although, doubtless, personal inspiration counted at Strassburg, as everywhere else, for more than circumstances—the genius of Claude Bernard and Pasteur spontaneously endowed biochemistry no less than other branches of biology in spite of circumstances—yet it is sure that those who occupied the Strassburg Chair during the years when their subject continued to gain in dignity were better able to impress their genius upon it than if they had been compelled to cultivate and expound, in addition to their own branch of physiology, others less

> suited to their powers and predilections. Though it is fifty years since Hoppe-Seyler printed his appeal for the recognition of biochemistry as an independent discipline, such recognition cannot be said to have yet arrived in his own country. In proportion to its outstanding academic equipment modern Germany provides but little institutional freedom for the discipline in question. It is under the circumstances difficult to see how she can continue to lead along the path she has trod almost alone.[66]

Opposition to Hoppe-Seyler's appeal for the establishment of additional independent institutes of biochemistry came not only from physiologists but also from some influential organic chemists. If during the 1850s Wöhler's views may have reflected those of other leading chemists of his time, by 1900 the transformation of the theoretical and practical style of organic chemical research, with emphasis on the proof of structure through synthesis, accentuated the separation of "pure" chemists from the "medical" chemists. In many instances, the work of physiological chemists did not meet the standards of the new organic chemistry, and a widely quoted saying was "Tierchemie ist Schmierchemie": Animal chemistry is messy chemistry. Although there were several leading physiological chemists (for example, Eugen Baumann) who not merit this disdain, the quip was not entirely unjustified. Indeed, some of the agricultural chemists who studied plant products (notably Heinrich Ritthausen and Ernst Schulze) came closer to the standards set by the professional organic chemists.

The disdain was especially marked in the case of Emil Fischer, who had entered the arena of biochemical investigation by establishing during 1880–1900 the structure of sugars and of purines in a brilliant series of synthetic operations, and who then embarked on a synthetic program aimed at the synthesis of proteins. In his papers and lectures dealing with proteins, he repeatedly disparaged the efforts in that field of physiological chemists whom he often denoted only as physiologists, presumably to indicate that, in his opinion, they were not proper chemists. In a letter to one of his academic colleagues in Berlin, Fischer wrote: "Regrettably, biological chemistry is that part of our science in which imprecise and incomplete experiments are often heavily padded with the dazzling ornamentation of so-called ingenious reflections to produce pretentious treatises."[67] I will deal at some length in later pages with Fischer's synthetic successes, as well as some of his failures, and only note here that as an adviser of the Kultusministerium, Fischer was in a position to express such views to Althoff and Schmidt-Ott and to influence their response to appeals from German biochemists for greater institutional recognition of their academic discipline. In fairness I must add that in January 1912, as a member of the governing body of the Kaiser Wilhelm Society, Fischer participated in a meeting to discuss proposals for the establishment of biological research institutes, and although he refrained from offering his personal opinions on scientific matters, he said that "since my own researches have frequently touched upon the field of biology, I must confess that my sympathies lie on the side of biochemistry, as recommended by Herr Kossel." Kossel had stated that "biochemistry is now at a stage of development in which it is no longer a part of physiology. It is now occupied in seeking new methods and general concepts which will benefit all parts of biology."[68]

As is evident from the variety of proposals discussed at that meeting, by 1912 what was called "biology" had become a somewhat disparate and often competitive set of research disciplines. The term had been introduced around 1800 and adopted by the German anatomist Karl Friedrich Burdach, the German physician and naturalist Gottfried Treviranus, and the French naturalist Jean Baptiste Lamarck. It was later popularized by Auguste Comte and William Whewell and had been intended to emphasize the unique attributes and continuity of animal and plant life. Despite its attractive brevity, *biology* was not readily accepted as the designation of an academic discipline and was often considered to be synonymous with *physiology*.[69] Moreover, one of the distinctive features of the institutional development of the biological sciences during the nineteenth century, especially at German-speaking universities, was the effort of botanists and zoologists to free themselves from the medical faculty. This separation appears to have come most clearly in the case of botany. In 1863 the botanist Hugo von Mohl became the dean of the newly established faculty of natural science at Tübingen, and in his inaugural address he called upon German universities "not to lag behind the times, to recognize the significance attained by the natural sciences, and to accord to their further development the necessary independent status."[70]

A feature of the development of botany as an independent discipline in Germany during the nineteenth century was increased emphasis on the study of plant anatomy and physiology at the expense of taxonomy. Although the microscopic observations of Robert Brown in London excited some interest, the impetus was provided by Matthias Jacob Schleiden (Jena), one of the proponents of the cell theory of biological structure and function, who also played a large role in the introduction of the microscope in biological research. After 1840 there was a succession of German university professors who contributed much to the study of plant development, metabolism, and inheritance. Among them were, in addition to Mohl, Wilhelm Hofmeister (Heidelberg, Tübingen), Carl Nägeli (Munich), Julius Sachs (Würzburg), Anton de Bary (Strassburg) Eduard Strasburger (Jena, Bonn), Oscar Brefeld (Münster), and Wilhelm Pfeffer (Tübingen, Leipzig). Special mention should be made of Ferdinand Cohn (Breslau), a botanist with an interest in algae and fungi, whose studies on bacteria, along with those of other nonmedical microbiologists (notably Martinus Willem Beijerinck and Sergei Vinogradsky), laid the groundwork for the expansion of biochemical knowledge during the twentieth century.[71]

In zoology, Mohl had his counterpart in Ernst Haeckel. Before becoming professor of zoology (1865–1909) at Jena, Haeckel stated in 1861 that "in its significance and scope, zoology or animal science *(Tierlehre)*, as it is now fairly widely understood, is the study of the outer form and inner structure of animals, of their general life phenomena, and their relationship to the rest of nature. Zoology is no longer, as it still was a few decades ago, a dry systematically arranged description of animals according to particular prominent features of their form, but much more a general animal physiology or biology."[72] Among the disciplines comprising the new zoology, Haeckel included animal chemistry *(Zoochemie)*, comparative histology, comparative physiology, and comparative embryology. After 1840 research in these areas of investigation was conducted in Germany both in institutes attached to the medical faculty and in independent university zoological institutes that retained a relation to a museum of natural history. Among the professors who headed such institutes were such former Haeckelschüler as Oscar Hertwig (Jena, Berlin), Richard Hertwig (Bonn, Munich), Wilhelm Roux (Innsbruck, Halle), and Hans Driesch (Leipzig). Other prominent investigators were Wilhelm Waldeyer (Strassburg, Berlin), August Weismann (Freiburg), Walther Flemming (Kiel), and Theodor Boveri (Würzburg).[73] By 1900 these biologists had provided much knowledge (and much speculation) about the changes in cell morphology in the processes of fertilization and nuclear division, and about embryonic development. In that year Oscar Hertwig claimed that work in his field of microscopic anatomy had shown that "the cell, this elementary keystone of living nature, is far from being a peculiar chemical giant molecule or even a living protein and as such is not likely eventually to fall prey to the field of an advanced chemistry. The cell is itself an organism, constituted of many small units of life."[74] That work was greatly furthered by important advances in microscopy, through the use of immersion objectives and the development of apochromatic lenses by Ernst Abbe, the improvement of microtomes for cutting sections, and the introduction of synthetic dyes as biological stains.[75] As will be evident, the techniques of microscopic anatomy also produced artifacts, much depended on the choice of the organism for study, and there was lively speculation and controversy about the significance of particular observations and experiments. The period before 1900 was also characterized by disputes among biologists who, like Haeckel, embraced Charles Darwin's theory of natural selection and those who rejected it, and by the appearance of variously named units of inheritance.[76] Nor had the centuries-old debate of mechanism versus vitalism subsided. As before, and as it continued during the twentieth century, the debate reflected the limits of the empirical knowledge of the time. Moreover, after 1900 German biologists became increasingly embroiled in controversy over the significance of the Mendelian "factors" in inheritance and evolution, with Theodor Boveri, Carl Correns, Erwin Baur, Richard Goldschmidt, and Fritz von Wettstein as participants.[77]

As was mentioned earlier in this chapter, during the eighteenth century and the first three decades of the nineteenth century, French and German pharmacists were major participants in the development of the interplay of chemistry and biology. After 1830, with the emergence of prominent German centers of research and instruction, not only in chemistry but also in botany and pharmacology, pharmacy lost much of its status as an academic discipline.[78] It was otherwise in France, where schools of pharmacy, though not raised to university status until 1900, had faculty members of considerable distinction as investigators, among them Emile Bourquelot and Marcel Delépine at the Ecole Supérieure de Pharmacie in Paris.[79]

Another difference between the nineteenth-century interplay of chemistry and biology in France and Germany was evident in the institutional development of experimental physiology. François Magendie was the chief animator of the movement to replace anatomical observation by the use of vivisection and chemical methods. It is noteworthy, however, that he gave private lectures and demonstrations in experimental physiology for about eighteen years before he was appointed professor of medicine at the Collège de France. His most famous pupil and successor, Claude Bernard, had many students after he began lecturing in 1853 and was later considered to have been the greatest physiologist of the nineteenth century. He emphasized vivisection, and, although he depended on the chemists Charles-Louis Barreswil and Jules Pelouze for help on chemical methods as the principal tools of experimental physiology, he tended to place less emphasis than his German counterparts on the use of physical instruments. This aspect of his style of research is evident in his 1872 report on the state of the development of general physiology in France.[80] Indeed, the lines of biochemical research opened by Bernard were not pursued as actively in France as in Germany and Great Britain. The later work of his French students Paul Bert and Arsène d'Arsonval was more biophysical than biochemical in nature, while Albert Dastre and Louis Ranvier turned to research in embryology and histology, respectively.[81] Among the foreign guests in Bernard's laboratory were the German Willy Kühne, who continued Bernard's work on pancreatic enzymes, and the Englishman Frederick Pavy, who then worked on diabetes, as well as the neurophysiologists Ivan Sechenov and Silas Weir Mitchell.

One of the most interesting personalities in the interplay of chemistry and physiology in nineteenth-century France was François Vincent Raspail, who chose not to identify himself with an established institution. Because of his political activities, he incurred the displeasure of influential scientists and government officials, but through his chemical and microscopic observations he became one of the pioneers in histology and contributed to the early development of the cell theory. Another early histologist of note was Charles Robin, a friend of Wurtz's.[82]

Between 1860 and 1914 some French biologists rejected both Darwin's theory of natural selection and the Mendelian theory of inheritance.[83] Despite such adherence to the ideas of Jean-Baptiste Lamarck, much outstanding research was done in France and in French Switzerland by the zoologists Laurent Chabry, Lucien Cuénot, Yves Delage, Maurice Caullery, Alfred Giard, and Emile Guyénot, and by the botanists Gaston Bonnier, Léon Maquenne, Alexandre Guillermond, Marin Molliard, and Henri Devaux.[84] Although some of these men (notably Maquenne and Devaux) were excellent chemists, I think it is fair to say that the emergence of biochemistry as an independent discipline in France was principally furthered not by biologists but by chemists like Pasteur and Wurtz, and also by the efforts of lesser-known men such as Antoine Béchamp, Paul Schützenberger, Emile Duclaux, and Gabriel Bertrand, some of whom were in schools of pharmacy.[85] It should be added, however, that the specialty of endocrinology, later to figure largely in the institutional development of biochemistry, had its pioneers in Charles Edouard Brown-Séquard and Eugène Gley.[86] Although, as might have been expected in the country of Lavoisier, Bernard, and Pasteur, what came to be called *chimie biologique* in France had acquired by 1914 greater institutional recognition than that accorded to Biochemie in Germany, the disparity in the quality and scope of the research in that field is undeniable.

In considering the differences in the course of the interplay of biology and chemistry in France, compared with that in Germany, it is necessary, I believe, to take account of the succession of political and military events that affected all aspects of French life after 1815—the revolts of 1830 and 1848, the coup d'état of 1851, the outcome of the Franco-Prussian War (the Paris Commune and the loss of Alsace-Lorraine), the Dreyfus affair, the expensive victory in 1918, and the disaster of 1940—and to wonder at the fact that the institutions for scientific research and instruction that had been established by 1830 had largely maintained their identity and purpose. Although these events influenced and, during the Nazi occupation, interrupted the development of the scientific enterprise, it would seem that the intellectual tradition of the eighteenth-century "philosophes" was too firmly embedded to be destroyed by adverse circumstances. Contrary to the views of some historians and sociologists that during the second half of the nineteenth century there was a "decline" in the prestige of French science, there was, as I have indicated above, continuity in the role of French scientists in the development of the interplay of chemistry and biology.[87]

There can be no doubt, however, that by the 1850s French preeminence in several scientific fields had been seriously challenged. Leading French chemists called attention to the more favorable institutional development of their discipline in the German states and to the greater productivity of German chemists. As Adolphe Wurtz put it in his 1864 report to the newly appointed French minister of education, Victor Duruy, "It is a matter of first-order importance, the future of chemistry in France. This science is French, and God will not permit it to be surpassed." Duruy responded favorably to such appeals, and among his many contributions to the reform of scientific research and higher education was the establishment of the Ecole Pratique des Hautes Etudes to coordinate and promote scientific and scholarly research in Paris. Although the liberal and anticlerical Duruy was removed from office in 1869, largely owing to the opposition to his reforms by the Catholic Church, which administered separate universities in France, many of his initiatives were taken up after 1871.[88] Apart from improvements in material facilities and the creation of new professorships, there still remained some heritages of the past. One of them was the continued dominance of the Parisian scientific establishment. Another was the practice, retained well into the twentieth century, of paying low salaries at individual institutions, thus obliging scientists in Paris, even those at the professorial level, to seek multiple appointments in order to obtain a total income (known as a *cumul*) sufficient for a comfortable life. A third was the retention of the multistep academic ladder (préparateur, répétiteur, licence, agrégation, doctorat) a scientist had to ascend in aspiring to a professorship.[89]

Of special interest for the institutional history of the interplay of biology and chemistry in twentieth-century France is the fact that what came to be called molecular biology arose principally at the Pasteur Institute, with its strong research tradition in microbiology and biochemistry. Mendelian genetics did not play a significant role before the work of Félix d'Hérelle on bacteriophage was followed by that of Eugène Wollman and André Lwoff on lysogeny. Indeed, except for Lucien Cuénot, who welcomed the rediscovery of Gregor Mendel, most leading biologists at French universities adhered, in varying degrees, to some form of Lamarckism. Later Philippe L'Héritier and Georges Teissier (at the Ecole Normale Supérieure) conducted important research on population genetics, and at the Paris Institut de Biologie Physico-Chimique (established by the Rothschild Foundation), Boris Ephrussi, in collaboration with George Beadle, did pioneering work in what was once called biochemical genetics. It was at the Pasteur Institute, however, where Jacques Monod studied the biochemical phenomenon of bacterial "enzyme adaptation," with particular reference to the biosynthesis and regulation of the enzyme beta-galactosidase, that genetics was brought into the mainstream French biological investigation. This development, which raised questions about the relation of the new molecular biology to the academic discipline of biological chemistry, was furthered by the establishment after 1945 of the Centre National de la Recherche Scientifique (CNRS). At that time, the prominent French professors of biochemistry included Claude Fromageot (Lyon, Paris), Pierre Desnuelle (Marseille), and Jean Roche (Collège de France).[90]

In Great Britain the institutional development of the interplay of chemistry and biology during the nineteenth century followed a course that differed in some respects from that in France or the German states. As has been mentioned, several prominent British chemists of the first decades of the century were physicians. William Prout and Alexander Marcet, for example, each of whom had taken his medical degree in Edinburgh, sought to develop methods for the quantitative chemical analysis of urine and other body fluids. But as Prout stated in 1816 (and reiterated in 1831), "It must be confessed, that hitherto physiology has not so much benefited by the application of chemistry as might have been expected. This has been owing in great measure to the cultivators of what is termed Animal Chemistry, having been generally mere chemists. . . . Chemistry, however, in the hands of the physiologist, who knows how to avail himself of its means, will doubtless prove to be one of the most powerful instruments he can possess."[91] During the nineteenth century, before the terms *physiological chemistry* and *chemical physiology* were widely used to denote the use of chemistry in medical practice, *animal chemistry* was, for a time, the preferred designation in Britain. The author of a three-volume work on animal chemistry, published in 1803, wrote in the preface that

> modern chemistry has already thrown great light on several parts of the animal system; it has within these last few years commenced an investigation of several of the functions of the body, and explained the manner in which they are carried on with some degree of success; the processes of respiration, of perspiration, of digestion, of animalization, and the action of oxygen upon vital organization, no longer remain in that state of darkness in which they were so lately enveloped, while the proficiency already attained

> in this department of the science, has established the animal analysis upon so firm and broad a basis, as to promise in future the happiest results.[92]

The title of Berzelius's first book (in Swedish), published in 1806 and 1808, may be translated as *Lectures on Animal Chemistry*, and in 1813 there appeared an English translation of his report on the progress of animal chemistry.[93] The great interest in the English translation (by William Gregory) of Liebig's *Die Organische Chemie in ihrer Anwendung auf Physiologie und Pathologie* under the title *Animal Chemistry or Organic Chemistry in its Application to Physiology and Pathology*, both published in 1842, attests to the continued attachment to the term *animal chemistry*.[94] In 1845, the Sydenham Society published an English translation of the medical chemist Johann Franz Simon's *Handbuch der angwandthen Chemie* under the title *Animal Chemistry with Reference to the Physiology and Pathology of Man*. Although after 1850 the term *physiological chemistry* appeared more frequently in titles of books on this subject, in 1878 the English analytical chemist Charles Kingzett published *Animal Chemistry, or the Relations of Chemistry to Physiology and Pathology*. Less attention was paid to the Austrian chemist Vincenz Kletzinsky's *Compendium der Biochemie* (1858), in which he proposed the separation of biology into three disciplines: biochemistry, biophysics, and biomorphology.[95]

It is, I think, fair to say that during most of the nineteenth century, British physicians who were devotees of "animal chemistry" played a larger role in the development of the interplay of chemistry and biology than did university scientists. Among these chemical physicians were Henry Bence Jones and Ludwig Thudichum (both had studied in Liebig's Giessen laboratory) and Charles Alexander MacMunn. As will be seen, the findings reported by Thudichum and MacMunn were not accepted during their lifetime, in part owing to the disapproval of Felix Hoppe-Seyler.[96]

Although physiology was taught to medical students in Britain before 1870, the kind of physiological animal experimentation practiced by François Magendie was abhorrent to the sensibilities of Victorian society, and vivisection was denounced (as it continued to be for decades) by people who viewed the desperation of human poverty, as made public by Charles Dickens, with Malthusian equanimity tempered by charity. Other reasons have also been offered for the "stagnancy" of British experimental physiology before 1870, when Michael Foster was brought to Cambridge, where he created a remarkable school and, within a generation, had placed Britain in the vanguard of the development of that field. A member of William Sharpey's department of anatomy and physiology at University College London, Foster had gained the attention of Sharpey's friend the biologist Thomas Henry Huxley, who recommended Foster's appointment at Cambridge when that university was making an effort to enhance its scientific stature.[97]

Foster's students included John Langley and Sheridan Lea, both of whom spent some months in Heidelberg with Willy Kühne. After his return to Cambridge, Lea taught "chemical physiology" and wrote *The Chemical Basis of the Animal Body,* which appeared in 1893 as an appendix to the sixth edition of Foster's *Textbook of Physiology.* Lea extended Kühne's studies on the enzymatic digestion of proteins but was obliged to curtail his teaching and research activities owing to progressive spinal disease. Langley, who succeeded Foster in 1903 as head of the laboratory, wrote in an obituary notice for Lea that physiological chemistry "as it was carried out in England in the '70s and '80s did not involve any profound chemical knowledge; it was mainly concerned with the experimental determination of simple but fundamental reactions of the fluids and tissues of the body. Lea's knowledge of chemical theory and methods was more than adequate for the conditions of the time."[98] This statement omits consideration of the work of Frederick William Pavy on carbohydrate metabolism and of Arthur Gamgee on hemoglobin. It also appears that, during the 1870s and 1880s, Kühne was more highly regarded in Britain than Hoppe-Seyler. Foster admired Kühne's work on visual purple ("rhodopsin") and sent Langley and Lea to Heidelberg rather than to Strassburg. Gamgee, who had also worked in Kühne's laboratory, came into conflict with Hoppe-Seyler and was a severe critic of Ludwig Thudichum. For a time (1873–1882), Gamgee was professor of physiology at Owens College, Manchester, but he then turned to the practice of medicine. According to Sharpey-Shafer, "Gamgee may certainly be described as a genius, but erratic. If he had kept to Physiology he might have been the first physiologist of his day, at any rate the first in biochemistry, which is the part of the subject he favored. But he divided his energies between science and clinical medicine, and thus failed to achieve that high distinction in either, to which his intellectual powers seemed to entitle him."[99]

In 1898 Foster brought to his department Frederick Gowland Hopkins, a chemist and M.D. at the Guy's Hospital Medical School, London. Hopkins had been bred in the tradition of analytical "animal chemistry," had worked on a pigment in butterfly wings, and had developed an improved method for the determination of urinary uric acid. There will be frequent occasion in later pages to describe his personal researches, but it should be recalled here that Hopkins was not appointed professor of biochemistry at Cambridge until 1914, when he was fifty-three years old, and that his department was not given adequate facilities for teaching and research until 1924.[100] In 1926 he described his view of the place of biochemistry in the university:

> It does not seem to me enough that the biochemist, hitherto housed in an Institute of Animal Physiology, and therefore mainly occupied with animal studies, should migrate to a separate building and retain his preoccupations. The greatest need of biochemistry at the moment in my opinion is equipment which shall make possible the study under one roof (of course, from its own special standpoint alone) of all living material. No full understanding of the dynamics of life as a whole, no broad and adequate views of metabolism, can be obtained save by studying with equal concentration the green plant and microorganisms as well as the animal. . . . Such Institutes of General Biochemistry would have to be equipped for teaching as well as research, for we have to prepare a future generation to bear burdens greater than our own. Modern physical chemistry, and modern organic chemistry, reacting as they already are—even the latter—to the newer concepts of Physics, are to provide entirely new concepts for biochemistry. There must be experts to understand and apply them; a task which will be impossible for those who must continue to concern themselves with the rapid growth on other sides of physiology and biology, as it will be for the equally preoccupied pure chemist. I do not undervalue the difficulties of equipping and staffing an Institute of general biochemistry such as I am picturing, but I must not stop to discuss them. I will only state that we have, in not too ambitious fashion, attempted the task in Cambridge.[101]

This statement reflects a breadth of scientific vision Hopkins appears to have acquired not only from his association with Michael Foster and with other members of Foster's group—William Bate Hardy, Walter Fletcher, and Joseph Barcroft—but also from his acquaintance in London with William Sharpey's former student William Halliburton and with Archibald Edward Garrod, Arthur Harden, Samuel Schryver, and R. H. A. Plimmer.[102]

Between 1924, when the new Dunn laboratory was opened, and 1943, when Hopkins retired from his professorship, several hundred people had worked there. The staff included Marjory Stephenson and Ernest Gale (microbial metabolism), J. B. S. Haldane (enzymes, genetics), Muriel Wheldale and Robert Hill (plant biochemistry), Dorothy Needham (muscle biochemistry), Joseph Needham (chemical embryology), Ernest Baldwin (comparative biochemistry), Malcolm Dixon (biological oxidation), and N. W. Pirie (protein chemistry). Among the many guest investigators from abroad were Albert Szent-Györgyi, Luis Leloir, and David Green. Hopkins also extended a welcome to émigrés from Nazi Germany, notably Hans Krebs, Ernst Friedmann, Ernst Boris Chain, Rudolf Lemberg, Hans Weil-Malherbe,

Stephen Bach, and Hermann Lehmann. These names will appear again in later pages of this book.

The distinction of Cambridge as a world center of biochemical research was further enhanced by the contributions of the nonmedical biologist David Keilin at the nearby Molteno Institute of Parasitology, opened in 1921 and directed by G. H. F. Nuttall. A distinguished parasitologist in his own right, Keilin had joined Nuttall in 1915 and was named his successor in 1931. A scientist of at least equally broad vision as Hopkins, Keilin became one of the leading biochemists of the twentieth century through his studies, beginning in the 1920s, on intracellular respiration.[103] Between 1918 and 1939, outstanding biochemical research was also conducted at Cambridge in the Department of Physiology (Joseph Barcroft, Gilbert Adair, Francis Roughton), the Low Temperature Station for Biochemistry and Biophysics (William Hardy), and the Dunn Nutritional Laboratory (Leslie Harris). Moreover, during the period 1870–1910, there had been a parallel effort by Sydney Vines and Joseph Reynolds Green to develop at Cambridge a botany school in which plant physiology would replace taxonomy as the principal area of study. Work was undertaken on plant enzymes, in large part stimulated by the appearance in 1875 of Charles Darwin's *Insectivorous Plants.* Later members of the Cambridge botany school included the plant physiologist Frederick Blackman, the biochemist Charles Hanes, and the geneticist David Catcheside.

Hopkins was succeeded in 1941 by the protein chemist Charles Chibnall, who resigned in 1949, and was followed by the endocrinologist Frank Young. Two important new centers of biochemical activity emerged in Cambridge during the 1940s. The first appeared in 1943 with the appointment of Alexander Todd as the successor of William Pope in the Department of Chemistry, where Todd's research group made decisive contributions to the chemistry of nucleic acids. The second came in 1947, with the creation of the MRC Unit for the Study of the Molecular Structure of Biological Systems, headed by Max Perutz; nine years later its name was changed to the MRC Laboratory for Molecular Biology. As we have seen, that laboratory was an offshoot of the one headed until 1937 at the Cavendish Laboratory for physics in Cambridge by the X-ray crystallographer John Desmond Bernal.[104]

In 1949 Cambridge was the site of the first International Congress of Biochemistry, and at that meeting steps were taken to establish an International Union of Biochemistry (IUB), independent of those for physiology or pure and applied chemistry. That effort was opposed by some biochemists attached to the latter body, which had set up a section of biological chemistry, but after extended discussion, IUB was formally admitted to the International Council of Scientific Unions in 1955.

I attempted earlier to summarize the institutional development of organic chemistry in Great Britain. I interpolate here a brief account of the emergence of physical chemistry as an academic discipline in British universities, because of its relationship to colloid chemistry, which attracted the attention of many British biologists, notably the physiologist William Bayliss. He made several important biochemical contributions, and his *Principles of General Physiology* (fourth edition, 1924) was an admired textbook. During the first two decades of the twentieth century, the Ostwald school of physical chemistry was represented most prominently by Frederick Donnan, who, after his association with Ostwald and van't Hoff, was professor of this subject in Liverpool (1904–1913) and University College London (1913–1937). Donnan's 1911 paper on what came to be called the Donnan equilibrium was taken by some general physiologists to explain the ionic equilibria within cells and between cells and their exterior. Also, fundamental colloid chemistry was developed at Cambridge by Eric Rideal and by Neil Adam at University College London.[105] It should be recalled, however, that in addition to the areas of physical chemistry of special interest to many biologists and biochemists of the 1920s, important contributions—whose relevance to the study of biological problems became evident only later—were then being made in Great Britain in such fields as X-ray crystallography, reaction kinetics, photochemistry, and physical-organic chemistry.[106]

I return to the development, after 1900, of the institutional status of biochemistry in Great Britain. For historians of science who have preferred to emphasize "discipline building" in the study of the emergence of research disciplines, the career of the incumbent of the first British professorial chair of biochemistry, founded at Liverpool in 1902, may perhaps serve as a reminder that more than entrepreneurship is involved in the development of an academic discipline. Apparently, Hopkins had been offered that post, but the Cambridge authorities were persuaded by Langley to keep him there and, at the instigation of Charles Sherrington, the professor of physiology at Liverpool, the appointment went to Benjamin Moore.[107] A man of modest scientific attainment, with an inclination toward idiosyncratic physico-chemical speculation, Moore was a busy entrepreneur. The only lasting mark he left on British biochemistry was to found in 1906, with the help of his wealthy friend Edward Whitley, the *Biochemical Journal,* whose primary purpose was to serve as an outlet for papers from his department. In 1912, after the formation of the Biochemical Club (shortly afterward changed to Biochemical Society), and prolonged bargaining, Moore sold the journal to the society for 150 pounds. When Sherrington moved to Oxford in 1913, he arranged with Whitley to finance the establishment of a chair of biochemistry there for Moore. It was not until 1923 (Moore died in 1922), with the appointment of Rudolph Peters as Moore's successor, and the subsequent tenure of the Whitley professorship by Hans Krebs (1954–1967), that the quality of biochemical research and instruction at Oxford began to match that in Cambridge. Moreover, the Dyson Perrins laboratory of organic chemistry, with Robert Robinson and Ewart Jones as successive professors, took on a more biochemical character, and research was conducted on peptide and enzyme chemistry. At the department of pharmacology, Harold Burn and Hermann Blaschko did important biochemical work, and, in her laboratory for chemical crystallography, Dorothy Hodgkin made memorable contributions to the study of the structure of vitamin B_{12} and insulin. Under Moore's successors in Liverpool, Walter Ramsden (1914–1931), Harold John Channon (1931–1943), and especially Richard Alan Morton (1944–1966), matters also improved greatly, in particular through collaboration with Ian Heilbron, who was professor of organic chemistry there from 1920 until 1933.[108]

In London, a great variety of biochemical research was conducted before 1940 in diverse institutions—King's College, University College, Imperial College, Birkbeck College, National Institute for Medical Research, Lister Institute, and the several hospital medical colleges. At King's, the professor of physiology (1889–1923) was William Halliburton, of whom Hopkins wrote: "He was the first in this country by his works and his writing to secure for biochemistry general recognition and respect." In 1904 the outstanding organic chemist Otto Rosenheim joined the faculty at King's, and later he became reader in biochemistry there. At University College, where Ernest Starling (professor of physiology, 1899–1922) and William Bayliss (professor of general physiology, 1912–1924) collaborated in research, the protein chemist R. H. A. Plimmer (1912–1919) was followed by Jack Drummond (professor of biochemistry, 1922–1945) and Ernest Baldwin (1950–1969). Other biochemists who were professors at University College were J. B. S. Haldane (genetics, 1933–1957) and Claude Rimington (chemical pathology, 1945–1968).[109]

At Leeds, where an independent Department of Biochemistry was established only in 1945, with Frank Happold as the professor, the most significant figure in the interplay of chemistry and biology during the 1930s was the X-ray crystallographer William Astbury, who came to the textile department of the university in 1928 and was made professor of biomolecular structure in 1945. Moreover, many years earlier, the professorship of chemistry at Leeds was held (1904–1924) by Julius Berend Cohen, who had received his D.phil. at Munich, for work with Pechmann in Baeyer's institute. An outstanding teacher, with a keen interest in biochemistry, Cohen inspired a number of his students—notably Henry Dakin, Stanley Raper, Arthur Wormall, Percival Hartley, and Harold Raistrick—to enter that field of research.[110]

The development of biochemical research at Manchester is of equal interest for the study of the institutional features of the interplay of biology and chemistry in Great Britain. There, with its succession of distinguished organic chemists, and with Archibald Vivian Hill professor of physiology for a time (1920–1923), a second chair in that discipline was established for the biochemist Raper, who, like Kossel in Heidelberg, was content to develop his research program without changing his departmental affiliation, even after Hill left for University College London in 1923. The same may be said about the biochemist John Beresford Leathes, who remained as professor of physiology at Sheffield from 1915 until 1933. The rise of Sheffield to biochemical prominence came after 1935, when Hans Krebs was appointed lecturer in pharmacology and headed the MRC Unit for Cell Metabolism there. He was made professor of biochemistry in 1945, was passed over in 1949 as a possible successor of Chibnall at Cambridge, and was named Whitley professor at Oxford in 1954 after he had been awarded a Nobel Prize the year before.[111]

Other variants in the institutional development of the interplay of chemistry and biology in Great Britain are presented by the examples of Edinburgh and Glasgow. At Edinburgh the professor of chemistry (1908–1928) was the physical chemist James Walker, who had received his D.phil. for work in Wilhelm Ostwald's institute in Leipzig. In 1919 Walker brought to his department the outstanding biochemist George Barger to be professor of "chemistry in relation to medicine." The professor of physiology (1899–1933), Edward Sharpey-Shafer (originally Shäfer), considered this action to be an invasion of his academic domain and insisted on continuing to have a course on "chemical physiology" taught in his department. In this collegial contest, the chemist prevailed over the physiologist, largely owing to Barger's superior talents as an investigator and teacher. In 1937 Barger moved to Glasgow, where he became Regius professor of chemistry.[112] Glasgow's biochemical reputation was further enhanced in 1948, when James Norman Davidson became professor of biochemistry there. During the preceding decades, when Noel Paton was Regius professor of physiology (1906–1928) and was succeeded by Edward Cathcart (1928–1947), their adherence to some of the vitalistic aspects of "chemical physiology" led them to dispute the vitamin concept, as espoused by Gowland Hopkins, with special reference to the cause of rickets.[113]

I conclude this sketch of the institutional development of the interplay of chemistry and biology in Great Britain with some remarks about the role of industrial enterprises. In the case of the pharmaceutical industry, the establishment in 1894 of the Wellcome Physiological Research Laboratories, initially to produce antitoxins, involved the efforts of numerous outstanding chemists and biologists.

Among them were Henry Dale (director, 1904–1914), John Mellanby (1902–1905), George Barger (1903–1908), Patrick Laidlaw (1909–1913), Harold King (1912–1919), and Percival Hartley (1919–1921).[114] Associated with the fermentation industry were several pioneers in the study of enzymes—Cornelius O'Sullivan, Horace Brown, and Adrian Brown. During World War I the chemist Chaim Weizmann developed his famous microbiological method for the manufacture of acetone, and during World War II such methods proved to be decisive in the manufacture of penicillin.[115]

The International Scene

Before proceeding to a discussion of the institutional aspects of the interplay of chemistry and biology in the United States, some mention, however brief, must be made of the developments in nations other than France, Germany, and Great Britain where significant work was done in that field. Indeed, in many cases, the results of that work provided stimuli to important investigations in these more prominent countries. I consider first some of the smaller European nations, with relatively few universities.

In the French-speaking part of Belgium, Liège was outstanding in the contributions of a series of noted biologists—the physiologists Theodor Schwann and Léon Fredericq, the embryologists Edouard van Beneden and Albert Brachet, the microbiologist André Gratia, the pharmacologist Zenon Marcel Bacq, and the biochemist Marcel Florkin. At Brussels the botanist Leo Errera and the biochemist Jean Brachet made important contributions to biochemical knowledge.[116] During his stay at Ghent (1858–1867), Kekulé left a lasting mark on the development of organic chemistry in Belgium. One of his students was Théodore Swarts, who succeeded him at Ghent. Another was Walthère Spring, who studied with Kekulé and became professor of chemistry at Liège.[117]

In the Netherlands, the principal center of chemical research during 1870–1940 was Amsterdam, where J. H. van't Hoff was professor from 1877 until 1896. Among his many students was Ernst Cohen, later professor of chemistry (1902–1939) in Utrecht and a noted historian of chemistry; a Polish Jew, Cohen was murdered by the Nazis in Auschwitz. Other prominent figures in the emergence of physical chemistry as the dominant chemical specialty in Holland were Johannes Diderik van der Waals, professor of physics in Amsterdam, the crystallographer Johannes Martin Bijvoet, and the colloid chemist Hugo Kruyt at Utrecht. Organic chemistry was represented by Arnold Holleman (Groningen, Amsterdam), Frans Maurits Jaeger (Groningen), and Fritz Kögl in Utrecht. Kögl's reputation was tarnished by disproof of some of his scientific claims and by his reputed collaboration with the German occupation forces during World War II.[118] After van't Hoff, the most famous Dutch physical chemist was Peter Debye, who was professor at Utrecht for only one year; his subsequent university posts were at Zurich, Göttingen, Leipzig, and Berlin. He left Germany in 1939 when the Nazis demanded that he give up his Dutch citizenship, and he joined the faculty at Cornell.[119] Among the noted biologists were Hartog Jacob Hamburger, professor of physiology at Groningen (1901–1924), who was a leading advocate of the application of the new physical chemistry to biological problems, and Cornelius Adrianus Pekelharing, professor of physiological chemistry at Utrecht (1888–1918), who made valuable contributions to the study of gastric digestion and of human nutrition. Other noted physiologists were Willem Einthoven at Leiden and Christiaan Eijkman at Utrecht. Of particular distinction were the botanist Hugo de Vries in Amsterdam and the microbiologists Martinus Willem Beijerinck and Albert Jan Kluyver in Delft.[120]

In Denmark, where scientific research was largely concentrated in Copenhagen, there was, at the university and the technical colleges, a sequence of distinguished physical chemists—Julius Thomsen, Johannes Nikolaus Brønsted, Niels Bjerrum, the inorganic chemist Sophus Mads Jørgensen, and the organic chemists Einar Biilmann and Stig Veibel. The biologists included the noted physiologists Christian Bohr, August Krogh, and Einar Lundsgaard, as well as the geneticist Wilhelm Johannsen.[121] Although biochemistry did not shine as an academic discipline, outstanding work in that research discipline flowered, as mentioned, in the Carlsberg laboratories.

In Sweden, Olof Hammarsten, the professor of medical and physiological chemistry at Uppsala from 1883 until 1906, had a sizable research group. He was an excellent teacher, and the several editions of his *Lehrbuch der physiologischen Chemie* were ranked among the best textbooks of the time on that subject. Among his students was Thorsten Thunberg, who became professor of physiology (1905–1938) at Lund, where he made notable contributions to the study of biological oxidation. Another leading figure was Hans von Euler (also known as Euler-Chelpin), a Bavarian who came to Sweden in 1897 to work with Svante Arrhenius, and in 1906 was made professor of chemistry in Stockholm. Although he had become a Swedish citizen in 1902, Euler joined the German armed forces during World War I. He became best known for his work on the chemical nature of "cozymase." One of the important early investigators of the chemical structure of nucleic acids was Einar Hammarsten, professor of physiological chemistry at Stockholm (1928–1957); his successor and former student was Erik Jorpes, an outstanding biochemist and historian of chemistry. The distinction of biochemical research in Sweden was further enhanced by the work of Hugo Theorell and Torbjörn Caspersson at the Nobel Medical Institute in Stockholm, and by that of Sune Bergström, at Lund and Stockholm. More luster to the fame of Swedish biochemistry was added by the experimental achievements of The Svedberg, professor of physical chemistry at Uppsala (1912–1949), and his colleague Arne Tiselius. As will be related, around 1910 Arrhenius made an unsuccessful foray into immunochemistry. Among the many noted Swedish biologists who contributed to the interplay of biology and chemistry were the physiologists Magnus Blix and Ragnar Granit and the pharmacologist Goeran Liljestrand.[122]

No sketch, however brief, of the institutional features of science in Sweden would be complete without mention of the foundation established under the will of Alfred Nobel, the inventor of dynamite. Apart from the support given within Sweden for the promotion of research in the chemical and biological sciences, the role of Swedish chemists, biologists, and physicians who since 1901 (except during the two world wars) have chosen the recipients of the Nobel awards in chemistry and in physiology or medicine has conferred on the electors a responsibility that has grown to worldwide institutional significance. That nearly all of the selections have been hailed within the knowledgeable scientific community attests to the care with which such selections were made, although the omission of some worthy candidates has been widely deplored. The brief public visibility of Nobel prize winners has had many consequences, not the least of which has been the enhancement of their institutional status.

Another fruitful national breeding ground for the interplay of biology and chemistry was Czarist Russia, where the stimulus came largely from the experiences of Russian physicians, biologists, and chemists in German and French laboratories. The physician Sergei Botkin, for example, who became in 1861 a professor at the Medical-Surgical Academy in St. Petersburg, had worked with Hoppe in Virchow's institute and had also studied with Claude Bernard. Botkin advocated the use of chemical methods and vivisection in experimental medicine. One of Botkin's students was Ivan Pavlov, famous for the work of his group on the physiology of digestion and brain function, who was professor of physiology in St. Petersburg (Leningrad) from 1896 until 1924. Pavlov had also worked with Ivan Sechenov, a devoted disciple of Carl Ludwig's and a teacher of other leading Russian scientists, such as the pharmacologist Nikolai Kravkov.[123]

Another physician, Aleksandr Danilevski, who had worked in Kühne's laboratory in Berlin, became professor of animal physiology at Kharkov (1885–1892) and of physiological chemistry at St. Petersburg (1892–1923). An especially prominent biochemist of prerevolutionary Russia was Aleksei Bakh (often written as Bach).

In 1884 he fled to Paris, where he worked in the laboratory of Paul Schützenberger, and then stayed in Geneva (1894–1917), where he did important biochemical research in collaboration with the noted Swiss botanist Robert Chodat. Also in Geneva was the Russian émigrée Lina Stern, who worked with the physiologist Federico Battelli on oxidative enzymes and in 1918 was appointed there as professor of physiological chemistry. In 1925 she returned to Moscow to become a professor of physiology.[124]

Among the approximately thirteen Russians who studied in Liebig's laboratory at Giessen was Aleksei Khodnev, later professor of chemistry at Kharkov (1848–1883), who was one of the first Russian chemists to adopt the theories of Laurent and Gerhardt and who published articles on many subjects, including physiological chemistry. Another was Nikolai Zinin, a member of the famous "Kazan school" of organic chemistry and later professor of chemistry at the Medical-Surgical Academy in St. Petersburg. One of his students there was Aleksandr Borodin, best known as a composer, but also a physician and a productive chemist. Still another former *Liebigschüler* was Carl Schmidt, the noted professor of chemistry (1852–1892) at the German-speaking Russian university in Dorpat (known in Czarist Russia as Yuriev, and later named Tartu). There he collaborated with Friedrich Bidder, the professor of physiology, in important studies on protein metabolism. Among Schmidt's students were not only such future famous chemists as Wilhelm Ostwald and Gustav Tammann but also the physiological chemist Gustav von Bunge, later professor of physiological chemistry (1886–1920) at Basle. It was at Dorpat that the young pediatrician Nikolai Lunin, one of Bunge's students, found evidence of the nutritional requirement for what were later called "accessory factors" in the diet of animals.[125]

In his successive laboratories in Berlin, Tübingen, and Strassburg, Hoppe-Seyler had approximately forty Russian associates. In addition to Botkin, his guests included Aleksandr Schmidt, later professor (1869–1894) at Dorpat, who gained fame for his studies on blood coagulation. Others were Grigori Sakharin, professor of medicine (1862–1896) in Moscow, Aleksandr Buliginsky and Vladimir Gulevich, successively professors of medical chemistry in Moscow, and Viktor Pashutin, professor of pathology (1875–1901) in St. Petersburg.[126]

In the postrevolutionary institutional development of biochemistry in the Soviet Union, the leading figures—Aleksandr Braunstein, Aleksandr Oparin, Vladimir Engelhardt, Andrei Belozerski, Sergei Severin, and Mikhail Shemyakin—taught and directed research in the university and in newly established special institutes in Moscow. Except during the disastrous war years (1941–1945), important biochemical contributions came from their laboratories that greatly influenced research in other countries.[127]

The groundwork for the development of plant physiology and biochemistry in Czarist Russia was laid largely by Andrei Famintzin (professor 1872–1918 in St. Petersburg), by Sergei Navashin (professor 1894–1915 in Kiev), and especially by Kliment Timiryazev in Moscow. Before becoming professor there (1877–1911), Timiryazev had studied with Bunsen, Bernard, Boussingault, and Wilhelm Hofmeister. He contributed much to the study of plant nutrition and was an adherent of Darwin's theory of natural selection. Among Timiryazev's students was Dmitri Prianishnikov, later professor at several agricultural academies in Moscow (1895–1948), and Vladimir Palladin, who, as professor in Kharkov, Warsaw, and St. Petersburg (1889–1921), greatly promoted research in plant biochemistry. During the first decades of the twentieth century there also emerged in Russia a remarkable group of plant geneticists, some of whom had begun as students of the above botanists. Among them were Sergei Chetverikov, Yuri Filipchenko, Nikolai Vavilov, and Boris Astaurov. The persecution of the Mendelian geneticists, and especially the martyrdom of Vavilov, owing to Stalin's favor toward Trofim Lysenko, can be explained but not forgiven.[128]

As in other countries, at the turn of the century the interplay of biology and chemistry in Russia was evident in the response to what has been called the bacteriological revolution. Here appear the names of Ilya (or Elie) Mechnikov, who, after some unhappy years in Odessa, joined the staff of the Pasteur Institute, as did Sergei Vinogradsky (often written as Winogradsky), one of the pioneers in the emergence of general microbiology. There was also Dmitri Ivanovski, professor at Warsaw (1901–1915) and Rostov-on-Don (1915–1920), who discovered filterable pathogenic agents.[129]

In neighboring Finland, Edvard Hjelt, an outstanding organic chemist and historian of chemistry who had studied with Baeyer, was professor at Helsinki (1882–1907) and was succeeded by Ossian Aschan (1908–1927). Among the biochemists were Henning Karström and Artturi Virtanen. Before Poland regained its independence in 1918, Leon Marchlewski became professor of medical chemistry in Cracow after he had worked in England with Edward Schunck, and Michael Tsvet (often spelled Tswett), a student of Robert Chodat, was professor of botany in Warsaw. Both men worked on the chemistry of chlorophyll, and Tsvet's name later became widely known as one of the early users of column chromatography to separate plant pigments. Between the two world wars, the leading Polish research center in biochemistry was that of Jacob Parnas in Lwow (1921–1941). He had received his Ph.D. in organic chemistry with Richard Willstätter and had been an associate of Franz Hofmeister; his research associates included Thaddeus Mann, Pawel Ostern, and Tadeusz Baranowski.[130]

In post–World War II Czechoslovakia, Francisek Sorm headed the Institute for Organic Chemistry and Biochemistry in Prague, which became an important center of research on peptides, especially through the work of his colleague Josef Rudinger. In Hungary the professor of organic chemistry (1913–1947) at Budapest was Géza Zemplén, who had worked with Emil Fischer and who made significant contributions in the field of carbohydrate chemistry. He was succeeded by Victor Bruckner. The best-known Hungarian biochemical laboratory was the one in Szeged, led by the remarkable Albert Szent-Györgyi and his principal associate Ferenc Bruno Straub, who did pioneering work on the chemistry of muscle proteins. In Greece, Leonidas Zervas, as professor of organic chemistry in Athens, made his productive laboratory a center for the training of peptide chemists who made their mark elsewhere in Europe and in the United States. In Israel a notable center of biochemical research was developed at the Weizmann Institute of Science, founded in Rehovoth in 1949; its staff included Ephraim Katchalski-Katzir and Michael Sela.[131]

In Italy, where much of the groundwork of modern chemistry had been laid by the physicists Alessandro Volta (professor at Pavia, 1778–1819) and Amedeo Avogadro (professor at Turin, 1834–1850), and by chemists such as Stanislao Cannizzaro (professor at Rome, 1871–1909) and Giacomo Ciamician (professor at Bologna, 1889–1922), the institutional development of the interplay of chemistry and biology did not lead to the establishment of centers of biochemical research comparable to those in Germany, Great Britain, or France. I can only offer the surmise that the attitude of leading physiologists and pathologists toward biochemistry may have been a factor. Among this group were Jacob Moleschott (professor at Turin 1861–1879), Angelo Mosso (professor at Rome, 1879–1910), Filippo Bottazzi (professor at Naples, 1905–1941?), and Pietro Rondoni (professor at Milan, 1924–1952). It should be added, however, that the Zoological Station in Naples served for many years as an important site for the biochemical research of visitors from abroad.[132]

In other parts of the earth, the institutional development of the interplay of chemistry and biology was influenced by political and social circumstances, and by the development of scientific thought and practice in Germany, Great Britain, and France. Thus the ties were close in nations that had been part of the far-flung British Empire (or Commonwealth), despite their geographical distance from London, Cambridge, or Oxford. The connection was most notable in the cases of Australia and Canada[133] but also evident in New Zealand, India, and South Africa. On the other hand, in the universities of some Latin American countries, the influence of French physiology and pharmacy appears to have been greater. The outstanding example is the University of Buenos Aires, where, despite the brutal vagaries of Argentine politics, Bernardo Houssay established a distinguished research school of physiology and biochemistry, which included men such as Luis Leloir and Alberto Sols. In Chile, the German influence may have been more prominent, as is suggested by the appointment of Adeodato Garcia-Valenzuela, a pupil of Hoppe-Seyler's, as the first professor of physiological chemistry at Santiago.[134]

China was once a major source of technical knowledge that was transmitted to the West by diverse agents, among them, during the thirteenth century, Mongol invaders and Venetian commercial travelers. In his multivolume treatise, Joseph Needham called attention to the achievements of Chinese craftsmen and raised the question why a "scientific revolution" comparable to the one historians have associated with Galileo did not occur in China.[135] It is beyond the scope of this chapter to enter that arena of historical discourse, and I only note that in 1917 the Rockefeller Foundation established the Peking Union Medical College (PUMC), whose faculty was initially largely composed of scientists and physicians from the United States. In 1920 Hsien Wu, a Chinese-born biochemical Ph.D. at Harvard, was added to the PUMC staff, and he later became the head of Department of Biochemistry. He gave up this post in 1942, when PUMC was taken over by the Japanese occupiers of Peking. Another member of the PUMC staff was the pharmacologist Ko Kuei Chen, who later became director of pharmacological research (1939–1963) at the Eli Lilly Company. The subsequent development of science in China was affected adversely not only by the brutal rule of the Japanese invaders but also by the civil war in which the Kuomintang was defeated by the Communist forces and by the later "cultural revolution" (1966–1976). Only recently has the fruitful interplay of chemistry and biology resumed in China. Among the institutions is a revived Peiping (now Beijing) Medical College, where Tsou Chen-Lu (Zhou Cheng-Lu) has conducted brilliant biochemical research.[136]

In the development of the interplay of chemistry and biology in Japan after the so-called Meiji restoration (1868) and the foundation of new national universities in Tokyo (1877), Kyoto (1898), Kyushu (1911), and Osaka (1931), several distinctive features may be discerned. In the early years scientists were imported from abroad (mostly from Germany and Great Britain); for example, the plant biochemist Oscar Loew taught at the Tokyo agricultural school (1893–1897, 1900–1907), and after World War I, Leonor Michaelis was at the medical school in Nagoya (1922–1925). Another feature was a steady flow of Japanese students to German and British laboratories before 1914; among the vanguard were Joji Sakurai, who after his studies in London became the first professor of physical chemistry in Tokyo; Muneo Kamagawa who, after his work with Salkowski in Berlin, became professor of physiological chemistry at Tokyo; and Torasaburo Araki, who became professor of medical chemistry at Kyoto after his stay with Hoppe-Seyler in Strassburg (1889–1895, M.D. 1891). By 1889 Japan had two independent departments of physiological chemistry, which, under the direction of Kamagawa and Araki, produced many of the leading investigators in this field. Among Emil Fischer's Japanese students during 1900–1910 were Umetaro Suzuki (who later discovered what came to be called vitamin B_1), Koichi Matsubara, and Tokuhei Kametaka. All three went on to become professors of organic, agricultural, or pharmaceutical chemistry at Tokyo, where Sakurai's successor as professor of physical chemistry was Kikunae Okeda, a former student of Wilhelm Ostwald (1899–1901).

The flow to Germany and England was interrupted by World War I but increased markedly during the 1920s and 1930s. It is noteworthy that it was not until about 1950 that laboratories in the United States became the preferred choice of Japanese postdoctoral students in the biochemical sciences. What I find to be most striking is that from 1880 until the recent past, most of these people returned to Japan and were given the opportunity to develop their talents as researchers or teachers in universities, research institutes, and industrial establishments. There were exceptions, most notably in the case of Jokichi Takamine, who, through commercial enterprise, gained considerable wealth in the United States. The financial success of Takamine also offers a glimpse of another feature of the development of the interplay of chemistry and biology in Japan, namely the importance in Japanese society of scientific problems related to the chemical constituents of ancient plant medicines, to the fermentation industry, and to other chemical industries, such as the manufacture of natural lacquers, or of ingredients of the Japanese diet. Examples are the isolation, by Nagayoshi Nagai, of ephedrine from a plant long used in China as a cough remedy, of squalene from shark liver oil by Mitsumaru Tsujimoto, or of tetrodotoxin from the puffer fish by Yoshisumi Tahara. The use of natural lacquers led to identification of an oxidative enzyme ("laccase") by Hikorokuro Yoshida; the brewing of sake (fermented rice or wheat) encouraged the study of microbial enzymes, especially those of *Aspergillus oryzae,* such as the "Taka-diastase" of Takamine; and Ikeda identified an age-old plant flavoring agent as monosodium glutamate ("Ajinomoto"). After World War II there emerged outstanding centers of research on protein chemistry in Osaka, led by Shiro Akabori, and on antibiotics, led by Hamao Umezawa in Tokyo.[137]

In addition to university laboratories, there have been important independent research institutes in the Tokyo region, such as the one created in 1914 by the famous bacteriologist Shibasaburo Kitasato, and new ones, among them the one supported by the giant Mitsubishi chemical company, which until recently was headed by the biochemist Fujio Egami. After about 1980, when recombinant DNA techniques became a prominent feature of biotechnology, the availability of many well-trained biochemists and microbiologists enabled the Japanese government to launch a policy of supporting research on genetic engineering of value to its burgeoning chemical industry.[138]

These sketches of the institutional features of the development of the interplay of chemistry and biology in nations other than Germany, France, and Great Britain, and of some of the people who were involved, are intended to remind the reader that these three countries, along with the United States, were not the only ones where significant and original contributions were made to the advancement of chemical and biological thought and practice. Much depended in other nations, of course, on the opportunities provided by the institutional organization of scientific education and research, but of comparable importance was the appearance of individuals with theoretical insight or technical skill, or both, who were successful as scientific investigators.

The United States and Canada

During the last decades of the eighteenth century, the leading American professors of chemistry were the physicians Benjamin Rush, at the College of Philadelphia (later the medical school of the University of Pennsylvania), and Samuel Mitchill, at King's College in New York (later Columbia University). Both men had received their M.D. degrees at Edinburgh and had been students of Joseph Black. Before returning to America in 1769, Rush also inspected British chemical factories and visited Pierre Joseph Macquer and other French chemists. After 1789, when Rush took over the professorship of medicine in Philadelphia, chemistry was taught there by a succession of his students, among them James Woodhouse and Woodhouse's student Robert Hare. The early years of the new American republic also marked the beginning of the emigration of scientists to the United States as a consequence of political upheavals in Europe. During the 1790s, the anti-Jacobin reaction in England brought the chemist Joseph Priestley and his friend Thomas Cooper. Despite Priestley's fame as a chemist, Mitchill and Woodhouse, as well as Cooper, became adherents of the antiphlogistic chemistry opposed by Priestley. Another émigré was John Maclean (M.D. Glasgow), who became professor of chemistry at the College of New Jersey (later Princeton University).[139] The ties of chemistry not only to medicine but also to other specialties are evident in the professional titles of the time. Mitchill, who founded a journal, *The Medical Repository*, was named professor of chemistry, natural history, and agriculture at King's College, while Cooper became professor of chemistry, geology, and mineralogy at the University of South Carolina.

After about 1820 Yale emerged as a leading center of instruction in chemistry, geology, and mineralogy, largely owing to the enterprise of Benjamin Silliman. Although he was an effective lecturer, his chemical contributions were modest, and as a geologist Silliman pleased his audiences by seeking to reconcile Genesis and geology. He also served as professor of chemistry and pharmacy at the newly established (1813) medical school, and in 1846 he persuaded the Yale administration to create professorships in applied chemistry and agricultural chemistry. The first went to his son Benjamin and the other to his student John Pitkin Norton. During the 1840s Liebig's writings on agriculture and animal physiology attracted much interest in the United States, but it would seem that the elder Silliman preferred the 1841 book on agricultural chemistry by James Johnston at Edinburgh and advised Norton to go there. Norton's own laboratory work at Yale was limited to soil analysis; he did no chemical research.[140] After Norton's untimely death he was succeeded by John Porter, a Liebig pupil, whose performance in teaching and research was undistinguished. But he persuaded his father-in-law, the railroad tycoon Joseph Sheffield, to support science at Yale, and by 1861 the funds had become large enough to establish the Sheffield Scientific School, with Porter as its first dean. It was appointed the land-grant school in Connecticut under the Morrill Land Act, and after 1874, when the state legislature established the first agricultural experiment station in the United States, that station became one of the leading American centers for the interplay of chemistry and biology. Its director (1877–1900) was Samuel William Johnson, professor of analytical chemistry at Yale, who had studied in Munich with Liebig and Pettenkofer. In 1882 the station was moved into its own quarters in New Haven, but its ties to Yale remained unbroken. Johnson brought to the staff of the station Wilbur Atwater, who made important contributions to the development of the Pettenkofer-Voit approach to the problems of animal nutrition, and his son-in-law Thomas Osborne, one of the pioneers in the study of the chemical composition of proteins.[141]

In addition to Porter, the American students in Liebig's Giessen laboratory included Eben Horsford and Wolcott Gibbs. Horsford, who became interested in problems of nutrition, was for a time (1847–1863) professor of chemistry and applied science at Harvard and was succeeded there by Wolcott Gibbs (1863–1877), who had previously taught (1848–1863) at the City College of New York. Gibbs's most significant scientific contribution was made in collaboration with Frederick Genth, who had worked with Bunsen and had emigrated to the United States at the time of the 1848 revolution in Germany. Genth set up a private analytical laboratory in Philadelphia and during 1872–1888 was professor of chemistry and mineralogy at the University of Pennsylvania. Apart from his important work on cobaltammines with Gibbs, most of Genth's numerous publications dealt with mineralogy and analytical chemistry. Other highly regarded chemists of the time were John Lawrence Smith, who, after receiving an M.D. degree, was a student of Liebig's and later worked largely on problems in mineralogy, and Charles Frederick Chandler, who headed the department of chemistry at Columbia University (1867–1911). Chandler also taught at the university's medical school (College of Physicians and Surgeons) from 1869 to 1910. He had received his Ph.D. for work with Wöhler, but his later scientific activities were largely related to sanitation and industrial chemistry.[142] It should also be mentioned that the nineteenth-century German emigration had a large effect on the development of pharmacy and the pharmaceutical industry in the United States.[143]

If, as this sketch suggests, the principal research activity of nineteenth-century American chemists was in the field of analytical chemistry as related to agriculture, medicine, pharmacy, geology, or mineralogy, there were also signs of interest in basic problems in inorganic, organic, and physical chemistry. At Harvard, Josiah Cooke studied atomic weights; among his students (and his successor at Harvard) was Theodore William Richards, whose subsequent work in this field won him the 1914 Nobel Prize in chemistry. The organic chemist James Crafts, who had worked with Bunsen, Wurtz, and Friedel, did much to chart the course of the Massachusetts Institute of Technology (MIT, president, 1897–1900) toward the German idea of the research university. Ira Remsen, who received his Ph.D. at Göttingen for work with Rudolph Fittig, became professor of chemistry at the new Johns Hopkins University and gained fame for his effectiveness as a teacher.[144]

At the turn of the century, modern organic chemistry had been introduced at several American universities by a few outstanding men, notably Arthur Michael (Tufts 1880–1907, Harvard 1912–1936), who had studied with A. W. Hofmann, Bunsen, and Wurtz, as well as three former students at Baeyer's Munich institute—John Ulric Nef (Chicago 1896–1915), William Albert Noyes (Illinois 1907–1926), and Moses Gomberg (Michigan 1902–1936). From their laboratories came many of the people who made the United States a world leader in both basic and industrial organic chemistry.[145]

The twentieth-century development of physical chemistry in the United States was in some ways even more striking. Josiah Willard Gibbs had been professor of mathematical physics at Yale (1871–1903) and had developed the application of thermodynamic theory to chemical problems. Although he was highly regarded by American colleagues (Gibbs was offered a professorship at Johns Hopkins), the significance of his contributions was more fully appreciated by Wilhelm Ostwald, who translated some of Gibbs's writings into German. Moreover, two of the future leaders in the institutional development of physical chemistry in the United States—Arthur Noyes and Gilbert Lewis—both studied in Ostwald's Leipzig institute. Noyes became professor of physical chemistry at MIT (1903–1919) and then moved to the Throop College of Technology (the precursor of the California Institute of Technology, Caltech) which he helped to make one of the leading research universities in the world. As professor of physical chemistry at the University of California, Berkeley (1912–1946), Lewis led a remarkable research group, wrote (with Merle Randall) one of the most influential chemical textbooks of the century (*Thermodynamics and the Free Energy of Chemical Substances,* 1923), and was one of the founders of physical-organic chemistry. In their organizational efforts, Noyes and Lewis were greatly aided by the financial support of the Carnegie Institution of Washington. Among their numerous scientific progeny was Charles Kraus, later professor of chemistry at Brown (1924–1951) and leader of an informal group of physical chemists on the eastern seaboard that included Lars Onsager, Herbert Harned, George Scatchard, Duncan MacInnes, John Kirkwood, and Raymond Fuoss.[146] I note with fond recollection that during the 1950s several of these outstanding physical chemists (Onsager, Harned, Kirkwood, Fuoss) were members of the Yale department of chemistry. The most famous chemist to come from the Noyes school was Linus Pauling, of whom much will be said later. An offshoot of the new physical chemistry that attracted many biologists and biochemists during the first decades of the twentieth century was colloid chemistry. Its principal advocate in the United States was Wilder Bancroft, who had also studied at Ostwald's Leipzig institute and was professor of physical chemistry at Cornell (1903–1937).[147]

During the nineteenth century the interaction of academic chemists and chemical manufacturers in the United States was limited to such industries as oil refining and the production of heavy chemicals (acids, soda, fertilizers) and explosives, largely because "fine" chemicals (dyes and drugs) were imported from Germany. If an American manufacturer ventured to enter that market, he was quickly undersold by one of his German competitors. At the end of World War I, however, the lawyer Francis Garvan successfully led a campaign for the confiscation by the U.S. government of the German chemical patents and established the nonprofit Chemical Foundation, which bought them from the government for $250,000 and issued licenses to American companies. As a consequence, manufacturers such as DuPont set up research divisions in organic chemistry and hired as consultants Roger Adams and other university chemists. One of Adams's students at Illinois had been Wallace Carothers, who later headed such a division at DuPont and pioneered in the field of synthetic polymers. The royalties that came to the Chemical Foundation were used effectively to support chemical research and education at several universities and also to publish, during the 1920s, a series of low-priced books, one of which was entitled *Chemistry in Medicine,* edited by Julius Stieglitz.[148] Financial support for chemical and biochemical research at universities also came from pharmaceutical and food companies, many of which had also set up their own research laboratories. Although the rise in such financial aid slackened for a time with the availability of funds from federal agencies, by the 1980s the interaction of chemical manufacturers with academic investigators in the sciences allied to medicine had assumed major importance in the United States.[149]

The nineteenth-century institutional development of the biological sciences in the United States, while deriving much stimulus from Europe, exhibited some distinctive features. Apart from the importance attached to the scientific problems related to agricultural practice, a prominent feature was the interest in the search for vertebrate fossils, with its connection to theories of biological evolution, as well as to geology and comparative anatomy. Louis Agassiz, born in Switzerland and also educated in Germany and France, came to Boston in 1846, and the acclaim accorded to his lectures there and in other American cities led to his appointment in 1847 as a professor at Harvard and the establishment in 1859 of its Museum of Comparative Zoology. The rapid western expansion of the United States after the Civil War was attended by geological surveys, one of whose consequences was the discovery of vertebrate fossils (especially of dinosaurs) by Edward Cope, a wealthy independent investigator, and Othniel Marsh, professor of paleontology at the Peabody Museum of Natural History at Yale. Agassiz, Cope, and Marsh all opposed Darwin's theory of natural selection.[150]

Although the report by the military surgeon William Beaumont on his study of gastric juice during the 1820s has placed him as a pioneer among American physiologists, experimental physiology did not begin to take root in the United States until the latter half of the nineteenth century.[151] Before sketching that development, it is to necessary to call attention once again to the difference between research disciplines and academic disciplines and also to the variety of meanings attached to the term *physiology* during the period 1850–1940. Some historians have included among the "physiological sciences" not only physiology but also physiological chemistry, biochemistry, and pharmacology. This classification is largely but not entirely justified on historical grounds if one considers only the emergence of independent units of instruction in schools of medicine. There has been, however, another tradition in which physiology is essentially synonymous with biology. It is expressed in Claude Bernard's *Leçons sur les phénomènes de la vie communs aux animaux et végétaux* (1878–1879), and in the form of "general physiology" it was advocated most prominently during the first half of the twentieth century by the physiologist Jacques Loeb. Much has been made by some historians of the failure of Loeb's effort as a discipline builder, but in view of the considerable influence of Loeb's writings I consider such interpretations to provide further evidence of the limitations of the institutional approach to the history of the biological sciences, with its emphasis on the political aspects of academic life.[152]

As in the case of Michael Foster in England, three men—the physician Henry Pickering Bowditch, the physiologist Henry Newell Martin, and the zoologist William Keith Brooks—may be said to have initiated the transition of biological investigation in the United States from natural history and morphology to functional biology. After receiving his M.D. at Harvard, Bowditch spent three years (1868–1871) in Europe, beginning in Paris and ending in Leipzig. He appears to have derived little stimulus from Claude Bernard but found his association with Carl Ludwig to be exhilarating. As professor of physiology at the Harvard medical school (1876–1906), Bowditch brought to the United States the experimental physiology of the Ludwig school and, in association with students, conducted studies on a wide variety of physiological problems. Martin and Brooks both became professors of biology at the new Johns Hopkins University. The Englishman Martin came in 1876, bringing to the United States some of the spirit of Michael Foster's physiological laboratory at Cambridge, and only his untimely death in 1896 cut short what might have been an even more brilliant scientific career. Brooks joined Martin in developing a program of graduate education in the biological sciences, out of which came four of the outstanding leaders in the development of the biological sciences in the United States—the cytologist Edmund Beecher Wilson, the geneticist Thomas Hunt Morgan, and the embryologists Edwin Grant Conklin and Ross Granville Harrison.[153] Among the other noted American biologists during the first decades of the twentieth century were the zoologists Charles Otis Whitman, William Morton Wheeler, Clarence Erwin McClung, and Herbert Spencer Jennings, the geneticists Edward Murray East and William Ernest Castle, and the numerous professors of physiology at American medical schools included William Henry Howell, Henry Plimpton Lombard, Anton Julius Carlson, Walter Bradford Cannon, and Joseph Erlanger. With this galaxy of talent in the United States, aspiring young American biologists were less impelled to seek postdoctoral inspiration abroad.[154] Many of these men and their scientific progeny gathered each summer at the Marine Biological Laboratory at Woods Hole, which became a counterpart of the famous experimental station in Naples.[155]

This brief account of the institutional development of chemistry and experimental biology in the United States during the period 1850–1940 invites consideration of the view, offered in 1948 by the medical historian Richard Shryock, that "basic science" was not highly regarded by Americans during the nineteenth century. That view has been discussed by other historians, most cogently by Nathan Reingold.[156] Here also, in my opinion, one should make a distinction between the institutional development of particular scientific disciplines, a process in which national (even local) differences in social and political circumstances have played a determining role, and the development of reliable knowledge in the corresponding research disciplines, a process dependent on the insight and skill of various kinds of people, whether they be defined as contributors to "basic," "pure," or "applied" science, and whatever their geographical location. Because the historical relation of technology and basic science is a reciprocal one, new theoretical knowledge in a research discipline has frequently been derived from advances in technical practice. From this point of view, the fact that the contributions of nineteenth-century American analytical chemists, geologists, mineralogists, or astronomers to basic science were more significant than those of their compatriots in organic chemistry or experimental biology should not be taken as evidence either of indifference to basic science or of a distinctive American style of scientific research but rather of the pressure of social and political problems in the United States during the first half of that century. Moreover, government policy toward basic science began to change during the Civil War, as was also later the case during the two world wars of the twentieth century. In that respect, the effect of social and political events on the development of the natural sciences in the United States was similar to that in Europe, where the military adventures of Napoleon or of an expansionist unified Germany played a large role in promoting some scientific disciplines at the expense of others.

I now turn to the institutional aspects of the interplay of chemistry and biology in the United States, with special reference to the emergence of academic disciplines successively known as physiological chemistry and biochemistry. The first person to be appointed professor of physiological chemistry in the United States (in 1882) was Russell Henry Chittenden (Yale Ph.B. 1875; Ph.D. 1880) in the Sheffield Scientific School at Yale. As an undergraduate, he served as a laboratory assistant in the chemistry course for premedical students and, according to his account, was placed in charge of instruction in physiological chemistry; in his later historical writings, he stated that "the laboratory of physiological chemistry at Yale . . . had its beginning in the Sheffield Scientific School in 1874."[157] During 1878–1879, and briefly in 1882, Chittenden worked in the laboratory of Willy Kühne in Heidelberg, and afterward (1883–1898) they collaborated by correspondence in extensive but inconclusive research on the products of the action of proteolytic enzymes on proteins. The Yale group also investigated problems in toxicology, especially as related to the physiological effects of arsenic and alcohol. Like his mentor Kühne, and unlike Kühne's adversary Hoppe-Seyler, Chittenden considered himself to be primarily a physiologist, and his field to belong to biology rather than to chemistry. A reading of his correspondence with Kühne (in the Yale archives) and of the scientific publications from Chittenden's laboratory hardly suggests that, as stated by Kohler, his program involved "rigorous practical training in chemistry." Such training was indeed available at Yale, but in the chemistry department, where the analytical chemist Frank Austin Gooch was a professor (1885–1918), and important work on the chemistry of nucleic acids and proteins was being done by the organic chemists Henry Lord Wheeler and Treat Baldwin Johnson (at Yale 1901–1943, professor 1917–1943), and on lipids by Rudolph John Anderson (professor 1927–1948). Moreover, the protein chemist Thomas Burr Osborne, at the Connecticut Agricultural Experiment Station in New Haven, drew his successive postdoctoral associates from the Yale chemistry department, and not from its laboratory of physiological chemistry.[158] In 1898 Chittenden became director of the Sheffield Scientific School ("Sheff") but retained the chairmanship of the department of physiological chemistry until his retirement in 1922, and his scientific writings dealt largely with problems of human nutrition. He was heavily engaged in the political struggle between the defenders of the interests of Yale College and of Sheff, a fight he lost in large part because of the straitened finances caused by increased costs and fixed income of the university during World War I. The duplication of instruction in Yale College and Sheff was abolished, and during the succeeding years nearly all vestiges of Sheff's former status vanished.[159]

In 1903 Chittenden's pupil Lafayette Benedict Mendel (Yale A.B. 1891; Ph.D. 1893) became a second professor of physiological chemistry in the Sheffield Scientific School and, because of Chittenden's larger administrative duties, was largely responsible for the further development of the department. An outstanding teacher, Mendel was held in affection by his many students, who worked on a variety of biochemical problems. Like other American physiologists and biochemists, Mendel was strongly influenced by the "intake-output" approach to animal metabolism of the Voit school in Munich.[160] Mendel's collaborative research with Osborne on the nutritive properties of highly purified proteins, which was aided by grants from the Carnegie Institution of Washington, greatly extended the earlier work of Gowland

Hopkins on the need for certain amino acids in the animal diet. It also led to the independent discovery of the dietary factors Elmer McCollum had named "fat-soluble A" and "water-soluble B."[161] Under Mendel's leadership, his department became, after World War I, a leading center of research on problems of nutrition. Among his students who entered academic life were Howard Bishop Lewis (professor at Michigan, 1922–1954) and William Cumming Rose (professor at Illinois, 1922–1955), and several others played a large role in the growth of the food industry in the United States.[162]

One consequence of the reorganization of Yale in 1921 was the transfer of Mendel's department from the college campus to larger quarters in the Yale School of Medicine. Chittenden's attitude to this action is evident from his later statement that "physiological chemistry should be considered as a biological science in the broadest sense of the term, as much as zoology, comparative anatomy or physiology, ready, however, to give its aid in any direction that might be called for but not linked indissolubly with the science or art of medicine."[163] In the context of the institutional development of the biological sciences at Yale, in which distinguished colleagues such as James Dwight Dana and Ross Granville Harrison played a large role, Chittenden's echo of Gowland Hopkins's claim for the place of biochemistry in the university was not welcome. Mendel died in 1935, before reaching emeritus status, and the medical faculty chose Cyril Norman Hugh Long, an endocrinologist, to be his successor; when Long moved to the Department of Physiology in 1951, I was appointed chairman of the Department of Physiological Chemistry, which was renamed the Department of Biochemistry soon afterward.[164]

The Mendel laboratory represented only one of the American centers for research and professional education in what has become the separate specialty of nutrition. By 1930 leadership in that branch of physiological chemistry, in which Carl Voit and Max Rubner had been preeminent, had passed from Germany to the United States. In addition to Chittenden, Samuel William Johnson of the Sheffield Scientific School and the Connecticut Agricultural Experiment Station had, as Yale students, William Olin Atwater and Henry Prentiss Armsby. With financial support from the U.S. Department of Agriculture, both men brought animal calorimetry to the United States. Moreover, the distinction of Mendel's collaboration with Osborne was soon exceeded by that of the work of the outstanding group assembled by Stephen Moulton Babcock and Edwin Bret Hart at the Experiment Station and the College of Agriculture in Madison, Wisconsin—Elmer McCollum, Harry Steenbock, William Peterson, Perry Wilson, and Conrad Elvehjem. It was that institutional entity at Wisconsin, rather than the one responsible for the instruction of medical students, which became after 1930 one of the principal American centers of research and training in biochemistry. Other prominent American nutritionists were Francis Gano Benedict, Carl Lucas Alsberg, Graham Lusk, John Raymond Murlin, and George Raymond Cowgill.[165]

By about 1930 the United States had also emerged as the leading nation in the development of clinical chemistry (now often termed laboratory medicine) as an independent biochemical discipline. Whereas research and education in the field of nutrition had been stimulated in the United States by the use of analytical chemistry in agricultural practice, the new clinical chemistry arose from the "medical chemistry" of the nineteenth century and involved the development of more reliable quantitative chemical methods for the analysis of human fluids and tissues. Among the pioneers were two American biochemists—Otto Folin and Donald Dexter Van Slyke. In 1896 the Swedish-born Folin completed work for his Ph.D. in organic chemistry with Julius Stieglitz at the University of Chicago, where he also came to know Jacques Loeb. Folin then went back to Sweden to work in the laboratory of Olof Hammarsten and also visited Ernst Leopold Salkowski and Albrecht Kossel. He returned to the United States in 1898 and two years later found employment at the McLean Hospital, a private mental institution affiliated with the Massachusetts General Hospital. The research Folin conducted there won him international recognition and an appointment at the Harvard medical school, where he became head of the department of biological chemistry (1909–1934). Among his many pupils were Philip Anderson Shaffer, Edward Adalbert Doisy, Walter Ray Bloor, Cyrus Hartwell Fiske, James Batcheller Sumner, Yellapragada SubbaRow, Hsien Wu, and George Herbert Hitchings, several of whom later made outstanding contributions to the development of various aspects of modern biochemistry.[166] The development of clinical chemistry in the United States was also furthered by Stanley Rossiter Benedict and Victor Caryl Myers, both pupils of Chittenden and Mendel.

Van Slyke's scientific career represents, in my opinion, one of the high points in the fruitful interplay of chemistry and biology during the twentieth century. After his Ph.D. (1907) in organic chemistry for work with Moses Gomberg at Michigan, Van Slyke joined Phoebus Aaron Levene's research group at the Rockefeller Institute for Medical Research in New York and investigated several problems in protein chemistry and metabolism. After a short stay in Emil Fischer's laboratory, in 1914 Van Slyke was moved to the recently established hospital of the Rockefeller Institute, where he developed new gasometric and manometric methods for the study of acid-base equilibria and CO_2 transport in blood, as well as the distribution of chloride and bicarbonate between erythrocytes and plasma. Then followed an extensive investigation of nephritis. During the course of these studies Van Slyke had, as postdoctoral associates, many future leaders in American academic medicine, among them John Punnett Peters, later at Yale. Out of that association there emerged a two-volume work *Quantitative Clinical Chemistry* (1931, 1932) by Peters and Van Slyke; the second volume, subtitled *Methods*, became the most widely used handbook in its field. It should be added that some of Van Slyke's work on blood complemented that of Lawrence Joseph Henderson at the Harvard medical school. Upon Van Slyke's retirement from the Rockefeller in 1948, he moved to the Brookhaven National Laboratory, where he worked until his death in 1971.[167]

The achievements of Van Slyke make it appropriate to emphasize at this point the role of the Rockefeller Institute in the institutional development of the interplay of chemistry and biology in the United States before 1940, especially as regards the emergence, in many medical schools, of independent departments named physiological chemistry, biological chemistry, or biochemistry. According to Robert Kohler,

> European-style bioorganic chemistry was favored in nonacademic contexts, most notably at the Rockefeller Institute, whose leaders were committed to European ideals and were not constrained by academic domestic politics. Phoebus Levene's department at the institute was the center of bioorganic chemistry in America. Trained in medicine, Levene gradually moved into pure organic chemistry. . . . A compulsively productive researcher, Levene was less successful as a discipline builder. Russian-born and trained in the German style, Levene ran his department autocratically. As a result, he had few disciples; only those who moved on, like Donald Van Slyke, became independent scientists.[168]

It is true that, as Corner stated in his account of the early history of the Rockefeller Institute,

> each of the five laboratories [not "departments"] was a little kingdom ruled over by the distinguished scientist for whom it was organized. How egocentric the command, how strict the direction, varied with the character and temperament of the Member in charge and with the nature and range of his interests. . . . Levene, too, used his young chemists largely for his own experiments. Flexner did not conceal from Levene his fear that they were not being trained for independent work. Against this charge Levene defended himself, with some justification, by pointing out that Van Slyke and Jacobs were leading independent research in the Institute, and three other former assistants were filling responsible positions elsewhere.[169]

As a member of the institute (1907–1939), Levene had about eighty-five postdoctoral associates, of whom the great majority were not on the institute staff but holders of fellowships; these included men such as Fritz Lipmann and Paul Bartlett. During my stay at the institute (1934–1945), I heard about Levene's "autocratic" style, but many years later my friend Robert Tipson wrote me that

> as regards my association with Dr. P. A. Levene, which extended over 9 years (1930–1939), I hasten to say that those were among the happiest years of my life, due in great part to him and his splendid attributes. He had a brilliant mind and boundless energy, and directed the research of his associates in an extremely unregimented way. When I arrived at the Rockefeller Institute, he suggested a research topic close to his heart, namely, determining the furanoid or pyranoid structure of the sugar moieties of the nucleosides; he then left me entirely to my own devices! He would make the rounds of the laboratories every day, to learn the progress of the work, but *he never interfered* and *never pressed for results* but curbed his desire for them.[170]

The explanation of the discrepancy between this testimony and public rumor may perhaps lie in the not uncommon favor of the head of a research laboratory shown to his abler associates and his sharp criticism of others whose work does not meet his standards.

In addition to Levene's wide-ranging program of research, dealing with the structure of nucleic acids, proteins, carbohydrates, and phospholipids, as well as stereochemistry, or Van Slyke's contributions to clinical chemistry, other senior investigators at the Rockefeller Institute played a large role in the development of the interplay of chemistry and biology during the first half of the twentieth century. Their contributions will be discussed later, and I only list here the names of some of them and those of some of their research associates: Walter Jacobs (Michael Heidelberger, Robert Elderfield, Lyman Craig), Jacques Loeb, John Northrop (Moses Kunitz, Roger Herriott), Oswald Avery (Michael Heidelberger, Walther Goebel, Colin MacLeod, Maclyn McCarty, Rollin Hotchkiss), Leonor Michaelis (Sam Granick), Karl Landsteiner, Wendell Stanley, Duncan McInnes (Lewis Longsworth, Theodore Shedlovsky), Alfred Mirsky, Wayne Woolley (Bruce Merrifield), and Max Bergmann (William Stein, Stanford Moore, Joseph Fruton). The contributions of the Rockefeller Institute to the advancement of knowledge in the biochemical sciences included not only new empirical discovery and theoretical insight, but also the invention of such valuable instruments as the countercurrent apparatus of Craig, Longsworth's modification of the Tiselius electrophoresis apparatus, Stein and Moore's amino acid analyzer, and Merrifield's peptide synthesizer. These contributions had a large effect on the course of the institutional development of the biochemical sciences in the United States not only in terms of providing new knowledge, instruments, and personnel to university departments but also in influencing academic politics by setting standards within several research disciplines. This influence is evident in the choices, after about 1925, of new department heads at several medical schools, notably Johns Hopkins and the College of Physicians and Surgeons (P&S) at Columbia, and it also reflected one of the deficiencies in the biochemical research programs at the Rockefeller Institute—namely, the absence of studies of the pathways of intermediary metabolism, then actively pursued in German laboratories.

A chemist who, like Levene, was not a member of a university faculty but played a significant role in the development of physiological chemistry in the United States was Henry Drysdale Dakin. Born and educated in England, a pupil of Julius Cohen at Leeds and of Albrecht Kossel at Heidelberg, Dakin came to New York in 1905 at the invitation of the wealthy physician Christian Archibald Herter, who had established a private research laboratory in his Madison Avenue residence and was a patron of biochemical research. After Herter's untimely death in 1910, his widow married Dakin, and in 1918 they moved their home and the laboratory to Scarborough-on-Hudson. Dakin gained fame during World War I for devising a solution of borate-buffered hypochlorite (the so-called Dakin solution), used as an antiseptic agent at the battlefront, and he was admired for his experimental work and his monograph *Oxidations and Reductions in the Animal Body* (1912, 1922), which introduced English-speaking students to the problems of intermediary metabolism. His scientific reputation, as well as his genial modesty and considered judgment, made him an influential adviser to administrators in American medical schools and research institutes.[171] To Kohler's comments about Levene, I add those about Dakin: "Although his work was widely admired, Dakin had little influence on the institutions of American biochemistry. He was a pathologically shy man; his phobia of public appearance was disabling. He never gave a public lecture, refused many academic calls, and declined to attend professional meetings, even to accept honors and prizes. Dakin was the antithesis of the research managers and discipline builders who shaped institutionalized biochemistry in the 1920s. How different American biochemistry might have been if Dakin had had the organizational flair of Mendel, Gies, or Folin."[172] I had come to know Dakin during the early 1930s, saw him many times until 1945, and corresponded with him until his death in 1952. The man I knew did not decline the honor of election to the Royal Society of London, and he was not, in my association with him, "pathologically shy." That he did not fit the model of the "research manager" or "discipline builder" overlooks the fact that, although Dakin did not seek personal advantage or public acclaim, he did, in his own courteous way, influence administrative decisions that affected the institutional development of biochemistry in the United States. I will say more below about William John Gies, cited in Kohler's remarks.

John Jacob Abel, professor of pharmacology at Michigan (1891–1893) and Johns Hopkins (1893–1932), merits a place at least equal to that of Chittenden, Mendel, or Folin in the institutional history of biochemistry in the United States.[173] Among the Americans who studied at German universities, Abel stands out because of the length of his absence abroad (seven years) and the variety of his experience. Studies in Leipzig, Heidelberg, and Würzburg were followed by a three-year stay in Strasbourg, where he received his M.D. degree (1888) and became a disciple of Oswald Schmiedeberg, the professor of pharmacology. Then Abel went to Vienna for some clinical training and concluded his tour in Marceli Nencki's laboratory in Bern, where he was attracted to physiological chemistry. Upon his return to the United States, and after a three-year stint at Michigan, Abel joined the medical faculty at Johns Hopkins. Until 1908 he was responsible for instruction in both pharmacology and physiological chemistry, but in that year a separate department in the latter subject was created, and Walter Jennings Jones, one of Abel's students, was made its head. Along with Chittenden, Folin, Mendel, and several others, Abel was active in the establishment in 1906 of the American Society of Biological Chemists and of its *Journal of Biological Chemistry*, whose foundation was financed by Christian Herter.[174] Abel's personal research achievements will be discussed later, and I only note here that among Abel's junior associates were the future professors of pharmacology Reid Hunt (Harvard), Arthur Loevenhart (Wisconsin), E. M. K. Geiling (Chicago), and E. K. Marshall (Johns Hopkins), and the future professors of biochemistry Vincent du Vigneaud (Cornell) and Earl Evans (Chicago). Those who entered the pharmaceutical industry included Thomas Aldrich (Parke, Davis) and K. K. Chen (Lilly), and Carl Voegtlin became a leading member of the U.S. Public Health Service.

Walter Jones, the first head (1908–1927) of an independent department of physiological chemistry at the Johns Hopkins medical school, had worked in 1898 in the laboratory of Albrecht Kossel and chose to devote his later research almost exclusively to the chemistry of nucleic acids. In this effort Jones was markedly less successful than Levene, for many of Jones's reports required correction. By all accounts, Jones was an effective teacher of medical students, some of whom were drawn into his research; among them were such future leaders of American academic medicine as George Whipple and Milton Winternitz.[175]

A sign of the dissatisfaction at Johns Hopkins with Jones's performance as an investigator and administrator, a feeling apparently shared by Abel, was the choice in 1927 of William Mansfield Clark as Jones's successor. That appointment marked a significant departure from the prevalent preference, in American medical schools, for men whose chemical talent was a secondary factor in their rise to professorships in physiological chemistry. After receiving his Ph.D. (1910) at Johns Hopkins for work with the physical chemist Harmon Northrop Morse, Clark worked in two government laboratories—the dairy division of the U.S. Department of Agriculture (1910–1920) and as chief of the Hygienic Laboratory of the U.S. Public Health Service (1920–1927)—where he made important contributions in both bacteriology and biochemistry through fundamental studies on the measurement of pH and of oxidation-reduction potentials. In addition to his brilliant program of research on metalloporphyrins at Johns Hopkins (1927–1952), Clark brought modern physical chemistry into the medical curriculum and, through his lectures and writings, inspired a generation of younger American biochemists.[176] His successor at Johns Hopkins was the outstanding biochemist Albert Lester Lehninger, whose work on biological oxidation will be discussed later; he was also the author of an important textbook of biochemistry. It should be noted that, in addition to the medical school Department of Physiological Chemistry, after 1945 there were at Johns Hopkins the Department of Biochemistry in the School of Hygiene and Public Health, headed by Roger Herriott, and the McCollum-Pratt Institute, headed by William McElroy.[177]

According to Robert Kohler, "Hans T. Clarke's department [at Columbia P&S] was the largest and most influential producer of biochemists in the 1930s."[178] That department had been organized in 1898 by Russell Chittenden, who served as head until 1905 on a part-time basis, with his pupil William John Gies as his full-time deputy. In 1905 Gies was made a full professor, and he continued to be the chairman of the department (renamed biological chemistry in 1908) until 1928, when he was relieved of the post (he retired as professor in 1937). During that thirty-year stretch, the research and teaching in the department was undistinguished, and Gies's principal activity was the organization of various professional societies and the promotion of dental education. His successor, the organic chemist Hans Clarke, had been born and educated in England. After a stay (1911–1913) in Emil Fischer's laboratory in Berlin, Clarke joined the research staff of the Eastman Kodak Company in Rochester and was responsible for developing its organic chemicals division, which after 1918 produced many synthetic compounds.[179] In this capacity he became a member of the influential group of American organic chemists that included Roger Adams and James Conant. His ambitions as a research scientist were modest, and the list of his publications was not impressive, even by the standards of the time. Nor was he an effective lecturer. As one of his graduate students (1931–1934), I was obliged to take the medical school course in biochemistry, and found it to be inferior to those I had attended as an undergraduate at Columbia. Indeed, the best lectures on biochemistry were then given at the 116th Street campus by John Maurice Nelson in his introductory course in organic chemistry and by Selig Hecht in his course in general physiology.[180] One of the medical students at P&S at that time was Robert Berliner, who later became a noted physiologist and dean of the Yale School of Medicine. He recently wrote me that "I particularly recalled first year biochem at P&S—one of the worst courses I was exposed to—and Hans Clarke's deplorable lectures!"[181]

Apart from his insistence that incoming graduate students in his department have a sound grounding in organic and physical chemistry, the secret of Clarke's success as a discipline builder was, in my opinion, his aversion to anti-Semitism and his admiration of German culture. As a consequence, before 1935 most of his students were New Yorkers of Jewish extraction. But that alone did not make for breadth and distinction in biochemical research, for the problems he suggested for our Ph.D. work did not give us a sense of participation in important areas of investigation. What made Clarke's department "influential" was the welcome he extended to exiles from Nazi Germany and Austria. With only a few notable exceptions, the heads of biochemical departments in other leading American universities were less generous in their response to the influx of these men and women, many of whom later played a large role in the institutional development of the biochemical sciences in the United States. Among the group who settled at P&S, Rudolf Schoenheimer and Erwin Chargaff occupy a special place in the recent history of these sciences.[182] Clarke arranged for the other émigré biochemists (Karl Meyer, David Nachmansohn, Zacharias Dische, and Heinrich Waelsch) to have their laboratories in other departments at P&S. To this talent must be added Michael Heidelberger, who had come to the Department of Medicine in 1927 and also held an appointment in Clarke's department.

It is true, as stated by Nathan Reingold, that "the German émigrés of the 1930s were simply an instance in a two-century-old process."[183] After Joseph Priestley and Thomas Cooper, the German uprisings in 1848 brought many chemists and pharmacists, and the offspring of the Jews who fled Russia, Poland, and Romania before 1914 included many future leaders in the biochemical sciences. Indeed, after World War II there was a wave of emigration from the United Kingdom (the "brain drain") to the United States,[184] and more recently there has been a large influx of promising scientists from countries in East Asia. I submit, however, that the outstanding chemists and biologists who came from German-speaking countries during the 1930s affected the further development of the biochemical sciences in the United States more profoundly than any previous or later immigration, and I consider the transformation of Clarke's department at Columbia to be a striking example of that change.

It should also be noted that in 1922 Carl and Gerty Cori came to the United States from Prague (where both received their M.D. in 1920), and they worked for about nine years at the New York State Institute for the Study of Malignant Disease (now the Roswell Park Memorial Institute) in Buffalo. There they initiated their memorable research on the pathways of carbohydrate metabolism. In 1931 Carl Cori was appointed professor of pharmacology at the Washington University medical school in St. Louis, and in 1946 he succeeded Philip Shaffer as head of the Department of Biochemistry. Although the significance of Gerty Cori's participation in their joint and separate effort was widely recognized, she did not receive the title of professor until 1947, the year in which she and Carl shared the Nobel Prize in physiology or medicine with Bernardo Houssay. The Cori laboratory in St. Louis became one of the principal Meccas for aspiring biochemists, and its importance in the institutional development of biochemistry in the United States cannot be exaggerated. Among their students were such future notables as Arthur Kornberg, Severo Ochoa, Luis Leloir, and Earl Sutherland.[185]

To this sketch of the institutional development of the interplay of chemistry and biology in the United States before about 1940 may be added the names of individuals at other universities, independent research institutes, and industrial companies who also played a significant role in that development. Among them were men such as the endocrinologists Philip Edward Smith and Herbert McLean Evans at the University of California Berkeley (Smith moved to Columbia P&S in 1927), George Washington Corner (Rochester, Carnegie Institution), and Edward Calvin Kendall (Mayo Clinic and Foundation), who laid much of the groundwork for later work on mammalian hormones.[186] This field also flowered in Canada, where the surgeon Frederick Banting, aided by Charles Best (a student assistant), and working in J. J. R. MacLeod's physiology department at Toronto, demonstrated the existence of insulin, and where the biochemist James Collip developed a method for its isolation. Although the preparation of a physiologically active pancreatic extract by Banting and Best had been claimed by previous workers, the availability of a method for the purification of insulin in a form suitable for clinical use made their contribution one of the most notable medical advances of the 1920s.[187] Because of its historical connection to Great Britain, and its proximity to the United States, English-speaking Canada gave to and received from both nations much scientific talent in the biochemical sciences.[188] Other notable individuals who played a significant role in the emergence, before 1940, of biochemistry as a prominent scientific discipline in the United States and Canada will be mentioned later.

After 1945 the distinction of some of the university departments of biochemistry mentioned above was outmatched by that of departments such as those headed by Arthur Kornberg at Stanford, Hans Neurath at the University of Washington (Seattle), and Harland Wood at Western Reserve. Moreover, several research institutions, most prominently the National Institutes of Health (NIH) at Bethesda, emerged as leading centers of research and postdoctoral education in the biochemical sciences. These developments in the United States were spurred by the public expectation that the victorious and wealthy nation that produced the atomic bomb could also provide cures for cancer and other hitherto incurable diseases.[189] Sizable financial aid for research and both predoctoral and postdoctoral education became available from such private agencies as the American Cancer Society and, in even larger measure, through the extramural grant programs of NIH and the National Science Foundation (NSF). One of the consequences of this development was the gradual withdrawal during the 1950s of the Rockefeller Foundation from the support of the research and fellowship programs it had established before World War II to promote the interplay of chemistry and biology. The greater dependence of these programs on agencies such as NIH (and later the Howard Hughes Medical Institute) contributed to the acceptance of the designation of biochemistry as one of the "biomedical" sciences. Until about 1980 the increase in the funds available from agencies of the U.S. government for research in these sciences at American universities kept pace with a roughly tenfold growth in the number of independent investigators, and generous support was given to approximately one-half of the applicants. Inevitably, the continued growth of the scientific population, the introduction of many costly instruments, and unfavorable economic factors obliged NIH to reduce the approval rate to about one-tenth of the applications. By the 1990s holders of Ph.D. degrees in the "biomedical sciences" encountered difficulty in finding good academic jobs, and although there were opportunities in the biotechnology industry, the future of many of the new enterprises was uncertain. The only salutary feature of the seeming overproduction of people in these disciplines was that, in response to the belated recognition of the rights of able women, they now suffer less discrimination than before. Some spokesmen for university scientists have deplored what they considered to be a breach of the "social contract" promised in Vannevar Bush's famous 1945 report

Science: The Endless Frontier. Other scientists, notably the physicist John Ziman, have suggested that although the search for new reliable knowledge about the natural world is certain to continue, the pace of discovery may now have reached a "dynamic steady state."[190]

Scientific Societies and Publications

The association of like-minded scientists to form societies and to establish new periodicals is a social phenomenon characteristic of the emergence of many specialties in the chemical and biological sciences.[191] During the course of the development of the interplay of chemistry and biology since the appearance in 1789 of the first issue of the *Annales de Chimie,* edited by Lavoisier and his colleagues, many new journals were founded that, as in the case of the *Annales,* marked both a change in scientific outlook in its field of endeavor and a response to the resistance of the editors of established periodicals to such change. In succeeding decades, the example of the *Annales de Chimie* (in 1816 changed to *Annales de Chimie et de Physique*) was followed by a succession of chemists and biologists, many of whom were entrepreneurs primarily interested in the rapid publication of papers from their own laboratories and of polemics in scientific disputes. Among them was Liebig, whose *Annalen der Pharmazie* (established in 1832) became in 1840 the *Annalen der Chemie und Pharmazie* and, after his death in 1873, *Justus Liebig's Annalen der Chemie.*[192] One of Liebig's admirers was Hermann Kolbe, who, after becoming editor in 1870 of Otto Erdmann's *Journal für praktische Chemie* (founded in 1834), used its pages for violent criticism of the new organic chemistry, as practiced by Kekulé, Baeyer, and Emil Fischer.[193] Among the consequences of the formation of national scientific societies, such as the Deutsche Chemische Gesellschaft (1867) and the establishment of its *Berichte,* was a reduction in such misuse of editorial privilege. The Chemical Society of London (1841), the Société Chimique de Paris (1857), and the American Chemical Society (1876) also issued journals for the communication of research reports, setting the standards for the editorial handling of such publications in their respective countries. During the middle of the nineteenth century, there were similar developments in the biological sciences, as in the formation of the Société de Biologie in Paris and the appearance of its *Comptes Rendus,* although in Germany the new influential journals were launched by individuals such as Virchow *(Archiv für pathologische Anatomie und Physiologie)* and Pflüger *(Archiv für die gesamte Physiologie des Menschen und der Tiere).* During the succeeding decades, many important periodicals appeared in response to the emergence of new trends in chemical or biological research, and the growth of the population in particular fields. An outstanding example was Ostwald's *Zeitschrift für physikalische Chemie* (1887), which complemented the organic-chemical nature of the *Berichte* and *Annalen.*[194]

In the case of biochemical journals, the organic-chemical standards set by Hoppe-Seyler for his *Zeitschrift für physiologische Chemie* were fairly high and were maintained by his successors,[195] but the growth of interest in physical chemistry and colloid chemistry among German biochemists led to the appearance in 1906 of the *Biochemische Zeitschrift,* edited by Carl Neuberg. There was also Emil Abderhalden's *Fermentforschung,* which he set up in 1916 because of the rejection of many of his papers.

In other countries, such as France and Japan, new biochemical journals were being established under the auspices of their respective professional societies.[196] As was mentioned above, the transfer of the ownership of the British *Biochemical Journal* from Benjamin Moore to the newly formed Biochemical Society led to an improvement in its quality, and in the United States the *Journal of Biological Chemistry,* published by the American Society of Biological Chemists, reflected the transition from emphasis on problems of nutrition and clinical chemistry to those in the chemistry of proteins, nucleic acids, enzymes, and intermediary metabolism.[197] Inevitably, some papers were rejected by these journals, and new journals such as *Enzymologia* (later transformed into *Molecular and Cellular Biochemistry*), *Archives of Biochemistry* (later *Archives of Biochemistry and Biophysics*), and *Biochimica et Biophysica Acta* began during the 1930s and 1940s to meet the rapidly growing demand for outlets for the communication of biochemical research reports. Among the most significant of the publications established after 1945 were *Biochemistry* (sponsored by the American Chemical Society) and the *Journal of Molecular Biology,* initiated by members of the MRC Unit at Cambridge, England. Abstract journals such as Richard Maly's *Jahresbericht über die Fortschritte der Thierchemie* (1871–1922; in 1882 the title was expanded to include *oder der physiologischen und pathologischen Chemie*), *Chemical Abstracts,* and *Biological Abstracts,* as well as annual reviews such as the *Annual Reports of the Chemical Society* (1903–) and the *Annual Review of Biochemistry* (1932–) initiated by J. Murray Luck, helped workers in the biochemical sciences to keep up with the growing literature, but the more recent further multiplication of new periodicals and the transformation of older ones (notably the *Proceedings of the National Academy of Sciences USA* and *Nature*) into favored media for the communication of research reports in these sciences has made it all but impossible for an individual scientist, even with the aid of modern computer technology, to keep abreast of advances in research outside his or her particular subspecialty. This publication explosion has placed a severe burden on librarians. There has also been a proliferation of articles on the chemical aspects of the historical development of biological subspecialties—for example, endocrinology and microbiology.[198]

After World War II there were important changes in the international organization of the biochemical sciences. During the nineteenth century the concept of "international science" was implicit in the presence of foreign scientists at the regular meetings of the Gesellschaft Deutscher Naturforscher und Aerzte and the British Association for the Advancement of Science, or at special conclaves like the Karlsruhe Congress of chemistry (1860). Except for the period 1914–1918, after 1900 many international scientific organizations were founded, exchange professorships were established between universities in Germany and the United States, and the grant programs of the Rockefeller Foundation played a large role in bringing together European and American scientists.[199] After World War II there emerged the Federation of European Biochemical Societies (FEBS), which sponsored the publication of the *European Journal of Biochemistry* and *FEBS Letters*, and the European Molecular Biology Organization (EMBO).[200] The first International Congress of Biochemistry in 1949 at Cambridge was followed, on a triennial basis, by meetings in many cities, including Moscow, Tokyo, Montreal, and Prague.

CHAPTER 3

Philosophy, Chemistry, and Biology

The recent sociological approaches to the study of the nature and historical development of scientific knowledge have raised many questions regarding the extent to which thought and action in the research conducted by scientists have been influenced (and perhaps determined) by the philosophical debates of their time and by the writings of past philosophers. Certainly, in the interplay of what we now call chemistry and biology, the problem of the relation of the phenomena exhibited by living organisms to the nature and properties of the matter of which such organisms are composed has been a compelling one throughout the history of recorded philosophical thought.[1] Moreover, many biologists who studied the form, function, and evolution of living organisms have been active participants in the philosophical discussion of the question whether the phenomena of life can ever be "fully" explained without invoking "vital forces" inaccessible to study by means of the methods of chemistry and physics, or whether biological processes are "teleological" in nature. During the twentieth century, the philosophical battleground shifted from "vitalism" versus "materialism" to the issue of "organicism" (or "holism") versus "reductionism."[2] For example, in 1922, after summarizing the knowledge provided by cytology, genetics, and chemistry about the parts of the living cell, the noted biologist Edmund Wilson asked:

> How shall we put it together again? It is here that we first fairly face the real problem of the physical basis of life; and here lies the unsolved riddle. We try to disguise our ignorance concerning this problem with learned phrases. We are forever conjuring with the word "organization" as a name for the integrating and unifying principle in the vital processes; but which one of us is really able to translate this word into intelligible language? We say pedantically—and no doubt correctly—that the orderly operation of the cell results from a dynamic equilibrium in a polyphasic colloidal system. In our mechanistic treatment of the problem we commonly assume this operation

> to be somehow traceable to an original pattern or configuration of material particles in the system, as is the case with a machine. Most certainly conceptions of this type have given us an indispensable working method—it is the method which almost alone is responsible for the progress of modern biology—but the plain fact remains that there are still some of the most striking phenomena of life of which it has thus far failed to give us the most rudimentary understanding. . . . We are ready with the time-honored replies: It is the "organism as a whole"; it is a "property of the system as such"; it is "organization." These words, like those of Goldsmith's country parson, are "of learned length and thundering sound." Once more, in the plain speech of everyday life, their meaning is: *We do not know*. . . . Shall we then join hands with the neo-vitalists in referring the unifying and regulatory principle to the operation of an unknown power, a directive force, an archeaus, an entelechy or a soul? Yes, if we are ready to abandon the problem and have done with it once for all. No, a thousand times, if we hope really to advance our understanding of the living organism. To say *ignoramus* does not mean that we must also say *ignorabimus*.[3]

As will be evident from the account in succeeding chapters, since those words were spoken the continued interplay of chemistry and biology has provided much understanding of "some of the most striking phenomena of life," but there remains also much about which "we do not know."

Likewise, in the interaction of chemists and physicists, the philosophical issues relating to such entities as atoms, corpuscles, heat, energy, or affinity were prominent in the discussion among scientists engaged in the experimental study of the properties of chemical substances and of the nature of chemical processes. After the discovery of electrons, protons, neutrons, radioactivity, and isotopes, the quantum-mechanical interpretation of chemical "structure" in terms of such concepts as "valence" gave a new dimension to the philosophical question whether chemistry had been "reduced" to physics.[4]

With the increasing professionalization and specialization of scientists after the middle of the nineteenth century, the philosophical debates at the fringes of the various biological and chemical disciplines had progressively less impact on the course of the interplay of thought and action. No doubt, through their general education, many twentieth-century biologists and chemists acquired philosophical attitudes that inclined them to consider the search for chemical or physical explanations of biological phenomena to be a worthwhile occupation. Once that commitment had been made, however, their work and thought followed a professional track in which a particular line of investigation was being developed, and their philosophical attitudes receded into the background. If these attitudes were given prominence in a scientist's later account of discoveries that made him or her famous, they tended to reflect philosophical views at the time of recollection. Perhaps in addressing a wider audience some chemists and biologists wished to place their research achievements within the framework of philosophical traditions familiar to nonscientists.[5] A nineteenth-century example is Claude Bernard's famous *Introduction à l'étude de la médecine experimentale* (1865), long regarded as a valuable description of the "scientific method." In his preface to the English translation, Lawrence Henderson wrote that "we have here an honest and successful analysis of himself at work."[6] Unfortunately for this judgment, the recent examination of Bernard's research notebooks and published papers has revealed inconsistencies with his later analysis, and it would appear that Bernard's skill in experimental surgery may have been more important in his achievements than the views he expressed later about the principles of the scientific method.[7]

The philosopher Arne Naess has offered the following slogan:

> "Whatever the proofs put forward for the impossibility of a kind of happening or process or thing, do not always reject an invitation to inspect an argument for its existence; it *may* exist. *Anything is possible!*" . . . When a scientist, in his actual work, takes something for granted, without any reservations whatsoever, this does not imply that he holds certain assumptions, principles, or laws to be true or highly confirmed. He is at the moment not concerned with the truth or certainty of *those* propositions. . . . The above *possibilistic* slogan is not so much directed at the individual scientist engulfed in his extremely limited research programme as [at] the "consumers of science," especially the philosophers and the "educated public," who take their cues from science and are attracted by theories said to be scientific and repelled by those said to be unscientific or not up to date. To take something for granted—absolutely, and beyond any question—differs vastly when it is done inside a research-problem situation from when it is done outside. The producer of science has a research programme in mind, the consumer may have anything in mind *except* just that kind of situation.[8]

I find this statement to be relevant to the writings not only of present-day philosophers but also of philosophically minded scientists who seek to popularize their specialties. Such popularizations, with their oversimplified and frequently overdramatized accounts of particular scientific investigations, have been used by philosophers as a basis for the evaluation of the intrinsic quality and social value of some contemporary lines of research, especially in molecular biology.[9]

If most of the leading twentieth-century chemists and biologists have evinced less interest in the philosophical aspects of their research traditions than their nineteenth-century forebears and have managed to make significant contributions to the growth of scientific knowledge without the guidance of the writings of philosophers, these contributions have had a considerable influence on the approaches of modern philosophers to age-old questions relating to the study of the natural world (ontology), of human reasoning, concepts, and knowledge (respectively logic, metaphysics, and epistemology), and of moral evaluation (ethics). Indeed, by the mid-twentieth century, the hope of philosophers from Francis Bacon to Karl Popper to develop methodological rules for scientific investigation were abandoned, and the efforts of present-day philosophers of science have been largely limited to the analysis of past and contemporary scientific theories in the hope of developing rules for the evaluation of the place of such theories in the growth of scientific knowledge. In their debates more attention has been given to the scientific thoughts of physicists and biologists than to those of chemists, and until recently relatively little attention was paid to the technical aspects of their research. The problem of the growth of scientific knowledge appears to have been mainly one of defining the "logic" in the "intellectual history" of such specialties as mathematical physics, genetics, or evolutionary biology.

In what follows I offer my current opinions about the various views expressed in the recent philosophical discourse about the questions raised by the growth of reliable chemical and biological knowledge. Although I held some of these views during the course of the laboratory work I did personally or that done by my research associates and students between 1931 and 1984, such views played a negligible role in the choice of problems for investigation or the manner in which the research was conducted. Nor should my present opinions be regarded as an expression of a "consensus" within the scientific tribe of which I have been a member.

I distinguish "reliable" scientific knowledge from what is commonly called belief (or ideology) and recognize that my opinions belong in the latter category. My beliefs, like those of anyone else, reflect a personal experience that, in my case, included efforts as a teacher and investigator in a sector of the scientific field known as biochemistry and as a historian of that discipline. What obliges me to claim that a distinction must be made between beliefs about the natural world and reliable knowledge of the objects and processes in that world is the historical fact that the age-old intimate interplay of thought and action in which human experience, as by observation of natural phenomena and through practice in everyday human activities, generated ideas (hypotheses, models, analogies, classifications) that led to sustained experimentation, the invention of new instrumental and mathematical tools, and the formulation of new concepts, many of which became "theories" or "laws." The observation and control of the fermentation of fruit juices preceded the concept of enzymes, the hybridization of plants and the breeding of animals preceded the modern theories of inheritance, and the invention of the steam engine and the boring of cannons preceded the formulation of the laws of thermodynamics. The "facts" and "theories" that have emerged since about 1600 from this interplay of thought and action constituted conceptual frameworks that underwent change. What may have been considered by members of a particular scientific specialty, at a particular time, to have been a reliable structure has been modified, rebuilt, or abandoned by later investigators in favor of others deemed likely to be more reliable in leading to the formulation of new theories and the discovery of new facts with the aid of new tools. In my view, the changes in the conceptual frameworks are intimately linked to changes in the methodology of thought and practice, and as the empirical and theoretical knowledge in an area of inquiry became more "reliable," new methodologies supplanted those based on everyday experience and commonsense reason. The same cannot be said of knowledge about the natural world derived solely from the thoughts of theologians or of speculative philosophers, whether of the "idealist" or "materialist" stripe, for the varieties of belief they have engendered have been continually confounded by the growth of knowledge produced by the members of the self-correcting scientific specialties. To admit that such "reliable" knowledge is partial or incomplete, and that a vast terra incognita still remains to be explored, is no more than to reassert the validity of the distinction between such knowledge and that based on religious faith or a priori philosophical ideas. To these opinions I add my agreement with Karl Mannheim's assertion that "the type of knowledge conveyed by natural science differs fundamentally from historical knowledge."[10]

I believe that the overall growth of reliable scientific knowledge has provided increasingly "true" representations of "facts" about objects and processes in a natural world whose existence is independent of the human mind, and that these representations therefore have "objective" validity (which, I suppose, makes me some sort of a "realist"). In my view, therefore, "ontology" should be kept separate from "epistemology" or "semantics" when philosophers, sociologists, or historians develop concepts about the ways in which scientists have studied natural objects and processes, and I reject the kind of "relativism" espoused by the "social constructionists." I do not accept the view that the aim of scientific investigation is (or should be) to attain some absolute exact "truth," for account must be taken of "chance" and "accident,"[11] as well as of the limitations of the available methods of investigation. Nor do I believe that the main task of scientific research is to determine the "truth-or-falsehood" of statements of fact or theory, but rather to examine the value of such claims in the further investigation of particular research problems. Although I believe that rational thought—as in formal logic, of which mathematics is a part, or as in commonsense logic, based on experience—is essential, any theory, however logical or nonlogical it may have appeared to be, can become part of reliable chemical or biological knowledge if it has proved its worth in guiding further observation and experiment (which, I suppose, makes me some sort of a "conventionalist" or "instrumentalist" or "pragmatist").

In the evaluation of scientific theories, I do not accept "simplicity" or "beauty" as valid criteria, for I find in the historical record of the interplay of chemistry and biology that, more often than not, simple and beautiful theories have been most valuable when they have led to the discovery of previously unknown complexity. I also question the criterion of the "predictive scope" of scientific theories, for I find in that record empirical reports that did not "fit" into prevailing theories, but whose theoretical ("retrodictive") significance was only recognized during the further study of particular scientific problems.

Although I recognize the importance in scientific thought of some kinds of "hypothetico-deductivism," I find in some kinds of "inductivism" much that describes the interplay of thought and action in the chemical and biological sciences, and I question the relevance, to actual scientific practice, of philosophical disputations about a "scientific method" that is independent of time and circumstance, or a "logic of scientific discovery (or justification)," based exclusively on "verification" or "falsification."[12]

The Unity of Science?

One of the features of the discourse about the methodology of scientific research has been advocacy of the idea of the "unity of science." This idea, with roots in antiquity, appeared in the writings of Francis Bacon and other seventeenth-century philosophers and of the eighteenth-century French Encyclopedists, notably Antoine Caritat de Condorcet. It was implicit in William Whewell's *History of the Inductive Sciences* (first edition, 1837) and more explicit in Auguste Comte's *Cours de philosophie positive* (1830–1842).[13] There was set up a hierarchy in which physics is the fundamental natural science and provides explanations of the subject matter of chemistry, which, in turn, is expected to provide explanations of the problems of biology, with psychology and sociology following in that order. During the first decades of the twentieth century, this physics-based "unity of science" was embraced by a group of philosophers and physicists which became known as the Vienna Circle of "logical positivists" or, as they later called themselves, logical empiricists. One of the projects of the group was the launching of a publication that eventually became the *International Encyclopedia of Unified Science.*[14] The members of the Vienna Circle drew inspiration from the writings of the physicist and psychologist Ernst Mach, who rejected the atomic theory and insisted that the knowledge of sensations is the only "true" scientific knowledge.[15] As the French chemist-philosopher Emile Meyerson put it,

> Principles of positivism, or, at any rate analogous principles, have since [Comte] been adopted, at least in appearance, by several scientists who have often felt bound to protest, like Comte, against atomic theories; but, in reality, and despite the support given to this tendency by the great and legitimate authority of Mach, it remains today, as it did throughout the nineteenth century, without the least influence on the progress of science. The scientists at the beginning of the twentieth century continue to build atomic theories just as their predecessors did. All of them see . . . there, for want of something better, a research instrument of great value, a "working hypothesis."[16]

The word *positivist* has had a checkered career since 1830. During the rest of the nineteenth century, it was used as a synonym for *scientific,* but

> later, as part of a general and difficult argument about *empiricism* and *scientific method,* [it acquired] its largely negative and now popular sense of naive objectivity. It is significant that it is not now used, as are both *scientific* and *empirical,* to describe and justify a criterion of reliable knowledge. Rather, it is used by opponents of this criterion as absolute. What they urge against it is not what positivists themselves argued against, whether faith or *a priori* ideas. . . . It has become a swear-word, by which nobody is swearing. Yet the real argument is still there. It is simply that it would be more uncomfortable to centre it on *scientific,* where the issues would be at once harder and clearer.[17]

If *positivism* and its attendant physics-based *reductionism* have become words of opprobrium (or condescending indulgence), the idea of the unity of science has retained its appeal in many quarters. For example, the biologist George Gaylord Simpson wrote that "biology . . . is the science that stands at the center of all science. It is the science most directly aimed at science's major goal and most definitive of that goal. And it is here, in the field where all the principles of all the sciences are embodied, that science can truly become unified," and his colleague Ernst Mayr expressed the hope that "biologists, physicists, and philosophers working together can construct a broad-based unified science that incorporates both the living and non-living world."[18]

By the 1980s the increasing complexity of the reliable knowledge about the "chemistry of life" led some philosophers of biology who had embraced one or another form of the reductionist thesis to write of the "disunity of science," with the use of new terms such as *promiscuous realism* to denote the idea that the things in the natural world can be classified in different ways ("epistemological pluralism"). Another new trend is the study of "organized complexity," intended to replace the ever more detailed analysis of the fine structure and function of living organisms.[19] As a skeptical biochemist of uncertain philosophical affiliation, I am doubtful about the fruitfulness of these new approaches but welcome the challenge to the idea of the "unity of science."

Philosophy and Chemistry

In its institutional classification at universities and learned societies, chemistry is now listed among the physical sciences, but as we have seen, its historical connection to medicine, botany, and mineralogy make that designation rather arbitrary, for it implies that, like astronomy, chemistry was more akin to physics ("natural philosophy") than to the branches of "natural history" that came to be called biology or geology.

For Immanuel Kant, writing in the years of the Chemical Revolution, chemistry was a "systematic art" rather than a science because it lacked a mathematical foundation comparable to that of Newtonian dynamics. Nevertheless, he was an admirer of the work of Georg Stahl, as is indicated in the preface to the second edition (1787) of the *Kritik der reinen Vernunft:* "When Galileo rolled his cannon balls, whose weight he had selected, down a steep surface, or Torricelli let air carry a weight he had previously estimated to be equal to a definite column of water, or later Stahl converted metals to calxes and these back into metals, whereby he withdrew or restored something; thus a light appeared for all investigators of nature."[20] As a "transcendental idealist," Kant thus appears to have appreciated the significance of experimentation, not only by physicists but also by chemists who had sought to develop methods to "separate the pure from the gross" and to study systematically such reactions as those between "alkalis" and "acids," as described in many seventeenth- and eighteenth-century chemical treatises before 1770. During the 1790s Kant also expressed his admiration of the contributions of Antoine Lavoisier, especially in regard to the "caloric" theories of heat and of gases, and declared his support of the antiphlogistic theory at a time when most of the leading "pneumatic" chemists (Scheele, Priestley, Cavendish) opposed it. Kant used Lavoisier's concept of caloric to develop his hypothesis of a *Wärmestoff* that "constitutes the aether as the basis of matter filling all the universe, whose inner motion, set into eternal vibrations by the first impact, constitutes a living force."[21] For Lavoisier, caloric was an "elastic fluid," and similar substantial media such as "electric fluid" or "luminiferous ether" formed part of the conceptual framework of nineteenth-century physics and chemistry. The physicists wrote of an imponderable "electromagnetic" ether, and Joseph John Thomson proposed a "vortex" theory of the atom, while some chemists (notably Dmitri Ivanovich Mendeleev) advocated the idea of a material ether. Indeed, at the turn of the century, the ether assumed a spiritual role, and led several noted British physicists, among them William Crookes, to engage in psychical research.[22]

Although Kant warned against the use of "pure reason" outside the field of possible experience, some of his followers, such as Georg Wilhelm Friedrich Hegel, and members of the Naturphilosophie group of "romantic" German philosophers (notably Friedrich Schelling), developed various kinds of metaphysical chemistries.[23] For Johann Trommsdorff, one of the leading German chemists of the time, however, such a priori chemistry had to be distinguished from "pure," or "scientific" chemistry, which he defined as follows: "The subject matter of chemistry are all the things in the world of the senses [*Sinnenwelt*]; its foundation is experience, which it attains through observations and experiments; from these it builds through induction and analogies general conclusions, and from these it derives a theory which binds the facts into a scientific whole."[24] Much has been written about the Chemical Revolution and whether it was indeed a revolution in the sense defined by Thomas Kuhn.[25] Certainly, between 1770 and 1800, common air was shown to be a mixture of separable gases, and Stahl's phlogiston had been removed as an entity lost or gained in the interconversion of calxes and their metals, but what stands out among Lavoisier's many achievements is his use of quantitative gravimetric and volumetric methods to determine the composition of natural substances and to study the chemical processes (such as combustion) in which other substances are formed. By 1770 a vast number of "things in the world of the senses" had been identified and characterized. Apart from the inheritance from practical chemists of many valuable laboratory operations, such as distillation, precipitation, or combustion, there were concepts denoted by such words as *purity, element, compound, principle,* or *affinity.* The concept of a "pure substance," inherited from antiquity in the form of *mixtio vera,* had acquired by the middle of the eighteenth century the meaning of a homogeneous "individual" or "species." The four "elements" in the matter-theory of the pre-Socratic materialist philosophers (air, fire, earth, water), adopted by Aristotle, as well as the three alchemical "principles" (sulfur, salt, mercury) of Paracelsus, had been discarded by Robert Boyle in favor of a "corpuscular" matter-theory that provided for the existence of many "elements," but it was not until analytical methods, such as those introduced by Lavoisier, became available that it was possible to speak of a chemical element as an entity that cannot be further decomposed. Likewise, the concept of what Boyle called a compound, which had been discussed for centuries in relation to Aristotle's criticism of the atomism of Democritos, was also transformed by the application of analytical methods and by the reclassification of known "pure" substances and some types of chemical reaction by Lavoisier and his colleagues.[26] As for the word *principle,* its usage underwent change during the eighteenth century in response to the preparation, from plant and animal fluids and tissues, of many new substances, some of which were denoted as "immediate principles."

The word *molecule,* discussed in a previous chapter, appeared during the eighteenth century in the form of Georges Louis Buffon's *molécules organiques* as fundamental units of living things, and as *molécules intégrantes,* which the mineralogist René Just Haüy proposed as the regularly arranged units of crystals. As in the case of *molecule,* during the nineteenth century the concept of the *atom,* as developed by chemists from John Dalton and Joseph Gay-Lussac to Stanislao Cannizzaro (aided by Amedeo Avogadro's hypothesis), was rather different from the one inherited from Boyle and Newton, or from that of James Clerk Maxwell in his article on *atom* for the *Encyclopedia Britannica.* If organic chemists such as August Kekulé considered atoms and molecules to be real, though invisible, entities, their existence was denied by some chemists who sought, without success, to "mathematize" chemistry.[27]

In my opinion, the emergence of organic chemistry as a discipline in which analysis, synthesis, and classification were successfully combined without recourse to Newtonian dynamics represents an exemplary case of the irrelevance of Kantian methodology to the growth of reliable knowledge in at least one important area of nineteenth-century investigation. In the midst of the theoretical confusion before 1850, it was the empirical discovery of such phenomena as isomerism and substitution reactions, or of the existence of a benzoyl "radical," that provided the clues for the formulation of the fruitful concepts of *valence, chemical bonding,* and three-dimensional *molecular structure.* Although in the eighteenth-century discussion of the concept of *affinity* some chemists invoked Newtonian dynamics, its initial formulation in 1718 by Etienne François Geoffroy in terms of the relative tendency *(rapports)* of various acids and alkalis to form salts was largely based on the empirical knowledge provided by practical chemists such as Agricola and Glauber. In the form of "elective attractions," this concept also played a significant role in the nineteenth-century development of thermodynamic theory.[28] Moreover, during the first part of the twentieth century, Niels Bohr's application of Max Planck's quantum theory to Ernest Rutherford's model of the hydrogen atom was followed at once by the electronic theories of valence offered by Walther Kossel and Gilbert Lewis, and after the development of quantum mechanics during the 1920s there emerged the "valence bond" and "molecular orbital" theories of chemical bonding. To these examples may be added the emergence of "polymer" or "macromolecular" chemistry through the organic-chemical approach of Carl Harries, Hermann Staudinger, and Wallace Carothers to the study of the substances that Thomas Graham had considered in the 1860s to represent the "colloidal state of matter."

This brief sketch of some of the philosophical aspects of the historical development of chemical thought, and of its interaction with physical theory, indicates a degree of methodological pluralism that is not readily accommodated by a single style of scientific reasoning or experimental practice. One of the few modern philosophers to appreciate fully the epistemological significance of the pluralism of chemical methodology was Gaston Bachelard.[29]

It has long been recognized that changes in the classification and the naming of chemical substances and processes, and the symbols used to denote them, reflect changes in the conceptual framework and practical methodology of the branch of chemistry to which they belong.[30] In recent years, particular emphasis has been placed on the historical significance of pictographic representations of chemical structure and reaction, an increasingly prominent feature of the chemical and biochemical literature since the appearance of Kekulé's hexagonal structure of benzene or of Krebs's ornithine cycle.[31] The variety of such representations during the course of the interplay of chemical and biological thought illustrates the plurality of the methodologies in the historical development of biochemical specialties.

Philosophy and Biology

The medieval alchemists based their speculations on a neo-Platonic version of Aristotle's matter-theory, according to which metals and minerals grew from seeds in the earth, in a manner similar to the generation of living things. Many of these alchemists were physicians who treated human diseases with an "essence" or "spirit" extracted from some mineral, plant, or animal material, and the transmutation of a sick person into a healthy one was considered to be comparable to the transmutation of a base metal into gold. The elaborate pictorial symbolism, in which astrology was wedded to alchemy, depicted the concepts underlying the efforts to prepare a "quintessence" which would cure all human ailments, and attracted the interest of many educated people after 1600, among them Isaac Newton, Johann Wolfgang von Goethe, and, most recently, Carl Gustav Jung.[32]

In their sprightly "philosophical dictionary of biology," Peter and Jean Medawar wrote disparagingly of Aristotle's biological works but granted that his "philosophical opinions—on teleology, for instance—command respectful attention."[33] Aristotle linked this concept of goal-directed activity, which implied both "function" and "purpose," to his definition of *psyche* (which came to be translated as *soul*) as the distinctive power possessed by living things.[34] Although the empirical evidence Aristotle offered in support of his arguments against the materialist interpretations of life by Leucippos and Democritos was repeatedly questioned by later investigators (notably Erasistratus), his vitalist philosophy, as spiritualized by medieval theologians and scholars, was not seriously challenged until the beginning of the seventeenth century. The challenges came, in the first place, from the experimental work of physicians such as Santorio Santorio, William Harvey, and Giovanni Borelli, who applied the principles of Galilean quantitative mechanics to the study of the physiological processes of nutrition, blood circulation, and locomotion. This "iatrophysical" approach was rejected by "iatrochemists" who accepted the extreme vitalism of Georg Stahl, and his concept of the "sensitive soul" *(anima)* that worked directly on the chemical processes in living things. There was also the "dualist" philosophy of René Descartes, who assigned to the "body," as distinct from the "mind," the properties of a "machine," and the "monadology" of Gottfried Wilhelm Leibniz, who combined mechanism with theological optimism. Later French philosophers paid more attention to Descartes than to Leibniz, and the reverse was the case in the writings of German philosophers. Although that discourse is enshrined in accounts of the history of philosophy, the development of reliable biological knowledge during the rest of the seventeenth century and most of the eighteenth century owed more to the contributions of John Ray and Carl Linneaus to the classification of plants, to the empirical discoveries made with new instruments, especially the microscopes of Antoni van Leeuwenhoek and Robert Hooke, and to such observations as those of Charles Bonnet on the parthenogenesis of aphids or of Abraham Trembley on a freshwater polyp *(Hydra)*.[35] Indeed, by the middle of the eighteenth century, biologists like Louis Buffon, Pierre Louis Maupertuis, and Albrecht Haller had rejected the extreme mechanistic philosophy, as then advocated in the *L'Homme machine* (1748) of Julien Offray de la Mettrie. They called attention to the complexity of the phenomena of life, and concluded that vital forces such as "sensibility," "irritability," or "formative drive" *(Bildungstrieb)* distinguished the living from the nonliving. The last of these terms had been introduced by Johann Friedrich Blumenbach in connection with his view that living things are generated by the internal transformation of a seed or an embryo ("epigenesis") and in his opposition to Bonnet's idea of the "preformation" in the "germ" of the pattern of the developed organism. During the course of that debate, Maupertuis offered a model of inheritance that some historians of genetics have found to be a prefiguration of the one developed by Gregor Mendel.

The debate over epigenesis versus preformation also gave rise to controversy about the "spontaneous generation" of living things and about the "origin of life." Buffon proposed that living things are formed by the aggregation of "organic molecules" (by Newtonian attraction) that had originally been generated from mineral matter, and he offered as evidence the microscopic observations of John Turberville Needham, who claimed to have seen such aggregation of "globules" present in extracts of plant and animal tissues. That evidence was disproved by Lazzaro Spallanzani (1765) by using strongly heated extracts that were protected from air. The issue was revived in the famous Pasteur-Pouchet debate a century later and resolved in a similar manner.[36]

The question of the origin of life on the earth has continued to excite interest up to the present. In some speculations, the problem was transferred to another astronomical body, and primordial living matter was thought to be brought to the earth by meteorites. Most of the other hypotheses have been based on the available geochemical and biochemical knowledge of the time. During the 1920s, Aleksandr Oparin advocated the view that the oxygen-free ("reducing") atmosphere of the primitive earth favored the synthesis of such simple organic compounds as amino acids. At that time, one of the prevailing biological concepts was that the essential components of the "protoplasm" in the cells of living organisms were the "colloidal" proteins. Thirty years later much was made of Stanley Miller's experiments, in which a mixture of methane, hydrogen, and water was subjected to a prolonged exposure to an electrical discharge, and the appearance of detectable amounts of some of the amino acids known to be units of protein structure was taken as evidence for the Oparin hypothesis. Also, by heating amino acids to 130°C, Sidney Fox obtained water-insoluble particles that he named proteinoids, and he suggested that such bodies may have been the precursors of the proteins of living cells. With the more recent emphasis on the intracellular role of nucleic acids and the discovery by Sidney Altman that a bacterial RNA is an enzyme, attention has now shifted from the primordial formation of the first protein enzymes to that of the first RNA enzymes in bacterial systems.[37] It should be noted that the successive hypotheses that have emerged from such studies are examples of historical reasoning in biology, analogous to that in cosmology and geology, and depend on empirical evidence that is somewhat different from the experimental evidence offered in support of hypotheses about functional mechanisms in biological systems. That style of biological thought was evident in the writings of Buffon and Lamarck, who provided the first nontheological hypotheses about biological evolution and the origin of terrestrial life, but the most impressive example still remains Charles Darwin's theory of the origin of plant and animal species through natural selection. In his famous 1859 book, Darwin only surmised that living things arose from some primordial nonliving matter.

The two parallel eighteenth-century debates, of mechanism versus vitalism and of epigenesis versus preformation, were brought to a focus in Kant's attempt to resolve the "antinomy" between the idea of finding, through observation and experiment, mechanistic explanations of the phenomena of life, and the idea that such explanations must be supplemented by teleological interpretations. In his *Kritik der Urteilskraft* (1790), Kant devoted a section to a defense of the value of teleological reasoning in biological thought, cited approvingly Blumenbach's idea of Bildungstrieb, and developed the concept of a living "organism" as a self-reproducing system in which purpose *(Naturzweck)* plays an important role.[38] The issue Kant raised has been debated continually during the past two hundred years, most recently in relation to what has been called teleonomic process or telecomechanism in biological adaptation and in the regulation of physiological processes.[39] As will be seen in later chapters, at each stage of the debate the arena shifted in response to new empirical discovery and theoretical insight.

During the final decade of the eighteenth century, the concept of "vital forces" as characteristic features of living organisms was advocated in a 1793 lecture by Carl Friedrich Kielmeyer, who added to Haller's *irritability* and *sensibility* the terms *reproductive force, secretory force,* and *propulsive force.* A professor of botany, zoology, and chemistry at Stuttgart (he later taught chemistry, botany, pharmacy, and materia medica at Tübingen), Kielmeyer attracted the attention of several prominent contemporaries. He does not appear to have made any significant empirical contributions, and his philosophical writings were soon overshadowed by those of Friedrich Schelling, one of the leaders in the Naturphilosophie movement, associated with the rise of "romanticism" among German intellectuals.[40] Schelling and other members of the movement saw the natural world as an organized whole, identical with "spirit" *(Geist, esprit),* and found in Kielmeyer's writings support for a fanciful and poetic form of idealism, rather different from that of Kant. Naturphilosophie acquired an English adherent in Samuel Taylor Coleridge (later also in Thomas Carlyle), Humphry Davy found in it some merit, and there were also echoes in Denmark and Sweden. By the 1830s, however, that particular form of German idealism had largely been rejected by German biologists.

During the first half of the nineteenth century, however, the concept of "vital force" *(Lebenskraft)* retained considerable appeal.[41] For example, at the beginning of that period, a widely held view was that "even though the gastric fluid, as a result of its chemical composition, is the dissolving agent of both the simple and composite foods and its action on food is a chemical one, digestion is still a vital process conditioned by the life of animals."[42] Both before and after this passage was written, the idea of a vital force was linked to emphasis on the inability of chemists to imitate, in the laboratory, the processes readily effected by living organisms. In 1786 John Hunter had written that "the action and production of actions, both in vegetable and animal bodies, have been hitherto considered so much under the prepossessions of chemical and mechanical philosophy, that physiologists have entirely lost sight of life. . . . No chemist on earth can make out of the earth a piece of sugar, but a vegetable can do it."[43]

Different meanings were attached to the idea of the "vital force" by various writers; the sense in which many of the nineteenth-century chemists used this term was probably first formulated by Johann Christian Reil:

> The characteristic nature of the matter of which animal bodies are composed provides the chief basis of the characteristic phenomena animals exhibit. The vital force, which we consider to be the cause of these phenomena, is not something different from organic matter; rather, the matter itself, as such, is the cause of these phenomena. Most of the animal phenomena can largely be explained on the basis of the general properties of matter. We therefore do not need any vital force as a unique primary force to explain them; we only use the word to designate concisely the concept of the physical, chemical, and mechanical forces of organic matter, through whose individuality and cooperation the animal phenomena are effected.[44]

In the later writings of many chemists there was considerable ambiguity in regard to the existence of a "unique primary force" in living organisms. Thus in the 1827 edition of his famous textbook of chemistry, Berzelius defined organic chemistry as the chemistry of the substances that are formed under the influence of the vital force. Earlier he had expressed the view that "the cause of most of the phenomena within the Animal Body lies so deeply hidden from our view, that it certainly will never be found. We call this hidden cause *vital force;* and like many others, who before us have in vain directed their deluded attention to this point, we make use of a *word* to which we can affix no idea."[45] Nor did Friedrich Wöhler's preparation (in 1828) of urea from ammonium cyanate sweep away the conviction that vital forces are operative in the chemical changes effected by living things.[46] For the physiologist Johannes Müller,

> the mode in which the ultimate elements are combined in organic bodies, as well as the energies by which the combination is effected, are very peculiar; for although they may be reduced by analysis to their ultimate elements, they cannot be regenerated by any chemical process. . . . Woehler's experiments afford the only trustworthy instances of the artificial formation of these substances; as in his procuring urea and oxalic acid artificially. Urea, however, can scarcely be considered as organic matter, being rather an excretion than a component of the animal body. In the mode of combination of its elements it has not perhaps the characteristic properties of organic products.[47]

At midcentury a distinction was drawn between such biological processes as digestion, which had been recognized to involve chemical reactions similar to those observed in the laboratory, and the metabolic transformations that lead to the formation of the chemical constituents of living organisms. Although the latter phenomena were considered to be "the manifestation of an admirable power that could not yet be reproduced by chemical means,"[48] the successes in the study of digestion were offered in evidence of the conviction that systematic research would lead to chemical explanations of biological phenomena then ascribed to vital forces. Indeed, the recurrent theme during 1800–1950 that a particular physiological process was "linked to life" was intended by many investigators to mean that it had not yet been imitated with isolated chemical constituents of the biological system under study, rather than to indicate their adherence to a vitalist philosophy that denied the possibility of such chemical reconstruction of that biological process.

However, as is evident from the statement by Edmund Wilson quoted earlier, the concept of the "organism" continued to occupy an important place in biological thought. In the recent past, "organismic biology" became a subject of contention in the discourse about the relation of the "parts to the whole," and of the regulation of physiological processes:

> Organismic biology and vitalism differ in one important respect: The latter holds (and the former denies) that the characteristic features of organic activity—all of which fall under the heading of "regulation"—are caused by

> the presence in the organism of a nonphysical but substantial entity. . . . The affinity between vitalism and organismic biology is more than an accident. In the history of biology it is difficult to disentangle vitalistic and organismic strands, since both schools are concerned with the same sorts of problems and speak the same sort of language. Organismic biology may be described as an attempt to achieve the aims of the murky organismic-vitalistic tradition, without appeal to vital entities. . . . Organismic biology is to be interpreted as a series of methodological proposals, based on certain very general features of the organism—namely, the existence in the organism of levels of organization with the ends of maintenance and reproduction. These features are sufficient to justify "a free, autonomous biology, with concepts and laws of its own," whether or not the higher levels are ultimately reducible to lower ones.[49]

The history of this aspect of the interplay of biological and chemical thought and action will be examined in the final chapter. As will be evident in the intervening chapters, after the emergence of the cell theory, and the associated concept of a "protoplasm," "vital forces" continued to be invoked, whether in the form of "living proteins" in metabolism or of "entelechy" in embryonic development. I postpone more detailed discussion of these topics and only call attention to some of the twentieth-century restatements of the extreme vitalist position. Among them were the writings of Henri Bergson, with his concept of the "élan vital," of Alexis Carrel, of Pierre Lecomte du Nouy, and of Pierre Teilhard de Chardin, with his concept of the "noosphere."[50]

CHAPTER 4

From Ferments to Enzymes

At the close of the nineteenth century, Eduard Buchner reported that a cell-free press juice prepared from brewer's yeast caused carbon dioxide and ethyl alcohol to form in aqueous solutions of various sugars (sucrose, glucose, fructose, maltose). He concluded that "the initiation of the fermentation process does not require so complicated an apparatus as is represented by the living cell. The agent responsible for the fermenting action of the press juice is rather to be regarded as a dissolved substance, doubtless a protein; this will be denoted *zymase*."[1] Although some yeasts failed to give an active extract, and the first step of grinding pressed yeast with fine quartz and kieselguhr (a diatomaceous earth) was rather tedious, several investigators soon confirmed Buchner's report.[2] The acceptance of his claim marked a decisive stage in a scientific controversy that had agitated many minds throughout the nineteenth century. In this chapter, I examine why many of Buchner's contemporaries attached great importance to his discovery, and why one of them wrote: "The ancient conflict over the question 'What is fermentation?' had ended: fermentation is a chemical process."[3]

The Problem of Fermentation in 1800

The process of fermentation had its origins in the kitchens of prehistory, probably with the rise of organized agriculture in Neolithic times. The long-known effect of leaven (a mass of yeast; Greek *zyme*) on a cereal dough, with the evolution of gas and a change in the texture of the solid matter, had its counterpart in the effervescence when honey or sweet fruit juices were kept in a warm place. Before 2000 B.C. the Egyptians knew that when crushed dates were stored, there was at first produced a pleasantly intoxicating material, but if the mixture was allowed to stand for a longer time, it turned sour to yield vinegar, the strongest acid known to antiquity. This souring of wine was considered to be comparable to the souring of milk. By 1500 B.C. the use of germinated cereals (malt) for the preparation of beer from bread, and the formation of wines from crushed grapes, were established technical arts in Mesopotamia, Palestine, and Egypt. The ancient artisans also observed that the formation of beer, wine, or vinegar was followed by changes that led to the liberation of noxious odors; this slow putrefaction of plant material was compared to the more rapid decay of animal and human tissues. The practical arts of preserving animal foods—drying, smoking, curing, pickling in brine, treatment with granular salt—were well developed in the prehistoric Near East and Europe, and dehydration with natron (a mixture of salts, largely sodium carbonate) was a key operation in the mummification procedures in Egypt.[4]

These ancient arts provided empirical data for the speculations of Greek philosophers, notably Aristotle, about the nature of fermentation (Greek *zymosis*) and putrefaction (Greek *sepsis*). Aristotle considered the changes undergone by inanimate things to be analogous to those seen in the biological world. Thus grape juice is the infantile form of wine, and fermentation is a process of maturation, while the further change to vinegar is the death of the wine. Moreover, the juice matures through a "concoction" *(pepsis)* promoted by heat, as in the transformation of foodstuffs through the innate heat of the animal body. In Aristotle's philosophy such natural change expresses the tendency of an object to function toward the attainment of a specific end; in a living thing this property *(psyche)* is inherent in the organism as a whole and arises from the integrated functions of its parts (see *De Anima* 412^{b}, 415^{a}). In the centuries after Aristotle, the idea of psyche as a nonmaterial principle of biological organization became mingled with the idea of *pneuma*, the ethereal stuff postulated by the Stoics as a principle of cohesion and activity in all matter, both living and nonliving.[5] Like *psyche, pneuma* is a word used in Greek literature for human breath, and the idea that pneuma is a subtle and volatile entity essential for life became popular among Greek physicians and natural philosophers before the time of Galen. In subsequent translation *pneuma* became the Latin *spiritus,* the French *esprit,* the German *Geist,* and the English *spirit.* This entity was considered to be a fifth essence (*quinta essentia,* in addition to earth, water, fire, and air)—the vital principle that determines the specific nature and activity of all material things.[6]

With the rise of alchemy during the first five centuries of the Christian era, the conviction grew that it should be possible to isolate the quintessence of things. Because pneuma was considered to be a volatile stuff, the early alchemists distilled various materials, saw in the expulsion of fumes and vapors the liberation of the spirit characteristic of each material, and identified these fumes and vapors with powerful, divine agents that gave specific life to each thing, whether it was an egg or a metallic ore. By about the twelfth century the art of distillation had developed to the stage where highly volatile distillates could be collected by cooling the receiving flask, and the distillation of wine was found to yield an inflammable liquid *(aqua ardens)*. By the next century the effects of drinking this "burning water" were well recognized, and it came to be called *aqua vitae*, whence all the familiar variants—akvavit, uisge beatha, uaquebaugh, whisky—are derived.[7] For the medieval alchemists, what had been isolated was the quintessence of wine, and it followed logically that in order to extract the proper quintessence of any other animal or vegetable material, one should allow it to ferment or putrefy, and then distill off a "water" that could be purified by redistillation. The persistence of this idea for about four centuries is indicated by the following extract from the popular treatise by the French pharmacist Jean Beguin, first published in 1610:

> The word quintessence refers to a substance that is ethereal, celestial, and extremely subtle. . . . Some call it *Médecine par excellence*, others *Elixir*, because of the signal virtues it possesses in the protection of the human body from various ailments; others call it Heaven for two reasons. First of all, because heaven is composed, not of the four elements, but of a certain ethereal matter, or fifth element, and is entirely incorruptible. . . . Secondly, because just as heaven acts powerfully on sublunary things, giving life to all things and conserving their vitality, so the quintessence conserves the health of man, prolongs youth, retards old age, and drives away all sorts of diseases.[8]

Beguin's book contains detailed recipes for the preparation of the quintessence of various animal and plant materials (for example, the quintessence of blood—"a remedy of sovereign power"). The first step was to imitate the process whereby crushed fruits are converted into wine. This process was termed *fermentatio;* when, as in the conversion of cereal dough to bread, an agent such as yeast was required, the agent was termed *fermentum*.

It should be noted, however, that medieval alchemists such as Albertus Magnus and Raymond Lull applied the word *fermentatio* to various kinds of natural change, in living and nonliving matter, and for them *fermentum* might be any reactive substance, or even the Philosopher's Stone. Definitions of this kind lingered up to the seventeenth century, as in the famous chemical treatise *Alchymia* of Andreas Libavius, published in 1597: "Fermentation is the exaltation of a substance through the admixture of a ferment which, by virtue of its spirit, penetrates the mass and transforms it into its own nature." Thus a small quantity of specially prepared "medicine" was mixed with a substance (such as a base metal) and brought it to life *(vivificatio)* or reawakened it *(resuscitatio)*. There was also the age-old idea that fermentation was akin to cooking: "The ferment acts through its inner heat, but this must be activated by an external heat of not too high a degree, lest the spirit be driven off."[9]

These ideas about fermentation occupied a central place in the writings of the "chemical physicians" whose revolt against the medicine of Galen and the materia medica of the first-century herbalist Dioscorides had been launched during the sixteenth century by Theophrastus Bombastus of Hohenheim, who called himself Paracelsus. The Galenic doctrine, built upon the medical writings attributed to Hippocrates as well as the physiological ideas of Aristotle and the Stoics, assumed that human health depends on a balance of forces specifically associated with various fluids ("humors") of the body: blood, yellow bile *(chole)*, black bile *(melanchole)*, and phlegm (nasal outflow). These humors were related to each other by affinities or antagonisms defined by their special relationship to the four elements: blood and fire (hot and dry), yellow bile and air (hot and wet), black bile and earth (cold and dry), and phlegm and water (cold and wet). The objective of good medical practice, according to this doctrine, was to restore the balance of the humors by such treatment as bleeding or by purgation with plant extracts.

By the seventeenth century there was a ready ear among propertied classes for revolt against the inadequacies of Galenic medicine. The rapid and continuous development of Europe after the Crusades was marked by exploration, conquest, and exploitation of non-European territory, by the more efficient use of wind and water power, and by the adoption of numerous inventions (for example, windmill, compass, rudder) imported from Islam or Byzantium. At the end of the sixteenth century, this development had produced great wealth, the growth of cities, and the emergence of a class of craftsmen skilled in the practical arts of shipbuilding, mining, and metallurgy, as well as in the preparation of such materials as leather, paper, ceramics, and gunpowder, needed for trade and war. Because of the rise of these practical men, Europe began to export goods instead of bullion and to assume the economic dominance that reached its peak during the nineteenth century. In this new prosperity, however, the medical problems arising from urbanization, long sea voyages, and armed conflict assumed larger social importance. The terror generated by such diseases as the bubonic plague (especially the Black Death of 1347–1348) and syphilis, and the inability of Galenic medicine to meet the challenge, made people receptive to calls for reform.

The revolt against Galenic medicine was also furthered, as were other uprisings of the sixteenth century, by the diffusion of books written by scholars. Among these books those dealing with the pyrotechnical arts and with metallurgy played a large role in the rise of chemistry; such works as Hieronymus Brunschwig's *Liber de Arti Distillandi* (1500), Georgius Agricola's *De Re Metallica* (1530), and Vanniccio Biringuccio's *Pirotechnia* (1540) were widely read and went through numerous editions.[10] Not only did the scholars turn to craftsmen for knowledge of the practical arts, but some of them also did not scruple to soil their hands in the chemical laboratory. In thinking about natural phenomena, however, the sixteenth-century chemical philosophers were inspired by the magic of Hermes Trismegistos and the Kabbalah, as popularized by Marsilio Ficino, Pico della Mirandola, and Cornelius Agrippa.[11] Thus in his attack on Galenic medicine, Paracelsus combined chemical technology and Hermetic mysticism to create a philosophical basis for medical practice. He assigned special importance to the use of mineral substances such as salts of antimony or mercury in the treatment of disease, not only externally (a standard practice in Arabic medicine) but also for internal administration. He advocated the medical virtues of vitriol and of various quintessences; one of the best known of the elixirs was *aurum potabile* (first described in the fourteenth century), prepared by circulating the spirit of wine (he termed it *alcool vini*) with gold leaves for four weeks. It is debatable whether Paracelsus's advocacy of chemical medicines improved later medical practice. That he stirred up the medical profession is unquestioned; he is reported to have told an irate group of Basel physicians: "If you will not hear the mysteries of putrefactive fermentation, you are unworthy of the name of physicians."[12] Nor can there be any doubt that his ideas influenced seventeenth-century physicians to reformulate questions about the nature of living things in the chemical language of their time, especially in relation to the process of fermentation. The most important of these followers of Paracelsus was Johannes Baptista van Helmont.[13] In his writings there is a large mixture of Christian piety and Hermetic mysticism, but in the mass of arcana there were reports of experimental observations which attracted the attention of physicians, apothecaries, and natural philosophers. Robert Boyle, for example, found van Helmont "more considerable for his experiments than many learned men are pleased to think him."[14]

Fermentation figured largely in van Helmont's natural philosophy. He wrote that if one burned sixty-two pounds of coal, and one pound of ashes was formed, "the 61 remaining pounds are the wild spirit. . . . I call this Spirit, unknown hitherto, the new name of Gas, which can neither be constrained by Vessels, nor reduced to a visible body, unless the seed being first extinguished. But Bodies do contain this spirit . . . a Spirit grown together, coagulated after the manner of a body, and is stirred up by an attained ferment, as in Wine, the juyce of unripe Grapes, bread, hydromel, or water and Honey."[15] For van Helmont, "gas" was different from "air" because it was derived from a substance by the action of a ferment, and different from "vapor" because, unlike water vapor, it was dry. The "wild spirit" *(spiritus sylvester, gas sylvestre)* also arose by the action of acids on "salt" (a word with many meanings at that time—in this case, a carbonate). In noting (by taste) the sour taste of gastric juice, he explained digestion in the stomach as a process in which the action of a ferment was aided by acid, and he wrote of other kinds of fermentation to which food is subjected in passing through the body (the duodenum, liver, heart); later, the terms *chylosis, chymosis, hematosis, pneumatosis,* and *spermatosis* were applied to such fermentations.

Among the contemporaries of van Helmont were several men whose writings represented a bifurcation of the Paracelsian doctrine, with increased emphasis on its mysticism or on its iatrochemistry. The mystics included Jakob Böhme, who combined Paracelsus with the Bible to develop a pantheism later reflected in the Naturphilosophie of the German romantics, and the Oxford Rosicrucian Robert Fludd, who stressed the Hermetic and Kabbalistic aspects of Paracelsism. The iatrochemists who rejected the mysticism included Daniel Sennert, who adopted a corpuscular theory of matter and considered fermentation to be a process in which substances are separated into their smallest indivisible parts, followed by their reunion to form new bodies. During the seventeenth century, van Helmont's ideas about fermentation became fused with the corpuscular theory in the writings of Franciscus Sylvius, and the renowned Oxford physician and anatomist Thomas Willis, a founder of the "clubb" that included Robert Boyle. In 1659 Willis defined fermentation as follows:

> Fermentation is the intestine motion of Particles, or the Principles of every Body, either tending to the Perfection of the same Body, or because of its change into another. For the Elementary Particles being stirred up into motion, either of their own accord or Nature, or occasionally, do wonderfully move themselves, and are moved; do lay hold of, and obvolve one another; the subtil and more active, unfold themselves on every side, and endeavour to fly away; which notwithstanding being entangled, by others more thick, are detained in their flying away. Again, the more thick themselves, are very much brought together by the endeavour and Expansion of the more Subtil, and are attenuated, until each of them being brought to their height and exaltations, they either frame the due perfection of in the subject, or compleat the Alterations and Mutations designed by Nature.[16]

By the end of the seventeenth century, the iatrochemical approach to medical practice had lost much of its earlier appeal. The famous French physician Guy Patin wrote in 1672 that "Descartes and the ignorant chemists strive to spoil everything, in philosophy as well as in good medicine,"[17] and Thomas Sydenham urged more careful observation and clearer description of individual human diseases. Moreover, the growth of physiological experiment and microscopic observation was furthered by the achievements of physicians such as William Harvey and Marcello Malpighi, and the writings of Johann Rudolph Glauber and Robert Boyle had charted a new course for the development of chemistry.

As has been mentioned, the development of eighteenth-century chemistry was influenced by Georg Ernst Stahl and Hermann Boerhaave, professors of medicine at Halle and Leiden. Both opposed iatrochemistry in their lectures but differed in their philosophical outlook. As part of his "animist" philosophy, Stahl sought to develop a coherent theory of chemical composition and reaction and, like Johann Joachim Becher, assigned central importance to the chemical principle of inflammability (Greek *phlogiston*). Boerhaave did not adopt the phlogiston theory and criticized the high-flown systems of chemistry advanced by the iatrochemists. If Boerhaave's chemical writings were highly valued during the eighteenth century as a guide to laboratory operations, Stahl's phlogistic philosophy offered a satisfying theoretical basis for the study of chemical reactions.[18]

In the writings of Becher and Stahl, and especially in Stahl's treatise *Zymotechnia Fundamentalis* (1697; German translation, 1748), the problem of fermentation occupied an important place. Becher sought to clarify the use of the word *fermentatio* by distinguishing between three kinds of fermentation: that accompanied by the evolution of gas, the alcoholic fermentation (which he called *fermentatio proprie*), and the fermentations leading to the production of acid; to these Stahl added putrefaction, which he considered to be no more than the final stage of fermentation. Going beyond Willis, Stahl proposed that the action of a ferment is to communicate the motion of its particles to the particles (he called them *moleculae*) of the fermentable body so as to accelerate the decomposition of the latter; the separated particles then recombine to form more stable compounds, with the release as a spirit of "oily" ("sulfureous") particles which he identified with phlogiston.

Boerhaave's principal contribution to the problem of fermentation was to discuss carefully the available chemical knowledge; the following extract from his *Elements of Chemistry* (first authorized Latin edition, 1732) gives some indication of his views:

> I say then, that in every Fermentation, there is an intestine motion of the whole Mass, and all the parts, so long as this physical action continues; and

> I call it an intestine one, because it chiefly depends upon the internal principles of the vegetable Substances that are fermenting. . . . But I add further, that this intestine motion can be excited only in vegetable Substances. . . . I know very well, that some famous Authors make no scruple to assert the contrary, and therefore to distinguish here as nicely as possible, I define a true and perfect Fermentation by its proper effect, and that is, that always terminates in the production of the Spirit, or Acid. . . . By the word Ferment, I shall mean any Substance, that being intimately mixed with the fermentable Vegetables . . . will excite, increase, and carry on the Fermentation describ'd. . . . Hence therefore it appears at one view that such a Ferment must belong to the Class of Vegetables.[19]

Boerhaave's restrictive definition was not adopted by later eighteenth-century chemists, and around 1800 it was considered that three kinds of fermentation were possible both in plants and in animals: vinous or spirituous fermentation, acid or acetous fermentation, and putrid or putrefactive fermentation; the last was usually associated with animal matter and, because of the release of ammonia, was also termed alkaline fermentation.[20]

For the development of reliable knowledge about fermentation, the decisive contributions came initially from the experimental work of the seventeenth-century "pneumatic" chemists who used new apparatus, notably the barometer (invented by Evangelista Torricelli in 1643, published in 1663) and the air-pump (invented by Otto von Guericke in about 1654, published in 1672). The empirical findings, and their Newtonian interpretation, of these English "virtuosi"—Robert Boyle, Robert Hooke, John Mayow—provided the background for the studies of Joseph Black. In a memorable paper on *Magnesia Alba, Quicklime, and Some Other Alkaline Substances* (1755), Black reported that when magnesia alba (basic magnesium carbonate) is heated, or when limestone (calcium carbonate) is treated with acid, a gas (which he called "fixed air") is liberated and is readily absorbed by quicklime (calcium oxide) in water, with the formation of a milky precipitate. In 1757 Black used this lime-water test to show that "fixed air" is the gas evolved from a fermentation brew, and the one formed on burning charcoal in air, thereby rediscovering the gas sylvestre of van Helmont.[21]

Black's work was followed by the quantitative studies of Henry Cavendish, whose first memoir (1766) was entitled *Three Papers Containing Experiments on Factitious Air* ("in general any kind of air which is contained in other bodies in an unelastic state, and is produced from thence by art"). In these papers he described "inflammable air" (hydrogen), produced by the action of acids on metals, and also reported on fixed air ("that species of factitious Air, which is produced from Alcaline Substances, by Solution in Acids or by Calcination"). In particular, Cavendish showed that when brown sugar in water was treated with yeast, all the gas discharged during the fermentation was absorbed by "sope leys" (aqueous sodium hydroxide) and had the same density, solubility in water, and action on flame, as the fixed air derived from limestone. Then came the report (1772) of Daniel Rutherford that after an animal had respired in a closed container, and then died, the residual *aer malignus* (after removal of fixed air with alkali) was common air highly charged with phlogiston. In the same year Carl Wilhelm Scheele established the separate identity of this "noxious air," and its inability to support life was indicated in the new French chemical nomenclature (1787) by the name *azote;* Jean Chaptal called it nitrogen in 1790 to denote its association with nitric acid. Among the gases of the atmosphere, the center of the stage was held by the "dephlogisticated air" or "vital air" identified by Joseph Priestley and by Scheele in 1772–1774. In 1781 Cavendish exploded two volumes of inflammable air with one volume of dephlogisticated air in a closed vessel and showed that the gases were completely converted to water. During the course of his studies, Cavendish greatly improved the technique ("eudiometry") for measuring the volume of gases; the most accurate eudiometer described around 1780 was said to be that of Felice Fontana.[22]

It was Antoine Lavoisier who provided the chemical basis for all further studies on the nature of alcoholic fermentation. At the core of his system of chemistry was the idea that, in combustion, an inflammable principle (which he called caloric) resides in the "vital air" liberated on the calcination of calxes (metal oxides), thus moving the principle from the inflammable body, where Stahl had placed phlogiston, to the gas that supports combustion. Moreover, Lavoisier considered vital air to be composed of an acidifying principle (first, *principe oxigine;* later *oxygène*) combined with caloric, and that when charcoal is burned in vital air, the air is decomposed with the absorption of the acidifying principle by the charcoal and the release of caloric in the form of heat. He named the fixed air produced by this combustion *acide crayeux* (1777) or *acide de charbon* (1778), and finally *acide carbonique.*[23] Although the concepts of caloric and acidifying principle proved to be inconsistent with later experimental observations, the method Lavoisier introduced for the elemental analysis of organic compounds for carbon and hydrogen marked a new stage in the development of quantitative chemistry. Moreover, as will be seen, Lavoisier's use of such analytical techniques in the experimental study of animal respiration had a decisive impact on physiological thought and on the interplay of chemistry and biology during the nineteenth century.

In his famous *Traité elémentaire de chimie* (1789; English translation in several editions: 1790, 1793, 1796, 1799, 1801), Lavoisier described his experiments on alcoholic fermentation as follows:

> I did not make use of the compound juices of fruits, the rigorous analysis of which is perhaps impossible, but made use of sugar, which is readily analyzed. . . . This substance is a true vegetable oxyd with two bases, composed of hydrogen and carbon, brought to the state of an oxyd, by means of a certain proportion of oxygen; and these three elements are combined in such a way, that a very slight force is sufficient to destroy the equilibrium of their connection. By a long train of experiments, made in various ways, and often repeated, I ascertained that the proportion in which these ingredients exist in sugar, are nearly 8 parts of hydrogen, 64 parts of oxygen, and 28 parts of carbon, all by weight, forming 100 parts of sugar.
>
> Sugar must be mixed with about four times its weight in water, to render it susceptible of fermentation, and even then the equilibrium of its elements would remain undisturbed, without the assistance of some substance to give a commencement to the fermentation. This is accomplished by means of a little yeast from beer; and, when the fermentation is once excited, it continues of itself until completed . . . I have usually employed 10 *libs.* of yeast, in state of paste, for each 100 *libs.* of sugar, with as much water as is four times the weight of the sugar.[24]

Lavoisier then gave the amount (by weight) of the oxygen, hydrogen, and carbon in the water, sugar, and yeast and also noted the content of "azote" (nitrogen) in the yeast. He conducted the fermentation with a "few pounds of sugar" and converted his results to the 100-pound scale mentioned above. According to his report, from the 95.9 lb. of sugar that had disappeared (corresponding to 26.8 lb. of carbon, 7.7 lb. of hydrogen, and 61.4 lb. of oxygen), there were formed 57.7 lb. of alcohol (obtained by distillation), corresponding to 16.7 lb. of carbon, 9.6 lb. of hydrogen, and 31.4 lb. of oxygen, as well as 35.5 lb. of carbonic acid (trapped in alkali) corresponding to 9.9 lb. of carbon and 25.4 lb. of oxygen. With the addition of a small amount (2.5 lb.) of acetic acid produced in the fermentation the total weight of the products was 95.5 lb., corresponding to 27.2 lb. of carbon, 9.8 lb. of hydrogen, and 58.5 lb. of oxygen. (All these figures were given to seven decimal places.)

This relatively satisfactory agreement is in accord with Lavoisier's principle of the conservation of matter but is all the more remarkable because the analytical data for sucrose and alcohol do not correspond to their actual elemental composition. The alcohol was probably contaminated with water, and some of the carbon dioxide was probably lost. In 1815 Joseph Louis Gay-Lussac reported that he had checked and corrected the figures given by Lavoisier and that, of 100 parts of sugar, 51.34 had been converted into alcohol and 48.66 into carbonic acid, thus indicating that sugar had been converted into equal parts of the two products.[25] This correction came from the development, by Gay-Lussac and Louis Jacques Thenard, of the first general method for the analysis of organic compounds for carbon, hydrogen, oxygen, and nitrogen. It involved oxidation with potassium chlorate and gave results that compare favorably with those obtained by means of better methods developed later. In particular, they established the elemental composition of cane sugar; the term *carbohydrate* came from their finding that the proportion of hydrogen to oxygen in that sugar (as well as in starch) is the same as in water.

As written in recent times, the "Gay-Lussac equation" for alcoholic fermentation is $C_6H_{12}O_6 = 2\ CO_2 + 2\ C_2H_5OH$. This is not the formulation given by Gay-Lussac but by organic chemists many years later. Jean Baptiste Dumas and Félix Polydore Boullay noted in 1828 that Gay-Lussac's calculations applied only to sugars having a formula which they wrote as $C^{12}H^{12}O^{12}$, in keeping with the convention of their time that assigned a relative combining weight of 6 to carbon and of 8 to oxygen. For them, cane sugar had the formula $C^{12}H^{11}O^{11}$, and they concluded that the fermentation of such sugars required the uptake of a molecule of water (then written as HO). Eighteen years later Augustin Pierre Dubrunfaut showed that cane sugar is cleaved by dilute acids to two fermentable sugars (later termed glucose and fructose) which he was able to separate from each other.[26] This important investigation depended on the use of polarimetry, of which more will be said later; an aqueous solution of cane sugar is strongly dextrorotatory, while the mixture of the split products is weakly levorotatory (hence the terms *invert sugar* for the mixture and *invertin* for the enzyme that catalyzes the hydrolysis).

Moreover, the designation of alcohol as C_2H_5OH required many years of effort in the development of the concepts of organic chemistry; the recognition of the ethyl group came only in the 1830s. The subsequent reformulation of atomic and molecular weights by Cannizzaro, of the ideas of molecular structure, and of the arrangement of atoms in space, advanced by Butlerov, Kekulé, and van't Hoff, were followed by Emil Fischer's brilliant work on the chemical structure of the sugars. Some facets of the efflorescence of organic chemistry during the nineteenth century are key elements in the history of the fermentation problem, and we shall return to them later.

Microorganisms as Agents of Fermentation

In 1810 there appeared a book entitled *Le Livre de tous les ménages ou l'art de conserver, pendant plusieurs années, toutes les substances animales et végétales* by Nicolas Appert, a French manufacturer of confectionery, distilled spirits, and food products. In it he described methods for preserving foods by putting them into tightly closed vessels that were then heated in boiling water. His success won him wealth and recognition, and his work marks the beginning of the canning industry.[27] Gay-Lussac examined Appert's results and found that the air left in the closed heated vessels lacked oxygen. Because fermentation of grape-must or the putrefaction of food products sets in when the containers were opened to let in air, but not if the contents were protected from air under mercury, Gay-Lussac concluded that atmospheric oxygen was required for these processes and suggested that "the absorbed oxygen produces a new combination which is no longer able to excite putrefaction or fermentation, or which becomes coagulated by the heat in the same manner as albumin."[28] Along with earlier reports by Giovanni Fabbroni, who claimed that the "material which decomposes the sugar is the vegeto-animal substance" (which he called gluten), and by Thenard, who noted that brewer's yeast loses nitrogen when it ferments sugar, Gay-Lussac's report was taken as evidence for the view that an essential feature of the fermentation process is the decomposition of albumin-like matter and that oxygen promotes this process.[29] This idea reappeared in the writings of Justus Liebig. The view that oxygen is necessary for fermentation was attractive because it was consistent both with the older identification of fermentatio with heating and with the antiphlogistic explanation of combustion. Less attention was given before 1830, therefore, to Charles Bernard Astier's claim that "the air is the vehicle of every kind of germs" and is the source of the ferment that "lives and nourishes itself at the expense of the sugar, whereby there results a disruption of the equilibrium among the elementary units of the sugar." Nor was much account taken of the opinion of Christian Erxleben that fermentation is a chemical process associated with vegetation, or of the microscopic observations of Jean Desmazières, who described globules of yeast as living organisms, which he named *Mycoderma cerevisiae* Desmaz. Moreover, Jean Colin reported that yeast can promote the fermentation of sugar in the absence of oxygen.[30]

It is perhaps understandable that there was some skepticism about a theory of fermentation that made living organisms the agents of the process. Such ideas had been advanced repeatedly in relation to the causation of contagious diseases, notably by Athanasius Kircher (in 1657) and by Marcus Plenciz (in 1762). In large part, these ideas were based on the observation that in all putrefying material there appeared "innumerable animalcules" widely believed to arise by spontaneous generation. Although Francesco Redi had already disputed this belief, it was greatly encouraged by Needham's report that was mentioned earlier. The contrary evidence given by Spallanzani swung some of the opinion the other way so that the issue was still in doubt at the end of the eighteenth century.[31]

I insert here some words about the excitement generated by Alessandro Volta's description in 1800 of his "electric pile." Very large batteries were constructed for Gay-Lussac and Thenard in Paris, and for Davy in London. Napoleon is said to have been greatly interested in the "pile," and the international competition that attended the construction of ever-larger Voltaic batteries may perhaps be compared to the more recent competition in the construction of particle accelerators. Gay-Lussac reported that the passage of an electric current through a fermentation mixture excited the process, but in 1843 Hermann Helmholtz showed that electricity had no effect.

The clear demonstration that the agents of fermentation are living organisms came in 1837, when three investigators (Charles Cagniard-Latour, Theodor Schwann, and Friedrich Kützing) independently, and almost simultaneously, reported their microscopic observations and experimental results. As so often in the history of scientific problems, this instance of multiple discovery[32] was an outgrowth of improvements in instrumentation, in this case the construction of achromatic compound microscopes.[33] The beginnings of microscopy in the seventeenth century (Francesco Stelluti, 1625) led to the instruments used by Malpighi, Hooke, and Grew, and, most important, Leeuwenhoek. Hooke observed pores (which he termed cells), and Leeuwenhoek saw spermatozoa, red blood corpuscles, many kinds of protozoa and bacteria (which he called "little beasts"), and globules of yeast. The simple microscope used by Leeuwenhoek permitted magnifications of several hundred diameters, an achievement not to be duplicated with compound microscopes until after 1800. Before then the limits of optical resolution were frequently exceeded by the imagination of the observers, and during the eighteenth century the use of the instrument fell into disrepute. By 1840, however, achromatic compound microscopes were widely used in several scientific specialties, including the new field of histochemistry, pioneered by François-Vincent Raspail.

Cagniard-Latour was a professor at the military school in Paris and a noted inventor. He first presented his results in 1835–1836, then in a paper before the Académie des Sciences in June 1837, published in 1838. He described brewer's yeast as composed of spherical particles that multiplied by budding, and he gave evidence for his view that alcoholic fermentation "resulte d'un phénomène de végétation." He also took special pains to exclude the possibility that the apparent multiplication of yeast during fermentation was a consequence of the precipitation of albuminous material described by Thenard. Cagniard-Latour concluded that "brewer's yeast, this widely used ferment, is a mass of small globular bodies which can reproduce themselves . . . and not simply an organic or chemical substance, as has been supposed."[34]

Schwann was associated with Johannes Müller when he published his report on alcoholic fermentation in 1837. The year before, Schwann had described his studies on pepsin, and two years later he presented his cell theory. In his 1837 paper he showed that Guy-Lussac's view of the role of oxygen in fermentation was incorrect, for to prevent fermentation or putrefaction it was sufficient to heat the air before it came into contact with a previously heated infusion of plant or animal material. This experimental approach was similar to that of Franz Schulze, who reported in 1836 that putrefaction of such infusions could be prevented by passing the air through concentrated sulfuric acid. Because atmospheric oxygen was not affected by these treatments, Schwann concluded that the agent responsible for the fermentation was carried by the air. He then turned to the microscopic observation of yeast, saw the same budding globules as did Cagniard-Latour, and named them *Zuckerpilz* ("sugar fungus," later *Saccharomyces*). He later wrote: "That this fungus is the cause of the fermentation follows, in the first place, from its constant occurrence in fermentation, secondly because the fermentation ceases under all conditions which visibly kill the fungus, namely boiling, treatment with potassium arsenite, etc., thirdly because the exciting principle in the fermentation process must be a material that is evoked and increased by the process itself, a phenomenon which applies only to living organisms."[35] Schwann considered the organism to be a plant rather than an animal, because it was resistant to the action of nux vomica, and he declared that "alcoholic fermentation must therefore be regarded as the decomposition effected by the sugar fungus, which extracts from the sugar and a nitrogenous substance the materials necessary for its own nutrition and growth, and whereby such elements of such substances (probably among others) as are not taken up by the plant preferentially unite to form alcohol."[36] This view was developed by Louis Pasteur twenty years later.

Kützing, first a pharmacist and after 1836 a science teacher at the Realschule in Nordhausen, reached the same conclusions as did Cagniard-Latour and Schwann, of whose work he knew when he wrote: "I gladly renounce a claim to priority, since it does not matter for science who first made the discovery." He also noted that yeast "is not a chemical compound, but an organized body, an organism. Unfortunately, too many truly organized structures are still being included among the chemical compounds, where they do not belong."[37]

Kützing did not identify any chemists, but he clearly must have had in mind opinions such as those expressed by Berzelius, who had written:

> The conversion of sugar into carbonic acid and alcohol, as it occurs in the process of fermentation, cannot be explained by a double decomposition-like chemical reaction between a sugar and so-called ferment, as we name the insoluble substance under the influence of which the fermentation takes place. This substance may be replaced by fibrin, coagulated plant protein, cheese and similar materials, though the activities of these substances are at a lower level. However, of all the known reactions in the organic sphere, there is none to which the reaction bears a more striking resemblance than the decomposition of hydrogen peroxide under the influence of platinum, silver, or fibrin, and it would be quite natural to suppose a similar action in the case of the ferment.[38]

In his annual report on the progress of chemistry in 1838, Berzelius wrote unfavorably about the work of Cagniard-Latour, Schwann, and Kützing, and it was clear that the most influential chemist of his time was unwilling to accept their evidence as proof of the thesis that fermentation is effected by living organisms.[39] While Berzelius had suggested the idea of a "catalytic force" (about which more later), other prominent chemists preferred the term *contact substance*, as defined by Eilhard Mitscherlich for reactions such as: "the breakdown of sugars to alcohol and carbonic acid, the oxidation of alcohol when it is converted to acetic acid, the reaction of urea and water to form carbonic acid and ammonia. As such, these substances undergo no change, but upon the addition of a small amount of ferment, which is the contact substance, and a definite temperature, these reactions take place at once."[40] On the other hand, Justus Liebig explained fermentation on the basis of the general hypothesis that it is a consequence of "the ability of a substance in decomposition or combination, *i.e.*, undergoing chemical reaction, to evoke in another substance with which it is in contact the same reaction, or to enable that substance to undergo the same changes that it undergoes itself."[41] Liebig also attached importance to the participation of atmospheric oxygen and suggested that the insoluble nitrogenous material which appears during the fermentation is derived from the soluble "gluten" by the action of oxygen. For Liebig, therefore, yeast was oxidized gluten in a state of putrefaction. These views were widely disseminated in his popular writings and, with some variation, were reflected in textbooks of the time.[42]

It is clear that the chemists were divided about the nature of fermentation, with Berzelius and Mitscherlich advocating a catalytic or contact action and Liebig resurrecting the ideas of transmitted vibration and oxidation. This disagreement is less important, however, than their agreement that the cause of fermentation is the decomposition of nitrogenous material. Mitscherlich accepted the idea that yeast is a living organism in 1841, and Berzelius admitted it in his annual report for 1848 (he died in that year), but they left open the question whether the fermentation of sugar is a process indissolubly linked with the life of the yeast cell. Despite the negative reaction of prominent chemists of the 1840s, the claims of Cagniard-Latour and Schwann were widely adopted and further developed.

In France, on behalf of a committee appointed by the Académie des Sciences to examine the validity of Cagniard-Latour's report, Pierre Turpin stated: "Fermentation must be considered to be the cooperative action of water and living organisms which develop and nourish themselves by absorption of a structural element of the sugar, and separating from it alcohol and acetic acid; a purely physiological action, which begins and ends with the existence of the small plant or animal infusoria, whose life ceases upon the complete utilization of the saccharine nutriment, whereupon it is deposited as a slimy precipitate, or yeast, at the bottom of the container."[43] A German abstract of Turpin's article appeared in Liebig's *Annalen* for 1839 and was followed by an anonymous satire (presumably written by Liebig or Wöhler, or both) in which yeast was elaborately described as a tiny organism shaped like a distilling flask; under the microscope this organism could be seen to swallow sugar, digest it in its stomach, and excrete alcohol through its digestive tract and carbonic acid through its bladder.[44] Another example of this kind of humor in Liebig's *Annalen* appeared in 1840 in the form of an equally unjust letter from "S. C. H. Windler" that satirized the work of Dumas on substitution reactions.

In addition to Turpin, during the 1840s and 1850s numerous investigators provided further evidence in support of the organismic theory of fermentation, among them Hermann Helmholtz, Théodore Auguste Quevenne, Heinrich Schroeder and Theodor von Dusch, Andrew Ure, and Robert Thomson. Moreover, at that time, this theory was generally favored by men concerned with yeast technology. The brewing industry in England, Germany, and Bohemia had expanded considerably during the first half of the nineteenth century, and several academic scientists were closely connected with this development.[45] Among these men was Karl Balling, whose influential treatise on fermentation (first edition, 1845–1847) advocated the organismic theory. Another was Friedrich Wilhelm Lüdersdorff, whose unsuccessful attempt to effect cell-free fermentation was a forerunner of Buchner's work. I mention these features of the debate before 1857, when Pasteur published his first paper on fermentation, in part because some of his biographers have tended to overemphasize his originality in recognizing that alcoholic fermentation is a process linked to the life of yeast cells.

It must be noted, however, that although before 1857 the idea that alcoholic fermentation is caused by a living organism was widely accepted, the same cannot be said of all the various other processes then considered to be fermentations. Thus in addition to alcohol, the products of the fermentation of sugar might be lactic acid (as in the souring of milk) or butyric acid (as in the souring of bread). In animal physiology the term *fermentation* was also applied to such processes as the formation of ammonia from urinary urea, as well as the transformation of proteins, starches, and fats by agents in the saliva, the gastric juice, and the pancreatic juice. Before 1850 some of these agents had been obtained in a soluble form and, although produced by living organisms, such "soluble ferments" were able to exert their action independently of the life of those organisms. For the chemist Marcelin Berthelot, writing in 1860, one of the urgent scientific questions of the day was the relation of the soluble ferments to the living organisms that cause the fermentation of sugar, and he believed that processes such as alcoholic fermentation are the consequence of the production by the living organisms of "insoluble ferments" similar to the known soluble ferments: "If a deeper study leads to the extension of the view I propose here, and to its application with certainty to the insoluble ferments, as well as to the soluble ferments, all the fermentations would be brought back to the same general concept, and they could be definitively assimilated to effects of acids provoked by contact, and of truly chemical reagents. That is an absolutely essential result. Indeed, in every fermentation, one must try to reproduce the same phenomenon by chemical methods and to interpret them by exclusively mechanical considerations. To banish life from all explanations relative to organic chemistry, that is the aim of our studies."[46] In pursuing this objective, Berthelot, an energetic entrepreneur, met a formidable adversary in his slightly older contemporary, Louis Pasteur. The remarkable transformation of a brilliant and equally enterprising chemical crystallographer into a famous bacteriologist and healer has made Pasteur's scientific career one of the great success stories in the annals of science. In that transformation, Pasteur's work on fermentation occupies a central place, and its historical background requires a brief account.

Pasteur and Molecular Dissymmetry

As a student at the Ecole Normale Supérieure in Paris during the 1840s, Pasteur came under the influence of three teachers: Auguste Laurent, Gabriel Delafosse, and Jean Baptiste Biot.[47] These men represented a French tradition in chemical crystallography that had its roots in the eighteenth-century mineralogy of Jean Baptiste Romé de Lisle and, most important, of René Just Haüy; in 1784 Haüy provided a quantitative theory of crystal structure based on the assumption that the shape of crystals is determined by the geometry of their molécules intégrantes.[48] This view was challenged in 1820 by the chemist Eilhard Mitscherlich, who showed that substances of different chemical composition can have the same crystal form ("isomorphism") and that substances having the same chemical composition can crystallize in two or more different forms ("dimorphism" or "polymorphism").[49] Laurent and Delafosse studied the problem of polymorphism from somewhat different points of view. For the chemist Laurent, crystallography offered an approach to questions posed by current theories relating to the arrangement of atoms in organic compounds. For Delafosse, who had studied with Haüy, the principal objective was to refine his teacher's theory, and he was particularly interested in the phenomenon of "hemihedrism" (the appearance of asymmetry in one half of a crystal). Such asymmetry had been studied in 1820 for "rock crystal" (a form of quartz) by John Herschel, who found a correlation between the sites of hemihedrism (he used Haüy's term *plagiedral faces*) and the direction of the rotation of polarized light by the two kinds of crystals.[50]

Herschel's report came only a few years after the memorable work (1812–1818) of Biot on the polarization of light, and his demonstration that several naturally occurring oils and solutions of various organic compounds such as camphor or cane sugar were "optically active" and turned the plane of polarized light to the right or to the left. From this work there emerged a new instrument—the polarimeter—whose importance in the subsequent development of the interplay of chemistry and biology cannot be overestimated.[51] Biot recognized that the optical activity of the organic compounds was a property of their molecules, and the idea of the right- or left- "handedness" (now termed chirality; Greek *cheir*, hand) as a feature of molecular structure has played a large role in the development of modern chemical thought.

Pasteur's first research efforts (1847–1848) dealt with the dimorphism of various substances, including salts of tartaric acid, whose crystal forms had been described in 1841 by Frédéric Hervé de la Provostaye and whose optical activity had been examined (1844) by Mitscherlich, who found one of the salts (a "paratartrate") to be optically inactive; this form also came to be called a racemate. Initially, Pasteur studied the relationship between the crystal forms of these salts and their chemical constitution, especially as regards their content of water of crystallization. When he began to use a polarimeter as well as a microscope, he discovered that the paratartrate (a sodium ammonium tartrate) was composed of two nonsuperimposable hemihedral crystal forms, which he separated by hand and which, in aqueous solution, rotated the plane of polarized light in opposite and approximately equal directions. That this outstanding achievement gained a legendary status was in large measure a consequence of the manner in which Pasteur reported the result in 1848 and described it afterward. In his subsequent papers Pasteur defended the originality of his work by emphasizing "molecular dissymmetry" as the basis for distinguishing the correlation of hemihedrism and optical activity for quartz from that for salts of tartaric acid, but he was also obliged to revise his initial conclusion about the generality of that correlation.[52] There can be little doubt, however, that Pasteur's subsequent studies on the optical activity of asparagine, aspartic acid, and malic acid led him to the view that if an organic compound is optically active it must have been formed by a physiological process. In an oft-quoted lecture delivered in 1860, when he was already engaged in studies on fermentation, he distinguished between organic substances prepared in the laboratory and those obtained from biological sources by stating that "the artificial products do not have any molecular dissymmetry; and I could not indicate the existence of a more profound separation between the products born under the influence of life and all the others." In 1883 Pasteur returned to this topic in another lecture before the Chemical Society of Paris, speaking of dissymmetry as a "chapter in molecular chemistry" and insisting that in order to produce dissymmetric compounds in the laboratory, it is necessary to apply dissymmetric forces.[53]

Apart from this seemingly vitalist attitude, Pasteur made significant contributions to practical organic chemistry, notably in his use (1852) of alkaloids as reagents for the separation of the components of racemic mixtures. It should be noted, however, that he appears to have been indifferent to the great chemical debates of his time, including the structure theory of Kekulé and the stereochemical theory of LeBel and van't Hoff. In my opinion, one does not diminish the importance of Pasteur's work on molecular dissymmetry by denying that he was, as some have stated, the "father of stereochemistry."[54]

From Pasteur to Nägeli

Pasteur's exceptional qualities, both as an experimenter and as a scientific adversary, were evident in his first publication (1857) on the fermentation problem, which dealt with the formation of lactic acid from sugar. Much was already known about this kind of fermentation. In 1780 Scheele had isolated lactic acid from soured milk, and the elemental composition of this acid had been established in 1833 by Gay-Lussac and Pelouze. It had been formed from sugar by means of "albuminoid" materials, and in 1843 Pelouze and Gélis (following up the earlier work of Boutron-Charlard and Frémy) reported that the addition of chalk to the fermentation mixture markedly increased the amount of lactic acid produced. In his paper, Pasteur stated: "I intend to establish in the first part of this work that, just as there is an alcoholic ferment, the yeast of beer, which is found everywhere where sugar is decomposed into alcohol and carbonic acid, so also there is a particular ferment, a lactic yeast, always present when sugar becomes lactic acid, and if all nitrogenous material can transform sugar into this acid, it is because it is a suitable nutrient for the development of this ferment."[55] In this paper, and a fuller one published in the following year, Pasteur reported that he had in fact seen, under the microscope, "little globules . . . much smaller than those of beer yeast," and he described their isolation "in a state of purity." He noted that "the purity of a ferment, its homogeneity, its free development, with the aid of nourishment well suited to its individual nature, these are among the essential conditions for good fermentations. Now, in this regard, one must know that the conditions of neutrality, alkalinity, acidity, or chemical composition of the liquid play a large role in the preferential development of individual ferments, because their life does not accommodate itself to the same degree to various states of the media."[56] This passage provides a succinct statement of the reasons for Pasteur's successes in the study of fermentation, and it may also be considered to mark the beginning of experimental microbiology.

During 1857–1860 Pasteur communicated, in rapid succession, a series of remarkable preliminary papers on alcoholic fermentation. In the light of the previous history of the problem, the most important of these reports was the one showing that ammonia is not a normal product of the fermentation of sugar by yeast; on the contrary, if one added to a solution of pure sugar an ammonium salt such as ammonium tartrate, the mineral components of yeast (containing phosphate), and a very small amount of fresh yeast, the yeast developed and fermented glucose, while the ammonia disappeared. He stated: "In other words, the ammonia is transformed into complex albuminoid material which enters the structure of the yeast, while at the same time the phosphate gives to the new globules their mineral principles. As to the carbon, it is evidently furnished by the sugar."[57] In other reports in this series, Pasteur denied that there is a chemical equation, such as that of Lavoisier and Gay-Lussac, for the fermentation of sugar to alcohol and carbonic acid, for he found succinic acid and glycerine to be normal products of alcoholic fermentation; he attributed the acidification in such fermentations to the succinic acid.

In 1860 Pasteur presented an extended and detailed report of his studies on alcoholic fermentation, and concluded that

> the variations in the proportions of succinic acid, of glycerine, and consequently of the other products of fermentation, should not be surprising in a phenomenon in which the conditions contributed by the ferment seem of necessity to be so changeable. What has surprised me, on the contrary, is the usual constancy of the results. . . . I am therefore much inclined to see in the act of fermentation a phenomenon which is simple, unique, but very complex, as it can be for a phenomenon correlative with life, giving rise to multiple products, all of which are necessary. . . . My present and most fixed opinion regarding the nature of alcoholic fermentation is this: The chemical act of fermentation is essentially a phenomenon correlative with a vital act, beginning and ending with the latter. I believe that there is never any alcoholic fermentation without there being simultaneously the organization, development, multiplication of the globules, or the pursued, continued life of globules which are already formed. The totality of the results in this article seem to me to be in complete opposition to the opinions of MM. Liebig and Berzelius. I profess the same views on the subject of lactic fermentation, butyric fermentation, the fermentation of tartaric acid, and many other fermentations properly designated as such [*fermentations proprement dites*] that I will study successively. Now, what does the chemical act of the cleavage of sugar represent for me, and what is its intimate cause? I confess that I am completely ignorant of it. Will one say that the yeast nourishes itself with sugar so as to excrete it in the form of alcohol and carbonic acid? Will one say that the yeast produces, during its development, a substance such as pepsin, which acts on the sugar and disappears when that is exhausted, since one finds no such substance in the liquids? I have no reply on the subject of these hypotheses. I do not accept them or reject them, and wish to constrain myself always not to go beyond the facts. And the facts only tell me that all the fermentations properly designated as such are correlative with physiological phenomena.[58]

This excerpt from Pasteur's lengthy article provides further indication of his skill as a scientific debater, a quality to be evident in even greater measure in his replies to the counterattacks by Liebig in 1870 and Berthelot in 1878, as well as the various exchanges with others (notably Félix Archimede Pouchet) on the subject of the spontaneous generation of ferments. Pasteur disproved spontaneous generation once again, by means of experiments similar to those performed by Schroeder and von Dusch in 1854, in showing that the ferments came from the air.

It is fair to say that Pasteur's most significant microbiological discovery was his finding that the organism responsible for the conversion of sugar into butyric acid not only lives without free oxygen but also is killed by oxygen; he applied the term *anaérobies* to such organisms. Moreover, he found that when the yeast of beer develops in the absence of oxygen, it can ferment sugar, but that it loses this ability when it is grown in air. From these and related observations, Pasteur concluded that fermentation is a consequence of anaerobic life: "In summary, besides all the hitherto known organisms which without exception (at least it is so believed) cannot respire and nourish themselves without assimilating free oxygen gas, there is a class of organisms whose respiration is sufficiently active so that they can live without the influence of air by taking the oxygen from certain compounds, which are thereby slowly and progressively decomposed."[59] In later writings, Pasteur repeatedly referred to this *vie sans air* as a process by means of which an organism takes oxygen from a suitable compound, such as sugar, and he demonstrated experimentally the capacity of various microorganisms to live in the absence of oxygen gas, whereupon (according to his definition) they become ferments. This discovery of anaerobic microbial life provided a fruitful basis for the subsequent work of others, notably Winogradsky, Beijerinck, and Kluyver, who modified and clarified the views expressed by Pasteur. Moreover, during 1870–1890, the impact of Pasteur's results and ideas on yeast technology was evident in the writings and practice of such men as Adolf Meyer, Max Delbrück, and Emil Christian Hansen.[60]

In their relation to the history of the fermentation problem, however, Pasteur's experimental findings and the conclusions he drew from them must be considered separately. There can be no doubt that his demonstration of the growth of yeast and other microorganisms in a medium devoid of albuminoid material demolished the theory popularized by Liebig, who in his counterattack of 1870 questioned the validity of Pasteur's claim. Pasteur's answer was a terse and scornful challenge that the matter be submitted to impartial inquiry; Liebig never replied (he died in 1873), and, according to his biographer, the impact of Pasteur's work and words on Liebig was shattering.[61] Nor can there be any doubt that Pasteur's insistence on the necessity of working with pure cultures of microorganisms, of taking into account their age, and controlling carefully the conditions of their growth, markedly furthered the development of a sound microbiology, or that the failure to consider these factors led many of his opponents into error.

In recognizing the significance of Pasteur's experimental achievements, it is also necessary to note that he chose his scientific language in a manner best calculated to accord with his preconceived ideas. His use of the term *fermentations proprement dites* at once narrows the ground of discourse, and to say that such fermentations are "absolutely dependent on the presence of living organisms" is a skillful debating point but is none the less a tautology. What about all the other phenomena termed fermentations by his contemporaries? Clearly, for Pasteur, they were not fermentations, at least in 1860. For example, in that year, Berthelot reported that he had prepared from yeast a water-soluble ferment that changed cane sugar into invert sugar. Pasteur had suggested that the inversion of cane sugar by yeast might be caused solely by the succinic acid produced upon fermentation. Berthelot showed that this explanation was incorrect and proposed that there is no fundamental difference between alcoholic fermentation and the inversion of cane sugar. Pasteur's prompt rejoinder was that Berthelot "here calls *ferment* substances soluble in water, and able to invert sugar. Now everyone knows that there are very many [*une foule de*] substances which enjoy this property, for example, all the acids. As for me, when it is a question of cane sugar and beer yeast, I only term ferment that which causes the fermentation of sugar, that is to say which produces alcohol, carbonic acid, etc. As for inversion, I have not occupied myself with it. In regard to its cause, I only proposed a doubt in passing, in a note in a memoir in which I summarized three years of study on alcoholic fermentation."[62] Apart from the personal rivalry between Pasteur and Berthelot, it is clear that the debate was being conducted with different scientific terminology. To this may be added a parenthetical note on the Franco-German antagonism evident in the debate about the nature of fermentation. Armand Gautier wrote of the "French vitalist theory" and the "German theory" of Liebig, while Cosmas Ingencamp complained about Pasteur's neglect of Schwann's contributions.[63]

Although some accounts of the nineteenth-century debate about the nature of fermentation have emphasized the Pasteur-Liebig exchange in 1870, with its overtones of nationalistic feeling, equal importance should be given to the later confrontation between Pasteur and Berthelot over some posthumously published experiments performed by Claude Bernard. Bernard died on 10 February 1878, and on 20 July of the same year there appeared in the *Revue Scientifique* an article submitted by Berthelot and containing notes regarding experiments on alcoholic fermentation conducted by Bernard during the fall of 1877. The notes were obviously not intended for publication, but it was clear that Bernard believed that he had shown that fermentation had occurred in the juice of rotting fruit without the participation of living cells and that the conversion of sugar into alcohol could be effected by agents separable from living yeast. The sponsorship by Berthelot of Bernard's work must not be taken to indicate that the two men held the same views regarding the organismic theory of fermentation. While Berthelot wished "to banish life from all explanations relative to organic chemistry," Bernard's work on carbohydrate metabolism during the 1840s and 1850s (which will be discussed later) led him to make a distinction between synthetic or "anabolic" processes, which he considered to be associated with life, and degradative processes, which he considered to be purely chemical processes, independent of intact cells. In his last book, published ten days after his death, Bernard wrote:

> In my view, there are necessarily two orders of phenomena in the living organism: 1) The phenomena of *vital creation* or *organizing synthesis;* 2) the phenomena of death or *organic destruction*. . . . The first of these two orders of phenomena is alone without direct analogues; it is specific [*particulier, spécial*] to the living being; this evolving synthesis is what is truly vital. . . . The second, namely vital destruction, is on the contrary of a physico-chemical order, most often the result of a combustion, of a fermentation, of a putrefaction, in a word of an action comparable to a large number of chemical decompositions or cleavages. These are the true phenomena of *death,* as applied to the organized being. And, it is worthy of note that we are here the victims of a habitual delusion, and when we wish to designate the phenomena of *life,* we in fact indicate the phenomena of *death.*[64]

It is clear from this excerpt that Bernard did not accept Pasteur's claim that alcoholic fermentation is a process correlative with life, and the notes published by Berthelot indicated that Bernard had set out to disprove it.

Pasteur's first reply came on 22 July, only two days after the *Revue Scientifique* article appeared, and there followed an acerbic but inconclusive debate before the Académie des Sciences. As for Bernard's apparent belief that he had observed the cell-free formation of alcohol from sugar, Pasteur stated: "I must add finally that it is always an enigma to me that one could believe that I would be disturbed by the discovery of soluble ferments in the fermentations properly designated as such, or by the formation of alcohol from sugar, independently of living cells. Certainly, I confess it without hesitation, and if one wishes, I am ready to explain myself on this point at greater length. I do not now see either the necessity for the existence of these ferments or the utility of their function in this kind of fermentation."[65] In the following year, Pasteur published a lengthy critique of Bernard's notes, based on new experiments showing that the fermentation of rotting fruits in air is not observed if one carefully shields the fruit from contact with microorganisms. It should be noted that in 1869 Lechartier and Bellamy had reported that various fruits can form alcohol from their sugar when kept in closed vessels; Pasteur confirmed this finding and showed that in the absence of oxygen and of microbial growth, cells of higher plants were able to form alcohol. In discussing Bernard's evidence for cell-free fermentation, Pasteur also questioned the reliability of Bernard's technique for determining the presence of alcohol, and indeed suggested that Bernard may have been so presbyopic that his visual observations were in serious error as a consequence of his poor eyesight. And in characteristic style, Pasteur stated: "As much as anyone, I attach importance to the substances which are called soluble ferments; I would not be at all surprised to see that yeast cells produce a soluble alcoholic ferment; I would understand that every fermentation could be caused by a ferment of this kind; but it is more difficult for me to imagine that such agents should be formed by cells given over to organic destruction in a fruit or a cadaver that is rotting."[66] Once again, Pasteur held the field against his opponents, but there appears to have been some modification in his attitude toward the question of a soluble alcoholic ferment, as compared with the one he evinced in 1860.

To see the changes in Pasteur's views in somewhat broader perspective, one may note that in 1858 the chemist Moritz Traube had written the following about the organismic theory of fermentation: "Even if all putrefactions depended on the presence of infusoria or fungi, a healthy science would not block the road to further research by means of such a hypothesis; it would simply conclude from these facts that the microscopic organisms contain certain substances which elicit the phenomena of decomposition. It would attempt to isolate these substances, and if they could not be isolated without changed properties, it would only conclude that all the separation methods had exerted a deleterious chemical effect on these substances."[67] Indeed, before Buchner's work, there had been several other reports of attempts to prepare from yeast a soluble ferment which would convert sugar into alcohol. In 1846 Lüdersdorff described experiments in which he ground yeast on a glass plate until no globules were visible under the microscope, and reported that the resulting material failed to ferment sugar. In the following year, Carl Schmidt repeated these experiments, with longer trituration, and explained the negative result as a consequence of the destruction of the ferment; during the course of this work, Schmidt anticipated Pasteur's finding of succinic acid as a regular by-product in alcoholic fermentation. Marie von Manassein claimed in 1872 to have demonstrated cell-free fermentation, and after the appearance of Buchner's first paper she reiterated this claim, but Buchner questioned its validity on the ground that intact organisms may have survived her treatment. In Pasteur's laboratory Denys Cochin attempted, without success, to extract a soluble alcoholic ferment, and Emile Roux later reported that Pasteur himself had tried to do so, although he "did not think it actually existed."[68]

In addition to his studies on the fermentation of sugar to lactic acid, alcohol, and butyric acid, Pasteur also explained the formation of acetic acid from alcohol as a process caused by an aerobic microorganism which he named *Mycoderma aceti.* In his controversy with Liebig, much attention was given to the question of whether such organisms are involved in the conversion of wine to vinegar, for Liebig attached great importance to the fact that this oxidation had been effected by purely chemical means, notably by Johann Wolfgang Döbereiner, with platinum as a catalyst. Another process considered by Pasteur to belong to the category of "fermentations properly designated as such" was the conversion of urinary urea to ammonium carbonate. During the course of his debate during the 1860s about spontaneous generation, Pasteur observed the presence in urine of a torulous organism and stated: "I am led to believe that this production constitutes an organized ferment and that there is never any transformation of urea into ammonium carbonate without the presence and development of this small plant."[69] In 1876, however, Frédéric Musculus obtained from ammoniacal urine a "soluble ferment" (named urease in 1889 by Bourquelot) that readily converted urea into ammonium carbonate, and this finding was promptly confirmed by Pasteur (with Jules François Joubert). In his discussion, Pasteur stated:

> Physiologists will no doubt notice that here is the first example of an autonomous organized ferment which can be grown in any media, provided only that they are suitable for its nutrition, and able to form, during its development, a soluble material that can cause the same fermentation as that effected by the microscopic organism. . . . The yeast of beer produces a soluble ferment which inverts cane sugar, but which is independent of the function of yeast, at least as it is expressed on glucoses properly designated as such, where there is no need for inversion. In other words, the function of the soluble inverting enzyme is not the same as the function of these yeasts. This is not so with the soluble ferment of urea. The soluble ferment and the organized ferment act in the same way on their fermentable material, *i.e.,* urea, because the soluble ferment presupposes the existence of an organized being and, inversely, because the little plant necessarily gives rise to the soluble ferment.[70]

That the finding of a soluble agent of urea fermentation did not fit into Pasteur's notions is indicated by his statement that this result "was not and could not have been foreseen. It is the first example of an autonomous organized ferment whose function merges with the function of one of its unorganized products. It is also a new example of a *diastase* produced during life and able to modify a substance by the fixation of water, in the same manner as for all the *diastases.*"[71]

These remarkable statements not only attest to Pasteur's skill as a scientific debater and evoke recollection of his earlier argument with Berthelot but also represent one of the many instances in the history of the interplay of chemistry and biology where an "imaginative preconception of what the truth might be"[72] has had to yield to new empirical discovery. Moreover, the phrasing of these statements suggests that, as has been mentioned, Pasteur was largely indifferent to the transformation of the organic chemistry of his time. It should also be noted that although during the nineteenth century urease was considered to be a microbial ferment (in 1890 it was obtained from various bacteria by Pierre Miquel), it was later found in plant seeds, notably jack beans, from which it was isolated in crystalline form by James Sumner in 1926.

As for Pasteur's assertion that the discovery of ferments such as urease "could not have been foreseen," it is perhaps appropriate to note, in hindsight, the views of the chemist Moritz Traube, who in 1878 reiterated those he had expressed twenty years earlier:

> 1) The ferments are not, as Liebig assumed, substances in a state of decomposition, and which can transmit to ordinarily inert substances their chemical action, but are chemical substances related to the albuminoid bodies which, although not accessible in pure form, have like all other substances a definite chemical composition and evoke changes in other substances through definite chemical affinities. 2) Schwann's hypothesis (later adopted by Pasteur), according to which fermentations are to be regarded as the expressions of the lower organisms, is unsatisfactory. . . . The reverse of Schwann's hypothesis is correct: Ferments are the causes of the most important vital-chemical processes, not only in lower organisms, but in higher organisms as well.[73]

Traube's views may have been attractive to organic chemists, but most biologists found greater merit in those of the influential botanist Carl Nägeli:

> The agent of fermentation is inseparable from the substance of the living cell, *i.e.*, it is linked to plasma [Nägeli's word for protoplasm]. Fermentation occurs only in immediate contact with plasma in so far as its molecular action extends. If the organism wishes to exert an effect on chemical processes in places or at distances where the molecular forces of living matter are without power, it excretes ferments. The latter are especially active in the cavities of the animal body, in the water in which molds live, and in the plasma-poor cells of plants. It is even doubtful whether the organism ever makes ferments that are intended to function within the plasma; since here it does not need them, because it has available to it in the molecular forces of living matter much more energetic means for chemical action.[74]

From Diastase to Zymase

In 1833 Anselme Payen and Jean François Persoz reported that the addition of alcohol to an aqueous extract of germinating barley (malt) precipitated flocculent material which, when dried and redissolved in water, could liquefy starch paste and convert it into sugar. They named this material diastase (Greek *diastasis,* making a breach) because they considered it to effect the bursting of the outer envelopes of starch granules *(fecula)*,[75] and they later identified it in germinating oats, wheat, corn, and rice. This work followed that of Antoine Augustin Parmentier and Antoine François Fourcroy during 1780–1800, suggesting that acids can convert starch into sugar (recognized by taste), and especially that of Constantin Kirchhoff, who reported in 1811 that upon being treated with hot dilute sulfuric acid, starch is converted to sugar without any apparent change in the acid. Kirchhoff also found in 1814 that an aqueous extract of dry malt could effect this conversion and attributed it to some property of the gluten in the malt. These results were interpreted as indicating that "the gluten, in combination with the starch, appears only to accelerate a decomposition that the latter would have suffered in a longer time without this influence."[76] A similar ability to convert starch into sugar was found in human saliva by Erhard Leuchs in 1831, and in 1845 Louis Mialhe precipitated a "salivary diastase" (later termed ptyalin) from this source. In the latter year, Apollinaire Bouchardat and Claude Sandras also reported the diastase activity of pancreatic juice. Such activity was subsequently found in a great variety of other plant and animal materials and at the end of the nineteenth century it came to be called amylase (Latin *amylum,* starch).

As regards the identification of the products formed by the action of the malt diastase on starch, it was Dubrunfaut who isolated a new crystalline sugar (maltose), which, upon treatment with acid, was converted to the dextrorotatory glucose (dextrose). At about the same time, he showed that "invert sugar" is a mixture of dextrose and the levorotatory "levulose" (fructose, isolated by Bouchardat upon the cleavage of inulin). These results were based largely on measurements of optical activity and the reduction of cupric salts (Fehling's test). Dubrunfaut's identification of maltose as a major product of the action of malt diastase was not generally accepted until 1872, when his work was confirmed and greatly extended by Cornelius O'Sullivan; since maltose and glucose give different optical rotations and reduce cupric salts to a different degree, serious errors were made because of the neglect of Dubrunfaut's finding. O'Sullivan also extended earlier observations on the formation of a second, less well defined product (dextrin) of the action of dilute acids or of diastase on starch; the recognition of the appearance of dextrin (a partially degraded starch) came from the use of iodine as a reagent for the detection of starch. This reaction was introduced in 1814, shortly after the discovery of iodine by Bernard Courtois.

By the 1830s chemists came to consider alcoholic fermentation and the cleavage of starch by acids and diastase to be phenomena similar to the decomposition of hydrogen peroxide by metals and "blood fibrin" (Thenard) or the effect of platinum in promoting the combination of hydrogen and oxygen (Davy, Faraday, Döbereiner), or the effect of sulfuric acid on the conversion of alcohol into ether. The last had been studied by Mitscherlich, who considered all these phenomena, whether or not caused by ferments, to be examples of "decompositions and combinations by contact." Berzelius introduced the idea of a "catalytic force" and defined "catalysis as the decomposition of substances by this force, just as one defines analysis as the decomposition of substances by means of chemical affinity." As has been mentioned, Liebig did not welcome this idea, nor Berzelius's suggestion that "thousands of catalytic processes take place between the tissues and the liquids and result in the formation of the great number of different chemical compounds, for the production of which from the common raw material, plant juice or blood, no probable cause could be assigned. The cause will perhaps be discovered in the future in the catalytic power of the organic tissues of which the organs of the living body consist."[77] Although the partial validity of Berzelius's suggestion was not established until many decades later, during the nineteenth century much attention was devoted to other "soluble ferments," especially those involved in the digestion of foodstuffs, in particular the albuminoid components of the animal diet.

The role of ferments in the digestion of food by animals had been the subject of speculation since the time of van Helmont, and the eighteenth-century experimental studies of Réaumur and Spallanzani were followed by the joint work of the chemist Leopold Gmelin and the anatomist Friedrich Tiedemann, as well as the famous investigation by William Beaumont.[78] In 1834 Johann Nepomuk Eberle showed that an acidic extract of gastric mucosa causes the dissolution of coagulated egg white. This finding was followed up by Theodor Schwann in a brilliant investigation in which he identified the active principle as a soluble ferment he named pepsin, which he compared to the agent in alcoholic fermentation. Schwann also described the relationship between pepsin and the hydrochloric acid identified some years earlier as the acid of gastric juice by William Prout and by Jean Louis Prévost and Augustin LeRoyer. Schwann precipitated pepsin with lead acetate; treatment of the precipitate with hydrogen sulfide brought the active principle back into solution. In succeeding years, other methods for the preparation of pepsin were developed in response to its medical use in gastric disorders.[79]

Some French investigators did not accept these conclusions. In 1825 François Leuret and Jean-Louis Lassaigne asserted that the principal acid of gastric juice is lactic acid, and this view was adopted during the 1840s by Claude Bernard and his chemical colleague Charles Louis Barreswil. They also suggested that gastric juice, pancreatic juice, and saliva contain the same active principle, the differential action on albuminoid substances or starch being largely determined by the acidity or alkalinity of the medium.[80] This opinion was contradicted by Louis Mialhe:

> Each of the ferments has an action appropriate to itself. One of them, salivary diastase, liquefies starch in less than a minute, and transforms it into dextrin and glucose; another, pepsin, which possesses no saccharifying action on starch, coagulates milk, fibrin, and gluten; then dissolves the coagulum, and subjects it to a very particular kind of molecular transformation. . . . It is not possible to agree with Liebig, Bernard, Barreswil, and others, that the ferments are instantly produced and destroyed as soon as the need for their action is felt, or that these ferments are one and the same principle which exhibits different qualities depending on the medium in which it is placed, and depending on the substance to which it is exposed. For us, these materials are special and distinct, each conserving its nature, its particular role, and its complete independence. . . . Up to the present, we know only two, diastase and pepsin, but there certainly exist others which also participate in the maintenance of life.[81]

During the nineteenth century, among the problems posed by the soluble ferments which caused the coagulation of albuminoid substances and their subsequent dissolution was the difficulty of defining their chemical action in a manner comparable to that worked out for diastase. For this reason, for example, it was not clear whether the "chymosin" (from calf stomach) described in 1840 by Jean Deschamps as the active principle which coagulates milk (*présure, Labferment,* rennet) is identical with Schwann's pepsin. Moreover, the "pancreatin" identified by Tiedemann and Gmelin was not listed as a separate ferment until 1857, when Lucien Corvisart demonstrated the presence, in pancreatic juice, of a comparable ferment active in alkaline media; it was later named trypsin by Willy Kühne. Subsequently, trypsin-like ferments were found in plant extracts; these "vegetable trypsins" included "papain," identified by Adolphe Wurtz in 1879. During the 1850s the findings of Graham and Dubrunfaut on the diffusion of various substances through animal membranes were applied to the study of the action of pepsin, and the rapidly diffusible products were named albuminoses by Mialhe or peptones by Carl Gotthelf Lehmann. Later work by Kühne involved the separation of the products by precipitation with inorganic salts and only emphasized further the difficulty of defining the action of the protein-degrading ferments.

We have seen in connection with the Pasteur-Berthelot controversy that in 1860 Berthelot had obtained from yeast a soluble ferment which cleaves sucrose to glucose and fructose, and that since this process involves an "inversion" of optical activity, the ferment was named *ferment inversif* or invertin (later, invertase); in 1864 Antoine Béchamp called it zymase, but this name faded from view until it was revived by Eduard Buchner. The fact that the process could be followed with some precision by measurement of the change in optical rotation was important in the early quantitative studies on the kinetics of ferment action.

Another of the soluble ferments identified during the 1830s came from work on the constitution of amygdalin, obtained from bitter almonds, and crystallized by Robiquet and Boutron-Charlard. In 1837 Liebig and Wöhler reported that an extractable albuminoid material, present in both sweet and bitter almonds, decomposes amygdalin with the formation of benzaldehyde, sugar and hydrocyanic acid; they named this active principle "emulsin" and compared its action to that of yeast in alcoholic fermentation. The name was criticized in 1838 by Robiquet, who preferred to compare the process to that effected by diastase, and he termed the ferment "synaptase" (the name Pasteur used in 1878). Emulsin was the first soluble ferment to be described as having an action on a well-defined crystalline compound whose structure had been largely elucidated during the nineteenth century and whose cleavage was shown to involve the addition of the elements of water to the products (hydrolysis). In addition to amygdalin, some other naturally occurring derivatives of glucose (salicin, phlorizin, helicin, and arbutin) were also shown to be hydrolyzed by preparations of emulsin. This knowledge served as a background for Emil Fischer's studies during 1894–1898, about which more shortly.

In all, by the end of the nineteenth century the list of soluble ferments had grown to about two dozen.[82] Except for a few, such as laccase and tyrosinase, which catalyzed oxidation reactions, all the known soluble ferments promoted the hydrolysis of their substrates (a term introduced by Emile Duclaux in about 1880). In addition to those mentioned above, a ferment (lipase) which promoted the hydrolysis of fats had been identified in pancreatic juice and in plant seeds, separate ferments had been implicated in the breakdown of glycogen, inulin, pectin, and cellulose, and the hydrolysis of maltose to glucose was attributed to a "maltase" different from invertase. These soluble agents were also termed unorganized (or unformed) ferments, to distinguish them from the "organized" (or "formed") ferments, namely the microorganisms that caused the fermentation of sugar to alcohol, lactic acid, or butyric acid. In 1876 Willy Kühne proposed that the "unformed or unorganized ferments, whose action can proceed without the presence of organisms or outside of them, be denoted enzymes," and two years later he expanded on this proposal by stating that the designations of "formed" and "unformed" ferments

> have not gained wide acceptance, in that on the one hand it was stated that chemical bodies, like ptyalin, pepsin, etc., could not be called ferments, since the name was already assigned to yeast cells and other organisms (Brücke), while on the other hand it was said that yeast cells could not be called ferment, because then all organisms, including man, would have to be so designated (Hoppe-Seyler). Without wishing to inquire further why the name has generated so much excitement from opposing sides, I have taken the liberty, because of this contradiction, of giving the name enzymes to some of the better-known substances, called by many "unorganized ferments." This is not intended to imply any particular hypothesis, but it merely states that in zyme [yeast] something occurs that exerts this or that activity, which is considered to belong to the class called fermentative. The name is not, however, intended to be limited to the invertin of yeast, but is intended to imply that more complex organisms, from which the enzymes pepsin, trypsin, etc., can be obtained, are not so fundamentally different from the unicellular organisms as Hoppe-Seyler, for example, appears to think.[83]

It has been stated that Kühne "coined the new word 'enzyme,'" but that word was used by medieval theologians in their own dispute whether the Eucharist should be celebrated with leavened or unleavened (enzyme or azyme) bread.[84] Kühne's proposal was rejected by Hoppe-Seyler, who had preferred to make a distinction between biological organisms and the catalytic ferments they elaborate and who insisted that, like the hydrolytic reactions, fermentations are chemical processes.[85] The term *enzyme* was adopted fairly readily in England, in part because of the similarity of the words *fermentation* and *ferment.* In German, *Gährung* and *Ferment* are sufficiently different, and although *Enzym* came to be widely used by German-speaking biochemists, some of them (notably Otto Warburg) preferred *Ferment.* In France, *diastase,* the term introduced by Payen and Persoz in 1833, was gradually replaced by *enzyme* after 1900. That usage was accompanied by a shift from its original feminine gender to the masculine;[86] according to *Le Monde* (3 June 1970), the Académie Française had ruled on 5 February 1970 in favor of the feminine. The abandoned term *diastase* provided the suffix *ase* to be added to some part of the name of the substrate, as in *proteinase* or *protease.* In my opinion, the most significant aspect of the shift from *ferment* to *enzyme* was that the meaning of the latter term came closer to the broad definition of *Ferment* offered by Traube and Hoppe-Seyler than to Kühne's restrictive definition of "unformed" ferments.[87]

As I have suggested, Eduard Buchner's preparation of "zymase" attracted particular attention at the turn of the century. More recently, that achievement has been considered to mark the rise of the "enzyme theory of life," and one distinguished scientist has written that "enzymes are the machine tools of the living cell. They were first discovered in 1897 by Buchner, who received a Nobel Prize ten years later for his discovery."[88] It seems necessary, therefore, to emphasize that Buchner's report, so significant in relation to the debates that preceded it and the experimental advances that followed, was a matter of improved technique rather than new theoretical insight, and that it owed much to the stimulus and encouragement of his older brother Hans, at that time professor of hygiene at Munich, and to the technical skill of Hans's associate Martin Hahn. After his Ph.D. (1888) for work on diazo compounds with Curtius in Baeyer's institute, Eduard Buchner held brief appointments there and at Kiel and Tübingen. During the 1890s Hans Buchner and Hahn were seeking to develop more effective means of disrupting microbial cells and to obtain cell-free protein preparations from such cells in order to test the idea that pathogenic microorganisms produce not only the toxins that cause disease but also the antitoxins found by Behring and Kitasato in the sera of immunized animals. As a former associate of Nägeli's, Hans believed that it might be possible to replace such serum therapy through the administration of the protoplasmic protein (he termed it plasmine) obtained by mechanical disintegration of bacterial cells. Martin Hahn developed a method of grinding cells with quartz and adding kieselguhr to give sufficient consistency to the resulting paste so that it could be safely subjected to the high pressure of a hydraulic press. When Hahn applied his method to brewer's yeast, the resulting press juice underwent rapid change, and various substances were added as possible preservatives. Among those tested was sucrose (in high concentration, as in the preservation of fruit juices), and its fermentation by the cell-free extract was observed.[89] Because of the prominence of the controversy about the nature of alcoholic fermentation during the preceding decades, the significance of this observation was readily apparent. Some noted physiologists were not convinced; for example, Max Rubner noted that the cell-free extract was much less effective in fermenting sugar than was intact yeast, and he maintained that although a small part of yeast fermentation might be caused by Buchner's zymase, the major role was played by the "living proteins" in the protoplasmic structure.[90] Many of Eduard Buchner's papers, including his last one (in 1914; he died in 1917 while on military service) were defenses of his claim that the cell-free extract did not contain "bits of protoplasm" as suggested by critics who accepted Nägeli's theory of fermentation. Moreover, in response to Richard Neumeister's suggestion that the fermentation effected by Buchner's yeast juice is caused not by a single substance but by a more complex set of cell constituents, Buchner believed it to be "provisionally expedient to adhere to the simpler assumption of a homogeneous zymase as the agent of fermentation."[91]

In the light of the subsequent history of the fermentation problem, it is clear that while Buchner's achievement brought the study of the process into the stream of enzyme research, most of his conclusions required later revision. Although he recognized that zymase differed in many respects from enzymes such as pepsin or invertase, most notably in its great lability, Buchner considered zymase to be a single catalytic agent similar to the known enzymes. As an organic chemist and a pupil of Adolf Baeyer, Buchner gave weight to his teacher's theory, published in 1870, which predicted that in the alcoholic fermentation of glucose, the two central carbon atoms of the hexose should appear in the carbon dioxide, and the other four carbon atoms in the ethyl alcohol. This prediction turned out to be correct but had little impact on the developments which led to the elucidation of the process, for the empirical research of numerous biochemists between 1900 and 1940 showed that what Buchner thought to be a reaction catalyzed by a single agent is a complex series of reactions effected by a dozen separate enzymes. Indeed, Buchner's experimental contributions to the study of the action of his cell-free extract were soon overshadowed by those of other investigators, notably Augustyn Wróblewski and Arthur Harden, of whom more later. Also, before 1930, Richard Willstätter, another renowned pupil of Baeyer's and a Nobel Prize winner, claimed that the action of Buchner's yeast juice on sugar is different from that in living yeast. This claim must be taken into account in assessing Willstätter's later recollection that after the appearance of Buchner's first paper on zymase, Baeyer said "This will bring him fame, even though he has no chemical talent."[92]

As we shall see in later chapters, despite these doubts and criticisms, the study of Buchner's zymase led to new insights into the nature of enzyme action and of metabolic pathways in living organisms. Moreover, as Jacques Loeb expressed it, "Through the discovery of Buchner, Biology was relieved of another fragment of mysticism. The splitting up of sugar into CO_2 and alcohol is no more the effect of a 'vital principle' than the splitting up of cane sugar by invertase. The history of this problem is instructive, as it warns against considering problems beyond our reach because they have not yet found their solution."[93]

The Specificity and Kinetics of Enzyme Action

During the last decade of the nineteenth century, Emil Fischer brought to enzyme chemistry the power of the recently emergent synthetic organic chemistry by preparing substrates of known structure and molecular configuration for invertase and emulsin. The work that formed the basis of that achievement represented an impressive demonstration of the fruitfulness of the new concepts of valence, molecular structure, and stereochemistry. One of Fischer's first successes as an independent investigator was the synthesis in 1874 of phenylhydrazine, and in 1883 he described its reaction with aldehydes. The work of several chemists, notably Heinrich Kiliani, had shown that the sugars glucose and galactose (the latter identified as a cleavage product of the milk sugar lactose) are aldehydes, whose structure was written as

$$CH_2(OH)\text{-}CH(OH)\text{-}CH(OH)\text{-}CH(OH)\text{-}CH(OH)\text{-}CHO,$$

and it was also known that fructose is a keto sugar of the structure

$$CH_2(OH)\text{-}CH(OH)\text{-}CH(OH)\text{-}CH(OH)\text{-}CO\text{-}CH_2(OH).$$

Fischer used phenylhydrazine for the study of the chemical constitution of these and other sugars and, in a series of brilliant degradative and synthetic operations, established the spatial arrangement about each of the asymmetric carbon atoms in the known 6-carbon and 5-carbon sugars (hexoses and pentoses). To describe the structure of such sugars, Fischer introduced projection formulas and used the symbols *d*- and *l*- to denote the configuration (not the direction of optical rotation) of related compounds.[94]

d–Glucose	*d*–Mannose	*d*–Galactose	*d*–Fructose
COH	COH	COH	CH_2OH
H—C—OH	HO—C—H	H—C—OH	CO
HO—C—H	HO—C—H	HO—C—H	HO—C—H
H—C—OH	H—C—OH	HO—C—H	H—C—OH
H—C—OH	H—C—OH	H—C—OH	H—C—OH
CH_2OH	CH_2OH	CH_2OH	CH_2OH

In 1894 Fischer reported that these four sugars were readily fermented by a variety of pure strains of yeast, whereas several closely related hexoses (*l*-mannose, sorbose) were not. This finding led him to conclude that "among the agents used by the living cell, the principal role is played by the various albuminoid substances. They are optically active, and since they are synthesized from the carbohydrates of plants, one may well assume that the geometrical structure of their molecules, as regards their asymmetry, is fairly similar to the hexoses. On the basis of this assumption, it would not be difficult to understand that the yeast cells, with their asymmetrically-constructed agent, can only attack and ferment those kinds of sugars whose geometry is not too different from that of glucose."[95] In other experiments, Fischer tested the action of an aqueous yeast extract (which he called invertin) and of an almond extract (containing "emulsin") on the two isomeric glucosides he had prepared by the reaction of glucose with methyl alcohol in the presence of hydrogen chloride; the less soluble one was named α-methyl glucoside and the other β-methyl glucoside. He wrote the structures as indicated below (later work led to revision of the size of

```
H—C—O · R                 R · O—C—H
  /  ·                          /  ·
 /  CHOH                       /  CHOH
O    ·                        O    ·
 \  CHOH                       \  CHOH
  \  ·                          \  ·
    CH                            CH
     ·                             ·
    CHOH                          CHOH
     ·                             ·
    CH2OH                         CH2OH
```

the oxygen-containing ring) and reported that the α-glucoside was hydrolyzed by invertin but not by emulsin, whereas the β-glucoside was cleaved by emulsin but not by invertin. From this result Fischer concluded that

> as is well known, invertin and emulsin have many similarities to the proteins and undoubtedly possess an asymmetrically constructed molecule. Their restricted action on the glucosides may therefore be explained on the basis of the assumption that only with a similar geometrical structure can the molecules approach each other closely, and thus initiate the chemical reaction. To use a picture, I would say that the enzyme and the glucoside must fit each other like a lock and a key, in order to effect a chemical action on each other. . . . The finding that the activity of an enzyme is limited by the molecular geometry to so marked a degree should be of some use for physiological research. Even more important for such research seems to me the demonstration that the difference frequently assumed in the past to exist between the chemical activity of living cells and of chemical reagents, in regard to molecular asymmetry, is nonexistent.[96]

Clearly, Fischer's oft-cited lock-and-key analogy does not imply, as he stated, a similarity but rather a complementarity of structure. Also, his idea that the yeast protein discriminated in favor of glucoselike sugars because it had been derived from glucose-containing carbohydrate is but another example of the limits of chemical speculation about biological processes. And on a more technical level, Fischer's use of the term *invertin* was rightly criticized by Emile Bourquelot: "Unfortunately, it has been found that the solution of invertin used by Fischer was not a solution of a single ferment. . . . This solution contained invertin, maltase, and perhaps other ferments. . . . The ingenious hypothesis he advances may perhaps correspond to the facts . . . but it will not be possible to study it until a means has been found to prepare a chemically pure soluble ferment, something that it has not been possible to do up to the present."[97] Although Fischer's work demonstrated how synthetic organic chemistry can be applied fruitfully to the study of enzyme action, and his general conclusion was confirmed and extended by later investigators,[98] the question of the homogeneity of the enzyme preparations used in such studies continued to bedevil the field of enzymology for many decades afterward.

At the root of the problem was the uncertainty whether enzymes are in fact proteins, as Buchner and Fischer thought. The analytical criteria for the identification of a substance as a protein included its nitrogen content and various color tests. As we shall see in the next chapter, the analyses of well-defined proteins such as egg albumin indicated that they contain about 16 per cent nitrogen. When elemental analysis was applied to preparations of soluble ferments, lower values were usually obtained and, as Bourquelot noted, "When it is higher, the elemental composition approaches that of albuminoid materials; in the cases where it is lower, the nitrogen content approaches zero. From this [have come] two diametrically opposite views regarding the possible results of a complete purification of the soluble ferments. For some people, this purification should lead to substances having the composition of albuminoids; for others, it should yield nitrogen-free compounds."[99] Among the color tests were the xanthoproteic test (the yellow color upon treatment with nitric acid, followed by ammonia, described by Fourcroy and Vauquelin in 1800), the purple color formed with alkaline copper sulfate, noted by Ferdinand Rose in 1833 (later named the biuret reaction), and the test introduced by Eugène Millon in 1849 (the red color formed with mercurous nitrate in acid solution). For example, in 1861 Ernst Brücke described a method for the purification of pepsin that yielded a product which rapidly liquified coagulated egg white but gave none of these color tests and was not precipitated by tannic acid (a reaction considered to detect 0.0001 per cent protein in aqueous solution).[100] Indeed, the uncertainty persisted into the 1920s, when the renowned organic chemist Richard Willstätter claimed that enzymes are small reactive molecules adsorbed on inactive colloidal material, including proteins.[101] At that time, some biochemists classed the enzymes with the hormones and vitamins as biologically active small molecules. It was not until the 1930s, after the crystallization of pepsin by John Northrop, that James Sumner's claim in 1926 for the protein nature of his crystalline urease was generally accepted. It also came to be appreciated that in a highly purified state, many enzymes are measurably active at concentrations below the limit of their detectability by means of the tests mentioned above. One consequence of the uncertainty about the protein nature of enzymes was the appearance before 1910 of theories that invoked the operation of "forces." Lambertus de Jager and Maurice Arthus proposed that the catalytic activity of enzymes is inherent in a large variety of chemical substances and that this force could even be exerted through a parchment membrane; this idea was disbelieved by William Bayliss, who suggested that "there may have been little holes." According to Hendrick Pieter Barendrecht, enzymes emit radiation absorbed by their substrates.[102]

It should also be noted that in his studies on enzymes, Emil Fischer relied on semiquantitative measurements of the rates at which various substrates underwent change. Nevertheless, his idea that an enzyme acts on its substrate by first combining with it in a specific manner greatly influenced the physical-chemical approach to the study of the kinetics of enzyme action and thus encouraged the development of methods for the quantitative definition of the specificity of individual enzymes.[103] Although the concept of an enzyme-substrate intermediate had been foreshadowed in 1880 by the report of Adolphe Wurtz that a precipitate was formed upon the addition of papain to a protein substrate, the idea was not generally accepted until after 1900. By 1914, however, useful mathematical equations had been developed for the analysis of quantitative kinetic data, obtained by means of polarimetric measurements, for the action of invertase on sucrose.

This development began in 1850, with the report of Ludwig Ferdinand Wilhelmy on his polarimetric studies on the rate of the hydrolysis of sucrose in the presence of various acids. He found that the amount of sucrose converted (dZ) in an element of time (dT) is proportional to the amount of sugar (Z) at time T and the amount of acid (S). Integration of the equation $-dZ/dT = MZS$ (M was denoted a "velocity coefficient") gave the "first-order" rate equation $\log Z_0 - \log Z = MST$, where Z_0 is the value of Z for T_0. A general rate equation was then derived by William Harcourt and William Esson (and others) for "second-order" processes such as the reaction: ethyl alcohol + acetic acid = ethyl acetate + water. These studies formed the background for the proposal by Cato Maximilian Guldberg and Peter Waage in 1864, usually denoted the law of mass action, and the subsequent formulation by van't Hoff of the concept of the equilibrium constant of a reversible chemical process in terms of the ratio of the velocities of the forward and back reactions. Then followed van't Hoff's expression for the temperature dependence of the equilibrium constant and Arrhenius's equation for the temperature dependence of the rate of a chemical reaction.[104] By about 1898 the theoretical groundwork of chemical kinetics was beginning to be well established, and the question arose whether the rates of enzyme-catalyzed reactions could be described by means of the equations developed for ordinary chemical reactions.

During the last quarter of the nineteenth century conflicting reports appeared regarding the applicability of Wilhelmy's equation to the rate of the hydrolysis of sucrose by invertase. Indeed, in 1899 Emile Duclaux concluded that enzyme action cannot be described by means of equations based on the law of mass action, and this view appears to have been widely held.[105] Moreover, although van't Hoff had affirmed that enzymes can catalyze the rates of reversible reactions in both directions, the report by Arthur Croft Hill that maltase can promote a synthetic reaction with glucose as the substrate was not accepted by more physiologically minded critics.[106]

Consequently, the adoption by Victor Henri of Emil Fischer's concept of an enzyme-substrate intermediate, and the assumption that its "active mass" determines the rate of a catalytic process, marked a new stage in the development of enzyme kinetics.[107] In the further elaboration of this approach, the contributions of S. P. L. Sørensen, of Leonor Michaelis and Maud Menten, and of George Briggs and J. B. S. Haldane played a large role in twentieth-century studies on the rates of the catalytic action of many individual enzymes.[108] The use of such data in the definition of the specificity and mode of action of enzymes other than invertase and emulsin was attended, however, by problems arising largely from the limitations of the available analytical methods for the measurement of the rates of other well-known enzyme-catalyzed processes, notably those involving the physiological breakdown of proteins.

The validity of Henri's formulation of the kinetics of invertase action was confirmed experimentally by Michaelis and Menten, whose famous equation was written as

$$v_0 = k_3(E_0)(S_0)/[K_S + (S_0)]$$

where v_0 is the *initial* velocity, (E_0) is the molar concentration of the enzyme, (S_0) is the initial molar concentration of the substrate, and the terms k_3 and K_S (the dissociation constant of ES = k_2/k_1) were defined by

$$E + S \rightleftharpoons ES \rightarrow E + \text{products}.$$

The choice of initial velocity made it possible to neglect the reversal of the hydrolytic reaction, since the concentration of the products (glucose and fructose) was small. It was assumed that (S_0) is much greater than (E_0) and that the equilibrium between ES and E + S is established much more rapidly than the conversion of ES to E +

products. A series of replicate determinations of v_0 at increasing values of (S_0) provides data for a graphical plot of v_0 versus (S_0) in the form of a rectangular hyperbola or, as shown later, of $1/v_0$ versus $(1/S_0)$ in a straight line. From such plots one may estimate, with fair precision, the magnitude of K_S (later termed K_M, the so-called Michaelis constant) and the maximal velocity V (= $k_{cat}(E_0)$). According to this formulation, the initial step in the catalysis is the reversible "productive" binding of the substrate by the enzyme, and the measured maximum velocity is that attained when all the enzyme molecules are bound to substrate molecules. As was shown by Briggs and Haldane, however, the Michaelis-Menten treatment is only a special case of a more general "steady-state" interpretation of the catalytic process. Although the kinetics of some enzyme-catalyzed reactions have been found to accord with the above assumptions, subsequent work showed that in many cases, K_M is not simply a dissociation constant. For this reason, in the use of kinetic data to define the specificity of an enzyme, the preferred parameter is now k_{cat}/K_M.[109]

Although the analytical methods used for the determination, under specified conditions of temperature and pH, of the "specific activity" (expressed in terms of some arbitrarily defined "enzyme unit") of enzymes such as invertase were based on considerable knowledge of the chemical structure and physical properties (for example, optical rotation) of the substrate and products, the methods used for enzymes such as pepsin were based on less clearly defined chemical knowledge. For example, in the measurement of the rate of pepsin action on protein substrates, samples of the reaction mixture were treated with trichloroacetic acid (a known precipitant of proteins such as hemoglobin), and the amount of soluble material formed in the catalytic process was determined by means of some colorimetric or other analytical method. Such procedures were useful in following the course of attempts to purify an enzyme[110] but gave little or no information about the nature of the reaction undergoing catalysis. For some years, much was made of the "law" proposed by Emil Schütz and confirmed by Julius Schütz, that in a given time the amount of soluble material produced from egg albumin varied with the square root of the amount of pepsin.[111] Indeed, in the case of pepsin, dependence on such methods during the 1920s contributed to the confusion about the structure of proteins. At that time, some chemists believed that proteins are long-chain molecules in which amino acid units are linked by amide or "peptide" (CO-NH) bonds hydrolyzed by pepsin, while others considered them to be noncovalently linked aggregates of small molecules and that the action of pepsin was solely to disrupt such aggregation; such a view was consistent with the colloid chemistry of the time.[112] The importance of using substrates of known chemical structure for enzyme assays has become widely appreciated by clinical chemists and by commercial manufacturers of biochemical products.[113]

With the gradual but incomplete acceptance during the 1930s of the view that the catalytic activity of all the then-known individual enzymes resides in their protein component, it became clear to many biochemists that further progress in the study of the mechanisms of enzyme action depended in large measure on the solution of several fundamental problems of protein chemistry. At that time, however, there was much debate about such matters as the peptide theory of protein structure, and the decisive experimental contributions did not come until after 1945, with the introduction of new techniques. For this reason I postpone the further discussion of the historical development of reliable knowledge about the specificity and mechanism of enzyme action until the end of the next chapter. These advances were also significant in the study of other biochemical problems, such as those in the field of immunology.

The Enzyme Theory of Life

Some historians of biochemistry have linked Eduard Buchner's achievement to the publication in 1901 of a lecture by Franz Hofmeister on the "chemical organization of the cell" and have suggested that these events marked the emergence not only of the "new science of biochemistry" but also of the " 'enzyme theory of life,' which gave rise to a rich new programme of biochemical researches, the isolation, purification, and physico-chemical characterization of enzymes implied [*sic*] in every biochemical function."[114]

The main theme of Hofmeister's 1901 lecture was the problem of the spatial organization of the various intracellular chemical processes and their associated enzymes, and he suggested that these "colloidal reagents" are separated by "impermeable partitions." Like Moritz Traube and Felix Hoppe-Seyler before him, Hofmeister expected that "sooner or later there would be found for every chemical reaction a corresponding specific ferment," and he also stated that "in the protoplasm synthesis and breakdown occurs by way of a series of intermediate steps, whereby it is not always the same kind of chemical reaction that is involved, but rather a series of reactions of different kinds. . . . A regular sequence of the chemical reactions in the cell presupposes, however, the separate activity of the individual chemical agents and a definite direction of movement of the products which are formed, in short, a chemical organization . . . that helps to explain the speed and certainty with which it functions."[115] It should be noted that Hofmeister referred to "protoplasm" of living cells and that he made no reference to Buchner's dispute with those who claimed that, in his preparation of zymase, there were "bits of protoplasm." Moreover, like many German physiological chemists and British chemical physiologists of his time, Hofmeister was influenced by the success of the colloid chemists. In a paper published in 1914, he wrote:

> In a lecture twelve years ago I expounded ideas which I had formed at that time about the organization of protoplasm, namely in regard to the spatial arrangements of the chemical processes which occur in the cell. These remarks elicited much approval, but also disagreement on several points. Since then molecular physics and especially its affiliated colloid chemistry, which is relevant to the conception of cell structure, have made such great progress, that a revision of the views I expressed at that time is indicated. . . . In my earlier lecture I emphasized the separation of the reactions in the protoplasm by means of partitions, because at that time the solubility of the cell ferments was generally assumed. As a consequence of the demonstrated colloid nature of most enzymes, they are not present in true solution, and that assumption is, in principle, no longer necessary.[116]

Much more will be said about the "colloid chemistry of protoplasm" in the next chapter. At this point I only deplore the facile generalization that "the controversy over zymase was in fact the last fight between the adepts of the fading myth of 'protoplasmic substance' and the adepts of the molecular viewpoint, the enzymologists."[117] That statement oversimplifies and overdramatizes the complex development between 1897 and 1930 of the study of the nature of enzyme action. To term that period as the "dark age of biocolloidology"[118] is to dismiss the importance of the physical-chemical approach to that problem at a time when the chemical nature of the catalytic agents in enzymatic reactions was in doubt. For example, in 1897, Gabriel Bertrand who (along with Emile Duclaux) welcomed Buchner's achievement, assigned to the manganese ion the catalytic activity of the diastase he called laccase (because he found manganese to accelerate the chemical reaction), and he called the accompanying proteinaceous material a co-diastase. A few years later, Arthur Harden and William John Young showed that ultrafiltration of a sample of Buchner's zymase yielded a substance whose presence was essential for alcoholic fermentation, and they called that substance a co-ferment, as distinct from the equally essential material retained by the filter.[119]

As has been mentioned, during the 1920s the famous organic chemist Richard Willstätter claimed to have shown that the protein component of enzyme preparations is a nonspecific "carrier" of low-molecular-weight catalytic agents adsorbed on proteins. Nor should it be forgotten that the concept of "adsorption" as a noncovalent interaction at the surfaces of molecular species was developed by colloid chemists and was not only shown to be important in such laboratory practices as "chromatography" but was also embraced by biologists. The first four chapters of William Bayliss's influential *Principles of General Physiology* (fourth edition, 1924), for example, were successively entitled "protoplasm," "energetics," "surface action," and the "colloidal state." And in 1929 Otto Warburg, who became during the 1930s one of the greatest "enzymologists" of the twentieth century, wrote as follows in regard to the many soluble oxidases identified in previous years: "If the extract-oxidases had been preformed in the cell, a single type of cell would contain innumerable oxidases. But the multiplicity of oxidases in the living cell would be in opposition to a sovereign principle in the living substance. . . . Therefore, if many different oxidases have been found in extracts of a cell type, these were not ferments which were already present in the living cell, but are rather products of the transformation and decomposition of a single homogeneous substance present in life."[120] I shall discuss in some detail Otto Warburg's outstanding contributions of the 1930s to the study of biological oxidations, and I only note here that his principal experimental achievements in that field came after a remarkable change in his research program.

In my opinion, the debate between "protoplasmists" and "enzymologists" was a less significant feature of the period 1900–1930 than was the continuing tension between biologists and biochemists who embraced the physical-chemical approach of the colloid chemists and the organic chemists and biochemists who adhered to the tradition represented by Emil Fischer. For example, the chemical physiologist William Bayliss wrote in regard to a statement made by Gustav von Bunge that "the less a physiologist knows about chemistry, the greater is he inclined to work on the most difficult chemical subjects—the proteins and the ferments":

> If the chemistry to which reference is made is purely statical, structural organic chemistry, as would appear, it is a remarkable fact that such a mode of attack has taught us practically nothing about the nature of enzymes, and has only led to a multiplication of names. . . . It is only since the question has been attacked from the kinetical standpoint of physical and colloidal chemistry that we are beginning to see light. It is, of course, far from my intention to undervalue the work of organic chemistry as one of our helps to the comprehension of difficult problems . . . but, in view of opinions sometimes expressed, it is necessary to point out that there are other bodies of doctrine of equal importance in the study of physiology.[121]

One of these "opinions" was no doubt that of Bayliss's colleague Frederick Gowland Hopkins that biochemists should deal "not with complex substances which elude ordinary chemical methods, but with simple substances undergoing comprehensible reactions."[122]

What requires emphasis in regard to the suggestion that Buchner and Hofmeister initiated the "enzyme theory of life" is that the dispute over that issue began during the nineteenth century and continued long after 1930. Before 1900 it was not only the problem of alcoholic fermentation but also that of biological oxidation that occupied center stage. If Buchner's success in preparing "zymase" opened the door to the fruitful investigation of the enzyme-catalyzed pathway of alcoholic fermentation, the point of view about the soluble "oxidases" expressed by Warburg in 1929 mirrored the attitude of the chemist Louis Pasteur and the biologist Carl Nägeli to the views of Moritz Traube, Marcelin Berthelot, and Felix Hoppe-Seyler on the fermentation problem. After 1930 many biochemists did not consider hydrolytic enzymes to be associated with "life" but only with the degradative processes of metabolism and with postmortem change, a view reminiscent of that of Claude Bernard. That attitude changed in the face of new empirical discovery, and it was recognized that enzyme-catalyzed biodegradation, as well as biosynthesis, is an important intracellular process. After 1950, with the emergence of the "information theory of life" and disparaging references to "bags of enzymes" involved in the "petty details of metabolic pathways," the discovery of many new enzymes involved in the biosynthesis, transformation, and degradation of nucleic acids and proteins, together with the finding of new biochemical agents which regulate the action of enzymes, has provided further evidence of the continuity of "the enzyme theory of life." The historical development of these areas in the interplay of chemical and biological thought and action will be examined in the subsequent chapters.

CHAPTER 5

The Nature and Function of Proteins

For more than two centuries from about 1750, the substances we now call proteins occupied center stage in the interplay of chemistry and biology. First examined as constituents of the human and animal diet and as components in the tissues and fluids of plants and animals, these substances presented to nineteenth-century chemists and biologists some of the greatest challenges in their search for chemical explanations of the phenomena exhibited by living things. During the course of the development of reliable knowledge about the nature and function of these substances, new chemical information led to new hypotheses regarding their physiological roles, and new biological discovery led to new chemical interpretations of the structure of living organisms and of the transformations associated with such processes as respiration, nutrition, and reproduction.[1]

We have seen that many of the nineteenth-century students of the fermentation problem attached great importance to what were then usually termed *Eiweisskörper, matières albuminoides,* or albuminous bodies. At the beginning of the century these materials were identified largely by means of such familiar phenomena as the coagulation of egg white by heat, the curdling of milk by acids, or the clotting of blood. Some of the names given to such coagulable materials had a long heritage; for example, *album ovi* had been termed *albumen* by Pliny the Elder. In 1801 Fourcroy wrote about blood serum: "The serum, when subjected to heat, coagulates and hardens like egg white. This property is one of its striking characteristics; it is attributed to a particular substance which is thereby readily recognizable, and which is named *albumine,* because it is the one present in egg white, termed *albumen*."[2] The heat-coagulable "albumin" in egg white and blood thus became a substance whose presence could be recognized, by virtue of this property, in other biological fluids, or extracts of animal tissues. Other characteristics assigned to this substance by 1800 included the presence of sulfur (Scheele had shown that when an alkaline solution of egg white is acidified, there arises a "hepatic smell," which blackens silver) or the appearance, on treatment with nitric acid, of a yellow color that turns orange upon the addition of ammonia. It was also known that albumin is precipitated by salts of lead and mercury as well as by tanning agents.

The separation of the serum from clotted blood gave a red, water-insoluble material; when thoroughly washed, the resulting product (named fibrine by Fourcroy) lost its red color. During the eighteenth century, the coagulable fibrous principle of blood was termed gluten (Latin *gluten,* glue) or gelatin because of its resemblance to the jelly formed upon the extraction of many animal tissues (skin, ligaments, cartilage) with boiling water, followed by cooling. A similar gluten had been described in 1747 by Iacopo Beccari, who obtained it by kneading wheat flour with water to remove the starch, and the tendency of this plant product to undergo putrefaction suggested to him a similarity to animal materials. The gluten of wheat and of other plant foodstuffs was intensively studied during the 1770s by Antoine Augustin Parmentier.[3]

The "red coloring matter" of the blood had been shown by Leeuwenhoek to be associated with "globules." His microscopic observations were followed by those of Giovanni della Torre, whose description was disbelieved. After the reports of William Hewson, who described them as "flat vesicles," the particulate nature of the coloring matter was generally accepted, and its red color was ascribed to iron, whose presence in blood had been reported by Vincenzo Menghini in 1747 and confirmed by Hilaire Martin Rouelle in 1773.[4] Another animal fluid subjected to considerable chemical scrutiny during the eighteenth century was milk, from which a curdy precipitate was obtained upon acidification, and named caseum; it was later called casein.

During the period of the Chemical Revolution, Claude Louis Berthollet found that when animal materials are treated with nitric acid they release large amounts of "azote" (nitrogen). He concluded that it is a characteristic constituent of animal organisms and explained the formation of ammonia during putrefaction as the combination of the nitrogen with the "inflammable air" (hydrogen) derived from the decomposition of "oils" or of water. In confirming this observation, Fourcroy urged "the savants who occupy themselves with animal physics to continue research on this important point, and above all to determine whence this principle comes, and how and in what organ it is fixed in animals."[5] He also called attention, however, to the presence in plants of nitrogenous materials that resemble the albumin of egg white and serum,[6] and later work by Nicolas Deyeux and Louis Nicolas Vauquelin extended the list of plant products similar to those described earlier. Because of these findings, the terms *albumin, fibrin,* and *casein* were soon applied to nitrogenous plant materials whose properties resembled those of the corresponding animal products. Thus when clarified juices from cauliflower, asparagus, or turnips were boiled, and the coagulum was indistinguishable from heat-coagulated egg white or serum, it was named a vegetable albumin. The "hordeine" obtained from barley (by Joseph Louis Proust, 1817) was classified as an albumin and the "legumine" from leguminous plants (Henri Braconnot, 1827) as a casein, while the water-insoluble material deposited when vegetable juices were allowed to stand was named a fibrin. These products all contained nitrogen and gave the color tests considered to be characteristic of animal albumin, casein, and fibrin.

The search for better chemical knowledge of these and other substances present in foodstuffs was spurred in England and France by the recurrent food riots both before and after 1789.[7] In eighteenth-century England, the enclosure of farmlands was followed by higher prices and shortages of flour and bread in the towns, many of which were acquiring a growing population of factory workers. In prerevolutionary France, the impact of taxation and the application of laissez-faire capitalism to agriculture had a similar effect; during the years of the French Revolution the food situation worsened and contributed to the rise of Napoleon. Later his policy of closing the European ports to British commerce, as well as the British blockage of French ports, stimulated research on new methods of agriculture (as in the manufacture of sugar from beets) and in such industries as the dyeing of cloth. In these efforts the leading chemists played a key role during Napoleon's Hundred Days. Jean Chaptal, who had emphasized agricultural chemistry in his writings, was appointed minister of agriculture, commerce, and industry. In England, Humphry Davy delivered a series of lectures that were published in 1813 under the title *Elements of Agricultural Chemistry*. In Germany, Sigismund Friedrich Hermbstädt founded the first journal devoted to this subject (*Archiv der Agrikulturchemie*, vol. 1, 1804). The famous "sprig of mint" experiment of Joseph Priestley, followed by the work of Jan Ingen-Housz and Jean Senebier, stimulated the research of Théodore de Saussure on the role of oxygen and carbon dioxide in the growth of plants; his *Recherches chimiques sur la Végétation* (1804) is a landmark in the history of agricultural chemistry.

After 1800 a major preoccupation of many chemists became the study of what had come to be called the "immediate principles" of animals and plants. Biological tissues were subjected to extraction with acids, alkalis, and alcohol, and various methods were used in efforts to isolate pure substances from such extracts. What was meant by the term *immediate principle* may be seen from the following statements by Michel Eugène Chevreul:

> In employing the least energetic methods of analysis, one reduces plants and animals to principles that are called *immediate*, because having been separated in the state in which they existed before the chemical operation, one

> is justified in attributing to them properties of the plant or animal to which they belonged, and to consider them as their essential or immediate constituents.[8]
>
> . . .
>
> Some savants think that the expression immediate principles is faulty because it is repugnant to reason to apply the word principle to compound bodies; I do not share this opinion, and here is why. If one considers in general the composition of a salt, as established by Lavoisier, it is evidently formed by the union of an acid with an alkali, rather than the union of the elements of the acid and those of the alkali; since, if you conceive these elements to be united in proportions other than those which constitute an acid substance or an alkaline substance, those elements will no longer give you an idea of the salt. Consequently, it seems logical to say that the acid and the alkali are the *two immediate principles of salts.* It is the same for sugar, gum, starch, lignin, etc. in relation to a plant; of fibrin, albumin, cellular tissue, etc. in relation to an animal; one should consider these substances as immediate principles, and characteristic of the plant or animal to which they belong, whereas oxygen, azote, carbon and hydrogen are their ultimate or elementary principles.[9]

For Chevreul, whose work during 1811–1823 on the nature of the animal fats represents the first significant success in the chemical analysis of a complex organic material,[10] it was important to include among the immediate principles only those compounds in which the elements are present in definite proportions. Thus, because he had found that sheep fat, upon treatment with alkali (saponification), gave rise, in addition to glycerine, to the fatty acids stearic acid, oleic acid, and "margaric acid" (later shown to be a mixture of stearic and palmitic acids), Chevreul concluded that the fats are composed of two or more immediate principles. In applying his definition to wheat gluten, Chevreul concluded from the fact that it had been separated into an alcohol-soluble component (named gliadin by Gioacchino Taddei) and an insoluble residue that "gluten should no longer be listed among the immediate principles."[11]

The problem of the purification of the chemical constituents of biological systems was to plague investigators throughout the nineteenth and twentieth centuries, up to the recent past, and we shall return to it repeatedly. In the early 1800s the available techniques of extraction and precipitation often led to the isolation of crystalline materials; these methods had been used with great art by Carl Wilhelm Scheele, and Chevreul's success in crystallizing the individual fatty acids formed upon saponification enabled him to draw fruitful conclusions about their constitution.

For Chevreul and his contemporaries, one of the important criteria of the purity of an immediate principle was an elementary composition in accordance with the laws of definite and multiple proportion that had emerged from the theories of Jeremias Benjamin Richter and Joseph Louis Proust. As was noted previously, it was not until after 1810 that reliable analytical procedures began to be available for the quantitative elementary analysis of organic compounds.[12] By 1820 the atomic theory was generally accepted among chemists as the basis for the conversion of the percentage composition (by weight) of each element in a compound (as found by oxidation to carbon dioxide and water) into a formula that indicated the relative proportion of the elements in that compound. The conversion of the analytical data into the so-called empirical formulas required knowledge of the relative combining weights of the elements; before 1860 the values for the chemical "equivalents" were in the ratio H = 1, O = 8, C = 6, N = 7 (or 14), S = 16, and for a time the molecular weights were calculated on the basis of a chemical equivalent for oxygen of 100, in accordance with the proposal of Berzelius. There was much confusion about the assignment of atomic weights to the elements, and the same compound might be denoted by different empirical formulas. It was agreed, however, that whatever formula was used, its validity depended on the homogeneity of the isolated substance, and the precision of the methods used to determine its elementary composition. An example of what a skilled experimenter could do before 1820 was provided by William Prout, who isolated a sample of urea, which he found to contain 19.99% carbon, 6.66% hydrogen, 26.66% oxygen, and 46.66% nitrogen; since the theoretical values for $CO(NH_2)_2$ are 19.99% carbon, 6.71% hydrogen, 26.64% oxygen, and 46.66% nitrogen, it is safe to say that no present-day analyst could have done better.

By 1835 the improvements introduced by Berzelius, Gay-Lussac, and Liebig had transformed the original analytical procedure for the determination of carbon and hydrogen into a more reliable method; the substance was now burned in a long tube, copper oxide was used in place of potassium chlorate as the oxidant, and the carbon dioxide was collected in an alkali trap. Dumas had developed a combustion method for the determination of the nitrogen content of organic compounds; it involved the measurement of the volume of nitrogen gas after the absorption of the water and carbon dioxide. At midcentury the Dumas method was largely replaced by that of Will and Varrentrapp (described in 1841), which involved decomposition of a substance with alkali, collection of the liberated ammonia in acid, and precipitation of the ammonia as its chloroplatinate for gravimetric determination. This procedure was in turn supplanted in 1883 by the Kjeldahl method.[13]

Although the analytical methods available between 1810 and 1830 had been applied by some chemists, notably Jean Baptiste Boussingault, to the study of albumin, casein, and fibrin, the systematic attack on the elementary composition of these substances was undertaken by Gerrit Jan Mulder in the mid-1830s. By that time their importance in the diet of animals was widely appreciated, and Prout's classification (in 1827) of foodstuffs into the three categories saccharinous, oleaginous, and albuminous was generally accepted. Moreover, since much had been learned about the chemical similarity of these constituents of animal and plant tissues and fluids, this knowledge contributed to the emergence of "biology" as the field of investigation of all living organisms, and with the rise of the cell theory and the associated concept of an albuminoid protoplasm, the chemical nature of what came to be called proteins assumed considerable importance in biological thought.[14]

The Albuminoid Nature of Protoplasm

Before the emergence of the cell theory, some of the microscopists who examined animal and plant tissues or small organisms ("infusoria") saw what they called globules associated with a jellylike material that resembled egg white in its coagulability. Although many of the observed structures were shown to be artifacts, owing to the limitations of the microscopes of the time as well as to the imagination of the observers, the discussion of the nature of the globules was influenced by the available knowledge of the properties of albumin, fibrin, casein, and gelatin.[15] For example, during the 1820s, Henri Dutrochet attempted to produce contractile fibers by treating solutions of albumin or gelatin by desiccation or by means of electricity, acid, or alkali.[16] The ancient idea, attributed to Thales, that solid bodies arise from fluids assumed scientific importance in the proposal of Félix Dujardin that the seemingly structureless biological fluid, which he named sarcode, is the material basis of life.[17]

In the formulation of their cell theory, Matthias Schleiden and Theodor Schwann emphasized the role of the nucleus ("cytoblast") in the formation of cells from a surrounding structureless "cytoblastema" in a "plastic" process analogous to the growth of inorganic crystals. Although their view that cells constitute the fundamental structural units of both animal and plant tissues was widely accepted, their theory of cell formation was soon replaced by the recognition that, as Rudolph Virchow put it, "omnis cellula e cellula."[18] The idea that a jellylike material resembling albumin constitutes the intracellular stuff associated with the phenomena of life was retained, however, and it was named plasma or protoplasma. The first term had been used in 1836 by Carl Heinrich Schultz to denote the "living" blood fluid, and it came to refer to fluid obtained after removal of the blood cells. It was also used by Carl Nägeli to denote what others (Jan Evangelista Purkyně, Ferdinand Cohn, Hugo von Mohl, Max Schultze) had termed protoplasma.[19]

By the 1860s the protoplasm theory had received the support of Ernst Haeckel and other noted German biologists. In England, Thomas Henry Huxley delivered a stirring lecture on the "physical basis of life," in which he announced that "the researches of the chemist have revealed . . . a striking uniformity of material composition in living matter. . . . All protoplasm is proteinaceous."[20] The chemical work to which Huxley referred will be considered later in this chapter, but in connection with the protoplasm theory, mention must be made of the concept of "living proteins" as sources of "vital energy" in biological systems. In his important 1845 paper on chemical changes in muscle, for example, Hermann Helmholtz raised the question of the role of proteins in muscular contraction "because we find the protein compounds everywhere as bearers of the highest vital energies."[21]

In succeeding decades, the idea that protoplasmic proteins are (in recent biochemical parlance) "energy-rich" substances was bolstered by Thomas Graham's concept of nondiffusible colloids (including fibrin, casein, and albumen) as representing "a dynamic state of matter, the crystalloidal being the static condition. The colloid possesses ENERGIA. It may be looked upon as the primary source of the force appearing in the phenomena of vitality."[22] In 1875, Eduard Pflüger published in his journal a long paper[23] in which he offered the view that protoplasmic proteins are labile structures whose instability underlies the phenomena of life, and he drew on the conclusions of Justus Liebig, Willy Kühne, and Ludimar Hermann about muscular contraction.[24] Pflüger also made much of the fact that Rudolph Clausius had stated that the difference between the total energy associated with the molecular motion of a system and the heat content of that system is greater as the number of atoms composing the molecules increases.[25] Pflüger used these ideas, along with contemporary organic-chemical knowledge, to develop a theory based on the assumption that protoplasmic proteins are polymeric structures of large molecular size, and he also offered a chemical explanation of their lability. Among the known products of the metabolic breakdown of proteins were compounds such as creatine, uric acid, or guanine, in which the carbon and nitrogen atoms are linked in a manner similar to that in the highly reactive compounds cyanogen (NC-CN) and cyanic acid (HO-CN), and Pflüger asserted that "living protein has its nitrogen not largely in the form of ammonia, but in the form of cyanogen." He claimed that, in the metabolic conversion of dead food proteins to living tissue proteins, there occurs a change in which nitrogen atoms combine with carbon atoms to form cyanogen groups and that, upon death, the living proteins return to amides. In Pflüger's view, the function of oxygen in tissue respiration (about which more in the next chapter) is to combine with the carbon of the CN groups of the living protein to generate CO_2 in "a series of small explosions whose impact increases the strength of the intramolecular vibrations." In summarizing his theory, Pflüger stated: "The life process is the intramolecular heat of highly unstable albuminoid molecules of the cell substance, which dissociate largely with the formation of carbonic acid, water, and amide-like substances, and which continually regenerate themselves and grow through polymerization."[26]

Pflüger's speculations were received respectfully by some of his noted contemporaries. For example, August Kekulé stated: "The hypothesis of chemical valency further leads us to the supposition that also a relatively large number of single molecules may, through polyvalent atoms, combine to *net-like,* and if one may say so, *sponge-like masses* in order thus to produce those molecular groups which resist diffusion, and which, according to Graham, are called colloids. The same hypothesis leads us, in a natural manner, to the view already pronounced by our eminent colleague, Pflüger, that such a cumulation of molecules may extend yet further, and thus build up the *formative elements* of living organisms."[27] During the succeeding decades, as growing attention was paid to the role of the cell nucleus as a site of both intracellular synthesis and the transmission of hereditary characters, Pflüger's ideas received recurrent attention. Russell Henry Chittenden, for example, in writing of the recently discovered adenine as a constituent of nuclein (of which more in a subsequent chapter), stated:

> Such being the nature of adenin, it is not to be doubted that bodies from this substance with strong affinities must be important actors in the chemical physiological processes, especially of a synthetical order, going on in all tissues. In this connection it is to be remembered that Pflüger on purely theoretical grounds ascribed great importance to the physiological role played by the cyanogen group with polymerization, etc. in the living albumin molecule. . . . In the discovery of adenin and its close relationship to the typical xanthin bodies we have added proof of the existence of cyanogen-containing radicals in the protoplasm of the cell, especially in the karyoplasm of the nucleus. In all of these xanthin bodies there is to be seen a particular combination of carbon, nitrogen, and hydrogen such as is not found in dead protein matter.[28]

Another imaginative chemical approach was that of Oscar Loew, a disciple of Carl Nägeli. Loew proposed in 1880 that protoplasmic proteins arise by the polycondensation of aminoaldehydes, and that what he called the active protein of protoplasm is a large molecule (he used the empirical formula $C_{72}H_{112}N_{18}O_{22}S$ proposed in 1851 by Nathaniel Lieberkühn) containing twelve aldehyde groups whose instability is the chemical source of vital phenomena. With Thomas Bokorny, Loew offered as evidence their observation that, in contrast to dead proteins, living proteins give a positive test for the reduction of alkaline silver nitrate.[29] Although this claim was severely criticized by Eugen Baumann, other noted physiological chemists took a more favorable view. Marceli Nencki, for example, wrote in 1885 about the instability of protoplasmic proteins that "if we wish to approach the phenomena associated with the word 'life,' research on the chemistry of the albuminoid bodies must take a new direction. As Pflüger stated over ten years ago, the protein of the living cell must have an entirely different molecular structure from that of dead tissues, and the evidence that the protein of living cells has a labile aldehyde structure increases daily."[30] This "increase" in evidence was principally due to the persistence of Loew and Bokorny in reiterating their claim well into the 1920s; indeed, the aminoaldehydes reappeared in 1941 as putative precursors in the biosynthesis of proteins.[31]

The theme that there is a chemical difference between living and dead proteins also appeared in Paul Ehrlich's 1885 *Habilitationsschrift* on biological oxidation. There he espoused Pflüger's concept of a "giant" protoplasmic molecule and suggested that it attracts to itself "side-chains," which act as agents of metabolic processes. Ehrlich noted that "speculations about the nature and origin of these binding groups would be premature . . . and I content myself with the indication that possibly the aldehyde groups, whose existence is assumed by Loew and Bokorny, may play a role in this connection." Such "side-chains" then appeared in Ehrlich's theory of immunity (of which more later in this chapter), and he claimed priority when in 1897 Max Verworn invoked the presence of protoplasmic side-chains in the hypothesis that "the metabolism of living matter . . . is determined by the existence of certain very labile compounds which belong to the group of proteins, and which because of their elementary significance for life can best be denoted 'biogens,'" possessing the kind of explosive properties that Pflüger had assigned to living proteins.[32]

During the period 1870–1910, physiological chemists such as Eugen Baumann were studying the fate of defined chemical compounds of low molecular weight in biological organisms by isolating the metabolic products and establishing their chemical structure. These experimental investigations will be discussed in a later chapter. At this point, I only note that these physiological chemists rejected explanations of life in terms of such concepts as those offered by Pflüger and Loew, largely because, like Claude Bernard, they viewed the living cell as a very complex organized assembly. That attitude was expressed in 1889 by Gustav Bunge: "The more closely, broadly, or fundamentally we attempt to study the phenomena of life, the more we come to the view that the processes we thought could be explained chemically and physically are much more complicated, and at present defy every mechanical explanation." In 1903 Richard Neumeister developed this theme further in direct opposition to Verworn's theory, and also to the optimism generated by Buchner's preparation of zymase that the chemical dissection of intracellular processes might be possible through the study of enzymes. In particular, Neumeister emphasized the fact that no biosynthetic processes had been shown to occur in cell-free systems.[33] It may perhaps suffice to quote the statement of Frederick Gowland Hopkins in 1913 on this subject:

> There is, I know, a view which, if old, is in one modification or another still current in many quarters. This conceives of the unit of living matter as a definite, if very large and very labile molecule, and conceives of a mass of living matter as consisting of a congregation of such molecules. . . . In my opinion, such a view is as inhibitory to productive thought as it is lacking in basis. It matters little whether in this connection we speak of "molecule" or in order to avoid the fairly obvious misuse of a word, we use the term "biogen," or any similar expression with the same connotation. Especially, I believe, is such a view unfortunate when, as sometimes, it is made to carry the corollary that simple molecules, such as those provided by foodstuffs, only suffer change after they have become in a vague sense a part of such a giant molecule of biogen.[34]

Although the debate about the "physical basis of life" has been discussed in philosophical terms as a conflict between mechanism (or reductionism) and vitalism (or teleological organicism), the main issue within the community of experimental biologists and biochemists was, in my opinion, one of strategy in the further study of the phenomena of life. At the turn of the century, in the most active branches of biology—cytology and embryology—investigators were sharply divided on the issue, with men such as Oscar Hertwig and Theodor Boveri insisting that protoplasm is a morphological concept and not a chemical one, and others, like Otto Bütschli and Jacques Loeb, seeking physical-chemical explanations of life processes. As was noted above in connection with the protein nature of enzymes, many physiologists and biochemists chose to study the physical-chemical and colloidal properties of proteins. Before returning to that subject, it is necessary to examine the historical development of the organic-chemical approach after about 1830.

The Organic Chemistry of Proteins, 1830–1860

In his textbook of physiological chemistry, Gerrit Jan Mulder wrote:

> In plants as well as in animals there is present a substance which is produced in the former, constitutes the part of the food of the latter, and plays an important role in both. It is one of the very complex compounds, which very easily alter their composition under various circumstances, and serves especially in the animal organism for the maintenance of chemical metabolism [*Stoffwechsel*], which cannot be imagined without it; it is without doubt the most important of all the known substances of the organic kingdom, and without it life on our planet would probably not exist. It is found in all parts of plants, in the roots, stems, leaves, fruits, and juices, as well as in very dissimilar parts of the animal body. In plants it occurs in three different forms, as water-soluble, water-insoluble, or alcohol-soluble; in animals it occurs in a large variety of forms, being sometimes soluble, sometimes insoluble in water, and in its insoluble form its structure is variable. It combines with sulfur or phosphorus, or both, and thereby exhibits differences in its appearance and its physical properties. The substance has been named *protein*, because it is the origin of very different substances and therefore may be regarded as a primary compound.[35]

The word *protein*, which first appeared in the chemical literature in a paper by Mulder in 1838, had been suggested to him by his former teacher Berzelius, in a letter dated 19 July 1838:

> I consider it sufficiently well established that the immediate organic substances either are oxides of compound radicals or are combinations of two or even several oxides of this kind. It is necessary first to look for this radical. The addition of nitrogen complicates matters a bit, but in general the difficulty is not great. . . . Now I presume that the organic oxide which is the base of fibrin and albumin (and to which it is necessary to give a particular name, e.g., *protein*) is composed of a ternary radical combined with oxygen. . . . The word protein that I propose to you for the organic oxide of fibrin and albumin, I would wish to derive from *proteios*, because it appears to be the primitive or principal substance of animal nutrition that plants prepare for herbivores, and which the latter furnish to the carnivores.[36]

During the years around 1840 these ideas about the chemical nature of a "radical" called protein and its place in the order of nature were widely accepted. Although they were abandoned soon afterward, the impetus Mulder gave to the study of such materials as fibrin and albumin was a lasting one, and the name he introduced remained long after the chemical and physiological ideas it was intended to connote had been discarded.

From his data for the elementary composition of egg albumin, serum albumin, and fibrin, Mulder concluded that they all had the same content of carbon, hydrogen, nitrogen, and oxygen, corresponding to an empirical formula which he wrote as $C_{40}H_{62}N_{10}O_{12}$. Further, he declared that this unit (termed protein) was combined with an atom of sulfur and an atom of phosphorus to form fibrin and egg albumin, or with two sulfurs and one phosphorus to form serum albumin. Mulder also reported that by treatment of the albumins and fibrin with dilute alkali, it was possible to remove the sulfur and phosphorus completely and to isolate the fundamental unit common to all three substances. Moreover, he claimed that a sample of plant gluten that had been treated with alkali gave the same analytical data as those found for the "protein" derived from animal sources, and he concluded that "animals draw their most important proximate principles from the plant kingdom. . . . The herbivorous animals are, from this point of view, no different from the carnivores. Both are nourished by the same organic substance, protein, which plays a major role in their economy."[37]

These conclusions were adopted by Justus Liebig on the basis of analyses performed by his students Johann Joseph Scherer and Henry Bence Jones; on 28 June 1841 he wrote to Wöhler that "I have been working on the legumin of leguminous plants and have obtained the remarkable result that it is casein in all its properties and its composition. We have therefore a complete analogy, we have plant albumin, plant fibrin, and plant casein, all identical with each other and with the animal proteins which bear their names."[38]

Mulder's work also brought into the field Jean-Baptiste Dumas, who disputed the identity of legumin and casein and found a higher nitrogen content in fibrin than in egg albumin but nonetheless concluded that "the animal receives and assimilates almost intact the neutral nitrogenous substances which it finds fully formed in the animals and plants that form its food."[39] Other French chemists who supported Mulder's protein theory included Apollinaire Bouchardat, who reported in 1842 that the unit (which he called albuminose) has the same optical activity no matter what its source (egg albumin, casein, gluten), but this claim was later disproved by Antoine Béchamp, and Charles Gerhardt stated: "The composition of the albuminoid substances is extremely complex; it appears to be the same for all of them. . . . If one considers that these substances behave in an identical manner under the influence of agents which transform them, one is led to attribute to impurities the small differences encountered in the results of analyses."[40] These ideas of a fundamental protein unit, and of its passage from plants to animals, were attractive in their simplicity and gained much attention when they were popularized by Liebig in his 1842 book on the application of chemistry to animal physiology. The book elicited from Berzelius the comment that "this easy kind of physiological chemistry is created at the writing desk, and is the more dangerous, the more genius goes into its execution, because most readers will not be able to distinguish what is true from mere possibilities and probabilities, and will be misled into accepting as truths probabilities that will require great effort to eradicate after they have become imbedded in physiological chemistry."[41] Indeed, among the ideas found wanting was the protein concept Berzelius had suggested to Mulder. In 1845 Liebig wrote Wöhler that "after so much has been prattled and written about protein and protein oxide, it is a source of despair to have to see that there is no such thing as protein." This verdict came from the work in Liebig's laboratory by Nicholas Laskowski, who reported that "since the acceptance of the substance described by Mr. Mulder as protein was based solely on the belief that it had been isolated free of sulfur,—the substance isolated by Mr. Mulder contains sulfur, and the one described by him cannot be isolated—there is no basis left for the assumption that protein is a hypothetical fundamental substance."[42] Laskowski's long paper had been preceded by a short note, several months before, in which Liebig stated this conclusion and asked "Would Mr. Mulder please describe his procedure in full detail?" This drew from Mulder an indignant reply in the form of a pamphlet entitled *Liebig's Question to Mulder, Tested by Morality and Science,* and a revised theory that albuminoid substances are combinations of a hypothetical protein that cannot be isolated (its empirical formula became $C_{36}H_{54}N_8O_{12}$) and linked to various amounts of "sulfamide" or "phosphamide."[43] At midcentury, the various forms of Mulder's protein theory were discounted: "It must rather excite our surprise that chemists should have hazarded any theory of their composition, than that nothing positive should as yet have been ascertained regarding their composition and mutual relations. Although we have the most accurate analyses of the protein compounds, it is impossible to form any decisive conclusion regarding their internal composition."[44] After the demise of Mulder's theory, German chemists referred to *Proteinkörper* or *Proteinsubstanz,* although *Eiweisskörper* continued to be used well into the twentieth century, and "protein bodies" or "proteids" was a common usage in English writings before Mulder's term acquired its current meaning.

By the 1850s the complexity of the composition of the proteins had become evident from studies on the products of their cleavage by strong acids and alkalis.

This approach, used successfully by Chevreul in his work on fats, had been applied in 1820 by Henri Braconnot, who isolated two crystalline substances (later called glycine and leucine) by treatment of protein preparations with sulfuric acid, and Liebig's pupil Friedrich Bopp later obtained leucine and a new substance ("tyrosine") in this manner from casein (Greek *tyros*, cheese). By 1850 leucine and tyrosine had been found as well-defined products of the cleavage of many proteins by means of sulfuric acid, hydrochloric acid, potassium hydroxide, and barium hydroxide. In Liebig's laboratory another student (Carl Gustav Guckelberger) subjected several proteins to the action of oxidizing agents (chromic acid, manganese dioxide) and identified various volatile aldehydes and acids among the products.[45] In characteristic fashion, Liebig wrote:

> The study of the products which caseine yields when acted upon by concentrated hydrochloric acid, of which as Bopp has found, tyrosine and leucine constitute the chief part, and the accurate determination of the products which the blood constituents, caseine, and gelatine, yield when oxidized . . . oil of almonds [benzaldehyde], butyric acid, aldehyde, butyric aldehyde, valerianic acid, valeronitrile, and valeroacetonitrile, have opened up a new and fertile field of research into the numberless relations of the food to the digestive process, and into the action of remedies in morbid conditions; discoveries of the most wonderful kind, which no one could have imagined a few years ago.[46]

During the 1850s oxidation with potassium permanganate gave benzoic acid, among other products, and in succeeding decades there were other reports that added further complexity to the problem of the composition of proteins.

The idea that the albuminoid substances are related to the substances named acid amides had been mentioned repeatedly since Dumas proposed in 1830 that the oxamide he had prepared (by treatment of ethyl oxalate with ammonia) might be a component of the nitrogenous constituents of the animal body. Since urea and uric acid were considered to be derived from the metabolic breakdown of albuminoid substances, the color reaction of these substances with alkaline copper sulfate, shown by Gustav Wiedemann to be a property of "biuret" (formed by prolonged heating of urea at high temperatures), was taken to suggest that ureido groups (-NH-CO-NH-) are significant elements of the constitution of proteins.

As has been noted, the first half of the nineteenth century was marked by a succession of empirical discoveries and theoretical insights from which there developed, by the 1860s, a fruitful conceptual framework for the subsequent emergence of organic chemistry as a prominent research discipline. A brief summary of the developments during 1830–1860 may perhaps suffice to indicate their importance in the study of protein chemistry and also to suggest why, in these studies, the successes of the new organic chemistry were only partial and left open many questions.

During the 1830s a key figure was Dumas, who proposed that organic compounds be considered to be unitary assemblies rather than the result of a binary combination of units of opposite charge, as advocated in the electrochemical theory of Berzelius. Dumas also emphasized the importance of the relative positions of the atoms of a substance, suggested by the recently discovered phenomenon of isomerism, and attempted to develop classificatory schemes based on experiments in which certain elements or groups were substituted by others during the course of chemical reactions; among his famous empirical findings was the conversion of acetic acid into trichloroactic acid by treatment with chlorine in sunlight. Of at least equal importance, albeit of lower academic status, was Auguste Laurent. A crystallographer endowed with a bold imagination, Laurent brought new insights into the problems presented by the growing number of known organic compounds.[47] In 1836 he advanced the hypothesis that every organic compound is derived from a hydrocarbon, constituting a fundamental unit in which hydrogen can be replaced by other elements or groups of elements. Laurent's ideas were not well received by Berzelius and Liebig, and Dumas stated in 1838 that he did not accept Laurent's extension of his substitution theory. A further source of difficulty for Laurent was his association with Charles Gerhardt, both in chemistry and in the 1848 revolution. Both of these outstanding men were ostracized by most of the established French chemists, with the notable exception of the aged Jean-Baptiste Biot and the younger Adolphe Wurtz. Although Laurent and Gerhardt were denied professorships in Paris, their writings had a large effect on young chemists elsewhere, especially in England.

By 1850 the efforts of Dumas, Laurent, and Gerhardt had provided a classificatory scheme of organic compounds according to "types," in which one or more hydrogen atoms were replaced by "compound radicals." The first of the types to be established experimentally was the "ammonia" type, by Wurtz and A. W. Hofmann through their studies on aliphatic amines and aniline derivatives. Alexander Williamson then showed alcohols and ethers to be of the "water" type. The type theory thus enlarged the scope of the "radical" theory (from Greek *radix*, root) based on the work of Liebig, Wöhler, and Bunsen, who identified the benzoyl, ethyl, acetyl, and cacodyl groups as units in chemical reactions.[48] The limitations of the type theory became evident when it was realized that many compounds could not belong to one of the established types, and in his treatise on organic chemistry August Kekulé was obliged to write of "mixed types." The formula for glycine (empirical formula $C_2H_5NO_2$; C = 12; N = 14; O = 16) thus became:

$$\left.\begin{matrix} H \\ H \end{matrix}\right\} N \qquad \left.\begin{matrix} C_2H_2O \\ H \end{matrix}\right\} O$$

to indicate that it belonged to both the ammonia and water types.[49]

It should be emphasized that for the adherents of the type theory, such designations were not structural formulas but shorthand representations of the reactions that a compound might be expected to undergo. The suggestion that formulas might be used to represent the arrangement of the constituent atoms in a compound, made tentatively by Williamson in 1852, was fully developed several years later by the Scotsman Archibald Scott Couper and the Russian Aleksandr Butlerov, both working in Wurtz's laboratory in Paris, and by the German August Kekulé, who had been associated with Gerhardt and Wurtz in Paris and with Williamson and William Odling in London. This group of young men, stimulated by the ideas of Laurent and Gerhardt, charted a new course for organic chemistry.[50]

It was Couper who expressed most forcefully the idea that the type theory should be abandoned in favor of the idea of the "combining power" of individual atoms and who stressed the necessity of writing structural formulas, which he considered to represent physical reality. Shortly after the appearance of his paper on the subject in 1858, Couper became mentally ill, and he lived in retirement until his death, but his formulas became widely known, especially through Lothar Meyer's *Modernen Theorien der Chemie* (1864). In assuming the combining power of carbon to be four, and the linking of carbon atoms with one another, Couper had independently developed ideas with which Kekulé's name is most prominently identified.[51] Although partly anticipated by Edward Frankland, whose work on organometallic compounds such as zinc methyl had led him to a similar view in 1852, it was Kekulé's influence that placed the idea of combining power (or valency, as it later came to be called) at the center of chemical thought. The term *Valenz* was introduced in 1868 by Hermann Wichelhaus to denote what had also been termed "saturating capacity" or "atomicity."

The "new" atomic weights proposed by Gerhardt (for example, C = 12, O = 16, S = 32) in place of "equivalents" were based on the reduction of the formulas of all volatile compounds to equal volumes. That proposal was adopted in the 1850s by Odling, Williamson, Couper, and Kekulé, but Hermann Kolbe and Marcelin Berthelot rejected it, and decisive support for Gerhardt's system only came in 1860 through the intervention of Stanislao Cannizzaro, who reminded the chemical world of the hypothesis advanced by Avogadro and Ampère, that at constant temperature and pressure the number of molecules in all gases is always the same for equal volumes. When nine years later the new atomic weights appeared in Mendeleev's Periodic Table, only a few stubborn chemists still refused to use them in writing formulas.

Although Kekulé wrote in his textbook that "the rational formulas . . . are in no way intended to express the constitution, *i.e.*, the arrangement of the atoms in the compound in question,"[52] he drew pictures that suggested such "arrangement." In 1861 Butlerov introduced the term *chemical structure*, and in 1864 Alexander Crum Brown and Lothar Meyer drew graphical formulas similar to those used a century later. After 1865, when Kekulé offered his hexagonal picture of benzene, structural formulas appeared with increasing frequency in chemical publications.

The capstone of the new structural chemistry was the explanation of the optical activity of organic compounds in terms of the asymmetric carbon atom. Although there had been anticipations of this discovery before 1874, when it was announced independently by van't Hoff and Le Bel, its impact was felt only after the formulation of the theory of valence and structure. Their description of isomerism in terms of differences in the arrangement of atoms in space provided a convincing argument for considering structural formulas as representations of physical reality. When in 1877 van't Hoff's book *La Chimie dans l'espace* appeared in a German translation, with an introduction by Johannes Wislicenus, it drew from Hermann Kolbe the memorable statement that "a certain Dr. J. H. van't Hoff, employed at the Veterinary School at Utrecht, has, it seems, no taste for exact chemical investigation. He has deemed it more convenient to mount Pegasus (evidently hired at the veterinary stables) and to proclaim in his "La Chimie dans l'espace" how during his bold flight to the top of the chemical Parnassus, the atoms appeared to him to have grouped themselves throughout universal space."[53] Van't Hoff's reply came in the following year, when he assumed his professorship at Amsterdam; his inaugural lecture was entitled *Imagination in Science.*

There is a large measure of irony in Kolbe's stubborn opposition to the concepts of the new structural chemistry, for after 1860 they stimulated the development of the art of organic chemical synthesis, a field in which Kolbe had been a pioneer. In 1845 he reported the synthesis of acetic acid, and in succeeding years other successes, such as Adolph Strecker's synthesis of lactic acid and alanine, had also been described. During the 1850s Kolbe made further contributions, as did others, among them Adolphe Wurtz and Antoine Béchamp.[54] After 1860 the number of "name reactions" in synthetic organic chemistry began to multiply,[55] and many of them played a role in the growth of the chemical industry. In addition to Emil Fischer's work on sugars, cited in the preceding chapter, his synthesis of purines and Ludwig Knorr's synthesis of pyrroles are prominent early examples of important contributions to biochemical knowledge. Indeed, it may be said that the concepts that Kolbe opposed were validated by the synthetic achievements of the men whose theories he had derided.[56] Another prominent figure in the mid-nineteenth-century literature on organic synthesis was Berthelot, whose 1860 book on this subject was less warmly received by his chemical colleagues than by nonscientists in the Parisian intellectual community. In his aspiration to "banish life from all explanations relative to organic chemistry," Berthelot overemphasized his own scientific work, denigrated the work of others, and employed a florid literary style not uncommon, even in recent times, in writings by prominent scientists about the general significance of their achievements.[57] One consequence of the emergence of the new structural chemistry was a lessened interest among leading chemists in the albuminoid substances, and an extreme statement of that attitude was that these substances "do not constitute, properly speaking, chemical species; they are organs or debris of organs whose history should belong to biology rather than to chemistry."[58] For Hoppe-Seyler, however, as spokesman for the independence of physiological chemistry,

> The upsurge experienced by organic chemistry during the past few decades enables it not only to analyze biological problems in the manner attempted before, but also to conduct searching experiments on the chemical processes in the living organism. The synthetic results, in providing insights, of which the recent past may be proud, into the structure of chemical substances and their transformation through chemical processes, have provided the means and directions to investigate, with hitherto unexpected assurance, the causes of vital phenomena in the structure and relationship of the substances that are active in biological organisms.[59]

On the other hand, the English chemist Charles Thomas Kingzett stated that

> of late years, there has been introduced into chemical teaching the so-called "structure" hypothesis of carbon-compounds, under which these latter are represented as structures comprised of constituent atoms arranged graphically. This graphical arrangement is arrived at from a knowledge of the ways in which substances decompose when subjected to particular processes. That is to say, if they yield by some process a particular substance, this is considered sufficient evidence that they contained in their structure a particular group of atoms, and so far so good. But many chemists go further

> than this, and say not only such and such a group is present, but also that it is present in a certain position. It is this last-named assumption that is so extremely unprofitable and unmeaning, particularly for physiological chemistry.[60]

Amino Acids and Peptones

In 1860 it was generally agreed that leucine and tyrosine are characteristic cleavage products of the albuminous substances, and in subsequent years many other such products were identified. It was often difficult to decide whether they were in fact constituents of an original complex structure in the sense that fats had been shown to be composed of glycerol and fatty acids, or starch to be made up of glucose units. Moreover, problems relating to the purity and identity of many of the new cleavage products were repeatedly encountered, and the methods used for their separation underwent continual change. As a consequence, many decades of much trial and error were to elapse before the salient features of the composition of the proteins were established. The following sketch may indicate the difficulties and uncertainties that attended the development of that line of investigation.[61]

Because of its sweet taste, Henri Braconnot named the substance he had obtained from gelatin *sucre de gélatine;* it then was called glycocoll, and later glycine. On the basis of the elementary composition of glycine and leucine, Laurent and Gerhardt considered the two substances to be members of a homologous series, later named amido acids or amino acids. The synthetic "alanine" prepared by Strecker became another member of the series. During the 1850s, glycine was identified as aminoacetic acid and alanine as aminopropionic acid. In 1888 Theodor Weyl isolated alanine as one of the products of the acid hydrolysis of silk, a source from which Emil Cramer had obtained in 1865 the substance "serine," later shown to be β-hydroxyalanine. Before 1900, however, neither alanine nor serine was regarded as a regular protein constituent. The same was true of the leucinelike substance isolated in 1856 by Eugen Franz Gorup-Besanez from tissue extracts; by 1878 this substance ("valine") had been shown to be aminovaleric acid, and thirteen years later leucine was proved to be aminoisocaproic acid. In 1881 Ernst Schulze and Johann Barbieri identified phenylalanine as a constituent of plant proteins; in the following year it was synthesized by the method Strecker had used to make alanine, and its structural relationship to tyrosine (*p*-hydroxyphenylalanine) became evident. The recognition of tyrosine as a phenol identified this protein component as the one responsible for the xanthoproteic and Millon color reactions.

Of special interest in the history of protein chemistry is the discovery by Heinrich Ritthausen that the long-known aspartic acid is formed upon the acid hydrolysis of plant proteins. This acidic compound had been the object of extensive chemical study, having been obtained in 1827 by Plisson from the substance (asparagine) isolated from asparagus juice by Vauquelin and Robiquet in 1806. By 1838 work by Pelouze and Liebig had established the elementary composition of the two compounds, and during the 1840s Piria showed that aspartic acid is converted into malic acid upon treatment with nitrous acid, and Dessaignes made aspartic acid by heating the ammonium salts of malic acid or fumaric acid. The latter observation interested Pasteur, as both the malic acid and aspartic acid derived from biological sources were optically active, whereas the synthetic aspartic acid was optically inactive.

Ritthausen's method for the isolation of aspartic acid in 1868 was, in principle, the one employed by Scheele in the previous century for the isolation of acids (lactic acid, malic acid, tartaric acid) from biological fluids, namely the precipitation with alcohol of their barium, calcium, or zinc salts. Two years earlier Ritthausen had obtained from plant proteins *Glutaminsäure* (glutamic acid), soon recognized to be homologous with aspartic acid.[62] By 1890 the structure of aspartic acid (aminosuccinic acid) and of glutamic acid (aminoglutaric acid) had been established by synthesis, and their widespread occurrence as products of the acid hydrolysis of plant and animal proteins was generally acknowledged.

As the number of amino acids identified as cleavage products increased, attempts were made to develop hydrolytic methods that would yield these products in amounts that might account for the weight of the protein materials subjected to analysis. Heinrich Hlasewitz and Josef Habermann claimed that hydrolysis of casein with hydrochloric acid in the presence of tin gave leucine, tyrosine, aspartic acid, glutamic acid, ammonia, and little else.[63] The appearance of ammonia led them to speculate that it might have been derived from asparagine or (hypothetical) glutamine units, and free glutamine was found in 1877 by Ernst Schulze, who established the importance of these two amides in plant metabolism, but a definitive answer came only many years later, from studies on the enzymatic cleavage of proteins, showing that they are indeed constituents of many proteins.[64]

Other investigators could account for only a fraction of the starting material, and it was realized that "in the cleavage of the albuminoid substances by means of concentrated hydrochloric acid, there arise other hitherto unidentified products."[65] By 1891, through the use of phosphotungstic acid (previously employed as a precipitant of alkaloids), and laborious fractionation, Edmund Drechsel and his associates were able to isolate, from an acid hydrolysate of casein, the hydrochloride of a basic substance (lysine), whose structure was shown in 1899 to be diaminocaproic acid. By 1896 two other basic amino acids (arginine and histidine) had also been found to be protein constituents. For a time after its discovery in 1877 by Max Jaffé, ornithine (diaminovaleric acid) was so considered, but its occasional appearance was later explained by the finding of the enzyme arginase, which cleaves arginine to ornithine and urea.[66]

Before the end of the nineteenth century still another compound was added to the list of protein amino acids. It had been found in 1810 by William Wollaston as a component of a kidney stone, and named by him cystic oxide (Berzelius renamed it cystine). During the 1880s Eugen Baumann showed that it can be reduced to a sulfhydryl (SH) compound he named cysteine, but the structural formulas he proposed for the two sulfur-containing amino acids were revised by Ernst Friedmann in 1903. When in 1899 Karl Mörner isolated cystine from horn, and a year later he and Gustav Embden obtained it from other protein materials, it was concluded that this amino acid provided the alkali-labile sulfur, whose presence had been detected by Scheele over a century earlier. Also, the presence of cysteine in some proteins was indicated by the rose color formed upon the addition of nitroprusside. It was also noted, however, that a portion of the sulfur present in proteins was not released as sulfide by alkali, but the explanation of this discrepancy did not come until the 1920s, when John Howard Mueller discovered the amino acid methionine during the course of his studies on the nutritional requirements of a haemolytic streptococcus, which requires protein hydrolysates for growth. Its chemical structure as a thioether was established in 1928 by George Barger and Frederick Coyne.[67] Then in 1935 William Rose showed that hydrolysates of fibrin contain another hitherto unknown amino acid, essential for the growth of immature rats; it proved to be a hydroxyamino acid related to serine and to the four-carbon sugar threose, whence its name threonine. Earlier additions had been proline, valine, and isoleucine.[68]

To this list was also added tryptophan, a compound with a long prior history. At intervals during the nineteenth century, there were reports (for example, by Tiedemann and Gmelin in 1826 and by Claude Bernard in 1856) of the appearance of a red color when chlorine water was added to albuminoid matter that had undergone decomposition. By 1875 it was known that indole, a product of animal putrefaction, gives this color reaction. In 1890 Richard Neumeister concluded that the chromogenic material arises when proteins undergo extensive degradation, and he named it "tryptophane." During the succeeding decade several investigators sought to isolate this substance, but without success. Another color reaction given by proteins had been found in 1874 by Albert Adamkiewicz; a violet color appeared when sulfuric acid was added to a solution of egg albumin in glacial acetic acid. Gowland Hopkins and Sidney Cole examined this color test, found that the reaction is caused by the glyoxylic acid in the acetic acid, and used it to isolate a substance that gave both the chlorine and glyoxylic reactions. In 1907 Alexander Ellinger synthesized the compound and showed it to be indolylalanine. Because tryptophan, like other indole derivatives, is decomposed by acids, Hopkins and Cole isolated it from a digest of casein by a preparation of what was then called trypsin.[69]

By 1935 it was agreed that purified proteins yield upon hydrolysis with acids or alkalis approximately twenty amino acids; the figure below shows the structural formulas of those now commonly listed in textbooks as the "repertoire" from which proteins are built.[70] It should be emphasized that this list only represents most of the "survivors" of the nineteenth- and twentieth-century search for the units of protein structure, and that the cleavage methods used frequently yielded a variety of products of uncertain identity and purity that were given names that have disappeared from the chemical vocabulary. Apart from the ill-defined products of acid or alkaline hydrolysis, those obtained by the oxidation of proteins with reagents such as potassium permanganate, chromic acid, hydrogen peroxide, or ozone, added to the speculations about protein structure. Although some chemists (for example, Niels Troensegaard) pursued this approach well into the 1920s, it was evident at the turn of the century that the hydrolytic methods were likely to be most fruitful ones. It should also be noted that some well-defined amino acids (for example, hydroxyproline or diiodotyrosine) were found only in particular proteins (collagen or thyroglobulin) and were later shown to be formed by metabolic processes from proline or tyrosine after these amino acids had been incorporated into such proteins. Several of these derivatives will be mentioned later in connection with the metabolic transformations of proteins. As will also be evident from the later discussion, since 1945 the introduction and refinement of new techniques (especially chromatography) for the purification and quantitative estimation of naturally occurring substances, and for the ready determination of their structure (as by nuclear magnetic resonance [NMR] or mass spectroscopy), have led to the discovery of previously unknown amino acids, some of which may be as yet unrecognized additions to the recently accepted repertoire for the biosynthesis of proteins, especially in plants and microorganisms.[71]

Except in glycine, the carbon atom linking the α-amino and α-carboxyl groups of the protein amino acids is "asymmetric" in the sense of van't Hoff and Le Bel. Around 1900 Emil Fischer examined the optical activity of the amino acids he

NH_2CH_2COOH

Glycine (Gly)

Three-letter abbreviations denote amino acid residues (–NHCH(R)CO–), not free amino acids

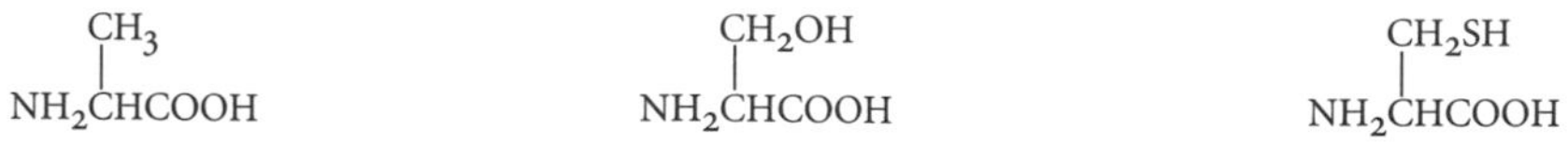

CH_3
|
$NH_2CHCOOH$

Alanine (Ala)

CH_2OH
|
$NH_2CHCOOH$

Serine (Ser)

CH_2SH
|
$NH_2CHCOOH$

Cysteine (Cys)

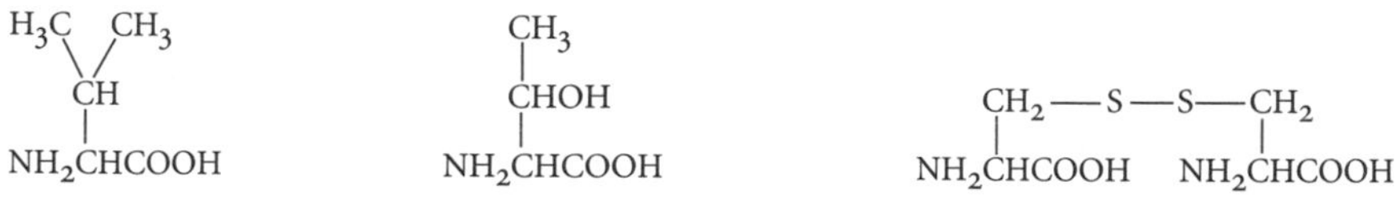

$H_3C \quad CH_3$
\ /
CH
|
$NH_2CHCOOH$

Valine (Val)

CH_3
|
$CHOH$
|
$NH_2CHCOOH$

Threonine (Thr)

$CH_2—S—S—CH_2$
| |
$NH_2CHCOOH \quad NH_2CHCOOH$

Cystine (Cys Cys)

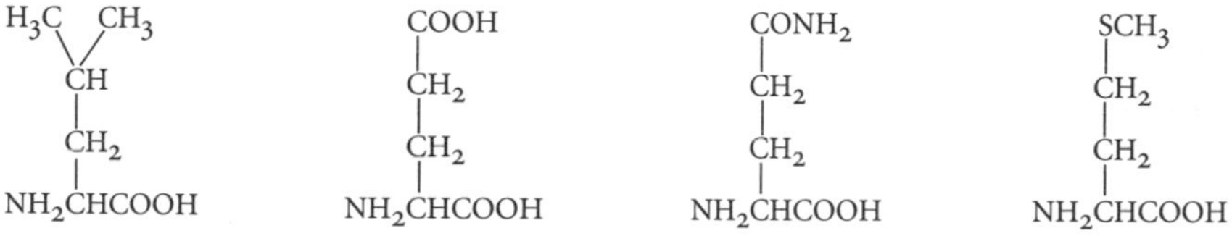

$H_3C \quad CH_3$
\ /
CH
|
CH_2
|
$NH_2CHCOOH$

Leucine (Leu)

$COOH$
|
CH_2
|
CH_2
|
$NH_2CHCOOH$

Glutamic acid (Glu)

$CONH_2$
|
CH_2
|
CH_2
|
$NH_2CHCOOH$

Glutamine (Gln)

SCH_3
|
CH_2
|
CH_2
|
$NH_2CHCOOH$

Methionine (Met)

CH_2CH_3
|
$CHCH_3$
|
$NH_2CHCOOH$

Isoleucine (Ile)

$COOH$
|
CH_2
|
$NH_2CHCOOH$

Aspartic acid (Asp)

$CONH_2$
|
CH_2
|
$NH_2CHCOOH$

Asparagine (Asn)

$CH_2—CH_2$
| |
$CH_2 \quad CHCOOH$
N
H

Proline (Pro)

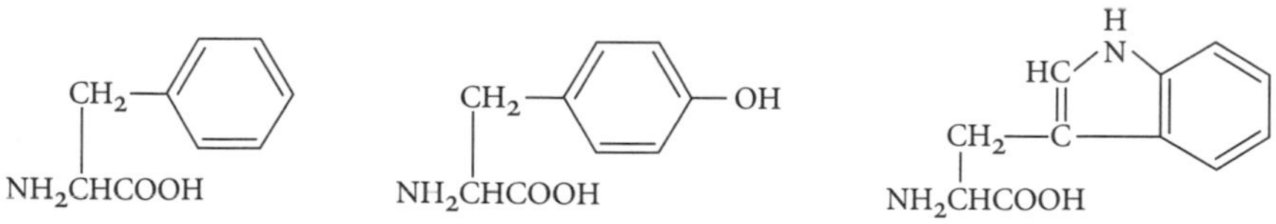

CH_2–(phenyl)
|
$NH_2CHCOOH$

Phenylalanine (Phe)

CH_2–(phenylene)–OH
|
$NH_2CHCOOH$

Tyrosine (Tyr)

H
N
HC
||
$CH_2—C$ (indole)
|
$NH_2CHCOOH$

Tryptophan (Trp)

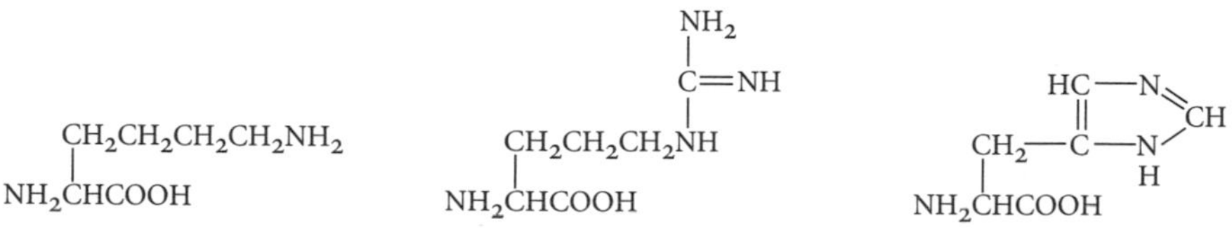

$CH_2CH_2CH_2CH_2NH_2$
|
$NH_2CHCOOH$

Lysine (Lys)

NH_2
|
$C{=}NH$
|
$CH_2CH_2CH_2NH$
|
$NH_2CHCOOH$

Arginine (Arg)

HC—N
|| CH
$CH_2—C—N$
H
|
$NH_2CHCOOH$

Histidine (His)

One-letter abbreviations: A = Ala; C = Cys; D = Asp; E = Glu; F = Phe; G = Gly; H = His; I = Ile; K = Lys; L = Leu; M = Met; N = Asn; P = Pro; Q = Gln; R = Arg; S = Ser; T = Thr; V = Val; W = Trp; Y = Tyr

knew about, and subsequent work led to the recognition that all twenty are characterized by the same type of arrangement of the four groups about the α-carbon—that is, they all have the same configuration or chirality (now usually designated with the prefix L-). Fischer also used the method devised by Pasteur to effect the resolution of the racemic forms of amino acids produced by organic synthesis. For a time, the isomeric D-forms were often termed the "unnatural" antipodes, but this practice was discontinued after 1945, as many biochemical substances were shown to contain D-amino acid units. In 1939 there was a brief flurry of excitement when Fritz Kögl reported that the proteins of tumor tissues contain D-glutamic acid units, but this claim was soon shown to be invalid.[72]

Before considering further the twentieth-century developments, it is necessary to return to the nineteenth century, when many physiological chemists were more interested in "albumoses" and "peptones" than in amino acids. After the discovery by Johann Nepomuk Eberle of the agent that Theodor Schwann called pepsin, during the 1840s the French pharmacist Louis Mialhe examined the action of gastric juice on a variety of protein materials. He was followed by Carl Gotthelf Lehmann, who expanded and corrected Mialhe's findings, and introduced the term *peptone,* which he defined as follows: "By the action of natural or artificial gastric juice on protein bodies . . . there are formed thoroughly new substances which, although they coincide in their chemical composition and in many of their physical properties, with the substances from which they are derived, essentially differ from them, not only in their ready solubility (in water, and even in alcohol), but in having now lost the faculty of forming insoluble combinations with most metallic salts."[73] Subsequently, it was recognized that the "trypsin" of the pancreatic secretion into the small intestine also produces peptones, with "albumoses" as intermediates in the process. In addition to the properties mentioned by Lehmann, peptones differed from proteins in being noncoagulable by heat and diffusible through membranes. For many years, peptones were considered to be the end products in the gastrointestinal digestion of proteins and to serve as precursors of blood proteins during their passage through the intestinal wall.[74]

During the period 1870–1900, extensive studies were conducted on these products by many investigators, including Franz Hofmeister and, most notably, by Willy Kühne and his associates Russell Chittenden and Richard Neumeister. Kühne's group performed laborious fractionations of the mixtures produced by the action of the gastric and pancreatic enzymes on proteins, using precipitation with salts (sodium chloride, ammonium sulfate, magnesium sulfate) under various conditions of temperature and acidity. They determined the elementary composition of the many products, to which they gave distinctive names. In their view, the action of pepsin on albumins first gave "syntonin," which was then converted to "protoalbumose" (soluble in water) and "heteroalbumose" (soluble only in salt solutions). From such albumoses, pepsin and trypsin produced "hemipeptones" and "antipeptones." By 1887 Neumeister had elaborated a scheme of considerable complexity, but since none of these products could be characterized in a manner acceptable to the chemists of his time, the scheme was of interest only to physiological chemists who, like Benjamin Moore, adhered to the idea that "albumoses and peptones are modified during their passage through the epithelial cells by the action of living protoplasm. What substances are formed from them is not known by direct experiment, but it is highly probable that the process is backward into coagulable proteid. It is known that coagulable proteid can be artificially obtained from peptone and albumose."[75] This idea, sometimes referred to as the peptone doctrine, gained public notice during the 1880s, when "peptonized" food materials became articles of commerce. The idea was laid to rest after 1901, when Otto Cohnheim showed that the intestinal mucosa elaborates an enzymatic activity ("erepsin") that converts peptones into amino acids, and Otto Loewi demonstrated that an extensively digested autolysate of pancreatic protein can replace intact protein in the animal diet.[76] Some years later, Otto von Fürth wrote: "There are few chapters in physiological chemistry that show more clearly the rapidity with which scientific ideas change than the doctrine of the albumoses and peptones. . . . How little of what I learned with great pains has importance today, and how many of the problems that the previous generation of physiologists fought about so passionately have lost for us any sense or significance."[77] The clear demonstration that amino acids are formed upon the enzymatic cleavage of proteins was of considerable chemical importance. The conditions under which these digestions were conducted were so much milder as regards temperature and acidity than those used by the protein chemists that the appearance of amino acids after extensive enzymatic digestion strengthened the conviction that these products were not artifacts of acid hydrolysis but were present in proteins in the form of "amide-like anhydrides." Such results also silenced some adherents of the idea of "living" protein molecules. Loew, for example, had stated that "we can conclude with a high degree of probability that the leucine radical does not exist as such in the proteids" and had compared the formation of amino acids upon the acid hydrolysis of proteins to the effect of such treatment on sugars to yield various products of decomposition.[78]

From Peptones to Peptides

Whatever appeal the peptone doctrine may have had for physiologists, the organic chemist Emil Fischer was convinced that the "various kinds of albumoses and peptones with which the physiologists deal are for the chemist only intractable mixtures. The next aim of research in this field must be directed to the isolation from them of definable homogeneous substances."[79] Neither Fischer nor the protein chemists who followed him could solve this problem before the introduction of chromatographic methods into protein chemistry after World War II.[80] Instead, Fischer chose to make peptonelike molecules by stepwise organic synthesis. Previous workers (for example, Edouard Grimaux and Hugo Schiff) had prepared materials that gave the biuret reaction for proteins by heating amino acids at high temperatures, but, as Fischer put it, "All the products described by them are amorphous substances that are difficult to characterize, and one can say as little about their structure as about the extent of their relationship to natural protein compounds. If one wishes to obtain secure results in this difficult field, one must first find a method that permits one to combine, in anhydride-like linkage, successively and with definable intermediates, the molecules of various amino acids."[81] The first such compound made in his laboratory was glycyl-glycine ($NH_2CH_2CONHCH_2COOH$), on the assumption that the amino acids of proteins and peptones are linked by amide (CO-NH) bonds. In 1902 Fischer proposed that "in analogy to the known designation of carbohydrates as disaccharides, trisaccharides, etc., the substances of the type glycyl-glycine be named dipeptides and anhydride-like combinations of a greater number of amino acids be denoted tripeptides, etc."[82] In subsequent articles, he began to use the term *polypeptides.*

This introduction of the word *peptide* appeared in Fischer's abstract of his talk during the afternoon of 22 September 1902 at the Karlsbad meeting of the Chemical Section of the Gesellschaft der deutscher Naturforscher und Aerzte. During the morning of the same day, Franz Hofmeister delivered a plenary lecture *(Hauptvortrag)* entitled "Ueber den Bau des Eiweissmoleküls," and the full text was published in the *Naturwissenschaftliche Rundschau.* The lecture was an adaptation of a review article, in which Hofmeister concluded that proteins "are formed by the condensation of alpha-amino acids bound through the regularly recurrent -CO-NH-CH= group." Among the hypotheses that Hofmeister judged to be less well founded was Paul Schützenberger's view (1875) that the amino groups of polyamino acid chains are linked by CO groups to form ureido linkages (-NH-CO-NH-), and that of Albrecht Kossel, who in 1900 considered complex proteins to be built around a central nucleus that resembles the protamines, relatively simple basic substances discovered in 1874 by Friedrich Miescher. Hofmeister attached special importance to the biuret reaction, and he noted that this test was given by the products made by Theodor Curtius during the 1880s either by the self-condensation of glycine ethyl ester or by the reaction of benzoyl chloride with silver glycinate. In the latter process one of the isolated products was benzoyl-glycyl-glycine, which, in retrospect, may be considered to have been the first well-defined synthetic peptide derivative. In favor of his theory, Hofmeister also offered evidence from physiological studies on the enzymatic cleavage of proteins and hippuric acid (benzoyl-glycine).[83]

In the abstract of his talk at the Karlsbad meeting, Fischer wrote: "Finally the speaker discussed the coupling of the amino acids in protein molecules. The idea that amide groups play the principal role is obvious, as Hofmeister also assumed in his general lecture this morning. I came to the same conviction one and a half years ago, when I initiated experiments to effect the synthetic linkage of amino acids."[84] In the general method developed in his laboratory after 1902, the synthesis began with the reaction of a halogenacyl halide (for example, chloroacetyl chloride) with an amino acid derivative to yield a product that was converted to a dipeptide derivative with ammonia. Treatment of the latter derivative with the halogenacyl chloride derived from another amino acid gave, after amination, the expected tripeptide derivative, and so on. By 1907 Fischer's associates had used this method to prepare more than a hundred peptides, including the octadecapeptide H-Leu-$(Gly)_3$-Leu-$(Gly)_3$-Leu-$(Gly)_8$-Gly-OH. In December 1905 he wrote to Baeyer: "Recently I have . . . prepared the first crystalline hexapeptide and hope to obtain a matching octapeptide before Christmas. Then we should be close to the albumoses. . . . My entire yearning is directed toward the first synthetic enzyme. If its preparation falls into my lap with the synthesis of a natural protein material, I will consider my mission fulfilled."[85]

At each stage of the investigation, Fischer called attention to the fact that, like the albumoses and peptones, the synthetic peptides gave the biuret reaction, and that most of them were cleaved by pancreatic enzymes. Indeed, he claimed that

> *l*-leucyl-triglycyl-*l*-tyrosine prepared artificially has all the properties of the albumoses. These observations are of importance in casting doubt on the view which formerly prevailed that, being intermediate products between proteins and peptones, the albumoses are substances of considerable complexity. . . . The synthesis of the higher terms [peptides] has been restricted hitherto to the combinations of glycine, alanine, and leucine; there is not a shadow of doubt, however, that all the remaining amino acids could be associated in complicated systems with our present methods. The knowledge

> of the artificial polypeptides thus acquired has opened up new ways of investigating the peptones and albumoses analytically.[86]

When Fischer reached the stage of the 14-amino acid compound and found that it gave a strong biuret reaction and was precipitated by protein reagents, he stated in January 1907 that "one cannot avoid the impression that this tetradecapeptide represents a product quite closely related to the proteins, and I believe that with the continuation of the synthesis to the eicosapeptide one will come within the protein group."[87] Later in the same year, however, he wrote: "From the experience gained thus far, I do not doubt that the synthesis can be continued by means of the same methods beyond the octadecapeptide. I must provisionally waive such experiments, which are not only very laborious but also very expensive."[88]

Clearly, for Fischer, the proteins isolated during the nineteenth century were aggregates of compounds having molecular weights below about 5,000, and one cannot escape the impression that in expressing the hope to synthesize a protein by combining peptides made by his synthetic method, Fischer was unwilling to admit publicly the possibility that he had failed in his ambitious objective. To his richly merited scientific prestige was added the great attention in the public press to his claims, and many people came to believe that, owing to Fischer's work, the preparation of synthetic proteins was around the corner. In the excitement, little notice was taken of the contributions of Franz Hofmeister and Theodor Curtius. In 1908 Hofmeister wrote: "That the end products of protein hydrolysis are largely linked to one another as amides was concluded by me on the basis of chemical and physiological evidence, and was securely established by E. Fischer through the preparation of such compounds—the peptides. . . . The rapid progress in the synthesis of polypeptides under the aegis of Fischer could lead to the view that one may expect the elucidation of protein structure to come entirely from this approach. This hope comes to naught if one considers the enormous number of synthetic possibilities."[89] As for Curtius, he resumed his work on peptides and in 1902 developed a coupling method involving the azides of acylamino acids. Some harsh words were exchanged with Fischer, and each of these great chemists felt that the other had not cited his work properly.[90] The public accolade went to Fischer, but except for Emil Abderhalden, later peptide chemists had little occasion to use halogenacyl halides, while the azide method continued to be useful. Although the strategy of Fischer's synthetic approach has guided further research in the peptide field, the tactical method he adopted was supplanted by newer procedures. During the 1920s valuable contributions were made by Rudolf Schoenheimer and Fritz Wessely, and in 1932 Max Bergmann and Leonidas Zervas introduced the "carbobenzoxy" method, which greatly broadened the scope of peptide synthesis to include amino acids that could not be handled by means of Fischer's method. It must be noted, however, that during the 1930s the incentive of making long-chain polypeptides had weakened, and it was not until the following decade, when it was recognized that some hormones and antibiotics are peptides, that there began an intensive effort to develop new synthetic methods in this field.[91] Many aspects of the more recent efflorescence of peptide chemistry will be mentioned later.

In embarking on his synthetic program in the protein field, Fischer realized that, as in the nineteenth-century development of knowledge about structure of organic molecules such as the fats and carbohydrates, one had to establish, by quantitative analysis, the complete unitary composition of well-defined proteins (in this case, of the amino acids). When Fischer began his work on proteins, about a dozen amino acids were known to be products of acid hydrolysis. By 1906, when Fischer presented a summary of the progress made in his laboratory, he had added to the list valine, proline, hydroxyproline, and diaminotrioxydodecanoic acid (he later withdrew the last one[92]). In 1901 Fischer introduced a general method for the separation of amino acids by the fractional distillation under reduced pressure of their ethyl esters. These were then converted to free amino acids, which were crystallized, weighed, and characterized. These analyses were largely done by Emil Abderhalden, or under his direction, and Fischer wrote to a Berlin colleague as follows: "Because of his unusual capacity for work, in a short time Abderhalden has become so adept in the difficult methods of organic chemistry that I was able to accept him as a collaborator in my private laboratory. . . . He is a good observer, and an enemy of all superfluous hypotheses."[93] This opinion led Fischer to turn over to Abderhalden the succession of post-M.D. students who flocked to Fischer's laboratory at that time; most of them worked on the application of the ester method to the analysis of a variety of protein preparations. Apart from the fact that Fischer's assessment of Abderhalden's chemical talent proved to be incorrect, the analytical data produced by the ester method were soon shown by Thomas Burr Osborne to require correction.[94] During 1920–1940 there were many other contributions to the quantitative analysis of the amino acid composition of protein hydrolysates, but as will be seen later in this chapter, the solution of the problem did not come until after about 1945, with important consequences for the study of protein structure.

In parallel with the development of the ester method of protein analysis, Fischer embarked on a systematic program to synthesize individual amino acids. As he noted in his first paper on this subject, "Whereas the hydrolytic cleavage of proteins usually produces optically-active amino acids, the artificial products are racemic. The complete synthesis of the natural compounds is only achieved when one can cleave the racemic mixture into the optically active compounds." In addition to improving older synthetic methods for making racemic amino acids, Fischer devised methods for their resolution into the optically active components. Because all the amino acids Fischer handled had only one asymmetric carbon, two optically active isomers were obtained, and comparison of the sign of optical rotation with that of the natural amino acid identified the desired isomer.[95]

The Multiplicity and Individuality of Proteins

After 1850, when some of the facile generalizations based on elementary analyses had been abandoned, it was increasingly recognized that there are many albuminoid substances with different properties. The impetus came principally from the use of neutral salts for their fractional precipitation ("salting-out"). Peter Ludwig Panum reported in 1852 that the addition of solid sodium chloride to blood serum gave a precipitate, but none was formed with egg white. In 1854 Rudolf Virchow described his extensive observations on the precipitation of albuminoid substances by means of a variety of inorganic salts and interpreted the failure of a salt to precipitate all of the protein in a biological fluid to indicate that the protein was present in the fluid in different states. Such observations were then followed by the systematic studies performed by Prosper-Sylvain Denis, who provided, during the 1850s, the first clear indications of the multiplicity of the serum proteins. From his work and the work of those who followed came the separation of a water-soluble material (serum albumin, which Denis termed serine) from a material soluble in dilute salt solutions but sparingly soluble in water. The latter subsequently came to be termed globulin,[96] a designation introduced by Berzelius to denote the material he believed to be associated with the iron-containing component of the "red coloring matter" (hematoglobulin) of red blood corpuscles. Denis also obtained from blood plasma a fraction ("soluble fibrin") that had the properties of a soluble precursor of fibrin (Virchow named it fibrinogen). The further study of the process of blood coagulation later came to a focus in the work of Alexander Schmidt and of Olof Hammarsten; by 1880 it was clear that the formation of the insoluble fibrin involves the action of a "fibrin ferment" (now called thrombin) on fibrinogen. This advance put the physiological process of blood coagulation within the scope of chemical investigation, and the ability of the circulating blood to remain fluid was no longer considered to be an expression of a vital force, as believed by John Hunter in the previous century. During the twentieth century, the fractionation of plasma proteins became a matter of military importance in connection with blood transfusion and was notably furthered by the work of Edwin Cohn's research group during World War II.[97] An important addition to the salts used by Denis was ammonium sulfate, introduced by Camille Méhu in 1878; its high solubility in water made it the protein precipitant of choice.[98]

In 1859 Willy Kühne reported the extraction from frog muscles of a coagulable globulin-like material (he termed it myosin), which he considered to be identical with Liebig's "muscle fibrin." Another indication of the rapid adoption of the method developed by Denis was Hoppe-Seyler's work on the albuminoid component of the yolk of hen's eggs. Hoppe-Seyler obtained a material (he named it vitellin) that had the solubility properties of a globulin. Substances similar to vitellin had been found in the eggs of other animals, notably fish and amphibia, and he grouped them in the class that included the ones identified in blood plasma and in muscle. By 1870 it was becoming evident that these various globulins might be different, and new names appeared to denote the proteins isolated from a particular tissue of an individual animal species; for example, the vitellin-like protein of fish eggs was named ichthulin.

Denis also showed that albuminoid substances present in plant seeds are extractable with 10 percent sodium chloride solutions. The decisive studies on these proteins were conducted later in the nineteenth century by the agricultural chemists Heinrich Ritthausen and Thomas Burr Osborne. In a report of the work he had done since 1860, Ritthausen concluded that despite the similar elementary composition of animal and plant proteins, there are significant differences between them, especially in their content in tyrosine, leucine, and aspartic acid, and that the metabolic transformation of plant proteins into animal proteins is more complex than had previously been assumed. As he put it, "Does not perhaps the circumstance that the plant proteins casein, albumin are distinguished from the animal materials by designating them plant casein, plant albumin conceal the idea that, despite their similar composition and properties, there are differences among them, and that it is necessary to consider the substances as distinct and not merely to regard them as identical?"[99] Ritthausen's initial protein preparations were largely obtained by extraction with acid or alkali, and Hoppe-Seyler considered them to be "not pure unchanged albuminoid substances, but more or less decomposed and insufficiently purified substances whose properties and composition tell nothing about the ones from which they were created."[100] This judgment was based on the studies of Theodor Weyl and Oswald Schmiedeberg on the crystalline seed proteins obtained by Denis's method. In rebuttal, Ritthausen published during 1880–1884 a series of papers to show that many of his original preparations had retained their solubility in saline solutions and, in the course of this work, he prepared crystalline globulins from the castor bean, hemp seed, and sesame seed in the manner that Edmund Drechsel (in 1879) and Georg Grübler (in 1881) had obtained crystals from the Brazil nut and squash seed. The procedure was simply to allow a warm sodium chloride solution saturated with the protein to cool slowly, whereupon well-developed crystals appeared.

The doubts about the validity of Ritthausen's results were settled by Osborne, who wrote:

> The fact that these proteid substances can be artificially crystallized is not only interesting in itself, but is important as presumably furnishing a means of making preparations of undoubted purity which will offer a secure basis for further study of their properties. The contradictory statements made by various investigators, not only in regard to the properties and composition of these bodies but also in respect to the value of the methods of solution and separation which have been employed hitherto, render an exact knowledge of all the facts relating to these substances a matter of the highest scientific and practical importance.[101]

Osborne's subsequent work, collected in his *Vegetable Proteins* (1909, 1924), established the importance of Ritthausen's contributions and counteracted the disfavor into which they had fallen owing to the verdict of Hoppe-Seyler. During 1890–1900, Osborne examined the seed proteins of more than thirty different plant species, and it became clear that the globulins previously grouped under the term *vitellin* were different. He assigned new names to the individual globulins; the one from Brazil nut became excelsin, the one from hemp seed edestin, and so forth. The alcohol-soluble proteins, which he termed prolamines—gliadin (from wheat), hordein (from barley), and zein (from corn)—were also included in Osborne's study. After 1900 he applied Emil Fischer's ester method for the separation of the amino acids formed upon acid hydrolysis, and he also used the Kossel-Kutscher method for the isolation of lysine, arginine, and histidine. Although these procedures did not give a complete account of the amino acid composition of the starting materials, it was clear by 1909 that proteins that resembled each other in solubility properties were distinctly different chemical substances. The most dramatic evidence of the individuality of the seed proteins came from Osborne's collaboration with the pathologist Harry Gideon Wells in a study of their immunological properties.[102]

It should be noted that during the 1850s Theodor Hartig, Otto Maschke, and Ludwig Radlkofer had studied the albuminoid granules in plant seeds, and that the influential botanist Carl Nägeli had questioned the crystalline nature of these structures. His microscopic observations led him to denote them as crystalloids, a term used by Ritthausen. Among the various hypotheses advanced by Nägeli during his scientific career one of the earliest was the idea that organic substances are arranged in the plant cell into a mosaic of submicroscopic particles (micelles) which had a crystalline appearance. This micellar hypothesis was adopted by leading plant physiologists, notably Wilhelm Pfeffer, but the beautiful crystals obtained in bulk by Osborne showed that the doubts of Nägeli and Pfeffer were not justified.

Even before the first recorded microscopic observations of crystalline plant proteins, it had been shown that the "red coloring matter" of animal blood could be brought to crystallization. In 1840 Friedrich Ludwig Hünefeld wrote that "I have occasionally seen in almost dried blood . . . rectangular crystalline structures which under the microscope had sharp edges and were bright red."[103] There were other isolated reports of the appearance of "blood crystals," and in 1851 Otto Funke described the phenomenon carefully. The red coloring matter had been called hematosine by Lecanu and hematoglobulin by Berzelius; in 1864 Hoppe-Seyler shortened the latter term to hemoglobin. By 1871, when Wilhelm Preyer collected the available knowledge about crystalline hemoglobin,[104] the blood from more than forty animal species had been examined, and it was seen that there were wide differences in the forms of the crystals, depending on their biological source. It was also known from the work of Hoppe-Seyler and George Gabriel Stokes during 1862–1864 that hemoglobin solutions take up molecular oxygen to form oxyhemoglobin, and Kühne had shown that oxyhemoglobin is less soluble in water and crystallizes more readily than does deoxygenated ("reduced") hemoglobin.[105] A decisive factor in the recognition of the reversible interconversion of the two forms of hemoglobin, and in the later elucidation of this phenomenon, was the use of the spectroscope, introduced into chemistry in 1860 by Kirchhoff and Bunsen.[106]

Striking evidence of the individuality of the hemoglobins was later provided by the painstaking crystallographic studies of Edward Reichert and Amos Brown. In a volume summarizing their results, they concluded that "the oxyhemoglobin obtained from the same blood crystallizes in the same form, with the same axial ratio, although often with different habit, when obtained by different methods of preparation. . . . The crystals obtained from different species of a genus are characteristic of that species, but differ from those of other species of the genus in angles or axial ratio, in optical characters, and especially in those characters comprised under the general term of *crystal habit*, so that one species can usually be distinguished from another by its hemoglobin crystals."[107] This evidence for the individuality of the hemoglobins was supplemented in 1923 by the immunological studies of Karl Landsteiner and Michael Heidelberger.[108]

By 1870 it was also known that treatment of hemoglobin with acid gives a colorless albuminoid constituent (globin) and a red iron-containing material named hematin. To denote nonalbuminoid constituents such as hematin, Hoppe-Seyler introduced the term *prosthetic group,* and he designated the conjugate with globin as a proteid.[109] Ludwik Teichmann had shown in 1853 that treatment of dried blood with sodium chloride and hot glacial acetic acid readily gives crystals of "hemin," and by 1857 this reaction had been introduced into forensic chemistry as a test for blood.[110]

Between 1880 and 1900 various methods for the preparation of hemin gave products that differed in their elementary composition, but in 1901 William Küster obtained a single product of the composition $C_{34}H_{33}O_4N_4ClFe$. The elucidation of the structure of hemin, and of its relationship to the pigment of hemoglobin, was one of the many great achievements of early twentieth-century organic chemistry as applied to the study of complex biological materials. The chief figure was Hans Fischer, who in 1929 established the structure of the tetrapyrrole unit (porphyrin) of hemin by synthesis; before his work, many important contributions to the solution of this problem had been made by Marceli Nencki and Richard Willstätter, among others. On the basis of the knowledge available in 1913, William Küster suggested that the four pyrrole rings of a porphyrin are linked by four single carbon atoms to form a 16-member ring system. Such large ring compounds were unknown at that time, however, and alternative formulas were proposed. In the face of this opposition, Küster withdrew his proposal in 1923, but Hans Fischer's synthesis proved that Küster's original idea had been correct. Fischer's work on hemin also set the stage for the elucidation of the structure of the chlorophylls; it had been known from earlier studies that they are porphyrin derivatives. The magnitude of Fischer's achievement was outmatched later in the twentieth century by the synthesis of vitamin B_{12} ($C_{63}H_{88}N_{14}O_{14}PCo$) by Albert Eschenmoser, Robert Woodward, and their associates.[111]

As we have seen, crystalline preparations of albuminoid substances were known throughout the latter half of the nineteenth century. It is therefore of interest to find in a lecture by Louis Pasteur the statement that "you know that the most complex molecules of plant chemistry are the albumins. You also know that these immediate principles have never been obtained in a crystalline state. May one add that apparently they cannot crystallize."[112] Apart from the studies on plant seed proteins, mentioned above, in 1889 Franz Hofmeister crystallized egg albumin from a half-saturated solution of ammonium sulfate, and a few years later his procedure was improved by Hopkins and Pinkus. In 1894 August Gürber crystallized horse serum albumin, and in 1899 Arthur Wichmann obtained a crystalline albumin from milk. By 1940 many other proteins had been crystallized, especially after James Sumner's crystallization of concanavalin and urease.[113] Although it was widely agreed that the crystallization of proteins is a valuable means of their purification, doubts were expressed about the homogeneity of such preparations. For example, it was remarked that "there is scarcely a crystalline substance which takes up dissolved substances, like a sponge, to such a high degree as does albumin," and, in speaking of crystalline oxyhemoglobin, Emil Fischer stated that "the existence of crystals does not in itself guarantee chemical individuality, since isomorphous mixtures may be involved, as is frequently the case for the silicates."[114]

In the history of the interplay of chemistry and biology, the phenomenon of the crystallization of chemical substances has repeatedly evoked the sense that "the beauty of crystals lies in the planeness of their faces" and that it provides an insight into the organization of biological systems.[115] More important, as was noted earlier, the availability of well-formed crystals of such proteins as oxyhemoglobin permitted the study of their intimate structure by means of the techniques of X-ray crystallography.

For many years after the discovery of X-rays by Wilhelm Röntgen in 1895 there was intensive but inconclusive experimentation to determine whether they were waves or particles. In 1912 Max von Laue suggested to his associates Paul Knipping and Walther Friedrich that, because the presumed wavelength of X-rays was shorter than the distance between the regularly spaced atoms in the lattice of a crystal, diffraction effects should be produced when these rays are passed through crystals. The striking symmetrical patterns of diffraction spots that were indeed found provided convincing proof of the wave nature of X-rays, and of their interaction with crystal structures. Moreover, the diffraction patterns accorded with the geometrical theory of crystal structure developed during the nineteenth century by Auguste Bravais and Evgraf Fedorov (among others), after René Just Haüy had transformed the theory of crystalline matter into a mathematical science of crystallography.[116] In a letter dated 2 October 1912, Fedorov wrote: "For us crystallographers this discovery is of prime importance because now, for the first time, we can have a clear picture of that on which we have but theoretically placed the structure of crystals and on which the analysis of crystals is based."[117]

The idea that X-rays might be used to determine the structure of a crystal by working back from the angles of reflection and the intensities of the diffracted rays was developed immediately afterward by William Henry Bragg and his son William Lawrence Bragg. By 1913 they had achieved striking success with several inorganic crystals, notably sodium chloride. Some years later, Henry Armstrong commented: "Prof. W. L. Bragg asserts that 'In sodium chloride there appear to be no molecules represented by NaCl. The equality in number of sodium and chlorine atoms is arrived at by a chess-board pattern of these atoms; it is a result of geometry and of a pairing off of the atoms.' This statement is more than 'repugnant to common sense.' . . . Chemistry is neither chess nor geometry, whatever X-ray physics may be. Such unjustified aspersion of the molecular character of our most necessary condiment must not be allowed to go unchallenged."[118] After World War I, work began on the structure of organic compounds. Among the pioneers were Roscoe Gilkey Dickinson (at Caltech) and Kathleen Lonsdale (at the Royal Institution); at the Cavendish Laboratory in Cambridge, Arthur Hutchinson was joined by John Desmond Bernal.[119] The first structures to be solved were those of hexamethylenetetramine and hexamethylbenzene. Although exact solutions of more complex organic structures were not yet possible, by 1932 X-ray diffraction played a decisive role in deciding between alternative structures, as in the case of the carbon skeleton of the sterol nucleus. During the 1920s this method was applied to the study of stretched natural fibers (silk fibroin, wool keratin, cellulose), and in 1934 the first satisfactory diffraction spectra of a crystalline protein (pepsin) were obtained by Bernal and Dorothy Crowfoot (later Hodgkin). Moreover, during the 1930s new approaches to the handling of X-ray data were developed (for example, the Patterson vector method and Beevers-Lipson strips), and new techniques such as the introduction of heavy metals by isomorphous replacement led to the solution of the structure of the porphyrin-like compound phthalocyanine.[120]

After 1945 the development of high-speed digital computers greatly increased the ability of X-ray crystallographers to solve complex organic structures.[121] The initial successes included Dorothy Hodgkin's complete solution of the structure of vitamin B_{12}. Max Perutz and John Kendrew undertook the study of horse oxyhemoglobin and of whale myoglobin, and by 1960 the three-dimensional structure of the simpler myoglobin molecule had been worked out. We shall return to the X-ray studies later in this chapter. It should be noted, however, that even during the 1990s, after many protein structures have been determined by means of X-ray crystallography, the rate-limiting step has often been the preparation of single crystals suitable for X-ray diffraction analysis. Moreover, it has been recognized that subjective factors in the analysis and refinement of the data have, on occasion, led to error.[122]

From Colloids to Macromolecules

In 1861 Thomas Graham described an instrument he called a dialyser, consisting of a bell jar closed at the large end with parchment paper and immersed in a reservoir of water, and he reported his studies on the diffusion across the membrane of substances in aqueous solution within the bell jar. He found that albuminoid substances dialyzed very slowly, whereas crystalline compounds such as cane sugar diffused rapidly. In writing of the slowly diffusible materials, Graham stated:

> Among the latter are hydrated silicic acid, hydrated alumina, and other metallic peroxides of the aluminous class, when they exist in the soluble form; and starch, dextrin and the gums, caramel, tannin, albumine, gelatine, vegetable and animal extractive matters. Low diffusibility is not the only property which bodies last enumerated possess in common. They are distinguished by the gelatinous character of their hydrates. Although often soluble in water, they are held in solution by a most feeble force. They appear singularly inert in the capacity of acids and bases, and in all the ordinary chemical reactions. But, on the other hand, their peculiar physical aggregation with the chemical indifference referred to, appears to be required in substances that can intervene in the organic processes of life. The plastic elements of the animal body are found on this class. As gelatine appears to be its type, it is proposed to designate substances of this class as *colloids* (*colla*, glue), and to speak of their peculiar form of aggregation as the *colloidal state of matter*. Opposed to the colloidal is the crystalline condition. Substances affecting the latter form will be classed as *crystalloids*. The distinction is no doubt one of intimate molecular constitution.[123]

Graham's instrument resembled the earlier "endosmometer" used by Henri Dutrochet in his studies on the movement of liquids through natural membranes ("osmosis"), a phenomenon described by Jean-Antoine Nollet in 1748.[124]

During the nineteenth century, several physiologists, notably Carl Ludwig, Adolf Fick, and Wilhelm Pfeffer, were greatly interested in osmosis. After the development of artificial semipermeable membranes (for example, copper ferricyanide) by Moritz Traube in 1867, quantitative studies on osmotic pressure were reported. In particular, the data presented by the plant physiologist Pfeffer in his book *Osmotische Untersuchungen* (1877) provided the basis for van't Hoff's formulation in 1885 of the osmotic pressure equation, and the development of a theory of solutions which connected osmotic pressure, freezing-point depression, and the lowering of vapor pressure as thermodynamic properties. This followed the work of François Raoult, published in 1882, on the freezing points of solutions and the use of this property for the determination of the molecular weights of dissolved substances.[125] When applied to proteins, values of 14,000 were reported during the 1890s for egg albumin and of 48,000 for hemoglobin. In addition, Picton and Linder examined the ability of solutions of crystalline hemoglobin to scatter light (Tyndall phenomenon) and concluded that "there is no hard and fast line between colloidal and crystalloidal solution."[126] Thus, although proteins were considered to be colloids, it was deemed appropriate to apply to their study the theory of solutions developed for crystalloids, and the molecular weights so obtained suggested that proteins are substances of considerable size. This conclusion was also indicated by the experiments of Charles Martin and of Heinrich Bechhold, who developed "ultrafilters" (for example, gelatin) that retained dissolved proteins but let salts and sugars through.

At the turn of the century, and for many years afterward, the problem of the molecular size of proteins was beset by debate. Apart from the question of the relation of the molecules postulated in the molecular-kinetic theory developed by Joule, Clausius, Maxwell, and Boltzmann to those considered by organic chemists to be involved in chemical reactions, there was debate about Wilhelm Ostwald's campaign to base all physical and chemical theory on energetics. As has been indicated, the philosophical aspects of this controversy enlivened German and Austrian intellectual life for many years. By 1910, however, the debate (at least among scientists) was over, and even Ostwald agreed that molecules existed. The developments that decided the issue were the formulation in 1905 and 1906 by Albert Einstein and by Marian von Smoluchowski of equations for Brownian motion, and their experimental verification, during 1907–1909, by The Svedberg and Jean Perrin. In 1828 the botanist Robert Brown reported that he had observed, under the microscope, an agitated movement of particles derived from pollen grains, and he attributed the motion to "active molecules."[127] Although the observation was confirmed repeatedly by others, the movement was usually attributed to convection currents. Ludwig Wiener recognized in 1863, however, that the agitation was caused by characteristic motions within the fluid, and subsequent work, notably by Lukasz Bodaszewski in 1881 and by Louis Georges Gouy in 1888, called attention to the possibility that the Brownian motion is a consequence of the kind of molecular movement postulated in the kinetic theory of heat. With the invention of the ultramicroscope in 1903 by Henry Siedentopf and Richard Zsigmondy it became possible to determine quantitatively the fluctuation in the number of colloidal particles in a given volume. The results were those predicted on the basis of the molecular-kinetic theory, and they gave a value for Avogadro's number of 6×10^{23}.[128]

Among chemists, the problem of the molecular size of proteins was of continuing interest throughout the latter half of the nineteenth century. Apart from the physical-chemical measurements mentioned above, advantage was taken of the fact that an accurate elementary analysis can give an estimate of the minimum molecular weight required to account for one atom of an element present in a compound in the smallest amount. For example, in 1872 Ludwig Thudichum reported analytical data (0.4 percent) for the iron content of crystalline oxyhemoglobin (he called it hematocrystalline) which indicated a molecular weight of about 13,000. Subsequent determinations gave values ranging from 0.335 (Oscar Zinoffsky, 1886) to 0.47 percent (Gustav Hüfner, 1884), corresponding to a minimal molecular weight of 16,700 to 12,000. From the capacity of hemoglobin to bind oxygen or carbon monoxide, a molecular weight of about 16,700 was estimated on the assumption that there was a 1:1 combination. Because it was known that the iron did not belong to the protein part of oxyhemoglobin and the sulfur did the sulfur analysis might have been a more reliable guide; Zinoffsky found a ratio of sulfur to iron of about 2:1 and suggested that two globin molecules are linked to one hematin, making the minimal molecular weight of globin about 7,000. From sulfur analyses for other proteins, as well as chemical substitution of proteins (for example, the iodination of crystalline serum albumin), values in the range 2,000 to 6,000 were obtained. In summarizing the evidence, Friedrich Schulz concluded that "on the whole the picture is not heartening. We are far from being able to state the molecular size of proteins with any degree of confidence. In this opinion, I disagree with the current widespread view which regards it almost firmly established that the molecular weight of oxyhemoglobin is about 15,000, that of crystalline egg albumin is 5–6,000, and so forth."[129] Similar doubts were expressed by Emil Fischer in 1907. In writing about the large number of amino acids that had been identified as products of the hydrolysis of proteins, he stated: "If they were really components of the same molecule, this must be a frighteningly large complex, and in fact older estimates of the molecular weight of several proteins gave values of 12–15,000. . . . I am however of the opinion that these calculations rest on a very insecure basis, principally because we do not have the slightest guarantee for the chemical homogeneity of the natural proteins; moreover, I believe that they are mixtures of substances whose composition is in fact much simpler than has been inferred hitherto from the results of elementary analysis and hydrolysis."[130] Fischer reiterated his doubts in 1916:

> Hofmeister based his considerations on a protein molecule of about 125 units (amino acids), as is thought to be the case for hemoglobin. In my opinion, however, the methods applied to the determination of the molecular weight of the hemoglobins are less certain than had been assumed previously. Although they crystallize beautifully, no guarantee of homogeneity is given, and even if one concedes this and accepts the validity of a molecular weight of 15,000–17,000 for several hemoglobins, it should always be remembered that the hematin, from all that we know of its structure, can bind

> several globin units. . . . On the other hand, I gladly concur in the view of Hofmeister and many other physiologists that proteins of molecular weight 4,000–5,000 are not rare. If one assumes an average molecular weight of 142 for the amino acids, this would correspond to a content of 30–40 amino acids.[131]

In 1916 the clear verdict of the most famous organic chemist of his time who had worked on proteins was that values above 5,000 for the molecular weights were not acceptable. Nor can there be any doubt, in my opinion, that Fischer's disappointment with the outcome of his efforts to synthesize a protein (perhaps even an enzyme) affected his attitude to the views of people like Franz Hofmeister, whom he termed only "physiologists." Small wonder that many of Fischer's biochemical contemporaries found more merit in the physical-chemical approach to the problems of protein chemistry!

Indeed, in the following year, the last of an important series of papers by S. P. L. Sørensen, entitled *Studies on Proteins,* reported measurements, under carefully controlled conditions, of the osmotic pressure of solutions of highly purified crystalline egg albumin and gave a value of 34,000 for the molecular weight of this protein. That this large value (by Fischer's standards) was in fact too low was demonstrated by Gilbert Adair, in whose hands the osmotic pressure method reached new experimental and theoretical refinement. In 1925 Adair startled protein chemists by reporting that the molecular weight of oxyhemoglobin is 66,800, or about four times the minimal value calculated from its iron content. Moreover, recalculation of Sørensen's data for egg albumin and serum albumin by Adair's method gave molecular weights of 45,000 and 74,000, respectively.[132]

Whatever doubts may have been entertained about the validity of Adair's values were stilled by the results obtained with the ultracentrifuge, invented by the colloid chemist The Svedberg during the 1920s. In this instrument, whose design was later extensively modified and improved, protein solutions were subjected to high centrifugal forces, and the rate of the movement of the protein outward from the axis of rotation was measured by means of ingenious optical techniques.[133] Not only were the rates of sedimentation in accord with the molecular weights determined by Adair, but the homogeneity of many proteins was evident from the single sharp boundary between the part of the solution containing the sedimenting protein and the solvent above it. Within about ten years a large number of purified proteins had been examined in the ultracentrifuge, and a range of molecular weights from 17,200 (for myoglobin) to 6,680,000 (for snail hemocyanin) was found. In the studies on some of the hemocyanins (blue copper-containing respiratory pigments), Svedberg observed that the one of highest apparent molecular weight was in equilibrium with smaller units that sedimented more slowly and whose molecular weights were approximately one-eighth and one-sixteenth of the largest value. These observations led Svedberg to propose that all proteins, regardless of apparent size, are aggregates of subunits having approximately the same molecular weight; at first he considered that value to be about 35,000, and later about 16,700: "Not only the molecular weights of the hemocyanins but also the mass of most protein molecules—even those belonging to chemically different substances—show a similar relationship. This remarkable regularity points to a common plan for the building up of the protein molecules. Certain amino acids may be exchanged for others, and may cause slight deviations from the rule of simple multiples, but on the whole only a very limited number of masses seems to be possible. Probably the protein molecule is built up by the successive aggregation of definite units, but only a few aggregates are stable."[134] This attractive generalization thus assigned a value of 17,600 to a fundamental protein unit, and the finding in 1938 that, under certain conditions, oxy-hemoglobin (molecular weight, 68,000) could dissociate into "half-molecules" appeared to support Svedberg's hypothesis. Later work showed, however, that although many well-defined proteins are indeed composed of subunits (hemoglobin has four subunits), the idea that all subunits have the same fundamental molecular weight was not correct. As will be discussed later, the case of hemoglobin was taken as a model for the "allosteric" regulation of the metabolic activity of multiunit enzyme proteins.

Although Svedberg's hypothesis invoked a minimum molecular size much larger than that of the largest synthetic polypeptides prepared by Fischer, the idea that natural proteins are aggregates or "micelles" composed of smaller units was a persuasive one for many decades. Graham had noted that "the inquiry suggests itself whether the colloid molecule may not be constituted by the grouping together of a number of smaller crystalloidal molecules" and, after the work of Adair and Svedberg, it was stated that "it is certain that, in almost all these cases, what is measured is the size of a *micelle* and not of a *molecule*."[135] Indeed, during the 1920s, most of the leading organic chemists were not ready to accept the evidence presented by Hermann Staudinger for the high-molecular character of such substances as rubber, starch, or cellulose, which he considered to be linear polymers (he introduced the term *macromolecule* in 1922), although such products had been prepared repeatedly since Agostinho Lourenço had condensed succinic acid and ethylene glycol in 1863 to yield a material he described as a long-chain substance. Instead, preference had been given to the view that rubber is composed of noncovalently linked aggregates of cyclic di-isoprene units, and starch and cellulose of cyclic disaccharide units.[136]

In the case of proteins, much attention was given during the 1920s to Emil Fischer's statement that "simple amide formation is not the only possible mode of linkage in the protein molecule. On the contrary, I consider it to be quite probable that . . . it contains piperazine rings, whose facile cleavage by alkali and reformation from the dipeptides or their esters I have observed so frequently with the artificial products."[137] In 1924 Emil Abderhalden advanced this suggestion to the status of a theory of protein structure. Diketopiperazines had been isolated repeatedly from protein hydrolysates since 1849, when Friedrich Bopp found "leucinimide" (a cyclization product of leucylleucine), and Abderhalden proposed that a protein molecule consists of a number of diketopiperazine-containing complexes that are held together by the kind of partial valency *(Nebenvalenz)* postulated by Johannes Thiele in 1899 for the interaction of unsaturated organic compounds. A variety of chemical evidence was offered, but to the protein chemists Vickery and Osborne it did not "seem that the enormous labors of Abderhalden have really furthered his fundamental view of the structure of the protein molecule." On the other hand, Ross Gortner, a leading colloid chemist, considered "that the evidence presented by Abderhalden for the presence of diketopiperazine rings in proteins amounts almost to proof."[138] Around 1925 Max Bergmann and Paul Karrer were also attracted to the idea that diketopiperazines formed structural units of proteins.

Apart from the respect shown to the opinions of Emil Fischer, during the 1920s X-ray diffraction studies by Reginald Herzog, Rudolf Brill, and Johann Rudolf Katz on natural fibers left open the question whether they are long-chain molecules or aggregates of small units. Thus Brill reported that silk fibroin gave an X-ray diffraction pattern consistent with either a long polypeptide chain consisting of glycyl and alanyl units, or an aggregate of glycyl-alanine diketopiperazine units. Among Herzog's collaborators was Michael Polanyi, who examined the X-ray diffraction pattern of cellulose in 1921. He later recalled that "I evaluated the elementary cell of cellulose and drew the conclusion that the structure of cellulose was either a straight giant molecule composed of a single file of linked hexoses, or else an aggregate of hexobiose anhydrides; both structures were compatible with the symmetry and size of the elementary cell—but unfortunately I lacked the chemical sense for eliminating the second alternative."[139] Independent work during the 1920s by Olenus Lee Sponsler favored the first alternative, which had indeed been suggested in 1913 by Richard Willstätter and Laszlo Zechmeister. Preference was given to the second alternative, however, because Bergmann and Karrer had reported molecular-weight data that indicated cellulose and starch to be aggregates of small molecules. By about 1930 the subsequent work of Kurt Meyer and Hermann Mark, together with the organic-chemical studies of Walter Norman Haworth, established the long-chain character of these polysaccharides. Haworth's contribution, in devising a method for the determination of the end groups of such chains, was especially important, and this approach was later used to good advantage in protein chemistry. Thus in the beginnings of X-ray studies on natural products, there was a close interdependence of the techniques of organic chemistry and the interpretation of the diffraction patterns in terms of molecular structure.

To this must be added that the idea of the synthesis of polyamides was offered during the 1880s by Edouard Grimaux and was fully accepted only after Wallace Carothers demonstrated the long-chain nature of such polymers.[140] The subsequent development of polymer chemistry, especially in the relatively recent invention by Giulio Natta of stereospecific polymerization, marks the emergence of a new chemical research discipline, with great impact on technology and influence on biochemical thought and practice.[141]

In addition to the problem of the molecular size of proteins, two other properties of colloidal solutions were actively discussed for many decades before 1940. One was coagulation (in Graham's terminology, the conversion of a "sol" to a "gel"), and the other was the adsorption of other substances by colloids. Among nineteenth-century biologists, interest in the coagulation of proteins was related to the question of the nature of protoplasm. Thus the botanist Carl Nägeli attached great importance to the idea that gels are aggregates of molecules (micelles), and a succession of investigators saw in the aggregation of proteins, as in the formation of fibrous material on shaking a solution of egg albumin, phenomena analogous to the deposition of microscopically visible intracellular structures.[142] During 1860–1880 leading physiological chemists (Kühne, Hoppe-Seyler, Halliburton) attempted to characterize individual proteins by the measurement of the critical temperature for heat coagulation. From their studies it became evident that the presence of neutral salts, or the addition of acid or alkali, markedly altered this property, in analogy to the effect of such agents on inorganic colloids. In 1888 Franz Hofmeister concluded that, whatever the nature of the protein, the salts fell in the same order in the relative concentrations required to cause incipient precipitation. This "Hofmeister series" was soon extended to the effect of salts on the ability of protein gels to swell by the imbibition of water.

Near the end of the century, a common feature of the explanations offered for the coagulation or precipitation of proteins (and other sols) was the emphasis on adsorption, the term given to the binding of substances at the surface of various materials. Adsorption phenomena had been described by Tobias Lowitz during the eighteenth century, and by 1812 the use of animal charcoal to decolorize sugar solutions was a well-known industrial practice. Experiments on the adsorption of various substances by filter paper were reported by Christian Friedrich Schönbein in 1861, but the fruitful exploitation of this method was not to come until some eighty years later. The use of adsorption techniques for the purification of enzymes was mentioned in the previous chapter, and some colloid chemists, notably William Bayliss, considered the interaction of an enzyme with its substrate to be an adsorption process. For Wilhelm Ostwald in 1885, phenomena of this type were consequences of mechanical affinity rather than chemical interaction, and his opinion was widely accepted. This point of view came to the fore at the turn of the century, when William Bate Hardy examined the effect of acidity or alkalinity on the direction of the movement of protein particles in an electric field.[143]

A biologist attracted to physical chemistry, and with little interest in organic chemistry, Hardy accepted the theory of Svante Arrhenius that the conductivity of electrolytes is an expression of their dissociation into ions. Also, the studies of Picton and Linder during 1892–1895 on the effect of ions on the migration of colloidal particles (arsenic sulfide) in an electric field (electrophoresis) indicated a correlation with precipitation phenomena. In 1900 Hardy reported that

> I have shown that the heat-modified proteid is remarkable in that its direction of movement is determined by the reaction acid or alkaline, of the fluid in which it is suspended. An immeasurably minute amount of free alkali causes the proteid particles to move against the stream while in the presence of an equally minute amount of free acid the particles move with the stream. In the one case therefore the particles are electro-negative, in the other they are electro-positive. Since one can take a hydrosol in which the particles are electro-negative and, by the addition of free acid, decrease their negativity, and ultimately make them electro-positive, it is clear that there exists some point at which the particles and the fluid in which they are immersed are iso-electric. This iso-electric point is found to be of great importance. As it is neared, the stability of the hydrosol diminishes until, at the iso-electric point it vanishes, and coagulation or precipitation occurs, the one or the other according to whether the concentration of the proteid is high or low, and whether the iso-electric point is reached slowly or quickly, and without or with mechanical agitation.[144]

Three years later, Hardy wrote as follows of the properties of globulins in solution: "They are not embraced by the theorem of definite and multiple proportions. Therefore they are conditioned by purely chemical forces only in a subsidiary way. A precipitate of globulin is to be conceived not as a compound of molecular aggregates but of particles of gel." In 1906, writing about proteins as the "material basis of life," he considered that "owing to their great size, to a certain extent [they] cease to be molecules at all in the physical sense, and possess the properties of matter in mass, [and] it is at any rate certain that in their chemical combinations they cease to follow the law of definite combining weights which is the basis of chemistry."[145] There followed some papers of which Gowland Hopkins generously wrote in an obituary notice that "These theoretical papers are important in showing the working of the author's mind at this time, but most find them difficult to understand and they have not counted for so much as the publications in which Hardy has described experimental work of his own."[146]

Indeed, the ability of amino acids and of proteins such as the albumins and globulins to act as both acids and bases had been clearly established by 1905; the term *amphoteric electrolyte* (later, *ampholyte*) was applied to such substances. In 1904 Hans Friedenthal recommended that the acidity or alkalinity of an aqueous solution be denoted by the numerical value of the hydrogen ion concentration; five years later Sørensen proposed that, for convenience, the negative logarithm of this value (pH) be used, and he described the use of buffer solutions to control the pH in experiments on enzymes. The English word *buffer* and the German *Puffer* are translations of the French *tampon*, introduced by Fernbach and Hubert in 1900. The importance of the pH concept was quickly recognized by Leonor Michaelis, who determined the pH values for the isoelectric points of several proteins. After World War I, William Mansfield Clark's book on the determination of pH was an essential item in every biochemical library.[147] Among the many consequences of the recognition of the importance of controlling the pH of protein solutions was the experimental evidence presented by Jacques Loeb against the view that proteins did not behave in accordance with the laws of classical chemistry. This work came at the close of a remarkable scientific career in which such diverse problems as animal tropisms, artificial parthenogenesis, and the diffusion of ions through biological membranes were studied in relation to an underlying mechanistic outlook.[148]

In the rejection of the colloidal approach, two other contributions, both in 1923, were especially important in the development of knowledge about the physical chemistry of proteins. One was the Debye-Hückel theory, which described the behavior of an ion in relation to its charge and radius, to the other ions in the solution, and the dielectric properties of the medium. The other was the demonstration by Niels Bjerrum that amino acids in their isoelectric state are not uncharged molecules (for example, NH_2CH_2COOH) but are dipolar ions *(Zwitterionen)* of the structure

$^{+}NH_3CH_2COO^{-}$. By the same token, the iso-electric forms of proteins and peptides are highly charged, with a net charge of zero. These important insights laid the groundwork for the systematic study of proteins as macromolecular polyelectrolytes, a field in which Edwin Cohn, John Edsall, and their associates played a leading role.[149] During the 1930s Arne Tiselius developed the technique of moving-boundary electrophoresis, subsequently improved by Lewis Longsworth, which made possible the separation of closely related proteins on the basis of differences in their electrical charge. One of the first important successes of the method was the separation of the proteins of blood plasma, and the demonstration that the globulin fraction contains multiple components, some of which (the so-called gamma-globulins) represent the antibodies.[150] After 1950 the moving-boundary method was replaced by other techniques, in particular zone electrophoresis in gels of agarose or polyacrylamide, now one of the most widely used analytical procedures in research on proteins and nucleic acids.

If by the late 1920s the macromolecular nature of proteins began to be generally acknowledged and many proteins had been shown to be chemical entities whose behavior as acid and bases obeyed the laws of stoichiometry, the property of coagulation, long considered to be their chief characteristic, had not been elucidated. As was noted above, the biological importance of this problem was repeatedly emphasized, with a variety of speculations about the chemical difference between "living" proteins and "dead" proteins. By 1910, however, the coagulation process had begun to be studied carefully with purified soluble proteins. Of special importance was the work of Harriette Chick and Charles Martin, who showed that heat coagulation of "native" egg albumin involves an initial process, termed denaturation, with an increase in the viscosity of the solution, followed by the precipitation of the coagulated protein.[151] Their work, together with that of Michaelis and of Sørensen, demonstrated that "under otherwise equal conditions, the rate of denaturation will be higher the higher the concentration of hydrogen ions in the solution, but the flocculation of the denatured protein only takes place at isoelectric reaction, or near it. The addition of salt extends the limits for complete flocculation."[152]

With the recognition that the precipitation of the denatured protein does not require heat but depends only on the pH and salt content of the solution, the problem became the cause of the denaturation. During 1910–1935, the effect of many chemical agents or physical treatments was studied; of special interest was Gowland Hopkins's finding that, at high concentration, urea is an effective denaturing agent. It was also noted that upon denaturation some proteins (for example, egg albumin) acquire the ability to react with nitroprusside, a reagent for sulfhydryl (SH) groups, indicating that denaturation had made cysteine units accessible to this reagent. The important studies of Mortimer Anson and Alfred Mirsky showed that under suitable conditions the denaturation of hemoglobin is a reversible process. After the isolation of enzymes in crystalline form, part of the evidence for their protein nature was the fact that denaturation (as measured by solubility) closely paralleled the loss of catalytic activity, and when the denaturation could be reversed both properties were restored together.[153]

In 1931, Hsien Wu offered the hypothesis that the essential feature of protein denaturation is the unfolding of tightly coiled chains, and five years later Mirsky and Pauling stated: "Our conception of a native protein molecule (showing specific properties) is the following: The molecule consists of one polypeptide chain which continues without interruption throughout the molecule (or in certain cases, of two or more such chains); this chain is folded into a uniquely defined configuration, in which it is held together by hydrogen bonds between the peptide nitrogen and oxygen atoms and also between the free amino and carboxyl groups of the diamino and dicarboxyl amino acid residues. . . . *The denatured protein molecule we consider to be characterized by the absence of a uniquely defined configuration.*"[154] The concept of the hydrogen bond, formed by the tendency of a hydrogen atom bound to one atom to share the electrons of another atom (as in -O-H ... O-H- in water or -N-H ... O=C- in a polyamide), had been formulated in 1920 by Wendell Latimer and Maurice Huggins.[155] Subsequent work forced a revision of the generalization that hydrogen bonding is the principal mode of noncovalent interaction in the formation of the internal structure of a globular protein. In addition to electrostatic interactions such as those between ammonium and carboxylate ions, disulfide (-SS-) bridges of cystinyl units, and nonspecific van der Waals interactions, much attention has been given to the role of what came to be called hydrophobic interactions involving nonpolar side-chains of amino acids such as leucine or phenylalanine. This usage does not mean that these side-chains repel water but that, in the presence of nonpolar molecules, water molecules attract each other more strongly.[156]

The Peptide Theory Revisited

During 1920–1940 doubt about the validity of the Fischer-Hofmeister theory of protein structure was usually based on the idea that "native" proteins are labile structures that, upon denaturation, are converted to polypeptides. As we have seen, some chemists gave serious consideration to Emil Fischer's suggestion that diketopiperazines might be important elements of protein structure. In a 1928 review of the various hypotheses of protein structure, Vickery and Osborne wrote: "Taken together this body of evidence is as nearly conclusive proof that the peptide bond occurs to a very considerable extent as could be wished. There are, however, certain facts which remain unexplained upon the view that the protein is essentially a single large polypeptide. In the first place the nature of the bonds that are attacked by pepsin is unknown. . . . Pepsin has never been found to have an effect upon any synthetic model substance whether it contained a peptide bond or not."[157] Similarly, Ross Gortner stated that "Fischer considered the chains of the polypeptides were not long enough for pepsin to act upon them, but it seems more probable that pepsin attacks some linkage other than the linkage in the peptide group," and J. B. S. Haldane wrote that "proteinases do not in general attack synthetic substrates."[158]

As was noted in the previous chapter, the crystallization of pepsin by John Northrop and his demonstration that its catalytic activity is a property of the protein marked a turning point for enzyme chemistry. Whereas James Sumner's earlier isolation of urease in the form of a crystalline protein was insufficient to sway opinion from Willstätter's contention that enzymes are small reactive molecules adsorbed on colloidal carriers, the evidence presented by Northrop left little doubt as to the protein nature of pepsin. Northrop's assay of pepsin activity was the traditional use of a protein substrate (he used denatured hemoglobin), and at that time many biochemists believed that, in contrast to the hydrolytic action of known enzymes that act on synthetic peptides, proteinases such as pepsin effect the "deaggregation" of protein substrates.

It was my good fortune to have the opportunity, while working in Max Bergmann's laboratory at the Rockefeller Institute, to find the first model peptides which were hydrolyzed by pepsin. These peptides were made by the new synthetic method developed by Zervas in Bergmann's Dresden laboratory and described by them in 1932. Their methods vastly enlarged the scope of peptide synthesis in making possible the inclusion of complex amino acids. The first synthetic substrates for pepsin, in which a Glu-Tyr bond was cleaved, were hydrolyzed very slowly, but later more sensitive peptide substrates were prepared.[159] Moreover, during the 1930s, Moses Kunitz had crystallized, from extracts of beef pancreas, two protein-splitting enzymes, which he named trypsin and chymotrypsin (both are components of Kühne's "trypsin"), and we also found synthetic substrates for these two enzymes. Whereas chymotrypsin acted preferentially at Phe-X or Tyr-X bonds, trypsin appeared to be specific for the hydrolysis of Lys-X or Arg-X linkages.[160]

Although the finding of synthetic peptide substrates for pepsin and other crystalline proteinases removed one of the objections to the peptide theory of protein structure, it was noted that

> it is well known that genuine proteins (at pH 7) are attacked by crystalline trypsin which, on the other hand, is able to split synthetic peptides. This has been taken as support for the view that these proteins contain peptide bonds in their molecules. We wish, however, to point out that in the light of the following consideration, this support loses a great deal, if not all, its importance. If, according to Anson and Mirsky, denaturation is reversible, then in a solution of a given globular protein there is an equilibrium between genuine and denatured protein, $G \rightleftharpoons D$. Hence it is sufficient that D and only D should contain peptide bonds open to fission by trypsin, because by removal of D by hydrolysis this process is forced in the direction from left to right and G will disappear as well.[161]

Experimental results on the temperature coefficients of the cleavage of a native ("genuine") and denatured substrate were consistent with this consideration, and led to the suggestion that they "provide sufficient basis for giving a warning against the conclusion that genuine proteins contain peptide bonds because they are split by proteinases like trypsin. They give a certain indication that peptide bonds are formed or 'appear' (like SH-groups) upon denaturation, but are not conclusive enough to decide whether or not some hydrolyzable peptide bonds are pre-formed in the molecules of the genuine globular proteins."[162] This cautious doubt was expressed at a time when other voices were raised in favor of the recurrent idea that linear polypeptides are merely the denatured forms of proteins. Considerable attention was given during the 1930s to possible ring structures in native proteins. From X-ray diffraction data on fibrous proteins (wool keratin, tendon collagen), William Astbury inferred the existence of hexagonal folds, and the physicist Frederick Charles Frank suggested a structure which was elaborated by the topologist Dorothy Wrinch, who offered a "cyclol" theory of protein structure. She proposed that a native globular protein exists as a honeycomb-like structure constructed of six-membered rings formed by the covalent union of the NH of one amino acid unit with the CO group of another amino acid to produce =N-C(OH)= bonds.[163]

Among the protagonists of the cyclol theory was the physicist Irving Langmuir:

> We may sum up the present position with regard to the structure of proteins as follows: A vast amount of data relating to protein structure have been collected by workers in a dozen different fields. No reasonable doubt remains as to the chemical composition of proteins. The original idea of native proteins as long-chain polymers of amino-acid residues, while consistent with the facts relating to the chemical composition of proteins in general, was

> not a necessary deduction from these facts. Moreover it is incompatible with the facts of protein crystallography, both classical and modern, with the phenomena of denaturation, with Svedberg's results which show that the native proteins have definite molecular weights, and with the high specificity of proteins discovered in immunochemistry and enzyme chemistry. All these facts seem to demand a highly organized structure for the native proteins, and the assumption that the residues function as two-armed units leading to long-chain structures must be discarded. The cyclol hypothesis introduced the single assumption that the residues function as four-armed units, and its development during the last few years has shown that this single postulate leads by straight mathematical deductions to the idea of a characteristic protein fabric which in itself explains the striking uniformities of skeleton and configuration of all the amino-acid molecules obtained by the degradation of proteins. The geometry of the cyclol fabric is such that it can fold round polyhedrally to form closed cagelike structures. These cage molecules explain in one single scheme the existence of megamolecules of definite molecular weights capable of highly specific reactions, of crystallizing, and of forming monolayers of very great insolubility.[164]

In the same year (1939), however, Linus Pauling and Carl Niemann wrote:

> We have carefully examined the X-ray arguments and other arguments which have been advanced in support of the cyclol hypothesis, and have reached the conclusion that there exists no evidence whatever in support of this hypothesis and that instead strong evidence can be advanced in support of the contention that bonds of the cyclol type do not occur at all in any protein. . . . It is our opinion that the polypeptide chain structure of proteins, with hydrogen bonds and other interatomic forces (weaker than those corresponding to covalent bond formation) acting between peptide chains, parts of chains and side chains, is compatible not only with the chemical and physical properties of proteins, but also with the detailed information about molecular structure in general which has been provided by the experimental and theoretical researches of the past decade.[165]

One of the arguments offered in support of the cyclol theory was that it explained the Svedberg unit of 35,000 in terms of a structural unit containing 288 amino acid residues, a number independently suggested by Bergmann and Niemann on the basis of their estimates of the amino acid composition of several proteins. According to Pauling and Niemann, "Considerable evidence has been accumulated suggesting strongly that the stoichiometry of the polypeptide framework of protein molecules can be interpreted in terms of a simple basic principle. This principle states that the number of each individual amino acid residue and the total number of all the amino acid residues contained in a protein molecule can be expressed as the powers of the integers two and three."[166]

The principle referred to was deduced by Bergmann and Niemann during 1936–1938 from Niemann's analyses for some amino acids in acid hydrolysates of several protein preparations (hemoglobin, fibrin, silk fibroin, gelatin), and from data reported by others for egg albumin. The total number of amino acid residues per "molecule" of protein appeared to fall into a series of multiples of 288; thus fibrin had 576 ($2^6 \times 3^2$) and silk fibroin 2,592 ($2^5 \times 3^4$) residues, respectively, corresponding to molecular weights of 69,000 and 217,000. Not only was the content of amino acid denoted by the formula $2^n \times 3^m$, but Bergmann and Niemann proposed, as a law of protein structure, that every amino acid residue occurs at a regularly periodic interval in the polypeptide chain of a protein.[167] For example, the polypeptide chain of silk fibroin, which has a preponderance of glycine (G) and alanine (A) residues (the others being denoted X), was considered to have the sequence

-G-A-G-X-G-A-G-X-G-A-G-X-G-A-G-X.

Periodicity hypotheses of this kind had been offered before. Albrecht Kossel concluded from analytical data that he and Henry Dakin had obtained in 1904 for the protamine clupein that arginine, the predominant amino acid, recurs repeatedly in the repeating triplet -Arg-Arg-X-. In 1934 William Astbury inferred from the available data for the amino acid composition of gelatin that every third residue could be a glycyl unit and every ninth a hydroxyproline unit.

Opinions regarding the Bergmann-Niemann hypothesis were divided, even in Bergmann's laboratory.[168] To the noted geneticist Richard Goldschmidt, the hypothesis suggested the operation of a mathematical principle that might link Mendel's laws to the structure of proteins, at that time considered by geneticists to be the most likely chromosomal constituents responsible for the transmission of hereditary characters.[169] Among those familiar with the limitations of the then available analytical techniques for the quantitative determination of the individual amino acids in protein hydrolysates, however, the reaction was largely negative and, when expressed publicly, was one of skepticism tempered with the respect due a chemist of Bergmann's distinction. During 1939–1941, such opinions came mainly from British protein chemists, among them Norman Pirie and Albert Neuberger. Charles Chibnall, who later succeeded Gowland Hopkins as professor of biochemistry at Cambridge, wrote: "To those of us interested in the subject it seemed like the dawn of a new era, as Bergmann himself so appropriately remarked: 'Everyone who is familiar with the history of protein chemistry may feel somewhat amazed on being confronted with a simple stoichiometry of the protein molecule.' Some of us, nevertheless, were not prepared to give this attractive generalization our immediate and unqualified support, for we questioned the reliability and completeness of the amino acid analyses on which it was based."[170] Indeed, by 1939 Bergmann was obliged to abandon the periodicity hypothesis, after his associate William Stein had produced incontrovertible analytical data that could not be accommodated by the $2^n \times 3^m$ rule.

One of the articles critical of the Bergmann-Niemann hypothesis was a 1941 paper in which the authors attempted to test its validity by analyzing the mixture of small peptides produced by the partial acid hydrolysis of proteins. They used various analytical methods, and although their work was not sufficiently conclusive "to form a rigid refutation of the Bergmann-Niemann hypothesis, it seems to suggest for proteins a considerably more complicated structure." It was evident that better methods were needed for the quantitative separation and analysis of the hydrolytic products derived from proteins. In an adjoining article, two of the authors, Archer Martin and Richard Synge, described an approach that set the stage for the solution of this problem, through the use of the technique known as chromatography, a term introduced by Mikhail Tsvet (Tswett) in his studies on plant pigments.[171]

Chromatography

> In 1906 Tsvet reported that if a petroleum ether solution of chlorophyll is filtered through a column of an adsorbent (I use mainly calcium carbonate which is stamped firmly into a narrow glass tube), then the pigments, according to their adsorption sequence, are resolved from top to bottom into various colored zones, since the more strongly adsorbed pigments displace the more weakly adsorbed ones and force them further downwards. This separation becomes practically complete if, after the pigment solution has flowed through, one passes a stream of solvent through the adsorbent column. Like light rays in the spectrum, so the different components of a pigment mixture are resolved on the calcium carbonate column according to a law and can be estimated on it qualitatively and also quantitatively. Such a preparation I term a chromatogram, and the corresponding method, the chromatographic method. It is self-evident that the adsorption phenomena are not restricted to the chlorophyll pigments, and one must assume that all kinds of colored and colorless chemical compounds are subject to the same laws.[172]

It should be noted that the phenomenon of the differential adsorption of solutes by filter paper had been described during the nineteenth century by several chemists, notably Schönbein and his pupil Goppelsroeder, and was extensively used in the dye industry and in the petroleum industry.[173] Although not unknown to organic chemists, the method does not appear to have been highly regarded, probably because of their understandable preference for crystallization and fractional distillation for the separation and isolation of compounds.

Tsvet's work received an unfavorable reception from Leon Marchlewski, then the leading expert on the chemistry of chlorophyll. As a consequence, credit for some of Tsvet's findings (for example, the existence in leaves of two forms of chlorophyll) went to Richard Willstätter, whose outstanding chemical work during 1906–1913 overshadowed all that had been done on the subject previously. Willstätter did not value the chromatographic method highly, but had a manuscript copy of a German translation of Tsvet's 1910 book (in Russian) on the theoretical aspects of the chromatographic method. During the 1920s this copy came into the hands of Willstätter's associate Richard Kuhn, who appreciated its significance and used adsorption chromatography in his studies on the carotenoid pigments, as did Leroy Palmer before him. By the 1930s Tsvet's method was used by numerous chemists, but not by those engaged in the study of proteins.

The chromatographic method developed by Archer Martin and Richard Synge in 1941 differed from adsorption chromatography in that it depended on the establishment of an equilibrium between two liquid phases, such as chloroform and water, one of which (water) is immobilized by being held by a solid support, such as silica gel, and the other (chloroform) flows through the column. This method thus involves the distribution of the molecules of a dissolved substance between two phases and was later termed partition chromatography. To apply this method to the amino acid analysis of protein hydrolysates, it was necessary to convert the amino acids (which are insoluble in chloroform) to acetylamino acids (for example, $CH_3CO\text{-}NHCH_2COOH$), which were found to be partitioned sufficiently differently between chloroform and water to permit the nearly quantitative separation of closely related acetylamino acids, such as those derived from valine and isoleucine. A disadvantage of this method was the necessity for the acetylation of the amino acids, for one could not be certain in the case of protein hydrolysates that the acetylation had been complete.

The appearance of the 1941 paper of Martin and Synge coincided with efforts in other laboratories to develop new methods for the quantitative analysis of the amino acids present in a protein hydrolysate. Apart from the impetus given this search by the Bergmann-Niemann hypothesis, other factors encouraged these efforts. During World War II, greater urgency was given to the need for accurate data on the amino acid composition of proteins, and there was much interest in antibacterial agents (gramicidin, penicillin) that had turned out to be related to polypeptides. Older chemical methods for the estimation of individual amino acids were subjected to renewed scrutiny and, in some cases, improved. In addition to studies on the adsorption chromatography of amino acids, their separation on the basis of ionic interaction with solid supports having charged groups ("ion-exchange chromatography") began to be examined, as was the possibility of electrophoretic separation of amino acids. The use of fastidious microorganisms—which, like the one whose study led to the discovery of methionine, required particular amino acids for growth—was explored. In 1945 Erwin Brand reported that by means of a combination of chemical and microbiological techniques, it was possible to account for 99.6 percent of the amino acid residues of β-lactoglobulin, and he concluded that "the constituent amino acids are primarily linked by typical peptide bonds."[174] As it later turned out, however, this complete accounting was the result of the compensation of errors.

One of the advances during this period was provided by Raphael Consden, Hugh Gordon, and Martin in the technique of "paper chromatography," which resembled "the 'capillary analysis' method of Schönbein and Goppelsroeder . . . except that the separation depends on the differences in partition coefficient between the mobile phase and the water-saturated cellulose, instead of differences in adsorption by the cellulose."[175] The immediate and widespread adoption of this method had a profound effect on many branches of biochemistry. For biochemists who were often obliged to work with very small amounts of a natural product, paper chromatography provided the solution to many intractable problems. During 1945–1955 numerous efforts were made to apply the method to the quantitative analysis of protein hydrolysates, but it became evident that its accuracy was too low for this purpose.

The next step toward the solution of this problem was the use of starch columns. In 1949 William Stein and Stanford Moore reported the complete amino acid analysis of beta-lactoglobulin, as determined by partition chromatography with the aid of an automatic fraction collector which they devised for this purpose. The slow flow rate through starch columns led Stein and Moore to use an ion-exchange resin (sulfonated cross-linked polystyrene, a polymer first described by Staudinger). The remarkable ability of this resin, which acts both by ion-exchange and adsorption, to separate not only free amino acids but also closely related polypeptides, made the new Moore-Stein method the procedure of choice.[176] An important factor in the ready acceptance of the method was the meticulous care with which it was described. Within a few years, Stein and Moore, together with Darryl Spackman, replaced the collection of individual fraction, as they emerged from the column, by an automatic recording assembly. This instrument measured and recorded the intensity of color produced by the emergent amino acids upon reaction with ninhydrin (described by Siegfried Ruhemann in 1910).

The development of the Spackman-Stein-Moore apparatus transformed the complete amino acid analysis of protein hydrolysates from a laborious operation, requiring much material, time, and skill, into a reliable and relatively rapid method, which has since been further accelerated and automated. Moreover, various new chromatographic methods became available, for example, high performance liquid chromatography (HPLC). The information so obtained is, however, only the equivalent of that given by the elementary analysis of an organic compound of low molecular weight. Just as for the organic chemist, the next step was the determination of the arrangement of the atoms in his compound through the study of its chemical transformations (he now has powerful physical methods, such as nuclear magnetic resonance spectroscopy, mass spectrometry, and X-ray crystallography), the challenge to the protein chemist was to determine the arrangement of amino acid units in the protein under study.

The Structure of Proteins

The challenge was taken up by Frederick Sanger even before the general methods of amino acid analysis had been developed, and within a ten-year period (1945–1955), he and his associates established the sequence of amino acid units in the polypeptide chains of the protein hormone insulin. It is noteworthy that during the course of this work, Sanger felt obliged to write in a review article on the arrangement of amino acids in proteins that "as an initial working hypothesis it will be assumed that the peptide theory is valid, in other words, that a protein molecule is built up only of chains of alpha-amino (and alpha-imino) acids bound together by peptide bonds between their alpha-amino and alpha-carboxyl groups. While this peptide theory is almost certainly valid . . . it should be remembered that it is still a hypothesis and has not been definitely proved. Probably the best evidence in support of it is that since its enunciation in 1902 no facts have been found to contradict it."[177] Although the experimental methods Sanger employed were new, for he depended heavily on the use of chromatographic and ionophoretic techniques and of proteinases of defined specificity, the basic approach was the time-honored one of nineteenth-century organic chemists—namely, the partial cleavage of the molecule, identification of the structure of the fragments, and the deduction from this information of the structure of the intact protein.

When Sanger began work on cattle insulin in the Cambridge Biochemical Laboratory, then headed by Charles Chibnall, this protein hormone had been shown to be composed of seventeen different amino acids and to contain an unusually large proportion of cystine units. Chibnall had selected insulin as a test protein for a critical assessment of the reliability of the available methods for the quantitative analysis of the amino acid composition of protein hydrolysates; by 1945 he had accounted for 96.4 percent of the amino acids in insulin and had concluded that its minimal molecular weight is about 12,000. Earlier estimates of the molecular weight of insulin had been 36,000 (X-ray crystallography) and 48,000 (ultracentrifugation), thus suggesting that subunits of 12,000 could aggregate to form larger particles.

Sanger's first step toward the elucidation of the amino acid sequence of insulin was to develop a method for the determination of the nature and number of the amino-terminal amino acid units in polypeptide chains of proteins. In 1945 he described the use of the reagent dinitrofluorobenzene for this purpose and reported that, per unit weight of 12,000, insulin contains two amino-terminal glycine and two amino-terminal phenylalanine units. Sanger concluded that "the presence of four alpha-amino groups suggests that the submolecule is built up of four open polypeptide chains bound together by cross-linkages, presumably chiefly -S-S- linkages. It is, of course, possible that other chains may be present in the form of a ring structure with no free amino groups."[178] It should be noted that cross-linking of separate peptide chains had been considered by Karl Freudenberg in 1935 and that ring structures had been encountered in some peptide antibiotics. Indeed, when Sanger's end-group method was applied to some proteins and no terminal amino acid was found, a cyclic structure was initially inferred, but later work showed that the amino-terminal group was blocked, as by an acetyl group.

The widespread adoption of Sanger's method and of paper chromatography was followed by the introduction of other procedures, the most important of which was Pehr Edman's development of a method for the successive chemical removal of individual amino acids from the amino-terminus of a peptide chain.[179] This method was subsequently automated and became a favored technique for the determination of the amino acid sequence of peptides. Sanger's agent was later replaced by diaminonaphthylsulfonyl chloride, and enzymes (aminopeptidases and carboxypeptidases) have been used as reagents to determine the amino-terminal and carboxyl-terminal units of peptides.

In 1949 Sanger described a method for the separation of the four polypeptide chains assumed to be present in the insulin molecule. The procedure involved the oxidative cleavage of the disulfide bridges, and the isolation of two fractions, one of which (Fraction A) had only glycine-terminal chains, the other (Fraction B) only phenylalanine-terminal chains. Further work indicated that each of the two fractions contained a single polypeptide, and Sanger (with Hans Tuppy) undertook the determination of the amino acid sequence of the B-chain by subjecting it to partial hydrolysis with acid or various enzymes (pepsin, trypsin, chymotrypsin). Their skill in the use of paper chromatography for the isolation of pure samples of the many fragments, and for the determination of the amino acid composition of each fragment, led to the amino acid sequence of the 30-member polypeptide constituting the B-chain of insulin. Two years later, Sanger and E. O. P. Thompson reported the sequence of the A-chain, which turned out to have 21 amino acid units, and by 1955 the Sanger group had determined the positions of the three disulfide bonds, two of which link the two chains, while the third forms a ring structure in the A-chain.[180] By 1952 it was known from the work of Lyman Craig that the minimal molecular weight of insulin is about 6,000. The structure is shown in the accompanying diagram.

As was noted above, the possibility of making peptide hormones and antibiotics by chemical synthesis provided a powerful spur to the development of the art of peptide synthesis. Thus after Paul Bell and his associates had determined in 1954 the complete 39–amino acid sequence of the adrenocorticotropic hormone (ACTH) of the anterior pituitary, not only had the entire chain been synthesized within ten years, but many structural analogues had been made. By 1965 chemists in

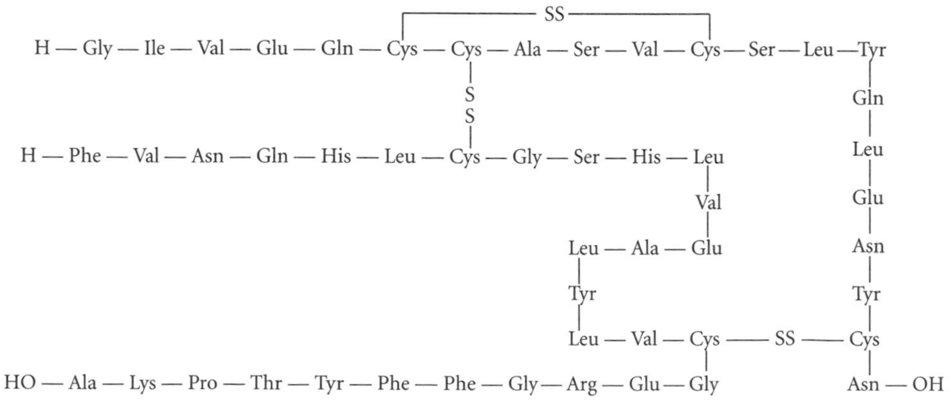

```
                                 ┌──────────── SS ────────────┐
H — Gly — Ile — Val — Glu — Gln — Cys — Cys — Ala — Ser — Val — Cys — Ser — Leu — Tyr
                                         |                                          |
                                         S                                         Gln
                                         S                                          |
                                         |                                         Leu
H — Phe — Val — Asn — Gln — His — Leu — Cys — Gly — Ser — His — Leu                 |
                                                                 |                 Glu
                                                                Val                 |
                                                                 |                 Asn
                                                 Leu — Ala — Glu                    |
                                                  |                                Tyr
                                                 Tyr                                |
                                                  |                                 |
                                                 Leu — Val — Cys ——— SS ——— Cys
                                                              |                  |
HO — Ala — Lys — Pro — Thr — Tyr — Phe — Phe — Gly — Arg — Glu — Gly            Asn — OH
```

Amino acid sequence of cattle insulin

three laboratories (in China, Germany, and the United States) had reported the preparation of synthetic insulin. Apart from providing evidence of the value of the classical route from analysis to synthesis for the determination of the structure of organic substances of biological interest, these achievements set the stage for the study of the processes whereby peptide hormones are generated in animal organisms.

Sanger's achievement was followed by the elucidation of the amino acid sequence of the 124-amino acid chain of the enzyme ribonuclease A by Stanford Moore and William Stein, in association with Werner Hirs and Derek Smyth.[181] Although the basic strategy was similar to that of Sanger, and the greater chain length presented special problems, the availability of new chromatographic techniques made it possible to derive the sequence in a shorter time. Since then, the complete amino acid sequences (often referred to as the primary structure) of many proteins have been determined, and new chemical methods, such as that devised by Edman, or physical methods (mass spectrometry) have made the task easier. The extent to which the new techniques in this field have influenced biological thought and experiment, from the treatment of human disease to the development of evolutionary theory, cannot be exaggerated.

An early example was provided by studies on the hemoglobin from normal adults (hemoglobin A) and that from human subjects suffering from sickle-cell anemia (hemoglobin S). In 1949 Harvey Itano and Linus Pauling found that the two proteins differ in their electrophoretic mobility, and a few years later Vernon Ingram showed that the mixture of peptides formed by action of trypsin, when examined by means of paper chromatography and electrophoresis ("fingerprint technique"), differed only with respect to the nature of a single peptide. The subsequent determination of the complete amino acid sequences of the two (α- and β-) peptide chains of hemoglobin A showed that a single glutamic acid unit, which bears a negative charge, is replaced in hemoglobin S by an uncharged valine unit in the β-chain. Since sickle-cell anemia was long known to be a hereditary trait, this finding was important in providing evidence for the genetic determination of the amino acid sequence of proteins.[182]

We have seen that although fibrous proteins such as silk fibroin or wool keratin were recognized to be long-chain macromolecules, and globular "native" proteins as high-molecular-weight substances, during the 1930s doubts were expressed about the polypeptide nature of the globular proteins. X-ray diffraction studies on crystalline pepsin, by Bernal and Crowfoot, led them to estimate from the dimensions of the unit cell that this protein had a molecular weight of about 40,000, and they concluded that

> not only do these measurements confirm such large molecular weights but they also give considerable information as to the nature of the protein molecules and will certainly give more when the analysis is pushed further. From the intensity of the spots near the centre, we can infer that protein molecules are relatively dense globular bodies, perhaps joined together by valency bridges, but in any event separated by relatively large spaces which contain water. From the intensity of the more distant spots, it can be inferred that the arrangement of the atoms inside the protein molecule is also of a perfectly definite kind, although without the periodicities characterising the fibrous proteins. . . . Peptide chains in the ordinary sense may exist only in the more highly condensed or fibrous proteins, while molecules of the primary soluble proteins may have their constituent parts grouped more symmetrically around a prosthetic nucleus.[183]

During the 1930s single-crystal X-ray photographs were also reported for other proteins (insulin, excelsin, lactoglobulin, chymotrypsin, hemoglobin). The one for hemoglobin, by Bernal, Isidor Fankuchen, and Max Perutz,[184] marked the beginning of Perutz's extended studies on this protein. Although the photographs obtained in this early work were of excellent quality, the very large number of diffraction spots generated by protein crystals presented problems that could not be solved by means of the methods that had been used in the study of small organic molecules. In 1954 Perutz solved one of the most troublesome problems by attaching heavy atoms, such as mercury, at specific sites on the surface of hemoglobin. This method ("isomorphous replacement") provided information about the phases of the individual X-ray reflections, and together with their intensities, such diffraction data were combined in a Fourier transform to give maps of electron density. In this way, and with computer analysis of the data, Perutz's associate John Kendrew succeeded in 1958 in establishing the general structure of whale myoglobin, a hemoglobin-like protein known to have a molecular weight of about 17,500. Subsequent high-resolution analysis permitted the construction of a three-dimensional model that not only revealed the convolutions of the single peptide chain of the protein and the position of the iron-porphyrin but also showed some of the amino acid units (such as tyrosine) as having a distinctive structure.[185] It was clear, however, that even at high resolution (1.5–2.0 Å), the uncertainties in the X-ray pattern required prior knowledge of the amino acid sequence for a complete model to be constructed, and this requirement has remained to this day now that very many protein structures have been "solved" by means of X-ray crystallography. In one of Kendrew's papers it was noted that "perhaps the more remarkable features of the molecule are its complexity and its lack of symmetry. The arrangement seems to be almost totally lacking in the kind of regularities which one instinctively anticipates, and it is more complicated than has been predicted by any theory of protein structure."[186] Eight years earlier, it seemed that "there appears to be a real simplicity of chain structure in myoglobin, which will perhaps be shown by other favourably built proteins, and which makes it particularly suitable for intensive X-ray investigation."[187]

Of particular interest was the finding that large sections of the peptide chain of myoglobin are twisted into a coil with 3.7 amino acid residues per turn (α-helix), which Linus Pauling and Robert Corey had proposed in 1951 as one of the general structural features of proteins, another being a "pleated-sheet" (later termed β-sheet) conformation, with hydrogen bonding between adjacent polypeptide chains. Such conformations have come to be termed the secondary structure of proteins. The formulation of these models involved the assumption, based on the theory of resonance and X-ray diffraction data obtained by Corey for the bond length and angles in a few amides, that all the atoms of a -CO-NH- group lie in a single plane.[188] Although the α-helix turned out to be a prominent feature of the structure of myoglobin, as more crystalline proteins were examined in later years, many of them exhibited much less helical content and some, like chymotrypsin, do not appear to have appreciable α-helix content. The pleated-sheet structure has been identified in parts of many proteins, but there are also stretches of the peptide chain in the form of irregular loops and turns. As Dorothy Hodgkin put it in 1979, "In protein molecules in general the proportions of α-helices and β-structures vary widely; both tend to be more distorted than we expected; β-sheets are usually much twisted. And within the more irregular strands of chains that run between the specific structures we can trace small stretches of chains in other folds, postulated in various papers and notebooks of those who thought about proteins in the 1930s and 1940s."[189]

It would seem, therefore, that despite the prominence assigned by some historians of molecular biology (as well as some philosophers of biology) to Pauling's model-building approach to protein structure, X-ray crystallography and (more recently) two-dimensional nuclear magnetic resonance (NMR) spectroscopy of dissolved proteins have revealed the limitations of that approach.[190] Apart from the greater complexity and individuality of the conformation of peptide chains (the so-called secondary structure) and the appearance of separate "domains," the specific assembly of the chains to form the "tertiary" structure and, in the association of separate peptide chains or subunits (as in hemoglobin) to form the "quaternary" structure of proteins, important structural features have been found in relatively short irregular peptide segments that form the location of the specific interaction of an enzyme protein with a substrate molecule or of a receptor protein with an effector such as a hormone. More will be said about these structural elements later in this chapter.

Indeed, at the present writing the most challenging problem in protein chemistry is presented by the biochemical process in the folding of peptide chains to form globular proteins. During the early 1960s it seemed that the problem had been solved by Christian Anfinsen and his associates by showing that if the disulfide bonds of pancreatic ribonuclease are reduced to form sulfhydryl groups in the presence of urea, the enzymatic activity is lost, but that removal of the urea and addition of mercaptoethanol ($HOCH_2CH_2SH$) leads to the reformation of the active enzyme.[191] This important finding supported the "thermodynamic" hypothesis that the intracellular formation of the tertiary structure of a globular protein is a spontaneous process whose outcome is solely determined by the amino acid sequence of the unfolded peptide chain. Subsequent research has indicated, however, the involvement of some enzymes that catalyze disulfide interchange and the isomerization of proline peptide bonds, as well as "chaperones," which include proteins elicited upon subjecting cells to heat shock.[192]

In the intracellular formation of particular proteins, the terminal amino or carboxyl group and one or more kind of amino acid side-chain may have undergone enzyme-catalyzed chemical change. For example, the OH group of serine, threonine, or tyrosine may have been phosphorylated, or the side-chain of lysine units may have been methylated, acylated, or oxidized.[193] Such so-called posttranslational modifications of protein structure confer upon these proteins properties that are important in their function, and they will be mentioned later in connection with some biochemical problems.

Proteins as Enzymes

In 1934 Wolfgang Pauli, a leading colloid chemist, wrote that "a comparative recent epoch has ascribed to the proteins the central position in the life process. Today, however, they appear rather as passive carriers of life phenomena, determining the important chemical and physico-chemical properties of the medium and the mechanical properties of the tissue or forming the base for certain attached active groups. The true directive participation in the maze of vital chemical reactions on the other hand devolves on scattered specific substances which, in general, belong to entirely different classes of bodies."[194] This statement reflects the view held at that time by many physiologists and biochemists that enzymes, hormones, and vitamins all belong to such "entirely different classes of bodies," and evidence was offered by

Willstätter's disciple Ernst Waldschmidt-Leitz and others, claiming that Sumner's crystalline urease and Northrop's crystalline pepsin could be separated into an active low-molecular weight material from its inactive protein "carrier." Before Abel's crystallization of insulin, the only known crystalline hormones were epinephrine (adrenaline), whose structure as a relative of tyrosine was established in 1904, and thyroxine, which Kendall identified as an amino acid in 1915. Secretin, described by Bayliss and Starling in 1904 (they introduced the word hormone, Greek *hormon,* arousing, exciting), was a dialyzable substance; its chemical structure as a 27-amino acid peptide was not elucidated until 1961. After the isolation of insulin, Allen and Murlin reported in 1925 that they had prepared a "biuret-free insulin" and concluded that "insulin *per se* is not a true protein in the sense of being composed of nothing but amino acids."[195] Abel's claim was the subject of prolonged controversy among German, Dutch, British, and American investigators before its validity was generally acknowledged during the 1930s. It is therefore not surprising that, given the state of protein chemistry during the 1930s, the idea that the catalytic activity of enzymes resides in some low-molecular compound associated with a protein "carrier" should have persisted even after the protein nature of urease and pepsin was accepted. For example, the "active group" of peroxidase (which catalyzes the oxidation of various organic compounds by hydrogen peroxide) was considered to be solely its ironporphyrin unit, and a similar role was assigned to other organic "prosthetic groups" of enzymes to be mentioned later. The first spectroscopic evidence for the existence of a rate-limiting enzyme-substrate complex in enzyme catalysis was provided by Britton Chance in his brilliant study of the rate of formation of peroxidase-H_2O_2 complexes, in which hydrogen peroxide is linked to the ironporphyrin.[196]

The first clear-cut studies on the nature of the "active sites" within enzyme proteins were not conducted until after 1945, with such enzymes as chymotrypsin, ribonuclease, lysozyme, pepsin, and carboxypeptidase, whose catalytic action does not involve such prosthetic groups. By that time the knowledge being gained by the use of new techniques had begun to transform the conceptual structure and experimental methodology of protein chemistry. In parallel with the development of methods for the determination of the amino acid sequence of peptide chains, and for the study of the three-dimensional structure of globular proteins by means of X-ray diffraction analysis, ultraviolet and fluorescence spectroscopy was increasingly used, and extensive studies were conducted on the specific chemical modification of proteins. A more recent addition has been the use of recombinant DNA technology to alter the amino-acid sequence of a protein in the method termed site-specific mutagenesis,[197] of which more later.

The identification of the amino acid units forming the active sites of protein enzymes provided a basis for studies on the physical-chemical mechanisms in their catalytic action. In the background of these efforts were important pre–World War II contributions, notably those of Jens Christianson to the theory of acid-base catalysis, of Henry Eyring to the concept of the transition state in the catalytic process, and of Louis Hammett on the factors affecting the rates of organic chemical reactions. After 1945 this knowledge was initially applied and further developed in studies on enzyme catalysis by several "physical-organic" chemists, mainly Myron Bender, Thomas Bruice, William Jencks, and Daniel Koshland.[198]

Beginning with the work in the laboratories of Arnold Kent Balls and of Brian Hartley during the early 1950s, and later of several other investigators, the hydrolytic cleavage of ester substrates (RCO-OR′) by chymotrypsin was shown to involve the direct participation of a single serine unit and a single histidine unit (identified as Ser-195 and His-57 in the linear sequence, with assistance from a single aspartate unit, Asp-102). The formation of an intermediate covalent O-acylserine-195 unit was demonstrated, evidence was presented for an analogous intermediate in the action of chymotrypsin at CO-NH bonds, and it was shown that the acyl-enzyme intermediate (or its kinetic equivalent) can react either with water (hydrolysis) or with the amino group of a peptide in a "transpeptidation" (or "transamidation") reaction. Pancreatic trypsin, as well as many other protein-splitting enzymes (among them elastase and subtilisin) have been shown to act by a similar mechanism, although with different specificity in respect to the nature of amino acid units forming the sensitive bonds, and are now termed serine proteinases (or proteases).[199]

An analogous catalytic mechanism appears to be operative in the action of a large group of enzymes (now denoted cysteine proteinases), among which papain (from *Carica papaya*) has been the most extensively studied representative. In this enzyme, a cysteine unit (Cys-25) and a histidine unit (His-159) are involved in a process in which the sulfhydryl (thiol) of the cysteine is acylated, and the intermediate thiol ester may either be hydrolyzed or, in the presence of an amino compound, participate in a transamidation reaction. Since the reactivity of thiol esters toward amines was long known to be greater than that of oxygen esters (as in the case of the serine proteinases), it was not surprising to find that papain is a more efficient catalyst of transamidation than is pancreatic trypsin in its action on the same substrate.[200] Other cysteine proteinases are the plant enzymes ficin and bromelain, the animal enzyme cathepsin B, and a streptoccocal proteinase.

A third large class of proteinases is represented by gastric pepsin, whose mechanism of action involves two aspartic acid units (Asp-32 and Asp-215), one of which acts as an acid and the other (in its carboxylate form) as a base, and the enzymes of this class have been denoted aspartic (or aspartyl) proteinases. Although pepsin catalyzes transpeptidation reactions, the available evidence argues against the formation of an acyl- (or imino-) enzyme intermediate in the catalytic process.[201] The aspartic proteinases include renin, the cathepsin D of lysosomes and many microbial enzymes, among which penicillopepsin and *Rhizopus* pepsin have been studied most extensively.

There is, in addition, a fourth group of enzymes that catalyze the hydrolysis of peptide bonds and whose proposed mechanism of action resembles that of pepsin, except for the involvement of a zinc ion in the catalytic process. These enzymes include pancreatic carboxypeptidase, which acts only at peptide bonds at the carboxyl-terminal end of peptide chains (a so-called exopeptidase) and thermolysin, a bacterial proteinase (an "endopeptidase"). In the case of carboxypeptidase, the zinc ion (linked to His-69, His-169, and Glu-172) is thought to act as a "Lewis acid" and a glutamate unit (Glu-270) as a base, and a similar active-site structure has been proposed for thermolysin.[202] In addition to leucine aminopeptidase, among the many other enzymes in which a metal ion has been shown to be involved in the catalytic process is urease, an enzyme prominent in the history of enzymology. In the case of urease, a nickel ion appears to act as a Lewis acid in the hydrolysis of urea (NH_2-CO-NH_2) to carbamate (NH_2-COO^-) and ammonium ion.[203]

In addition to these various enzymes, which act at amide or ester bonds as catalysts of the transfer of acyl (RCO-) groups, pancreatic ribonuclease, a catalyst of phosphoryl transfer, and egg-white lysozyme, a catalyst of glucosyl transfer, played a significant role in the post-1945 development of knowledge about the relation of protein structure to biological function. In the case of ribonuclease, the cleavage of the internucleotide 3′,5′-phosphodiester bonds of RNA involves the action of His-12, His-119, and (probably) Lys-41 of the protein, with a preferential action at bonds involving the 3′-hydroxyl of pyrimidine nucleosides and the intermediate formation of a 2′,3′-cyclic phosphate. In the case of lysozyme, the crystallographic data of David Phillips and his associates indicated the participation of Asp-52 and Glu-35 in the catalytic process.[204] Moreover, numerous enzymes that act at interior bonds of natural polymers have extended catalytic sites, as has been shown by the kinetics of their action on oligomeric synthetic substrates. In particular, crystallographic studies on proteinases have indicated the presence in such enzymes of a flexible "active site cleft" composed of approximately five amino acid units. It has also been found that enzymes such as pepsin or papain are strongly inhibited by natural or synthetic compounds (for example, pepstatin or acyl-Phe-glycinal) whose structure mimics the presumed transition state in the reaction.[205] These findings have been adduced in support of the hypothesis that in addition to the particular catalytic groups mentioned above, enzymes utilize the binding energy from the interaction of an oligomeric substrate with other groups in the extended active site to lower the free energy of the transition state.[206]

The introduction of new methods for the determination of the amino acid sequence of protein chains also provided chemical insights into the physiological role of many enzymes. It was long known that gastric pepsin is derived from an inactive precursor (a "zymogen," in this case pepsinogen) and that pancreatic trypsinogen is converted to trypsin by an intestinal "enterokinase." Pepsin, trypsin, and their zymogens were crystallized during the 1930s,[207] but only after their amino acid sequences had been determined was it possible to define the activation processes as instances of limited proteolysis at particular peptide bonds in the zymogens.[208] Such limited proteolysis was then shown to be an essential feature of the conversion of plasma fibrinogen to insoluble fibrin in the clotting of mammalian blood, with a series of reactions in which certain plasma proteins are successively converted to serine proteinases.[209] Moreover, after Donald Steiner discovered a precursor of insulin in 1967 and found that the formation of the two-chain hormone involves the enzymatic cleavage of two interior peptide bonds of the single-chain "proinsulin," other investigators showed that similar limited proteolysis is involved in the formation of many peptide hormones from precursor proteins.[210] To these examples of current interest in the various physiological roles of proteinases may be added recent studies on the mechanisms of the intracellular breakdown of proteins, involving cell elements termed lysosomes, multienzyme "proteosomes," and an accessory small protein named ubiquitin.[211] Most of the individual proteinases operative in these various physiological processes have not yet been characterized to the same degree as in the case of pepsin or trypsin, and it may be expected that this area of biochemical research will remain active.

Proteins as Antibodies

In significant respects, the nineteenth-century discussion of the nature of the response of a biological organism to foreign agents (microbes, toxic proteins) resembles the contemporary debate about the nature and role of enzymes.[212] Indeed, some of the notable pioneers in the emergence of immunology as a distinctive field of research, including Louis Pasteur, contributed to both areas of controversy. After the demonstration, by Pasteur and Robert Koch, of the specificity of infectious agents, Emil von Behring reported in 1890 that resistance to bacterial infection is attributable to "antibodies" in the serum, thus disputing Ilya Mechnikov's theory, based on microscopic observation, that mobile cellular elements ("phagocytes") are the principal agents of the immune response.[213] Among the alternative theories that attracted attention during the 1890s was the proposal by Hans Buchner (a disciple of Carl Nägeli and brother of Eduard Buchner) that toxic agents, among which he included peptones, are attacked by plasma "alexins" and converted into antitoxins. This theory was soon refuted.

In 1900 Paul Ehrlich advanced a theory according to which a toxin has two different combining groups, a chemically stable haptophore group and an unstable toxophore group, which is responsible for the toxic action. He compared the haptophore group of toxic agents to normal metabolic substances such as sugars and stated that

> we may regard the cell quite apart from its familiar morphological aspects, and contemplate its constitution from the purely *chemical* standpoint. We are obliged to adopt the view that the protoplasm is equipped with certain atom groups, whose function consists in fixing to themselves certain food-stuffs, of importance to the cell-life. Adopting the nomenclature of organic chemistry, these groups may be designated *side-chains*. . . . The relationship of the corresponding groups, *i.e.*, those of the food-stuffs, and those of the cell, must be specific. The groups must be adapted to one another, *e.g.*, as a male and female screw (PASTEUR), or as lock and key (E. FISCHER).[214]

Although Ehrlich offered organic-chemical analogies in support of his theory, and his first scientific paper (1877) dealt with the use of aniline dyes in histological staining, he espoused the ideas of Eduard Pflüger about "living protoplasm" as a giant molecule, cited approvingly Oscar Loew's claims about "living proteins," and asserted his priority for Max Verworn's "biogen" concept. In Ehrlich's theory the serum antibodies "represent nothing more than the side-chains, reproduced in excess during regeneration and therefore pushed off from the protoplasm—thus coming to exist in a free state."[215]

Ehrlich's side-chain theory of immunity grew out of an international effort during the 1890s to develop methods for testing the potency of antidiphtheria sera, and Ehrlich headed a state institute set up in 1896 for this purpose in Steglitz. From his studies, Ehrlich concluded that

> all observations, especially since it is possible to titrate the antibodies rather accurately (in favorable circumstances, the present method works with an error of 1 per cent), indicate that the reaction between toxin and antitoxin takes place in accordance with the proportions of a simple equivalence. *A*

> *molecule of toxin combines with a definite and constant quantity of antibody.* One must assume that this capacity to bind antibodies is attributable to the presence in the toxin complex of a definite group of atoms which exhibits a maximum specific relationship, and easily joins one to the other, as a key does a lock, according to Emil Fischer's well-known comparison.[216]

The theory was further developed in papers with Julius Morgenroth, and Ehrlich introduced many new words (for example, *amboceptor* and *haptine*). In 1899 Ehrlich moved to his new institute in Frankfurt, where research on chemotherapy, for which Ehrlich gained much public fame, soon replaced the experimental work on immunological problems.

Ehrlich's side-chain theory, and its implications for a sharply defined ("absolute") immunological specificity and tight binding of the antigen to the antibody, elicited several lively controversies during 1900–1905. In addition to a polite exchange between Ehrlich and Jules Bordet, there was an acerbic one with Max von Gruber (a disciple of Carl Nägeli) regarding the sharpness of immunological specificity. From those who had embraced the new colloid chemistry—some physiological chemists, notably Wolfgang Pauli (a pupil of Franz Hofmeister), and some physical chemists, for example Wilhelm Biltz—came insistence on the importance of the relatively nonspecific surface adsorption of protein antigens on antibodies.[217] Also, Svante Arrhenius, one of the founding triumvirate of the new physical chemistry, entered the fray and, with the experimental help of the bacteriologist Thorvald Madsen, claimed that the antigen-antibody interaction is a reversible process, analogous to the reaction of an acid and a base. The term *immunochemistry* first appeared in Arrhenius's writings. His idea was soon dismissed, even by the noted physical chemist Walther Nernst.[218] So far as I have been able to discover, none of the leading organic chemists of the time chose to enter the arena of this controversy. There were, however, some younger biologists with organic-chemical inclinations who supported Ehrlich's concept of the specificity of immunological reactions. Among them was Leonor Michaelis, whose remarkable scientific career began in the field of embryology and ended in the field of magnetochemistry. His important contributions in the enzyme field have been mentioned; they were preceded by a brief association with Ehrlich at the latter's institute in Steglitz. Michaelis's assignment there was to continue Ehrlich's research on the use, for histological staining, of organic-chemical dyes that had been developed for the textile industry. During the course of this work, Michaelis discovered the specific vital staining by Janus Green of intracellular granules later to be named mitochondria.[219] Support for the organic-chemical interpretation of immunological processes also came from the experimental work of the clinical chemist

Friedrich Obermayer and the physiological chemist Ernst Peter Pick (another former pupil of Franz Hofmeister's). They showed that the introduction of various substituent groups into bovine serum albumin altered the specificity of the immune response of rabbits to the injection of such chemically modified forms of the protein.[220] This approach was subsequently taken up by Karl Landsteiner, who, having been a pupil of Emil Fischer, Eugen Bamberger, and Arthur Hantzsch during the early 1890s, was exceptionally well equipped to develop it. As a former associate of Gruber's, Landsteiner inclined toward the concept of a relative rather than absolute specificity of immunological reactions, and he also had made important contributions, through his study of isoagglutinins in human blood, to the recognition of what came to be called the ABO groups and to the general problem of the existence of natural antibodies. Like other biologists of his time, Landsteiner was attracted to colloid chemistry, and in 1908 he collaborated with Wolfgang Pauli in studies on the electrophoresis of proteins, but unlike Pauli, Landsteiner later abandoned the colloid-chemical approach. Instead, during the succeeding two decades Landsteiner showed that many organic substances that, when injected alone, do not elicit an immunological response, do so when linked chemically with a protein by means of an azo (-N=N-) group. He termed such substances haptenes (a word borrowed from Ehrlich). He also made further important contributions to the study of human blood groups.[221]

Despite his wide-ranging achievements as an investigator in Vienna, Landsteiner did not receive there the professional recognition he merited, such as the directorship of an independent institute. This neglect has been attributed to his critical attitude toward Ehrlich's theories. He left Vienna in 1919 for a modest post in The Hague but was rescued by Simon Flexner two years later. Flexner brought Landsteiner to the Rockefeller Institute, where his research program was largely restricted to studies on blood groups and the specificity of immunological reactions. It has been reported that at the institute Landsteiner was "difficult to deal with" and "did not interact socially . . . by having lunch together" with members of other research groups.[222]

At this point it should also be recalled that during the early 1920s it was generally believed that for a foreign agent to be an antigen it had to be a protein or attached to a protein, but there was no consensus as to the chemical nature of the serum constituents involved in the various known immunological reactions (neutralization of toxins, the lysis or agglutination of bacterial cells, the precipitation of proteins, anaphylactic sensitization, or complement fixation). As the immunologist Harry Gideon Wells put it:

> A pure antibody is as yet as unattained as a pure enzyme; indeed antibodies and enzymes resemble one another in many respects, especially in that they are so intimately associated with the proteins, although we do not know whether they actually are proteins with a special molecular structure or with some special radical attached. We do not even know whether each of the several reactions described depends on a special sort of antibody (*i.e.*, antitoxins, agglutinins, precipitins) or whether there is only a single type of antibody which is demonstrable by these several different reactions. As far as chemical properties are concerned, we merely know that they are found in a certain part of the proteins of the serum, but we do not know whether the immune bodies are serum proteins modified by the process of immunization, or specific proteins formed and secreted by the cells to unite with the antigen, or specific chemical radicals either attached to or forming part of the protein molecule.[223]

The essay containing this excerpt was evidently written some time before 1928, for some years earlier, James Sumner had crystallized urease, and Michael Heidelberger and Oswald Avery had shown that a protein-free preparation of the type-specific capsular polysaccharide of the pneumocccus is an effective antigen, thus disproving the view that only proteins could serve in that capacity.[224]

Some words are needed about Heidelberger, whose role in the development of immunochemistry has, in my opinion, been insufficiently appreciated by several historians. A postdoctoral pupil of Richard Willstätter's, Heidelberger joined Walter Jacobs at the Rockefeller Institute in 1912 and, during World War I, participated in the synthesis of a trypanocidal agent (Tryparsamide). In 1921 he joined the section headed by Donald Van Slyke at the hospital of the institute, and shortly afterward he was drawn into Avery's research on the pneumococcus.[225] That collaboration lasted until 1927, when Heidelberger moved to the College of Physicians and Surgeons (P&S) at Columbia. There he embarked on a program to develop quantitative methods for the determination of the antibody content of sera. It had been the practice of immunologists (including Landsteiner) to denote the extent of a precipitin reaction by means of plus signs (+ to ++++), tr (trace) or ±, and a minus sign or zero. I do not think it an exaggeration to suggest that the introduction of the reliable methods devised by Heidelberger, and his associates Forrest Kendall and Elvin Kabat, marks the beginning of quantitative immunochemistry.[226] To this I must add that Heidelberger was a modest and undemanding person, who was not made a full professor at P&S until 1945, three years after his election to the National Academy of Sciences. Although the significance of his experimental achievements was widely recognized, Heidelberger's avoidance of participation in the immunochemical speculations of the 1930s may have been a factor in the place accorded him by some historians of immunology in their accounts of the development of that field.

In response to the seemingly infinite capacity of a biological organism to make a specific antibody to almost any haptene, several variants of what came to be called the direct template hypothesis were offered during the 1930s; according to this view, the presence of the antigen at the site of serum globulin synthesis leads to the formation of a peptide sequence complementary in structure to that of the antigen.[227] A modification of this idea was proposed in 1940 by Linus Pauling, who suggested that the antigen influences only the conformation of preexisting globulin molecules; in 1942 he issued a press release announcing the artificial formation of antibodies, in accordance with his theory. This claim proved to be irreproducible and was quietly forgotten in the face of the scientific prestige of its proponent.[228] The failure of these direct-template hypotheses to account for many known facts about antibody production led Macfarlane Burnet to suggest that an antigen modifies the proteinases within globulin producing cells to make a specific antibody. In support of this idea, he cited Max Bergmann's extrapolation of the finding in his laboratory that proteinases can effect peptide bond synthesis.[229] Subsequently, Burnet and others, including Jacques Monod, offered modifications of this so-called indirect template hypothesis, based on the available knowledge about what was then termed enzyme adaptation.[230] These various speculations faded into the background, however, after experimental achievements on two fronts: the determination of the chemical structure of immunoglobulins, made possible by newly available methods of protein chemistry, and biological studies, such as those of Peter Medawar and his associates, on tissue transplantation and immunological tolerance.[231]

From the electrophoretic examination of serum proteins, by Arne Tiselius and Elvin Kabat,[232] it was clear that antibodies are gamma-globulins, and subsequent work showed this fraction of the proteins in animal sera to be a mixture separable by means of chromatographic techniques. The purified protein initially subjected to the most intensive study was the one denoted as immunoglobulin G (IgG) of the rabbit. In the work of the groups associated with Rodney Porter and with Gerald Edelman, limited proteolysis of IgG by crystalline papain or pepsin or reduction of the disulfide bonds linking separate peptide chains yielded fragments whose amino acid sequences were then examined. The results led to the formulation during the 1960s of a remarkable chemical structure, with great explanatory power.[233] Briefly, it turned out that the IgG molecule (molecular weight, c. 150,000) is composed of four peptide chains, two longer ones (the "heavy" chains, c. 50,000, denoted H) and two shorter ones (the "light" chains, c. 20,000, denoted L) to form a structure (L-H-H-L) linked by disulfide bonds. Limited proteolysis by papain yielded an antigen-binding fragment (Fab) and a crystallizable fragment (Fc) with no antibody activity. Because of the immunological heterogeneity of IgG, a unique amino acid sequence could not be determined for the Fab fragment, found to contain the four "hypervariable" amino-terminal regions (or "domains") of IgG, in two of which the amino-terminal unit is pyrrolidone carboxylic acid (a cyclic derivative of glutamic acid), while the crystallizability of the Fc fragment indicated that it represents a homogeneous set of peptides (linked by means of disulfide bonds) with largely "constant" amino acid sequences. X-ray crystallographic studies have given a three-dimensional picture and have shown the presence of a cleft, formed by the hypervariable regions of the H and L chains, which can serve as a complementary site for interaction with the antigen.[234]

An answer to the chemical problems posed by the heterogeneity of the isolated rabbit immunoglobulins came from an unexpected source, namely the examination of the serum proteins in patients suffering from the disease known as multiple myeloma, a tumor of the bone marrow. Such individuals excrete a urinary protein described in 1848 by Henry Bence Jones (hence named accordingly) and found during the 1960s to be an immunological counterpart of the L chains of IgG. In contrast to IgG, the serum immunoglobulins in the myeloma B-lymphocytes are chemically homogeneous, and the finding that each myeloma patient makes a different immunoglobulin has served as the basis for important advances in immunological theory and practice.[235]

During the 1950s, in parallel with the initial chemical work that led to the elucidation of the structure of immunoglobulins, there appeared a series of stimulating papers offering a new view of the biological process of antibody formation. The first was by Niels Jerne, and it was quickly followed by those of Burnet, David Talmage, and Joshua Lederberg.[236] As gradually modified, what came to be called the clonal selection theory of antibody production rejected the so-called instructional theories, which assigned to the antigen a "direct" or "indirect" template role in determining the specificity of the elicited antibody. Instead, attention was drawn to the recently emergent view that the amino acid sequence of a protein is determined by a sequence of nucleotides in the DNA of the protein-forming cell, and it was suggested that a single immunocompetent cell produces a single antibody. Although experimental support for this view was provided by Gustav Nossal and Lederberg, and the validity of the clonal selection theory was widely accepted during the 1960s, its great practical utility was not made evident until 1975, through the work of Georges

Köhler and Cesar Milstein.[237] They described a technique in which B-lymphocytes from the spleen of a mouse immunized with an antigen are fused with myeloma cells (from a cell line that has lost the ability to make its own immunoglobulin) to produce hybrids with a capacity for making relatively large amounts of the specific "monoclonal" antibody. This advance in cellular immunology has had a great impact on the development of other areas of the interplay of chemistry and biology. One example is the successful effort of several organic chemists (notably Richard Lerner and Peter Schultz) who prepared specific "catalytic antibodies" by using as an antigen a compound which mimics the presumed transition-state intermediate in an enzyme-catalyzed reaction. Thus in the enzymatic hydrolysis of esters (RCO-OR′) the transition state structure has been formulated as $RC(O)(O^-)$-OR′; by using the phosphonate analogue $RP(O)(O^-)$-OR′ as an antigen, esterase-like antibodies were obtained.[238] Another line of research in the new immunology has been the study of the structure and function of the protein factors ("lymphokines") that are produced by T-lymphocytes and control the activity of the immune system.[239]

The history of immunology has been divided by some historians into "three distinct periods: the 'physiological' period (roughly 1880–1910), the 'chemical' period (roughly 1910–1950), and the 'biological' period (from 1950 to the present)."[240] It will, I hope, be clear from the above sketch of the development of immunology that I do not accept this division. If I did not mention in my account the names of the many biologists who made important experimental contributions during the so-called chemical period, I wished to emphasize the dependence of immunology on the contemporary state of knowledge about the chemistry of proteins. Moreover, I do not consider the biological development of immunology since 1950 to be merely a reaffirmation of the biological ideas of Mechnikov or Ehrlich but rather a model example of a complex and fruitful interplay of biological and chemical thought and practice, which merits closer scrutiny by future historians.

I am certain that they will also find in the biochemical literature of the 1990s the beginnings of similar important developments in which the study of old biological problems was transformed by the use of the concepts and methods of modern protein chemistry. An outstanding example of such a recent development was provided by Stanley Prusiner's discovery, in mammalian brain tissue, of protein particles he named prions.[241]

Protein-Protein Interaction

The development of reliable knowledge about the chemical structure and biological role of individual proteins has brought to prominence the fundamental problem of the factors that determine their specific interaction with one another.

The importance of this problem was appreciated before the 1950s, but studies directed to the identification of the chemical groups and of their spatial arrangement in the regions of such interactions were possible only after the unqualified acceptance of the polypeptide theory of protein structure, the development of chemical methods for the determination of the amino-acid sequence of polypeptide chains, and knowledge provided by X-ray crystallography of the three-dimensional structure of such chains. Thus the outstanding studies of Max Perutz's group on hemoglobin (a tetramer composed of two α-chains and two β-chains) showed that it is held together largely by a set of ion pairs, among them a "salt bridge" linking Lys^+-40 of an α-chain to the terminal COO^- of a β-chain.[242] Another of the many examples that can be cited was provided by the elegant work of Michael Laskowski Jr. and his associates on the interaction of trypsin with an inhibitory protein that binds at the active site of the enzyme.[243] As mentioned above, in the interaction of enzymes such as lysozyme or pepsin with oligomeric substrates or specific inhibitors, the binding site has been found to be an extended region enclosing the catalytic site, and this is also a feature of the interaction of immunoglobulins with haptene molecules.[244] A general feature of these various specific interactions is that they involve the formation of noncovalent bonds that are relatively weak in nature individually, but that reinforce each other.

As will be evident from the discussion in subsequent chapters of the role of multiprotein complexes in a great variety of intracellular processes, the study of the principles of protein-protein interaction has assumed considerable biological importance and has become a major area of theoretical and experimental effort.[245]

CHAPTER 6

Chemical Energy of Biological Systems

In a famous memoir dated 1789, Armand Seguin and Antoine Lavoisier wrote of the respiration of animals that

> in general, respiration is nothing but a slow combustion of carbon and hydrogen, which is entirely similar to that which occurs in a lighted lamp or a candle, and . . . from this point of view, animals that respire are true combustible bodies that burn and consume themselves. . . . The proof of this identity of effects in respiration and combustion are immediately deducible from experiment. Indeed, upon leaving the lung, the air that has been used for respiration no longer contains the same amount of oxygen; it also contains not only carbonic acid gas but also much more water than it contained before it had been inspired. Now since the vital air can only convert itself into water by the addition of hydrogen; since this double combination cannot occur without the loss, by the vital air, of a portion of its specific caloric, it follows that the effect of the respiration is to extract from the blood a portion of carbon and hydrogen, and to deposit there a portion of its specific caloric which, during circulation, distributes itself with the blood in all parts of the animal economy, and maintains that nearly constant temperature observed in all animals that breathe.[1]

This memoir appears to have been read to the Académie des Sciences on 17 November 1791 and was published in 1793, a time of increasing desperation and cruelty in the French Revolution; Lavoisier was guillotined on 8 May 1794. The final paragraph reads:

> We conclude this memoir with a consoling reflection. To be rewarded by mankind and to pay one's tribute to the nation, it is not essential to be called to those public and brilliant offices that contribute to the organization and regeneration of empires. The physicist may also, in the silence of his laboratory and his study, perform patriotic functions; he can hope, through his

> labors, to diminish the mass of ills that afflict mankind, to increase its happiness and welfare; and if he has only contributed, through the new avenues he has opened, to prolong the average life span of human beings by a few years, even by a few days, he could also aspire to the glorious title of benefactor of mankind.[2]

The conclusions regarding the chemical events in animal respiration were an outgrowth of Lavoisier's explanation of combustion as an uptake of oxygen. In 1777 he had concluded that during respiration, as in the combustion of charcoal, oxygen (he then called it *l'air éminemment respirable*) is removed from the air and is converted to carbon dioxide *(acide crayeux aériforme)*. He suggested that this process occurs in the lungs and that oxygen combines with the blood to give the latter its red color. By 1793 he and Pierre Simon Laplace had performed their celebrated experiments on the amount of heat released when equal amounts of carbon dioxide were produced in the combustion of charcoal and in the respiration of a guinea pig. Laplace devised for these studies an ice calorimeter, and the amount of ice melted in the two processes was compared. They found that during ten hours the animal had produced a quantity of heat sufficient to melt 13 ounces of ice, whereas the burning of charcoal melted 10½ ounces of ice. No account was taken of the fact that the two processes had occurred at different temperatures, and the difference of 2½ ounces was attributed largely to the cooling of the limbs of the animal. Lavoisier and Laplace stated that "the conservation of animal heat is due, at least in large part, to the heat produced by the combination of the pure air respired by animals with the base of the fixed air furnished by the blood."[3] A contemporary and complementary theory of animal heat was offered in 1779 by Adair Crawford, a pupil of Joseph Black's and William Irvine's; he applied their studies on latent and specific heats and suggested that the inspired air contains a great amount of "absolute heat," which is released in the pulmonary blood.[4]

In the studies of Lavoisier and Seguin, gas analysis (eudiometry) was employed, and they concluded that the operation of what Lavoisier called *la machine animale* is controlled by three processes: respiration, a variable transpiration through the skin depending on the need to remove heat, and digestion, which provides the blood with "chyle," derived from the food and composed principally of carbon and hydrogen, for the replacement (by combustion) of the heat lost in the other two processes. These studies served as the starting point of the discussion, throughout most of the nineteenth century, of two general questions of animal physiology: in what parts of the organism does the conversion of oxygen into carbon dioxide and water occur, and in what form does the carbon and hydrogen exist?

As for the first question, Lavoisier's surmise that respiratory combustion occurs principally in the lungs elicited the criticism that "M. de la Grange considered that if all the heat which distributes itself in the animal economy were released in the lungs, it would necessarily follow that the temperature of the lungs would be increased to such an extent that one might continually fear their destruction, and if the temperature of the lungs were so considerably different from the other parts of animals, it was impossible that this had never been observed. He believed it to be very probable that all the heat of the animal economy is not solely released in the lungs, but also in all parts where the blood circulates."[5] Indeed, in 1792 Vauquelin reported that insects release carbon dioxide, although they have no lungs or capillary system and oxygen is supplied directly to the tissues through tracheal tubes. In 1807 Jean Senebier called attention to the findings, during the 1780s, of Lazzaro Spallanzani on the formation of carbon dioxide by a variety of animal tissues, and Théodore de Saussure showed in 1804 that in the dark, plants take up oxygen and release carbon dioxide. Many of these observations were confirmed and extended in 1824 by William Edwards and during the 1840s by Victor Regnault and Jules Reiset.[6]

During the first four decades of the nineteenth century, several investigators (notably Eilhard Mitscherlich, Leopold Gmelin, and Friedrich Tiedemann) tried and failed to extract gases from the blood by means of an air pump, and it was not until 1837 that Gustav Magnus, using the higher vacuum of a mercury pump, established the presence of dissolved oxygen, carbon dioxide, and nitrogen in the blood. His analytical data indicated that arterial blood has a higher proportion of oxygen than venous blood and he concluded: "It is very probable from these [results] that the gaseous oxygen is absorbed in the lungs and carried by the blood around the body, so that it serves in the so-called capillary vessels for an oxidation, and probably for the formation of carbonic acid. . . . It follows necessarily that, during the circulation of the blood, the carbonic acid is formed there, or is taken up by it."[7] Although criticized by Gay-Lussac and Magendie, Magnus's results were soon confirmed in several laboratories and drew increasing attention to the role in animal respiration of the formed elements of the blood, because whole blood appeared to be able to hold more oxygen than serum.[8]

Most of the quantitative experiments on respiration during 1780–1840 were performed by men who identified themselves as physicists or chemists, and who studied animal organisms in the sense of the machine animale. An early critic of this approach was the brilliant young clinician Xavier Bichat, who disavowed belief in the ideas of eighteenth-century vitalists but insisted on considering animal organisms in terms of the physiological functions of their tissues. In 1801 he wrote: "Attempts have been made recently to determine precisely what amount of oxygen is inhaled, what amount is required to produce the water of respiration, what amount of carbonic acid gas is formed, what amount of heat is released, etc. . . . Chemists and physicists, accustomed to study the phenomena over which physical powers preside, have carried their mathematical spirit into the theories they have imagined to apply to the laws of vitality but it is no longer the same thing. The mode of theorizing about organized bodies must be quite different from that of theories applied to the physical sciences."[9] The claim that quantitative physiological measurements are pointless was criticized by François Magendie, although he and his students accepted Bichat's view that the phenomena exhibited by animal organisms must be studied in their own right and in relation to the specific structure and composition of individual tissues. This attitude was to find its clearest expression in the writings of Claude Bernard later in the century, and Bichat's negative reaction to Lavoisier's work on respiration was reflected in that of Bernard to the physiological theories of Justus Liebig and his followers.

Liebig and Biological Oxidation

Liebig's enormous public reputation at midcentury, especially in Germany, England, and the United States, placed his ideas at the center of discussion and has colored many historical accounts of the development of biochemical thought. His ideas about alcoholic fermentation and the nature of proteins, mentioned in previous chapters, formed part of a larger design that he developed in successive editions of his two books on agricultural and animal chemistry and his bestseller *Familiar Letters on Chemistry*. They were presented with such assurance, often without acknowledgment to his immediate predecessors, that their truth and originality seemed incontrovertible to the reading public. To some leading physiologists of Liebig's time, however, in the words of Emil du Bois-Reymond, "his physiological fantasies [are] worthless and pernicious, because he entirely lacked the necessary factual knowledge and critical training."[10]

Liebig's theory of respiration was a modernized version of the one offered by Seguin and Lavoisier. By the 1820s, however, that theory had been subjected to serious criticisms. Apart from the finding that the ice calorimeter is an unreliable instrument, evidence was presented for the involvement of the nervous system in heat production, and there were inconsistent reports about the ratio of carbon dioxide output to oxygen intake and about the relation of oxygen consumption to heat production.[11] Liebig dismissed the doubts raised by the inconsistencies in the available data, and for his theory of the oxidation of food materials he adopted the classification of William Prout (saccharine, oily, albuminous). Liebig divided them into (1) the "respiratory" elements, which included the nonnitrogenous materials of the diet (fat, starch, sugars, as well as wine, beer, and spirits), and (2) the nitrogenous components, represented by plant albumin, casein, and fibrin (eaten by herbivorous animals) and animal flesh and blood (eaten by carnivores), which he called the plastic elements of nutrition, convertible into blood. In Liebig's view the fats and carbohydrates are the fuel of the animal body. Carbohydrates are not deposited at all, because no starch or sugar had been detected in arterial blood, and fats are "unorganized" substances that are deposited only when there is insufficient oxygen for their complete combustion. The central place in the machinery of the animal body is occupied by the "vital" proteins of the blood and the tissues, especially muscle. During muscular exercise or starvation, the respiratory oxygen attacks these proteins, with the release of nitrogenous products such as urea, uric acid, and ammonia, which are excreted in the urine. These speculations, which appeared in the first English edition (1842) of Liebig's *Animal Chemistry,* were derived from comparisons of elementary composition, were severely criticized, and were omitted from the third (1846) edition.[12] If Liebig's ideas were dismissed by physiologists, they were hailed by his chemical friends. Friedrich Mohr wrote to Liebig in 1842: "Your discoveries on the economy of the animal organism are tremendous [*kolossal*], the physicians watch dumbstruck the approach of a period in which they will be shaken from their customary torpor." In his reply Liebig wrote: "Chemistry, the noblest and most beautiful of all the sciences, will and must gain the victory."[13]

Liebig was not alone among the leading chemists of his time in believing that important physiological conclusions could be drawn from comparisons of the elementary composition of chemical constituents of biological organisms. During 1840–1844 he conducted a lively argument with Jean-Baptiste Dumas, who had offered his ideas on the subject.[14] Early in his scientific career, Dumas had worked with the physiologist Jean-Louis Prévost; they showed in 1823 that after surgical removal of the kidneys in dogs, cats, and rabbits, urea accumulates in the blood. As for Liebig, his only recorded physiological investigation with living organisms was the determination, in November 1840, of the amount of carbon in the food and excreta of a company of soldiers. During the course of his debate with Liebig, Dumas stated: "We are also convinced that if the views we have summarized in this lecture retain in the future the importance accorded them today, it will have been shown that the principal studies upon which these views were founded were performed in France, and that their synthesis into a general formula goes back to Lavoisier, who gave to physiology an impetus we have followed."[15] Although the Liebig-Dumas debate largely involved matters of priority, there was an important difference, for Liebig went far beyond Lavoisier's conception of the role of oxygen in the animal economy.

Indeed, Liebig appears to have considered oxygen to be harmful to the tissues of the animal organism. In writing about the role of fats and carbohydrates in providing carbon and hydrogen for respiratory combustion, he stated that "there can hardly be any doubt that this excess of carbon alone, or carbon and hydrogen, is used for the production of animal heat and for resistance against the external action of oxygen."[16] In his earlier writings (1839) about fermentation, Liebig had termed such action of oxygen on the tissues, when it occurred outside the body, *Verwesung* or *eremacausis* (Greek *erema,* quietly, plus *causis,* burning) and distinguished such slow oxidative decomposition from putrefaction, in which oxygen did not take part, and from mouldering, which occurred under conditions of low oxygen and moisture. Thus in the first edition of *Animal Chemistry,* Liebig wrote: "If we consider the transformation of the constituents of the animal body (the utilization of matter by the animal) as a chemical process, which proceeds under the influence of the vital force, then their decomposition outside the animal body into simpler compounds is one in which the vital force does not participate. The action in both cases is the same, only the products are different."[17]

Liebig's use of the term *Lebenskraft* (the word *Kraft* can be translated not only as "force" but also as "power" or "energy") has invited the attention of historians interested in the nineteenth-century mechanism-vitalism debates.[18] He certainly was not a vitalist in the sense of the Naturphilosophie movement. His importance in the development of the interplay of chemistry and biology was rather that, although most of his physiological speculations turned out to be wrong, the public position he had won through his achievements in organic chemistry, and the reception accorded his popular writings on agriculture and nutrition, assured a respectful hearing to his bold hypotheses. In seeking to prove or disprove Liebig's ideas, his defenders and critics were spurred to examine more closely biochemical substances and processes. In particular, his writings inspired the efforts of a succession of German and American physiologists to develop what came to be called energy metabolism. The background of that approach, with its roots in the thermochemistry of Lavoisier, and its relation to the newly emergent specialty of chemical thermodynamics, will be considered later in this chapter. I now return to the nineteenth-century debate about the site of biological oxidation in the animal organism. Although, as was mentioned above, Spallanzani and others had previously shown that the tissues of lower animals release carbon dioxide, during the 1840s it was widely supposed that the respiratory oxidation occurs principally in the red cells of the blood as it circulates through the capillaries. For example, Robert Mayer recalled musing as a young physician in 1840 about the source of animal heat:

> Starting from Lavoisier's theory, according to which animal heat is a combustion process, I considered the double color change undergone by the blood in the capillary vessels of the greater and lesser circulation as a perceptible indication, as a visible reflection, of an oxidation process proceeding the blood. . . . Not the hundredth part of the oxidative process occurs outside the walls of the blood vessels. . . . The blood corpuscles take up the atmospheric oxygen in the lungs, and the vital chemical process accordingly depends essentially on the combination of the oxygen absorbed by the blood corpuscles with the combustible constituents of the blood to form carbonic acid and water.[19]

The Role of Hemoglobin

In the first edition of *Animal Chemistry,* Liebig wrote:

> From the invariable presence of iron in red blood it must be concluded that it is absolutely essential for animal life, and since physiology has shown that the blood corpuscles do not participate in the nutritional processes, there can be no doubt that they assume a role in the respiratory process. . . . Blood corpuscles of the arterial blood contain an iron compound that is saturated with oxygen, and in which the living blood loses its oxygen during the passage through the capillary vessels; the same occurs when blood is taken from the body and begins to decompose (begins to putrefy); the oxygen-rich compound is transformed by the loss of oxygen (reduction) into a less oxygenated compound. One of the resultant products is carbonic acid.[20]

Little was known, however, about the nature of the iron compound except that blood iron had been shown by Louis-René Lecanu to be associated with the pigment Berzelius had named hematin, and there was disagreement about the reason for the difference in the color of arterial and venous blood.[21] In lectures given in 1857, Claude Bernard described a series of experiments from which he concluded that the conversion of bright red arterial blood to dark blood is a consequence of an increased proportion of carbon dioxide relative to the amount of oxygen.[22] By the end of the decade, some of the uncertainties had been resolved, and it was recognized that the difference in color arose from the oxygenation and deoxygenation of hemoglobin. In part, the answers came from physiological experiments in Carl Ludwig's laboratory with improved mercurial gas pumps, showing that oxygen is carried in the blood almost entirely in the form of a readily dissociable compound. Also, in 1857 Bernard and Hoppe-Seyler independently discovered that carbon monoxide (CO) can displace oxygen from arterial blood; this finding was applied during the 1860s to the quantitative determination of the oxygen-binding capacity of the blood.

The decisive advances in settling the question of the nature of the iron-containing compound that combines with oxygen in the respiratory process came during the 1860s from the chemical and spectroscopic studies of Felix Hoppe-Seyler and George Gabriel Stokes. Their work established the identity of the crystalline material previously described by Otto Funke and others with the respiratory pigment of mammalian blood. Hoppe-Seyler showed that the isolated crystals contain oxygen in a loosely bound form, and in 1864 he named the pigment oxyhemoglobin. In 1862 he reported that in aqueous solution, oxyhemoglobin exhibits two absorption bands in the visible region of the spectrum, and two years later Stokes found that these bands disappear upon the addition of what later came to be called Stokes's solution (an ammoniacal solution of ferrous sulfate and ammonium tartrate). In place of these absorption bands, there appeared a single more diffuse band, with a change in color from that of "scarlet cruorine" to that of "purple cruorine," the names Stokes gave to oxyhemoglobin and "reduced" hemoglobin respectively. When the solution was shaken with air, the two-banded spectrum was restored. Stokes considered the loss of oxygen by scarlet cruorine to be a "simple reduction," comparable to the known behavior of indigo.[23] Later work showed this comparison to be inappropriate, but the fact that it was made calls attention to extensive work on indigo (the coloring matter of woad) and Chevreul's finding in 1808 that "indigo-blue" can be reduced to "indigo-white." In 1823 Walter Crum purified indigo-blue by sublimation, and in 1841 Auguste Laurent laid the foundation for its subsequent synthesis by Adolf Baeyer.[24] As will be seen later in this chapter, the reduction of natural pigments such as indigo and the reoxidation of the reduced form were important reactions in pioneering nineteenth-century studies on biological oxidations.

Stokes also found that treatment of blood with acid changes its absorption spectrum, and that, upon reduction of the resulting pigment in alkaline solution, there appear two very intense absorption bands whose positions differ from those of the oxyhemoglobin bands. He named this two-banded species reduced haematine; in confirming Stokes's finding, Hoppe-Seyler changed its name to hemochromogen. In these studies, the possible fate of the protein component of hemoglobin was overlooked, and later work (after 1910) showed that hemochromogens are complexes of iron-porphyrins with various kinds of nitrogenous bases. When Hoppe-Seyler showed that treatment of the hemochromogen derived from hemoglobin with strong acid gave an iron-free pigment, which he named hematoporphyrin, the complexity of the nomenclature in this field was further increased; this finding had been anticipated in 1867 by Ludwig Thudichum, who named the pigment cruentine. The elucidation of the chemical relations among hemoglobin, oxyhemoglobin, hemochromogen, hematin, and hematoporphyrin required many decades of research, with numerous wrong turnings. Also, during the 1870s, the question of the valence of the iron in the various iron-porphyrin compounds was open, although Hoppe-Seyler surmised that hematin contains ferric (Fe^{3+}) iron. The brown product he named methemoglobin, obtained by oxidation of hemoglobin, was not clearly identified to be the ferric form of hemoglobin until the 1920s.

Despite the shortcomings of the knowledge provided by Hoppe-Seyler and Stokes, their work not only provided the basis for subsequent chemical studies but also marked the introduction into biochemical research of the powerful new technique of absorption spectroscopy. One consequence was the emergence of a group of British investigators (Sorby, Lankester, Gamgee, MacMunn) who conducted spectroscopic studies on biological pigments that could undergo the kind of reversible oxidation and reduction that Stokes believed he had shown for hemoglobin.

By about 1875, therefore, physiological, chemical, and spectroscopic studies had indicated that the interaction of hemoglobin with oxygen is not an oxidation but an oxygenation, and that the physiological function of hemoglobin is that of an oxygen carrier. The recognition of this function obliged physiologists to consider anew the many experiments, from Spallanzani to Regnault and Reiset, on the production of carbon dioxide by animal tissues. Before examining the development of the concept of intracellular biological oxidation, it is appropriate to summarize briefly the results of important later research on the interaction of hemoglobin and oxygen.

This research reached a high point in 1904, when Christian Bohr described his quantitative studies on the proportion of blood hemoglobin (Hb) in the form of oxyhemoglobin (HbO_2) and demonstrated the remarkable capacity of the animal organism to regulate the Hb/HbO_2 ratio in response to changes in the O_2 and CO_2 content of the blood. The curve obtained by plotting the percent of HbO_2 against oxygen pressure was found to be a sigmoid curve, not a rectangular hyperbola, as would have been expected from the equation $Hb + O_2 = HbO_2$. Bohr also showed that, with increasing CO_2 pressure, the percent of HbO_2 at a given O_2 pressure is reduced. These two regulatory properties are of physiological advantage in unloading the oxygen of HbO_2 in the tissues, where the oxygen pressure is lower than in the lungs and the CO_2 is liberated from the tissues. Later work by John Scott Haldane showed that, conversely, oxygenated blood takes up less CO_2 than does deoxygenated blood. Various theories were offered to explain the sigmoid nature of the saturation curve of hemoglobin. In relation to later developments, of special interest is the contribution in 1911 of Archibald Vivian Hill, who assumed a reversible aggregation of hemoglobin units and a cooperative interaction among them, such that the conversion of the first one to HbO_2 increases the affinity of the others for oxygen. At that time, the molecular weight of hemoglobin was considered to be about 16,000, but during the 1920s it became evident that it is a tetramer of such units. More recently, the development of knowledge of the detailed chemical structure of hemoglobin has permitted a closer study of the dissociation of these subunits and of the relation of their specific interaction to the regulatory properties of the protein.[25] As we shall see in a later chapter, this knowledge has provided a model for general theories of the regulation of biochemical processes.

Intracellular Biological Oxidation

In the gradual acceptance, after about 1850, of the idea that the principal sites of the biological utilization of oxygen are the cellular elements of animal tissues, of particular importance were experiments on muscular contraction performed by Hermann Helmholtz during the 1840s, and those by Liebig's son Georg in 1850. In 1845 Helmholtz described his studies on the chemical changes in isolated frog muscle when it contracts upon electrical stimulation. The paper began as follows: "One of the most important physiological questions, intimately related to the nature of the vital force, namely whether the life of organic bodies is the effect of a self-generated purposeful force, or the result of forces also active in inanimate matter but specifically modified through the nature of their cooperation, has recently been given a much more concrete form in Liebig's effort to derive physiological conclusions from known chemical and physical laws, namely whether or not the mechanical force and the heat produced in the organism can be deduced completely from metabolic changes [*Stoffwechsel*]."[26] In this study Helmholtz determined the weights of solid material extractable from stimulated muscle by means of water or alcohol and found that, in comparison to nonstimulated muscle, the water-soluble material was decreased and the alcohol-soluble material was increased. He also noted, echoing Justus Liebig, that "in this article I must leave unresolved, however, another of the most important problems; whether the muscle fiber participates in the decomposition. A priori this would be quite probable, because we find in the protein compounds everywhere as carriers of the highest life energies."[27]

The 1845 paper was followed two years later by Helmholtz's memorable *Ueber die Erhaltung der Kraft,* of which more later, and a 1848 paper on his study of heat production during muscular contraction, an investigation conducted with remarkable experimental ingenuity and craftsmanship. Georg Liebig then reported that isolated frog muscles contract for a longer period in oxygen than in other gases and that, after they have ceased to contract in the absence of oxygen, carbon dioxide is still released. He concluded that "the formation of carbonic acid . . . proceeds in the body not within the capillary vessels, but outside them in the muscle tissue." Georg Liebig's findings were soon confirmed and extended by Gabriel Valentin and Carlo Matteucci.[28]

These observations provided the background for Moritz Traube's criticism in 1861 of the idea that the principal site of respiratory oxidation is the blood, and of the theory that the energy for muscular contraction is derived solely from the oxidative breakdown of proteins. After citing the reports of Helmholtz, Regnault and Reiset, and Georg Liebig, Traube concluded that "the released oxygen passes in a dissolved state through the capillary walls and forms with the muscle fiber a loose combination that is able to transfer the oxygen to other substances, dissolved in the muscle fluid, and [the muscle fiber] can then take up new oxygen. . . . The muscle fiber thus behaves toward reducing substances on the one hand, and to oxygen on the other, just as do indigo, indigo sulfonate."[29] With regard to the site of biological oxidation, Traube wrote:

> The fact that all organs of the animal body require arterial blood indicates that not only the blood, but all organs of the body respire. . . . What we call respiration is therefore a very complex process. It represents the sum of the consumption of all those quantities of oxygen needed by each organ, either for its nutrition or for its maintenance. Thus, there can be an increase in the respiration of the brain, or liver and spleen, or indeed any individual groups of muscles, nerves, spinal cord, without an accelerated respiration in other organs of the body. . . . The motive forces, however, which oxygen elicits in the muscles, nerves, spinal cord, and brain are a consequence of the characteristic construction and chemical nature of the apparatus in which the oxidative processes occur, so that these forces do not appear in the form of heat, but in the form of their specific, as yet inexplicable, vital functions.[30]

Such views were not generally accepted during the 1860s, and some physiologists preferred the opinion of Claude Bernard that "it is infinitely probable that the carbonic acid of the venous blood arises from an oxidation which is effected in the blood corpuscle itself." In 1867 Ludimar Hermann dismissed Georg Liebig's results as having no relevance to muscular contraction in the intact animal, and in 1870

Bernard's disciple Paul Bert stated that "the quantitative importance of the metabolic activity of our tissues has been greatly exaggerated, and it is in the blood that there occurs the greater part of the changes which ultimately produce carbonic acid, urea, etc."[31] At that time, the most convincing experimental evidence for this opinion appeared to be that provided by Carl Ludwig, who believed that muscular activity is largely controlled by the rate of blood flow.

Among physiologists, the chief protagonist of the view that the tissues represent the principal site of biological oxidation was Eduard Pflüger. In a series of papers during 1866–1877, frequently dotted with acerbic comments about the work of his opponents, he and his associates provided important experimental data in favor of that view. In characteristic style, Pflüger wrote: "Here lies, and I want to declare this once and for all, the real secret of the regulation of the oxygen consumption by the intact organism, a quantity determined only by the cell, not by the oxygen content of the blood, not by the tension of the aortic system, not by the rate of blood flow, not by the mode of cardiac action, not by the mode of respiration."[32] In stressing that the oxidative activity is localized in cells, Pflüger cited examples from comparative physiology, such as the work with insects mentioned above. The most impressive evidence was provided by his associate Ernst Oertmann, who determined the respiration of a frog whose blood had been replaced by a saline solution *(Salzfrosch),* and reported that "the oxidative processes of the frog are not altered by exsanguination, since the bloodless frog has the same metabolism as the one that contains blood. The site of the oxidative processes is therefore the tissues, not the blood."[33]

Pflüger's conviction regarding the importance of intracellular respiration was related to his advocacy of the idea of "living" protein molecules: "The life process is the intramolecular heat of highly unstable albuminoid molecules of the cell substance, which dissociate largely with the formation of carbonic acid, water, and amide-like substances, and which continually regenerate themselves through polymerization."[34] From ill-founded chemical considerations, he affirmed that the "living protein contains nitrogen, not in the form of ammonia, but in the form of cyanogen," and he attributed to the cyanogen radical the "explosive" quality of living matter, which allowed it to react with the chemically passive oxygen brought to the tissues by the circulation.

To conclude this section, it should be added that other leading physiologists did not accept Pflüger's views about intracellular respiration. At the close of his life, Claude Bernard came to the opinion that Lavoisier's and Liebig's identification of animal respiration with combustion was incorrect. Bernard wrote that chemists

> believed that organic combustion had as its counterpart the combustion effected outside living things, in our furnaces, in our laboratories. On the contrary, in the organism there is probably not a single case of these supposed phenomena which is effected by the direct fixation of oxygen. All of them involve the participation of special agents, for example, the ferments. . . . The production of carbonic acid, which is so general a phenomenon of life, is a consequence of a true organic destruction, of a decomposition analogous to those produced by fermentations. Moreover, these fermentations are the dynamic equivalent of the combustions; they fulfill the same purpose in the sense that they generate heat and therefore are a source of energy necessary for life.[35]

As for the role of oxygen in the animal organism, Bernard stated that

> it is not in direct combination that this gas is used. The usual formula repeated by all the physiologists that the role of oxygen is to support combustion is not correct, because there is really no true combustion at all in the organism. What is true is that the precise role of oxygen, which we thought we knew, is still unknown to us, one can hardly guess at it. . . . It is quite certain that this gas is fixed in the organism and it thereby becomes one of the elements of organic structure or creation. But it is not all through its combination with the organic matter that it incites vital function. Upon contact with the tissues, it makes them excitable; they cannot exist without this contact. It is therefore as an agent of excitation that it would participate intimately in most of the phenomena of life.[36]

Before considering the further efforts to explain the role of oxygen in intracellular respiration, and in such processes as muscular contraction, it is necessary to summarize the nineteenth-century development of theory and practice in thermochemistry, and of its relation to what came to be called chemical thermodynamics.

Thermochemistry and Thermodynamics

In his famous 1847 treatise *Ueber die Erhalting der Kraft,* Helmholtz presented a general concept of the interconversion of forms of energy and enunciated the principle that "the sum of the available living and potential forces *(lebendigen und Spannkräfte)* [in a system] is always constant." He discarded the idea of "caloric" as a material substance and considered heat to be a form of motion. He had been preceded during the 1840s by Robert Mayer (1842), James Prescott Joule (1843), and Ludvig Colding (1843), but Helmholtz did not refer to Mayer, and Colding had presented his theory in the form of a lecture that was not published until 1856. Owing to an error in computation, Helmholtz did not appreciate fully the importance of Joule's experimental determination of the mechanical equivalent of heat.[37] Many years later, Helmholtz apologized for his seeming ignorance in 1847 of Mayer's 1842 paper and for his misjudgment of Joule's work, but added:

> Very recently, the adherents of metaphysical speculation have attempted to put the stamp of an *a priori* principle on the law of the conservation of energy, and therefore hail R. Mayer as a hero in the field of pure thought. What they regard as the peak of Mayer's achievements, namely the metaphysically-formulated evidence for the *a priori* necessity of the law, will seem to every investigator accustomed to strict scientific method just the weakest aspect of his explanations, and has unquestionably been the reason for the fact that Mayer's work remained unknown so long in scientific circles. Only after the road had been opened from another direction, namely through the masterly researches of Mr. Joule, was attention given to Mayer's writings.[38]

At the close of his 1847 article, Helmholtz applied the law to biological systems:

> In plants, the processes are chiefly chemical ones and, in addition, there occurs, in some cases, a slight development of heat. Mainly, there is deposited in them an enormous quantity of potential energy [*Spannkräfte*], whose equivalent is provided to us as heat in the burning of plant substances. So far as we know at present, the only living energy [*lebendige Kraft*] absorbed during plant growth are the chemical rays of sunlight. . . . Animals take up oxygen and complex oxidizable compounds made by plants, release largely as combustion products carbonic acid and water, partly as simpler reduced compounds, thus using a certain amount of chemical potential energy to produce heat and mechanical forces. Since the latter represent a relatively small amount of work in relation to the quantity of heat, the question of the conservation of energy reduces itself roughly to whether the combustion and transformation of the nutritional components yields the same amount of heat as that released by animals. From the experiments of Dulong and Despretz, this question can be answered at least approximately in the affirmative.[39]

In his 1847 article Helmholtz had only a little more to say about the conservation of energy in chemical processes, and he merely referred to the work of Germain-Henri Hess, who in 1840 had offered the "law of heat summation," according to which "the heat developed in a chemical change is constant, whether the change occurs directly, or indirectly in several steps." In his physiological speculations, Liebig had implicitly assumed this law, and in his 1845 article Helmholtz noted that accurate calorimetric data were needed for the combustion of foodstuffs, and he questioned the procedure of equating the heat of combustion of a substance with the sum of heats for its carbon and hydrogen.[40] In this sense, Helmholtz's statement that the heats of combustion of nutrients approximate the heat released by animals provided support to those inspired by Liebig to develop the calorimetric approach to the "intake-output" method of studying the energy changes in animal metabolism.

Hess's work was followed by that of Thomas Andrews, Pierre-Antoine Favre, Julius Thomsen, and Marcelin Berthelot. Major advances were made during the second half of the nineteenth century in instrumentation; for example, the bomb calorimeter, introduced by Berthelot, continued to be used for many decades afterward. During the 1870s Berthelot enunciated the "principle of maximum work," according to which "all chemical change accomplished without the intervention of external energy tends to the production of a substance or a system of substances which liberate the most heat." A slightly different statement of the same principle had been previously put forward by Thomsen.[41] Berthelot also proposed that chemical reactions be denoted as exothermic or endothermic, depending on whether they were accompanied by the release or uptake of heat.

The application of thermochemistry to physiology reached a peak in 1904, when Max Rubner accounted for the heat production of a dog by measuring its carbon and hydrogen balance in a respiration calorimeter and by determining the heats of combustion of the dog fat, dog protein, and the organic constituents of dog urine. Because he found that the heat produced by the animal equaled the heats of combustion of the fat and protein, minus that of the urinary matter, Hess's law was taken to apply to the animal organism.[42]

This development of thermochemistry was paralleled by the interest of physicists in the problem of the relation of heat to motion, as defined in 1824 by Sadi Carnot in his *Réflexions sur la puissance motrice du feu.* Carnot's analysis of the working of a steam engine as a cycle of operations, and his proof that the most efficient engine is one in which all the steps are reversible, were put into mathematical form ten years later by Benoit Pierre Emile Clapeyron to produce what came to be called the Carnot-Clapeyron equation. Both men adhered to the caloric theory of heat, and Carnot compared the flow of heat from a higher to a lower temperature to a waterfall. With the replacement of the caloric theory by the concept of heat as a mode of motion, the significance of the Carnot-Clapeyron equation was appreciated more fully during the 1840s by William Thomson (later Baron Kelvin of Largs) and by Rudolph Clausius. In 1850 Clausius pointed out that if a Carnot cycle is operated reversibly between two temperatures (T_1 greater than T_2), the heat (Q_1) released from the heat reservoir may be divided into two parts, one of which is converted to work (W) and the other (Q_2) "sinks" to the lower temperature. From this formulation, in 1854 Clausius drew the conclusion that for the complete cycle there is a *Verwandlungsinhalt* [bound transformation content], which he denoted by the integral of dQ/T between T_1 and T_2, and which he later termed entropy.[43] A similar but less precise statement of what came to be known as the second law of thermodynamics was offered by Kelvin, who wrote in 1852: "1. There is present in the material world a universal tendency to the dissipation of mechanical energy. 2. Any *restoration* of mechanical energy, without more than an equivalent of dissipation, is impossible in inanimate material processes, and is probably never effected by means of organized matter, either endowed with vegetable life or subjected to the will of an animated creature."[44] A more famous statement is Clausius's 1865 dictum: "Die Energie der Welt ist konstant. Die Entropie der Welt strebt gegen ein Maximum." The question whether "organized matter" can flout the inexorable demand of the second law has been discussed repeatedly, up to the recent past, by philosophically minded scientists and by scientifically minded philosophers. As will become evident, this question has occupied an important place in the development of biochemical thought.

In its relation to the thermochemistry of Thomsen and Berthelot, the thermodynamics of Clausius and Kelvin predicted that the "principle of maximum work" is strictly applicable only to systems at absolute zero. In conducting a lively debate with Berthelot on matters of priority, Thomsen admitted that their "law" was only approximately valid and that it did not account for the spontaneous occurrence of some endothermic chemical reactions. Moreover, the thermochemical approach to the problem of chemical affinity was severely criticized first by several German and British chemists (notably Lothar Meyer and Pattison Muir) and then by members of the Ostwald group of physical chemists. In France, Berthelot was attacked by Pierre Duhem in writings that reflected differences not only in scientific opinion but also in political and personal attitudes. Like previous French scientists who had gained high prestige, Berthelot blocked the professional advancement of his scientific adversaries, and Duhem was one of his victims.[45]

The alternative to thermochemistry—chemical thermodynamics—emerged during the 1870s in large part owing to the experimental work of August Horstmann and the theoretical contributions of Josiah Willard Gibbs, who was followed during the 1880s by Helmholtz and van't Hoff. Horstmann showed that Clausius's formulation of the Carnot-Clapeyron equation accords satisfactorily with the measured heats of the vaporization and decomposition of ammonium chloride, and he also derived a thermodynamic interpretation of the dissociation of calcium carbonate, phosphorus pentachloride, and urea. Gibbs's papers appeared in the *Transactions of the Connecticut Academy of Arts and Sciences,* and reprints were sent to noted European scientists, including Helmholtz. With a few exceptions, their reception of Gibbs's treatment was, to put it most generously, ambiguous; some were outright dismissive. It was not until Gibbs's writings were translated into the scientific language of his European contemporaries, at the behest of Wilhelm Ostwald, that their importance was generally acknowledged. In his lengthy 1878 paper entitled "On the Equilibrium of Heterogeneous Substances" Gibbs developed the relation between energy and entropy in a reversible system as it approaches its equilibrium state. He proposed a new equation, now written as G = H – TS, where G is the "free energy" or "Gibbs energy," H is the "heat content" or "enthalpy," S is the entropy, and T is the absolute temperature. In its various forms, as presented in successive textbooks, this equation represents one of the theoretical foundations of present-day chemical thermodynamics.[46]

Gibbs's writings were unknown, or neglected, or used without acknowledgment in Europe during the 1880s, and an equation of similar form was presented by Helmholtz. He had moved in 1871 from his professorship of physiology at Heidelberg to the professorship of physics in Berlin, where he embarked on a study of electrochemical processes and returned to thermodynamics. In 1882 he developed a mathematical treatment, also based on Clausius's formulation of the Carnot-Clapeyron equation, in which the portion of the total energy (U) available for conversion to other forms as "free energy" was termed A, and the residual "bound energy" as TS. According to Helmholtz's treatment, a chemical reaction can proceed spontaneously only when there is a decrease in the quantity of A. The difference between the Gibbs and the Helmholtz equations, and those derived from them, is that the Gibbs free energy applies to processes at constant pressure, while the Helmholtz free energy refers to those proceeding in a constant volume.[47]

Gibbs and Helmholtz were mathematical physicists whose theoretical contributions were relevant to the problems of chemistry, but it was the chemist van't Hoff who, during the 1880s, combined physical theory and chemical experiment to produce the most lasting formulation of the thermodynamics of reversible chemical reactions. He defined the equilibrium of such reactions as the state at which the velocities of the forward and reverse reaction are equal, and he put the "law of mass action" and the concept of chemical equilibrium (proposed by Cato Maximilian Guldberg and Peter Waage during the 1860s) on a more satisfactory theoretical basis. Of particular importance in the subsequent development of chemical thermodynamics, and its application to biological problems, was van't Hoff's formulation of the relation between the equilibrium constant (K) of an ideal reversible chemical reaction operating at constant temperature and the maximum work that can be derived from that reaction. Moreover, van't Hoff identified the elusive concept of chemical "affinity" as the maximum amount of such work. This theoretical treatment was later modified by Gilbert Newton Lewis for nonideal solutions, and "concentrations" were replaced by "activities."[48] It is now customary in biochemistry to write the van't Hoff equation for the reaction A + B = C + D as $\Delta G = \Delta G° + RT\ln K$, where ΔG is the actual change in Gibbs (free) energy, $\Delta G°$ is the "standard" energy change (where all reactants are at unit molar concentrations), R is the gas constant, T is the absolute temperature, and $K = [C][D]/[A][B]$, the brackets indicating concentrations rather than activities. The preferred unit of energy is now the "joule" (J) or "kilojoule" (kJ) instead of "calorie" (cal) or "kilocalorie" (kcal) generally used in the past (2.3RT = 1.37 kcal/mol = 5.72 kJ/mol). Biochemists usually refer to reactions at a particular pH value, and ΔG or $\Delta G°$ is frequently denoted $\Delta G'$(pH 7) or $\Delta G°'$(pH 7).

Although biochemists soon appreciated the importance of the new chemical thermodynamics, they continued for many decades to assume the equivalence of work and heat in biological systems, largely because calorimetry readily gave values for the heat change in any chemical reaction, while the determination of free-energy changes was possible only for freely reversible reactions for which the equilibrium constant could be measured experimentally. Thus when Otto Warburg and Otto Meyerhof began their studies on the energetics of cellular metabolism, they assumed that "for the oxidation of hydrogen or carbon, it is permissible to set, with some degree of approximation, the heat liberation as equal to the decrease in free energy."[49]

It seemed for a time possible to calculate equilibrium constants from thermal data by means of Walther Nernst's "heat theorem" (often termed the third law of thermodynamics), and his theory was applied by Julius Baron and Michael Polanyi to the calculation of the free energy change in the combustion of glucose; they obtained a value (at 37°C) about 13 percent higher than the heat measured in a calorimeter.[50] It is also noteworthy that in 1912 A. V. Hill called the attention of his fellow physiologists to the fact that, in addition to Nernst's approach,

> in many cases it is already possible to calculate the equilibrium constant K of a reaction and hence also . . . its free energy. The development of ferment chemistry, especially in the case of reversible changes carried out by ferments, may make it possible to calculate directly the equilibrium constants of many breakdowns of organic material. If, *e.g.*, the bio-chemist can decide what chemical reactions go on in the process of carbohydrate breakdown, and if we can determine directly the values of K_1, K_2, K_3 . . . for these several

> reactions, then it will be possible not only to give the total free energy . . . but the free energy of every stage.[51]

It was only after several decades of experimental work, with some wrong turnings before the identity of the chemical participants in the "several reactions" had been firmly established and the "ferments" had been identified and isolated, that the validity of this statement was confirmed.

In addition to the applications of the second law to the determination of the free energy and equilibrium constants for freely reversible chemical reactions, of particular importance for the later study of biological oxidation was the nineteenth-century development of electrochemistry.[52] The effect of the passage of electricity through aqueous solutions of chemical substances had been examined by many investigators before 1830, but it was the memorable series of studies by Michael Faraday on "electrolysis" (in the sense of electrochemical decomposition) that mark most significantly the beginning of electrochemistry as a specialty. He showed that the amount of a substance deposited at an "electrode" depended only on the product of the current and the time, and that the laws of definite and multiple proportions hold not only for the chemical elements but also for electricity. He also introduced terms suggested to him by William Whewell and used today—anode, cathode, ion, anion, cation.[53] During the 1850s Clausius questioned Faraday's idea that the electric current decomposes a substance and suggested that some charged units exist in the solution. The validity of this view was supported by the experiments of Wilhelm Hittorf and Friedrich Kohlrausch on the electrical conductivity of solutions. Together with van't Hoff's theoretical treatment of osmotic pressure, these studies formed the background of Svante Arrhenius's formulation in 1886 of the theory of electrolytic dissociation. Moreover, by that time Gibbs and Helmholtz had developed a quantitative theory of the electromotive force of a galvanic cell. In 1881 Helmholtz stated: "Now the most startling result of Faraday's law[s] is this. If we accept the hypothesis that elementary substances are composed of atoms, we cannot avoid the conclusion that electricity also, positive as well as negative, is divided into definite elementary portions, which behave like atoms of electricity."[54] Johnstone Stoney had advanced a similar idea, and in 1891 he called the unit of negative electricity electron, but it was not until six years later, when Joseph John Thomson reported his studies on the particles produced in a cathode ray tube, that this term could be applied to a physically demonstrable entity.

In 1889 Nernst (then an associate of Wilhelm Ostwald) proposed a relation between electromotive force and ionic concentration. As now written for an oxidation-reduction reaction, the Nernst equation has the form $\Delta G° = -nF\Delta E_O$, where n is the number of electrons transferred, F is the faraday (23.1 kcal/volt equivalent or 96.5 kJ/volt equivalent), and E_o is the "normal" oxidation-reduction potential (in volts).[55] Among the first determinations of such potentials was the one performed by Ostwald's associate Rudolf Peters, who measured the difference in potential between one cell containing variable proportions of ferric and ferrous salts and a reference cell containing mercury and mercuric chloride ("calomel electrode"). His data fitted the equation $E_h = E_o + 0.06 \log [(Fe^{3+})/(Fe^{2+})]$, where E_h is the measured difference in potential and the terms in the brackets are the concentrations of the ferric and ferrous ions.[56] Consequently, the determination of the value of E_o for any oxidation-reduction system that equilibrates rapidly with a metallic electrode indicates the free energy change and equilibrium concentrations to be expected when the oxidant of that system reacts with the reductant of another oxidation-reduction system of known potential. With the emergence of the electronic theory of valency,[57] organic chemists began to consider reactions in their domain as involving electron transfer, and after about 1916 several investigators (Einar Biilmann, William Mansfield Clark, John Maurice Nelson, James Conant, Leonor Michaelis) reported electrometric studies on organic oxidation-reduction systems such as the quinone-hydroquinone system or on those involving dyes such as methylene blue and indigo sulfonate.[58] In these processes, hydrogen ions are involved, and the normal potential varies with pH. The convention was adopted to refer the potential of any system to that of the hydrogen electrode; this oxidation-reduction system ($H_2 = 2H^+ + 2e$) provided the basis for the electrometric determination of pH, and in succeeding years many technical improvements were made in the instruments for such measurements, including the use of glass electrodes.

The use of such electromotive systems in the study of biological oxidations and reductions will be considered later in this chapter, as will the thermodynamics of muscular contraction and of the transfer of chemical substances across biological membranes. After 1930 it was recognized that a prominent feature of these processes is the "energetic coupling" of biochemical reactions. It is therefore of considerable historical interest that in 1900 Wilhelm Ostwald called attention to the difference between the transfer of chemical energy and the transfer of heat or mechanical energy and stated that "a direct reciprocal transformation of chemical energies is only possible to the extent that chemical energies can be set in connection with each other, that is, within such processes which are represented by a stoichiometric equation. Coupled reactions of this kind may be distinguished from those that proceed independently of each other; their characteristic lies in the fact that they can be represented by a single chemical equation with definite integral coefficients."[59] At the time, little was known about the reactions that might be operative in the utilization, by energy-requiring processes, of the chemical energy made available by intracellular oxidation. The chemical nature of the reactants, labile intermediates, and products in each "stoichiometric equation" had to be identified, together with the enzymes that catalyze the coupled reactions. The succeeding sections of this chapter offer an account of the search for such knowledge.

Some readers may deplore my omission of mention of the contributions of Clerk Maxwell and Ludwig Boltzmann to the statistical interpretation of Clausius's thermodynamics, or those of the founders of quantum theory and quantum mechanics in the further development of thermodynamic theory. The significance of these contributions to theoretical and experimental physics is undeniable, but thus far their role in the interplay of chemistry and biology has been rather limited, with more impact in the realm of discourse than of experimentation. In particular, much has been made in recent decades of the relation of entropy (or "negentropy") to "disorder" or of "information" to the question "what is life?"[60]

The Activation of Oxygen

For mid-nineteenth-century chemists and chemically minded physiologists who concerned themselves with the problem of biological oxidation, an important question was whether this process depends on changes in the "chemical affinity" of oxygen, or of the substances undergoing oxidation, or the "activation" of both. From chemical experiments such as those of Humphry Davy and Johann Wolfgang Döbereiner on the effect of porous platinum in accelerating the oxidation of organic compounds it had been inferred by Robert Mayer and others that biological oxidations occur at porous surfaces in the organism. With the growing acceptance of the primacy of the tissues, it was suggested that such oxidations are effected at intracellular "catalytic membranes."[61] The finding that aqueous solutions of glucose or pyrogallol are altered by atmospheric oxygen only after alkali had been added led to speculations about the role of alkali in promoting the oxidation of glucose in the animal body and to the idea that human diabetes is a consequence of a deficiency of alkali in the blood.[62] The known tendency of aldehydes to take up oxygen spontaneously led to experiments such as those of Friedrich Wöhler and Theodor Frerichs, who fed benzaldehyde to a dog and found an increased urinary excretion of benzoic acid in the form of hippuric acid (benzoyl-glycine).[63]

During the 1860s there was great interest in the possibility that biological oxidations involve a particular activated form of oxygen. The leader in this development was Christian Friedrich Schönbein.[64] In 1840 he had assigned the name ozone to the peculiar odor emitted during the electrolysis of sulfuric acid, and four years later he noted the same odor when moist phosphorus was exposed to air. Such "ozonized" air appeared to have a strong oxidative capacity, as judged from the liberation of free iodine from a solution of potassium iodide, and Schönbein concluded that ozone represents an active form of O_2. Among the chemical changes effected by ozonized air, but not by O_2, was the rapid bluing of an alcoholic extract of guaiac. This resin from the West Indian plant gouyaca had been used in Europe as a medicine since the sixteenth century (under the name *lignum vitae*), and its chemical constituents had been studied extensively.[65] Its importance in our story arises from the fact that in 1820 Louis Antoine Planche had described the bluing of guaiac by various plant materials, including horseradish, and had attributed this action to an unstable plant constituent he denoted as cyanogen. In the same year, Gioacchino Taddei confirmed these observations and noted a requirement for air, and during the 1840s Schönbein concluded that some constituent of plants such as horseradish or potato can combine with atmospheric oxygen to generate ozone or hydrogen peroxide. The latter oxidant had been prepared by Thenard in 1818 and shown to be decomposed to water and oxygen in the presence of finely divided platinum. Berzelius had already named such reactions as "catalytic" processes, and Schönbein accordingly suggested that the bluing of guaiac involves the participation of a catalyst.[66] In many of his experiments, Schönbein exposed paper strips impregnated with the guaiac tincture to ozonized air and then to reducing agents. After 1850, when ozone became for a time a popular item of medical investigation, such paper strips (usually impregnated with iodide and starch) were widely used to detect its presence in air. In 1861 Schönbein extended the use of paper strips to the separation of chemical substances, but such "paper chromatography" did not gain the favor of the chemists of his time.

For some years Schönbein's chemical contemporaries were skeptical about ozone. His lively imagination brought forth the idea that nitrogen is composed of ozone and hydrogen, and then he made ozone negatively charged oxygen and hydrogen peroxide a compound of water and positively charged oxygen. After Jacques Soret showed that ozone is O_3, it was isolated as a blue liquid by Hautefeuille and Chappuis in 1882 and became a valuable oxidizing agent in organic chemistry. From Schönbein's ebullient letters to Faraday, Berzelius, and Liebig, he emerges as a delightful person, in the "romantic" spirit of early nineteenth-century Naturphilosophie, whose spokesman was Schönbein's teacher Friedrich Schelling. A less romantic aspect of Schönbein's career was his invention of guncotton, announced in 1849.[67]

Despite his shortcomings in quantitative experimentation, Schönbein's empirical observations are part of the groundwork for the subsequent study of biological oxidations. In 1855 he described the bluing of the interior of the mushroom *Boletus luridus* after it had been cut and reported that a colorless alcoholic extract turned blue upon the addition of a press juice of the plant. The similarity of this behavior to that of guaiac suggested to him that plants contain a material that activates oxygen to form ozone and transfers it to oxidizable substances such as the chromogen of Boletus. Schönbein tested aqueous extracts from a wide variety of plant and animal tissues and found many of them to effect the bluing of guaiac in the presence of hydrogen peroxide and to decompose hydrogen peroxide with the liberation of O_2. Because this catalytic activity was lost when the extracts were heated, Schönbein concluded that he was dealing with ferments of an albuminoid nature, and he considered these ozone-producing ferments to be responsible for biological oxidations in both plants and animals.[68]

Schönbein's ozone theory received support from some of the physiologists interested in the problem of animal respiration. In his widely discussed book *Ueber Ozon im Blut* (1862), Alexander Schmidt adopted the theory, and a few years later Willy Kühne stated that "oxyhemoglobin really contains oxygen which gives the ozone reaction, and is a more energetic oxidant than ordinary oxygen."[69] Efforts to demonstrate the presence of ozone or hydrogen peroxide in blood were unsuccessful, however, and after 1870 physiologists gave more attention to other theories of biological oxidation. A prominent opponent of the ozone theory was Eduard Pflüger, who dismissed the claims of Schönbein and Schmidt and insisted that "it is not oxygen, but the protein which undergoes change, when it becomes an integrating constituent of the organism. . . . As soon as this incorporation has occurred it has lost its indifference to oxygen; in other words, it begins to respire, to live."[70]

This idea of "living protein" was offered repeatedly during the second half of the nineteenth century as a ready explanation of biological oxidation. Before Pflüger, Ludimar Hermann attempted to account for the effect of oxygen in promoting the recovery of exhausted excised muscle by postulating the regeneration of a labile albuminoid material. Carl Nägeli extended his theory of fermentation to biological oxidations, while Wilhelm Pfeffer wrote that "The affinities developed in the functional plant suffice completely to cleave the molecule of neutral oxygen, and the maintenance of normal respiration does not require the participation of an active oxygen." As has been noted, the views of Pflüger and Nägeli reappeared several times in new dress, notably in the writings of Nencki and Sieber, Loew, and Verworn. Although these speculations elicited a lively interest among biologists, their experimental fruit was meager.[71] Less favor was shown to an alternative theory, initially offered by Moritz Traube, based on the activation of molecular oxygen by intracellular enzymes.

Whereas Schönbein approached the problem of biological oxidation from his chemical studies on ozone, Traube began with a theory whose formulation shows its debt to the ideas of Justus Liebig: "The ferments concerned with putrefaction and slow combustion [*Verwesung*] are definite chemical compounds formed by the transformation of proteins (perhaps with the cooperation of oxygen). . . . Among the ferments made within and outside the organism there are a) those which can take up free oxygen easily and bind it loosely [*Verwesungsfermente*]; b) those which can accept oxygen that is already bound, that is, can easily deoxidize other compounds [*Reduktionsfermente*]; c) the ferments which can decompose water directly, with the liberation of free hydrogen."[72] In 1874 Traube challenged Pasteur's interpretation of the finding that yeast fermentation occurs in the absence of oxygen: "Upon exclusion of air the fermentation stops after a small part of the dissolved sugar has been decomposed, whereas upon the admission of air . . . the fermentation of the same sugar-containing solution proceeds with the disappearance of the sugar. . . . If, as Pasteur states, the yeast were able to take from the sugar the oxygen it needs for its growth, why does it cease to grow in the absence of air, when most of the sugar is still present?"[73] Traube's answer to this question was indecisive, but he took the opportunity to restate the theory he had advanced in 1858:

> I have shown by means of numerous examples that just as there are substances which . . . can transfer free oxygen to other substances and to effect their oxidation (oxygen carriers, oxidation ferments), there are also substances which can transfer bound oxygen, that is, they can effect reduction of one part and oxidation of the other. If we imagine the sugar molecule to be composed of 2 atomic groupings, a reducible A and an oxidizable B, then the cleavage by the yeast ferment is effected in such a manner that it extracts oxygen from group A (the deoxidized product is alcohol) in order to transfer it to group B, which is thereby burned to carbonic acid.[74]

Two years later, Traube restated his views once again after Hoppe-Seyler had proposed a theory similar in some respects to that of Traube, but which assumed that "all reductions occurring in putrefying fluids are secondary phenomena elicited by nascent hydrogen. . . . Instead of the reduction, there appears during the putrefaction oxidation, which can have as its cause in nothing but the cleavage of the oxygen molecule by the nascent hydrogen . . . whereby the oxygen is converted to an activated state and can then act as a powerful oxidizing agent."[75]

In his paper Hoppe-Seyler did not refer to Traube, and there ensued a lively exchange between them. During 1882–1886, Traube showed that the formula of hydrogen peroxide is HO-OH, that it arises during the oxidation of organic compounds by the addition of hydrogen atoms to O_2, and that in such "autoxidation" there is no cleavage of O_2 to activated atomic oxygen but rather the addition of O_2 to the organic molecule in a loosely bound form, with the formation of what he called a holoxide. Traube's interpretation was adopted during the 1890s by Carl Engler and Aleksei Bakh, as well as by Guido Bodländer, and it received strong support from subsequent organic-chemical studies. In particular, the work of Baeyer and Villiger in 1900 on the autoxidation of benzaldehyde to benzoic acid gave clear evidence for the intermediate formation of a benzoyl peroxide.[76]

During the 1880s another participant in the discussion about biological oxidation was Paul Ehrlich. In extending his studies on the staining of animal tissues by synthetic dyes, Ehrlich injected into mice such dyes as methylene blue, alizarin blue, or indophenol blue, which differed in the ease with which they are reduced to their leuco forms by chemical reagents. From the microscopic examination of the extent to which the various tissues were stained, Ehrlich concluded that they differed in their "oxygen saturation," with heart muscle and brain at a high level (indophenol blue remained oxidized), liver, lungs, and fatty tissues at a low level (even alizarin blue was reduced), and skeletal muscle at an intermediate level (indophenol blue was reduced, but not alizarin blue). Ehrlich also injected rabbits with a mixture of alpha-naphthol and dimethyl-*p*-phenylene diamine (later known as the Nadi reagent) and observed that indophenol blue was synthesized in the tissues and deposited as either the oxidized or reduced form of the dye. His interpretation of the results was in line with the prevailing ideas about the nature of protoplasm, to which he added his "side-chain" theory; in his view, there is present in "the living protoplasm a [chemical] nucleus of special structure which determines the specific, characteristic function of the cell, and to this nucleus there attach themselves as side-chains molecules [*Atome*] and molecular complexes which are of subsidiary importance for the specific cellular activity, but not for life as a whole. Everything points to the view that it is indeed the undifferentiated side-chains which represent the points of initiation and attack in physiological combustion, in that some of them mediate the combustion by the transfer of oxygen, and that the others consume it."[77] Ehrlich also suggested that biological oxidations depend on the alkalization of the tissues, because the leuco forms of the synthetic dyes are oxidized more readily in alkaline solution than in an acidic medium.

In contrast to the medical physiologists of the 1880s, some of the plant physiologists were more receptive to the ideas of Schönbein and Traube. For example, Johannes Reinke endorsed Traube's views and proposed that biological oxidations depend on the autoxidation of specific organic substances with the formation of hydrogen peroxide, which in the presence of intracellular enzymes oxidizes the compounds undergoing metabolic transformation.[78] Indeed, the first oxidative enzymes to be identified as such were derived from plants. In 1883 Hikorokuro Yoshida described the preparation of a "diastatic matter" that promoted the darkening and hardening of the latex of the Japanese lacquer tree, and he showed that this material catalyzes the autoxidation of the plant constituent urushiol. This work was extended during the 1890s by Gabriel Bertrand, who named the enzyme laccase. By 1897 Bertrand had established the existence of several individual members of a group of enzymes, which he called oxidases and which exhibited specificity with respect to the substrate undergoing oxidation by molecular oxygen.[79]

After Oswald Schmiedeberg's student Alfred Jaquet showed in 1892 that aqueous extracts of animal tissues contain catalysts for the oxidation of substances such as benzyl alcohol or salicylaldehyde,[80] medical physiologists who had been skeptical about Traube's claims began to consider seriously the view that discrete enzymes are involved in biological oxidations. A consequence of this change in attitude was the recognition that the oxidation of the chemical constituents of the diet or of the tissues to carbon dioxide and water is not a single explosive combustion, as envisaged by Pflüger, but that it may be a stepwise process, with partially oxidized intermediates. This idea had been advanced by Dumas during the 1840s and later by Hoppe-Seyler, but as was noted earlier, it was stated more explicitly in 1901 by Franz Hofmeister.[81]

Several decades elapsed before the validity of this view, as applied to the intracellular utilization of oxygen, was established. The limitations of the available methods for the isolation and characterization of enzymes made the task difficult, and the extensive literature on biological oxidations before 1914 is full of uncertainties and disputes.[82] For example, the intracellular role of hydrogen peroxide was unclear. The bluing of guaiac in the presence of H_2O_2, observed by Schönbein with plant extracts, had also been observed with leucocytes, and in 1898 it was attributed by Georges Linossier to a "peroxidase." In 1903 Chodat and Bach proposed that oxidase preparations contain an "oxygenase" that binds O_2 to form an organic peroxide, and that this substance participates in a peroxidase-catalyzed oxidation of substrates such as the polyphenols, but attempts to isolate such an oxygenase were unsuccessful.[83] A related problem was presented by the wide distribution of an enzyme that catalyzes the decomposition of H_2O_2 to water and O_2, named catalase by Oscar Loew; the existence of this enzyme provided an explanation for the failure of H_2O_2, a toxic substance, to accumulate in the tissues, but did not clarify matters a great deal.[84]

The preoccupation with hydrogen peroxide as an active participant in biological oxidations led to chemical studies designed to mimic such oxidations in model reactions. Advantage was taken of the observations of Henry Fenton and Wilhelm Manchot that ferrous salts catalyze the oxidation of organic substances by H_2O_2; the Fe^{2+} ion was considered to act as an oxygen carrier through oxidation by H_2O_2 to Fe^{3+}, from which the ferrous form was regenerated through reduction by the organic substance. Examples of such model experiments were provided during 1903–1910 by Henry Dakin and Carl Neuberg on the oxidation of fatty acids and amino acids. Comparison of the products formed in the chemical oxidations with those that appeared in the urine or formed in the tissues provided some useful clues as to the nature of the intermediates formed in the oxidative breakdown of these compounds in the animal organism.[85]

At this point it should be noted that many of the puzzles presented by this early work on the intracellular utilization of O_2 and H_2O_2 were only clarified after 1950, with the recognition that in addition to the partial reduction of O_2 to H_2O_2 and total reduction to H_2O, there is the formation of the oxygen radical O_2^- ("superoxide") as an intermediate in the stepwise reduction of O_2 or the stepwise oxidation of H_2O_2. Some of the biochemical reactions in which the superoxide ion participates will be considered later in this chapter; of particular importance was the discovery during the 1970s by Irwin Fridovich and his associates of the enzymes he called superoxide dismutases, which catalyze the reaction $O_2^- + O_2^- + 2H^+ = H_2O_2 + O_2$.[86]

By 1914 the roster of oxidative enzymes identified in animal tissues had grown considerably. Wilhelm Spitzer claimed to have isolated the catalytic agent responsible for the "Nadi" reaction mentioned above, and he reported that its action was strongly inhibited by cyanide; Joseph Kastle named it indophenol oxidase. In 1905 Richard Burian identified a "xanthine oxidase" that catalyzes the conversion of hypoxanthine to xanthine and of xanthine to uric acid. Battelli and Stern attempted to determine the extent to which the respiratory activity of animal tissues could be accounted for in terms of the catalytic activity of known oxidative enzymes. During the course of this work they found that while some of them, such as xanthine oxidase, could be extracted with aqueous solvents, the ability to oxidize substrates such as succinic acid or citric acid appeared to be bound to labile water-insoluble material. Of particular interest to later investigators were the observations of Battelli and Stern, and also of Horace Vernon, indicating a correlation between the oxygen uptake of various tissues and their apparent content of indophenol oxidase or succinoxidase. For example, the rapidly respiring heart muscle was found to be richest in indophenol oxidase, and one of the poorest in the other intracellular oxidases then known.[87] It is at this stage of the development of knowledge about intracellular respiration that the dominant twentieth-century figure—Otto Warburg—enters our story.

The Role of Iron in Intracellular Respiration

Although the idea that iron is involved in intracellular oxidations had been advanced before 1910—for example, by Spitzer in 1897—it was Warburg's work during the succeeding two decades that established the validity of this view. In 1908 he began a series of researches (interrupted by his service in the German army during 1914–1918) characterized by experimental ingenuity and precision, whose results he presented with an assurance that commanded respect.[88] Warburg's initial quantitative studies, stimulated by the work of Jacques Loeb, dealt with the oxygen uptake of sea urchin eggs and of red blood cells. At first, Warburg used a small-scale version of the manometric apparatus (devised by Thomas Brodie) described by Joseph Barcroft and John Scott Haldane in 1902 to measure the binding of gases by hemoglobin; in time, it came to be known as the Warburg apparatus and was one of the most widely used instruments in biochemical laboratories between the two world wars.[89]

Warburg observed that the mechanical disintegration of red blood cells reduced the respiration to a very low value, while with unfertilized sea urchin eggs such disruption reduced the oxygen uptake only slightly. He examined the effect of ethyl urethane (a narcotic) on the respiration of whole cells and, in line with the theory of Ernst Overton that such agents act as poisons by virtue of their solubility in the lipid-containing cell membranes, Warburg concluded that "the oxidative processes stand in closest connection with the physical state of the lipids." Soon afterward, Warburg abandoned Overton's theory and considered that the effect of narcotics is on "a more physical-chemical catalysis, associated with a specific arrangement of the substances of the cell, on the structure of the cell." Further, he stated, "The chemical aspect of the oxidative process in the disintegrated [sea urchin] egg substance is elucidated to the extent that we can say that it involves the oxidation of lipids in the presence of iron salts."[90] The experimental basis of this assertion was that chemical analysis showed iron to be present in the egg substance, that addition of ferrous or ferric salts to the suspension increased the rate of oxygen uptake, and that extraction with ether gave a material (containing lecithin) that was oxidized by O_2 in the presence of iron salts. Torsten Thunberg had already reported the catalytic effect of iron salts on the autoxidation of lecithin, and Warburg found that this also applied to linolenic acid, a constituent of lecithin. To these observations Warburg added a study of the effect of iron salts in promoting the autoxidation of various aldehydes and thiol compounds such as cysteine. In large part these results confirmed those obtained earlier by chemists (notably Manchot) who had adopted Traube's theory of autoxidation during 1900–1910. Warburg also found, as had others before him—for example, Albert Mathews and Sydney Walker—that cyanide inhibits the autoxidation of cysteine. In offering the theory that "the oxygen consumption in the egg is an iron catalysis; that the oxygen consumed in the respiratory process is taken up initially by dissolved or adsorbed ferrous ions," Warburg did not include these previous findings as evidence.[91]

After World War I, when Warburg resumed work on biological oxidations, the inhibitory action of cyanide, carbon monoxide, or ethyl urethane was a central feature of his theory. During 1921–1924, he reported the results of model experiments in which oxidations were catalyzed by iron-containing charcoals prepared by the incineration of blood, hemin, or impure aniline dyes containing iron salts. In a paper entitled "On iron, the oxygen-transferring constituent of the respiratory enzyme [*Atmungsferment*]," showing that amino acids (cystine, tyrosine, leucine) are extensively oxidized in the presence of such charcoals, and that this action is inhibited by cyanide and ethyl urethane, Warburg concluded that "molecular oxygen reacts with divalent iron, whereby there results a higher oxidation state of iron [that] reacts with the organic substance with the regeneration of divalent iron. . . . Molecular oxygen never reacts directly with the organic substance." In justifying the use of data obtained with the charcoal models for a theory of physiological oxidation, Warburg stated that "the experiments are more than model experiments if one succeeds with the help of iron in transferring the oxygen to the combustible substances of the cell," and he then reiterated the view he had expressed ten years earlier: "Thus there arises that remarkable interplay of unspecific surface forces and specific chemical forces, characteristic for the hemin-charcoal as well as for the living substance. Both systems behave on the one hand like unspecific surface catalyses, on the other as specific metal catalyses. The specific anticatalyst is hydrocyanic acid, the unspecific anticatalysts are the narcotics. . . . The catalytically active substance in hemin-charcoal is therefore iron, but not iron in any form whatever, but iron bound to nitrogen."[92] Warburg offered no suggestion as to the nature of the nitrogenous intracellular material to which iron is bound in biological systems, and his reference to "hemin-charcoal" cannot be interpreted to imply the involvement of heminlike compounds in physiological oxidations, for the hemin used to make the charcoal had been completely incinerated. As will be noted, in 1925 David Keilin called attention to the work of Charles MacMunn on intracellular iron-porphyrin compounds, and assigned to them (Keilin renamed them cytochrome) a significant role in biological oxidations.

After Keilin's first paper on cytochrome, Warburg turned his attention to the inhibitory action of carbon monoxide (CO), long known to be toxic to animals because it displaces O_2 from hemoglobin, but believed to be nontoxic to cellular respiration. John Scott Haldane had stated during the 1890s that "apart from its action in putting the red cells out of action as oxygen carriers carbonic oxide would thus appear to be a physiologically indifferent gas like nitrogen." With Lorrain Smith, Haldane had also found that, like the iron carbonyls (for example, $Fe(CO)_4$) studied by Ludwig Mond, the CO-hemoglobin described by Claude Bernard and Felix Hoppe-Seyler was dissociated by light. This knowledge provided the background for Warburg's most impressive experimental achievement of the 1920s, for which he received a Nobel Prize in 1931. By measuring the efficiency of various wavelengths of light in counteracting the inhibition by CO of the respiration of a yeast, Warburg and his assistant Erwin Negelein determined the absorption (or more accurately, the photochemical action) spectrum of the so-called *Atmungsferment*, which showed a strong band near 420 nm, characteristic of porphyrins (the Soret band), and a weak band associated with the aromatic amino acids of proteins, as well as distinctive bands in the visible region of the spectrum, thus indicating a chemical relation to iron-porphyrin ("heme" or "haem") compounds such as hemoglobin.[93]

Warburg's physicochemical orientation during the 1920s, with its emphasis on "unspecific surface forces," reflected the popularity of colloid chemistry among cell physiologists but neglected the organic-chemical problem of the specificity toward the molecules undergoing oxidation. This aspect of the problem of intracellular respiration, raised in the earlier work of Battelli and Stern, was developed during 1910–1925 by Heinrich Wieland and Torsten Thunberg and focused attention on the specific "dehydrogenases" that catalyze the removal of hydrogen atoms for reaction with molecular oxygen. During the 1920s Warburg disputed the significance of this approach and reiterated his view that "molecular oxygen never reacts directly with the organic substance." Thus after Malcolm Dixon and Sylva Thurlow reported that the action of xanthine oxidase is not inhibited by cyanide, Warburg stated that "the life processes that are specifically inhibited by hydrocyanic acid are heavy-metal catalyses. If the oxidation of hypoxanthine is not inhibited by hydrocyanic acid, then it represents a system that plays no role in the respiration of the liver, a system that is *not* a respiration system."[94] Warburg's comment brings to mind Pasteur, who used the phrase "fermentation proprement dite" in his debates of the 1860s.

In the introduction to a selected collection of his papers on biological oxidation, published in 1928, Warburg stated that "since experience teaches that the catalysts of the living substance—the ferments—cannot be separated from their inactive accompanying material, it is appropriate to forgo the methods of preparative chemistry, and to study the ferments under the most natural conditions of their activity, in the living cell itself."[95] After 1930, however, there was a striking change in Warburg's experimental approach to the problem of intracellular oxidation, for he turned to "preparative chemistry" and thereby made his most decisive contributions to the study of this problem. They will be described shortly, and at this point it is, I believe, appropriate only to call attention to the fact that Warburg was visiting the United States in 1929 and, in connection with his contacts at the Rockefeller Foundation, had undoubtedly learned of John Northrop's crystallization of pepsin in the form of a protein, and probably of James Sumner's earlier crystallization of urease. Before about 1930, however, the views of Richard Willstätter regarding the nature of enzymes held sway. Apart from the general impact of these views, Willstätter entered the debate on biological oxidation through his work during 1918–1925 on the purification of horseradish peroxidase. It was thought from the earlier work of Jules Wolff and of Walther Madelung showing that oxyhemoglobin can act as a peroxidase, though more weakly than the plant enzyme, that peroxidase might be an iron-porphyrin associated with a protein. When Willstätter applied his adsorption techniques to the purification of peroxidase, he found no correlation between the iron content of various preparations and their enzymatic activity and concluded that "iron compounds are closely associated with peroxidase, but the enzyme does not contain iron as an integral constituent."[96] At that time many biochemists shared Hopkins's view that "the production of hydrogen peroxide in the course of dehydrogenations and the simultaneous activity of peroxidases are together responsible for prominent oxidations in the living cell."[97] Consequently, when Willstätter announced that he had obtained a nearly iron-free peroxidase, a blow appeared to have been struck at Warburg's theory of the essential role of iron compounds in biological oxidations.[98] As in the case of other enzymes—for example, invertase—that Willstätter had purified, the chemical tests he used for the detection of iron or protein were much less sensitive than the test for enzyme activity. After the establishment of the protein nature of enzymes during the 1930s, the peroxidases were shown to be iron-porphyrin proteins, in which the nonprotein group is the same as that in methemoglobin.[99]

Cytochrome and Cytochrome Oxidase

David Keilin's famous 1925 paper on cytochrome began as follows: "Under the names myohaematin and histohaematin MacMunn (1884–1886) described a respiratory pigment, which he found in muscles and other tissues of representatives of almost all the orders of the animal kingdom. He found that this pigment, in the reduced state, gives a characteristic spectrum, with four absorption bands occupying the following positions: 615–593/567.5–561/554.5–546/532–511. When oxidized, the pigment does not show absorption bands."[100] As was noted earlier, Charles Alexander MacMunn belonged to the group of British biologists who were stimulated by Stokes's observation on blood to use the spectroscope for the study of animal and plant pigments. In 1873 Henry Sorby, who had invented a microspectroscope for this purpose, called their pursuit chromatology. The instrument was used by Ray Lankester in a systematic search for hemoglobin-like pigments in a large variety of organisms. One such pigment had already been identified by Albert Kölliker and Willy Kühne in mammalian muscle and had been named muscle hemoglobin, but its identity as a distinct heme protein was established later, through the work of Karl Mörner in 1897 and of Hans Günther in 1921; the latter gave it its present name, myoglobin. During the course of his studies, Lankester found a similar pigment in mollusks devoid of blood hemoglobin, and a new respiratory pigment (which he named chlorocruorin) in some marine worms.[101] In such investigations, the recognition of a respiratory pigment was based on the observation of spectroscopic changes when reducing agents were added, and the reappearance of the original "oxidized" form when the solution was shaken with air. It was this approach that led MacMunn to find the four-banded spectrum of the presumed respiratory pigment to which Keilin referred. In a detailed account of his finding, MacMunn wrote:

> In no animal have I succeeded as yet in isolating the histohaematins. In them the coloured constituent occurs united to a proteid in all probability, and hence the difficulty attending attempts at isolation. On the other hand, oxidations and reductions can be brought about in the *solid* organs and tissues. . . . I have proved to my own satisfaction that the *banded condition belongs to the reduced state and the bandless to the oxidised;* but . . . this oxidation and reduction are not as simple as the reduction and oxidation of haemoglobin, for example, the oxygen being apparently more firmly fixed than in the case of oxyhaemoglobin. Thus, from echinoderms to man throughout the animal kingdom, we find in various tissues and organs a class of pigments whose spectra show a most remarkable resemblance to each other. . . . Their bands are intensified by alkalis and enfeebled by acids,

> intensified by reducing agents, and enfeebled by oxidizing agents; they accordingly appear to be capable of oxidation and reduction, and are therefore *respiratory*. . . . Hence the histohaematins are concerned in the *internal* respiration of the tissues and organs of invertebrates and vertebrates. . . . Why a *coloured* constituent should be more useful than a *colourless* one is not clear, but in haemoglobin, hemocyanin, and my echinochrome and Professor Lankester's chlorocruorin, as well as Sorby's aphidein, we have colouring matters which are respiratory pigments.[102]

MacMunn considered myohematin to be related to hemoglobin but different from it, and he offered spectroscopic data for products derived from the muscle pigment in support of his claim that they were different from the hemochromogen and hematoporphyrin identified by Hoppe-Seyler as derivatives of hemoglobin. This claim was strongly criticized by Ludwig Levy, a student of Hoppe-Seyler's, and there was a brief exchange between MacMunn and Hoppe-Seyler. In a footnote appended to MacMunn's final note on the subject, in the journal edited by Hoppe-Seyler, the latter terminated the discussion by reiterating the view that "the investigations that have been reported provide no basis for the assumption that special pigments are present in fresh pigeon muscle."[103] Subsequent printed references to MacMunn's work, until 1925, stressed the probable identity of his myohematin with myoglobin, and no account was taken of his evidence for the presence of histohematins in animal tissues where hemoglobin or myoglobin were unlikely to be present. In a book published posthumously MacMunn wrote: "The Histohaematins and Myohaematins have not found their way into text-books because they do not belong to the ordinary pigments. . . . A good deal of discussion has taken place over this pigment, and the name of Hoppe-Seyler has prevented the acceptance of the writer's views. The chemical position is undoubtedly weak, but doubtless in time this pigment will find its way into the text-books."[104] It was not only Hoppe-Seyler but also MacMunn's fellow chromatologists in England who found it difficult to accept his claims. The reasons have been discussed carefully by Keilin.[105] One was that no hemoglobin derivative was known to exhibit a four-banded spectrum in the reduced state and only faint absorption in the oxidized state. It remained for Keilin to resolve this difficulty in 1925 and to provide evidence justifying the inclusion of MacMunn's hemoglobin-like pigment in "the text-books."

It is important to recall that during the forty years between the work of MacMunn and of Keilin, significant progress had been made not only in the study of intracellular respiration but also in the elucidation of the chemical structure of the porphyrins and of their metal complexes. Thus in addition to proposing in 1913 what was shown during the 1920s by Hans Fischer to be the correct cyclic tetrapyrrole structure of the porphyrins, William Küster had also suggested (in 1910) that in both hemoglobin and oxyhemoglobin the iron is in the divalent (ferro) state, whereas the iron of methemoglobin is in the ferri state. By 1925 the validity of this suggestion had been demonstrated experimentally, largely through the work of James Conant and Louis Fieser.[106] The quotation above from MacMunn's paper is a typical example of the confusion generated in the early literature on respiratory pigments by the use of the terms *oxidation* and *reduction* in the sense in which Stokes had applied them to hemoglobin. Another chemical advance was the clearer definition in 1910, by Walter Dilling and Richard Zeynek, of the term *hemochromogen* as referring to a compound in which a ferroporphyrin is combined with two additional nitrogenous groups; this was firmly established by the work of Robert Hill.[107] Finally, it should be recalled that in the background of these and related studies on the iron-porphyrin compounds was the new inorganic chemistry formulated from 1893 onward by Alfred Werner, who broadened the concept of valency to include the "coordination" of atomic groups about a metal ion.[108] As applied to heme, Werner's work led to the recognition that the ferrous ion is hexacoordinate, with four valencies satisfied by the nitrogens of the pyrroles in the porphyrin ring and the other two available for other interactions with ligands such as O_2 or CO, or nitrogenous compounds such as pyridine or imidazole.

A noted parasitologist with wide-ranging biological interests, Keilin had begun work in this field in the laboratory of Maurice Caullery in Paris. In 1915 he joined George Henry Falkiner Nuttall in Cambridge, where Keilin remained the rest of his life; in 1931 he succeeded Nuttall as the Quick professor of biology and director of the Molteno Institute for Parasitology, located near Gowland Hopkins's Biochemical Laboratory. In the course of his work on the life cycle of the horse botfly *(Gasterophilus intestinalis)*, Keilin observed in 1924, with the aid of a microspectroscope, the appearance of a four-banded absorption spectrum in the muscles of this and other insects, and he also found a similar spectrum with a suspension of yeast. The absorption bands of the "cytochrome" (as Keilin called it) disappeared when the suspension was shaken with air and reappeared shortly afterward. Cyanide inhibited the oxidation of the reduced form, and ethyl urethane inhibited the reduction of the oxidized form, suggesting that the pigment acted as a mediator in the oxidation of cell metabolites by molecular oxygen: "Cytochrome acts as a respiratory catalyst, which is functional in oxidized as well as in partially reduced form. The oxygen is constantly taken up by the pigment and given up to the cells. In the living organism the state of the cytochrome as seen spectroscopically denotes only the difference between the rates of its oxidation and reduction."[109] From the effect of various chemical treatments, Keilin concluded that the four-banded spectrum is associated with three separate hemochromogens, which he named cytochromes a, b, and c. Each of the three bands at the longer wavelengths (605 nm, 565 nm, and 550 nm) was assigned to a different hemochromogen, and the second bands of all three cytochromes were considered to be fused into the single band near 520 nm. Only cytochrome c was stable to heat and extractable with water. Ethyl urethane inhibited the reduction of cytochromes a and c, and the oxidation of cytochrome b. Keilin considered the oxidized forms to contain ferric iron, and he denoted them as parahaematins.

Although it was clear from these observations that the oxidation and reduction of the cytochromes, as seen spectroscopically with insect muscle or yeast suspensions, was intimately related to the intracellular respiration of these biological systems, in contrast to Warburg's Atmungsferment none of the cytochromes appeared to be autoxidizable. Keilin was therefore obliged to assume the existence of an oxidase that catalyzes the oxidation of the ferrocytochromes by molecular oxygen, and he turned to the indophenol oxidase system previously studied by Vernon and by Battelli and Stern. On the basis of the light-sensitive inhibition of yeast indophenol oxidase by carbon monoxide, Keilin concluded in 1927 that "Warburg's respiratory ferment is a polyphenol or indophenol oxidase system . . . [and] the oxidase systems revealed by the indophenol test belong to the respiratory catalysts essential for the oxygen uptake of the living yeast cells."[110] In subsequent experiments with a heart-muscle preparation that effected the cytochrome-dependent oxidation of succinate by molecular oxygen, Keilin found that cyanide and carbon monoxide inhibited both the conversion of the "Nadi" reagent to indophenol blue and the oxidation of cytochrome. He therefore considered indophenol oxidase to be the enzyme responsible for the oxidation of reduced cytochrome by oxygen. He later modified this view and used the term *cytochrome oxidase* to denote his equivalent of Warburg's Atmungsferment.[111]

Warburg's response was to question the significance of Keilin's findings: "Whether the MacMunn hemins lie on the normal path of respiration . . . is a question we cannot answer today. The available spectroscopic observations are also consistent with the view that the MacMunn hemins are only reduced in the cell when, as a consequence of oxygen deficiency, the concentration of the activatable substances exceeds the physiological level."[112] Similar doubt had been expressed by Anson and Mirsky:

> Some properties of cytochrome are known qualitatively; it is known, for instance, that it can be oxidized and reduced. But what cytochrome actually

> does accomplish in normal respiration still remains to be demonstrated. . . . It is of course conceivable that the ferment which combines with CO is only one catalyst in a complicated catalytic system of which cytochrome and glutathione are essential parts. . . . The chromatologists have always called every pigment which can be oxidized and reduced a respiratory pigment—why, it is not clear. There are many colorless substances in the cell which can be and probably are oxidized and reduced. This property is not taken as proof that they are respiratory substances.[113]

Within a few years, not only the cytochromes but also some "colorless substances" were to be established as participants "in a complicated catalytic system," while glutathione (about which more later) was to fall from the favor it had enjoyed during the 1920s. Moreover, at the end of the 1930s, Keilin and his associate Edward Francis Hartree demonstrated the existence of a CO-sensitive, autoxidizable cytochrome ("cytochrome a_3") whose distinctive absorption band near 600 nm lies close to that of cytochrome a. The absorption spectrum of the CO-complex of cytochrome a_3 was found to agree with the photochemical action spectrum of Warburg's Atmungsferment.[114] It was not until the 1950s, however, that Britton Chance showed the compound of cytochrome a_3 and CO to be dissociated by light.[115]

Like the cytochromes a and b, cytochrome oxidase (clearly related to cytochrome a_3) was not extractable from animal tissues with ordinary aqueous solvents, and it was not until the 1950s that methods began to be devised, using detergents, to disperse such water-insoluble cell constituents. By that time, investigators of the problem of intracellular respiration in animal tissues had begun to focus their attention on the long-known mitochondria, described by Richard Altmann in 1890 and later studied by the cytologist Robert Bensley. During the 1940s Albert Claude had isolated them by means of differential ultracentrifugal centrifugation, and George Hogeboom, his colleague at the Rockefeller Institute, found that mitochondria are the principal sites of cell respiration.[116] Soon afterward, Eugene Kennedy and Albert Lehninger showed that several known processes of oxidative metabolism (the citric acid cycle, fatty acid oxidation, and oxidative phosphorylation, about which more later) are effected in isolated mitochondria from animal tissues.[117]

More recent work has demonstrated that cytochrome c is the substrate of the cytochrome oxidase identified by Keilin and Hartree and that the oxidase is an aggregate of a molecule of cytochrome a, a molecule of cytochrome a_3, two copper ions, several proteins, and bound phospholipids.[118] At the present writing, the investigation of the structure and mode of action of this complex assembly is still in active progress; of particular interest is the involvement of a copper protein in the electron transfer from reduced ferrocytochrome c in the overall reaction $4H^+ + 4e^- + O_2 = 2H_2O$.[119]

As was noted above, Keilin found cytochrome c to be extractable by means of aqueous solvents. During the 1930s and 1940s the work of Karl Zeile and especially of Hugo Theorell led to the purification of this cytochrome, and its characterization as a heme protein having a molecular weight of about 13,000, with the porphyrin unit joined by covalent linkage to two cysteine units of the protein. After 1955 cystalline preparations were obtained from a large variety of biological sources, and their amino acid sequences were determined. This new knowledge permitted the closer study of the oxidation-reduction behavior of cytochrome c, and comparisons of the amino acid sequences for different biological forms have provided new insights into problems of biological evolution.[120] In this connection, the following statement by Ray Lankester in 1871 is worthy of notice: "The chemical differences among various species and genera of animals and plants are certainly as significant for the history of their origins as the differences in form. If we could define clearly the differences in molecular constitution and function of different kinds of organisms, there would be possible a more illuminating and deeper understanding of the question of the evolutionary relations of organisms than could ever be expected from morphological considerations."[121] As regards the finding that a copper protein is a component of cytochrome c oxidase, it should be noted that for a time Keilin considered this possibility, and that copper proteins had long been known to be present in living organisms. Examples are hemocyanin (the respiratory pigment of some marine animals), hemocuprein (found in blood red cells), polyphenol oxidase (from potato), and azurin (from bacteria). Moreover, nutritional studies showed that dietary copper promotes the incorporation of iron into hemoglobin.[122] As has been mentioned, in addition to iron and copper ions, other metals (cobalt, manganese, magnesium, molybdenum, nickel, selenium, vanadium, zinc) play significant roles in biological processes when bound to proteins. In the coordination complexes of metal ions with proteins, the amino acid side-chain most frequently encountered has been the imidazolyl group of one or more histidine units, although other groups, such as the sulfhydryl group of a cysteine unit, also serve as metal-binding ligands.[123] The development of knowledge about the chemical nature and physiological function of several of these other kinds of metalloproteins will be considered later.

Two other enzymatic oxidation-reduction systems related to the cytochromes described by Keilin, and of more recent vintage, should be mentioned. During the 1940s a cytochrome of the c type was identified (first named cytochrome e, later c_1), and subsequent work showed that it participates, with cytochrome b and an "iron-sulfur" protein, in a mitochondrial complex that catalyzes the reduction of cytochrome c. In this process, a widely distributed derivative of the quinone-hydroquinone ("ubiquinone-ubiquinol") system is involved; more will be said about the role of that oxidation-reduction system.[124] The other group of cytochromes which has received much attention is that of the heme proteins present in microorganisma and in animal tissues that are termed cytochrome P-450 because they form derivatives with carbon monoxide whose principal (Soret) absorption band is near 450 nm. Whereas cytochrome c oxidase catalyzes an autoxidation reaction in which both atoms of O_2 are reduced to water, cytochrome P-450 is a "mixed-function" oxidase that utilizes one of the two atoms of O_2 to perform reactions such as the introduction of a hydroxyl group into a substrate, as well as a remarkable variety of other biochemical reactions.[125]

The Dehydrogenases

In 1912 Heinrich Wieland published a series of papers on the catalysis by finely divided palladium of the oxidation of various organic compounds. He offered evidence for the view that the oxygen entering the compound (as in the oxidation of an aldehyde RCHO to the corresponding carboxylic acid RCOOH) is not derived from molecular oxygen, and that there was no need to postulate the activation of O_2 in such processes. Instead, he proposed that oxidations of this type involve the catalytic labilization of the organic compound, with the activation of its hydrogen atoms, which are transferred to O_2 to form hydrogen peroxide. Thus, in the case of the oxidation of an aldehyde to an acid, Wieland assumed an initial addition of water, followed by the catalytic dehydrogenation *(Dehydrierung)* of the aldehyde hydrate:

$$RCHO + H_2O \rightarrow RCH(OH)_2 \xrightarrow{+O_2} RCOOH + H_2O_2$$

In support of this theory of "hydrogen activation," as it came to be called in the ensuing debate, Wieland reported that in the absence of O_2, substances known to be readily reduced in presence of palladium or platinum (for example, quinone, methylene blue) can serve as acceptors in the palladium-catalyzed oxidation of organic substances and can be converted into their hydrogenated forms (hydroquinone, leucomethylene blue).[126]

The reports of these chemical studies were soon followed by an important paper in which Wieland stated that he had begun work on biological oxidations "with the aim of testing whether the dehydrogenation theory could contribute to the understanding of the mechanisms of these largely unexplained reactions. As is well known, the almost universally accepted view of biologists and chemists who have worked on this subject is that the intracellular oxidations and combustions owe their rapid rate to the participation of oxygen-activating ferments."[127] He proceeded to show that his model system can effect, in the apparent absence of O_2, the oxidation of glucose to CO_2, the conversion of lactic acid to pyruvic acid [CH_3-CH(OH)-COOH → CH_3-CO-COOH], or the oxidation of polyphenols such as pyrogallol. To demonstrate the relevance of his theory to oxidations in biological systems, Wieland reported that the oxidation of ethanol (CH_3CH_2OH) or of acetaldehyde (CH_3CHO) to acetic acid (CH_3COOH) by "acetic acid bacteria" *(Acetobacter)* can proceed anerobically provided that quinone or methylene blue is present as a hydrogen acceptor. Wieland's attempt to prepare from this organism a cell-free press juice which could effect this process, however, was unsuccessful.

Wieland's work brought greater clarity to a problem that had concerned physiological chemists for several decades—namely, the question of the existence of separate "reduction ferments" as counterparts of the "oxidation ferments." As noted earlier, the nineteenth-century chemists and biologists who were concerned with the problem of animal respiration tended to emphasize either the activation of oxygen or some special property of protoplasmic protein that made it more reactive toward oxygen. By 1861 the work of Louis Pasteur on anaerobic fermentation had shown that the long-known formation of hydrogen-rich compounds such as methane (CH_4) during putrefaction is effected by living microorganisms, and such reductions had to be viewed as biological processes. Moreover, by 1876 several metabolic transformations, such as the conversion of hemoglobin into bile pigments, were recognized to be reductions, and Felix Hoppe-Seyler was led to assume the occurrence in animal tissues of "processes in which organic substances are changed and cleaved by the action of water in a manner similar to that found in the process of putrefaction."[128] To explain such reductions, he called attention to the earlier studies of Gottfried Osann and Thomas Graham.

Osann found that the hydrogen gas released at a platinum electrode during the electrolysis of dilute sulfuric acid had a stronger reducing action than that of ordinary hydrogen. He considered this "active" hydrogen to be analogous to Schönbein's ozone and named it "ozone-hydrogen."[129] Graham later reported that "hydrogen (associated with platinum) unites with chlorine and iodine in the dark, reduces a persalt of iron to the state of protosalt [ferric to ferrous], converts red prussiate of potash to yellow prussiate [potassium ferricyanide to the ferrocyanide], and has considerable deoxidizing powers. It appears to be the active form of hydrogen, as ozone is of oxygen."[130]

We saw earlier that Hoppe-Seyler's exchange with Traube on the relative importance of hydrogen activation and oxygen activation in biological oxidations had led, by the end of the nineteenth century, to the preference for Traube's theory. Also, as little interest was shown in the active hydrogen theory of Osann and Graham as in the ozone of Schönbein. Nevertheless, the problem of the mechanism of biological reductions was a continuing subject of discussion during 1880–1910. Ehrlich's studies, reported in 1885, on the intracellular reduction of various synthetic dyes were followed by a succession of others indicating the presence, in animal tissues, of substances that catalyze reductions. For example, Joseph de Rey-Pailhade observed the reduction of sulfur to hydrogen sulfide by extracts of yeast and of animal tissues and gave the name *philothion* to the substance he thought to be responsible for this reaction. He concluded that "Acting like an enzyme [*diastase*], it serves to add one more proof of M. Berthelot's theory of fermentation. It is the first known example of a substance extracted from a living organism and possessing the property of hydrogenating sulfur."[131]

In the face of the uncertainty about the nature of enzymes, this claim was ill-founded and was soon disproved, but similar reports came from other investigators. At the end of the century, Emile Abelous and Ernest Gérard reported the presence, in extracts of animal tissues, of soluble ferments that catalyze the reduction of nitrate to nitrite, of nitrobenzene to aniline, and of methylene blue to its colorless leuco form. They concluded that "in the aqueous maceration from horse kidney there co-exist a soluble reducing ferment and a soluble oxidizing ferment,"[132] but the view that such reductions are enzymatic in nature was disputed by Arthur Heffter, who considered these reactions to involve the sulfhydryl (SH) groups of protoplasmic proteins.[133] The SH group of cysteine was known to be autoxidizable, with the formation of the disulfide (-SS-) group of cystine, and Heffter thought that this group was reduced by suitable metabolites without enzymatic catalysis, thus making the assumption of the existence of reducing enzymes unnecessary. By 1909, however, it was clear from the work of Franz Schardinger and Richard Trommsdorff that milk contains a heat-labile factor that catalyzes the rapid reduction of methylene blue or indigo sulfonate to their leuco forms, provided that an aldehyde such as acetaldehyde is added. Also, Georg Bredig described a model for this enzymatic catalysis; in the presence of colloidal palladium, methylene blue was reduced by aldehydes.[134] Once again the ideas of Osann, Graham, and Hoppe-Seyler about the activation of hydrogen came to the fore, principally in the writings of Aleksei Bach. He proposed, in analogy to his earlier theory of the nature of oxidases, that the reducing system of animal tissues consists of the Schardinger enzyme (the counterpart of peroxidase) and a tissue component replaceable by aldehydes (the counterpart of hydrogen peroxide). In Bach's scheme the function of the enzyme was to activate the hydrogen of a hypothetical "perhydride" form of water.[135]

The studies on the Schardinger enzyme, as well as other enzymes to be mentioned later, thus called attention to the possibility of the enzyme-catalyzed transfer of hydrogen atoms. Evidence for this view came from the studies of Battelli and Stern and of Hans Einbeck on the enzyme-catalyzed oxidation of succinic acid, with the demonstration that this compound (HOOC-CH_2-CH_2-COOH) is converted to fumaric acid (HOOC-CH=CH-COOH). Moreover, Jacob Parnas reported that an "aldehyde mutase" present in animal tissues can catalyze the conversion of an aldehyde into the corresponding alcohol and acid, indicating that this chemical process (described by Cannizzaro in 1853) involves a "dismutation" in which one molecule of the aldehyde acts as a hydrogen acceptor and another molecule of the aldehyde (plus H_2O) acts as a hydrogen donor.[136] Thus when Wieland advanced his dehydrogenation theory, he could write:

> If oxidative processes are considered to be dehydrogenations, as has been definitely shown for a few important cases, then they include at the same time a reductive process, since the hydrogen activated by the ferment must be taken up by some acceptor. . . . Because of this relationship, the so-called reduction ferments that have frequently appeared in the literature lose their separate status if it can be shown that their characteristic reductive action, for example, the decolorization of some dye by some substrate, can also be effective in the hydrogenation of oxygen, [or] in the sense of the views held up to now, the "reductase" can also function as an oxidase.[137]

After offering evidence to show that this criterion was met by the Schardinger enzyme, Wieland concluded: "One can easily imagine that the course of a dehydrogenation process in the direction of the hydro product will depend on the nature of the available acceptors, so that the dehydrogenating ferments, which undoubtedly will have the most varied specificity with respect to both the dehydrogenating and hydrogenating agents, sometimes will find the most suitable hydrogen acceptor in the substrate itself (mutase), at other times in molecular oxygen (oxidase), or finally in a dye, in nitrate, or similar substances (reductase)."[138] In writing of the "dehydrogenating ferments," Wieland used the term *Dehydrase;* this was later replaced by *dehydrogenase,* so as to avoid confusion with enzymes that catalyze the dehydration (removal of the elements of water) of their substrates.

Comparison of Wieland's theory with that of Warburg, as they were expressed in 1914 and during 1922–1924, when their debate reached its peak, provides a striking example of the way the chemical mechanism of a biological process could be formulated in entirely different terms at a time when the extent of the chemical complexity of the process was unknown. As was suggested above, Warburg's theory reflected the influence of the new colloid chemistry on the general physiologists of that period; although trained in organic chemistry (he had worked with Emil Fischer on amino acids), Warburg's interpretation of his observations on cellular respiration was a physicochemical theory that emphasized catalysis at surfaces. On the other hand, Wieland's formulation was in the tradition of the best organic chemistry of his time, but the limitations of the available knowledge about the nature of enzymes and of the essential participants in biological oxidations made his theory no more convincing. The choice was between two kinds of artificial models of the respiratory apparatus of living cells: the iron-containing charcoals of Warburg as models of his hypothetical Atmungsferment or the palladium system of Wieland as a model of the activation of hydrogen in the presence of specific enzymes. The post–World War I debate between these two great scientists was largely about the shortcomings of the other's models. Wieland disputed Warburg's finding that amino acids are oxidized in the presence of iron-charcoal by noting that "in the presence of oxygen, amino acids can be dehydrogenated, with the release of ammonia, by palladium black," while Warburg disputed Wieland's contention that the cyanide inhibition of respiration was a consequence of the inhibition of catalase.[139] As before and since, in debates of this sort both models were inadequate, although the discussion about them enlivened the scientific scene and stimulated fruitful experimentation. Indeed, later work showed that if impurities were removed from palladium black, the catalytic dehydrogenation of hydroquinone reported by Wieland could no longer be observed.[140] This blemish on a brilliant scientific career must be seen in the context of Wieland's elucidation of the structure of the bile acids and of the pigments (pterins) of butterfly wings; the latter subsequently turned out to be functional units of the widely distributed and physiologically important folic acids.

For our story, the most significant consequence of Wieland's formulation of the dehydrogenation theory was perhaps the stimulus it provided to the work of Torsten Thunberg during 1917–1920. To examine the validity of the theory, Thunberg devised a technique in which thoroughly washed minced tissue (such as frog muscle) was suspended in a solution containing methylene blue, which was not decolorized by the washed tissue. The suspension was placed in a special test tube from which air could be evacuated (later known as a Thunberg tube) with a side bulb into which a solution of an organic compound was placed. After removal of air, the side-bulb contents were tipped into the suspension, and the time required for the decolorization of the dye was noted. In this manner, Thunberg found that several organic acids (among them lactic acid, succinic acid, malic acid, citric acid, α-ketoglutaric acid, glutamic acid, alanine) promoted the rapid reduction of methylene blue, and concluded that these substances were dehydrogenated by specific dehydrogenases, in accordance with Wieland's theory.[141] Thunberg offered some evidence, based on differences in the apparent stability of the various dehydrogenase activities to freezing or to heat, indicating that the dehydrogenation of the individual substrates was attributable to separate enzymes, and other investigators added supporting evidence in favor of this view, but it was not until many years later that its validity was established through the crystallization of the dehydrogenases as well-defined catalytic proteins.

Despite the deficiencies in the characterization of the dehydrogenases, many investigators accepted the Wieland-Thunberg theory during the 1920s and sought to reconcile it with the evidence in favor of the activation of oxygen. Thus, from the fact that cyanide inhibits the dehydrogenation of succinic acid by washed muscle preparations when molecular oxygen is the hydrogen acceptor, but not when methylene blue is used in the Thunberg method, Alfred Fleisch and Albert Szent-Györgyi concluded that "activation of both hydrogen and oxygen is necessary at least for the oxidation of succinic acid" and that "in cellular oxidation the *activated hydrogen* is burned by the *activated oxygen*, but in stages. In the terminology of the hydrogen activation theory this means that molecular oxygen is not a hydrogen acceptor; the biological hydrogen acceptor is the oxygen activated in Warburg's system."[142] Also, Gowland Hopkins suggested that in biological oxidation "hydrogen is transported, not directly from primary donators to oxygen, but by stages. It would appear that the path may in a given case be smoothened, and so the velocity of transport increased, by the intervention of a substance which can act alternately as an intermediate acceptor and donator. Such a substance acts therefore catalytically as a carrier of hydrogen."[143] There had been anticipations of this idea. For example, Vladimir Palladin concluded from his studies on plant respiration that certain polyphenols, acting as "respiratory pigments," remove hydrogen from substances undergoing oxidation, and the reduced pigments are then reoxidized under the catalytic influence of the oxidases.[144]

One of the sources of confusion among biologists who debated the question of the separate existence of "reductases" and "oxidases" was the absence of an agreement on the chemical meaning of the terms *oxidation* and *reduction*. By 1914 biological *oxidation* came to include the removal of hydrogen atoms from a molecule, and the ancient term *reduction* (originally used in the sense of the "revivification" of a metal from its ore), could mean not only the removal of oxygen from a compound but also the addition of hydrogen. It was not until the 1920s, after the emergence of the electronic theory of valency and the application of electrochemistry, as enriched by thermodynamic theory, that the intracellular oxidation of a metabolite came to be considered as the loss of electrons (with or without the gain of oxygen or loss of hydrogen), and reduction as the gain of electrons (with or without the loss of oxygen or gain of hydrogen). The background of this development has been sketched earlier, and mention was made of the studies of Biilmann, Clark, Conant, and Michaelis on the oxidation-reduction potentials of several dyes that were then used as "indicators" for the determination of the potentials of metabolite systems (such as succinate-fumarate) with which the dyes reacted under the catalytic influence of a suitable enzyme preparation. Apart from the availability of indicators having widely different normal potentials (according to the convention employed, a much more positive potential for system A than for system B indicates that the reaction $A_{ox} + B_{red} = A_{red} + B_{ox}$ has a tendency to go far to the right), these studies during the 1920s emphasized the limitations inherent in the formulation of Wieland's theory in terms of hydrogen transfer. As Clark put it, the reduction of the dyes "consists essentially in the transfer of an *electron pair* accompanied or not accompanied by hydrogen ions according to the state of the acid-base equilibrium in the solution."[145]

Among the leading biochemists of the 1920s who were somewhat skeptical about this approach to the study of biological oxidation was Gowland Hopkins:

> All students of the subject must indeed be grateful to Mansfield Clark and his colleagues for their successful endeavour to provide a number of reducible dye-stuffs for which the oxidation-reduction potentials have been accurately determined electrometrically. These can be arranged in a series and used for determining the reducing power of a given solution or system, and for stating it in definite terms. Again, however, discrimination is necessary. Whether a given transference of hydrogen can occur, or cannot occur, is a matter determined by potentials; but because in the study of tissue oxidations we seldom deal with equilibria, observing, rather, relative velocities, we must remember that though thermodynamics decide that a given reaction may, or must occur, yet it may proceed with a velocity too slow to observe. Kinetics may then, as Wieland has somewhere said, appear to leave thermodynamics in the lurch![146]

This comment gains interest from the fact that Juda Quastel and Margaret Whetham, working in the Cambridge Biochemical Laboratory, showed in 1924 that, under anaerobic conditions, a resting suspension of *Escherichia coli* catalyzes the reaction

$$\text{succinate} + \text{methylene blue} \rightleftharpoons \text{fumarate} + \text{leuco-methylene blue}$$

in both directions to the same equilibrium point. Shortly afterward, Thunberg reported a similar experiment, using the succinic dehydrogenase present in washed heart muscle as the catalyst; from Clark's value for the oxidation-potential of the methylene blue system, Thunberg estimated a potential for the succinate-fumarate system that was consistent with the equilibrium data of Quastel and Whetham. In these studies the extent of the reduction of methylene blue was determined colorimetrically. The subsequent important work of Jørgen Lehmann showed that it is possible to determine the oxidation-reduction potential of the succinate-fumarate system at a metallic electrode, provided there is present, in addition to succinic dehydrogenase, a small amount of methylene blue acting as a "mediator" between the electrode and the electromotively inactive metabolite system. Moreover, Henry Borsook and Hermann Schott calculated the free-energy change in the reduction of fumarate to succinate by a totally independent method, using thermal and supplementary data, and obtained a value in good agreement with those calculated from the earlier data of Quastel, Thunberg, and Lehmann.[147] It was clear, therefore, that the dehydrogenase accelerates the attainment of equilibrium in an oxidation-reduction reaction but does not change the equilibrium constant, in agreement with the principle stated thirty years earlier by van't Hoff for enzymic catalysis in general. Although it was recognized that "kinetics may appear to leave thermodynamics in the lurch," the data obtained after 1930 on the oxidation-reduction potentials of systems considered to play a role in biological oxidations, and on the free-energy changes in the reactions catalyzed by enzymes, profoundly influenced thought about the nature of intracellular respiration. To this should be added the important finding in 1931, independently by Bene Elema and by Leonor Michaelis, that in the oxidation-reduction of some organic compounds the two-electron transfer specified by Mansfield Clark occurs in one-electron steps, with the intermediate formation of a "semiquinone."[148] This discovery provided a clue to the mechanism whereby the oxidation of two-electron metabolite systems is linked to the one-electron heme-protein systems and molecular oxygen.

Intracellular Electron Carriers

We have already considered some of Keilin's achievements, through his work on the cytochromes, in the identification of links in intracellular respiration. During the 1920s other substances were also thought to be possible participants in this process. One of them was isolated from yeast by Hopkins, who used the nitroprusside test to follow the course of the purification. The substance appeared to be identical with the "philothion" of Rey-Pailhade and to be a peptide composed of glutamic acid and cysteine. Hopkins wrote: "Until the constitution is finally established it may be premature to suggest a name for the substance. But, provisionally, for easy reference, the name *Glutathione* will perhaps be admissible. It leaves a link with the historic *Philothion,* has the same termination as in *Peptone,* which has long served as a name for the simpler peptides, and is a sufficient reminder that the dipeptide contains glutamic acid linked to a sulfur compound."[149] Glutathione (GSH) was found to be autoxidizable in the presence of metal ions, and the resulting cystine derivative (GSSG) was readily reduced to GSH by various tissues. The addition of GSSG to washed muscle tissue promoted the reduction of methylene blue, and the glutathione system thus appeared to "possess what are essentially catalytic properties."[150] Although the tissue constituent that caused the reduction of GSSG turned out to be stable to heat, indicating that it was not an enzyme, Hopkins nevertheless concluded that "while there is much that is obscure in the phenomena involved, the facts in our opinion fully justify the claim that a non-enzymic oxidation-reduction system represented by the thermostable residue *plus* the sulfur grouping of glutathione actually functions in the cell."[151]

Hopkins's evidence for the structure of glutathione as γ-glutamyl-cysteine was questioned by George Hunter and Blythe Eagles, and a reinvestigation led Hopkins to conclude that it is a tripeptide composed of glutamic acid, cysteine, and glycine;[152] its structure as γ-glutamyl-cysteinyl-glycine was confirmed after 1930 by synthesis. Despite extensive work in many laboratories, notably that of Alton Meister, the role of this widely distributed peptide in cell metabolism remained obscure for many years, and although it has been assigned a variety of biochemical functions,[153] participation in the electron transfer from metabolites to molecular oxygen is not one of them.

Another candidate of the 1920s was discovered by Albert Szent-Györgyi in the form of a "reducing factor" whose oxidation "is a reversible one, and under the given conditions this factor plays the role of a catalytic hydrogen carrier between the peroxidase and other oxidising or reducing systems."[154] Although the factor was readily reduced by glutathione or by the thermostable "fixed SH" of animal tissues, it was not reduced by succinate in the presence of succinic dehydrogenase, and it was therefore unlikely that the factor is a participant in the process of intracellular respiration. These studies, however, led Szent-Györgyi to the isolation of the reducing factor in crystalline form from extracts of adrenal cortex and of plants (orange, cabbage), and he identified the compound as an acidic carbohydrate (a "hexuronic acid"). He also noted that "the reducing properties of plant juice have repeatedly attracted attention, especially from students of vitamin C. . . . The reducing properties of lemon juice have been the object of a thorough study by Zilva, . . . who established interesting relations between vitamin C and the reducing properties of the plant juice. The main reagent employed by Zilva (1928) was phenol indophenol. Indophenol blue is readily reduced by the hexuronic acid, so that it is probable that it was this substance which has been studied by Zilva."[155] Some years later Szent-Györgyi wrote about the naming of his newly found substance that "I called it ignose, not knowing which carbohydrate it was. This was turned down by my editor. 'God-nose' was not more successful, so in the end 'hexuronic acid' was agreed upon. Today the substance is called ascorbic acid."[156] The identity of hexuronic acid and vitamin C was not established until about 1933, when Charles Glen King reported the isolation of the antiscorbutic substance in crystalline form from lemon juice, and Szent-Györgyi tested his preparation for vitamin activity. He found the Hungarian red pepper to be an exceptionally rich source of the substance. Also, by 1933 its chemical structure had been completely determined and its synthesis effected, largely by Walter Norman Haworth and his associates.[157] Later work showed that in the presence of O_2 and Fe^{2+}, ascorbic acid activates the enzyme (prolyl hydroxylase) that catalyzes the hydroxylation of peptide-bound proline units in the precursor (procollagen) of connective tissue collagen and of plant cell–wall proteins.[158]

The determination of the nature of vitamin C came during an exciting decade, when organic chemists exhibited the power of their analytical and synthetic techniques in the rapid elucidation of the chemical structure of several vitamins. The term *vitamine* (the final e was later dropped) was suggested in 1912 by Casimir Funk to denote what Hopkins had named accessory factors; the term *food hormones* had also been proposed. Interest in these dietary factors grew out of the work of Christiaan Eijkman during 1890–1897 on the effect of a constituent of rice husks in curing a neurological disease of birds (polyneuritis) that resembled the human disease known as beri-beri. In 1906 Hopkins stated that "no animal can live upon a mixture of pure protein, fat and carbohydrate, and even when the necessary inorganic material is carefully supplied the animal still cannot flourish. . . . In diseases such as rickets, and particularly in scurvy, we have had for long years knowledge of a dietary factor; but though we know how to benefit these conditions empirically, the real errors in the diet are to this day quite obscure."[159] An anticipation of these views appeared in a paper by Nikolai Lunin, who studied the effect of milk on the survival of mice fed artificial diets. He concluded that "since as shown by the above experiments they were not able to survive on albuminates, fat, sugar, salts, and water, it follows that in milk there are other substances besides casein, fat, lactose, and salt that are essential for nutrition. It would be of great interest to track them down and to study their significance in nutrition."[160] It was not until after the work of Eijkman, however, that systematic studies were conducted along this line.[161] Thus when Axel Holst and Alfred Fröhlich found in 1907 that the guinea pig is susceptible to scurvy, a test animal was provided for the study of the antiscorbutic factor in plant juices, whose curative value was known since James Lind's *A Treatise on the Scurvy,* published in 1753.

An account of the events leading to the identification of the various vitamins is beyond the scope of this chapter. Briefly, by 1925 five classes of vitamins had been identified and labeled alphabetically. The water-soluble antineuritic factor was termed vitamin B, vitamin A being the fat-soluble factor that prevented eye disease (xerophthalmia) and promoted the growth of the rat. Because the water-soluble antiscorbutic factor differed from both of these, it was named vitamin C. When it was shown that the fat-soluble antirachitic factor is different from vitamin A, it was denoted vitamin D. A distinct fat-soluble antisterility factor (for the rat) appeared soon afterward and was termed vitamin E. As work progressed on the so-called vitamin B, it was recognized to be composed of many factors; after 1927 the antineuritic factor was known as vitamin B_1. This vitamin, along with other members of the B complex, will be mentioned again. Two general comments may be made here, however: (1) The availability of a suitable biological system (animal or microbial) for testing the activity of fractions obtained in the purification of a vitamin usually led to its isolation in pure form (or to its disappearance from the list) and to the elucidation of its chemical structure through controlled degradation and synthesis in the laboratory; apart from the highly developed state of organic chemistry, technical advances such as the microanalytical methods introduced by Fritz Pregl played a major role, for at first only small amounts of the isolated vitamins were available for chemical study. (2) The elucidation of the chemical structure of the vitamins came at a time when some of them were recognized to be components of electron-transfer systems in enzyme-catalyzed biological oxidations. In this development the achievements of Otto Warburg during the 1930s were decisive. By 1940 his work had brought together into a single sequence of electron transport the dehydrogenases of Wieland and Thunberg, the cytochromes of Keilin, and his own autoxidizable Atmungsferment.

The starting point of this aspect of Warburg's work was the report by Guzman Barron and George Harrop that in the presence of glucose the normally very slow oxygen uptake by mammalian red cells is greatly increased by the addition of small amounts of oxidation-reduction indicators such as methylene blue, and that the glucose is oxidized in this process. As will be seen later in this chapter, it was known during the 1920s that the metabolic breakdown of glucose involves the formation of hexose phosphate, and Barron and Harrop suggested that in their oxidation system "the principal point at which methylene blue acts is upon the oxidation of hexose phosphate. . . . As to the exact nature of the methylene blue effect little may be said. It is conceivable that it acts as a coenzyme or a catalyst, rendering the substrate (hexosephosphate?) more sensitive to the action of molecular oxygen. On the other hand, one might consider that methylene blue plays in this system the role ascribed to iron in the oxidations produced by Warburg with his charcoal model."[162] Warburg proceeded to study the mechanism of the methylene blue effect. First, he examined the possibility that the dye oxidized the ferrous iron of oxyhemoglobin to the ferric iron of methemoglobin and, in line with his views of the 1920s he concluded that "in the methylene blue respiration the oxidation of sugar is nothing but an oxidation by the hemin iron, namely the iron of methemoglobin," that the reaction between methylene blue and glucose is "a surface reaction. Methylene blue, which is adsorbed on the surfaces of the blood cells, forms methemoglobin on the surfaces—that is, at the reaction sites—and therefore a small methemoglobin concentration suffices during methylene blue catalysis to cause a large oxidative effect," and that closer investigation had shown that here also "there is heavy-metal catalysis that closely resembles the normal catalytic actions of the living substance."[163]

Shortly thereafter, however, Warburg found that although disruption (cytolysis) of the erythrocytes abolished their ability to oxidize glucose in the presence of methylene blue, the so-called Robison ester (glucose-6-phosphate) was readily oxidized by such cell-free preparations. Upon fractionating the constituents of the fluid from the cytolyzed red cells (after removal of the cell debris by centrifuging the suspension), Warburg concluded that "the reaction in the blood cells between methemoglobin and hexose monophosphate *or between methylene blue and hexose monophosphate* occurs by the cooperation of at least two substances, of which we name one 'ferment' and the other 'coferment.' "[164] Thus methemoglobin was not an obligatory participant in the oxidation catalyzed by methylene blue, heavy-metal catalysis was not essential, and "cell structure" was not required for the oxidation of glucose-6-phosphate by methylene blue, provided that a heat-labile, nondialyzable "ferment" and a heat-stable, dialyzable "coferment" were present.

After examining various biological sources of these new factors, Warburg and his associate Walter Christian proceeded to isolate from yeast a yellow-red protein which they termed "oxygen-transporting ferment," whose pigment was decolorized by a reducing system composed of glucose-6-phosphate, the "coferment," and an additional "ferment" they found in yeast. They proposed that in aerobic cells the following pathway was operative:

$O_2 \rightarrow$ hemin-$Fe^{2+} \rightarrow$ hemin-$Fe^{3+} \rightarrow$ leuco form of pigment $\rightarrow$ pigment $\rightarrow$ reducing system

In the presence of cyanide, the iron-containing components are inhibited, and molecular oxygen oxidizes the leuco form of the pigment directly, with the formation of hydrogen peroxide. Because in the absence of oxygen the reduced pigment is oxidized by methylene blue, they concluded that "the yellow ferment is therefore not only an oxygen-transporting ferment but also a ferment of 'oxygen-less respiration.' . . . It is probable that in life, the yellow ferment does not transfer molecular, but 'bound' oxygen. Probably, in life, it is not an oxygen-transporting ferment but an oxidation-reduction ferment."[165] This statement marks a significant shift in Warburg's approach to the problem of biological oxidation. Despite his use of the terms *bound oxygen* and *oxidation-reduction ferment,* implying a stubborn refusal to refer to "dehydrogenases" or, for that matter, to "electron transfer," it is clear that by 1933 Warburg had come closer to Keilin in his conception of the mechanism of intracellular respiration. Whether this change in research strategy merits classification as a "paradigm shift" is a question I leave for the consideration of others.

The discovery of the yellow enzyme was soon followed by Warburg's chemical studies that showed that the pigment is a small molecule that is released from the protein when the latter is denatured, and that the pigment belongs to the class of substances named flavins, one of which appeared to be identical with vitamin B_2. In 1926 Joseph Goldberger had shown that the human disease pellagra is caused by a deficiency of a dietary factor belonging to the B-complex but different from the antineuritic vitamin, and the antipellagra factor was accordingly labeled vitamin B_2. Several laboratories embarked on its isolation, and by 1933 Richard Kuhn, Paul György, and Theodor Wagner-Jauregg had obtained from milk an orange pigment, which they named lactoflavin. Although the pigment was active as a vitamin in promoting the growth of immature rats, it had no antipellagra activity, and the Goldberger factor was then named vitamin B_6. (Numbers 3, 4, and 5, which had been assigned to other presumed members of the B-complex, later disappeared from the list.) Pigments similar to lactoflavin had been described by several investigators since 1879, when Wynter Blyth reported the presence in milk of "lactochrome," but the chemical nature of these materials was uncertain. During the early 1930s Ilona Banga and Albert Szent-Györgyi found in animal tissues a yellow pigment ("cytoflave") that could undergo reversible oxidation-reduction, and Philipp Ellinger and Walter

Koschara were led to study of such pigments, which they named lyochromes, by the green fluorescence of some animal tissues upon irradiation with ultraviolet light.[166]

During the course of their chemical study of the yellow pigment, Warburg and Christian found that in alkaline solution it is decomposed by light to yield a product that turned out to be a member of the class of compounds termed alloxazines by Otto Kühling some thirty years earlier. This finding, together with the spectroscopic studies of Kurt Stern and Ensor Holiday, gave valuable clues for the synthesis of an extensive series of the flavins in the laboratories of Richard Kuhn and Paul Karrer, and vitamin B_2 was recognized to be the substance named riboflavin. Also, Kuhn determined the oxidation-reduction potentials for a series of flavin derivatives.[167] Then followed the decisive work of Hugo Theorell, who succeeded in separating reversibly the pigment and protein of the yellow enzyme without denaturation and showed the pigment to be a flavin phosphate. After the chemical synthesis of riboflavin-5′-phosphate (see formula) by Kuhn, Hermann Rudy, and Friedrich Weygand, and their demonstration that the synthetic material combines with the separated protein to regenerate the catalytic agent, it was clear that a derivative of a vitamin is a "prosthetic group" of a catalytic conjugated protein, in this case a "flavoprotein."[168] The prosthetic group of this enzyme protein came to be called flavin mononucleotide (FMN); in the biochemical nomenclature of the time, the term *nucleotide* was not limited to the constituents of nucleic acids but denoted a substance in which a nitrogenous compound is linked to a sugar phosphate (in FMN the sugar is ribitol, a reduction product of ribose). As will be evident, it is matter of some physiological importance that although FMN is bound to the protein by noncovalent interaction, the binding is rather tight.[169]

In continuing their investigation of the factors involved in the aerobic oxidation of glucose-6-phosphate, Warburg and Christian undertook the chemical examination of the additional "coferment" and "ferment" mentioned above, which they named *Zwischen-Co-Ferment* and *Zwischenferment*, respectively, "because their

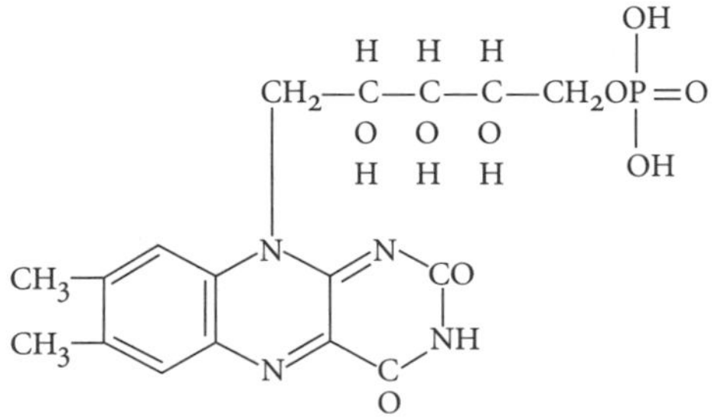

Riboflavin phosphate (flavin mononucleotide)

area of action is between the oxygen-transporting ferments and the substrates."[170] This investigation brought the study of biological oxidation into connection with the parallel study, largely by Hans von Euler, of the nature of the "cozymase" in alcoholic fermentation, discovered by Harden and Young. By 1930 Euler had obtained from yeast a preparation that resembled the adenylic acids from yeast nucleic acid (adenosine-3′-phosphate) and from muscle (adenosine-5′-phosphate) but was not identical with either.[171] Although Euler provided evidence for his view that cozymase promotes the action of some of the dehydrogenases studied by Wieland and Thunberg, its intracellular role remained uncertain so long as its chemical nature had not been established and its relationship to well-defined enzymes was unknown. It was the work of Warburg and Christian on their so-called *Zwischen-Co-Ferment* that clarified the chemical nature of Euler's cozymase and explained its function in the anaerobic fermentation of glucose.

Warburg and Christian found that, on acid hydrolysis, their product released not only adenine but also the amide of nicotinic acid.[172] Nicotinic acid (pyridine-3-carboxylic acid) had been known since the 1870s as an oxidation product of the plant alkaloid nicotine, and in 1912 it had been isolated from rice by Umetaru Suzuki in his search for the anti-beri-beri vitamin; in 1926 Hubert Vickery found it in yeast. The chemical analyses of the Zwischen-Co-Ferment (which was renamed the hydrogen-transporting coferment) showed it to be composed of adenine, nicotinamide, pentose, and phosphate in the ratio 1:1:2:3, and Warburg concluded that "the pyridine component of the co-ferment is its active group, because the catalytic action of the co-ferment depends on the alternation of the oxidation state of its pyridine part."[173] Warburg and Christian also isolated what they called the fermentation coferment (Euler's cozymase), showed it to differ in composition from their "hydrogen-tranporting coferment" in having only two phosphate groups, and named the two cofactors diphosphopyridine nucleotide and triphosphopyridine nucleotide. Euler proposed that the diphospho- and triphosphocompounds be named Cohydrase I and II, respectively, and that the substrate-specific proteins with which they are associated be denoted Apodehydrasen. On the other hand, Warburg continued to eschew the terms introduced by Wieland and Thunberg, and wrote of "specific colloids" he named *Gärungs-Zwischenfermente,* which he considered to be necessary for the reduction of the "fermentation coferment." For several years afterward most biochemists referred to the two cofactors as codehydrogenases I and II, or coenzyme I and II. After 1945 preference was given to the use of abbreviations (DPN and TPN) of the terms introduced by Warburg. The structure of DPN, as proposed by Fritz Schlenk and Euler in 1936, is shown in the accompanying formula.

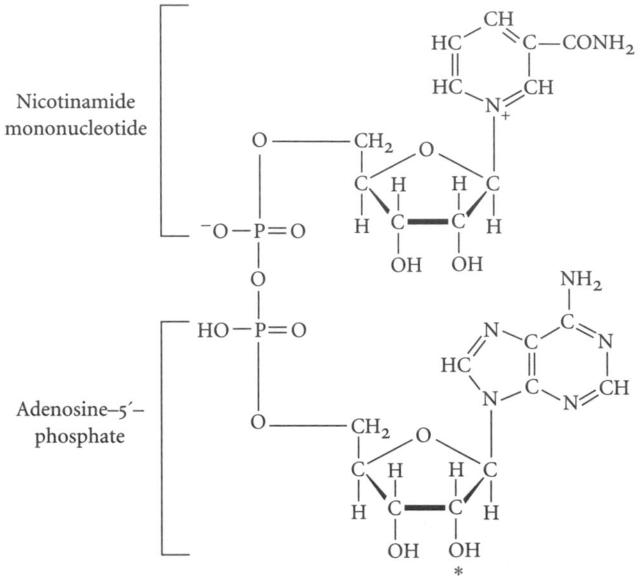

Nicotinamide mononucleotide

Adenosine–5′–phosphate

Diphosphopyridine nucleotide (DPN)

The location of the third phosphate of TPN (at the hydroxyl group marked by an asterisk) was established in 1950 through the work of Arthur Kornberg, as well as the characterization of the adenosine-2′- and adenosine-3′-phosphates derived from the alkaline hydrolysis of ribonucleic acids.[174] Since about 1960 the terms DPN and TPN have been displaced by NAD (nicotinamide adenine dinucleotide) and NADP (nicotinamide adenine dinucleotide phosphate), in accordance with a recommendation of an international commission on enzyme nomenclature. Warburg referred to the reduced forms as Dihydropyridin, and later workers denoted them as CoH_2 or $DPNH_2$, although it was evident from the chemistry of the reactions that only one hydrogen is added to the ring in a process involving the transfer of two electrons from the substrate (the other hydrogen atom derived from the "hydrogen donor" appears in the solution as a hydrogen ion); the reduced forms are now written as NADH and NADPH. It was first thought that in the reduced forms the hydrogen atom added to the pyridine ring is at the 2-position, but later work showed that it is the 4-position. In 1954 Birgit Vennesland and Frank Westheimer demonstrated that in a dehydrogenase-catalyzed reaction, the added hydrogen is transferred directly from the substrate, without mixing with the hydrogen ions of the solution.[175] This important discovery influenced the further study of the mechanism of dehydrogenase-catalyzed reactions.

A striking property of the nicotinamide portion of the pyridine nucleotides is the appearance of a new absorption band near 340 nm upon their enzyme-catalyzed reduction. The discovery of this property permitted Warburg and his associates to develop a rapid quantitative assay for the pyridine nucleotide-dependent enzymes and to use this method for their purification. After 1945, when reliable photoelectric quartz spectrophotometers (especially the Beckman DU instrument) became available, they displaced the use of Warburg manometers and variants of the Thunberg methylene blue technique in studies on the dehydrogenases. Although by 1936 numerous dehydrogenases had been identified on the basis of their specificity toward individual metabolites, the enzyme preparations used in earlier studies represented only partially fractionated extracts. These crude preparations from animal or plant tissues, or from microorganisms, were named alcohol dehydrogenase or lactic dehydrogenase and were those used, for example, in the estimation of the apparent oxidation-reduction potentials of such systems as alcohol-acetaldehyde or lactate-pyruvate. Many investigators agreed with the view reiterated by Thunberg that these dehydrogenases represent separate catalytic entities, but the failure to emulate the success of Sumner, Northrop, and Kunitz in the crystallization of enzymes was a repeated source of uncertainty.[176] There were echoes of ideas of the 1920s about the nature of enzymes, as in the statement by Richard Kuhn, a former associate of Richard Willstätter, that "R. Willstätter thought an enzyme consists of a colloidal support and an active group. The explanation that O. Warburg and H. Theorell give for the structure of the yellow enzyme illustrates exactly this conception."[177] The situation began to change, however, with the crystallization in 1937 of the alcohol dehydrogenase from yeast by Warburg's associates Erwin Negelein and Hans Wulff. According to Warburg's terminology, they had isolated "this colloid as a crystalline protein. The protein combines with diphosphopyridine nucleotide to form a dissociating pyridino-protein, the reducing fermentation ferment, which reduces acetaldehyde to alcohol."[178] Moreover, in 1938 Warburg and Christian obtained from a purified preparation of D-amino acid oxidase (an enzyme identified by Hans Krebs in 1935) a flavin derivative whose structure was later shown to be that of riboflavin phosphate joined to adenosine-5′-phosphate. This "flavin adenine dinucleotide" (FAD) was also found to be the prosthetic group of a flavoprotein ("new yellow enzyme") isolated from yeast by Erwin Haas. While it catalyzed the same reaction as the "old yellow enzyme," their protein portions were different. Thus although the reaction $FAD_{red} + O_2 \rightarrow FAD_{ox} + H_2O_2$ is catalyzed by both the D-amino acid oxidase and the "new yellow enzyme," in the first case the reduction of FAD is specifically effected by D-amino acids in the presence of a particular protein, whereas in the second case a different protein promotes the specific reduction of FAD by the "reducing system" with glucose-6-phosphate as the substrate.[179]

After 1938 there came what I consider to have been the most significant reports from Warburg's laboratory—namely, the work of Erwin Negelein and Heinz Brömel on what was called the oxidizing fermentation ferment, which catalyzes the oxidation of glyceraldehyde-3-phosphate, in the presence of DPN and inorganic phosphate, to 1,3-diphosphoglyceric acid; the crystallization of this enzyme; and the demonstration by Theodor Bücher that the oxidation of the aldehyde to 3-phosphoglyceric acid is coupled to the conversion of adenosine diphosphate (ADP) to adenosine triphosphate (ATP). These discoveries opened a new chapter in the history of biochemistry, in providing a well-defined chemical route for the coupling of the energy released in an oxidative reaction to the enzymatic synthesis of ATP.[180] The immediate impact of these achievements was dampened by the onset of World War II but became clearly evident after 1945. We shall consider their historical background and their consequences later in this chapter.

It must be added that despite the elegance and significance of the work of Warburg's group during the 1930s, his nomenclature and his views about the role of the protein components in the pyridine nucleotide-dependent reactions were a source of some confusion. For example, it was noted that

> the question of how the coenzyme functions in the catalytic system has now become a matter of great dispute. The concept of "Zwischenferment" introduced by Warburg implies that the coenzyme combines with the dehydrogenase to form the catalytically active complex. What is ordinarily referred to as a dehydrogenase is considered by Warburg to be merely a highly specific protein with no catalytic properties apart from its prosthetic group—the coenzyme. Euler and his school have accepted this view but they prefer to call the active complex the "holoenzyme." There is a good deal of evidence in favour of the view that the dehydrogenase is the seat of catalytic activity. . . . The classical conception of the dehydrogenase as the actual activating mechanism seems to be in fair agreement with the facts. No doubt the coenzyme combines with the dehydrogenase in the same way that the substrate does. But the function of the coenzyme seems to be that of highly specific hydrogen acceptor which cannot be replaced by any other substance.[181]

This point of view was developed further by other investigators and continued to be a focus of discussion for many years.[182] After the crystallization of numerous dehydrogenases, and their study as catalysts and as proteins, there was no longer any doubt that the dehydrogenase protein is the "seat of catalytic activity."

To sum up this story, before 1945 the electron transfer systems considered to constitute the chemical apparatus of intracellular respiration were iron-porphyrins, flavin nucleotides, and pyridine nucleotides. Other candidates, such as glutathione and ascorbic acid, as well as adrenochrome (an oxidation product of adrenaline), had been considered but not accepted. As stated by Edward Charles ("Bill") Slater many years later, "When I first started working on the respiratory chain under David Keilin in 1946, it was assumed that electrons flowed in a linear chain from substrate to oxygen, first in a two-electron step from substrate to the cytochrome system (via NAD(P) and flavoprotein), and then in one electron steps through the cytochrome system until oxygen was reached."[183] After 1945 the picture of the respiratory chain became more complex, notably with the recognition that cytochrome c oxidase represents a tightly knit assembly that includes copper proteins. Also, the ubiquinol-ubiquinone (Q) system (in the form of the semiquinone) has been added as link between cytochrome b and cytochrome c, with the participation of an iron-sulfur protein. Moreover, new electron transfer systems, such as the pyrrolo-quinoline quinone (PQQ) system, were found in some microorganisms.[184] Among the many remarkable oxidations performed by bacteria, and whose mechanisms still remain to be elucidated, is that performed by "hydrogenase," which converts H_2 to $2H^+ + 2e^-$ and can reduce methylene blue or fumarate.[185] It is not unlikely that the further investigation of this and other biological oxidations may reveal the existence of hitherto unknown electron transfer agents attached to catalytic proteins. Later in this chapter mention will be made of the iron-sulfur protein ferredoxin and the copper-protein plastocyanin, both electron transfer agents important in photosynthesis.

Alcoholic Fermentation as Oxidation-Reduction

In the background of Eduard Buchner's discovery of cell-free alcoholic fermentation was the suggestion offered in 1870 by his teacher Adolf Baeyer that the sugar molecule is converted to an unstable linear six-carbon compound which is split in the middle into two three-carbon units; these yield lactic acid (in lactic fermentation) or are cleaved to produce alcohol and CO_2.[186] During the succeeding decades, as the structure of glucose and related hexoses was elucidated, there were several reports (by Hoppe-Seyler, Nencki, and others) on the chemical cleavage of glucose to lactic acid in alkaline solution, and these reports were cited in support of Baeyer's hypothesis. Moreover, by the 1870s lactic acid had begun to assume importance in the study of muscle physiology. After the discovery by Emil du Bois-Reymond in 1859 that upon muscular contraction or after death acid appears in the muscles, Ludimar Hermann concluded in 1867 that this acid production is anaerobic in character, and ten years later Claude Bernard wrote that "this lactic ferment occurs in the blood, in the muscles, even in the liver, since I have found that muscles do not become acid after death unless they contain sugar or glycogen [*la matière glycogène*], which rapidly undergoes a lactic fermentation."[187] This "fermentation" in animal tissues later came to be called glycolysis; the historical development of that topic will be considered later in this chapter.

What needs to recalled at this point is that when Pasteur studied lactic fermentation in 1860 there was uncertainty about the relation between the product formed in soured milk and the muscle lactic acid found by Berzelius in 1807. On the basis of elementary analysis of salts of the two acids, Liebig concluded in 1847 that "the nitrogen-free acid which occurs in the animal organism is identical with the acid that arises in souring milk."[188] There were indications, however, from differences in the solubility of comparable salts that the two acids might be different, and for a time muscle lactic acid was thought to be a structural isomer of the fermentation lactic acid. In 1873 Johannes Wislicenus showed that both acids have the structure CH_3-CH(OH)-COOH but differ in their optical activity, the muscle acid being dextrorotatory, whereas the other is optically inactive. In the following year van't Hoff and Le Bel explained the difference between the two lactic acids as a consequence of the two modes of arranging the four groups CH_3, H, OH, and COOH in defined positions about the tetravalent α-carbon atom, thus founding the specialty known as stereochemistry. (Later work showed that the fermentation lactic acid is initially levorotatory and is then racemized to an optically inactive mixture of the two stereoisomers). Although Buchner's contemporaries knew, or should have known, that the two acids differed in their stereochemistry, and Emil Fischer's work had shown the importance of stereochemical configuration for enzymatic catalysis, hypotheses were developed after Buchner's discovery that made lactic acid an intermediate in alcoholic fermentation.

Indeed, Buchner believed for a time that "lactic acid plays an important role in the cleavage of sugar and probably appears as an intermediate in alcoholic fermentation."[189] He then proposed that the term *zymase* be applied to the enzyme that cleaves glucose to lactic acid, and he attributed the conversion of lactic acid to another enzyme, "lactocidase." The lactic acid theory was questioned by Arthur Slator, who found that lactic acid is not fermented at an appreciable rate by yeast; earlier findings of alcohol in animal tissues were attributed to bacterial contamination, and it was concluded that "the post-mortem formation of lactic acid in animal tissues is not in any way connected with alcoholic fermentation."[190] The demise of the lactic acid theory, which Buchner also abandoned, invited speculation about the possible role of other three-carbon compounds (see formulas) that had been found

CH_2OH | H—C—OH | CH_2OH — Glycerol ⇌ (±2H) CHO | H—C—OH | CH_2OH — Glyceraldehyde ⇌ CH_2OH | C=O | CH_2OH — Dihydroxyacetone

$COOH$ | C=O | CH_3 — Pyruvic acid → ($-CO_2$) CHO | CH_3 — Acetaldehyde

$COOH$ | H—C—OH | CH_2OH — Glyceric acid ⇌ (±O) Glyceraldehyde ⇌ (±H_2O) CHO | C=O | CH_3 — Methyl glyoxal ⇌ $COOH$ | H—C—OH | CH_3 — Lactic acid ⇌ (±2H) Pyruvic acid

Acetaldehyde ⇌ (±2H) CH_2OH | CH_3 — Ethanol

upon the alkaline degradation of hexoses. Alfred Wohl modified Baeyer's scheme to include glyceraldehyde and methyl glyoxal as intermediates,[191] and several investigators, including Buchner, tested them, along with the closely related dihydroxyacetone, for their fermentability by yeast juice. Because this compound was found to be fermented more rapidly than the other two, Buchner inserted it into his theory in place of lactic acid, and assumed the existence of some enzyme other than "lactocidase." This proposal was not accepted by his contemporaries and, after a brief debate on the subject, in 1912 Buchner stopped writing about the mechanism of alcoholic fermentation.

Another line of investigation was opened in 1903, when Arthur Harden reported that the addition of blood serum to Buchner's yeast juice increased the rate of CO_2 formation and attributed this result to an "an inhibitory effect which the serum exerts on the proteolytic enzyme of the press-juice; one may therefore infer that the agent for alcoholic fermentation is active for a longer time."[192] In continuing this research, Harden tested boiled and filtered solutions of autolyzed yeast juice and concluded that the stimulatory effect was not related to an inhibition of proteolysis. Instead, he showed (with William John Young) that the effect was attributable to the presence of phosphates and, in addition, that "the fermentation of glucose by yeast juice is dependent upon the presence of a dialyzable substance which is not destroyed by heat," which they termed a co-ferment.[193] Later work showed that the material they had denoted as a substance included several heat-stable compounds, all of which are involved in the conversion of glucose to alcohol and CO_2. One of them was the compound named by Euler and Myrbäck as cozymase, whose chemical nature and intracellular function were not elucidated, as was sketched above, until the 1930s.

Of more immediate impact was the discovery by Harden and Young of the effect of inorganic phosphate. Although the stimulation of zymase by phosphate had been reported earlier by Augustyn Wróblewski, he attributed the effect to the protection of zymase from inactivation by acids or alkalis.[194] A noteworthy feature of the work of Harden and Young is that they measured the CO_2 production volumetrically rather than gravimetrically (as Buchner had done), thus permitting frequent determinations and enabling them to find that, upon the addition of successive equal amounts of phosphate, "The extra amount of carbon dioxide evolved after addition is the same, and is equivalent . . . to the phosphate added." They accordingly concluded that a portion of the phosphate had been linked to organic material and suggested the bound phosphate "exists in combination with glucose, probably in the form of a phosphoric ester,"[195] and they then isolated from the fermentation mixture a substance that they identified as a hexose diphosphate.

The determination of the chemical structure of this "Harden-Young ester" was achieved two decades later by Phoebus Levene and Albert Raymond.[196] In the intervening years, two other hexose phosphate esters were isolated, and the uncertainty about the structure of these compounds was a source of some confusion in the study of alcoholic fermentation. In 1918 Carl Neuberg showed that mild acid hydrolysis of the Harden-Young ester gave a hexose monophosphate ("Neuberg ester") which turned out to be different from the one isolated four years earlier by Harden and by Robert Robison. Later Robison concluded that his product was a mixture of isomeric hexose monophosphates, probably those of glucose and fructose, and it was not until 1931 that he and Earl King obtained the glucose monophosphate ("Robison ester") in a pure state.[197] In the determination of the structure of these compounds, the new methods developed in the carbohydrate field during the 1920s, notably by W. N. Haworth, played a decisive role. The Harden-Young ester turned out to be fructose-2,6-diphosphate, the Neuberg ester fructose-6-phosphate, and the Robison ester glucose-6-phosphate.

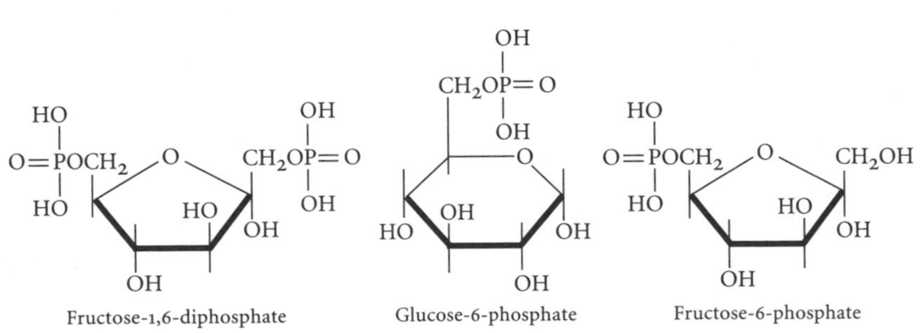

Fructose-1,6-diphosphate

Glucose-6-phosphate

Fructose-6-phosphate

One of the important consequences of the discovery of the Harden-Young ester was the demonstration by Gustav Embden and Fritz Laquer that its addition to a press juice from muscle tissue caused a large increase in the production of lactic acid. This finding led Embden to suggest that the "lactacidogen" he had proposed in 1912 as a precursor in glycolysis is a substance related to hexose diphosphate.[198] To this evidence of the possible similarity of the chemical pathways in glycolysis and in alcoholic fermentation was added Otto Meyerhof's report in 1918 that the "co-ferment" discovered by Harden and Young is also present in muscle tissue, and his later finding that such a coferment is required for muscle glycolysis.[199] These results called attention to the possibility that similar pathways are present in other biological systems, and gave support to the concept of the "unity of biochemistry."[200]

There were, however, doubts about the role of hexose diphosphate as a direct intermediate in alcoholic fermentation because it was not fermented by living yeast. Harden himself stated that "it is not impossible that the hexose phosphate is formed by combined synthesis and esterification from smaller groups produced by the rupture of the sugar molecule," and a detailed mechanism of this sort had been proposed in 1912 by Aleksandr Lebedev.[201] Moreover, during the years immediately after 1910, attention was drawn away from hexose diphosphate to pyruvic acid as an intermediate in alcoholic fermentation.

During the period 1900–1910, the technique of perfusing animal organs such as liver with saline fluids (Ringer-Locke solution) containing substances of biological interest gave valuable data on the metabolic conversion of these substances. Embden and Otto Neubauer applied this method to the study of the fate of α-amino acids and found them to be deaminated to the corresponding α-keto acids (R-CO-COOH). Neubauer and Konrad Fromherz then suggested that the deamination of alanine by yeast would yield pyruvic acid (CH_3-CO-COOH), whose decarboxylation would yield acetaldehyde (CH_3-CHO), whose reduction would produce ethanol, and since they found pyruvic acid to be fermented by yeast concluded that it might be an intermediate in alcoholic fermentation. Shortly afterward Auguste Fernbach and Moise Schoen reported the appearance of pyruvic acid during alcoholic fermentation, and Carl Neuberg announced the discovery of a new enzyme ("carboxylase") that catalyzes the conversion of pyruvic acid to acetaldehyde and CO_2. By 1913 Neuberg had developed a theory of alcoholic fermentation in which glucose is first cleaved to two molecules of methyl glyoxal, one molecule of which is reduced to glycerol and the other is oxidized to pyruvic acid; decarboxylation of pyruvic acid to acetaldehyde is followed by the reduction of the latter to ethanol, balanced by the oxidation of the second molecule of methyl glyoxal to pyruvic acid.[202] There was no place in this theory, with its roots in the ideas of Baeyer and Wohl, for phosphorylated sugars of the kind studied by Harden, Young, and Robison.

This theory held sway, with minor modifications, for more than fifteen years. The fact that methyl glyoxal was not readily fermented by yeast was explained by assuming that one of its possible isomers is the "true" intermediate. An argument offered in favor of methyl glyoxal was the discovery, by Henry Dakin and Harold Dudley, of a widely distributed enzyme ("glyoxylase") that catalyzes the interconversion of methyl glyoxal and lactic acid.[203] The most attractive features of Neuberg's theory were that separate enzyme-catalyzed reactions were provided for the generation of CO_2 and the formation of ethanol, and that the overall process was formulated as a set of coupled oxidation-reduction reactions. Moreover, Neuberg's theory received support from its successful application in the industrial manufacture of glycerol in Germany. He showed that the addition of sulfite to a fermentation mixture blocks the reduction of acetaldehyde by forming an aldehyde-sulfite compound, and that the process becomes: Glucose = Glycerol + Acetaldehyde + CO_2; he explained this result in terms of a disturbed balance of oxidation-reaction reactions: "Glycerol is the reduction equivalent of pyruvic acid, which decomposes to carbonic acid and acetaldehyde. If the reduction of the latter is blocked, the only remaining possibility is the increased correlative formation of glycerol."[204] The sulfite process was patented in 1914 but "could not be published earlier because, during the war, the German army administration had an interest in keeping the experiments and results secret. Our work arose from the necessity of the time and owes its origin to the expectation that the supply of glycerol available to the European Central Powers would soon be insufficient, because of the blockade."[205] It should also be added that the inhibition of alcoholic fermentation by sulfite had been reported in 1874 by Jean Baptiste Dumas.[206]

If in demonstrating the enzymatic decarboxylation of pyruvic acid, Neuberg followed the lead of Neubauer, the conception of the coupling of biochemical oxidation-reduction reactions had already been stated by Jacob Parnas, in connection with his discovery of the enzymatic catalysis of the Cannizzaro "dismutation" of aldehydes: "In the Cannizzaro rearrangement of the aldehydes we have come to know a simple system of coupled reactions, in which through oxygen transfer and hydrogen uptake there occur simultaneous oxidation and reduction. Through an enzyme from the liver the reaction is catalyzed to such an extent that it leads to the complete disappearance of the aldehyde. . . . Aldehydes may be regarded as general reductants for the reduction of carbonyl groups in the animal organism. Through specific ferments the Cannizzaro reaction between two reacting substances is accelerated, and there are formed an alcohol (or a hydroxy acid) and a fatty acid."[207] This view of the oxidation-reduction of aldehydes was accepted after 1910 as a key step in alcoholic fermentation and in glycolysis, and by 1925 there was general agreement that pyruvic acid is an intermediate in both processes, but the role of the hexose phosphates was still unclear and the status of methyl glyoxal was hypothetical. Within about ten years, however, the situation had changed dramatically, and the efforts of several research groups had led to the formulation of a clear-cut sequence of enzyme-catalyzed reactions, from glucose-6-phosphate to pyruvic acid. The impact of this development was far-reaching, not only because it represented the elucidation of a complex metabolic pathway but also because it provided a new basis for the understanding of the intracellular mechanisms for the utilization of the energy made available by oxidative processes in biological systems. In this development the new attitude toward enzymes as individual catalytic proteins, as well as the contributions of organic chemistry in determining the structure of the intermediates and cofactors, played decisive roles.

During the mid-1920s Otto Meyerhof described the preparation of a cell-free extract of frog muscle that, like the yeast maceration juice of Lebedev, was relatively free of carbohydrate and could therefore be used to test the role of suspected intermediates in the conversion of muscle glycogen to lactic acid.[208] This achievement ended the discussion whether the anaerobic breakdown of carbohydrate in muscle is linked to the integrity of cell structure, and marks the beginning of the study of anaerobic glycolysis as an enzymatic process. Meyerhof showed that, as in the case of alcoholic fermentation, the addition of inorganic phosphate promoted the process, and that the extract converted hexose diphosphate into lactic acid. Glucose, however, was not utilized effectively unless there were present, in addition to Harden's heat-stable coferment, a heat-labile factor in yeast juice that Meyerhof named hexokinase.[209] Before Meyerhof's work, Gustav Embden had shown that "lactacidogen" accumulates in muscles poisoned by fluoride and had suggested that this effect is a consequence of the decomposition of hexose diphosphate. The inhibition of glycolysis and of alcoholic fermentation by fluoride had long been known, and Embden cited his finding in support of the view that hexose diphosphate is a direct intermediate in carbohydrate breakdown.[210]

The establishment of the structure of the Harden-Young ester as fructose-1,6-diphosphate invited more modern formulations of the chemical hypotheses regarding alcoholic fermentation than those advanced earlier by Baeyer, Wohl, and Lebedev. Heinz Ohle suggested that the anaerobic breakdown of glucose begins with its conversion to glucose-6-phosphate and then (via fructose-6-phosphate) to fructose-1,6-diphosphate, which he thought to undergo a series of reactions leading to glyceraldehyde-3-phosphate and dihydroxyacetone phosphate.[211] This proposal stimulated the synthetic efforts of Hermann Fischer and Erich Baer, who made DL-glyceraldehyde-3-phosphate, whereupon Carl Smythe and Waltraut Gerischer (in Warburg's laboratory) showed this compound to be fermented by yeast.[212] Because the synthetic material was a racemate, only 50 percent was fermented, and later work showed the D-isomer to be the reactive compound.

Although the phosphorylated three-carbon sugars (triose phosphates) had appeared in successive earlier hypotheses, it was only in 1933 that the accumulated knowledge permitted their inclusion in a scheme that proved to be fruitful for later research. In that year, the last of his life, Embden proposed that, in glycolysis, fructose-1,6-diphosphate is cleaved directly to glyceraldehyde-3-phosphate and dihydroxyacetone phosphate (see the formulas of the unphosphorylated three-carbon compounds on p. 291), and that these products underwent a Cannizzaro (mutase) reaction to yield 3-phosphoglycerol and 3-phosphoglyceric acid, with the conversion of the latter product into pyruvic acid and phosphate. A second mutase reaction involving pyruvic acid (which is reduced to lactic acid) and phosphoglycerol (which is oxidized to glyceraldehyde-3-phosphate) completed the balance of the equations, and accounted for the complete conversion of glucose into lactic acid. In the background of this proposed scheme were not only Parnas's concept of dismutations as coupled oxidation-reduction reactions, but also the identification, by Embden's research group, of 3-phosphoglyceric acid in fluoride-poisoned muscle extracts undergoing glycolysis.[213]

In 1934 Meyerhof and Karl Lohmann reported that fructose-1,6-diphosphate is cleaved by muscle extracts to two molecules of triose phosphate, and they named the enzyme supposedly responsible for this action zymohexase, later shown to represent two enzymes, one named aldolase (after the aldol reaction described by Wurtz during the 1870s), which effects the cleavage of the six-carbon chain, the other named triose phosphate isomerase, which catalyzes the interconversion of the two triose phosphates.[214] Thus the three-carbon compounds postulated on purely chemical grounds around 1910 as intermediates in alcoholic fermentation reappeared during the 1930s as their phosphorylated derivatives. Apart from the elucidation of their place in the increasingly complex metabolic pathway from glucose to lactic acid or to alcohol, the subsequent studies on the mechanisms of the catalytic action of these two enzymes, notably by Irwin Rose and by Jeremy Knowles, have provided important knowledge about the nature of enzymatic catalysis.[215]

Whereas as late as 1932 it was still possible to state that "the trend of opinion seems to be that methyl glyoxal is an intermediate between hexosephosphate and lactic acid,"[216] after 1933 methyl glyoxal disappeared from the generally accepted pathway and was replaced by glyceraldehyde-3-phosphate and its oxidation product 3-phosphoglyceric acid. In 1935 Meyerhof and Kiessling showed that it is this oxidation of the aldehyde to the acid that balances the reduction of pyruvic acid in anaerobic glycolysis and, by inference, of acetaldehyde to alcohol in alcoholic fermentation.[217] The catalysts of these oxidation-reduction reactions were, as has been recounted, the substrate-specific dehydrogenases, with the pyridine nucleotide NAD as the electron-transfer agent. We will return to these reactions shortly.

Before concluding this section, more should be said about the relation of enzymes to vitamins, made evident by Warburg's work on the catalytic flavoproteins. Shortly after the discovery that nicotinamide is the reactive component of the pyridine nucleotides, Elvehjem and his associates reported that nicotinic acid is the active agent in liver concentrates that are effective in curing black tongue, a deficiency disease in dogs, and other investigators demonstrated the curative value of nicotinic acid in human pellagra.[218] The designation of the antipellagra factor as vitamin B_6 was abandoned, and this term was applied to a group of pyridine derivatives related to a coenzyme denoted pyridoxal phosphate, of which more will be said in the next chapter. Moreover, by 1937, owing to the work of André Lwoff, B. C. J. G. Knight, and Howard Mueller, nicotinic acid was recognized to be an essential growth factor for a variety of microorganisms. The subsequent development of research on vitamins has depended heavily on the use of such microbial species for the assay of fractions obtained during the course of purification. For example, the problem of the isolation and characterization of the antianemia factor (vitamin B_{12}, cobalamin) proved to be intractable until a microbiological assay was found in 1947.[219] One of the important consequences of the convergence of studies on the nutritional requirements of humans and microbes has been the tacit assumption that a new vitamin or microbial growth factor will turn out to be part of a coenzyme. Since 1940 there have been several instances in which this expectation has been fulfilled; we will refer to some of them later.

At this point it must be noted that during the 1930s the antineuritic vitamin B_1 was shown to be a constituent of a coenzyme important in alcoholic fermentation. In 1932 Ernst Auhagen found that the Harden-Young coferment contained, in addition to Euler's cozymase, a heat-stable organic substance that is a cofactor for yeast carboxylase, the enzyme that converts pyruvic acid into acetaldehyde and CO_2. The studies of Rudolph Peters and his associates around 1930 had shown that vitamin B_1

is involved in the metabolism of pyruvic acid in the animal body, because lactic acid accumulated in the brain tissue of B_1-deficient pigeons, and the addition of the vitamin (then available in crystalline form) and pyrophosphate to the minced tissue caused the rapid disappearance of pyruvic acid without the accumulation of lactic acid. After the elucidation of the chemical structure of vitamin B_1 in 1935, through the efforts of several research groups, notably those led by Robert Williams, Adolf Windaus, and Alexander Todd, it was recognized to be composed of a thiazole linked to a pyrimidine and named thiamine (for a time it was also termed aneurin). Then followed the report of Karl Lohmann and Philipp Schuster on the isolation of crystalline cocarboxylase from yeast, and the demonstration that it is thiamine pyrophosphate (TPP). Subsequent studies on the pyruvate oxidation system in animal tissues and some microbes showed it to be an integrated assembly of several enzymes, acting in cooperation with TPP and other cofactors and yielding in place of acetaldehyde the key metabolite acetyl-coenzyme A, to which we shall return shortly.[220]

Chemical Energetics of Muscular Contraction[221]

Some facets of the nineteenth-century research on muscle have been considered, and we have noted Liebig's view that the energy for muscular contraction came from the oxidative breakdown of proteins. This idea was challenged by the physiologist Adolf Fick and the chemist Johannes Wislicenus in their celebrated experiment performed in 1865. After seventeen hours on a nitrogen-free diet, they walked up the Faulhorn, an Alpine peak near Grindelwald, and they determined the amount of nitrogen excreted in their urine. From their calculations they concluded that no more than about a third of the energy expended during the climb could have been derived from the combustion of body protein.[222] Experiments by Max Pettenkofer and Carl Voit during the 1860s also showed that muscular work did not lead to increased nitrogen excretion, but they concluded that "through the oxygen uptake of the organs and through the commensurate protein decomposition there accumulates a tension [*Spannkraft*] which is gradually used up even during rest and which we can convert at will into mechanical work."[223] A similar view was expressed by Liebig in 1870, and Ludimar Hermann postulated the existence of a labile albuminoid material (later termed inogen) that undergoes anaerobic breakdown to myosin, "fixed acid" (lactic acid) and CO_2. Although Otto Nasse offered chemical evidence to suggest that the appearance of lactic acid during muscular contraction paralleled a decrease in glycogen content, it was uncertain whether these changes reflected a vital process or merely the death of the animal.[224]

On the other hand, Fick insisted that muscular contraction is a "chemical-dynamic" process not comparable to the operation of a heat engine, which depends on differences in temperature, and more like the process of fermentation, in which a complex compound (not protein) is decomposed.[225] As A. V. Hill put it in 1912, "Unfortunately, there have been, even among physiologists, many and grievous misconceptions as to the application of the laws of thermodynamics; these have been due partly to the desire to make over-hastily a complete picture of the muscle machine, partly to the completely erroneous belief that the laws of thermodynamics apply only to heat engines, and not to chemical engines. . . . The muscle fibre has been treated as a heat engine when it is inconceivable that there are finite differences of temperature in it. . . . The muscle is undoubtedly a chemical machine working at constant temperature."[226] This statement appeared a few years after the studies of Walter Fletcher and Gowland Hopkins, who showed that lactic acid is formed during muscular contraction and disappears upon the admission of oxygen.[227] Through the use of a sensitive microcalorimeter, Hill was able to correlate these chemical changes with two phases of heat production by surviving muscle, an "initial heat" and a "recovery heat." In 1913 Hill estimated that, under optimal conditions, the total energy measured as initial heat is approximately equal to the potential energy of the tension when a contracting muscle is made to do work; it was later shown that such high mechanical efficiency is, in fact, not attained.

The experiments of Fletcher and Hopkins marked a new stage in the study of the chemical energetics of muscular contraction, and their success was due in large part to the use of ice-cold alcohol into which the frog muscles were plunged before the chemical determinations were made. As Hopkins described it later, "We found that the confusion in the literature as to the quantitative relations of lactic acid in muscle was wholly due to faulty technique in dealing with the tissue itself. When the muscle is disintegrated as a preliminary to extraction for analytical purposes, the existing equilibrium is entirely upset. . . . Fletcher and I, however, found it quite easy, by means of a simple method, not only to avoid starting the changes which led to the formation of lactic acid, but to arrest them at any point in their progress, and thus establish their time-relations."[228] Fletcher and Hopkins supposed that, during oxidative recovery, the lactic acid was converted to CO_2 and H_2O, but Hill's measurements indicated that the total heat liberated in a complete cycle of contractions was only about one-fifth of that expected from such complete oxidation of the lactic acid produced during the anaerobic phase. He concluded that "lactic acid is therefore built up in the body into some, at present unknown, chemical combination of greater energy than glucose. It is suggested that this body is one containing a large amount of 'free energy,' and that it may be able to account for the mechanical work done by a muscle simply by the process of breaking down into lactic acid."[229]

The next decisive advances were provided by Otto Meyerhof. In a series of papers during 1920–1922, he established that in surviving muscle the lactic acid is wholly derived from glycogen, and that during oxidative recovery about four-fifth of the lactic acid is reconverted to carbohydrate. In 1925 he summarized his view of the process as follows:

> The first phase, the formation of lactic acid from carbohydrate, is anaerobic and spontaneous. This process is the immediate source of muscular force. In the second phase, with the expenditure of oxidation energy, the lactic acid is reconverted to carbohydrate. This second process corresponds to the recovery or restitution of the muscle. . . . The heat liberated at the instant of contraction corresponds to the breakdown of glycogen to lactic acid and to a physical-chemical change, induced by the presence of lactic acid, of the muscle protein directly associated with the contractile process. The oxidative heat, on the other hand, represents the excess of oxidation energy over the endothermic reactions that occur during the reversal of the processes during the work phase.[230]

Although the importance of Meyerhof's experimental results cannot be exaggerated, his conclusion that the formation of lactic acid is the cause of muscular contraction had to be abandoned a few years later. This came about through the discovery of other chemical constituents in muscle, and the demonstration that contraction can occur in the absence of the formation of lactic acid. There ensued what Hill called "the revolution in muscle physiology."[231] These findings also brought new light to the problem of the anaerobic breakdown of carbohydrates, not only in muscular contraction but in alcoholic fermentation as well.

The "revolution" began in 1927, with reports of the independent discovery of creatine phosphate in muscle extracts, and two years later came the announcement of the independent discovery in such extracts of adenosine triphosphate (ATP). In 1927 Cyrus Fiske and Yellapragada SubbaRow reported that upon applying to muscle extracts the method they had devised for the determination of inorganic phosphate, they had found an acid-labile substance which yielded creatine (known since 1832 to be constituent of muscle) and phosphate upon hydrolysis; further, they reported, this "phosphocreatine" is hydrolyzed during muscular contraction and resynthesized during the recovery phase. At about the same time, Philip and Grace Eggleton reported the finding of a similar acid-labile material, which they called phosphagen and which they first thought to be a hexose monophosphate related to

Embden's "lactacidogen." Its identity with creatine phosphate, as the substance came to be called, was soon established, and it was recognized that its acid lability arises from the fact that the phosphoryl group is linked to a nitrogen atom in the guanidino group of creatine.[232]

The discovery of creatine phosphate focused attention on acid-labile phosphate compounds. Meyerhof and Lohmann showed that the "phosphagen" in the muscles of invertebrates is arginine phosphate, in which the phosphoryl group is also linked to a nitrogen of a guanidino group, and that the hydrolysis of the N-P bond of both organic phosphates is strongly exothermic. Moreover, Lohmann found the acid-labile inorganic pyrophosphate in muscle and in yeast, but after the report of Harold Davenport and Jacob Sacks that fresh muscle does not contain inorganic pyrophosphate, this substance was soon shown both by Fiske and SubbaRow and by Lohmann to be derived from a new acid-labile organic phosphate related to what had been called muscle adenylic acid, and the new compound was named adenylpyrophosphate. Of the several structures that were proposed, adenosine-5′-triphosphate (ATP) was shown by chemical synthesis to be the correct one.[233]

The second stage in the "revolution in muscle physiology" came in 1930, when Einar Lundsgaard discovered that muscles poisoned with iodoacetic acid (ICH_2COOH) can contract without the formation of lactic acid and that in the process creatine phosphate (in crustacean muscle, arginine phosphate) disappears. He suggested that "phosphagen is the substance directly supplying the energy for contraction, while lactic acid formation in the normal muscle continually provides the energy for its resynthesis."[234] After Lundsgaard's work, it was recognized that Embden had been correct when he claimed in 1924 that much of the lactic acid is formed after the contraction.

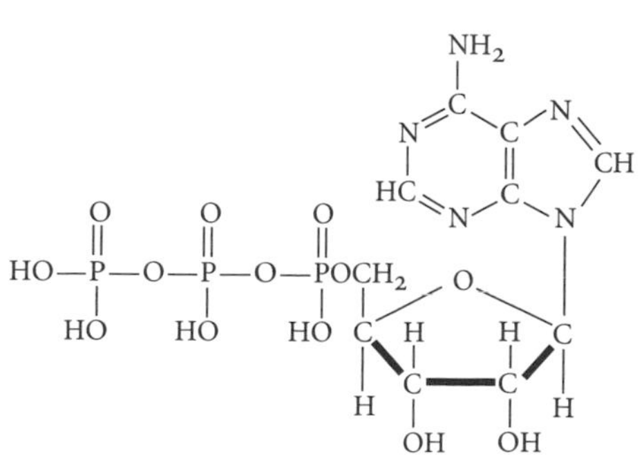

Adenosine-5′-triphosphate (ATP)

With the seeming demise of the lactic acid theory of muscular contraction, there was extensive debate during 1930–1934 about the nature of the chemical reaction most immediately related to the physical process. An important feature of the discussion was noted in a review article of the time: "It has been tacitly assumed, and this view is still very widespread, that the greatest significance must be assigned to that partial process, within the totality of the chemical processes associated with muscular contraction, which has undergone the greatest change after a muscle twitch." Indeed, much attention was given to Gerhard Schmidt's discovery in muscle extracts of an enzyme that catalyzes the deamination of muscle adenylic acid.[235] This line of investigation was abandoned, however, after it was recognized that the central chemical processes in muscular contraction are the cleavage and resynthesis of the pyrophosphate bonds of ATP.

In 1934 Lohmann showed that the apparent hydrolysis by dialyzed muscle extracts of creatine phosphate (CP) to creatine (C) and inorganic phosphate (P) is promoted by the addition of ATP, which is cleaved to adenylic acid (adenosine monophosphate, AMP) and two equivalents of P. He concluded that ATP acts as a "coenzyme" in the hydrolysis of CP in the following manner: (1) AMP + 2CP $\rightarrow$ ATP + 2C; (2) ATP + $2H_2O$ $\rightarrow$ AMP + 2P. The sum of the two reactions is 2CP + $2H_2O$ $\rightarrow$ 2C + 2P.[236] Calorimetric measurements in Meyerhof's laboratory had already shown that the hydrolysis of a molecule of ATP to AMP and 2P is attended by a large heat liberation, approximately double that found for the hydrolysis of one molecule of CP. These data were consistent with the conclusion that the ATP-C reaction is readily reversible, on the assumption that the free-energy changes parallel the changes in heat content, and were offered in support of the view offered in 1931 that the role of ATP "appears to consist in the fact that the esterification of phosphate, which precedes the cleavage of carbohydrate to lactic acid, occurs with the simultaneous cleavage of adenylpyrophosphate, which is resynthesized during the further cleavage [of carbohydrate]. In this manner, the adenylpyrophosphate cycle maintains the lactic acid formation. The synthesis of phosphagen is therefore made possible . . . by the cleavage energy of the adenylpyrophosphate, while the energy of lactic acid formation (from phosphate esters) serves to resynthesize the cleaved pyrophosphate."[237] Lohmann therefore concluded that "as regards the question of the chemical and energetic relation of the breakdown and resynthesis of adenylpyrophosphate to the fundamental process of muscular contraction, it may be assumed that there is no direct relation."[238]

Within five years, however, this conclusion was challenged by the report of Vladimir Engelhardt and Militsa Lyubimova that myosin, the fibrous protein long identified with the process of muscular contraction, has the properties of an enzyme that specifically catalyzes the hydrolysis of ATP, and that the contractile process is directly linked to this energy-yielding chemical reaction.[239] Myosin had been described (and named) by Willy Kühne in 1864, and the subsequent work during 1880–1910 of Aleksandr Danilevski, William Halliburton, and Otto von Fürth confirmed and extended Kühne's observations.[240] Before 1930 the available data were usually interpreted in terms of theories of colloidal behavior, but subsequent physical-chemical studies dealt with myosin as a macromolecular species. Thus Alexander von Muralt and John Edsall, whose viscosity studies showed that myosin solutions exhibit a strong double refraction of flow, concluded that this fibrous protein is a highly asymmetric molecule.[241] Also, William Astbury included myosin in his X-ray diffraction studies on fibrous proteins, and Hans Weber found that a purified preparation sedimented in the untracentrifuge as a monodisperse species.[242] In later work purified myosin from skeletal muscle was found to have a particle weight of about 540,000 and to consist of two "heavy" peptide chains a large portion of which represents a tightly coiled "thick filament" and a "head" (amino-terminal) portion, constituting a chain-length of about 2100 amino acid units per heavy chain, as well as four "light" peptide chains (particle weight, about 20,000 each).

After the announcement of the discovery that myosin catalyzes the hydrolysis of ATP, Albert Szent-Györgyi and his associates at Szeged examined the effect of ATP on the contraction of myosin threads derived from various kinds of muscle extracts. Among their numerous important findings was that the contraction depends on the presence of an additional factor, which was isolated by Bruno Straub and named actin. It was also found that the complex formed between myosin and actin ("actomyosin") is dissociated by ATP.[243] This work was done during the war years (1941–1944) and did not become widely known until after 1945. Many significant contributions appeared soon afterward, notably those of Kenneth Bailey, who discovered "tropomyosin" as an additional muscle component and pioneered in the chemical characterization of myosin, and of Andrew Szent-Györgyi, who used proteinases to cleave myosin into smaller units ("meromyosins") and to show that the ATPase activity is localized near the amino-terminal end of the long molecule. Tropomyosin was found to be associated with the actin-containing "thin" filaments of striated muscle and involved in the regulation by calcium ions of the ATPase activity.[244]

Of particular importance in the post–World War II development of the study of the relationship between the chemical and mechanical events in muscular contraction have been the contributions during the 1950s of Andrew Huxley, Hugh

Huxley, and Jean Hanson. Their use of light and electron microscopy, as well as X-ray diffraction, for the study of muscle structure led them to develop a fruitful version of the "sliding mechanism" of the contraction of vertebrate striated muscle. In their theory the enzymatically active site of the myosin filament and the actin filament in the actomyosin complex oscillates between two positions (as in a mechanical spring) on the contractile myosin filament, with the ATP system as a cross-bridge between the two filaments.[245] In the subsequent development of the theory, evidence has been presented for the view that in the cycle of muscular contraction the products of the hydrolysis of ATP (ADP and P) are not released from the assembly, thus markedly reducing the magnitude of negative free-energy change in the hydrolytic process and making the process freely reversible. As will be seen shortly, during the 1940s much was made of the "irreversibility" of the intracellular hydrolysis of "energy-rich bonds," such as the pyrophosphoryl bonds of ATP.

It can only be added here that the recent advances in the understanding of the chemistry and mechanics of vertebrate striated muscle have greatly influenced the study of the alternative relations between ATPase action and the structural components of vertebrate smooth muscles and of the muscles of invertebrates, as well as the study of the cilia and flagellae of motile small organisms like *Paramecium*, where the counterpart of myosin is an ATPase named dynein.[246]

The Energetics of Anaerobic Glycolysis

I return to the early 1930s, when ATP was considered to be a "coenzyme" in glycolysis, but its function was not clear. It had been shown to be a component of the Harden-Young coferment, in addition to magnesium ions, Euler's cozymase, and Auhagen's carboxylase. Thus the increasing heterogeneity of the heat stable dialyzable "substance" identified by Harden and Young as a cofactor in alcoholic fermentation matched the growing complexity of the zymase that Buchner had considered to be a single enzyme.

In 1934 a clue to the role of ATP in glycolysis and fermentation was provided in a paper by Jacob Parnas, Pawel Ostern, and Thaddeus Mann, who stated that "the resynthesis of phosphocreatine and adenosine triphosphate is not linked to glycolysis as a whole, but to definite partial processes; and this leads further to the conclusion that the resynthesis does not involve a relationship that might be called 'energetic coupling,' but more probably involves a transfer of phosphate residues from molecule to molecule."[247] In writing this passage, Parnas may have intended to refer to the conclusion drawn by Meyerhof from his measurements of the relation between oxygen uptake and glycogen resynthesis: "Oxidation and resynthesis do not represent a chemically-coupled process, for which one can give a stoichiometric equation, but an energetically coupled one." This statement was criticized by Amandus Hahn, who reiterated the definition of coupled reactions given by Wilhelm Ostwald in 1900, cited earlier.[248]

To define which molecules might be involved in the phosphoryl transfer postulated by Parnas, a more detailed identification of the intermediates in glycolysis and alcoholic fermentation was needed; this came quickly, largely through work in Meyerhof's laboratory. According to Embden's theory, the 3-phosphoglyceric acid formed by the oxidation of glyceraldehyde-3-phosphate is the precursor of pyruvic acid, and this conversion was found to involve the "phosphoglyceromutase"-catalyzed migration of the phosphoryl group from the 3-position to the 2-position, followed by the dehydration of 2-phosphoglyceric acid to phosphoenolpyruvic acid (PEP) by an enzyme denoted enolase, which was later crystallized in Warburg's laboratory. This enzyme was found to be activated by magnesium ions and to be strongly inhibited by fluoride, thus explaining the long-known effect of fluoride on alcoholic fermentation and glycolysis.[249] Later work (in Carl Cori's laboratory) showed that in the enzyme-catalyzed isomerization of the two phosphoglyceric acids, catalytic amounts of 2,3-diphosphoglyceric acid are required; this substance had been isolated in 1925 from erythrocytes by Isidor Greenwald, but no physiological role could be assigned to it until the enzyme involved in the reaction had been purified.[250]

Moreover, in 1935 Meyerhof and Kiessling found that in iodoacetate-poisoned muscle extracts, the phosphoryl group of PEP is transferred to glucose via ATP to form hexose phosphates and pyruvate. Also, calorimetric measurements showed that the hydrolysis of PEP is a strongly exothermic reaction. It became clear therefore that two of the individual phosphoryl-transfer reactions in which ATP or its cleavage products participate are the entry of glucose into the glycolytic sequence (the hexokinase reaction) and the dephosphorylation of PEP to form pyruvic acid. The stoichiometry of the overall process was not clarified, however, until Lohmann had isolated adenosine diphosphate (ADP) and after Herman Kalckar had shown that there is a widely distributed enzyme ("myokinase") that catalyzes the reaction 2ADP $\rightleftharpoons$ ATP + AMP.[251] It was then recognized that in the hexokinase reaction ATP

$$\begin{array}{ccccc} \mathrm{COOH} & & \mathrm{COOH} & & \mathrm{COOH} \\ | & & | & & | \\ \mathrm{HCOH} & \longrightarrow & \mathrm{HC{-}OPO_3H_2} & \longrightarrow & \mathrm{C{-}OPO_3H_2} + \mathrm{H_2O} \\ | & & | & & \| \\ \mathrm{CH_2OPO_3H_2} & & \mathrm{CH_2OH} & & \mathrm{CH_2} \end{array}$$

D-3-Phosphoglyceric acid — D-2-Phosphoglyceric acid — Phosphoenolpyruvic acid reacts with glucose to form glucose-6-phosphate and ADP, that in the "creatine kinase" reaction the equation is C + ATP → CP + ADP, and that pyruvic acid + ATP are formed in the reaction between PEP and ADP.

Earlier in this chapter it was noted that the importance of coupled dehydrogenase reactions had been emphasized by Euler during the 1920s and that Meyerhof and Kiessling had shown that the oxidation of glyceraldehyde-3-phosphate is balanced by the reduction of pyruvic acid to lactic acid (in muscle) or of acetaldehyde to alcohol (in yeast). Dorothy Needham and R. K. Pillai then reported in 1937 that this "dismutation" between triose phosphate and pyruvic acid in muscle extracts is accompanied by the esterification of inorganic phosphate. David Green and associates showed this process to be reversible and to depend on the presence of DPN. In the same year Zacharias Dische reported similar findings with red blood cells. These results focused attention on the role of ATP in the DPN-dependent dehydrogenation of glyceraldehyde-3-phosphate (G-3-P) to 3-phosphoglyceric acid (3-PG), and in 1938 Meyerhof, Paul Ohlmeyer, and Walter Möhle showed that the process G-3-P + DPN + ADP + P $\rightleftharpoons$ 3-PG + reduced DPN + ATP is reversible and that it is the one inhibited by iodoacetate in the anaerobic breakdown of carbohydrate.[252] This reagent was known to react with the cysteine-sulfhydryl groups of proteins, and Louis Rapkine proposed in 1938 that such a group is involved in the catalytic action of glyceraldehyde-3-phosphate dehydrogenase. For a time during the 1950s it was thought that glutathione provided this group, but later work showed it to be the sulfhydryl group of a particular cysteine unit of the catalytic protein.[253]

As was mentioned earlier, in 1939 Warburg and Christian reported the crystallization from yeast juice of G-3-P dehydrogenase (which they called the protein of *das oxydierenden Gärungsferment*). In the purification procedure, they used the spectrophotometric determination of reduced DPN as the assay method, and they also took advantage of the results of Needham and Pillai on the effect of arsenate on the oxidation of triose phosphate to phosphoglyceric acid. This reagent, which resembles phosphate in its chemical properties, had been found by Harden and Young to accelerate the fermentation by yeast juice, and they interpreted the effect in terms of a faster hydrolysis of hexose diphosphate. After Needham and Pillai showed that arsenate completely inhibited the coupled phosphorylation of ADP but did not affect the oxidation-reduction process, it was clear that arsenate was replacing phosphate in the reaction catalyzed by the dehydrogenase. The nature of the chemical reaction involving arsenate was not clarified, however, until Negelein and Brömel isolated the immediate product of the reaction in the presence of phosphate, and showed it to be 1,3-diphosphoglyceric acid. It followed, therefore, that in the presence of arsenate and ADP, the corresponding 1-arseno-3-phosphoglyceric acid only undergoes rapid hydrolysis. With the later recognition that the key step in the catalytic action of G-3-P dehydrogenase is the NAD-dependent formation of a thiolester (RCO-SR′), which then reacts with phosphate to form the diphosphoglyceric acid, Warburg's original interpretation of the process was found to be incorrect.[254]

What had been achieved in these studies was the demonstration of a chemically defined enzymatic mechanism whereby the energy released by the oxidation of the aldehyde group of G-3-P by NAD is utilized for the synthesis of a pyrophosphate bond of ATP. The key discovery was the identification of an intermediate in which the carboxyl group formed in the oxidation is combined with a phosphoryl group, that is, an anhydride of a carboxylic acid and phosphoric acid (an "acyl phosphate"). This discovery was made possible by the availability of a crystalline enzyme preparation, relatively free of other enzymes that act on the participants in the dehydrogenase-catalyzed reaction. When seen against the background of the earlier uncertainties regarding the chemical events in glycolysis and fermentation, most of which arose from the use of yeast and muscle juices or of crude enzyme preparations, the work of Otto Warburg and his associates marks the recognition by biochemists that the chemical dissection of complex intracellular processes depends on the isolation and characterization of individual catalytic proteins. That recognition also marked a change in the attitude of biochemists engaged in the study of such processes to the contributions of James Sumner, John Northrop, and Moses Kunitz to the study of such hydrolytic enzymes as urease, pepsin, or trypsin. If the view that the physiological function of enzymes of this kind was somehow different from that of those associated with intracellular processes still prevailed during the 1930s, the techniques for the crystallization of enzyme proteins became essential components of the methodology of biochemical research, and were later refined in response to the emergence of X-ray crystallography as a powerful tool in the study of protein structure. Below is a diagram of what came to be called the Embden-Meyerhof pathway of the anaerobic breakdown of glucose in yeast, as formulated at the end of the 1930s; the investigation of the fate of pyruvic acid in muscle and other microorganisms will be considered later in this chapter.

Apart from this general influence on biochemical thought and practice, the work leading to the discovery of an acyl phosphate as a participant in the phosphorylation of ADP had an immediate impact on the work of two investigators—Fritz Lipmann and the younger Herman Kalckar. Early in 1939 Lipmann reported that the oxidative decarboxylation of pyruvic acid by *Lactobacillus delbrückii* in the presence of inorganic phosphate and thiamin pyrophosphate, both of which were required for

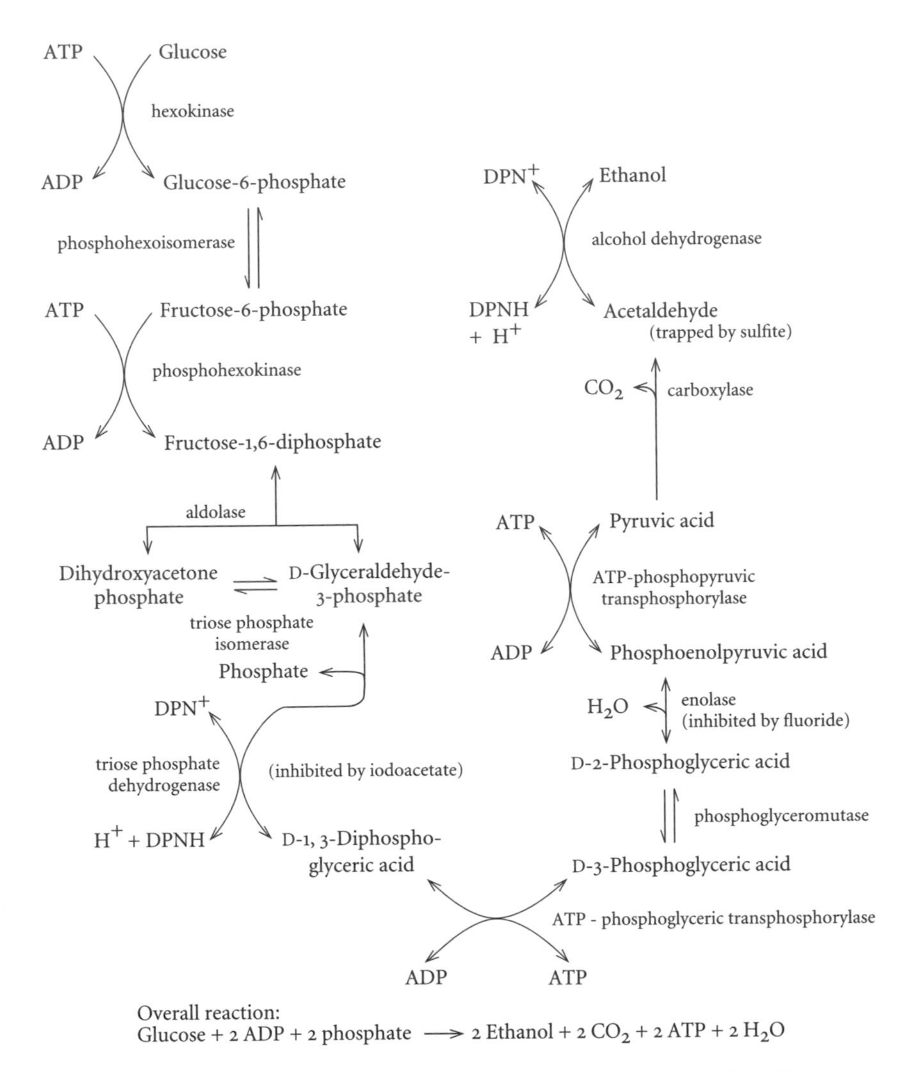

Pathway of anaerobic breakdown of glucose to ethanol and carbon dioxide in yeast

the oxidation, was accompanied by the phosphorylation of muscle adenylic acid. He first suggested that the coupled phosphorylation might involve the reduction of the thiazole ring of TPP, in analogy to the reduction of the pyridine ring of DPN. After the appearance of the report by Negelein and Brömel, however, Lipmann offered evidence for the view that acetyl phosphate (the anhydride of acetic acid and phosphoric acid) is an intermediate in the process.[255] Kalckar's work dealt with oxidative phosphorylation in kidney extracts.[256] In 1941 both men published important review articles in which the coupling of oxidation reactions to the generation of ATP was considered in the light of the discoveries of the 1930s.[257] Both articles rendered great service in collecting and organizing the available knowledge on this subject, but Lipmann went further in his generalizations by introducing the concepts of "group potential" in transfer reactions and of the "energy-rich phosphate bond." He used the latter term to denote linkages in compounds such as creatine phosphate, ATP, phosphoenolpyruvate, 1,3-diphosphoglyceric acid, or acetyl phosphate, whose hydrolysis is associated with a relatively large negative free-energy change, whose value he took to be roughly equal to heats of hydrolysis determined in Meyerhof's laboratory. As was indicated earlier, such large $-\Delta G$ values mean that the equilibria in such hydrolytic reactions lie very far in the direction of the cleavage products. In enzyme-catalyzed phosphoryl-transfer reactions, such as Glucose + ATP → Glucose-6-phosphate + ADP, there is no hydrolysis of the "energy-rich" bond of ATP, and the fact that the equilibrium lies very far to the right is, in Lipmann's terminology, associated with the higher "group potential" of the terminal phosphoryl group in ATP than the one in glucose-6-phosphate. Thus the numerical difference between the ΔG in the hydrolysis of ATP to ADP and P (Lipmann used the value of about –10 kcal) and of glucose-6-phosphate (–3 kcal) makes the transfer reaction a strongly exergonic process.[258]

Lipmann proposed that in the presence of suitable catalysts, acetyl phosphate might be an acetylating agent, and he went on to state:

> A broader application of this metabolic principle can be visualized if it is remembered that the formation of a mixed acyl phosphate anhydride can be effected through phosphorylation from adenyl pyrophosphate. The major part of the constituents of protoplasma are compounds which contain the ester or the peptide linkage. In the routine procedure of organic chemistry for synthesis of compounds of this type the acyl chloride of the acid part is first prepared and then brought into reaction with the hydroxyl or amino group of the other part. In an analogous procedure the cell might first prepare the acyl phosphate with adenyl pyrophosphate as the source of energy-rich phosphate groups. The acyl phosphate of fatty acids might then condense with glycerol to form the fats. The acyl phosphate of amino acids likewise might condense with amino acids to form the proteins.[259]

These views reflected the emerging knowledge about enzyme-catalyzed transfer reactions, as well as the growing conviction that ATP plays a central role as a funneling agent of the chemical energy derived from the intracellular breakdown of carbohydrates to "drive" endergonic processes. Although, as we shall see later, the chemical mechanisms Lipmann envisaged required correction, his attractive concept of the "high-energy phosphate bond" (and the "squiggle" he used to denote it), took firm root in biochemical textbooks. For the biochemist Daniel Atkinson, however, "Such concepts as the storage of energy in a peculiar kind of bond, the release of energy on breaking a bond, and the movement of a phosphoryl group, complete with a squiggle bond, to another site have no chemical meaning. They and related statements have been presented in textbooks and reviews, and have had the unfortunate effect of confining metabolic chemistry behind a jargon curtain that has tended to isolate it from simple and direct chemical considerations, and to confuse and confound the relations between thermodynamics and metabolic processes in the minds of many students."[260] As for myself, upon reading Lipmann's 1941 article, I could not accept his sharp distinction between energy-rich and energy-poor bonds, and I questioned the estimates of the ΔG values, nor could I "unlearn the dicta that thermodynamic data alone cannot predict the pathway or the mechanism of a chemical process," and that in an open system "a strongly endergonic (and hence apparently irreversible) reaction can be made to proceed if a chemical mechanism is available for coupling the reaction to an appropriate exergonic process."[261]

Among the enzyme-catalyzed group-transfer reactions discovered during the 1930s, the process whereby the glucose units of glycogen enter the pathway of anaerobic glycolysis was also of general importance. In 1935 Jacob Parnas and Thaddeus Baranowski found that in muscle extracts glycogen and inorganic phosphate react to form hexose monophosphate under conditions where the oxidation had been blocked by iodoacetate and no ATP is being generated. This "phosphorolysis" of glycogen was elucidated by Carl and Gerty Cori through the isolation of glucose-1-phosphate as the product, and the subsequent crystallization, by Arda Green and Gerty Cori, of the enzyme ("phosphorylase") that catalyzes this readily reversible reaction. Cori and Cori also showed that a separate enzyme catalyzes the interconversion of glucose-1-phosphate and glucose-6-phosphate, thus linking the product of the hexokinase reaction to glycogen through two reversible reactions. We will return to these important findings in the next chapter, but it should be mentioned here that they also influenced the thought of biochemists who were interested in the enzymatic mechanisms in the biosynthesis of polymeric cell constituents. After the emergence of the concept of the "energy-rich phosphate bond," the reaction catalyzed by glycogen phosphorylase provided an example of the formation of a long-chain molecule in a process involving an "energy-poor phosphate bond," indicating that a sharp distinction could not be made between the two kinds of bonds.[262]

The Pasteur Effect

The great achievements of the 1930s in the study of the energy-yielding pathways of electron transfer from metabolites to molecular oxygen and of the anaerobic breakdown of glucose left unanswered, however, a problem that had been posed by Pasteur's observation that much less glucose is metabolized by yeast in the presence of oxygen than in its absence: "Fermentation by yeast appears to be essentially linked to the property this little plant possesses to respire, in some manner, with oxygen bound in sugar. . . . If one gives it a quantity of free oxygen sufficient for its requirements for life, nutrition, respiratory combustions, . . . it ceases to be a ferment. . . . On the other hand, if one suppresses all influence of air on the yeast, makes it develop in a sugar medium devoid of oxygen gas, it multiplies there just as if air were present, although less actively, and it is then that its character as a ferment is most pronounced."[263] Pasteur did not speculate extensively about the cause of this phenomenon but suggested that free oxygen "gives to the yeast a great vital activity"[264] whereby it can assimilate nutrients and grow as rapidly as ordinary aerobic organisms. As we have seen, Claude Bernard considered the principal role of oxygen in respiration to be one of stimulating the "vital activity" of animal tissues. Also, at that time, much attention was given to Pflüger's views about "living proteins" as energy-rich cell constituents that react directly with molecular oxygen to cause the combustion of metabolites. These views encouraged the formulation of general hypotheses about the connection of fermentative and oxidative processes, such as those offered by Wilhelm Pfeffer and Julius Wortmann.[265]

Many years later, when Otto Meyerhof found that the Harden-Young coferment also acts as a respiratory catalyst, he concluded that "the coenzyme of fermentation is at least partly identical with the respiratory substance of killed yeast and muscle tissue. This result supports the already old hypothesis, for which there has been hitherto little evidence, that the initial phases of respiration and fermentation are closely related." Subsequently, after he had explained the chemical changes during the aerobic recovery after muscular contraction in terms of a coupling of the oxidation of about 20 percent of the lactic acid and the resynthesis of the rest to glycogen, Meyerhof stated that "this connection between carbohydrate breakdown and oxidation is a general phenomenon of vital metabolism, and in the final analysis underlies the inhibition of the cleavages and fermentative processes, assumed by Pasteur."[266]

Also, during the 1920s Otto Warburg showed that cancer cells (Jensen sarcoma) exhibit high rates of glycolysis even in the presence of oxygen. From the finding that ethyl isocyanide (C_2H_5-NC), which inhibited heavy-metal catalysis of oxidation in his charcoal model systems, also abolished the inhibition of glycolysis by free oxygen, Warburg concluded that "respiration and fermentation are therefore linked by a chemical reaction which I term the 'Pasteur reaction' after its discoverer."[267] In his later writings, he continued to insist that when the respiration of normal cells is irreversibly damaged in some manner their metabolism becomes anaerobic in character. Warburg defended this view, with its implications for the search for a means of preventing and curing cancer, well into the 1950s, and, as in his debate with David Keilin or (as we shall see later, about photosynthesis), he treated his critics with the arrogance of a Prussian army officer.[268]

During the 1930s and 1940s several investigators sought to identify the cell component presumably associated with the "Pasteur reaction" (or Pasteur effect, as it came to be called). For example, in 1933 Lipmann found that treatment of muscle extracts with chemical oxidizing agents inhibits glycolysis and suggested that the Pasteur effect might be a consequence of the oxidation of the "glycolytic ferment" by an electron carrier such as one of the cytochromes. Some years later, Kurt Stern and Joseph Melnick reported the photochemical action spectrum of the CO derivative of the "Pasteur enzyme" in retina and yeast. In 1943 Vladimir Engelhardt offered evidence for the view that the Pasteur effect is due to the inhibition, by oxygen, of the conversion of hexose monophosphate to hexose diphosphate.[269] With the growth of knowledge in this field after 1970, however, the Pasteur effect has turned out to be a more complex phenomenon than had been thought.[270] Before considering the newer studies, it is necessary to sketch the development of several earlier lines of investigation.

The Citric Acid Cycle

During the 1930s more conclusive information about the link between the anaerobic breakdown of carbohydrate and the respiratory utilization of oxygen was provided by the search for carbon compounds that might be intermediates in the intracellular oxidation of pyruvate to CO_2 and H_2O. Before 1930 particular attention was focused on the chemically plausible sequence of four-carbon dicarboxylic acids: succinate → fumarate → malate → oxaloacetate; the decarboxylation of oxaloacetate gives pyruvate, whose oxidative decarboxylation gives acetate (then considered to be a product of the metabolic breakdown of glucose and of fatty acids). This sequence was based on the work of Battelli and Stern, Einbeck, Wieland, and especially Thunberg, who proposed that if there were a mechanism for linking two molecules of acetate to form succinate, a cyclic process would be available for the oxidation of one molecule of acetate to CO_2 and H_2O.[271] The assumption of a reductive condensation of two molecules of acetate to form succinate was entirely hypothetical, and no evidence was found for the occurrence of the reaction in biological systems. Similarly, in 1930 Erich Toenniessen proposed a cycle involving the reductive condensation of two molecules of pyruvate to form diketoadipate (HOOC-CO-CH_2-CH_2-CO-COOH), but other investigators soon provided evidence against a role of this compound in the oxidation of pyruvate by animal tissues.[272]

A more fruitful approach to the problem proved to be that of Albert Szent-Györgyi and his associates during 1934–1937. They observed that the oxygen uptake by suspensions of minced pigeon-breast muscle (selected because of the rapid respiration of this tissue) is stabilized at a high rate by the addition of small amounts of fumarate. Since the fumarate did not disappear, it was clearly acting as a catalyst. From the ratio of the CO_2 produced and the O_2 consumed (the "respiratory quotient") in the process catalyzed by fumarate, they inferred that carbohydrate, or its product ("triose"), was being subjected to oxidation. These results were confirmed and extended by other investigators, notably Frederick Stare and Carl Baumann, and it was also found that the addition of the other four-carbon dicarboxylic acids (succinate, malate, oxaloacetate) produced an effect similar to that shown by fumarate. Moreover, malonate (HOOC-CH_2-COOH), known to inhibit the action of succinic dehydrogenase (which was believed to react directly with the cytochrome system), did not abolish the catalytic action of fumarate. Szent-Györgyi therefore proposed that the cozymase-dependent dehydrogenation of metabolites is linked to oxygen by the sequence of reactions: "hydrogen donor" → oxaloacetate → malate → fumarate → succinate → cytochrome → oxygen.[273]

In a memorable paper published in 1937, Hans Krebs and William Arthur Johnson reported that the six-carbon tricarboxylic citric acid also exerts a catalytic effect on the respiration of minced pigeon breast muscle, that citrate is successively converted to α-ketoglutarate and succinate, and that citrate is formed from oxaloacetate by the addition of two carbon atoms from an unidentified source, provisionally denoted as triose.[274] Citric acid had been known to be a major constituent of plant tissues since its isolation by Scheele (in 1785) and Proust (in 1801) and its study by Liebig during the 1830s contributed to the formulation of his theory of polybasic acids. Later, citric acid was identified as product of the fermentation of sugar by some molds, and also was found in small amounts in animal tissues.

Shortly before the appearance of the paper by Krebs and Johnson, Franz Knoop and Carl Martius drew attention to the chemically plausible condensation of oxaloacetate and acetate but stated that their efforts to effect this reaction by chemical means had been unsuccessful. They showed, however, that if one uses pyruvate instead of acetate and subjects the mixture to oxidation by hydrogen peroxide in alkaline solution, citrate is thereby produced. (I mentioned earlier that this kind of chemical treatment had been used by Henry Dakin and Carl Neuberg in experiments designed to mimic biological oxidations). Martius and Knoop then showed that Thunberg's "citric dehydrogenase" (from liver) converts citrate into α-ketoglutarate anaerobically, and they suggested the following sequence of reactions: citrate → *cis*-aconitate → isocitrate → oxalosuccinate → α-ketoglutarate.[275] The structural formulas of these compounds are given in the scheme on p. 316. Isocitric acid had been synthesized by Rudolf Fittig in 1889, and oxalosuccinic acid had been studied by Emile Blaise and Henry Gault in 1908. Although *trans*-aconitic acid was described by Ludwig Claisen in 1891, the less stable *cis* isomer was prepared only in 1928 by Roman Malachowski.

Krebs and Johnson also showed that in agreement with the theory of Knoop and Martius, isocitrate and *cis*-aconitate are oxidized by muscle preparations as rapidly as citrate. With the addition of the four-carbon acids of the earlier schemes, and the key step of the conversion of oxaloacetate into citrate, they accordingly proposed a "citric acid cycle": citrate → isocitrate → oxalosuccinate → α-ketoglutarate → succinate → fumarate → malate → oxaloacetate → citrate, and they left open the question whether *cis*-aconitate is a component of the direct pathway. In the operation of this cycle, two carbon atoms are derived from "triose," and two carbon atoms are released as CO_2 in the steps from oxalosuccinate to succinate.

The decisive experiment in support of the theory was the demonstration of the aerobic formation of succinate from fumarate in the presence of malonate, which blocks the interconversion of these compounds. In a paper with Leonard Victor Eggleston, Krebs wrote:

> The following reaction takes place in the presence of malonate: (1) Fumarate + pyruvate + O_2 = succinate + 3 CO_2 + H_2O. In the absence of malonate this action is followed by the oxidation of succinate; (2) Succinate + ½O_2 = fumarate + H_2O. The net effect of 1 and 2 is the complete oxidation of pyruvate: Pyruvate + 2½O_2 = 3CO_2 + 2H_2O. In reaction 1 the succinate does not arise from fumarate by anaerobic reduction. It is formed from fumarate and pyruvate by a series of oxidative processes formulated in the theory of the "citric acid cycle." This theory is supported by the fact that a change in experimental conditions directs reaction 1 in such a way as to yield citrate (about 15%) or α-ketoglutarate (about 50%) instead of succinate and CO_2. Up to the present the citric acid cycle is the only theory accounting for experimental observations in pigeon breast muscle.[276]

Although the citric acid cycle was originally proposed for pigeon breast muscle, work in many laboratories soon showed it to be operative in other animal tissues, in plants, and in aerobic microorganisms. For some years during the 1940s, the term *tricarboxylic acid cycle* was preferred because the use of isotopes in metabolic studies (about which more in the next chapter) had given results which appeared to be inconsistent with the fact that citric acid is a symmetrical molecule, and it was thought that citric acid is therefore not on the direct pathway of the cycle. In 1948 Alexander Ogston pointed out that if a symmetrical molecule interacts with an asymmetric center at three points, the reaction product will be labeled in an asymmetric manner. Ogston's idea was found by Van Potter and Charles Heidelberger to apply to the conversion of citric acid, which was then restored to the direct pathway. As was noted by Ronald Bentley, Ogston did not cite the earlier introduction of the general concept by Leslie Easson and Edgar Stedman in 1933 or the similar "polyaffinity theory" offered by Max Bergmann in 1935 in their writings about the stereochemistry of enzyme-substrate interactions.[277]

After 1945 the operation of the citric acid cycle was intensively studied in many laboratories, through the isolation and crystallization of the enzymes that catalyze the individual steps and the identification of the cofactors involved in the action of these enzymes, and by the use of radioisotopes to trace the path of labeled carbon atoms through the cycle. The most significant contributions were those relating to the process whereby a two-carbon unit is derived from pyruvate and condensed with oxaloacetate to form citrate. It was recognized that the oxidation of pyruvate leads to the intermediate formation of a reactive two-carbon fragment ("active acetate") and that thiamine pyrophosphate is a cofactor in the oxidative decarboxylation, but the chemical nature of this intermediate was unknown.

In Fritz Lipmann's experiments acetyl phosphate did not live up to his earlier expectations, and he turned to the study of a model system, the acetylation in liver extracts of the amino group of the drug sulfanilamide. James Klein and Jerome Harris had previously shown, with liver slices, that this process is linked to respiration and during the course of his studies Lipmann not only showed that in the presence of added ATP the extracts also effected this reaction, but he also discovered the requirement for a heat-stable dialyzable cofactor. A similar cofactor was independently found by David Nachmansohn to be required for the ATP-dependent acetylation of choline in brain tissue. Lipmann named the cofactor coenzyme A (A for acetylation), subsequently abbreviated to CoA. Because of the numerous instances in which coenzymes had already turned out to be related to known vitamins, partially purified samples of CoA were tested for vitamin activity and found to contain a

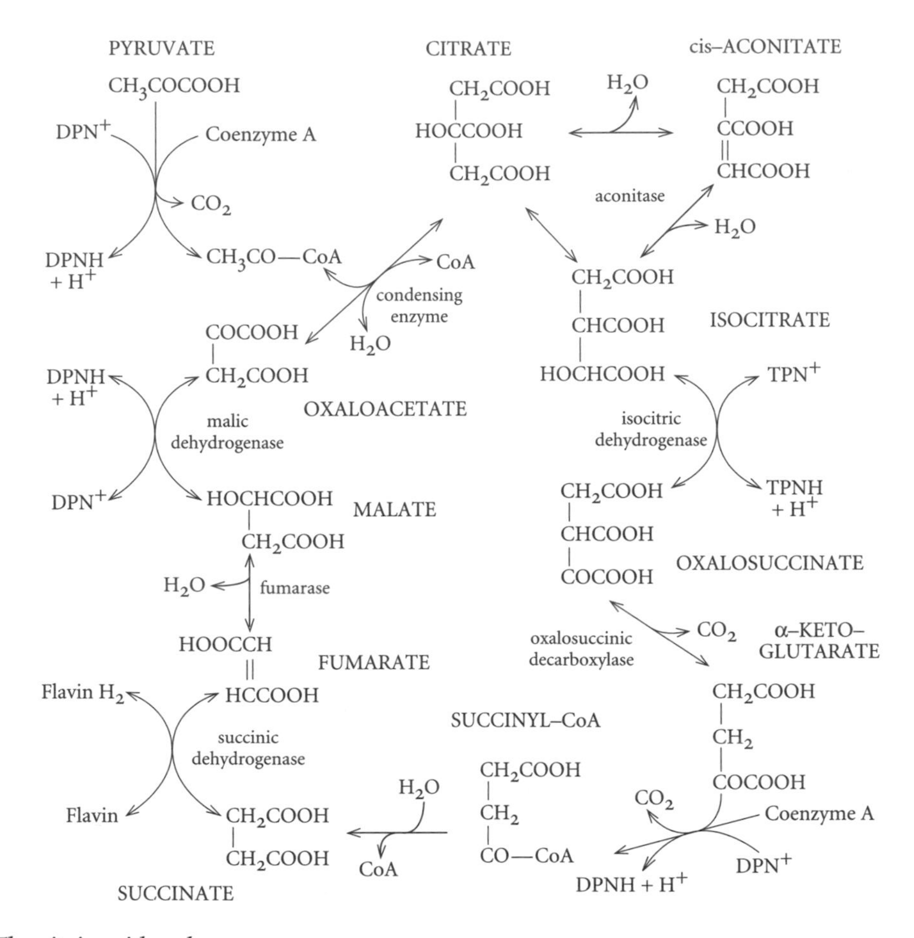

The citric acid cycle

vitamin that Roger Williams had discovered in 1933 and named pantothenic acid (Greek *pantothen,* from every side) because of its wide distribution.[278] By 1940 its chemical structure had been determined and, before its identification as a constituent of CoA, Williams wrote: "The presumption, in view of the known function of thiamin, nicotinamide, and riboflavin, is that pantothenic acid fits into some enzyme system (or systems) which is essential for metabolism. What this enzyme system is or what they are is not known. There are some facts which suggest that pantothenic acid may be concerned with carbohydrate metabolism, but this is not certain."[279] The structure of CoA was deduced in 1952 from the chemical studies in Lipmann's laboratory, as well as by James Baddiley, Esmond Snell, and their associates; its synthesis was effected by John Moffatt and Gobind Khorana.[280] A major achievement was Feodor Lynen's isolation of an acetylated CoA from yeast, and his demonstration that the acetyl group is linked to the sulfur atom to form a thiol ester. This acetyl-CoA was then shown to be the substance which reacts with oxaloacetate to form citrate in the presence of the "condensing enzyme" (now termed citrate synthase) crystallized in Severo Ochoa's laboratory.[281]

As for the formation of acetyl-CoA from pyruvate, the process was found during the 1960s to be effected by a widely distributed organized multienzyme complex consisting of three enzyme proteins and six cofactors. In addition to TPP, NAD, FAD, CoA, and Mg^{2+}, "lipoic acid" (a newly discovered disulfide) was required. In this remarkable investigation, notably in the laboratories of Lester Reed and of Irwin Gunsalus, not only were the components of the complex dissociated and the individual steps in the process characterized, but it was also shown that the protein components reassociate spontaneously to reform the active complex. I will not discuss the mechanisms of the chemical reactions in this process, for which I refer the reader to a modern biochemical textbook.[282] It should be noted that, apart from the significance of these investigations in elucidating a key step in the generation of biochemical energy, they brought to the fore the importance of specific protein-protein interactions in the organization of the functional units of living cells. We have already mentioned cytochrome c oxidase; other examples will appear later.

To conclude this section, it is appropriate to note that the establishment of the citric acid cycle as the principal intracellular mechanism for the oxidation of pyruvate in animal tissues has influenced the study of similar processes in other biological systems. For example, some plants and bacteria that can utilize acetate for

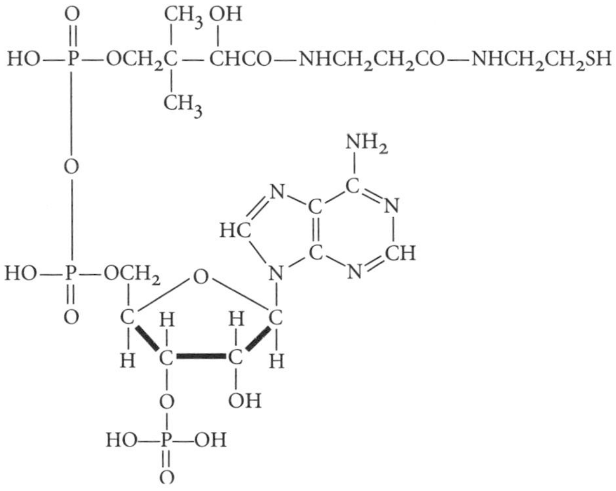

O　　CH_3 OH
HO—P—OCH_2C—CHCO—$NHCH_2CH_2CO$—$NHCH_2CH_2SH$
CH_3
NH_2
O
N　C　N
HC
C　CH
HO—P—OCH_2　O　N　N
O　C　H　H　C
H　C　C　H
O　OH
HO—P—OH
O

Coenzyme A

growth have been found to generate succinate in a "glyoxylate cycle" in which isocitrate is cleaved enzymatically to succinate and glyoxylate [HC(O)-COO$^-$]; the enzyme-catalyzed addition of the acetyl group of acetyl-CoA to glyoxylate produces malate, which is converted to oxaloacetate and citrate.[283]

Oxidative Phosphorylation

The acceptance of the citric acid cycle as a general mechanism for the aerobic oxidation of pyruvate to CO_2 and H_2O had provided by 1940 the long-sought link between anaerobic glycolysis or fermentation and the respiratory utilization of oxygen. In 1940, however, the available experimental knowledge did not permit the formulation of coherent schemes for the transfer of the five electron pairs, to oxygen, from any of the five metabolites undergoing dehydrogenation in the citric acid cycle. Although the great achievements of the 1930s had revealed the role of such electron carrier systems as the flavin and pyridine nucleotides as links in a respiratory chain from metabolites such as lactate to the cytochromes and cytochrome oxidase, the synthesis of this knowledge into a conceptual model of the chain in a defined biological system such as muscle proved to be difficult. It could only be surmised, on thermodynamic grounds, that the chain represents a sequence based on the order of their oxidation-reduction potentials. The estimated values (in volts, at pH 7 and 37°) were: DPN (or TPN) –0.3, flavoprotein –0.1, cytochrome b –0.04, cytochrome c +0.26, cytochrome a +0.29. On the basis of such relations, Eric Ball concluded that "The energy liberated when substrates undergo air oxidation is not liberated in one large burst, as was once thought, but is released in stepwise fashion. At least six separate steps appear to be involved. The process is not unlike that of locks in a canal. As each lock is passed from a lower to a higher level a certain amount of energy is expended. Similarly, the total energy resulting from the oxidation of foodstuffs is released in small units or parcels, step by step. The amount of free energy released at each step is proportional to the difference in potential of the systems comprising the several steps."[284] During the 1950s Britton Chance developed rapid and sensitive spectrophotometric methods for the measurement of the oxidation-reduction state of intracellular electron carriers and provided direct experimental evidence in favor of this sequence of reactions.[285]

After the discovery of ATP and Lundsgaard's work on the phosphorylation of creatine during oxidative recovery in muscle, many investigators recognized the general importance of phosphorylations in the biological transformations of energy. Among the first to study the coupling of respiration and phosphorylation in animal tissues were Vladimir Engelhardt, John Runnström and Leonor Michaelis, Herman

Kalckar, and especially Vladimir Belitser and Elena Tsibakova.[286] Belitser and Tsibakova reported that the phosphorylation of creatine in heart muscle is coupled to the oxidation of any of a variety of metabolites (citrate, α-ketoglutarate, succinate, fumarate, malate, pyruvate, lactate), and that at least two molecules of creatine phosphate are formed per atom of oxygen consumed. Their findings were soon confirmed and extended, most notably by Severo Ochoa, who provided evidence for a "P/O ratio" of three in the aerobic oxidation of a substrate such as pyruvate.[287] This result was used in thermodynamic calculations (based on assumptions of doubtful validity) of the efficiency of oxidative phosphorylation, and encouraged renewed studies on the free energy change in the hydrolysis of ATP to ADP and P. As a consequence, the accepted value of $\Delta G°$ (pH 7) dropped from about 13 kcal/mole to about 7 kcal/mole.[288] It should also be recalled that although it was known that the generation of ATP from ADP and P in the dehydrogenation of glyceraldehyde-3-phosphate is a consequence of the intermediate formation of the "energy-rich" acyl phosphate 1,3-diphosphoglyceric acid, no comparable mechanism could be invoked for the coupling of electron transfer by the respiratory chain to the phosphorylation of ADP.

By 1950 it was clear from the work of George Hogeboom, Alfred Lehninger, and their associates that mitochondria isolated from animal tissues contain the complete enzymatic apparatus of the citric acid cycle, whereas the glycolytic enzymes are largely present in the "cell sap," namely the supernatant fraction obtained upon high-speed centrifugation. In 1949 Kennedy and Lehninger wrote: "The striking fact is that all the individual enzymes concerned in these complex systems should be found in one species of morphological element. These findings in some measure justify the early views of Altmann that these bodies are fundamental biological units and possess a certain degree of autonomy and certainly, together with the considerable work already done by the Rockefeller school, [Walter] Schneider, and others, provide considerable basis for the apt designation 'intracellular power plants' conferred on the mitochondria by Claude."[289] Some words may be inserted at this point about the contribution of Richard Altmann, mentioned in this quotation. During the latter half of the nineteenth century, there was a remarkable development of cytology, aided by the availability of improved microscopes and staining techniques. The most notable achievements were in the study of the role of the nucleus in cell division, fertilization, and heredity (about which more in a later chapter). During this period there were recurrent reports about supposed structural elements in the cytoplasm, among them the *microzymas* of Antoine Béchamp and Altmann's *Elementarorganismen,* which were thought to be fundamental units of life and homologous with bacteria.[290] Many of Altmann's contemporaries considered such cytoplasmic structures to be artifacts produced by the reagents, such as osmic acid, used as fixatives in preparing tissue slices for staining and microscopic observation. For example, William Bate Hardy wrote: "It is notorious that the various fixing agents are coagulants of organic colloids, and that they produce precipitates which have a certain figure or structure. . . . Altmann demonstrates his granules by the aid of an intensely acid and oxidizing mixture."[291] Among the cytoplasmic granules of Altmann were rodlike structures to which Carl Benda gave the name *mitochondria* in 1898. Two years later, while working in Paul Ehrlich's laboratory, Leonor Michaelis described the "vital" staining of these structures with the dye Janus green; this technique, which avoided the use of fixatives, had been introduced by Ehrlich in 1886, when he used it to stain nerves with methylene blue.[292] In giving reality to the mitochondria, and in providing a specific dye for their microscopic observation, Michaelis laid the groundwork for an extensive series of later cytological studies, notably those of Robert Bensley and Normand Louis Hoerr and of Albert Claude, who separated mitochondria from suspensions of mechanically disintegrated tissues ("homogenates") by differential centrifugation. This method for isolating intact mitochondria was subsequently improved by dispersing the homogenate in concentrated sucrose.[293]

It should also be noted that by the 1940s biochemists were well prepared to accept the idea that intracellular oxidation is associated with particulate cytoplasmic elements. Although Altmann had suggested that his granules are responsible for this process, most of his contemporaries placed the site of respiratory activity in the nucleus. In retrospect the first definite indications that cell respiration might be associated with such water-insoluble elements may be discerned in the work of Battelli and Stern and of Warburg before 1914. Later, as new knowledge was acquired, this idea was reiterated by several investigators, notably Keilin and Szent-Györgyi during the 1930s.

After 1945 commercial electron microscopes became available for biological research, and this instrument was used by George Palade (at the Rockefeller Institute) and Fritiof Sjöstrand (at the Karolinska Institute, Stockholm) to examine the internal structure of mitochondria. They used different methods for preparing the material for observation, and they differed somewhat in their interpretation of the results.[294] In the model reproduced in present-day biochemical textbooks, mitochondria have an outer membrane and a convoluted inner membrane system, which is the site of oxidative phosphorylation, with the enzymes of the citric acid cycle in the space between the folds of the internal ridges (Palade named them cristae). I have already mentioned the fruitful use of electron microscopy in the study of the components of multiprotein systems and will discuss in a later chapter its role in the development of knowledge about the structure of DNA and bacteriophages.[295]

During the 1950s other new techniques, in particular the use of the radioactive isotope ^{32}P, were applied to the study of oxidative phosphorylation. In this manner Albert Lehninger and his associates showed that mitochondria effect the coupling of the phosphorylation of ADP to the electron transfer from reduced NAD to oxygen.[296] Parallel work at the laboratory of David Green led to the discovery by Frederick Crane of coenzyme Q and to the study of its place in the chain in relation to other electron carriers (flavin, iron-sulfur proteins, cytochrome b).[297] Britton Chance and his associates made brilliant use of spectroscopic and polarographic methods in identifying three steps in the respiratory chain as loci of oxidative phosphorylation: NADH → flavinH_2, → cytochrome b → cytochrome c, cytochrome a → oxygen.[298] Another valuable technique was the use of compounds found to "uncouple" phosphorylation from electron transfer.

If by 1960 there was general agreement that three molecules of ATP are generated per atom of oxygen consumed, and that the approximate location of the three coupling sites in the respiratory chain had been determined, the mechanism of the processes that link electron transfer to the phosphorylation of ADP was unknown. Valuable clues had come from the isotope experiments of Mildred Cohn and Paul Boyer on the exchange reactions between ATP and inorganic phosphate labeled with ^{18}O or ^{32}P in the presence of mitochondria, but the interpretation of the results was uncertain.[299] In 1960, however, Efraim Racker and his associates made a decisive contribution in achieving a partial resolution of the mitochondrial apparatus that effects ATP synthesis. This approach, whose strategy had proved to be successful in the study of anaerobic glycolysis, led to the separation, from the inner membrane of beef heart muscle mitochondria, of a soluble coupling factor (F_1) that the Racker group identified as an ATPase.[300] The discovery that a "degradative" enzyme was intimately involved in the intracellular catalysis of the reversal of what had been pronounced to be a physiologically "irreversible" process marked a significant change in biochemical thought and a return to van't Hoff's insistence on the reversibility of enzyme-catalyzed reactions. It was recognized that, given coupling to a sufficiently effective exergonic process, a "hydrolase" can also act as a "synthetase."[301] Moreover, it was not always remembered that for the maintenance of a "steady state" concentration of ATP in a biological system, a mechanism is required for its removal by hydrolysis, or by other means.[302] After the identification of the F_1 coupling factor, it was shown to be a component in a multiprotein complex, now usually termed ATP synthase. In this complex, F_1 is closely associated with a membrane structure denoted F_O; recently, the three-dimensional structure of F_1 has been determined by X-ray diffraction analysis.[303] The Racker group also showed that the activity of F_1 is inhibited by 2,4-dinitrophenol, long known as a stimulant of respiration, and identified as an "uncoupler" of oxidative phosphorylation.[304] In extending their investigation, Racker and his associates also made significant progress in reconstituting the mitochondrial apparatus for oxidative phosphorylation.[305]

During the 1960s there began a spirited debate about the mechanism of mitochondrial oxidative phosphorylation. As has been mentioned, an attractive possibility was the intermediate formation of an "energy-rich" phosphate, comparable to 1,3-diphosphoglycerate or phosphoenolpyruvate. Such a theory was put forward by Edward Charles ("Bill") Slater in 1953, but the subsequent search for that kind of intermediate proved to be unavailing. In response to such hypotheses, in 1961 Peter Mitchell offered a theory that invoked no chemical intermediates, and he considered the problem of the oxidative phosphorylation of ADP to be closely linked to the problem of the transport, across the mitochondrial membrane, of protons (H^+) generated by the respiratory chain.[306] Mitchell's interest in biological membranes was derived in large part from his brief association with James Danielli during the 1940s, as a student at the Cambridge Biochemical Laboratory, where he began work (with Jennifer Moyle) on the transport of ions through the osmotic barrier of bacteria. This line of research was continued in the Department of Zoology at Edinburgh until 1963, and Mitchell then set up a private laboratory (in Cornwall) for his further studies. In the formulation of what he called the chemiosmotic coupling mechanism, Mitchell postulated that the energy transfer from the membrane enzymes of the respiratory chain to the ATP-synthesizing system is effected by a transmembrane electrochemical proton gradient ("proton-motive force") composed of a membrane potential and a difference in pH across the membrane. Mitchell then extended his approach to other intracellular processes and contrasted it with that in which "metabolic pathways . . . were intended to represent the sequence of the chemical transformations, catalyzed by enzymes and catalytic carriers that could be dissolved or dispersed in homogeneous aqueous solution, according to the classical bag-of-enzymes view of metabolism." Some years later he wrote: "Fifty years ago, the bag-of-enzymes view of cell metabolism was prevalent, and the chemical actions of metabolism were generally looked upon purely as processes of primary chemical transformation."[307] This may have been Mitchell's view of the attitude of the enzyme group at the Cambridge Biochemical Laboratory at the time he was there, but it cannot, in my opinion, be considered to be an accurate or unbiased assessment.

The initial response to Mitchell's theory was largely unfavorable, and there ensued a lively debate that stimulated experimental research in this field. Evidence was presented for the importance of proton translocation, but in its original form the chemiosmotic theory was shown to be incorrect as regards the stoichiometry of the process, and alternative hypotheses were considered. The idea that intramembrane proton fluxes might mediate the coupling of electron transfer and ATP synthesis had already been proposed by R. N. Robertson and by R. J. P. Williams, and Paul Boyer invoked the occurrence of conformational changes arising from the direct interaction of the enzymes of the respiratory chain and those of the ATP-synthesizing system. The debate reached its peak during the 1970s, with the appearance of a set of articles by six of the leading participants, including Mitchell.[308] It seems likely that important features of some of the other hypotheses will be found in whatever mechanism emerges from future investigation, although the award of the 1978 Nobel Prize in Chemistry to Mitchell alone has tilted some historical accounts of the recent development of knowledge about oxidative phosphorylation, with comparisons to "paradigm shifts" during past scientific "revolutions."[309]

Active Membrane Transport

Since about 1970 membrane research has become one of the most vigorous areas of the interplay of chemical and biological thought and action. This development was spurred not only by the discussion of the chemiosmotic theory but also by the availability of new techniques introduced after 1945. In addition to those mentioned above, these techniques included the use of detergents to disperse membrane proteins, of limited proteolysis, of phospholipid vesicles, of acrylamide gel electrophoresis, and of new spectroscopic methods.[310]

In the nineteenth-century background of this upsurge of interest in the phenomena associated with the movement of water and of dissolved solutes across biological membranes are the investigations of Henri Dutrochet during 1827–1832, the introduction of artificial semipermeable membranes by the chemist Moritz Traube during the 1860s, the measurements of osmotic pressure by the plant physiologist Wilhelm Pfeffer during the 1870s, and the pharmacologist Ernest Overton's proposal during the 1890s that cell membranes are largely lipoidal in nature. Dutrochet used the terms *exosmose* and *endosmose* in describing his finding that certain biological membranes allow the passage of water but hold back dissolved solutes, so that when two solutions of different concentration of a solute are separated by such a "semipermeable" membrane, water passes from the less concentrated to the more concentrated solution.[311] Traube's gelatinous copper ferricyanide membrane, extensively used by later investigators, was freely permeable to water but not to dissolved sucrose.[312] As we have seen, Pfeffer's quantitative data were used by van't Hoff to develop a theory of the osmotic pressure for dilute solutions.[313] Overton's theory was widely adopted and later found a modern expression in the "fluid mosaic model" of the cell membrane.[314]

During the first three decades of the twentieth century, important contributions were made by numerous investigators, notably S. C. Brooks, J. F. Danielli, E. Gorter, M. H. Jacobs, P. R. Collander, H. G. Lundegardh, and W. J. V. Osterhout. Much work was done on blood red cells, which were shown to possess a bilayer lipoidal membrane.[315] Also, Lars Onsager had developed a thermodynamic treatment of the flow of ions, solvent, and current through a membrane.[316] After about 1950, with the knowledge of cell structure provided by electron microscopy and the study of the mechanism of oxidative phosphorylation, the results of this earlier work on ion transport began to be interpreted in biochemical terms. Among the first notable achievements was Jens Skou's discovery, in 1963, of the role of an ATPase in the transfer of sodium and potassium ions across cell membranes. This Na^+,K^+-ATPase has been the object of intensive biochemical study and, like the mitochondrial H^+-ATPase, is a part of a multiprotein assembly. Its physiological function is to use the energy derived from the cleavage of ATP to drive the pumping in of K^+ and the pumping out of Na^+. An important feature of the mechanism of this process is that, in the presence of Na^+ (and Mg^{2+}), the terminal phosphoryl group of ATP is transferred to a protein aspartyl unit to form an acyl phosphate, whose cleavage requires the presence of K^+.[317] The finding that this process is inhibited by steroids present in digitalis, used since the eighteenth century to treat congestive heart failure, is another example of the practical fruits of the detailed chemical investigation of medical problems. Another ion pump driven by the cleavage of ATP involves the action of a Ca^{2+}-ATPase in the regulation by calcium ions of muscular contraction. The mechanism is similar to that in the case of the Na^+,K^+-ATPase, with the intermediate formation of a phosphorylated enzyme protein.[318] A third ATP-ase of this type is the one (K^+,H^+-ATPase) that pumps acid into the stomach. In addition, transport mechanisms have been identified that are not directly dependent on the cleavage of ATP. Among these processes are the transport, in some animal cells, of sugars and amino acids driven by a Na^+ gradient and, in some bacteria, by a proton gradient generated by the flow of electrons through the respiratory chain.[319] The known membrane proteins involved in such natural transport are considered to form channels, and this also appears to be the case in the action of the linear peptide antibiotic gramicidin A in the transport of potassium ions. Another antibiotic, the cyclic peptide valinomycin, acts instead as a carrier of K^+ ions by forming a complex with the ion in the center of the ring.[320]

Light as a Source of Biochemical Energy

In his *Vegetable Staticks,* published in 1727, Stephen Hales raised the question of whether the effect of light on plant life is comparable to the respiration of animals, by deriving "some part of their nourishment from the air." With the identification of the various "airs" later in the eighteenth century, there also came Joseph Priestley's famous observation that the "fixed air" exhaled by mice is converted into "dephlogisticated air" by "a sprig of mint." Soon afterward, Jan Ingen-Housz showed that this action of green plants occurs only in sunlight, and Jean Senebier confirmed and extended this conclusion.[321] From the quantitative experiments of Théodore de Saussure, it seemed that the function of light was not only to cleave carbon dioxide with the retention of the carbon and release of oxygen, thus "revivifying" the atmosphere, but also that water is not the source of the oxygen.[322] Indeed, during the nineteenth century and some years afterward, the released oxygen was generally considered to come from CO_2, with the "assimilation" of the carbon for the formation of carbohydrate.

After the work of Joseph Pelletier and Joseph Caventou on the green pigment ("chlorophyll"), during the 1830s its absorption spectrum and fluorescence were described by David Brewster, and its chemical properties were further examined by Berzelius. In the succeeding years of the nineteenth century, many investigators, including Gerrit Mulder, François Verdeil, George Gabriel Stokes, Felix Hoppe-Seyler, Edward Schunck, and Leon Marchlewski, provided valuable data indicating the relation of chlorophyll to the iron-porphyrin of hemoglobin. After 1900 the work of Michael Tswett, Richard Willstätter and his associate Arthur Stoll, Hans Fischer, James Conant, and Robert Woodward established the structure of chlorophyll.[323] From these contributions it was learned that chlorophyll exists in two forms, designated *a* and *b,* which differ in a substituent on a porphyrin ring ("pheophytin") with magnesium ion at its center, and with a long-chain aliphatic alcohol ("phytol") as a ring substituent.

For the study of the role of chlorophyll in what came to be called photosynthesis, of special importance were the nineteenth-century contributions of Julius Sachs on the light-dependence of starch formation, and of Theodor Wilhelm Engelmann, who showed that the process is effected by intracellular structures (chloroplasts).[324] Unsuccessful attempts to prepare from green leaves cell-free juices that effect photosynthesis reinforced the view that it depends on the integrity of cell structure.[325] Important clues regarding the chemical process in photosynthesis were provided through work on the purple bacteria discovered by Engelmann during the 1880s, and the finding that these organisms effect the light-dependent synthesis of cell material without the release of O_2, as well as the parallel discovery by Sergei Winogradsky of "chemosynthetic" bacteria, which contain no chlorophyll and can assimilate CO_2 in the dark.[326] It was not until the studies of René Wurmser and Cornelis van Niel during the 1920s and 1930s, however, that the general significance of these clues became evident. From their work, it appeared that the overall chemical change in the conversion of CO_2 and H_2O into hexoses is

$$6\,CO_2 + 12\,H_2A \rightarrow (CH_2O)_6 + 6\,H_2O + 12\,A$$

where A is the oxidized form of a hydrogen donor H_2A. This formulation, based on the dehydrogenation theory of Wieland and Thunberg, separated the photosynthetic process into a photochemical reaction in which light energy is used to effect the endergonic reduction of water (or H_2A) and a nonphotochemical reaction in which the endergonic synthesis of carbohydrate from CO_2 and H_2O is driven by the energy transferred by the photochemical reaction.[327]

Decisive evidence in support of the view that the action of light is to effect the photolytic reduction of water was provided by Robert ("Robin") Hill, who showed that in the absence of CO_2 isolated chloroplasts release O_2 when illuminated in the presence of a hydrogen acceptor such as ferric oxalate, quinone, or one of the oxidation-reduction dyes. In addition to the discovery of what came to be called the "Hill reaction," Samuel Ruben and his associates gave further support by the use of H_2O and CO_2 labeled with ^{18}O, and showed that the released O_2 comes from the water while the oxygen of CO_2 enters into organic compounds.[328]

I interrupt the continuity of the historical account at this point to consider the development of knowledge about the efficiency of the photosynthetic process. During the nineteenth century there were numerous studies on the action spectrum of the process.[329] Just before the outbreak of World War I, Otto Warburg undertook to determine the quantum efficiency of photosynthesis, and in 1919 he described new methods for this purpose. In addition to his modification of the Brodie manometer, he introduced the use of the unicellular green alga *Chlorella,* which was later used by other investigators. With Erwin Negelein, Warburg then reported that upon illumination with red light (660 nm), where chlorophyll has a strong absorption band, three to four light quanta are sufficient for the production of one molecule of O_2.[330] He considered the released oxygen to come from CO_2 and stubbornly adhered to this view long after the evidence to the contrary became available. Moreover, he continued to insist on the validity of his estimate of the efficiency of photosynthesis, in the face of the data presented by others, notably Robert Emerson, for a much higher quantum requirement per molecule of CO_2 utilized.[331] As in his debates about the Atmungsferment or the cause of cancer, Warburg dismissed the work of others who presumed to criticize his theories, most of which proved to be wrong, and the esteem he had earned for the brilliance of his experimental achievements was diminished by the arrogance of his behavior in these debates.

The question of the quantum efficiency of photosynthesis faded into the background after 1940, when the isotope technique was applied to the study of the photosynthetic process. Earlier work, involving model experiments and much speculation, was mainly based on Adolf Baeyer's suggestion that the first product of CO_2 assimilation is formaldehyde (H_2CO), which is polymerized to hexoses.[332] Among the first experiments conducted with radioactive carbon isotopes were those on the fate of $^{11}CO_2$ in Chlorella; Samuel Ruben, Martin Kamen, and Zev Hassid found the incorporation of labeled carbon into organic compounds to be the same in the dark as in the light, thus showing that the utilization of CO_2 is indeed a nonphotochemical process and offering further support for the earlier recognition, by Frederick Blackman, of separate "light" and "dark" reactions in photosynthesis. Because of the long-held view that formaldehyde is an intermediate in the formation of carbohydrates, they attempted to isolate labeled H_2CO in various ways, without success.[333] By 1940, as we have seen, the outlines of the sequence of enzyme-catalyzed reactions in the Embden-Meyerhof pathway and the Krebs citric acid cycle had been formulated, and the coupling of oxidation-reduction processes to the cleavage and formation of ATP had been demonstrated. It was against the background of this new knowledge that, in the last year of his short life, Ruben offered the perceptive hypothesis that in photosynthesis the reduced pyridine nucleotide and ATP needed to reverse the glycolytic pathway are generated by coupling to the light-dependent reduction of water.[334]

After 1940, with the availability of the long-lived radioactive ^{14}C,[335] it became the carbon isotope of choice for metabolic studies. In particular, the organic chemist Melvin Calvin and his associates applied the new chromatographic techniques to the separation of the radioactive substances formed from $^{14}CO_2$ by Chlorella either in the dark or after different periods of exposure to light.[336] The identification of the chemical nature of the individual substances required several years of sustained effort. It was found that after brief illumination (five seconds), most of the ^{14}C that is incorporated into organic material appears in the carboxyl group of 3-phosphoglyceric acid, long known as a key intermediate in glycolysis.

Because the radioactive hexoses which appeared after longer illumination were largely labeled in the two central carbon atoms (carbons 3 and 4), it was reasonable to conclude that the hexoses arose from 3-phosphoglyceric acid by a reversal of the glycolytic pathway. The solution of the problem of the rapid formation of 3-phosphoglyceric acid came from parallel studies on the metabolic fate of 6-phosphogluconic acid, the product of the NADP-dependent dehydrogenation of glucose 6-phosphate studied by Warburg. After 1945 it was shown that 6-phosphogluconic acid is oxidatively decarboxylated to a five-carbon sugar phosphate (ribulose-5-phosphate) that turned out to be a key intermediate in an "alternative" pathway of glucose breakdown in various plant and animal tissues, as well as in some microorganisms. By 1955 the main features of this "pentose phosphate pathway" had been worked out through the efforts of the research groups associated with Frank Dickens, Seymour Cohen, Efraim Racker, and especially Bernard Horecker.[337] These advances led Calvin to conclude that ribulose-1,5-diphosphate, made from ribulose-5-phosphate by enzymatic phosphorylation with ATP, is the immediate CO_2 acceptor in photosynthesis, and that the resulting intermediate is rapidly cleaved to yield two molecules of 3-phosphoglyceric acid, in a process denoted the reductive pentose phosphate cycle.

During the 1950s, in addition to the achievement of the Calvin group, of equal importance was the demonstration by Daniel Arnon and others that isolated chloroplasts can, upon illumination, form carbohydrate from CO_2 and H_2O, and that particulate elements ("grana") obtained upon disintegration of chloroplasts can effect the phosphorylation of ADP upon illumination.[338] Arnon's group also identified an iron-containing protein ("ferredoxin") as a component of the electron transfer system in the photosynthetic process.[339] They also provided experimental evidence for Robert Emerson's conclusion that two photosystems are involved in the process, in showing that photosystem I (700 nm) is involved in the light-dependent reduction of pyridine nucleotide, and is linked to photosystem II (680 nm), which accepts electrons generated by the reduction of H_2O and the release of O_2. Later work in numerous laboratories has led to the formulation of an intricate mechanism for the conversion of light energy into the chemical energy required for the reduction of $NADP^+$. In this mechanism, oxidation-reduction systems analogous to those found to be operative in the mitochondrial transfer of electrons from NADH to O_2 use light energy to reverse that oxidative process. In brief, the cleavage of water and release of O_2 was found to involve the catalytic action of a manganese-containing multiprotein complex coupled to chlorophyll photosystem II, which, upon illumination, is converted into a strongly reducing system; from that system electrons are transferred, via several electron carriers (plastoquinone, the cytochrome bf complex, plastocyanine) to photosystem I, whose illumination produces an even more strongly reducing species; finally, from that species electrons are transferred to $NADP^+$ via ferrodoxin and flavoprotein. An alternative cyclic mechanism, involving only photosystem I, leads to the phosphorylation of ADP without the formation of NADPH or O_2.[340]

Of particular importance was the demonstration, by Andre Jagendorf and Ernest Uribe, that in the dark spinach chloroplast grana can effect the phosphorylation of ADP, provided an artificial pH gradient has been established across the membrane.[341] This achievement provided some of the first experimental support for Mitchell's chemiosmotic hypothesis and the impetus to study the role of proton gradients in the various steps of electron transfer in the above pathways of the utilization of light energy. These studies included microorganisms that contain, as light harvesters, bacteriochlorophyll (purple sulfur bacteria), linear tetrapyrrole units named bilins (cyanobacteria), or bacteriorhodopsin (halobacteria). Bacteriorhodopsin is closely related to the rhodopsin of the retina, about which more will be said shortly. These microbial light receptors are components of multiprotein assemblies, such as the "phycobilisomes" of the cyanobacteria, and electron micrographs have shown that in most cases they are associated with membranous "thylakoids." Recent studies, notably by Alexander Glazer, Richard Henderson, and Walther Stoeckenius, have provided much knowledge about the structure and operation of the photosystems in such organisms, and Hartmut Michel succeeded in crystallizing the reaction center for photosystem II in a purple bacterium, thus permitting a determination of its three-dimensional structure by X-ray diffraction analysis.[342]

During the course of these studies it became evident that the mitochondrial H^+-ATPase discovered by Efraim Racker has its counterpart in the photosynthetic apparatus of microorganisms, for he and Stoeckenius showed that the combination of purple membrane vesicles with preparations of this enzyme effected light-driven proton uptake and ATP formation.[343] If one seeks to find in the available experimental evidence a generalization about the key enzymatic process in the conversion, by living organisms, of light energy into chemical energy, first place must now be given to the reversal of the hydrolytic action of such ATPases. At a time when historians of the biological sciences have devoted much attention to the impact of the Watson-Crick model of DNA, it is gratifying to note that at least one of them has called attention to the equally exciting and fruitful development of knowledge about the chemical mechanisms in the process upon which, in the last analysis, all living organisms depend for their survival.[344]

I turn next to the interplay of chemistry and biology in the study of the physiological process of vision. Other important biological responses to light include phototropism, as in the bending of plants; or phototaxis, as in the free movement of purple bacteria; or photoperiodism, as in the flowering of plants; or indeed the phenomenon of bioluminescence, in which the light emitted by an organism is received by members of the same or different species.

In 1842 Ludwig Moser suggested that, as in photography, light might induce chemical reactions in the retina of the eye, and some years later Heinrich Müller described the red color of the rods in the frog retina.[345] It was Franz Boll, however, who concluded that the pigment (he called it *Sehrot* or *erythropsin*) is localized in the outer segments of the rods and that its bleaching by light and the return of color in the dark is associated with the visual process.[346] Immediately after the publication of Boll's 1876 paper Willy Kühne repeated and extended Boll's findings, changed the name of the pigment to *Sehpurpur* or *rhodopsin,* and extracted it from retinas by means of a solution of bile salts.[347] There inevitably was a priority dispute between Boll and Kühne, but it was short-lived, because Boll died in 1879. Boll's friend Hermann Helmholtz wrote him on 28 October 1877: "Do not be too upset by Kühne; he is opposed by impartial observers of the dispute, all the more as he becomes more disagreeable. I considered him to be more sensible."[348]

In the subsequent development of knowledge about the mechanism of the visual process, before 1950 the contributions of Selig Hecht and of his student George Wald were especially decisive.[349] During the course of his research, initially (1917–1926) performed at the Marine Biological laboratory in Woods Hole, Hecht demonstrated the validity of what Boll and later workers had only surmised—the essential role of visual purple—by determining the action spectrum for human rod vision and its close similarity to the absorption spectrum (maximum at 500 nm) of the extracted pigment. Some twenty years later, as professor of biophysics in the zoology department at Columbia University, Hecht and his associates showed that only one quantum of light is needed to trigger the amplification of the photochemical reaction involving rhodopsin, with the production of an electric signal transmitted to the brain.[350] In the intervening years, Wald established the chemical nature of the chromophore in rhodopsin as the aldehyde derivative of vitamin A, long known as an antixerophthalmic (fat-soluble) constituent of the human diet, and whose carotenoid structure had been determined around 1930 by several organic chemists, notably Paul Karrer, Richard Kuhn, Edgar Lederer, and Laszlo Zechmeister. Wald showed that, in rhodopsin, this compound ("retinine" or "retinal") is the prosthetic group of a protein ("opsin"). When rhodopsin is bleached by light, the chromophore is released from the protein, and converted to vitamin A ("retinol"); during dark adaptation, the alcohol is oxidized to the aldehyde, which recombines with opsin.[351] As Wald and his associate Ruth Hubbard later showed, this oxidation-reduction involves catalysis by a NADP-dependent dehydrogenase. Of particular importance was the identification of the photochemical reaction as the multistep conversion of the protein-bound isomer of retinal in which one of the double bonds in the carotenoid chain has a *cis* conformation into an isomer with a fully extended (all-*trans*) chain.[352] During the 1960s Wald and his associates also identified the photoreceptors of color vision, in which the cone cells of the retina are involved.[353]

More recent work in this field has been spurred by the development of new techniques, the upsurge of interest in the mechanisms of active membrane transport, and the discovery of the role of guanosine nucleotides (GTP, cyclic GMP) in the visual process. Among the many striking findings was that the chromophore in the "bacteriorhodopsin" discovered by Stoeckenius is retinine, and that its isomerization by light is the same as for rod rhodopsin. The "visual cycle" formulated by Wald during the 1930s has been transformed into a sequence of reactions beginning with the activation, by the photoexcited rhodopsin, of a multiprotein "transducin" that includes a GTPase, followed by a series of enzyme-catalyzed and membrane-transfer processes.[354]

To conclude this chapter I offer some comments about the problem of the intracellular organization of the chemical components of biological systems. If, as recounted in the preceding chapters on the development of reliable knowledge about enzymes and proteins, much depended on the isolation of chemically homogeneous preparations of these substances and their characterization as individual molecular entities, studies on the intracellular utilization of the energy derived from respiration or light absorption brought this problem to the attention of biochemists long before the advent of electron microscopy. Indeed, as we have seen, the question of the intracellular organization of enzymatic activity was raised in the early years of the twentieth century by Franz Hofmeister, and one of the reasons for the appeal of colloid chemistry is that it appeared at that time to offer more promise than the organic-chemical approach advocated by Felix Hoppe-Seyler and Gowland Hopkins. If during the 1930s the notable success of that approach to the study of glycolysis, of the structure and biochemical role of vitamins, and of the action of crystalline enzymes, swung the opinion in its direction, to my knowledge no leading biochemist of the time denied the validity of the long-held view of cytologists expressed in 1924 by Ralph Lillie: "It is evident that a mere random mixture of cell constituents, in the same proportions and concentrations as in protoplasm, would not give a system having the properties of a living cell. The constituents must have a definite arrangement, and must be present in a definite physical state; only under these conditions is it possible to conceive of any kind of ordered interaction as that underlying the life of the cell."[355] What changed during the 1930s and 1940s was the attitude toward statements such as that in 1925 of the noted microbiologist Albert Kluyver that "it is no longer necessary to have recourse to the assumption of a large number of separate enzymes to explain the partial reactions of the dissimilation process. . . . The changes which are ascribed to separate enzymes such as catalase, reductase, zymase, lactozymase, alcoholoxidase, carboxylase, carboligase, glyoxylase, aldehydomutase, Schardinger's enzyme, etc., are actually manifestations of a definite degree of affinity of the protoplasm for hydrogen."[356] A similar point of view was expressed in 1926 by Sergei Kostychev and by Juda Quastel.[357] Despite the acceptance during the 1930s of the idea that all enzymes, whatever their physiological function, are individual proteins, and that the understanding of that function required their isolation in as pure a state as possible, there were recurrent statements of the view that the enzymatic apparatus within cells represents an organized assembly. For example, Hans Euler urged the search for cell constituents that have multiple catalytic functions, and Kurt Stern proposed that intracellular electron transport might involve a very large protein particle in which "the active groups of the catalysts are arranged . . . in an orderly fashion."[358]

This view was expressed before the post-1945 work of Frederick Sanger on insulin, of William Stein and Stanford Moore on ribonuclease, and of John Kendrew on myoglobin. The methods they applied and developed were further improved, and the chemical structure of proteins could be studied with the same assurance as in the case of organic substances of lower molecular weight. It became possible to study the cooperative chemical interaction of proteins, increasingly recognized to be of crucial biological importance. Multienzyme assemblies, such as the mitochondrial cytochrome c oxidase or the ATP synthase, were studied with knowledge of the properties of the individual components, and examined by means of X-ray diffraction analysis and electron microscopy. It should also be emphasized that in narrowing the gap between the reliable knowledge about cell structure and biochemical function, the structural models and reaction schemes derived from such studies have depended on the results of efforts to dissect the assembly, to isolate and characterize the individual components, and to reconstitute all, or at least part, of the assembly. In this respect, resolution and reconstitution are roughly analogous to the traditional chemical procedures of analysis and synthesis.

CHAPTER 7

Pathways of Biochemical Change

With the rise of iatrochemistry, the ancient "coctions" in animal nutrition, based on the analogy to cookery, were replaced by the "fermentations" akin to those operating in the wine vat. By 1800 the theories of van Helmont, Sylvius, and Willis had declined in prestige, and the "assimilation" of nutrients was considered to be a process in which the triturated and partially dissolved food ("chyme") is transformed in the stomach and intestinal canal, under the influence of air and heat, into a fluid ("chyle") that serves as the source of the blood, whose constituents are converted into materials characteristic of the individual tissues. Thus it was believed that the "fibrin" of muscle is formed by solidification of the dissolved "fibrin" of the blood. The physiological experiments of René Antoine Réaumur, described in 1752, and their later confirmation and extension by Lazzaro Spallanzani, focused attention on digestion in the stomach. A contemporary report by Edward Stevens stated that "it is not the effect of heat, trituration, putrefaction, or fermentation alone, but of a powerful solvent, secreted by the coats of the stomach, which converts the aliments into a fluid, resembling the blood."[1] The task of the new chemistry, based on the work of Lavoisier and his contemporaries, was seen to be the definition of this conversion in terms of the elements (carbon, hydrogen, oxygen, nitrogen) found in animal substances and those (carbon, hydrogen, oxygen) found in plant materials.

Among the early attempts was that of Jean Noel Hallé, who suggested that under the influence of oxygen, foodstuffs become "animalized" through the loss of carbon (as carbon dioxide) and the gain of nitrogen; he proposed that these chemical changes occur successively in the digestive organs, the lungs, and the skin.[2] Several years later, that view was modified by Antoine François Fourcroy, who considered the general process of assimilation in the animal body to represent "a complete transformation of the primitive alimentary substance into each particular organic substance; . . . this assimilation, begun in digestion, continued in respiration, almost achieved in the courses of the circulation, and entirely completed upon entry into each organ to be nourished, consists mainly in the loss of carbon and hydrogen, in an increase of azote, and a sort of transmutation hitherto termed animalization."[3] This view reappeared in the writings of leading medical investigators during the first half of the nineteenth century. Its limitations became increasingly apparent, however, after the new methods of elemental analysis of organic substances had been applied to the "immediate principles" obtained from animal organisms.[4] Although in 1810 it could be stated that "a little oxygen or nitrogen more or less; therein, at the present state of science, lies the only apparent cause of these innumerable products of organic bodies,"[5] the interplay of chemistry and animal physiology showed, within a few decades, that the chemical processes in assimilation are more complex than had been envisaged by Hallé and Fourcroy. This complexity was not accepted by all medical physiologists; for example, in his famous book on digestion, William Beaumont wrote:

> The ultimate principles of nourishment are probably always the same, whether obtained from animal or vegetable diet. It was said by Hippocrates, that "there are many kinds of aliments, but that there is at the same time but one aliment." This opinion has been contested by most modern physiologists; but I see no reason for scepticism on this subject. . . . The perfect chyle, or assimilated nourishment, probably contains the elements of all the secretions of the system; such as bone, muscle, mucus, saliva, gastric juice, etc., etc., which are separated by the action of the glands, the sanguiniferous and other vessels of the system.[6]

Foremost among these "modern physiologists" were François Magendie and Friedrich Tiedemann.[7]

By 1815 many immediate principles of animal organisms had been characterized as well-defined substances, including crystalline compounds like urea (from urine) or uric acid (from kidney stones) as well as a variety of nonnitrogenous organic compounds such as milk sugar, the sugar of diabetic urine, and several acids—for example, benzoic acid. In addition, many noncrystalline materials (albumin, fibrin, casein, olein, stearin, and others) were also included in the list. The recognition of this chemical heterogeneity influenced the new experimental physiology promoted by Magendie and led to studies on the nutritional role of individual components of the animal diet. Thus, from his pioneer nutritional experiments with dogs fed controlled diets, which showed that nitrogenous food is essential for their survival, he concluded that his results "make it very probable that the azote of the organs has its original source in the aliments; they also throw light on the causes of gout and gravel [kidney stones]. . . . People suffering from these diseases are usually heavy eaters of meat, fish, milk products, and other substances rich in azote. Most of the kidney stones, bladder calculi, and arthritic tophi are composed of uric acid, a principle which contains much azote. In reducing the proportion of azotized elements in the diet, one succeeds in preventing these diseases."[8] This specific example of the application of the new chemical knowledge to medical practice was offered against the background of a general view of the assimilatory process in which "all parts of the human body undergo an internal motion, which has the double effect of expelling the molecules no longer needed as components of the organs, and of replacing them by new molecules. This internal motion constitutes nutrition."[9] To denote such "internal motion," Tiedemann used the word *Stoffwechsel,* whose English equivalent was taken to be "metamorphosis." Later in the nineteenth century the terms *progressive metamorphosis* and *regressive metamorphosis* were often used to distinguish between assimilation as a constructive process and "disassimilation" as a degradative one. After 1880, with the recognition that both kinds of processes occur in normal cells, the term *metabolism* came into favor among British physiologists; *anabolism* denoted the creation of cell constituents, and *catabolism* their degradation to simpler chemical compounds.[10]

According to William Hardy, the view held in active nineteenth-century English schools of physiology was as follows: "Metabolism—that is the whole chemical cycle—consisted . . . of a phase of increased molecular complexity in which protein, fats, and carbohydrates with oxygen were built up into the living substance, and a phase of decreasing chemical complexity and liberation of energy. The picture of two operations, anabolism and katabolism—or loading and discharge—was based in the first instance on the properties of muscle, especially on its capacity for doing work when not supplied with food or even oxygen, but it received immense support when the processes of loading and discharge of gland cells were discovered in the late eighties."[11] In German writings the terms *Stoffwechsel* and *Assimilation* continued to be widely used well into the twentieth century. As Ewald Hering stated, "the most significant feature that distinguishes living from dead matter is its *Stoffwechsel, i.e.,* the internal chemical processes whereby on the one hand materials are formed which have become foreign to the living substance and either accumulated nearby or transferred to the circulating fluids, and on the other hand, and concurrently, food materials are taken up and appropriated by the living substance and converted into its own constituents."[12] In the background of these statements was the work, before 1840, of Tiedemann and his chemical colleague Leopold Gmelin, of Eberle and Schwann, and of Apollinaire Bouchardat and Claude Sandras, which had led to acceptance of the view that animal digestion is a chemical process in which ferments and bile effect a transformation of food material into products used by the organism.[13] As the chemist Thomas Thomson put it, however, "Over the problem of assimilation the thickest darkness still hangs; there is no key to explain it, nothing to lead us to the knowledge of the instruments employed"; it appears, according to Thomson, to be effected by the action of "a *living* or *animal* principle [that] does not act according to the principles of chemistry."[14] In the formulation of his version of the cell theory, Theodor Schwann discussed the problem of assimilation and offered the hypothesis that "organisms are nothing but the form under which substances capable of imbibition crystallize."[15] In his view the transformation of an undifferentiated extracellular matter *(blastema),* through the "metabolic power" of existing cells, is analogous to the formation of crystals in a supersaturated solution. The word *metabolic* (Greek *to metabolikon,* disposed to cause or suffer change) was introduced by Schwann to denote the chemical changes undergone by cell constituents and by the surrounding material. Although his hypothesis, based on the presumed specific chemical attraction of molecules by existing cellular membranes, contributed to the philosophical discussion that enlivened biological discourse during the nineteenth century, it gave little hint of the chemical experimentation needed to test its validity.[16]

During the early 1840s there were also specific problems that indicated the complexity of the process of assimilation. One of them came from the work of a French gelatin commission, headed by Magendie. The government had sought to respond to the social turmoil arising from the price of food by introducing gelatin as a dietary staple, because it was readily available from bones, and therefore cheap. By 1830, however, there was sufficient resistance among the poor to encourage doubts about the nutritive value of gelatin. Magendie's group used dogs to study the problem from 1832 to 1841, and his report stated that life could not be sustained with gelatin as the sole source of nitrogen; indeed, other "immediate principles" (albumin, fibrin, casein) considered to be good nutrients also were unsatisfactory when fed alone. Magendie could do no more, therefore, than to report the results and to emphasize the obscurity surrounding the problems of animal nutrition.[17]

If Magendie was cautious, Justus Liebig was not. During the 1840s Liebig offered specific chemical answers to many questions relating to the problem of assimilation. As we have seen, he assigned to carbohydrates and fats the role of "respiratory" substances, whose oxidation produces heat, and he considered proteins to be "plastic" substances whose decomposition causes the chemical transformation (Stoffwechsel) of body constituents. These views were presented against the background of the careful studies of Jean-Baptiste Boussingault, who had used Dumas's new method of nitrogen analysis to show that the apparent increase in the nitrogen content of animal tissues, compared with that of vegetable matter, could be largely accounted for by the loss of nitrogen in the animal excreta and secretions. This quantitative "intake-output" approach to the problem of assimilation ended the discussion about "animalization" and marked the beginning of studies on the "nitrogen balance" in animal nutrition.[18]

The course charted by Liebig for the study of the Stoffwechsel of body constituents was indicated in the third edition of his *Animal Chemistry:* "Urea and uric acid are products of changes undergone by the nitrogenous constituents of the blood under the influence of water and oxygen, [and] the nitrogenous constituents of the blood are identical in composition with the nitrogenous composition of the diet. The relation of the latter to uric acid, urea, the oxygen of the atmosphere, and the elements of water, [and] the quantitative conditions for their formation, are expressed by chemistry by means of formulas and, within the limits of its domain, [explain] them thereby."[19] Although Liebig's physiological speculations spurred the further development, after 1850, of Boussingault's approach, by 1875 the site of Stoffwechsel had been transferred from the blood to the cellular elements of the tissues, the organic chemical "formulas" were quite different from those of Liebig in 1846, and most of his speculations had turned out to be wrong. In view of the extremes of adulation and rejection accorded these ideas at midcentury, it may be appropriate to quote the opinion of a leading French biologist:

> I am far from wishing to say that the speculative views of Mr. Liebig on the transformations of organic matter in the interior of the animal economy, and the use he has made of equations to show how it would be possible to conceive the formation of the various products of chemical-physiological work, have been useless for the progress of science. On the contrary, I believe that in giving a precise form to his argument, he has rendered a real service and has accustomed physiologists to a mode of thought that is very useful for the study of the phenomena of nutrition. It is only necessary to take care in accepting these hypotheses as the expression of what is actually occurring in the organism, where the intermediate reactions are very complex and very important to know.[20]

These lines were published twenty years after the appearance of the first edition of Liebig's *Animal Chemistry*. During the interval the grand design that he and Dumas had based on elementary analysis had been shown to be inadequate by new discoveries which indicated the existence, in the animal economy, of "intermediate reactions that are very complex." The most dramatic of these discoveries was Claude Bernard's demonstration that glucose is manufactured in the liver.

Glucose and Glycogen in the Animal Body

In 1853 Bernard published his celebrated experimental report in which he stated: "I will establish in this work that animals, as well as plants, have the ability to produce sugar. Furthermore, I will show that this animal function, hitherto unknown, is localized in the liver."[21] There followed an account of experiments begun ten years earlier, whose high point came in 1848, when Bernard discovered the presence of sugar in the blood leaving the liver of animals that had received no carbohydrate in their diet.[22] Bernard exhibited exceptional surgical skill in preparing the experimental animals (mostly dogs) for these studies, and he largely used a yeast fermentation method as his test for sugar. In this work he was assisted by the chemist Charles Barreswil.

Bernard wrote that he had been impelled to examine the fate of sugar in the animal body in the hope of understanding the nature of the wasting human disease diabetes mellitus and "was led to think that there might be in the animal organism phenomena still unknown to chemists and physiologists, and able to give rise to sugar from something other than starchy substances."[23] Before 1840 the chemical identity of the sugar of diabetic urine with grape sugar had been demonstrated by Chevreul and by Peligot, and Ambrosioni showed that the sugar of diabetic blood is glucose, as judged by its alcoholic fermentation with yeast. That the ingestion of carbohydrates is followed by the appearance of fermentable sugar in the chyme and blood of normal animals had been shown repeatedly since the work of Tiedemann and Gmelin, and the salivary conversion of dietary starch into sugar had been established during the 1830s. It seemed, therefore, that the pathological defect in diabetes might be a consequence of the inability of the organism to utilize the glucose derived from the diet. In *Animal Chemistry,* Liebig wrote: "In some diseases the starchy materials do not undergo the transformations that allow them to support the respiratory process or to be changed into fat. In diabetes mellitus, starch is not converted beyond sugar which is not utilized and is removed from the body."[24]

After Bernard had found that the liver can produce sugar, he concluded that this discovery "must necessarily change the hitherto prevalent ideas about the nature of diabetes, based on the belief that the sugar found in the organism is exclusively derived from the food. . . . Since this glycogenic function has been found to be localized in the liver, it is evident that it is in this organ we must now seek to place the seat of the disease."[25] In showing that the physiological function of the liver is not limited to the production and "external secretion" of bile, as had been thought previously, but rather included the "internal secretion" of sugar, Bernard asked whether "the albuminoid substances of the blood, upon coming into contact with the hepatic cells, are cleaved into two products, a hydrocarbonated one which becomes the sugar, the other a nitrogenous one which becomes the bile?"[26]

To this Liebig-like speculation must be added Bernard's insistence for many years that the liver was the sole organ capable of secreting sugar into the blood. During the 1850s he vigorously disputed the statements by Louis Figuier and Auguste Chauveau that sugar is present in all parts of the circulating blood, and the controversy led to the establishment by the *Académie des Sciences* of a commission headed by Dumas to adjudicate the matter. They decided in favor of Bernard on the stated ground that the yeast fermentation test for glucose was considered to be more reliable than the copper-reduction method used by Figuier, which had been successively improved during the 1840s by Carl Trommer, Carl Fromherz, and especially by Hermann Fehling.[27] In hindsight, it is clear that the commission's verdict was influenced by the fact that Figuier's status, as a well-known popularizer of science, was lower in the upper circles of French science than that of Bernard.

To meet the objection that the sugar emerging from the liver might have been produced in the blood flowing through the tissue, Bernard submitted an excised dog liver to thorough lavage in order to remove the blood and showed that by the next day sugar had again been formed; if the washed liver was boiled, the formation of sugar was abolished. Bernard therefore concluded that the glucogenesis is a property of the hepatic tissue, and that in addition to glucose there is present in the liver "another substance, sparingly soluble in water . . . [which] is slowly changed to sugar by a kind of fermentation."[28] Two years later, Bernard reported the isolation of this *matière glycogène* and was aided by the chemist Théophile Jules Pelouze in its characterization as a starchlike substance *(amidon animal)*. This important discovery was immediately confirmed by Victor Hensen and André Sanson, but Sanson called attention to the presence of "glycogen" in the muscle, a claim that Bernard also disputed for some years.[29]

By 1880 the formation of glucose in the liver was widely acknowledged to be a normal physiological function, and the objections of Frederick Pavy (among others) that it represented a postmortem process had been largely discounted.[30] The existence of glycogen as a chemical entity had been firmly established, and its starchlike character had been confirmed by a series of investigators, notably August Kekulé, Joseph von Mering, Frédéric Musculus, and Eduard Külz. Moreover, a starchlike material was found in other organs, although muscle glycogen did not appear to be a source of blood sugar, and Bernard considered it to undergo a "lactic fermentation," with glucose as an intermediate.[31]

As has been mentioned, Bernard saw the glycogenic role of the liver as involving two distinct processes: "The first entirely vital action, so termed because it is not effected outside the influence of life, consists in the creation of the glycogenic material in the living hepatic tissue. The second entirely chemical action, which can be effected outside the influence of life, consists in the transformation of the glycogenic material into sugar by means of a ferment."[32] Bernard reiterated this view during the succeeding twenty years, and in 1877 he described the extraction (with glycerol) of the "diastatic ferments" of liver and barley and reported that both convert starch and glycogen into glucose. He believed that he had demonstrated "the identity of the mechanism of formation of sugar in animals and plants, since we have not only seen that glycogen is identical with starch, but that the diastase of the seed is also identical with the diastase of the liver. . . . The mechanism of the formation of the amylaceous material [starch or glycogen] is on the contrary completely unknown for plants as well as for animals, and it is the problem which now presents itself for the studies of chemists and physiologists."[33] In the last book he wrote, Bernard stated that he had been drawn thirty years before to the study of the transformations of sugar in the animal body "by the conviction that the phenomena of nutrition should not be considered by the physiologist from the same point of view as that of the chemist. Whereas the latter seeks to determine a nutritional balance, *i.e.*, to establish the balance between the substances that enter and those that leave, the physiologist should set himself the task of following them step by step during their transit, and to study all their successive transformations in the interior of the organism."[34] It may be questioned whether this account accurately describes Bernard's thought during the 1840s, when the physiological ideas of Dumas and Liebig were in fashion. At that time he wrote: "In order to avoid error and to render all the services of which it is capable, chemistry must never venture alone into the study of animal functions; I think that in many cases it alone can resolve the difficulties that block physiology, but cannot anticipate them, and finally I think that in no case can chemistry consider itself authorized to restrict the resources of nature, which we do not know, to the limits of the facts or processes which constitute our laboratory knowledge."[35] Thirty years later, he restated this attitude toward chemistry in this fashion: "The chemist can make the products of a living thing, but he will never make its tools, because they are the result of the organic morphology which . . . is beyond the scope of properly defined chemism; and in this respect it is no more possible for the chemist to make the most simple ferment than to make the entire living thing."[36]

Whatever may be said about Bernard's attitude toward the role of chemists in physiological research, or his own shortcomings as a chemist, his experimental achievements in the study of carbohydrate metabolism and other physiological problems, and his emphasis on the need to replace the "intake-output" approach of Dumas and Liebig by the search for intermediates in the "step-by-step" transformations of body constituents, justify his rather immodest claims (which he did not put into print) that "I am the first who has studied the intermediate. One knew the two extremes and made [from them] a probability physiology with the rest," and that "I have pushed physiology forward by my discoveries and I have enabled it to make advances which it could not have achieved in a century if I had not been there."[37]

The Biosynthesis of Glycogen

After 1870 much attention was given to the proposal of Baeyer and Nencki that the loss or gain of the elements of water might be important in metabolic processes.[38] To Friedrich Wöhler's finding in 1842 that the ingestion of benzoic acid (C_6H_5COOH) by a human subject led to the excretion of hippuric acid (benzoylglycine, $C_6H_5CONHCH_2COOH$) were added other examples of the seeming reversal of the "hydratation" (hydrolysis) effected by acids or enzymes. During the 1850s Adolf Strecker had shown that the glycocholic and taurocholic acids of the bile represent condensation products of cholic acid with glycine and taurine, respectively. In 1875 Eugen Baumann discovered that phenol (C_6H_5OH) appeared in the urine in the form of an ester of sulfuric acid extractable with ether ("ethereal sulfate"), and three years later Oswald Schmiedeberg reported that camphor was excreted in the form of a condensation product with glucuronic acid, a new sugar. The formation of such amides, phenyl sulfates, or glucuronides was interpreted by Hoppe-Seyler as indicating that

> the materials of which the organs are composed, from which they build and regenerate themselves, belong to a class of substances that collectively represent *anhydrides,* and which . . . are converted or cleaved with the uptake of the elements of water by treatment with alkalis and acids, and in many cases also by ferments. . . . Upon passage of oxygenated blood through the living kidney, the combination of glycine and benzoic acid—an anhydride formation—has been demonstrated. The living organ has also been found to effect the reverse process, namely the cleavage of hippuric acid and similar compounds by the uptake of water. It may be expected that the latter process will also occur in the absence of oxygen.[39]

This anhydride theory figures largely in the discussion during 1880–1910 about the biosynthesis of glycogen. Owing to the work of Emil Fischer, much was known about the chemical structure of glucose, other monosaccharides, and oligosaccharides such as maltose. Although the detailed structure of glycogen and starch was still a matter of conjecture in 1910, there was no doubt that they represent polyglucoses in which the monosaccharide units are joined by glycosidic linkages to form anhydrides analogous to the fats and proteins.

There was uncertainty, however, about the nature of the metabolic precursors of glycogen. In 1891 Carl Voit summarized the situation as follows: "According to one theory, glycogen arises by the loss of water from dietary carbohydrates, that is, glucose present or made in the intestinal canal. It is the theory of the anhydride formation of glycogen. It has been based on the fact that the greatest accumulation of glycogen in the liver occurs upon the uptake of certain carbohydrates, especially glucose. According to the other view, glycogen arises from protein breakdown, and the only role of the readily-decomposable sugar derived from the diet is to be burned instead of glycogen, thus protecting it from decomposition; this has been called the sparing theory."[40] Voit favored the anhydride theory but granted the possibility that glycogen can arise from noncarbohydrate sources. Eduard Pflüger, in a massive paper, denied this possibility: "If it were true that the liver can make the same glycogen from the most varied molecules, whose atomic groupings have no relation whatever to glycogen, one would have to assign to the liver quite mysterious synthetic capacities."[41] Shortly afterward, however, Hugo Lüthje showed clearly that a diabetic dog on a carbohydrate-free diet excreted much more sugar than could have arisen from the total carbohydrate in the body, and some years later Pflüger confirmed this finding and accepted the view that "the liver can effect the synthesis of glycogen from protein."[42]

Lüthje induced experimental diabetes by the method introduced in 1889 by Joseph von Mering and Oskar Minkowski, who found that surgical removal of the pancreas of a dog produced a large increase in blood sugar (hyperglycemia) and a concomitant urinary excretion of glucose (glycosuria).[43] This experimental production of a condition resembling human diabetes was followed by Eugene Opie's demonstration in 1901 that in diabetic patients, special pancreatic cells (the islets described by Paul Langerhans in 1869) were damaged, and in 1909 Jean de Meyer suggested that these cells elaborate a hormone which he termed insuline. The search for this hormone encouraged the development of micromethods for the determination of blood sugar, so that many samples could be taken from the same animal, and with the help of such analytical procedures in 1922 Frederick Banting and Charles

Best succeeded in establishing the existence of insulin as a chemical entity.[44] I have already referred to the important contribution of James Bertram Collip in the purification of insulin, and to John Jacob Abel's subsequent crystallization of this hormone.

It should be added that three years before his collaboration with Minkowski, Mering had found that the injection of the plant glucoside phlorizin causes the urinary excretion of glucose, without an increase in the level of blood sugar. This artificial diabetes, arising from an increased permeability of glucose in the kidney, was used for many decades to examine the effect of various nutrients on the metabolic production of glucose. For example, Graham Lusk employed this method to show some amino acids (glycine, alanine, glutamic acid, aspartic acid) are converted into glucose in the animal body.[45] By about 1910 it was widely agreed that the immediate precursor of liver glycogen is glucose, but that the metabolic capacities of the liver permitted it to convert into glucose noncarbohydrate substances such as lactic acid, pyruvic acid, and some of the amino acids derived from the gastrointestinal breakdown of dietary proteins.

During the succeeding two decades important advances were made in the study of the structure of glycogen and starch, notably by the British chemists Thomas Purdie, James Irvine, and Walter Norman Haworth, who developed a method involving the methylation (with dimethyl sulfate) of the hydroxyl groups of polysaccharides. Such substituted products were then hydrolyzed to yield the monosaccharide units in which the hydroxyl groups involved in glycosidic linkages had not been methylated. The application of this method resolved a problem arising from the fact that during the 1920s glycogen, cellulose, and starch (known to be composed of two different polysaccharides named amylose and amylopectin) were thought to be aggregates of relatively small polyglucoses held together by noncovalent linkages. As mentioned earlier, X-ray data on cellulose fibers were taken to support this view, and Hermann Staudinger's ideas about the existence of long-chain polymers were not yet accepted. Although osmotic pressure measurements had indicated particle weights for glycogen (as prepared by Pflüger) in the order of several hundred thousand, there was reluctance to consider such values as indicating the size of a covalently bound molecule. Hydrolysis of methylated glycogen produced, as the major product, 2,3,6-trimethyl glucose and minor products as 2,3,4,6-tetramethyl glucose (for nonreducing ends) and 2,3- dimethyl glucose. These results indicated that glycogen has a highly branched structure in which straight-chain arrays of eleven to eighteen glucopyranose units [in α(1–4) glycosidic linkage] are linked by means of α(1–6) glycosidic bonds (see formula).[46] A similar structure was found for amylopectin,

Segment of glycogen

which, like glycogen, gives a red-brown color with iodine; the amylose portion of starch is a straight-chain α(1–4) polysaccharide, with 100–200 glucose units, and gives the blue color reaction with iodine.[47] These chemical advances were significant for the subsequent elucidation of the biological formation of glycogen and starch, but before they were made, the metabolic aspects of the problem had been redefined by Carl and Gerty Cori.

In a series of striking experiments beginning in 1925, the Coris studied the metabolic relations among lactic acid, glucose, and glycogen in intact rats, and in 1929 they were able to conclude that

> formation of liver glycogen from lactic acid is thus seen to establish an important connection between the metabolism of the muscle and that of the liver. Muscle glycogen becomes available as blood sugar through the intervention of the liver, and blood sugar in turn is converted into muscle glycogen. There exists therefore a complete cycle of the glucose molecule in the body which is illustrated in the following diagram:
>
> 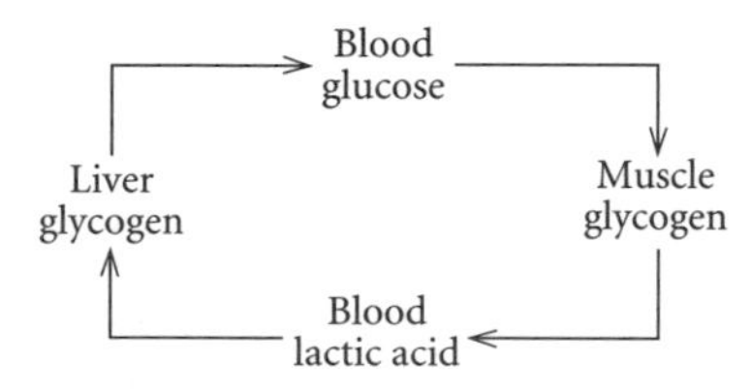
>
>
> Epinephrine was found to accelerate this cycle in the direction of muscle glycogen to liver glycogen and to inhibit it in the direction of blood glucose to muscle glycogen; the result is an accumulation of sugar in the blood. In-

> sulin, on the other hand, was found to accelerate the cycle in the direction of blood glucose to muscle glycogen, which leads to hypoglycemia and secondarily to a depletion of the glycogen stores of the liver. . . . There is also the possibility that other hormones besides epinephrine and insulin influence this cycle.[48]

The paper in which this statement appears, coming as it did at a time when the outlines of the Embden-Meyerhof pathway of anaerobic glycolysis and of the respiratory electron-transfer chain were beginning to emerge, also marks the beginning of a new stage in the development of knowledge about the regulation of metabolic processes. As regards the last sentence in the above quotation, in the same year Bernardo Houssay showed that, upon surgical removal of the pituitary of depancreatized animals, the severity of the diabetic state was markedly ameliorated, and the animals became very sensitive to insulin, indicating that the pituitary elaborates a factor (or factors) that counteract the action of insulin. One of these factors later turned out to be the adrenocorticotropic hormone (ACTH), made in the anterior pituitary; this hormone stimulates the adrenal cortex to release steroid hormones that antagonize the action of insulin. Indeed, in 1936 Hugh Long reported that surgical removal of the adrenal gland of diabetic animals had an ameliorating effect similar to that demonstrated by Houssay.[49] These observations revealed the existence of an intricate set of regulatory mechanisms in the pathways of biochemical change linking muscle and liver glycogen, blood glucose, and the substances that either give rise to them or are formed upon their breakdown. By the mid-1930s, the dissection of the enzymatic apparatus in glycolysis had developed to a stage where there was hope of identifying the individual steps in carbohydrate metabolism that are under hormonal control. We will return to the development of this line of investigation in a later chapter, and will now consider the important contributions of Carl and Gerty Cori to the study of the enzymes involved in the interconversion of glucose and glycogen.[50]

During the 1930s, in the course of their studies on the metabolism of hexose phosphates in frog muscle, the Coris discovered the formation of a nonreducing sugar phosphate, which they identified as glucose-1-phosphate, and they showed that it arises by the reversible phosphorolytic cleavage of glycogen, catalyzed by a "glycogen phosphorylase" present in various animal tissues (muscle, liver, heart, brain). They also demonstrated the enzymatic interconversion of glucose-1-phosphate and glucose-6-phosphate by a "phosphoglucomutase" as well as the presence in liver of a phosphatase that hydrolyzes glucose-6-phosphate.[51] These achievements not only ended the decades-old speculations about the action of a "diastase" in the formation of blood glucose from liver glycogen, or the reversal of that action in the synthesis of glycogen, but they also represented the first *in vitro* synthesis of a natural polymer of high molecular weight. Indeed, it may be said that these findings, and those that followed soon afterward, profoundly influenced biochemical thought about the processes in hitherto unexplored metabolic pathways. Shortly after the discovery of the animal phosphorylases, Charles Hanes described the reversible phosphorolysis of starch by enzyme preparations from higher plants (potatoes, peas), and showed the cleavage product to be glucose-1-phosphate.[52]

The product of the polymerization reaction catalyzed by the plant phosphorylases behaved like the amylose (straight-chain) fraction of starch. After the elucidation of the branched-chain structure of amylopectin and glycogen, additional enzymes were found in plant and animal tissues which catalyze the "branching" of straight-chain α(1–4) polyglucose, with the transfer of pieces of the chain to the 6-hydroxyl of glucose units in the main chain. In this respect, the crystallization of muscle phosphorylase by Arda Green and Gerty Cori was especially important,[53] because it made possible a more precise definition of the catalytic properties of the enzyme. In particular, it became evident that the enzyme uses the numerous nonreducing ends of preexisting glycogen molecules as sites of attachment of the glucosyl units derived from glucose-1-phosphate. Glycogen synthesis was thus viewed as the elongation of the branches of the "primer" glycogen, followed by further branching; conversely, the enzymatic breakdown of glycogen was seen to involve the successive shortening of the branches and cleavage of the α(1–6) branch points until a "limit dextrin" is reached.

An unexplained feature of the Coris' initial results was the stimulation of the phosphorolysis of glycogen by adenosine-5′-monophosphate (AMP). In a paper accompanying the one on the crystallization of the enzyme, Gerty Cori and Arda Green showed that in addition to the crystalline form, which exhibits activity in the absence of added AMP, muscle extracts contain another form that is inactive in the absence of AMP; the two forms were denoted phosphorylase *a* and phosphorylase *b* respectively. They concluded that the *a* form contains AMP as a prosthetic group that can be removed by a "PR enzyme" present in muscle extracts. This inference proved to be incorrect, but the subsequent investigation of the interconversion of the two forms of the enzyme was to have unexpected and important consequences. Among them was the recognition that the *a* form is a dimer of the *b* form, that the so-called PR enzyme is a phosphoprotein phosphatase, and that the conversion of the *b* form to the *a* form is effected by a "phosphorylase *b* kinase," with ATP as a phosphoryl donor and a serine unit of the protein as the phosphoryl acceptor.[54] Another chemical feature of glycogen phosphorylase, found by Carl Cori and his associates during the 1950s, is the participation of pyridoxal phosphate in its catalytic action. A plausible explanation of the role of this cofactor was offered during the 1980s.[55]

It seemed during the early 1950s that the work of Carl and Gerty Cori had established the role of glycogen phosphorylase as the catalytic agent in the biosynthesis of glycogen. This view had to be abandoned for several reasons, the most compelling of which was the discovery by Luis Leloir of a cofactor in which glucose-1-phosphate is linked by a pyrophosphate bond to uridine-5′-phosphate (uridine diphosphoglucose, UDPG), and his demonstration that this compound is a participant in the intracellular formation not only of disaccharides such as sucrose or lactose but also of glycogen.[56] That reaction, catalyzed by an enzyme now called glycogen synthase, differs from the one catalyzed by phosphorylase in the fact that the glucosyl unit of UDPG is transferred to the nonreducing ends of preexisting glycogen branches (with the formation of UDP) and the equilibrium in this process is much farther in the direction of polysaccharide synthesis than in the phosphorylase-catalyzed process. The resynthesis of UDPG is effected in an enzyme-catalyzed reaction involving uridine triphosphate (UTP) and glucose-1-phosphate. Nucleoside diphosphate sugars analogous to UDPG have been shown to be donors of sugar units in the biosynthesis of a large variety of polysaccharides, including starch and the "peptidoglycans" of bacterial cell walls.[57]

By the 1960s it was generally agreed that the physiological role of glycogen phosphorylase is a degradative one.[58] Support for this view came from biochemical studies of patients suffering from a relatively benign type of glycogen storage disease (McArdle's disease) in which there is a deficiency of phosphorylase but only a moderate increase in tissue glycogen. Such studies had been initiated during the 1950s by Gerty Cori (she had trained in her youth in pediatrics), who showed that in the case of the familial von Gierke's disease, where there is a massive enlargement of the liver and low blood sugar, there is a deficiency in the liver of glucose-6-phosphatase, thus blocking normal glycogen breakdown to glucose.[59] This finding appears to represent the first identification of a defective enzyme as being responsible for an inherited disease in human subjects. In a subsequent study Gerty Cori found that in another type of glycogen storage disease the defect lay in the "debranching" enzyme amylo-1,6-glycosidase. Later work by others led to the attribution of the causes of such diseases to enzymatic defects, and the knowledge so gained has served as a guide in attempts to develop new therapeutic agents.[60]

To these consequences of the line of research initiated by Carl and Gerty Cori during the 1920s must be added the discovery of adenosine-3′,5′-phosphate (cyclic AMP, cAMP) by their student Earl Sutherland. As was noted above, the Coris had shown before 1930 that insulin promotes glycogen formation in the liver, while epinephrine stimulates glycogen breakdown in muscle. The work in their laboratory during the succeeding two decades had identified many of the enzymes in glycogen synthesis and degradation, and the question of the relation of the hormonal effects to the catalytic activity of these enzymes was a matter of continuing interest. The insulin effect proved to be difficult to explain in biochemical terms and it is only recently, with new knowledge gained from the study of the physiological role of cAMP, that progress has been made in elucidating the mechanism of insulin action.[61] Sutherland's study of the epinephrine effect, begun in the Cori laboratory, as well as the effect of glucagon (a 29-amino acid peptide hormone found to stimulate glycogen breakdown in liver), led to the unexpected discovery of cAMP.[62] We will return to the historical background and subsequent development of research on the hormonal control of carbohydrate metabolism in the final chapter.

We have seen that Claude Bernard's discovery of glycogen in 1857 led to a more explicit formulation of the problem of assimilation in terms of the metabolic synthesis of a complex molecule from simpler ones. About twelve years later a metabolic process long considered to be a disassimilation—namely, the formation of urea in the animal body—also was recognized to be a synthetic one.

Protein Metabolism and Urea Synthesis

At the beginning of the nineteenth century, urea and uric acid were the only nitrogenous "immediate principles" of the animal organism that had been isolated in crystalline form. Through the work of several investigators, notably Fourcroy and Vauquelin (who introduced the term *urée*), urea was recognized to be the principal nitrogen-containing constituent of human urine and to yield much ammonia on being heated.[63] It was widely believed that "since urea and uric acid are the two animal substances richest in azote, one may presume that the secretion of urine is intended to remove from the blood the excess azote supplied by the food, just as the respiration removes the excess carbon."[64] Indeed, as we have seen, in 1823 Prévost and Dumas demonstrated the accumulation of urea in the blood upon surgical removal of the kidneys of experimental animals. They stated that "all chemists know that the urine of patients with chronic hepatitis contains little or no urea, which would seem to prove that the functions of the liver are necessary for its formation."[65] This view does not appear to have won much favor among physiologists (including Claude Bernard) during 1830–1870.

In 1828 Friedrich Wöhler reported his famous synthesis of urea by the reaction of lead or silver cyanate with ammonia. In his oft-quoted letter of 22 February 1828 to Berzelius, he wrote, "I cannot, so to speak, hold my chemical water and must tell you that I can make urea without the aid of kidneys or even an animal, whether man or dog. . . . This artificial formation of urea, can it be regarded as an example of the formation of an organic substance from inorganic materials?"[66] Although, as I have noted, this discovery aroused great enthusiasm among chemists, the response of physiologists was more reserved, and Wöhler's synthesis of urea did not sound the death knell of nineteenth-century vitalism.

The physiological theories offered by Dumas and Liebig around 1840 invoked the oxidative breakdown of proteins as the source of urinary urea, and Liebig suggested that uric acid, which he and Wöhler had oxidized chemically to urea and alloxan, represented an intermediate stage in that process. Moreover, Liebig insisted that "there can be no greater contradiction than to assume that the nitrogen of the food could pass into the urine as urea without first becoming a constituent of the organized tissues, since albumin, the only blood constituent whose amount justifies consideration, cannot have undergone the slightest change upon its passage through the liver, because we find it in all parts of the body in the same condition and with the same properties."[67] This view was challenged by Theodor Frerichs, by Carl Lehmann, and especially by Friedrich Bidder and Carl Schmidt, whose extensive measurements of the respiratory exchange and nitrogen excretion led them to conclude that in fed animals only a small part of the urea in the urine could be derived from the tissue proteins owing to what they called *Luxusconsumtion*.[68] According to Theodor Bischoff, an adherent of Liebig's views, these claims had "dealt a death blow to the doctrine of metamorphosis, as it has sprung from the comprehensive labors and genius of Liebig, and had forever robbed us of the hope of finding a measure of metabolism."[69] A few years later Bischoff and his associate Carl Voit published results that, in their opinion, disproved the idea of Luxusconsumtion, and they attributed the lowering of the rate of the oxidation of tissue proteins to the more ready oxidation of dietary fats and carbohydrates. In reaffirming the Liebig doctrine, they wrote: "It will and must remain forever true that only the nitrogenous substances are producers of energy, that is, only they determine the energy effects, the phenomena of motion, in the animal body; and equally it will remain incontrovertible that, upon their transformation, fat and the so-called carbohydrates produce only heat and no effects of motion."[70] As we have seen, this doctrine crumbled soon afterward, especially after the famous mountain climb of Fick and Wislicenus. Also, during the late 1860s Voit turned against the ideas that he and Bischoff had defended. From experiments on dogs brought into nitrogen balance at different levels of protein intake, he concluded that there are two kinds of protein in the animal body: a variable pool of *circulierendes Eiweiss* related to the level of dietary protein, and *Organeiweiss*. Of these, according to Voit, only the circulating protein is subjected to oxidation, whereas the tissue protein is stable, and only a small portion of it is continually regenerated from the circulating protein.

There ensued a bitter exchange between Voit and Liebig, during the course of which Voit wrote that Liebig "still stands on the ground he created 25 years ago; from chemical experiments he attempts to draw analogies and conclusions about the processes in the animal body. . . . But he has forgotten, to the sorrow of of those who know and value his high services to science better than his flatterers, that these are only ideas and possibilities, whose validity must first be tested through experimentation on animals."[71] Voit's theory was widely accepted among students of human nutrition for several decades, and in 1908 it was elaborated by his disciple Max Rubner. In opposing the theory Hoppe-Seyler noted that Voit's concepts had become commonplace among agricultural chemists, physiologists, and pathologists; he asked "How in the world . . . can Voit have reached any conclusion whatever about the stability or instability of protein in the lymph or organs by weighing dogs, their diet, etc.?" After drawing attention to his own work in the 1860s showing that the oxygen of oxyhemoglobin cannot oxidize proteins, sugars, or fats, and reviewing the new knowledge about metabolic transformations in the organs (notably glycogen in the liver), Hoppe-Seyler asserted that "these organs are not stable, but remarkably variable, with the most active metamorphosis, and consequently wholly dependent on the supply of foodstuffs."[72] In a characteristically combative paper, Pflüger subjected Voit's theory to severe criticism, offered evidence in favor of the stability of circulating protein, and insisted that food protein must first be assimilated by the tissues before it can be metabolized.[73]

The apogee (or nadir) of the nitrogen balance studies came in 1905, when Otto Folin reported an elaborate series of analyses of the urine of human subjects maintained on diets of varying nitrogen content. After criticizing adversely the techniques offered in support of Pflüger's view, Folin used the "laws governing the composition of the urine" that he had deduced from his data to conclude:

> To explain such changes in the composition of the urine on the basis of protein katabolism, we are forced, it seems to me, to assume that katabolism is not all of one kind. There must be at least two kinds. Moreover, from the nature of the changes in the distribution of the urinary constituents, it can be affirmed, I think, that the two forms of protein katabolism are essentially in-

> dependent and quite different. One kind is extremely variable in quantity, the other tends to remain constant. The one kind yields chiefly urea and inorganic sulphates, no kreatinin and probably no neutral sulphur. The other, the constant katabolism, is largely represented by kreatinin and neutral sulphur, and to a less extent by uric acid and ethereal sulphates. . . . It is clear that the metabolic processes resulting in the end products which tend to be constant in quantity appear to be indispensable for the continuation of life; or, to be more definite, these metabolic processes probably constitute an essential part of the activity which distinguishes living cells from dead ones. I would therefore call the protein metabolism which tends to be constant, *tissue* metabolism or *endogenous* metabolism, and the other, the variable protein metabolism, I would call the *exogenous* or intermediate metabolism.[74]

This formulation of the problem dominated the field for about thirty years, especially among American nutritionists and clinical investigators; we will consider later the developments which brought about its downfall. At this point one may recall Claude Bernard's comment, made near the beginning of the long series of nutritional studies on protein metabolism, from Bidder and Schmidt to Folin:

> One can undoubtedly establish the balance between the food consumed by a living organism and what it excretes, but these would be nothing but purely statistical results unable to throw light on the intimate phenomena of the nutrition of living things. It would be, according to the phrase of a Dutch chemist [Mulder], like trying to tell what happens in a house by watching what goes in the door and what leaves by the chimney. One can determine exactly the two extreme limits of nutrition, but if one wishes to construe the intermediate between them, one finds himself in an unknown region largely created by the imagination, and all the more easily because numbers often lend themselves admirably to the proof of the most diverse hypotheses.[75]

Many years later, two investigators chiefly responsible for the downfall of the Folin theory offered a different analogy, reflecting a higher state of technology, in comparing nutritional balance studies with "the working of a slot machine. A penny brings forth one package of chewing gum, two pennies bring forth two. Interpreted according to the reasoning of balance physiology, the first observation is an indication of the conversion of copper into gum, the second constitutes proof."[76]

Among the "proofs of the most diverse hypotheses" that Bernard may have had in mind was the espousal, by Dumas, of Béchamp's claim to have produced urea by the oxidation of albumin by permanganate. Dumas welcomed this report because it provided support for his view that urea is formed in the blood by the combustion of protein, and he stated that "the combustible constituents of the blood therefore yield, as the main products, carbonic acid, water, and urea, unless the last is replaced by products of less extensive combustion."[77]

The Biosynthesis of Urea

It will be recalled that throughout most of the latter half of the nineteenth century the chemical nature of the albuminoid substances was unclear, shrouded by physiological speculations about their relationship to "living protoplasm." Moreover, the metabolic fate of the products of the action of pepsin and trypsin on dietary proteins was also in doubt, and the physiologists studying protein metabolism paid more attention to "albumoses" and "peptones" than to the amino acids. Thus Otto Schultzen and Marceli Nencki noted, "It is striking that hitherto no detailed investigations have been conducted on the behavior of these cleavage products in the animal body, indeed that nobody has suggested that these substances might be the natural intermediates between protein and urea."[78] When they fed glycine or leucine to a dog, they observed an increase an urea excretion. Because these amino acids contain only one nitrogen atom, whereas urea contains two, Schultzen and Nencki concluded that the metabolic formation of urea must involve a synthetic process. A few years later, Woldemar von Knieriem showed that the administration of ammonium salts also gives rise to an increased output of urea, and he added aspartic acid and asparagine to the amino acids whose ingestion had this effect.[79]

Of particular importance was the work of Woldemar von Schroeder, who reported in 1882 that perfusion of a dog liver with ammonium salts led to the extensive formation of urea, thus adding a third function, in addition to the production of glycogen and of bile, to this organ, and providing decisive evidence for the surmise offered sixty years earlier by Prévost and Dumas.[80] The technique of the perfusion of isolated organs had been developed during the 1860s in Carl Ludwig's laboratory and subsequently played an increasingly important role in the study of metabolic processes in animal tissues. Schroeder's results were confirmed by numerous workers, but they left open the question of the intermediates between dietary proteins and the ammonia used by the liver for urea synthesis. The prevailing view was expressed by Hoppe-Seyler: "Since acid albumin and peptone are produced in the stomach and by the pancreatic secretion, it appears to be the function of the epithelial cells of the intestine to transform these substances into serum albumin and the fibrin-generating materials."[81] Apparent support for this view was offered by Hofmeister, on the ground that peptones disappeared when in contact with the intestinal wall.[82] Given the choice between resynthesis to proteins and cleavage to amino acids, the latter possibility was discounted because, as stated by Gustav von Bunge in successive editions of his textbook: "It may be *a priori* doubted on teleological grounds, whether under normal conditions the amount of amido-acids formed in the intestine is a large one. It would be a waste of chemical potential energy, which would serve no purpose when converted into kinetic energy by their decomposition, and a reunion of the products of such a profound decomposition is highly improbable."[83] This statement is a sobering example of the seductive power of both teleological and thermodynamic reasoning, especially because a year before its appearance in the 1902 English translation of Bunge's book, Otto Cohnheim had shown that "the disappearance of peptones in contact with the intestinal wall does not arise from their assimilation or their reconstitution into protein, but their further cleavage to simpler products. . . . This cleavage is effected by a special ferment elaborated by the intestinal mucosa, erepsin, which acts on peptones and some albumoses, but not on true proteins."[84] Shortly afterward, Cohnheim demonstrated the formation of several amino acids upon the action of "erepsin" on peptones, and this finding was confirmed by Emil Abderhalden in Emil Fischer's laboratory. Moreover, in 1902 Otto Loewi reported that he had fed an extensively digested "biuret-free" autolysate of pancreatic protein to dogs and that "the sum of the biuret-free end products replaces food protein, that is, can substitute for all parts of the body protein that is degraded in protein turnover."[85] The same result could not be obtained with acid hydrolysates of food proteins, but Gowland Hopkins called attention to the fact that the acid-labile tryptophan is indispensable for the maintenance of the nitrogen balance of animals.[86] The decisive experiments on the question were performed during 1912–1913 by Donald Van Slyke and Gustave Meyer, with the aid of an apparatus Van Slyke had devised to determine the volume of nitrogen gas produced by the action of nitrous acid on amino acids and peptides. Their data showed that "ingested proteins are hydrolyzed in the digestive tract setting free most, if not all, of their amino-acids. These are absorbed into the blood stream, from which they rapidly disappear as the blood circulates through the tissues."[87]

Although the metabolic source of the nitrogen in urea was known, the mechanism for urea synthesis was still a matter of chemical speculation. Among the early hypotheses were the suggestions in 1877 by Ernst Salkowski that "nascent" cyanic acid (HNCO) is an intermediate, and by Oswald Schmiedeberg that urea formation involves the dehydration of ammonium carbonate [$(NH_4)_2CO_3$]. Edmund Drechsel preferred a process in which the ammonium salt of carbamic acid (NH_2-$COONH_4$) is involved.[88] Later, Drechsel proposed that urea is formed by the hydrolysis of "lysatine," a material he obtained from an acid hydrolysate of casein. Subsequent work by Ernst Schulze showed that the urea-producing component of "lysatine" was arginine, which he had isolated in 1886, and by 1898 he had demonstrated that arginine yields, on alkaline hydrolysis, urea and the amino acid ornithine (α,δ-diamino valeric acid).[89] The latter substance had been obtained in 1887 from bird urine in the form of its dibenzoyl derivative (ornithuric acid) by Max Jaffé.[90]

Albrecht Kossel, who had found protamines to be rich in arginine, then suggested that, in addition to the amide linkage (-CO-NH-) postulated by Franz Hofmeister and Emil Fischer as a general structural feature of proteins, and the guanidino group [NH_2-C(=NH)-NH-] identified in arginine by Schulze, there might be combinations of the two [that is, -CO-NH-C(=NH)-NH-]. This idea was related to Kossel's hypothesis that the protamines represent fundamental structural units of all proteins. Accordingly, with Henry Dakin, he subjected a protamine to the action of Cohnheim's "erepsin" and found that, according to the conditions of the test, either arginine or ornithine was formed. They concluded that the enzyme preparation contained, in addition to the enzyme that cleaved CO-NH bonds, another enzyme that hydrolyzes arginine to ornithine and urea. They named it arginase and found it to be especially abundant in the liver.[91] This discovery explained the observations by Charles Richet during the 1890s on the production of urea in macerated liver; together with the parallel demonstration by Ernst Salkowski and Martin Jacoby that animal tissues such as liver contain "autolytic" enzymes, which can degrade proteins to peptones and amino acids, it seemed possible that the combined action of these enzymes and arginase could give rise to urea.

Subsequent work, especially by Antonio Clementi, strengthened the view that arginase plays a role in the metabolic production of urea. He noted the apparent absence of this enzyme from the livers of birds and of reptiles, which excrete uric acid, in contrast to its presence in the livers of mammals, amphibians, and fishes, which excrete urea.[92] Clementi named the first group of animals uricotelic and the second ureotelic. His finding was largely confirmed during the 1920s by Siegfried Edlbacher and Andrew Hunter. By that time the quantitative estimation of urea (and of arginase activity) had been made more reliable through the use of urease as a specific reagent to convert urea into CO_2 and NH_3, either of which could be readily determined. Indeed, it was the availability of this method that led James Sumner, who had worked in Otto Folin's laboratory on the problem of urea formation, to begin work in 1917 on the purification of urease.[93]

Although it seemed by 1930 that arginase is a participant in the metabolic formation of urea, the massive conversion of ammonia to urea could not be accounted for solely in terms of the hydrolytic cleavage of the arginine derived from proteins. The long-discarded cyanate theory reappeared in new guise and, what was more important, Wilhelm Loeffler had concluded that "urea formation in the surviving liver is bound to the integrity of the cell structure; oxidative processes play an integrating role." This view was supported by the failure of efforts to demonstrate the formation of urea from ammonia in minced liver preparations.[94]

In 1932 entirely new light was thrown on the problem of urea synthesis in a paper by Hans Krebs and his student-assistant Kurt Henseleit. Their work showed that: "the primary reaction of urea synthesis in the liver is the addition of 1 molecule of ammonia and 1 molecule of carbonic acid to the δ-amino group of ornithine, with the elimination of 1 molecule of water and the formation of a δ-ureido acid citrulline. . . . The second reaction of urea synthesis is the combination of 1 molecule of citrulline with an additional molecule of ammonia, with the loss of a second molecule of water and the formation of a guanidino acid, arginine. . . . The third reaction is the hydrolytic cleavage of arginine to ornithine and urea."[95] The achievement summarized in these sentences marked a new stage in the development of biochemical thought. Not only was an explanation of an intracellular synthesis offered for the first time in terms of chemical reactions identified in the appropriate biological system and not merely inferred from the known chemical behavior of presumed reactants, but the formulation of the "ornithine cycle" also provided a clue to the organization of metabolic pathways. This became more evident in 1937, with the appearance of the "citric acid cycle." Krebs later generalized the concepts that had emerged from these advances in terms of the physiologically unidirectional character of metabolic cycles, because of the "irreversibility" of some of the intermediate steps.[96] Apart from its importance, the Krebs-Henseleit paper has several interesting features. As Krebs later wrote: "When I set out to study the mechanism of urea formation, I had no preconceived hypothesis. My idea was that the in vitro system of liver slices, incubated in a saline medium, provided a novel and easy method of measuring accurately the rate of urea synthesis under a variety of conditions."[97] A large variety of substances, including all readily available amino acids, were tested for their effect on the rate of the process $2NH_3 + CO_2 = NH_2\text{-}CO\text{-}NH_2 + H_2O$. The rate data were expressed as the amount of "urea-CO_2" (liberated by urease, and measured in a Warburg manometer) per unit weight of tissue per hour. The action of most amino acids could be explained by their deamination to produce ammonia, but unexpectedly ornithine was found to promote urea synthesis in a catalytic manner.[98] Since the arginase reaction was well known, the effect of ornithine appeared to indicate that it is used, together with CO_2 and NH_3, in the energy-requiring synthesis of arginine, and it was reasonable to ask whether the δ-ureido derivative of ornithine is an intermediate in this synthesis. Only two years earlier, Mitsunori Wada had shown that a substance isolated in 1914 from watermelon juice (and hence named "citrulline") is δ-ureido ornithine.[99] When liver slices were found to produce urea rapidly in the presence of citrulline and ammonia (but not citrulline alone), the experimental evidence for the operation of the ornithine cycle was in hand.

Another interesting item in the Krebs-Henseleit paper was the conclusion that the steps leading from ornithine to arginine are "linked to life" and dependent on the integrity of the cellular structure.[100] As we have seen, attempts to demonstrate the synthesis of urea from ammonia in disintegrated tissues were unsuccessful, and the small urea formation found there was attributable to the action of arginase on the arginine formed by the autolysis of liver proteins. The view expressed by Krebs was consistent with the decades-old conviction that "syntheses are reserved to life" and that "the synthetic formation of urea from its precursors has not been imitated outside the living cell, any more than it has been successful for other vital syntheses."[101]

By 1946, however, a cell-free liver "homogenate" had been shown to make urea from NH_3 and CO_2. During the years after 1932, the metabolic importance of such substances as cytochrome c, NAD, and ATP had been clarified, in large part through the work of Krebs's mentor Otto Warburg who during 1910–1930 had insisted on the dependence of intracellular respiration on the integrity of cellular structure. By adding these three substances to a cell-free liver preparation, Philip Cohen and Mika Hayano showed that earlier unsuccessful efforts to demonstrate urea synthesis in such systems were in large part a consequence of the dilution and destruction of essential cofactors in the process.[102]

Shortly afterward, Sarah Ratner showed that in the conversion of citrulline into arginine, the added nitrogen does not come directly from ammonia but by enzyme-catalyzed and ATP-dependent transfer from aspartic acid, with the intermediate formation of argininosuccinic acid, which is cleaved to arginine and fumaric acid.[103] This important discovery thus linked the biosynthesis of urea to the metabolic processes involving aspartic acid and fumaric acid, a component of the citric acid cycle. Then followed the work of Mary Ellen Jones and Leonard Spector, in Fritz Lipmann's laboratory, showing that the formation of citrulline involves the enzyme-catalyzed reaction of ornithine with carbamoyl phosphate, which is made in a separate enzymatic reaction: $NH_3 + CO_2 + 2ATP \rightarrow NH_2CO\text{-}P + 2ADP + P$.[104]

Before these contributions of the 1940s and 1950s in support of the validity of the ornithine cycle, Krebs had replied to some critics as follows:

> The ornithine cycle is a theory; it is a supposition designed to account for a number of observations and to serve as a starting point for further investigations. . . . These observations "prove," i.e., establish as fact, the occurrence of each step of the cycle as postulated in the theory. This of course is true only for the conditions of the experiments and it can be said with some justification that the reactions have not been proved to occur under "physiological" conditions, i.e., in the intact liver in situ. In all experiments some facts were "unphysiological." The liver was taken out of the body and sliced or artificially perfused. The concentrations of the reactants were artificially raised. Hence the statement that the ornithine cycle is operative in the liver in situ implies the assumption (which some do not wish to make) that the cycle occurs under conditions different from those of the experiments. In this sense it may be formally correct to say that the cycle has not been proved to occur in "normal" liver, but such a statement has little scientific value for it applies to every intermediary reaction mechanism. Experiments can only prove that a tissue has the ability to perform certain chemical reactions, but if these reactions are intermediary stages, the "normal" occurrence cannot be proved.[105]

Another metabolic role of arginine will be considered in a succeeding chapter. At this point I only mention the discovery during the late 1980s of an enzyme system ("nitric oxide synthase") that catalyzes the oxidative cleavage of arginine to citrulline and nitric oxide (NO); the multiple physiological function of NO is now an active area of investigation.

The Metabolism of Amino Acids

Many of the questions encountered in the early studies on the formation of urea in animal organisms were considered by plant physiologists who studied the formation of asparagine in seedlings. Because of its ready crystallization from concentrated aqueous solution, asparagine was isolated (and so named) from asparagus juice by Vauquelin and Robiquet in 1806. Later Raffaele Piria obtained it from vetch seedlings but reported that it appeared to be absent in ungerminated seeds or in mature plants exposed to light, and suggested that "it is probable that there exists, in the seeds, an azotized material (perhaps casein) which is transformed into asparagine and other products during germination."[106] In 1868 Boussingault concluded, from the changes in carbon and nitrogen content in germinating seedlings, that there is a respiratory combustion of seed protein, and he compared this process to what was then believed to be the oxidative conversion of protein into urea in animals. This comparison was also made by Wilhelm Pfeffer, who proposed that the asparagine formed during germination is derived from the seed protein and cannot be excreted but diffuses in the growing plant to sites where it is used for the resynthesis of protein.[107]

Although Pfeffer's theory was widely accepted among botanists, the work of Ernst Schulze after 1876 made it untenable. Schulze showed that during the first stages of plant growth, a variety of amino acids (leucine, tyrosine, and others) as well as peptones are formed by the enzymatic breakdown of seed proteins, in a process comparable to the gastrointestinal digestion of proteins in animals. Because asparagine appeared later, it clearly represented a secondary product of protein breakdown, possibly derived from ammonia formed by the decomposition of amino acids.[108] Schulze's approach was taken up by his student Dmitri Prianishnikov, who wrote: "Since asparagine contains two NH_2-groups in the molecule, and the primary amino acids (leucine, aminovaleric acid, tyrosine, etc.) contain only one, it must be assumed that in the formation of an asparagine molecule the residues of two molecules of some primary amino acid (or two different amino acids) are involved."[109] Twenty years later, his studies on the nitrogen metabolism of plants, and the parallel work of others on the metabolic fate of amino acids in animal tissues, led Prianishnikov to emphasize anew the similarity of the roles of asparagine and urea:

> Both amides are not primarily products of the hydrolytic cleavage of proteins, but are formed in a secondary manner, involving above all ammonia formed by the oxidation of amino acids; both amides are formed equally easily at the expense of externally supplied ammonia, with the obvious difference that for the formation of asparagine a part of the unoxidized carbon chain is necessary, whereas ammonia and carbonic acid suffice for the production of urea. Not only the mechanism, but also the physiological purpose of the two amides is entirely similar—it consists in the detoxification of ammonia. . . . The further fate of asparagine and urea is quite different: urea is excreted by the organism, since the animal does not need to economize on nitrogen, because it is abundantly supplied in the food; . . . asparagine, on the other hand, remains in the cells as a reserve form of nitrogen which, with a more abundant supply of carbohydrate, can later contribute to various components of the protein molecule.[110]

By the early 1930s it was clear from the work of Krebs and of Prianishnikov that much more detailed knowledge was needed about the metabolism, in both animals and plants, of the amino acids that constitute protein molecules.

Earlier work by Otto Neubauer and by Gustav Embden had shown that amino acids are oxidatively deaminated in animal tissues to the corresponding keto acids, and that perfusion of dog liver with α-keto acids and ammonia gives rise to the reversal of this process, with the formation of the corresponding amino acids.[111] Also, Franz Knoop had reported that if the synthetic keto acid C_6H_5-CH_2-CH_2-CO-COOH is fed to a dog, the corresponding amino acid (in the form of its N-acetyl derivative) appears in the urine.[112] The use of such unphysiological phenyl compounds in metabolic studies was introduced by Knoop in connection with his important studies on the oxidation of fatty acids, to which we will return shortly. During the 1920s, with the rise of the Wieland-Thunberg dehydrogenation theory, Knoop also offered a chemical model of amino-acid synthesis by the reaction of α-keto acids and ammonia in the presence of platinum or palladium catalysts.[113]

During the 1930s Hans Krebs showed that the kidney contains separate oxidases for L- and D-amino acids and that, in the case of L-glutamic acid, which markedly enhanced the oxygen uptake of kidney slices, there was a reduction in the rate of ammonia formation. He explained this phenomenon by showing that the kidney converts ammonium glutamate into glutamine and that this energy-requiring process is also effected in brain tissue.[114] Glutamine had been isolated from beet juice in 1883 by Ernst Schulze, who considered it to be equivalent to the less-soluble asparagine in plant metabolism.[115] By the 1930s, with the development of analytical methods for the separate determination of the two amides, it was recognized that both occur in all plants. After 1945 the formation of glutamine from glutamic acid and ammonia was shown to be effected in animal and plant tissues, as well as microorganisms, by an ATP-dependent synthetase, and a similar enzyme system was identified in plants and microorganisms for the synthesis of asparagine. Moreover, glutamine has been identified as the metabolic donor of its amide NH_2 group in reactions catalyzed by "amidotransferases" involved in the synthesis of purine and pyrimidine nucleotides, amino sugars, and some amino acids.[116] Several of these metabolic processes will be described later.

It was already known before 1940 that the reversible dehydrogenation of glutamate and aspartate is effected in biological systems by pyridine nucleotide-dependent enzymes, with α-ketoglutarate and oxaloacetate as the respective products of oxidative deamination. These two keto acids were also known to be components of the citric acid cycle, thus linking the metabolism of the two amino acids to the metabolism of pyruvate and the carbohydrates from which it is derived. Moreover, in 1937 Aleksandr Braunstein and Maria Kritzmann showed that minced pigeon or rabbit muscle can effect the reversible transfer of the amino group of glutamate to pyruvate (to form alanine) or to oxaloacetate (to form aspartic acid).[117] This discovery of "transamination" reactions not only gave further evidence of enzyme-catalyzed group transfer in metabolic processes but also emphasized the central position of glutamate and aspartate in protein metabolism. After 1945 Fritz Schlenk and Esmond Snell found that the tissues of vitamin B_6-deficient rats have low transaminase (now termed aminotransferase) activity. The discovery of the natural occurrence of the aldehyde (pyridoxal) and amine (pyridoxamine) forms of vitamin B_6 led to the identification of their 5′-phosphoryl derivatives as interconvertible cofactors in the enzyme-catalyzed reaction. Later work showed that the aldehyde group (R-CHO) of the cofactor is bound to a lysine unit of the catalytic protein by "Schiff-base" linkage (-CH=N-), and is converted to the amine ($R\text{-}CH\text{-}NH_2$) by reaction with the amino-group donor to form the corresponding keto acid.[118]

Intermediate Steps in Metabolic Processes

During most of the nineteenth century, the term *Stoffwechsel* or *metabolism* was taken to refer to the transformation, in animal and plant organisms, of nutrients into excretory or storage products. Although Claude Bernard wrote about intermediate steps in what he called circulation chimique, it was this "intake-output" concept that was dominant in physiological thought and experiment. By about 1900, however, considerable empirical evidence had become available to warrant closer attention to intermediate chemical events in metabolic processes. After the discoveries made during the first decade of the twentieth century, this point of view was expressed by Henry Dakin: "A true knowledge of metabolic processes can only be obtained by the tedious unraveling of the complex system of biochemical changes into individual chemical reactions. At the present time only a few of these simple reactions have been recognized and studied, but even now it requires little imagination to realize that in the future it will be possible to construct an accurately itemized account of the animal body's chemical transactions both anabolic and catabolic. The value of such knowledge for the advancement of biology and medicine is sufficiently obvious."[119] Shortly afterward, in a famous lecture, Gowland Hopkins stated that "in the study of the intermediate processes of metabolism, we have to deal, not with complex substances which elude ordinary chemical methods, but with simple substances undergoing comprehensible reactions. By simple substances I mean such as are of easily ascertainable structure and of a molecular weight within a range to which the organic chemist is well accustomed."[120] These opinions of two noted British biochemists reflected not only their negative attitude to the colloid chemistry of their time but also their adherence to the view expressed by Emil Fischer some years earlier: "The ultimate aim of biochemistry is to gain complete insight into the unending series of changes which attend plant and animal metabolism. To accomplish a task of such magnitude complete knowledge is required of each individual chemical substance occurring in the cycle of changes and of analytical methods that permit of its recognition under conditions such as exist in the living organism. As a matter of course, it is the office of organic chemistry, especially of synthetic chemistry, to accumulate this absolutely essential material."[121] However prescient one may now judge these statements to have been, at a time when the chemical nature of high-molecular weight "complex substances" was uncertain, the fact remains that there was then only fragmentary evidence for the conviction that metabolic processes involve "a complicated series of chemical reactions following upon each other in definite sequence."[122]

Thus when Franz Hofmeister wrote in his eulogy of Willy Kühne that the latter had foreseen that "the answers to the immediate and fundamental physiological questions depend above all on three fields accessible only to the chemist, those of protein chemistry, of the action of ferments, and of intermediary metabolism,"[123] Hofmeister referred primarily to Kühne's work on the role of peptones as intermediates in the digestive breakdown of proteins. Similarly, the dextrins were considered to be intermediates between glycogen and glucose, and uric acid had long been thought to be an intermediate in the metabolic oxidation of proteins to urea. In addition to these doubtful items of evidence, however, there had been a series of findings, many of them haphazard, that later led to fruitful research. Among them was the isolation of several curious organic substances from mammalian urine and the application of the new organic chemical methods to the determination of their constitution.

Two of these compounds were constituents of dog urine; one was found by Justus Liebig, who named it kynurenic acid (Greek *kynos*, dog + *ouron*, urine), the other by Max Jaffé, who called it urocanic acid (Greek *ouro* + Latin *canis*, dog). The latter was more elusive, as it did not appear in all dogs, and the one that produced it for Jaffé ran away, but his report was confirmed by Andrew Hunter, and urocanic acid was shown to be imidazolylacrylic acid.[124] Later work, during the 1950s, led to the identification of urocanic acid as the initial intermediate in the metabolic breakdown of histidine in animal tissues, and to the finding that the imidazole ring of urocanic acid is then opened to yield formiminoglutamic acid (NH=CH-Glu), whose further conversion to glutamic acid is an enzyme-catalyzed process that involves the participation of a cofactor related to the member of B-group of vitamins known as folic acid.[125] More will be said about this cofactor shortly.

As for kynurenic acid, by about 1900 the work of Zdenko Hans Skraup and Rudolf Camps had shown it to be a derivative of quinoline and some biochemists took this finding to suggest the presence of "quinoline-like radicals in proteid molecules."[126] After the discovery of tryptophan by Hopkins and Cole, Alexander Ellinger reported that the administration of this amino acid to dogs or rabbits induced the excretion of kynurenic acid, and subsequent work by Yashiro Kotake and by Adolf Butenandt established its chemical structure as well as that of kynurenine, formed by the opening of the indole ring of tryptophan, its metabolic precursor.[127] We will return to kynurenine in the next chapter, in connection with studies on the genetic control of its utilization for the synthesis of insect eye pigments.

A third unusual chemical compound, named homogentisic acid, was found in the urine of human subjects with the condition known as alcaptonuria, described in 1828 by Alexander Marcet and in 1859 by Carl Boedeker. Such subjects excrete a urine that darkens on exposure to air; in 1891 Mikhail Volkov and Eugen Baumann showed the compound responsible for this property to be 2,5-dihydroxyphenylacetic acid and demonstrated that its excretion by an alcaptonuric subject is promoted by the ingestion of tyrosine. Although they inferred from clinical reports that the disease represented a "metabolic abnormality that lasted throughout the patient's life," they also concluded that "the formation of homogentisic acid from tyrosine is not effected by an inexplicable abnormal function in the tissues, but is to be regarded as the action of a particular kind of microorganism."[128] The latter statement reflected the contemporary fashion of attributing disease to the action of microbial agents, and there was much evidence for the degradation of some amino acids by intestinal bacteria. Extensive studies by Marceli Nencki and others during 1870–1890 on the fate of aromatic compounds in the animal body had indicated that urinary products such as phenol or indole were formed by the action of putrefactive bacteria on substances derived from food proteins.

In 1903 Wilhelm Falta and Leo Langstein showed that phenylalanine also gives rise to the excretion of homogentisic acid in alcaptonuria, and a year later Otto Neubauer and Falta obtained a similar result with the α-keto acids corresponding to tyrosine and phenylalanine. They concluded that "in the normal organism the combustion of the aromatic amino acids proceeds by way of the alcapton acids and the disorder in alcaptonuria consists only in the fact that, as a consequence of the inhibition of metabolism, the degradation stops at this point."[129] As we shall see, Archibald Garrod had already noted the hereditary character of this disease and, after the report by Neubauer and Falta, he and Thomas Hele showed that the homogentisic acid excreted by alcaptonuric subjects corresponded to the "whole of the tyrosine and phenylalanine broken down."[130] Subsequent work, largely by Neubauer, led to the recognition that the metabolic pathway from phenylalanine to homogentisic acid involves (1) oxidation of Phe to Tyr, (2) deamination of Tyr to *p*-hydroxyphenylpyruvic acid; (3) a rearrangement to produce 2,5-dihydroxyphenylpyruvic acid; (4) decarboxylation to homogentisic acid. Although this pathway had been formulated and largely accepted by the 1920s, the enzymes and cofactors that catalyze the successive reactions were not identified and studied until after 1945. Thus the conversion in the liver of Phe to Tyr, demonstrated in perfusion experiments by Gustav Embden in 1913, was shown to be effected by a "phenylalanine hydroxylase" catalyzed reaction in which O_2 is the oxygen donor and in which reduced NADP is an electron donor, and another oxidation-reduction (dihydropterin-tetrahydropterin) system, related to the vitamin folic acid, is an essential participant.[131] In another hereditary human disease, known as phenylketonuria, associated with severe mental retardation, this process was found to be blocked. The second step is catalyzed by a transaminase, and the resulting α-keto acid is converted to homogentisic acid in a reaction catalyzed by a "dioxygenase" with the incorporation of both atoms of O_2 into the product.[132]

The Metabolism of Fatty Acids

The finding of the curious urinary products mentioned above and the recognition that they were derived from protein amino acids were typical of the haphazard development of the study of intermediate metabolism before about 1900. They reflected the rise of "pathological chemistry" and the entrance of the new structural organic chemistry into clinical practice and research. In this application of chemistry to medicine, special attention was given to diabetes mellitus, considered by some to be a "purely chemical derangement of health,"[133] and chemical observations on diabetic patients opened one of the most fruitful avenues of research on intermediate metabolism.

During 1850–1880 several investigators identified the odorous material exhaled by diabetic patients as acetone (CH_3-CO-CH_3), which was found in the blood and urine, and the coma in the advanced stages of the disease was associated with the appearance of this ketone. In 1883 Rudolf Jaksch found in diabetic urine the closely related acetoacetic acid (CH_3-CO-CH_2-COOH), whose structure had recently been determined and which was known to give rise to acetone by loss of CO_2. One year later, Oscar Minkowski and Eduard Külz reported the finding in such urine of an optically active (levorotatory) β-hydroxybutyric acid [CH_3-CH(OH)-CH_2-COOH], whose chemical relation to acetoacetic acid had already been established by

Wislicenus.[134] Acetone was first considered to be the product of the incomplete oxidation of sugar, a view consistent with the long-held opinion that diabetes is associated with the inability of the body to convert glucose to CO_2 and H_2O. By 1885 it was known, however, that these "ketone bodies" are formed in normal human subjects kept on a diet free of carbohydrate. The discovery of the two acids in diabetic urine shifted attention to the carbon side-chains of protein amino acids, and during the 1890s a widely held view was that the ketone bodies are "undoubtedly . . . products of protein degradation, since their amount in the urine increases unmistakably with the increased excretion of nitrogen."[135] Adolf Magnus-Levy dissented from this view, and in his *Habilitationsschrift* at Strassburg (where he was associated with Bernhard Naunyn and Franz Hofmeister) he provided strong evidence for the opinion that the appearance of β-hydroxybutyric acid in the blood of diabetic patients is a consequence of the incomplete metabolic oxidation of fatty acids.[136] In the background of nineteenth-century studies on fat metabolism was the work of Michel Eugène Chevreul on the constitution of fats and their hydrolytic cleavage ("saponification") to glycerol and fatty acids, as well as the subsequent demonstration by Johann Eberle and Claude Bernard of the emulsification of fats by pancreatic juice.[137] In 1896 Maurice Hanriot gave the name *lipase* to this enzymatic activity.

It will be recalled that before 1890 many physiologists shared Pflüger's view of biological oxidation as "explosive" combustions effected by "living" protoplasmic proteins. From that point of view, the idea of a stepwise chemical process in the metabolic breakdown of a fatty acid such as stearic acid [$CH_3(CH_2)_{16}COOH$] was superfluous. With the gradual acceptance among physiological chemists that such intracellular processes are effected by specific enzymes, and with the formation of intermediate products, metabolic pathways became subjects of experimental study rather than only matters for chemical speculation. This approach found its first full expression in the work of two chemically minded junior members of Franz Hofmeister's Strassburg laboratory—Franz Knoop and Gustav Embden—on the problem of the metabolic oxidation of fatty acids and the formation of ketone bodies.

Knoop fed dogs the sodium salts of various straight-chain fatty acids in which the carbon farthest from the carboxyl group is linked to a phenyl group, as in phenylbutyric acid [$C_6H_5(CH_2)_3COOH$]. The rationale of this approach was that, whereas naturally occurring fatty acids such as butyric acid and stearic acid were known to be completely oxidized in normal animals, earlier work had also shown that aromatic acids such as benzoic acid (C_6H_5COOH) or phenylacetic acid ($C_6H_5CH_2COOH$) are resistant to metabolic oxidation in the animal body and are excreted in the urine in the form of conjugates with glycine (hippuric acid or phenylaceturic acid). Knoop found that, after phenylbutyric acid had been fed, the urinary product was phenylaceturic acid, while with phenylpropionic acid [$C_6H_5(CH_2)_2COOH$] or phenylvaleric acid [$C_6H_5(CH_2)_4COOH$] the product was hippuric acid.[138] From these results, Knoop concluded that the metabolic breakdown of fatty acids occurs by successive oxidation at the β-carbon, according to the scheme:

$$CH_3(CH_2)_nCH_2CH_2COOH \rightarrow CH_3(CH_2)_nCOCH_2COOH \rightarrow$$
$$CH_3(CH_2)_{n-2}COOH + C_2 \text{ unit.}$$

This "beta-oxidation theory" received strong support from Embden's perfusion experiments with surviving liver during 1904–1908; with unsubstituted straight-chain fatty acids, those with even-numbered chains (C_4, C_6, C_8, C_{10}) gave acetone, which must have been derived from acetoacetic acid, whereas those with odd-numbered chains (C_3, C_5, C_7, C_9) did not, but gave rise to the formation of glucose.[139] Further support came from the parallel studies of Henry Dakin, who also showed that the liver contains an enzyme which interconverts acetoacetic acid and *l*-β-hydroxybutyric acid.[140] By 1910, therefore, the formation of ketone bodies appeared to be a process in which long-chain fatty acids are shortened by the successive removal of -CH_2-COOH units (released as acetic acid [CH_3COOH]) from the β-keto acids formed by beta-oxidation, and it was thought that, in diabetes, this process is blocked when acetoacetic acid is reached. In addition, from the studies of Naunyn and others on the "antiketogenic" action of glucogenic substances, it seemed that in normal animals there is a link between the oxidation of glucose and and the complete oxidation of ketone bodies, and much was made of the idea that "fats burn in the fire of the carbohydrates."

Doubts about the validity of the Knoop theory were raised by Ernst Friedmann, who found in perfusion experiments that the liver can convert acetic acid to acetoacetic acid and concluded that "the pairwise degradation of normal saturated fatty acids does not proceed with the liberation of acetic acid. It also follows that it is incorrect to assume that normal saturated fatty acids are degraded by the cleavage of β-keto acids having two fewer carbon atoms."[141] Later, after the perfusion method had largely been replaced by the tissue-slice technique in such studies, Maurice Jowett and Juda Quastel reported that the oxidation of fatty acids with 6, 8, or 10 carbon atoms gave rise to more acetoacetic acid than was expected on the basis of the successive beta-oxidation theory, and Eaton MacKay proposed that the two-carbon units join to form acetoacetic acid.[142] Moreover, before 1945 Maurice Edson and Luis Leloir found a relation between the formation of ketone bodies and the oxidation of pyruvate, and subsequent work in Leloir's laboratory showed that the two-carbon units enter the citric acid cycle.[143] These findings, and Albert Lehninger's report in 1945 that the oxidation of fatty acids to acetoacetic acid by a rat liver homogenate requires the participation of ATP, marked a new stage in the study of the metabolic breakdown of fatty acids. During the 1950s, an outstanding series of investigations on the beta-oxidation pathway, largely in the laboratories of Earl Stadtman, Feodor Lynen, Severo Ochoa, and David Green, showed this process to involve the following enzyme catalyzed reactions: (1) ATP-dependent conversion of the fatty acid (RCH_2CH_2COOH) into its acyl-CoA derivative; (2) FAD-dependent dehydrogenation to form an "enoyl" compound (RCH=CHCO-CoA); (3) hydration of the double bond to form the β-hydroxy compound [$RCH(OH)CH_2CO$-CoA]; (4) NAD dependent oxidation to the β-keto compound ($RCOCH_2CO$-CoA); (5) "thiolysis" by the sulfhydryl group of CoA to form RCO-CoA and CH_3CO-CoA.[144] Carnitine (trimethyl-γ-ammonium-β-hydroxybutyrate), once considered to be a B-vitamin, is a dipolar substance which can exchange CoA with acetyl-CoA, and appears to be involved in the transfer of long-chain fatty acyl-CoA across the mitochondrial membrane into the interior site of biological oxidation.[145]

As regards the relation between the metabolisms of fatty acids and carbohydrates, during the early years of the twentieth century there was a lively but inconclusive debate whether animals, like plants, can manufacture glucose from fatty acids. There was no debate, however, about the reverse process, whose occurrence in the animal body had been well established for more than fifty years. By 1845 Boussingault had acknowledged that he and Dumas were wrong in denying that animals can convert sugar into fat (as Liebig had stated). The later balance experiments of Lawes and Gilbert provided decisive evidence on this question and also showed that fat can arise from dietary protein.[146] Among the many chemical speculations about the mechanism whereby fatty acids are synthesized were those of John Norman Collie and of Henry Stanley Raper, who suggested that the -CH_2CO- group may be involved in polymerization reactions and that fatty acids "are produced by the condensation of some highly reactive substance containing two carbon atoms and formed in the decomposition of sugar."[147] It was not until the 1950s that any expectation that the metabolic pathway of fatty acid synthesis is simply the reversal of their oxidation to acetyl-CoA was dispelled by discoveries made by Salih Wakil and Roy Vagelos. Wakil, working with liver extracts, showed that the initial step in the synthetic process is the conversion of acetyl-CoA into malonyl-CoA (HOOC-CH_2CO-CoA) in an ATP-dependent reaction in which the vitamin biotin is a participant. At the time, such carboxylation reactions in animal tissues were not considered to be unusual, but as we shall see, the earlier recognition of the importance of CO_2 fixation by such tissues is a matter of considerable historical interest. Vagelos, working with a bacterial *(Escherichia coli)* system, showed that in the elongation of the fatty acid chain, the growing acyl group is attached to a sulfhydryl group of an "acyl carrier protein" (ACP).[148] The sequence of steps was shown to be: (1) the conversion of acetyl-CoA to malonyl-CoA and the transfer of the malonyl group to ACP and the parallel transfer of the acetyl group of acetyl-CoA to ACP; (2) the reaction of malonyl-ACP and acetyl-ACP to form acetoacetyl-ACP; (3) the reduction of this intermediate by NADPH to hydroxybutyryl-ACP, followed by dehydration to form crotonyl-ACP ($CH_3CH{=}CHCOACP$); (4) reduction by NADH to butyryl-ACP, which then reacts with another molecule of malonyl-ACP to begin another cycle in the elongation reaction. Although the sequence of reactions appears to be the same in the bacterial and animal systems, in *E. coli* the enzymatic catalysis of the individual steps is effected by separable proteins, while the "fatty acid synthase" of higher organisms is a single huge protein in which separate domains are combined to form a multienzyme complex.[149]

To conclude this section—and in hailing the exceptional distinction of the most recent achievements in the study of fatty acid metabolism—it is appropriate to recall that Franz Knoop's experiments on the fate of phenyl-substituted fatty acids marked the introduction of the labeling technique into the study of metabolic processes. The limitations of his approach were obvious, for the chemical modification he made was not generally applicable to a large variety of metabolites. The availability during the 1930s of the stable isotopes of hydrogen (^2H) and nitrogen (^{15}N), and after 1945 of the radioactive isotopes of carbon (^{14}C) and other elements, made the labeling technique an indispensable tool in the study of the pathways of biochemical change in living organisms. Together with the advances made in enzyme chemistry, in chemical genetics, and in the chromatographic separation of organic compounds, the isotope technique made possible the transformation during the 1950s of biochemical thought and practice.

Isotopes in Metabolic Research

In August 1935 there appeared a short paper which began as follows:

> Many attempts have been made to label physiological substances by the introduction of easily detectable groups such as halogens or benzene nuclei. However, the physical and chemical properties of the resulting compounds differ so markedly from those of their natural analogues that they are treated differently by the organism. The interpretation of metabolic experiments involving such substances is therefore strictly limited. We have found the hydrogen isotope deuterium to be a suitable indicator for this purpose.

> The fact that it occurs in the same proportion (1 atom of deuterium to 5,000 atoms of protium) in the hydrogen of ordinary water and of organic matter is in itself evidence that the living body is unable to distinguish the few organic molecules which contain deuterium from those that do not. Were the reverse the case, organic matter of biological origin would display differences in isotopic ratio. We have prepared several physiological compounds (fatty acids and sterol derivatives) containing one or more deuterium atoms linked to carbon, as in methyl or methylene groups. Their physical properties are indistinguishable from those of their naturally occurring analogues by the methods commonly employed. As, however, the deuterium content of these substances or of their physiological derivatives can readily be determined from the properties of the water formed on combustion, their fate in the body can be followed even after considerable dilution.[150]

During the remaining six years of Schoenheimer's life, a succession of outstanding papers appeared from his laboratory, illustrating the power of the isotope method in the study of intermediate metabolism and laying the groundwork for even more exciting later research.

Schoenheimer's use of deuterium as a metabolic tracer was not the first application of isotopes in biological studies. This approach had been initiated by George de Hevesy, who used radioactive thorium B to study the transport of lead in the bean plant.[151] Ten years earlier, Hevesy and Friedrich Paneth had performed the first chemical tracer experiments, in which they demonstrated the electrochemical interchangeability of lead and the radioactive radium D. These results provided additional evidence for the views developed earlier in 1913 by Kasimir Fajans and by Frederick Soddy regarding the relation of the newly discovered radioactive elements to Mendeleev's periodic table. In a paper on this subject, Soddy wrote: "The same algebraic sum of positive and negative charges in the nucleus, when the arithmetical sum is different, gives what I call 'isotopes' or 'isotopic elements,' because they occupy the same place in the periodic table. They are chemically identical, and save only as the relatively few physical properties which depend upon atomic mass directly, physically identical also."[152] The existence of isotopes was definitely established in 1914 by Theodore William Richards and Max Ernest Lembert,[153] and subsequent work by Francis William Aston showed that the occupancy of a place in the periodic table by more than one element is not limited to the heaviest ones. Aston identified the stable neon isotopes of mass 20 and 22, and in connection with these studies he invented a "mass spectrograph" for the identification of the isotopic elements.[154] This instrument was later superceded by the "mass spectrometer," which permitted quantitative estimation of the relative abundance of stable isotopes.[155]

During the 1920s and 1930s Hevesy also studied the partition of labeled lead between normal and tumor tissues in mice, and the distribution of bismuth in animals, with radium E as a tracer, because of the clinical interest in the replacement of arsenic by bismuth in the treatment of syphilis. Many years later, Hevesy wrote: "The investigation of the partition of lead was carried out in the Institute of Physical Chemistry of the University of Freiburg. Feeling the need of an assistant trained in biology, I approached Dr. Aschoff, the director of the Institute of Pathology. He selected a Japanese scientist for this work, who however proved to have difficulties in coping with this task. It was then that Dr. Schoenheimer, associate of the Institute of Pathology, was asked by Dr. Aschoff to step in. It was in the course of these investigations that Schoenheimer became familiar with the method of isotopic indicators."[156] With the rise of Hitler, Schoenheimer left Freiburg in 1933 and came to Columbia University, where, the year before, Harold Urey and his associates had separated and identified deuterium, the hydrogen isotope of mass 2.[157]

Of the elements present in biological systems, the first for which isotopes had been identified was oxygen (by William Francis Giauque in 1929), but deuterium was the first to become available for biological studies. After Urey's group had developed a practical method for preparation of D_2O ("heavy water"), many investigators at once performed a variety of experiments to determine its effect in animals, plants, and microorganisms.[158] It should be emphasized that in none of these initial studies was deuterium introduced at specific loci in organic molecules by means of known chemical reactions, so as to permit one to determine the metabolic fate of the label through the isolation of well-defined biochemical substances from the organism under study. Schoenheimer, as a biochemist trained in organic chemistry, was the first to do this and he thereby transformed the labeling technique devised by Franz Knoop into a powerful method for the study of intermediate metabolism. In the initial stages of this line of research, Schoenheimer was aided by David Rittenberg, who had received his doctorate for work with Urey. Mildred Cohn, another Urey student, also entered biochemistry and soon made important contributions in the laboratories of Vincent du Vigneaud and of Carl Cori. The Schoenheimer research group at Columbia included, as graduate students or research associates, Konrad Bloch, Sarah Ratner, and David Shemin.

Within a few years after the discovery of deuterium, Urey and his associates developed methods for concentrating the naturally occurring isotopes of nitrogen

(^{15}N), carbon (^{13}C), and oxygen (^{18}O). Moreover, in 1934 Irène Curie and Frédéric Joliot discovered that radioactivity can be induced by bombardment of a natural element with α-rays (helium nuclei) from radium. Thus a radioactive isotope of phosphorus (^{32}P) could be prepared from sulfur (^{32}S), and phosphate labeled in this manner was used by Hevesy during the mid-1930s in studies on the turnover of phosphorus in animal tissues. After the development of the cyclotron by Ernest Lawrence, the preparation of radioactive isotopes by bombardment with nuclear particles was simplified, and by 1938 a variety of additional tracers, such as ^{35}S, became available for biochemical studies. Some of the others, such as ^{11}C, decayed so rapidly that biological experiments had to be of short duration and could only be performed in laboratories near a cyclotron. As we have seen, the more valuable long-lived isotope of carbon, ^{14}C (half-life, 5570 years), had been prepared by Samuel Ruben and Martin Kamen in 1940, but it did not become generally available for metabolic studies until after 1945, when the cyclotron was replaced by the nuclear reactor ("atomic pile") as the source of artificial radioisotopes. For many years, improved versions of the instrument devised in 1908 by Hans Geiger were used to measure radioactivity; such "Geiger counters" were replaced by scintillation spectrometers. The introduction of ^{14}C into metabolic research, together with the development of chromatographic techniques, were decisive factors in the explosive growth of biochemical knowledge after 1945.[159]

In the mid-1930s it was known that deuterium atoms linked to oxygen or nitrogen (as in carboxyl, hydroxyl, or amino groups) can exchange readily with protium atoms in water and that the label is lost during the course of metabolic experiments. Compounds were required, therefore, that had been labeled with formation of stable linkages such as the carbon-hydrogen bond. For his studies on the metabolism of fatty acids, Schoenheimer prepared deuterated stearic acid chemically by reduction, in the presence of D_2, of the two double bonds of the C_{18} linoleic acid [$CH_3(CH_2)_3(CH_2CH{=}CH)_2(CH_2)_7COOH$]. The labeled compound was included in the diet of mice for twelve days, the animals were killed, and their fatty acids were isolated and analyzed for their deuterium content. Among the labeled fatty acids were the saturated C_{16} palmitic acid and the monounsaturated C_{18} oleic acid. When the oleic acid so obtained was fed to mice, their fatty acids included labeled stearic acid. These pioneer studies demonstrated the metabolic interconvertibility of the three principal fatty acids of animal fats and showed that the body fats are in a "dynamic state" even when adequate fat is supplied in the diet.[160]

The use of deuterium, however, could not answer the questions discussed above about the formation of ketone bodies. After ^{13}C became available, Sidney

Weinhouse and his associates showed that the labeling of the acetoacetic acid produced from labeled octanoic acid [$CH_3(CH_2)_6{}^{13}COOH$] is inconsistent with the successive beta-oxidation theory, and in accord with the view that the ketone bodies are formed by random condensation of pairs of the four two-carbon units derived from the metabolic cleavage of the C_8 acid.[161] Later, by the use of ^{14}C and tissue homogenates, Crandall and Gurin demonstrated a difference in the metabolic fate of the terminal CH_3CH_2unit of a fatty acid and its other two-carbon units.[162] These findings form the background for the elucidation of the enzyme-catalyzed steps in the oxidation of fatty acids, after the discovery of acetyl-CoA.

In 1937, ^{15}N became available for metabolic studies, and during 1939–1941 Schoenheimer and his associates published a series of important papers on protein metabolism. In this research various amino acids enriched with ^{15}N in their amino group were administered to animals, and amino acids were then isolated from hydrolysates of the tissue proteins for analysis of their isotope content. In some cases stably bound deuterium was also introduced into the ^{15}N labeled amino acids to serve as an indicator of the fate of the carbon chains. A major consequence of these experiments was the demise of the theory of endogenous ("wear-and-tear") and exogenous protein metabolism offered by Otto Folin and adopted by Max Rubner. Schoenheimer and his associates wrote:

> It is scarcely possible to reconcile our findings with any theory which requires a distinction between these two types of nitrogen. It has been shown that nitrogenous groupings of tissue proteins are constantly involved in chemical reactions; peptide linkages open, the amino acids liberated mix with others of the same species of whatever source, diet, or tissue. This mixture of amino acids, while in the free state, takes part in a variety of chemical reactions: some reenter directly into vacant positions left open by the rupture of peptide linkages; others transfer their nitrogen to deaminated molecules to form new amino acids. These in turn continuously enter the same chemical cycles which render the source of the nitrogen indistinguishable. Some body constituents like glutamic acid and aspartic acid and some proteins like those of the liver, serum, and other organs are more actively involved than others in this general metabolic pool originating from interaction of dietary nitrogen with the relatively larger quantities of reactive tissue nitrogen.[163]

The Folin-Rubner theory had already been challenged during the 1930s. Henry Borsook and Geoffrey Keighley concluded from nutritional experiments that there is a "continuing" metabolism of protein and that tissue proteins are constantly being synthesized from amino acids.[164] The studies of Sidney Madden and George Whipple on the replacement of plasma proteins after animals had been bled led them to postulate a dynamic state of body protein.[165] After the work of the Schoenheimer group, the issue no longer seemed in doubt, and the existence of a "metabolic pool" of nitrogen in the animal body was generally accepted. During the 1950s, however, Jacques Monod questioned whether protein turnover occurs at the intracellular level, and he suggested that in mammalian tissues protein is released only by secretion or by cellular disruption. This view was based on the apparent lack of protein turnover in microbial cells, but subsequent work by Joel Mandelstam showed that the breakdown as well as synthesis of proteins occurs in intact bacteria.[166]

Apart from the demolition of the Folin-Rubner theory, and in ending the era of "intake-output" nutritional studies, the work of the Schoenheimer group confirmed the special metabolic role of glutamic acid and aspartic acid, as well as the general importance of transamination reactions, and, what is perhaps most significant, it initiated the study of the detailed pathways in the breakdown and synthesis of the individual amino acids, and their metabolic relation to other body constituents. In addition to their finding that the labeling of arginine accorded with Hans Krebs's formulation of the pathway of urea synthesis, the isotope technique provided a new approach to the problem of creatine synthesis.

The metabolic origin of creatine (methylguanidinoacetic acid), and of the urinary creatinine derived from it by the loss of the elements of water, had been studied extensively but inconclusively for decades.[167] After the identification of arginine as a protein constituent at the turn of the century, its metabolic relationship to creatine was suspected on the ground that both compounds have a guanidino group [$-NHCH(=NH)NH_2$], and various chemically plausible hypotheses were advanced for the conversion of arginine to creatine. The isotope experiments of Konrad Bloch showed that the metabolic synthesis of creatine involves the transfer of an amidino group [$-C(=NH)NH_2$] from arginine to glycine, with the formation of guanidinoacetic acid, which is then methylated in a process involving methionine.[168] The "transamidination" reaction had been effected chemically in 1927 by Max Bergmann and Leonidas Zervas,[169] and the possibility that the methyl group of creatine might come from methionine had been suggested after the discovery of that amino acid, and its identification during the 1920s as a methylthio compound.[170] Although nutritional studies during the 1930s indicated a metabolic relation among methionine, choline [$(CH_3)_3NCH_2CH_2OH$], and creatine, these anticipations did not win wide acceptance until the isotope technique had been applied to the study of the problem.[171] In addition to the report of Bloch and Schoenheimer, and the parallel results of Henry Borsook and Jacob Dubnoff with tissue slices,[172] extensive studies in the laboratory of Vincent du Vigneaud delineated the metabolic pathway and laid the groundwork for the later study of the enzymatic mechanisms in transmethylation reactions. Of particular importance was the demonstration, by means of the isotope technique, that the methyl groups of methionine and choline are metabolically "labile" but that of creatine is not.[173] Other instances of biological methylation had been found earlier. During the latter years of the nineteenth century, the younger Wilhelm His and Franz Hofmeister studied examples of such reactions, and by 1940 many other biological transmethylations had been identified.[174]

Among the many notable biochemical achievements of the 1950s was Guilio Cantoni's finding of what proved to be the key intermediate (S-adenosylmethionine) in the transfer of the methyl group to acceptor molecules (see scheme below).[175] Later work showed that in many other methyl transfer reactions the methyl-tetrahydrofolate system and methyl-cobalamin are necessary participants.[176]

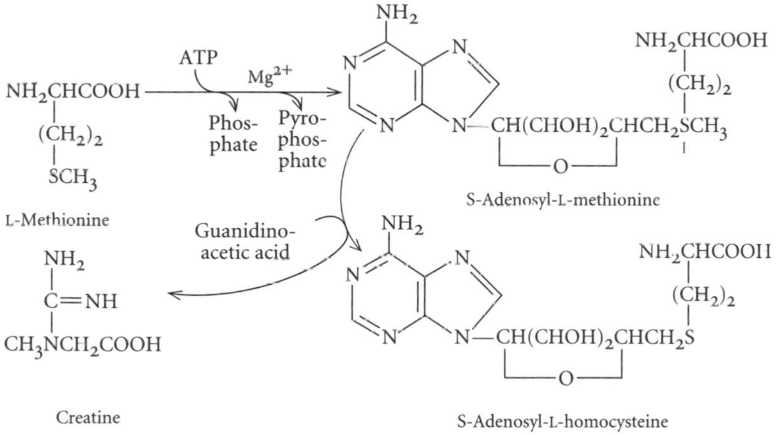

$NH_2CHCOOH$ — $(CH_2)_2$ — SCH_3

L-Methionine

ATP, Mg^{2+} → Phosphate, Pyrophosphate

$NH_2CHCOOH$ — $(CH_2)_2$ — N—CH(CHOH)$_2$CHCH$_2$SCH$_3$ (adenine, NH_2)

S-Adenosyl-L-methionine

Guanidinoacetic acid →

NH_2 — C=NH — CH_3NCH_2COOH

Creatine

$NH_2CHCOOH$ — $(CH_2)_2$ — N—CH(CHOH)$_2$CHCH$_2$S (adenine, NH_2)

S-Adenosyl-L-homocysteine

The Fixation of Carbon Dioxide

Before about 1935 it was widely believed that the metabolic incorporation of CO_2 into organic compounds is limited to photosynthetic plants or "autotrophic" (also termed "chemosynthetic") organisms, and that in "heterotrophic" organisms (those that require some organic compounds as nutrients) CO_2 is an inert end product of cellular respiration. When Harland Wood, working in the laboratory of Chester Werkman, found that the formation of succinic acid by propionic acid bacteria is related to the uptake of CO_2,[177] relatively little attention was paid to that discovery by biochemists concerned with problems of animal metabolism. Soon afterward, however, work by Earl Evans in Hans Krebs's laboratory indicated that animal tissues can effect CO_2 fixation in the reactions leading from pyruvate to α-ketoglutarate, and the use of the available carbon isotopes (^{11}C and ^{13}C) established the validity of this surmise.[178] The belated recognition of the importance of the discovery of the "Wood-Werkman" reaction also drew the attention of biochemists to earlier studies on a large variety of microorganisms and to the contributions of general microbiology to the study of fundamental biochemical problems.[179] Moreover, research on the metabolic utilization of labeled CO_2 revealed many previously unknown biochemical reactions, whose examination led to the discovery of new enzymes, of new cofactors related to known vitamins, and of new intermediates that were also important in the metabolism of animals.

Wood's active interest in the mechanisms of CO_2 fixation continued throughout his scientific life, although he occasionally strayed into other pastures, as in his studies on lactose synthesis in the cow. His career, which began with Werkman at Iowa State College and then continued in the Department of Physiology at the University of Minnesota, reached its peak after 1946, when Wood became head of a new Department of Biochemistry in the medical school of the Case-Western Reserve University, to which he brought his former colleagues Merton Utter, Vincent Lorber, Warwick Sakami, and Lester Krampitz. Wood was also a leading participant in an experiment to reform the preclinical education of medical students.[180]

In 1945 Utter and Wood reported that with extracts of pigeon liver ATP is essential for the exchange reaction:

$$HOOCCH_2COCOOH + {}^{14}CO_2 \rightleftharpoons HOO^{14}CCH_2COCOOH + CO_2$$

and subsequent work by Utter led to his finding in 1960 that, in addition to ATP, the vitamin biotin is involved in the conversion of pyruvate to oxaloacetate, because CO_2 fixation was inhibited by the egg-white protein avidin, which binds biotin strongly.[181] Indeed, the identification of biotin and the determination of its structure came from studies on the cause of the toxic effect on rats of raw egg white, and the widely distributed curative factor had been named vitamin H.[182] After 1945 several other biotin-dependent carboxylases were identified and purified, among them the acetyl-CoA carboxylase in fatty acid synthesis, another involved in the metabolic breakdown of leucine in bacterial and animal systems (β-methylcrotonyl-CoA carboxylase), and a third (methylmalonyl-CoA-pyruvate transcarboxylase) found in propionic acid bacteria.[183] It was found that biotin (a bicyclic sulfur ureido compound with a long fatty acid side-chain) is linked to lysine side chains of protein subunits in a multiunit catalytic assembly, and that, as Feodor Lynen showed in 1959, CO_2 is first bound in an ATP-dependent reaction to one of the ureido nitrogen atoms of biotin in one kind of subunit, and then transferred to the substrate by a different kind of subunit.[184]

The Biosynthesis of Steroids

As we have seen, the use of isotope techniques transformed such questions as the role of "active formaldehyde," prominent after 1870 in chemical speculations about biosynthetic mechanisms, into large chapters of present-day biochemistry dealing with the metabolic interconversion and utilization of such one-carbon units as CH_3, CH_2OH, and CHO in defined enzyme-catalyzed reactions involving, as cofactors, compounds chemically related to various vitamins. I shall discuss next the use of the isotope technique to study the metabolic utilization of acetic acid (CH_3COOH) in the biosynthesis of cholesterol and of its chemical relatives.

Before I do so, it should be recalled that during the early years of the twentieth century, the ingenuity of organic chemists was challenged by the structural complexity of many natural products, in particular plant alkaloids such as morphine, strychnine, and quinine, which had been isolated during the early 1800s and had acquired importance in medical practice. By 1890 the new organic chemistry was being intensively applied to the determination of their structure, and methods were sought for their synthesis in the laboratory. For example, Richard Willstätter's first experimental success, during the 1890s, was in the study of the alkaloids related to cocaine. Moreover, speculations were offered about the mode of biosynthesis of such natural products. A high point in this development was the paper by Robert Robinson, who had effected the synthesis of the alkaloid tropinone from succinaldehyde, methylamine, and acetone under relatively mild conditions, and who considered that this synthesis "on account of its simplicity, is probably the method employed by the plant."[185] Robinson also developed a theory that assigned to certain amino acids (ornithine, lysine) and to presumed products of carbohydrate breakdown, for example, acetone dicarboxylic acid ($HOOCCH_2COCH_2COOH$) the role of intermediates in the biosynthesis of a large number of known alkaloids. Not only were such theories intellectually satisfying, but they also proved to be of practical value in the determination of the structure and in the laboratory synthesis of many alkaloids, as in the case of Robert Woodward's proposal that the complex structure of strychnine can be derived from tryptamine (the decarboxylation product of tryptophan), phenylalanine, acetic acid, and a one-carbon unit. During the period 1920–1945, some organic chemists considered such "biogenetic" hypotheses to reflect the processes in the plant and sought to effect the artificial synthesis of alkaloids under "physiological" conditions, but afterward, as Robinson stated, it was generally agreed that "the comparison of structures *per se* gives no information about the details of the mechanisms of biosynthetic reactions. It is the task of the biochemist to determine these by appropriate experiment."[186] Although later "appropriate experiments," using isotopes, showed some aspects of these hypotheses to be correct, for the plant biochemist Ray Dawson, writing in 1948, they "can be regarded as little more than kinetic and thermodynamic possibility of occurrence of otherwise highly improbable *in vivo* events."[187]

Like the plant alkaloids, the substance to which Chevreul gave the name *cholestérine* (Greek *chole,* bile + *stereos,* solid) attracted much interest during the nineteenth century. He identified it as an "unsaponifiable fat" to distinguish it from the fats that could be hydrolyzed to glycerol and fatty acids. The chemical purification and elementary analysis of cholesterol presented some difficulty, and it was not until the 1890s that its empirical formula was definitely set at $C_{27}H_{46}O$. This elementary composition indicated a chemical relation to what Adolf Strecker called in 1848 cholic acid ($C_{24}H_{40}O_5$), found in bile in the form of conjugates with glycine or taurine. Another bile acid ("deoxycholic acid") was found by Olof Hammarsten in 1881. Cholesterol turned up as a constituent of gallstones and nearly all animal tissues, and a similar substance ("phytosterine") was also found in plant tissues. All these compounds were later grouped under the collective term *sterols,* and together with their chemical relatives as steroids.

The determination of the structures of cholesterol and cholic acid was not accomplished until 1932, after more than thirty years of intensive chemical effort by numerous investigators, notably Adolf Windaus and Heinrich Wieland. From their empirical formulas, with less hydrogen per carbon atom than required for a straight-chain hydrocarbon, it was evident that cholesterol and cholic acid are ring compounds, and by 1920 it had been established that they are derivatives of the same $C_{27}H_{46}$ cyclic molecule. In 1896 Stanislaw Bondzynski reported that the ingestion by animals of cholesterol leads to the excretion of a substance ("coprosterol"), and later chemical work by Julius Mauthner and by Heinrich Wieland showed that the $C_{27}H_{48}$ hydrocarbon ("coprostane") derived from this sterol can be converted to a derivative of cholic acid ("cholanic acid").[188] The formulation of the structure of the ring system of the steroids presented some difficulty, however, and the structure proposed during the 1920s by Windaus and Wieland proved to be incorrect. In one of the first demonstrations of the contribution of X-ray crystallography to the solution of difficult organic-chemical problems, Bernal reported in 1932 that the diffraction pattern of ergosterol (a plant sterol known to be related to vitamin D) was not consistent with this proposal. In the same year, Otto Rosenheim and Harold King offered an alternative ring structure, based on the work of Otto Diels on the dehydrogenation of cholesterol to produce a $C_{18}H_{16}$ hydrocarbon.[189] Before the year was out, Wieland had accepted that formulation as the correct one, shown in the formula below.

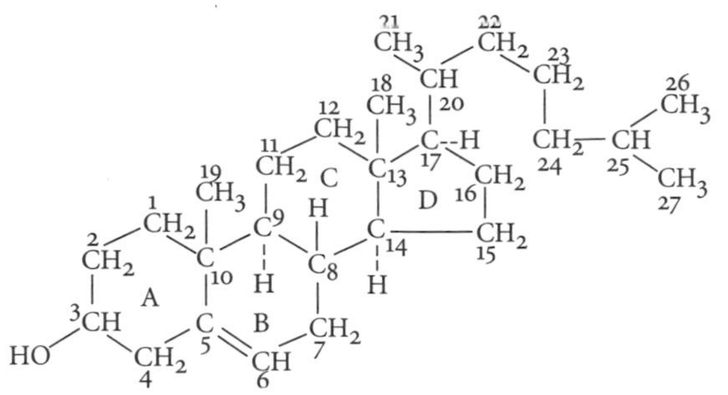

Cholesterol

Before the structure of cholesterol had been established, important studies had been conducted on its formation in the animal body. In particular, during the 1920s Harold Channon had provided strong evidence favoring the view that animals can synthesize cholesterol, thus settling a lively but inconclusive debate that had proceeded for several decades, largely because of inadequacies in the analytical methods for the determination of cholesterol. During the course of his experiments, Channon also showed that, when squalene (a nonsaponifiable lipid isolated in 1916 from shark liver) is fed to rats, there is an increase in the cholesterol content of the tissues.[190] This test of squalene was a direct consequence of the parallel chemical work of Ian Heilbron, who studied its structure because of his interest in the lipid components of fish-liver oils as sources of vitamin A.[191]

In the determination of the structure of squalene, a long-chain $C_{30}H_{50}$ hydrocarbon, and of its metabolic relation to cholesterol, a chemical hypothesis—the "isoprene rule"—played a significant role. The origins of the rule lie in the early decades of the nineteenth century, when there was much interest in the volatile products produced upon the dry distillation of rubber. In 1826 Michael Faraday isolated a low-boiling material later identified as an unsaturated C_5H_8 hydrocarbon related to isopentane, and named isoprene. This product also appeared upon heating turpentine oil and other plant oils, and during the 1880s Otto Wallach developed the idea that natural $C_{10}H_{16}$ "monoterpenes" are composed of isoprene units. In 1897 Vladimir Ipatieff showed the structure of isoprene to be $CH_2{=}C(CH_3){-}CH{=}CH_2$. The hypothesis that isoprene constitutes a fundamental structural unit of the terpenes became the "isoprene rule" only after 1920, when Leopold Ruzicka showed its general validity for various C_{15} terpenes ("sesquiterpenes" such as farnesol), and subsequently for "diterpenes" and "triterpenes" with 20 or 30 carbon atoms, respectively.[192] The synthesis of squalene by Paul Karrer in 1931 established the details of its structure as a triterpene composed of six isoprene units. After the acceptance of the Rosenheim-King formula for the ring system of steroids, Robert Robinson called attention to the fact that squalene contains the intact skeleton of cholesterol and proposed a chemical mechanism for the conversion of the linear squalene molecule into the tetracyclic sterol nucleus.[193]

These exciting developments came at a time when Schoenheimer and his colleague Friedrich Breusch were conducting studies on the effect of changes in the diet of mice on the formation and breakdown of cholesterol; from the results they concluded that cholesterol is synthesized when needed and destroyed when present in excess.[194] Thus when Schoenheimer came to the United States (Breusch went to Turkey) and began to use isotopic tracers, his first experiments dealt with the uptake of deuterium from D_2O into the tissue cholesterol of mice. The extensive incorporation of the isotope suggested that the sterol had been synthesized by the condensation of small molecular units. Parallel studies by others on the utilization of deuterium-labeled acetate by yeast indicated that "the sterols of yeast arise from acetic acid by a rather direct route."[195]

After the carbon isotopes became available, Konrad Bloch showed that in rat-liver slices, acetic acid can supply all 27 carbon atoms of cholesterol, and he then embarked on the determination of the pattern of the incorporation of the carbon atoms of acetic acid labeled either in the methyl group or the carboxyl group ($^{14}CH_3COOH$ or $CH_3{}^{14}COOH$) or both. A similar effort was undertaken by John Cornforth and George Popjak, and by 1957 the carbon-by-carbon dissection had been completed.[196] Also, by that time important information had been obtained on the biosynthesis of squalene from isoprenoid units.

The story of isoprene also begins with nineteenth-century studies on the decomposition of rubber, and the later study, by Barbarin Arreguin and James Bonner, of the utilization of acetate by the guayule plant for the biosynthesis of rubber,[197] influenced the early interpretation of the labeling pattern of the cholesterol formed from ^{14}C-acetate. After Robert Langdon and Konrad Bloch had shown that labeled squalene is converted to labeled cholesterol in animals,[198] the metabolic precursor of the isoprene units of squalene appeared from an unexpected quarter; in 1956, a research group led by Karl Folkers isolated from "distiller's solubles" a new compound (mevalonic acid), which replaced acetate in the growth medium for lactobacilli and whose chemical structure they determined; the compound turned out to be an excellent metabolic precursor of cholesterol in the liver.[199] This discovery opened the way for the study of the reactions in the formation of mevalonic acid and in its conversion to squalene. Subsequent work revealed a complex metabolic pathway, involving the following enzyme-catalyzed reactions: (1) the conversion of

$$HOOCCH_2{-}\underset{OH}{\overset{CH_3}{C}}{-}CH_2CH_2OH \xrightarrow{-CO_2} \left[CH_3\overset{CH_3}{C}{=}CHCH_2CH_2\overset{CH_3}{C}{=}CHCH_2CH_2\overset{CH_3}{C}{=}CHCH_2{-}\right]_2$$

Mevalonic acid Squalene

acetyl-CoA to acetoacetyl-CoA; (2) the formation, from these two compounds of 3-hydroxyl-3-methylglutaryl-CoA (HMG-CoA); (3) the reduction of HMG-CoA to mevalonate; (4) the conversion of mevalonate, in a series of ATP-dependent reactions, to the pyrophosphate of isopentanol [CH_2=C(CH_3)-CH_2-CH_2OH]; (5) the successive elongation of this C_5 pyrophosphate, via C_{10} and C_{15} intermediate isoprenoid pyrophosphates, to the C_{30} squalene.[200] In addition to its role as a key intermediate in the synthesis of steroids, in plants mevalonic acid is a metabolic precursor of carotenoids (including vitamin A), the phytyl side-chains of chlorophyll, the hormone gibberellic acid, and other compounds.[201]

As regards the metabolic conversion of the linear C_{30} squalene to the tetracyclic C_{27} cholesterol, the elucidation in 1952 by Leopold Ruzicka's group of the structure of the C_{30} lanosterol (from wool fat) led to the imaginative hypothesis of Woodward and Bloch, who suggested how the squalene chain might be folded to yield lanosterol as an intermediate, followed by the removal of the three CH_3 groups at C-4 and C-14 of that sterol. The Woodward-Bloch hypothesis, which predicted a different labeling pattern from that expected for the arrangement Robinson had proposed in 1934, was supported by subsequent experimental data.[202]

Among the many consequences of this remarkable series of organic-chemical and enzymatic studies has been the discovery during the 1970s, by Michael Brown and Joseph Goldstein, that the cholesterol in the so-called low-density lipid of the blood inhibits the activity of HMG-CoA reductase.[203] This finding was followed by the development of synthetic inhibitors of this enzyme, and such drugs are now widely used in clinical practice to control blood cholesterol levels and thus to reduce the incidence of atherosclerotic plaques.

The Biosynthesis of Porphyrins

By the 1930s the work of Hans Fischer had established the detailed chemical structure of the porphyrin (protoporphyrin IX) of blood hemoglobin, as well as of many other naturally occurring members of the group of tetrapyrrole compounds that differ from protoporphyrin IX in the nature and arrangement of the eight side-chains attached to the four pyrrole rings.[204] It was also known that animals can maintain a normal level of hemoglobin when kept on a diet free of porphyrins, and there were speculations about the metabolic precursors. As was customary, the guiding principle in such hypotheses was one of structural similarity, and the amino acids proline and tryptophan (which contain rings similar to the pyrrole nucleus) and glutamic acid (which can be converted chemically into a derivative of proline) were considered as possible biological sources of the porphyrins. With the application of the isotope technique, these hypotheses were discarded, for it became evident that protoporphyrin IX is made in the animal body from small molecules. In 1945 David Shemin reported that the administration of ^{15}N-glycine to human subjects leads to the labeling of the nitrogen atoms of the hemin (ferriprotoporphrin) isolated from the blood, while ^{15}N-labeled glutamic acid was ineffective in this respect.[205] Later, advantage was taken of the ability of the nucleated erythrocytes of birds to synthesize porphyrins, in contrast to the inability of mammalian (nonnucleated) red cells to do so. After ^{14}C became available, experiments by Shemin, Albert Neuberger, and their associates, with labeled glycine ($^{15}NH_2CH_2COOH$, $NH_2{}^{14}CH_2COOH$, $NH_2CH_2{}^{14}COOH$), showed that all four nitrogen atoms of the tetrapyrrole ring are derived from glycine, whose CH_2 group provides eight of the 34 carbon atoms of protoporphyrin, and that the COOH group is lost during porphyrin synthesis.[206] Moreover, upon the systematic chemical degradation of the hemin obtained after the incubation of duck erythrocytes with labeled acetate ($^{14}CH_3COOH$, $CH_3{}^{14}COOH$), it was found that the other 26 carbon atoms of protoporphyrin are derived from acetate.[207] In the diagram, the metabolic sources of the carbon atoms of protoporphyrin IX are indicated as follows: asterisks denote atoms derived from the CH_2 carbon of glycine; solid circles denote atoms derived from the CH_3 carbon of acetate; open circles denote atoms derived mainly from the

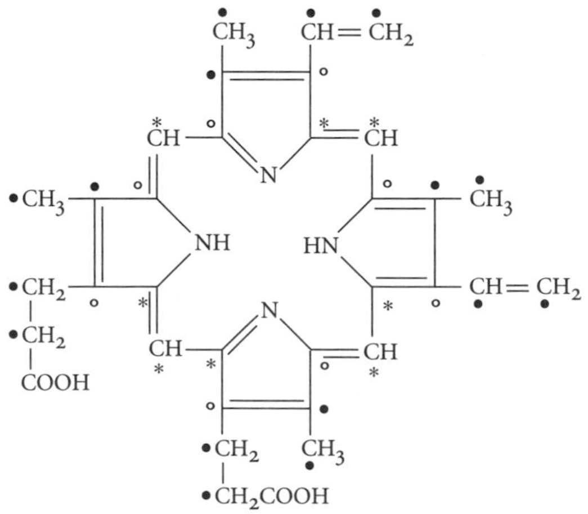

Metabolic sources of atoms of protoporphyrin IX, as shown by isotope experiments

CH_3 carbon of acetate and in small part from the COOH carbon of acetate; the unmarked carbon atoms are derived solely from the COOH carbon of acetate.

From this labeling pattern, Shemin concluded that "in the biosynthesis of protoporphyrin a pyrrole is formed which is the common precursor of both types of pyrrole structure found in protoporphyrin . . . [and] in each pyrrole ring the same compound is used for the methyl side of the structure and for the vinyl [$CH{=}CH_2$] and propionic acid [CH_2CH_2COOH] sides of the structure . . . and that the common precursor originally formed contained acetic acid and propionic acid side chains."[208] Shemin proposed that the hypothetical precursor might arise by the condensation of a succinyl compound ($HOOCCH_2CH_2CO$-X) derived from acetate through the operation of the Krebs citric acid cycle, and later work showed it to be succinyl-CoA. The condensation reaction postulated by Shemin had precedents in synthetic organic chemistry and was analogous to the one described by Ludwig Knorr in 1902 for the synthesis of pyrroles. Also, in 1944 Hans Fischer drew attention to the ready condensation of glycine with formylacetone (CH_3COCH_2CHO) under "physiological" conditions as a possible route of pyrrole synthesis in biological systems. Among the other speculations whose validity could not be tested experimentally before the use of the isotope technique was the proposal by William Turner in 1940 that all four pyrrole rings of protoporphyrin are derived from a common precursor pyrrole having acetic acid and propionic acid chains.[209]

The question of the existence and chemical nature of a common precursor pyrrole was solved through a development in the study of the hereditary human disease known as acute porphyria (or porphyrinuria). As its name implies, this condition is associated with the appearance of porphyrins in the urine. The clinical manifestations include periodic attacks of abdominal pain and mental symptoms, which have occasionally led to wrong diagnoses of schizophrenia or paranoia. A strong case has been made for the view that the unstable mental state of King George III was a consequence of porphyria, whose recurrence has been traced in several royal houses of Europe.[210] During the 1930s Jan Gosta Waldenström observed that the urine of porphyric patients contains a colorless material ("porphobilinogen") which is converted into porphyrins when the urine is acidified and heated.[211] The prophyrins so produced were identified as uroporphyrins I and III, already shown by Hans Fischer to have only acetic acid and propionic acid side-chains; the two porphyrins differ in the way the two kinds of side-chains are arranged, uroporphyrin I having a symmetrical structure. The nature of porphobilinogen (PBG) was unknown, but it was isolated in crystalline form in 1952, and the structure determined soon afterward by Gerald Cookson and Claude Rimington[212] immediately suggested its synthesis from δ-aminolevulinic acid (ALA, H_2N-CH_2-CO-CH_2-CH_2-COOH), two molecules of which could be expected to condense by the Knorr reaction to produce PBG. Neuberger and Shemin showed independently that ALA labeled with ^{14}C at the CH_2 group next to the amino group is an excellent precursor of protoporphyrin and that the position of the label (denoted with an asterisk in the above diagram) corresponded to that previously found with methylene-labeled glycine. Subsequent work in several laboratories led to the characterization of the enzymes involved in the pyridoxal phosphate-dependent reaction of succinyl-CoA and glycine to form ALA, and in its condensation to PBG, but the enzymatic mechanisms in the combination of four PBG molecules to form the asymmetrically substituted type III porphyrins (to which protoporphyrin IX is related by modification of the acetyl side-chains) proved to be more difficult to unravel.[213] It has become clear, however, that a key intermediate in the biosynthesis, not only of protoporphyrin IX but also of the chlorophylls in plants and photosynthetic algae, as well as of vitamin B_{12}, is the reduced form of uroporphyrin III derived from ALA.[214]

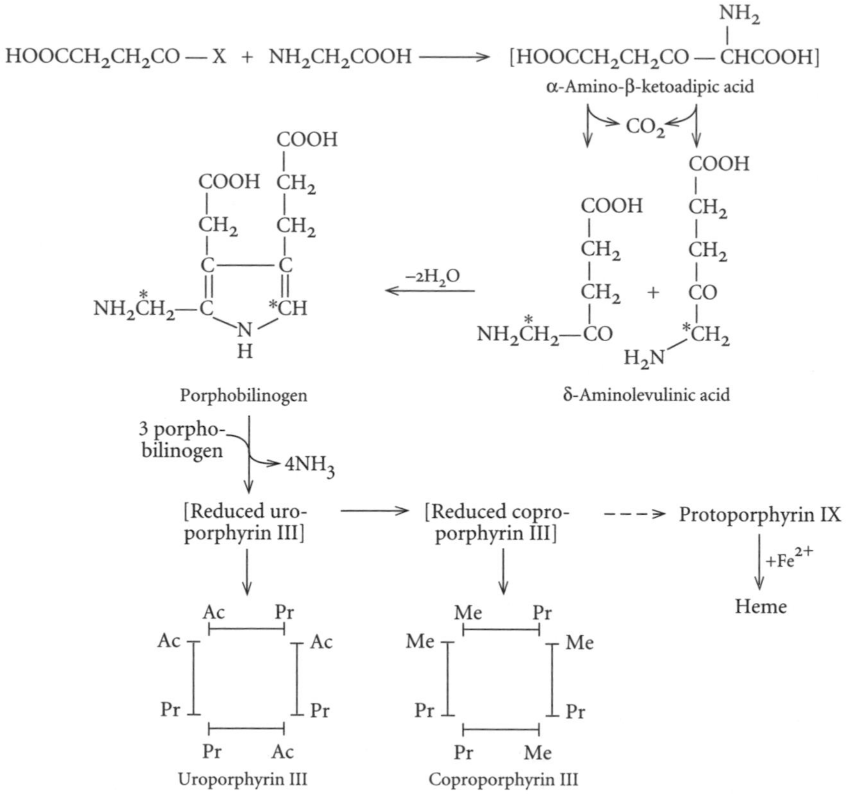

$$HOOCCH_2CH_2CO—X + NH_2CH_2COOH \longrightarrow [HOOCCH_2CH_2CO—CH(NH_2)COOH]$$

α-Amino-β-ketoadipic acid

$\downarrow -CO_2$

δ-Aminolevulinic acid (2 molecules: $NH_2\overset{*}{C}H_2—CO—CH_2CH_2COOH$)

$\xleftarrow{-2H_2O}$

Porphobilinogen

3 porphobilinogen, $-4NH_3$

[Reduced uroporphyrin III] → [Reduced coproporphyrin III] - - -> Protoporphyrin IX

↓ Uroporphyrin III (Ac, Pr substituents) ↓ Coproporphyrin III (Me, Pr substituents) ↓ $+Fe^{2+}$ Heme

Role of aminolevulinic acid and of porphobilinogen in the biosynthesis of heme

The achievements during the quarter-century after 1945 in the study of such metabolic processes as photosynthesis, the utilization of CO_2 in animal tissues, and the biosynthesis of complex natural products such as the steroids and porphyrins reflect the role of the isotope labeling technique in the transformation of biochemical thought during that period. As we shall see, the use of the technique was also decisive in the study of the mechanisms in other metabolic pathways, such as the fixation of inorganic nitrogen, the synthesis of nucleic acids and proteins, and the regulation of physiological processes. Before considering some of these other applications, brief mention must be made of the fact that the use of the hydrogen isotopes deuterium and tritium in studies of enzyme-catalyzed reactions in which there is a transfer of carbon-bound hydrogen—as in those effected by dehydrogenases or isomerases—has provided valuable information about the intimate mechanism of such reactions.[215] Also, in parallel with the isotope technique, a new method emerged during the 1940s from genetics for the investigation of the chemical pathways of metabolism.

Mutants in Metabolic Research

In 1902 the clinician Archibald Garrod wrote as follows about the human condition known as alcaptonuria: "All the more recent work on alkaptonuria . . . [has] strengthened the belief that the homogentisic acid excreted is derived from tyrosin, but why alkaptonuric individuals pass the benzene ring of their tyrosin unbroken and how and where the peculiar change from tyrosin to homogentisic acid is brought about, remain unsolved problems. There is good reason for thinking that alkaptonuria is not the manifestation of a disease but rather of the nature of an alternative course of metabolism, harmless and usually congenital and lifelong."[216] In his book *Inborn Errors of Metabolism,* published in 1909, Garrod summarized the available evidence on the breakdown of tyrosine and phenylalanine in the mammalian organism and concluded that the metabolic error lies "in the penultimate stage of the catabolism of the aromatic protein fractions. . . . We may further conceive that the splitting of the benzene ring in normal metabolism is the work of a special enzyme, and that in congenital alkaptonuria this enzyme is wanting."[217]

This idea reflected the convergent influence during the first decade of the twentieth century of the emergence, on the one hand, of the study of Mendelian inheritance, with William Bateson as its chief proponent in England, and, on the other, of the study of intermediate metabolism by such physiological chemists as Franz Knoop, Otto Neubauer, and Gustav Embden. The importance of Garrod's contribution was recognized by Gowland Hopkins: "Extraordinarily profitable have been the observations made upon individuals suffering from those errors of metabolism which Dr. Garrod calls 'metabolic sports, the chemical analogues of structural malformations.' In these individuals Nature has taken the first essential step in an experiment by omitting from their chemical structure a special catalyst which at one point in the procession of metabolic events is essential to its continuance. At this point there is arrest, and intermediate products come to light."[218] As J. B. S. Haldane later noted, however, "alkaptonuric men are not available by the dozen for research work."[219] This difficulty, as well as the limited knowledge regarding enzymes and metabolic pathways, was not conducive to an intensive study of the biochemical aspects of genetics. Nor did the research conducted before about 1935 on the relation of tyrosinase to the inheritance of coat color in animals or on the genetic control of the pigmentation of flowers by investigators such as Muriel Wheldale, Sewall Wright, or Rose Scott-Moncrieff encourage many others to enter this field.[220] A significant change came during the late 1930s, as a consequence of the work of George Beadle and Boris Ephrussi on the inheritance of eye color in Drosophila,[221] and the subsequent studies by Beadle and his biochemical associate Edward Tatum on the mold *Neurospora crassa*.[222]

The power of the Beadle-Tatum technique was first made evident in a series of studies on the metabolism of various amino acids that Neurospora can make from sucrose as a source of carbon and from ammonia as a source of nitrogen. For example, among the mutant strains of the organism found among the survivors after irradiation with X-rays were several that had lost the ability to make tryptophan. There had been indications, from the work of Paul Fildes and Esmond Snell, that anthranilic acid and indole might be intermediates in the microbial biosynthesis of this amino acid.[223] When the two compounds were tested, it was found that one of the mutant strains could use either anthranilic acid or indole in place of tryptophan for growth, while another strain grew on indole but not on anthranilic acid.[224] Subsequent use of this technique by David Bonner, William Jakoby, and Herschel Mitchell showed that in Neurospora tryptophan is a metabolic precursor of nicotinic acid, with kynurenine and 3-hydroxyanthranilic acid as successive intermediates.[225] These findings of the 1940s and 1950s were followed by intensive studies on the enzymes involved in that metabolic pathway.[226] I note in passing that work during the 1950s on the insect eye pigments studied twenty years earlier by Beadle and Ephrussi led to the determination of their chemical structure, and the demonstration, by use of the isotope technique, that what has been named xanthommatin is formed from tryptophan, with kynurenine and 3-hydroxykynurenine as metabolic intermediates.[227]

Shortly after the initial reports on the biosynthesis of tryptophan, further evidence of the value of Neurospora mutants for the study of intermediate metabolism came from the work of Adrian Srb and Norman Horowitz on several "arginineless" auxotrophs.[228] Three groups of mutants were found: (1) those that grew on arginine, citrulline, or ornithine; (2) those that used arginine or citrulline, but not ornithine, for growth; (3) those that grew only in the presence of arginine. No mutants appeared that could use arginine and ornithine but not citrulline. Because Neurospora contains arginase, it was concluded that the Krebs ornithine cycle is operative in this organism and, because it also contains urease, that the urea is decomposed to CO_2 and NH_3.

After 1945, Neurospora mutants were increasingly displaced by those of other microorganisms, especially *Escherichia coli*.[229] Work by several investigators on the microbial metabolism of phenylalanine, tyrosine, and tryptophan revealed a branched metabolic pathway from carbohydrate to each of these aromatic amino acids, with a common intermediate in a new compound named chorismic acid, and leading via anthranilic acid to tryptophan or via "shikimic acid" and "prephenic acid" to phenylalanine or tyrosine.[230]

In terms of its physiological outcome, the block of a step in a metabolic process because an enzyme is absent owing to an inherited genetic anomaly is equivalent to the inhibition of the action of that enzyme by means of a chemical agent. By the late 1930s several examples of such chemical effects had been recognized. For example, the discovery by Gerhard Domagk of the antibacterial action of sulfanilamide led to the concept offered by Paul Fildes that such agents interfere with the utilization of "essential metabolites." In particular, Fildes's associate Donald Woods provided experimental evidence for the view that "the enzyme reaction involved in the further utilization of *p*-aminobenzoic acid is subject to competitive inhibition by sulphanilamide, and that this inhibition is due to a structural relationship between sulphanilamide and *p*-aminobenzoic acid (which is the substrate of the enzyme in question)."[231] Shortly after Woods's discovery, *p*-aminobenzoic acid was identified as a naturally occurring metabolite, and in 1945 it was shown to be a constituent of the vitamin folic acid. It should be recalled that the idea of the competitive inhibition of an enzyme by a structural analogue of its substrate had been anticipated during the 1920s by Juda Quastel in connection with the inhibition of succinic dehydrogenase by malonate. It was not until after the work of Fildes and Woods, however, that biochemists (notably Wayne Woolley) undertook the systematic study of such "antimetabolites."[232] The intensive search for new chemotherapeutic agents produced many inhibitors of bacterial growth and the enzyme-catalyzed reactions blocked by some of them were identified as steps in important metabolic pathways. In the case of penicillin, the most famous successor of sulfanilamide, during the 1960s Jack Strominger and James Theodore Park showed that it inhibits a "transpeptidase," which catalyzes the cross-linking reaction leading to the assembly of the polymeric peptidoglycan of bacterial cell walls.[233]

Biological Nitrogen Fixation

In parallel with the active nineteenth-century discussion among chemists and biologists about the mechanisms in the utilization of atmospheric oxygen by animal organisms and of atmospheric carbon dioxide by plants, there was debate about the biological utilization of atmospheric nitrogen, termed *azote* because of its inability to support animal life. It was known from the work of Berthollet that this element is present in the chemical constituents of plants and animals, and that a "volatile alkali" (ammonia) is formed upon the decomposition of animal matter. Since herbivorous animals received their nitrogen from plants, what was the source of plant nitrogen? Beginning in 1836, Jean Louis Boussingault conducted an extensive series of experiments to determine the elementary composition of plants, and he used the analytical methods developed by Dumas for this purpose. In particular, he noted in 1838 that some plants, such as clover, raised the nitrogen content of the soil, and he asked whether "under some circumstances plants find in the atmosphere a part of the nitrogen which contributes to their organization."[234] He mentioned the possibility that the source might be ammonia, and two years later this suggestion was raised by Justus Liebig in his book on agriculture to the level of the "ammonia hypothesis," supported solely by his finding of some ammonium carbonate in rain water. Although Boussingault confirmed this observation, that hypothesis was challenged by Thomas Way and by John Bennet Lawes and Joseph Gilbert, whose analytical data showed that the amount of ammonia in rain water (even in England) is insufficient to supply the nitrogen requirements for the growth of plants.[235] Moreover, during the 1850s the young French chemist Georges Ville offered evidence for the view that plants can fix atmospheric nitrogen, and after some discussion his conclusion was accepted, despite the difficulty of explaining how such a seemingly inert element can be utilized by biological organisms.[236] As has happened often in biochemical research, the puzzle was a consequence of the use of a wholly reasonable technical procedure. In this case, it was the fact that the experiments had been conducted with soil subjected to intense heat ("calcination"), which destroyed all organic matter.

It was not until the late 1870s, under the influence of Louis Pasteur's insistence on the role of microorganisms in fermentative processes, that answers to the puzzle began to appear. First, Jean Jacques Schloesing and Achille Müntz, and the later work of Schloesing's son Achille, demonstrated that soil bacteria are involved in the biological fixation of atmospheric nitrogen.[237] The most convincing evidence came in 1886, when Hermann Hellriegel showed that this process was effected by leguminous plants in symbiosis with a *Rhizobium* present in root nodules.[238] Also, after 1880, Beijerinck and Winogradsky isolated many "chemosynthetic" microorganisms, including the nitrogen-fixing *Rhizobium,* as well as the free-living aerobe *Azotobacter* and the anaerobe *Clostridium* (which were also found to be able to fix N_2), and developed valuable methods for the preparation of pure cultures of these organisms.[239] At the turn of the century, therefore, the problem of biological nitrogen fixation by leguminous plants had turned into the study of the symbiotic relationship between these plants and the microbes present in their root nodules, and during the next four decades there was much chemical speculation about the mechanism of this process.[240]

After 1930 the most active laboratory in this field was the one at the College of Agriculture at the University of Wisconsin, with Edwin Fred, Perry Wilson, and Robert Burris as successive leaders of productive research groups.[241] When $^{15}N_2$ became available, the group associated with Burris made considerable progress in the study of nitrogen fixation by the free-living *Azotobacter vinelandii* and *Clostridium pasteurianum.* From the finding that the isotope appeared first in glutamic acid and aspartic acid it was inferred that the key intermediate is ammonia, rather than hydroxylamine (NH_2OH), as had been proposed by Artturi Virtanen.[242] The biochemical dissection of the process of nitrogen fixation was impeded, however, by the failure to obtain active cell-free extracts. This difficulty was not overcome until the 1960s, by the group associated with James Carnahan and Leonard Mortenson at the DuPont research laboratory, who also succeeded in purifying the "nitrogenase" of *Clostridium* cells.[243] This enzyme system was found to consist of two proteins, a "dinitrogenase" with an iron-molybdenum center at the active site, and a "dinitrogenase reductase" (an iron-sulfur protein), which uses reduced ferredoxin as the electron donor in a reaction coupled to the hydrolysis of ATP. The binding and reduction of N_2 is effected by the reduced dinitrogenase. The overall process for the reduction of N_2 to ammonia is now formulated as $N_2 + 8H^+ + 8e \rightarrow 2NH_3 + H_2$.[244]

This necessarily brief sketch of recent achievements in the study of biological nitrogen fixation should be viewed against the background of the interest of nineteenth-century agricultural chemists in the general problem of the utilization by plants of soil nitrate. By the 1880s, in large part owing to the work of Ulysse Gayon and Gabriel Dupetit, it was recognized that soil microorganisms can convert nitrate to less oxygenated compounds such as nitrous oxide (N_2O).[245] Before the importance of nitrogen fixation in the "nitrogen cycle" of nature was fully appreciated, this "denitrification" of the soil was a source of some concern. Later work by numerous investigators, more recently by Victor Najjar and Alvin Nason, among others, has identified the enzymes in various bacteria, molds, and plants that catalyze the successive reduction of nitrate to nitrite, hydroxylamine, and ammonia, with pyridine nucleotides and flavin nucleotides as electron carriers. The finding that molybdenum is involved in the reduction of nitrate marked the first appearance of this metal ion in an enzyme catalyzed reaction.[246] It should also be noted that the presence of nitrate in the soil inhibits the formation of the root nodules of leguminous plants.[247]

To conclude this section, I call attention to the use of the term *symbiotic* in referring to the relation between a leguminous plant and the microorganisms in its root nodules. The concept of symbiosis came to the fore during the 1870s among botanists after Simon Schwendener described his microscopic studies on the lichens and advanced the view that they represent the association of green algae and "parasitic" fungi.[248] His work on lichens was followed by that of Johannes Reinke and Ernst Stahl, and in 1877 Bernhard Frank proposed the term *symbiosis,* which was given prominence by Anton de Bary.[249] The validity of this concept, as applied to lichens, was strengthened by the 1886 report of Gaston Bonnier, who described the "synthesis" of a lichen from algae and fungi.[250] It was recognized that the fungi ("mycobionts") are more dependent on the symbiotic relationship than the algae ("phycobionts"), as many of the algae are also free-living organisms, while the fungi sporulate only when part of a lichen. In contrast to our present biochemical knowledge about the symbiosis involving nitrogen fixation, the metabolic pathways in the physiology of the lichens are less clearly defined, especially in relation to the formation of the many unusual fungal metabolites, including the lichen acids, characterized by Harold Raistrick and his associates during 1930–1960.[251] In recent years, the concept of symbiosis has been extended to the relation between mitochondria or chloroplasts and the animal or plant cells of which they form a part, and there has been much discussion of the evolutionary significance of that relation.[252]

CHAPTER 8

The Chemical Basis of Heredity

In 1919 Thomas Hunt Morgan began his book *The Physical Basis of Heredity* as follows: "That the fundamental aspects of heredity should have turned out to be extraordinarily simple supports us in the hope that nature may, after all, be entirely approachable. Her much-advertised inscrutability has once more been found to be an illusion due to our ignorance. This is encouraging, for, if the world in which we live were as complicated as some of our friends would have us believe we might well despair that biology could ever become an exact science."[1] These lines were written at the end of a decade-long investigation on the genetics of the fly *Drosophila* by Morgan and his associates Alfred Sturtevant, Calvin Bridges, and Hermann Muller, about whose work much will be said in this chapter. As an introduction, I add two other quotations. In 1970 Warren Weaver recalled that upon his association with the Rockefeller Foundation he "started with genetics, not because I realized in 1932 the key role this subject was destined to play, but at least in part because it is a field congenial to one trained in mathematics."[2] In 1990 Robin Holliday wrote that around 1960 "geneticists tended to think in formal rather than in biochemical or molecular terms. . . . DNA was a physical structure which could mutate and recombine and it was not necessary to worry at this stage about the underlying biochemistry."[3] In this chapter, I sketch the interplay of biological and chemical thought and action, with recurrent attempts to find physical or mathematical simplicity in the face of growing complexity in the chemical knowledge about the biological mechanisms in the transmission of hereditary characters.

Fertilization, Development, and Inheritance

As we have seen, during the first half of the nineteenth century the stimulus for the interplay of biology and chemistry came largely from the study of physiological processes such as respiration and the changes undergone by ingested food materials. With the acceptance of the cell theory, after 1870 important cytological studies broadened the scope of the interplay to include the phenomena observed upon cell division, in the fertilization of eggs by sperm, and in the development of the fertilized egg. Also, after the publication of Charles Darwin's *Origin of Species* in 1859 and his *Variation of Animals and Plants Under Domestication* in 1868, there was a succession of hypotheses in which material factors were assumed to be bearers of the specific characters transmitted from parents to offspring.[4]

The cytological studies were greatly furthered by technical advances in microscopy, through the use of immersion objectives and the development by Ernst Abbe of apochromatic lenses, the improvement of the microtome for cutting tissue sections, and the use of better fixatives and of synthetic dyes as biological stains. As has been mentioned, the appearance of the aniline dyes, after William Perkin's discovery of mauve in 1856, had considerable economic impact in Europe, as it marked the emergence of a powerful German chemical industry with strong ties to university chemists. That development also influenced the microscopic observation of cell structures, especially after 1877, when Paul Ehrlich reported the first systematic study of the biological staining properties of the new synthetic dyes.[5] By the late 1870s it was known that when natural coloring matters such as carmine or hematoxylin, or synthetic dyes such as methyl green, are applied to biological tissue sections treated with such fixatives as acetic acid or osmic acid, the cell nuclei commonly appear deeply stained, whereas the cytoplasm remains relatively pale. Such dyes were recognized to be basic in chemical character, and Ehrlich referred to the cell elements that took them up as basophilic. The cytoplasm was preferentially stained by dyes such as eosin or acid fuchsin, which are acidic in nature, and the cell elements made visible in this manner were termed oxyphilic.

Among the notable cytological achievements of the 1870s was Oscar Hertwig's observation that in the sea urchin *Toxopneustes lividus* fertilization involves the union of the nucleus of the spermatozoon with the egg nucleus to form a single "cleavage nucleus," which is the precursor of all the nuclei of the embryo. This discovery was soon confirmed and extended by Hermann Fol and was found to apply to other animals and to plants.[6] In 1879 Walther Flemming described the transformation of the threadlike strands in cell nuclei during cell division in a process he later called mitosis. Flemming wrote of the structural framework of the nucleus as follows: "The framework owes its refractile character, the nature of its reactions, and particularly its affinity for dyes to a substance which I have tentatively named chromatin because of the last-named property. Possibly this substance is identical with the nuclein bodies; at any rate it follows from the work of Zacharias that it is their carrier, and if it is not nuclein itself, it consists of substances from which nuclein can be split off. I retain the word chromatin until a decision about it is made by chemical means, and I use it empirically to denote the substance in the cell nucleus which is stained by nuclear dyes."[7] After some doubts, the botanist Eduard Strasburger agreed with Flemming's observations, and Edouard van Beneden reported that in the case of the worm *Ascaris megalocephala* there occur during fertilization remarkable changes in the chromatin of the egg and sperm. The inequality of the size of their nuclei disappears and, before or during their union, each of them is transformed into a definite number of rodlike structures that appear to be of the same shape, size, and number in the two sexes.[8] Confirmation of this discovery, and its extension to other animals and to plants, was provided by numerous workers, notably Theodor Boveri and Eduard Strasburger. The rodlike chromatin segments were given a variety of names, but the term *chromosomes*, proposed by Wilhelm Waldeyer, won preference.[9] During the 1880s Wilhelm Roux and August Weismann advanced the view that the linear structure and apparent longitudinal division of the chromosomes argued for their role as bearers of the hereditary material, and Weismann offered his theory that "the nature of heredity is based upon the transmission of nuclear substance with a specific molecular constitution. This substance is the specific nucleoplasm of the germ-cell, to which I have given the name of germ-plasm."[10] Weismann also assumed that in the maturation of egg and sperm cells there occurs a "reduction division" (meiosis), in which the chromosome number is halved; the validity of this prediction was soon established.

These studies influenced the discussion of the numerous theories of inheritance that had been proposed during the latter half of the nineteenth century.[11] Among the key terms that epitomized these precursors of Weismann's *germ-plasm* were Herbert Spencer's *physiological units* (1864), Charles Darwin's *gemmules* (1868), the *Anlagen* of Wilhelm His (1874), Louis Elsberg's (and Ernst Haeckel's) *plastidules*, and Francis Galton's *stirps* (1876). Special attention was given to Carl Nägeli's *idioplasm* (1884), thought to be a micellar network providing the physical basis of heredity and set in a matrix of structureless "trophoplasm," which represents the nutritive components of cells. The proliferation of new terms to denote presumed fundamental units of life continued into the 1890s, with Hugo de Vries's *pangens* (1889), Richard Altmann's *bioblasts* (1890), and Weismann's *biophors* (1892). According to Weismann, the germ-plasm is a nuclear material with a specific molecular constitution, and the units of heredity are "determinants" composed of biophors needed to form cells of a given type: "*The germ-cell of a species must contain as many determinants as the organism has cells or groups of cells which are independently variable from the germ onwards,* and these determinants must have a definite mutual arrangement in the germ-plasm, and must therefore constitute a definitely limited aggregate, or higher vital unit, the 'id.' "[12] To these terms of the 1890s may be added the *plasomes* of Julius Wiesner and the *idioblasts* of Oscar Hertwig. Of the various particulate units of inheritance postulated before 1900, the idioblast approximated most closely the *gene* as defined by Wilhelm Johannsen in 1909.

Nuclein and Heredity

When, in 1869, Friedrich Miescher reported his preparation in Hoppe-Seyler's laboratory of a material he called nuclein, it was not uncommon for disputes to arise about the identity and homogeneity of supposedly new chemical constituents of biological systems. For example, a few years earlier, another of Hoppe-Seyler's students, Oscar Liebreich, isolated from brain tissue a material he called protagon. As was his practice, Hoppe-Seyler then turned the matter over to his student Konstantin Diakonov, who concluded that protagon was merely an impure preparation of the well-known lecithin. Ludwig Thudichum and Arthur Gamgee then initiated a lively discussion whose principal consequence was the disappearance of protagon.[13] Thudichum also questioned the validity of Willy Kühne's work on rhodopsin, and his own work on the isolation of new chemical constituents, such as sphingosine, from brain tissue was not accepted by leading physiological chemists.[14] It was a time in which names given to new biochemical substances soon lapsed into obscurity together with those of their creators. Sometimes, as with Thudichum and, as we saw before in the case of Charles MacMunn, later research with improved methods established the validity of a portion of the disputed work and brought to the attention of historians the vindicated scientist.

Miescher received his M.D. degree at Basle in 1868, and on the advice of his uncle Wilhelm His (at that time professor of anatomy and physiology there) went to Tübingen, first to study organic chemistry with Adolf Strecker, then to work with Hoppe-Seyler on the chemical constitution of the cells present in pus. Miescher obtained the cells by washing used bandages from the surgical clinic, and, as he wrote his uncle in February 1869, "After the pus cells had been prepared, there was the question of the aims and methods of the research. . . . There presented itself the task of giving as complete accounting as possible of the characteristic chemical constituents from whose variety and arrangement there results the structure of the cell."[15] After describing some of his experimental difficulties, Miescher continued:

> In the experiment with weakly alkaline fluids, I obtained, by neutralization of the solutions, precipitates which were insoluble in water, acetic acid, very dilute hydrochloric acid, or sodium chloride solutions; consequently, they

> could not belong to any of the known albuminoid substances. Where did this substance come from? By prolonged action of very dilute hydrochloric acid on the cells, a point is reached when the acid does not take up anything more. The residue consists of partly isolated, partly shrunken nuclei. . . . Very weakly alkaline fluids (even 1/10000 sodium carbonate) cause the nuclei to become pale and to swell considerably. . . . From this fact, known to histologists, the substance could belong to the nuclei, and therefore gripped my interest. The most reasonable next step was then to prepare pure nuclei.[16]

The method he eventually used involved extraction of the cells with alcohol to remove fatty material, followed by treatment of the alcohol-insoluble residue with an acidified extract of swine gastric mucosa (containing pepsin). This procedure had been described by Willy Kühne: "The protoplasm of the cells of all edible glandular structures, such as liver, etc., largely dissolves, leaving behind small crumbs and greatly shrunken nuclei."[17] Miescher's preparation contained 14 percent nitrogen, 2 percent sulfur, and 3 percent phosphorus. The relatively high content of phosphorus was of special interest, for at that time this element was associated with proteins such as casein or with alcohol-soluble substances such as lecithin, known to be a fatty acid derivative of glycerophosphoric acid, and under active study in Hoppe-Seyler's laboratory. Because the resistance to pepsin and insolubility in acid set his product apart from these substances, Miescher concluded that he had isolated a new cell constituent, which he named nuclein.[18]

Miescher's paper was published in Hoppe-Seyler's house journal along with four others: two were by his fellow students Pal Plósz (who used Miescher's method to demonstrate the presence of nuclein in the erythrocytes of birds and snakes) and Nikolai Lubavin (who examined the action of pepsin on casein to produce nuclein-like products). The fourth was by Hoppe-Seyler himself, who confirmed Miescher's claim and also reported the presence of nuclein in yeast: "Just as the pus cells, the elementary yeast organisms contain in addition to cholesterol, lecithin and fat, a phosphorus-containing substance which resembles albuminoid bodies, is indigestible by gastric juice, is colored brown by iodine, also belongs to the nuclei, and perhaps may play a highly important role in all cell development."[19] The fifth paper, by Miescher, was based on work he had done after he left Tübingen in 1869 and described the application of his method for the isolation of nuclein to the chemical examination of the microscopic particles in the yolk of the hen's egg. He chose this material because His, among others, considered these yolk platelets to be cell nuclei because of their strong light refraction and because they stained red with carmine (an insect coloring matter), which had been found during the 1850s to stain the nuclei of plant and animal cells. Miescher obtained from the yolk platelets a material that contained organic phosphorus, was soluble in alkali and precipitated by acid, gave the biuret, Millon, and xanthoproteic tests for protein, and was resistant to pepsin. He concluded, therefore, that "a substance having such a composition and such properties clearly can only be ranked with the nuclein of pus."[20]

This work was a prelude to Miescher's most important biochemical investigation, namely the study of the chemical composition of the spermatozoa of the Rhine salmon, during 1871–1873, upon his return to Basle after a stay in Carl Ludwig's laboratory. As this fish, having fed in salt water, travels upstream to spawn, there is a massive conversion of skeletal muscle to gonadal tissue, and it was known that the sperm heads are essentially equivalent to cell nuclei. The ready availability of large amounts of salmon sperm (Basle was a center of the salmon fishing industry) made it an attractive starting material. In 1874 Miescher reported that the sperm heads are largely composed of "an insoluble salt-like combination of a very nitrogen-rich organic base with a phosphorus-rich nuclein body which assumes the role of the acid."[21] He crystallized the organic base, which he named protamine, in the form of its hydrochloride, and after reprecipitation the acid-insoluble component was found to contain 13.1 percent nitrogen and 9.6 percent phosphorus, was free of sulfur, and failed to give the color tests for protein. In spite of the differences between this acidic material and the nucleins he had described earlier, Miescher named it a nuclein as well. To explain the variable composition of the nucleins from various biological sources (he also examined bull spermatozoa), especially as regards their sulfur content, Miescher suggested that "there exists a sulfur-containing nuclein which is cleaved by warm alkali into sulfur-free nuclein and a compound containing sulfur in an unoxidized state. This compound is not an albumin; the sulfur content is too high for that; rather one might consider an atomic grouping like the one involved in the structure of the keratin substances. The cleavage occurs readily with bull sperm, with greater difficulty with pus nuclei. It is most probable that both nucleins occur together in the nuclei. This may be regarded as certain for the yolk platelets of the hen's egg."[22] As we shall see shortly, about twenty years of work was required to sort out the confusion generated by these views. It should be added, however, that later investigators confirmed Miescher's observation that the sperm nuclein did not diffuse through a parchment membrane and therefore appeared to be colloidal in nature. It also behaved as a polybasic acid, and Miescher suggested that the saltlike "nucleoprotamine" might exist in various forms, in which the organic base is replaced by such inorganic components as sodium.

The contemporary chemical judgment as to validity of Miescher's claims was not entirely favorable. In Germany his conclusions were questioned by Jacob Worm-Müller, and, after carefully summarizing the results, the French chemist Adolphe Wurtz concluded that they "appear to be somewhat vague from a chemical point of view and seem to require new research." In England, Charles Kingzett analyzed nuclein and stated that it "is nothing but an impure albuminous substance," and Ludwig Thudichum cited this report in writing that "nuclein was exploded, in this country at least."[23]

As I have indicated, the uncertainties about the identity and homogeneity of many newly discovered chemical constituents of biological systems led to disputes. Miescher declined to defend his claims for nuclein, and this attitude was evident at the start of his work. In a letter of February 1869 to His, he wrote: "There is nothing more difficult than sharp separation in the field of the albuminoid substances. I can well understand why their definition is so variable and controversial; and it is indeed the curse of the amorphous substances that one has no assurance as to the purity of one's preparation. That is why the real chemists shun them so much."[24] That attitude was evident five years later, after Miescher's work on the nuclein of salmon sperm, when he wrote to His that "physiological chemistry consists of such a heap of unconnected facts that it makes little sense to wish to add still more chaff."[25] After 1874 Miescher's published papers dealt largely with purely physiological problems, but shortly before his death in 1895 he returned to the study of the nucleins. A posthumous report of this work was published in 1896 by Oswald Schmiedeberg.

In separating himself from the physiological chemistry of his time, Miescher also reflected the influence of his uncle's biological thought about the nature of fertilization. In 1874 Wilhelm His defended the idea that "it is neither the form, nor the form-building material that is transmitted, but the excitation to form-developing growth."[26] This physicalist view was shared by Miescher in 1874 when he wrote that many authors

> incline to the idea that the spermatozoa might be carriers of specific substances which act as fertilizing agents by virtue of their chemical properties. ... To the extent that we wish to assume at all that a single substance, acting as a ferment or in another manner as a chemical sensitizer, is the specific cause of fertilization, one must unquestionably think above all of nuclein. ... If we consider the sperm to be only a carrier of a specific fertilization substance, how do we explain the variations in action, from species to species, from genus to genus, from individual to individual? ... A variety of reasons

> speak against a decisive role of chemical phenomena. . . . There are no specific fertilization substances. The chemical phenomena have only a secondary significance; they are subordinate to a higher explanation. If we seek an analogy to explain all the available knowledge, it seems to me that there remains nothing but a picture of an apparatus that evokes or transforms some kind of motion.[27]

As we have seen, Walther Flemming and several other cytologists of the 1870s and 1880s found in Miescher's nuclein a "specific fertilization substance." In 1881 Eduard Zacharias applied to various kinds of cells the tests Miescher had used—namely, resistance to pepsin, solubility in alkali, and swelling in salt solutions—and obtained isolated nuclei that were readily stained by suitable dyes.[28] A few years later, Oscar Hertwig wrote: "I believe that I have made it at least highly probable that nuclein is the substance which is responsible not only for fertilization but also for the transmission of hereditary characteristics and thus corresponds to Nägeli's idioplasm. . . . Nuclein is in an organized state before, during and after fertilization, . . . [that] is not only a physico-chemical process, as was usually assumed by physiologists, but . . . is also a morphological process."[29] In 1889 Richard Altmann gave the name *nucleic acid* to an apparently protein-free preparation of Miescher's nuclein, and this term was adopted by Edmund Wilson in summarizing the view held by many cytologists during the 1890s:

> The precise equivalence of the chromosomes contributed by the two sexes is a physical correlative of the fact that the two sexes play, on the whole, equal parts in hereditary transmission, and it seems to show that the chromosomal substance, the *chromatin,* is to be regarded as the physical basis of inheritance. Now chromatin is known to be closely similar to, if not identical with a substance known as nuclein, . . . a tolerably definite chemical compound of nucleic acid (a complex organic acid rich in phosphorus) and albumin. And thus we reach the remarkable conclusion that inheritance may, perhaps, be effected by the physical transmission of a particular compound from parent to offspring.[30]

Miescher's reaction to the development of these ideas, from Zacharias to Weismann, was wholly negative. In his letters to His during 1890–1895, Miescher inveighed against the "guild of dyers who insist that there is nothing but chromatin (nuclein)" and characterized the "speculations of Weismann and others [as] afflicted with half-chemical concepts which are partly unclear, and partly correspond to an obsolete state of chemistry." Instead, he claimed that nuclein forms a structural envelope for an iron-containing protein which he named karyogen. He wrote to His that "it is not the nuclein, but the iron-containing phosphorus-free substance which gives the chromatin reactions (methyl violet, safranin, etc.)" and offered the view that "the key to sexuality lies in stereochemistry. . . . With the enormous molecules of the albuminoid substances . . . the many asymmetric carbon atoms permit such a colossal variety of stereoisomers that the entire wealth and multiplicity of hereditary transmissions can find as good an expression as can the words and concepts of all languages in the 24–30 letters of the alphabet. It is therefore entirely superfluous to make of the egg or sperm cell, or the cell in general, a repository of innumerable chemical substances, of which each is supposed to be a carrier of a particular hereditary trait (de Vries pangenesis)."[31] Miescher did not publish these views during his lifetime (he died in 1895), but their impact on the discussion of the mechanism of heredity seems to have been negligible after they appeared posthumously in 1897.

During the 1890s doubts arose among biologists about the role of chromatin and nuclein because, as Weismann wrote, "The affinity of the chromosomes for colouring matter varies markedly at different periods, and this indicates that slight changes, which are beyond our control, take place in the constitution of this substance, and are sufficient to cause its most striking reaction with regard to colouring matters to disappear for a time."[32] Martin Heidenhain suggested that the nucleic acid (which he termed basichromatin) of chromosomes may change its affinity for dyes by the "uptake or release of phosphorus," and based this view on the chemical studies of Hans Malfatti, who reported that the basophilic affinity of nuclein preparations was related to their phosphorus content.[33] It was also found that combinations of a nucleic acid with egg albumin stained less intensely with basic dyes than did the nucleic acid itself, and Edmund Wilson wrote: "We may infer that the original chromosomes contain a high percentage of nucleic acid; that their growth and loss of staining power is due to a combination with a large amount of albuminous substance to form a lower member of the nuclein series, perhaps even a nucleo-albumin; that their final diminution in size and resumption of staining power is caused by a giving up of the albumin constituent, restoring the nuclein to its original state as a preparation for division."[34] This explanation was also considered to be inadequate when cases were reported where the basichromatin had apparently disappeared completely. After 1900 Eduard Strasburger spoke out "against every conception, even partial ones, of true fertilization as a purely chemical process, therefore against every chemical theory of heredity"; he also wrote that "the chromatin cannot itself be the hereditary substance, as it . . . leaves the chromosomes, and its amount is subject to considerable variation in the nucleus, according to its stage of development."[35]

Doubts about the nuclein theory of heredity were also expressed in 1906 by Jacques Loeb and Richard Burian,[36] and it seems that by 1910 the views expressed earlier by Hertwig and Wilson had lost favor. Indeed, after summarizing the available evidence in the third edition of his great treatise on the cell, published in 1925, Wilson concluded that "these facts offer conclusive proof that *the individuality and genetic continuity of chromosomes does not depend upon a persistence of 'chromatin' in the older sense (i.e., basichromatin),*" and he also inferred that the loss in affinity for dyes "seems to indicate a progressive accumulation of protein components and a giving up, or even a complete loss, of nuclein."[37]

The decline of the nuclein theory of heredity during the first three decades of the twentieth century is also indicated by the fact that the three most significant monographs on the chemistry of nucleic acids written at that time say little about their presumed biological role. For example, Levene and Bass noted that by virtue of their acidic character the nucleic acids "may serve as regulators of the physicochemical properties of the cell or at least of the nucleus of the cell. This role of nucleic acids was especially emphasized by Einar Hammarsten."[38] Hammarsten had performed a thorough study of carefully prepared calf thymus nucleic acid, and the inferences drawn from his work were influenced by the considerable interest during the 1920s in the physiological regulation of acid-base equilibria and of osmotic pressure. In particular, much importance was attached to the "Donnan equilibrium," which describes the effect of a colloid on the distribution of diffusible ions on either side of a membrane.

Chemistry of Nucleic Acids and Associated Proteins

The investigator who began to unravel some of the chemical problems presented by Miescher's nuclein was Hoppe-Seyler's associate Albrecht Kossel, whose entry into that field appears to be related to the appearance of a paper by Carl von Nägeli and Oscar Loew in which they concluded that the nuclein of yeast "did not differ from albumin with a slight admixture of potassium and magnesium phosphates. In view of the considerable phosphate content of yeast, a slight contamination with 'phosphorus,' from whose presence Hoppe-Seyler inferred the existence of yeast nuclein, cannot be surprising."[39] After refuting this assertion, Kossel provided a new basis for distinguishing between nucleic acids and proteins, by showing that while the nucleins from yeast, pus, and erythrocytes contain "hypoxanthine" as a structural unit, the material from egg yolk does not.[40] The latter was designated a paranuclein and turned out to be a pepsin-resistant phosphorus-rich protein subsequently named phosvitin. Similar phosphoproteins were later found, with phosphoryl groups attached to serine or tyrosine units.

Miescher's failure to distinguish the material from egg yolk from the one from salmon sperm was not a consequence of ignorance about the test for hypoxanthine, for in his 1874 paper he reported that upon treatment of the protamine preparation with nitric acid there appeared a yellow color that turned red upon the addition of alkali. This color reaction was known to be given by uric acid and several closely related animal products. The isolation of uric acid from human urine by Scheele in 1776 and from bird excrement (guano) by Fourcroy and Vauquelin in 1805 was followed in 1818 by the work of Brugnatelli and of Prout, who showed that nitric acid converts uric acid into a product (later named alloxan) that gives a red ammonium salt ("murexide"). The elementary composition of uric acid was established in 1834 by Liebig and by Mitscherlich, and the subsequent joint studies of Liebig and Wöhler on the chemical degradation of this compound evoked the admiration of their contemporaries. This work laid the basis for that of Strecker around 1860; by 1870 the relation of uric acid to xanthine (found in kidney stones by Alexander Marcet in 1817), to guanine (isolated from guano in 1845 by Unger), and to sarkine (later named hypoxanthine) obtained in 1850 by Scherer from pancreas had been clearly established, although the definitive proof of their structure did not come until later through the synthetic work of Emil Fischer. All these compounds gave the murexide reaction, and Miescher, as a former pupil of Strecker, could not have failed to recognize the possibility that some of the relatives of uric acid might be constituents of sperm. He did not follow up his observation but asked his colleague Jules Piccard to do so.

Piccard found that both guanine and sarkine were formed upon treatment of salmon nuclein with acid, and that the positive murexide test Miescher had obtained with protamine was probably a consequence of its coprecipitation with these bases.[41] There was still uncertainty, however, whether hypoxanthine is a constituent of albuminoid substances. In 1878 Georg Salomon stated that it is present in fibrin, and this report was confirmed in the following year by Russell Chittenden, working in Kühne's laboratory. Shortly afterward Kossel showed that carefully prepared fibrin does not yield hypoxanthine, which had probably come from leucocyte material.[42]

During the 1880s Kossel provided further evidence for the separate identity of the nucleins not only by the isolation of the "xanthine bodies" guanine and hypoxanthine after acid hydrolysis but also by the discovery of another member of the group (adenine) as a cleavage product of the nucleins. Moreover, in 1882 Emil

Fischer began work on the chemistry of uric acid and its relatives, and during the next fifteen years he established their chemical structure as being derived from a bicyclic unit he named purine.[43] Through unequivocal synthesis, Fischer showed that uric acid is 2,6,8-trioxypurine (as had been suggested by Ludwig Medicus in 1875), that xanthine is 2,6-dioxypurine, hypoxanthine is 6-oxypurine, adenine is 6-aminopurine, and guanine is 2-amino-6-oxypurine. As in his studies on the sugars and his later research on polypeptides, Fischer's work in the purine field was elegant in its execution and placed this aspect of nuclein chemistry on a firmer chemical foundation. With the knowledge of the structure of the various purines, it was possible around 1905 to show that the hypoxanthine found by Kossel upon the degradation of the nucleins was a secondary product formed by the deamination of adenine.[44] By that time it was generally agreed that the two purines guanine and adenine represent authentic constituents of the nucleins.

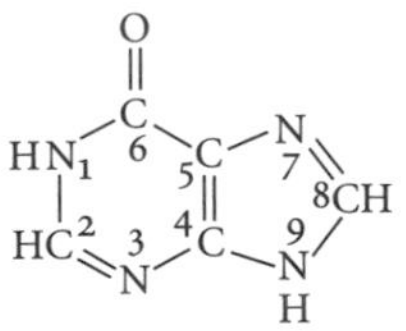

Hypoxanthine

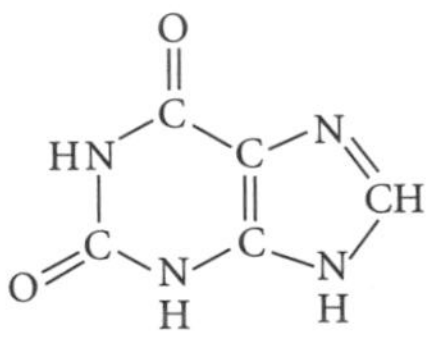

Xanthine

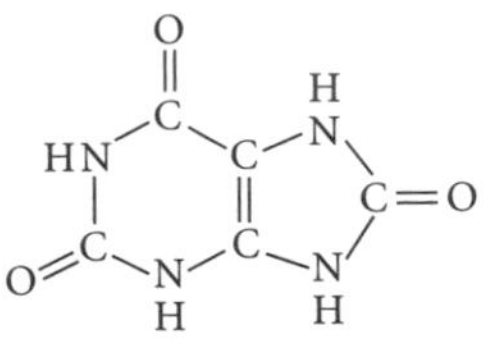

Uric acid

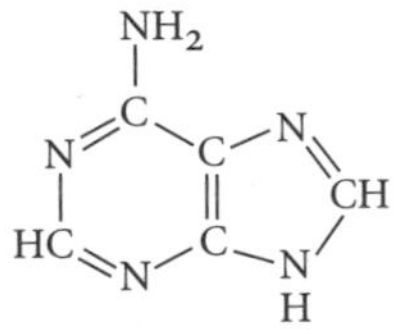

Adenine

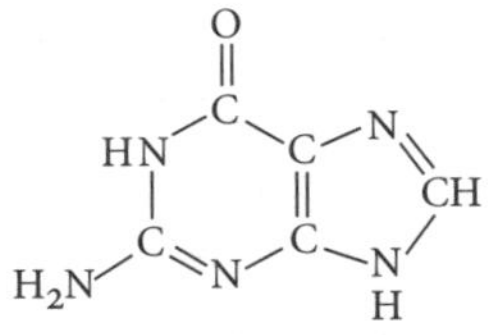

Guanine

Kossel's initial evidence for the separate identity of the nucleins was not accepted by some physiological chemists. During the period 1888–1892, Leo Liebermann, Julius Pohl, and Hans Malfatti reiterated the claim that these products were merely phosphorylated derivatives of ordinary proteins. Much of the uncertainty arose from the difficulty in obtaining nuclein preparations which were relatively free of protein, and (as we have seen) some progress was made by Richard Altmann, who improved Miescher's isolation method to yield what Altmann named nucleic acid. Also, in 1884 Kossel isolated from goose erythrocytes a peptonelike substance that he named histone and that readily combined with nuclein to form a "nucleohistone." Similar histones were prepared from various biological sources (for example, calf thymus) and differed from the protamines by their less basic character and the fact that, upon acid hydrolysis, the histones yielded a larger variety of amino acids. It seemed likely, therefore, that the protein-containing nuclein preparations, such as the one Miescher had obtained from pus cells, belonged to a class of conjugated proteins in which a nucleic acid is combined with a proteinlike substance, just as in hemoglobin an iron-porphyrin is combined with globin. However, since isolated nucleic acids were found to precipitate a variety of proteins, it was uncertain whether the "nucleoproteins" obtained from biological sources were preformed cell constituents or artifacts of isolation. Indeed, Altmann concluded that "until now no fact is known to establish the authenticity of the so-called nucleins as compounds of nucleic acids and proteins."[45] Despite this uncertainty, before 1910 there was the assertion by Wilhelm Spitzer that an iron-nucleoprotein is involved in intracellular respiration, while Cornelis Pekelharing suggested that pepsin might be a nucleoprotein, and Gustav Mann concluded that "the nucleoproteids are the agencies by which amino acids are built into the cell-plasm."[46] As will be seen, the possible enzymatic nature of nucleoproteins continued to be discussed for many years afterward in relation to theories of heredity.

The availability of nucleic acid preparations relatively free of protein was soon followed by the identification of degradation products other than guanine and adenine. By subjecting thymus nucleic acid to hot concentrated acid, Kossel and Neumann found among the products a new substance they named thymine; this compound appears to have been also isolated by Miescher during the 1890s.[47] Thymine was recognized to be a new member of a known class of organic ring compounds, the pyrimidines, and Kossel's associate Hermann Steudel showed it to be 5-methyl-2,6-dioxypyrimidine (the numbering of the ring was the one then used); this structure was established by synthesis in Emil Fischer's laboratory. Kossel and Neumann also isolated from the acid hydrolysate of thymus nucleic acid another pyrimidine ("cytosine") which turned out to be 6-amino-2-oxypyrimidine; this was later synthesized by Henry Wheeler and Treat Johnson.[48] In 1900 Alberto Ascoli, working in Kossel's laboratory, found that a similar hydrolysate of yeast nucleic acid contains the well-known uracil (2,6-dioxypyrimidine), and a few years later Phoebus

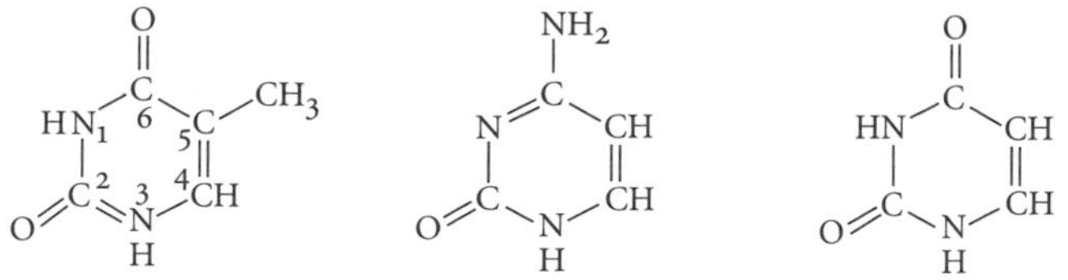

Thymine

Cytosine

Uracil

Aaron Levene showed that this nucleic acid yields, as pyrimidines, only cytosine and uracil, and no thymine. The relatively rapid elucidation of the chemical structure of these pyrimidines is noteworthy, because it marks another of the many early fruits of the new synthetic chemistry as applied to the study of a complex biochemical constituent. In this case, the work of Adolf Pinner and Robert Behrend during the 1880s was decisive. Moreover, by 1900 the work of Albrecht Kossel and Olof Hammarsten had shown that, in addition to the two purines and three pyrimidines, the nucleic acids contain a carbohydrate component. Kossel identified it as a 5-carbon sugar because, as with known pentoses, heating with strong acid produced furfural; with thymus nucleic acid, similar treatment gave levulinic acid, a product of the decomposition of hexoses. In 1903 Levene summed up the status of the nucleic acids of animal tissues as follows: "It remains to establish the nature of the carbohydrate present in their molecule, and further to establish the proportions of the different components in the acids of different tissues. It seems probable from our present experience that the proportions of the purine bases to the pyrimidine bases, as well as the proportion of the different bases of each group, varies considerably in acids of different tissues."[49] Much of Levene's later scientific life was devoted to the study of these questions.

The next substantial contribution was made by Levene and his associate Walter Jacobs in their study of "inosinic acid," isolated by Justus Liebig from meat extract and examined some fifty years later by Franz Haiser, who found that, in addition to phosphate, it yields hypoxanthine upon hydrolysis. In 1907 Friedrich Bauer reported that inosinic acid also contains a pentose (arabinose), Carl Neuberg identified it as xylose, and Haiser as lyxose.[50] During the next two years, Levene and Jacobs put out a set of papers on inosinic acid, questioned the validity of all these proposals, and, with the work of Emil Fischer on the stereochemistry of the sugars as a background, showed that the pentose is *d*-ribose (now denoted D-ribose), a new sugar that is the optical antipode of the known *l*-ribose.[51] They also established the arrangement of the components of inosinic acid as hypoxanthine-ribose-phosphate, with a glycosidic bond linking the purine to the sugar-5-phosphate.

The elucidation of the structure of inosinic acid, though not in all detail, led Levene and Jacobs to apply their methods to the examination of a substance isolated by Olof Hammarsten upon the hydrolysis of a pancreatic nucleoprotein with dilute alkali and named guanylic acid because it contained phosphorus and guanine. Steudel had found in 1907 that this product was composed of guanine, pentose, and phosphate. Levene and Jacobs established the sequence of these components as the same as in the case of inosinic acid, and showed that mild hydrolysis gave the guanine riboside ("guanosine"). Compounds such as inosinic acid and guanylic acid were named nucleotides and the corresponding inosine and guanosine as nucleosides.[52]

Some years before this work in Levene's laboratory, Thomas Osborne and his associate Isaac Harris reported the analysis of an acid hydrolysate of a nucleic acid preparation from wheat germ ("triticonucleic acid").[53] Their data were based on the weights of isolated products, and therefore of uncertain validity, but they noted that their nucleic acid gave equivalent amounts of adenine and guanine upon hydrolysis. Shortly afterward, a similar finding was reported by Hermann Steudel for thymus nucleic acid and by Levene for yeast nucleic acid, not only for the two purines but for the two pyrimidines as well.[54] This apparent equivalence of the four bases led to the hypothesis that the intact nucleic acids are composed of tetranucleotides, and speculations were offered about the mode of linkage of the four units.[55] Apart from the limitations of the analytical methods then available, an assumption implicit in the formulation of the tetranucleotide hypothesis was that each of the nucleic acid preparations they handled represented a single chemical entity. The multiplicity and individuality of proteins, made evident by 1910, does not appear to have been considered to apply to the nucleic acids, presumably because of the greater simplicity of the tetranucleotide structure compared with that of the polypeptide chains of proteins. As has been noted, the search for simplicity in the arrangement of the amino acid units in a protein molecule attracted noted investigators, from Albrecht Kossel to Max Bergmann, and the demise of their hypotheses, based on analytical data of uncertain validity, came principally from the acceptance of the evidence for the macromolecular nature of proteins and the invention of new methods for the determination of the amino acid sequence in the polypeptide chains of proteins. Likewise, the demise of the tetranucleotide theory of the structure of nucleic acids was a consequence of the recognition of the fact that they are long-chain polymers and the invention of new analytical methods for the study of their composition. Before retelling that part of our story, it is necessary to return to the organic-chemical work earlier in the twentieth century.

By 1912 several important chemical differences were evident between the nucleic acid preparations from calf thymus and from yeast. In addition to the appearance of thymine in place of uracil among the products of acid hydrolysis and the presence of a sugar different from D-ribose, thymus nucleic acid was resistant to hydrolysis by aqueous ammonia, which cleaved yeast nucleic acid to nucleosides and, under milder conditions, to nucleotides. Moreover, during the 1890s Kossel had observed that thymus nucleic acid preparations can differ in their ability to form gels, and denoted the gel-forming species the a-form, as distinct from a nongelating b-form. His associate Albert Neumann then reported the conversion of the a-form to the b-form by means of hot alkali and considered the process to be a "depolymerization" although, as later noted by Robert Feulgen, it was attended by considerable destruction.[56] One consequence of Neumann's work was an effort in 1903–1905 by several investigators (Torasaburo Araki, Leonid Ivanov, Fritz Sachs) to study the action of enzymes on thymus nucleic acid.[57] Such "nuclease" preparations were subsequently used by Siegfried Thannhauser, Levene, and others and proved to be valuable in establishing the nature of the sugar unit in thymus nucleic acid.

For many years that unit was considered to be a hexose, but repeated efforts to isolate it and identify its structure were unsuccessful. In 1914 Feulgen reported that a neutralized acid hydrolysate of thymus nucleic acid restores the color of a fuchsin solution previously decolorized by means of sulfur dioxide.[58] This reaction, introduced by Hugo Schiff in 1866, is given by ordinary aldehydes but not by glucose. Feulgen named it the nucleal reaction and suggested that the carbohydrate of thymus nucleic acid is "glucal" (a glucose derivative described by Emil Fischer in that year), but this idea was abandoned when it was shown in 1920 that the original preparation of glucal was contaminated with aldehyde material and that the pure substance does not give the Schiff reaction.[59] The situation was not clarified until 1929, when Levene succeeded in isolating the nucleosides from an enzymatic digest of thymus nucleic acid and found that the sugar component, which he named thyminose, belongs to a group of pentoses, then denoted as desoxypentoses (later, deoxypentoses), lacking an oxygen atom. Such sugars had been studied by Heinrich Kiliani around 1910, and Max Bergmann had shown during the 1920s that 2-deoxysugars are readily formed from members of the glucal family.[60] This knowledge served as the background for Levene's identification of thyminose as 2-deoxy-D-ribose, which gives the aldehyde reaction. An analytical method for the quantitative determination of thymus nucleic acid, using diphenylamine as an aldehyde reagent, was described by Zacharias Dische in 1930.[61]

With the completion of the establishment of the nature of the components of the nucleotides yielded by yeast nucleic acid and thymus nucleic acid, there was a change in nomenclature. In 1920 Walter Jones summarized the view then held by

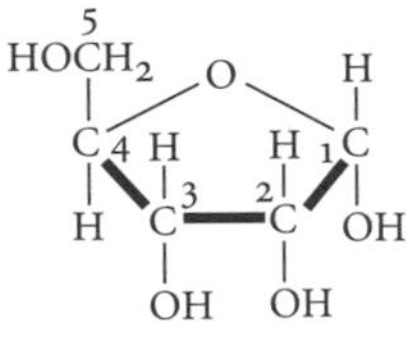

D-Ribose

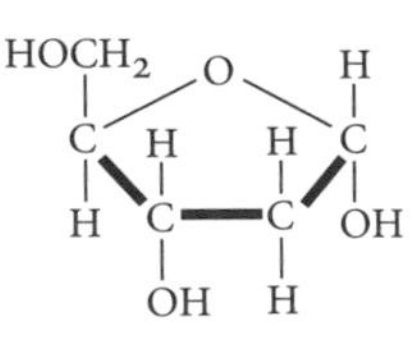

2-Deoxy-D-ribose

most of the nucleic acid chemists in stating that "there are but two nucleic acids in nature, one obtainable from the nuclei of animal cells, and the other from the nuclei of plant cells" and that the designation of the animal or plant from which either is derived "is as superfluous as would be the application of a similar nomenclature to lecithin."[62] Thus the term *thymus nucleic acid* was frequently applied to any nucleic acid preparation from an animal source, and *yeast nucleic acid* often denoted a material from an organism other than yeast. After about 1930 the distinction was made between ribonucleic acids and desoxyribonucleic (later deoxyribonucleic) acids, and during the 1950s the abbreviations RNA and DNA came into general usage. Also, it had become evident from the use of Feulgen's reaction in cytochemical research that both plant and animal cells contain DNA and that RNA is present in animal tissues (for example, beef pancreas).[63]

The widespread acceptance of the tetranucleotide hypothesis led to speculation about the mode of linkage of the nucleotides to each other. A variety of structures were proposed: in some, the sugar units were joined by means of ether linkages, in others, the nucleotides were joined solely through the phosphoryl groups, and various combinations of these two types of linkage were also suggested. Most of these ideas were discarded, however, when the presumed partial degradation products (dinucleotides) on which the ideas were based proved to be mixtures of mononucleotides. The structure that was generally accepted during the 1930s was that proposed by Levene for thymus nucleic acid in 1921; in his formulation the individual units were joined to each other by phosphodiester bridges.

```
phosphate-sugar-base
              |
          phosphate-sugar-base
                        |
                    phosphate-sugar-base
                                  |
                              phosphate-sugar-base
```

The application of the techniques of carbohydrate chemistry had shown by 1935 that the deoxyribosyl unit was present in nucleotides in the form of a five-membered "furanose" ring, leaving only the hydroxyl groups at carbons 3 and 5 for joining the nucleotide units. On the basis of the tetranucleotide formula, Levene concluded that "in *desoxyribose nucleic acid* the position of the phosphoric acid radicles are carbon atoms (3) and (5) of the desoxyribose."[64] This view was confirmed in subsequent research, but the problem of the mode of linkage of the mononucleotide units in ribonucleic acids proved to be more difficult because the furanose form of ribose has three free hydroxyl groups (at carbons 2, 3, and 5). During the 1930s preference was given to a phosphodiester bridge linking carbons 2 and 3,[65] and the alkali-lability of ribonucleic acids was attributed to this mode of linkage, but later observations were incompatible with this idea. The solution of this problem did not come until after 1950, from the study of the degradation of yeast nucleic acid to mononucleotides with the enzyme ribonuclease A.

As we have seen, this heat-stable enzyme occupies an important place in the development of knowledge about protein structure and the mechanism of enzymatic catalysis. Its presence in extracts of swine pancreas was reported in 1920 by Walter Jones, who found that it did not cleave thymus nucleic acid and noted that "this is curious. An active agent is present in animal pancreas which is specifically adapted to plant nucleic acid. It suggests evolutionary matters."[66] This finding was confirmed by René Dubos and Robert Thompson, who named the enzyme ribonuclease, but Levene (with Gerhard Schmidt) reported that mononucleotides are not formed, as claimed by Jones, and renamed it a ribonucleodepolymerase, whose action is "limited to the dissociation of the tetranucleotides of high molecular weight into those of lower molecular weight."[67] This view is similar to that held for the action of pepsin on proteins before the first synthetic peptide substrates for that enzyme were described in 1939.[68] The crystallization of pancreatic ribonuclease had been reported by Moses Kunitz in 1940,[69] but the first synthetic substrates were not available until ten years later, during the course of the work of Alexander Todd and his associates in the nucleic acid field.[70]

Through unequivocal chemical synthesis of nucleosides, Todd established the validity of earlier conclusions that the sugar is linked to the nitrogenous base by a β-glycosidic bond and that the site of attachment is the nitrogen in position 9 of the purine ring or in position 3 in the pyrimidine ring. Then followed the synthesis of nucleotides in which the phosphoryl group was attached to the hydroxyl at positions 2, 3, or 5 of the sugar unit of various nucleosides.[71] By demonstrating the identity of particular synthetic products with those obtained by enzymatic degradation of nucleic acids, Todd's group provided a sounder basis for the consideration of mode of linkage between the mononucleotide units, and their work contributed much to the now accepted formulation of ribonucleic acids as 3′,5′-linked polynucleotides. During the early 1950s, the examination of alkaline hydrolysates of yeast nucleic acid by Charles Carter, using paper chromatography, and by Waldo Cohn, with the more valuable ion-exchange method, showed the appearance of two forms of each of the four nucleotides,[72] which were identified by the Todd group as the 2′- and 3′- isomers. Roy Markham and John Smith, as well as the Todd group, then explained this result by demonstrating the intermediate formation of cyclic nucleoside-2′,3′-phosphates, in which the phosphoryl group is linked to the hydroxyls at both the 2′ and 3′ positions. When synthetic cyclic phosphates of this kind were tested as substrates of pancreatic ribonuclease, only the cytidine or uridine compounds were cleaved, and the sole product was the nucleoside-3′-phosphate. Moreover, when the 2′- and 3′-forms of synthetic phosphodieters of these pyrimidine nucleosides (the phosphoryl unit also bore a benzyl group) were tested, only the 3′-form proved to be a ribonuclease substrate.[73] Together with the parallel finding by Waldo Cohn and Elliot Volkin that a phosphodiesterase from snake venom cleaves yeast nucleic acid with the formation of nucleoside-5′-phosphates,[74] these findings strongly supported the idea of the 3′,5′ mode of internucleotide linkage.

After the identification and purification of ribonucleases from biological sources other than beef pancreas, the substance crystallized by Kunitz was named ribonuclease A. In 1947 Kunitz also crystallized a deoxyribonuclease A from beef pancreas, and many other deoxyribonucleases were found later.[75] As in the study of protein structure, the use of these enzymes was decisive in the nucleic acid field after the acceptance of the idea that the nucleic acids are long-chain polynucleotides.

In 1931 Levene and Bass stated that "it must be borne in mind that the true molecular weight of nucleic acids is as yet not known. The tetranucleotide theory is the minimum molecular weight and the nucleic acid may well be a multiple of it."[76] During the 1930s, however, organic-chemical studies on nucleic acids were based on the assumption that they are simple linear tetranucleotides, not polymeric macromolecules. For example, when Katashi Makino reported in 1935 that, per atom of phosphorus, yeast nucleic acid behaved as a monobasic acid, he supported the idea advanced by Hitoshi Takahashi that it is not a linear but a cyclic tetranucleotide.[77] This idea is similar to that entertained during the 1920s about cyclic dipeptide units as structural elements of proteins, mentioned earlier.

Miescher had found that his salmon nuclein did not diffuse through a parchment membrane but, when nucleic acids were thought to be tetranucleotides this colloidal behavior was attributed to the association of small units to form nondiffusible aggregates. The recognition that nucleic acids are macromolecules came during the 1930s through studies of the physical properties (viscosity, rate of diffusion, rate of sedimentation) of carefully prepared samples of calf thymus DNA, such as those described in 1924 by Einar Hammarsten. With Rudolf Signer and Torbjörn Caspersson, he reported in 1938 that this DNA behaved like thin rods, whose length was about 300 times its width and whose molecular weight was between 500,000 and 1,000,000.[78] Later DNA preparations were found even larger and more asymmetric, but around 1950 there was no agreement whether these enormous numbers represent a single "molecule" or an aggregate of relatively small polynucleotide chains.[79]

As Miescher had recognized qualitatively during the course of his work on salmon nuclein, the physical properties of DNA preparations were found to depend on the methods used to isolate them, and a succession of studies during the 1940s on the acid-base properties of nucleic acids, notably by John Gulland and Denis Jordan, showed that extremes of acidity or alkalinity could lead to degraded DNA preparations. In particular, Gulland likened the degradation of calf thymus DNA to protein denaturation, and concluded that in its undegraded form the polynucleotide chains are held together by hydrogen bonds whose rupture by acid or alkali causes a decrease in molecular size.[80]

By 1945 it was also evident that the older RNA preparations represented partially degraded polyribonucleotides, owing to the use of alkali and the presence of ribonucleases in the tissue extracts. Earlier estimates of the molecular weight of yeast RNA ranged from about 1,500 (close to the theoretical value for a tetranucleotide) to about 23,000 (corresponding to about seventy nucleotide units). In 1935 Wendell Stanley reported that he had isolated, from the juice of Turkish tobacco plants infected with tobacco mosaic virus (TMV), a crystalline protein that caused the appearance of the disease in healthy plants.[81] It had been known since 1899, through the work of Dmitri Ivanovski and Martinus Beijerinck, that the characteristic leaf mottling is associated with a factor that can pass through the filter designed in 1884 by Charles Chamberland to retain bacteria.[82] During the succeeding decades, many diseases of plants, animals, and man were found to be caused by such filterable factors, and a variety of names were assigned to this class of infective agents: *contagious living fluid* (Beijerinck), *filterable viruses* (Wolbach), *ultraviruses* (Levaditi), among others. By the 1930s these terms had been replaced by *viruses,* defined as submicroscopic, filterable pathogenic agents that could multiply only within living cells. It was also recognized that the "bacteriophage" discovered by Frederick Twort (1917) and Félix d'Herelle (1919) represents a class of viruses that infect bacterial cells.[83]

In his isolation of the apparently crystalline TMV, Stanley used methods like precipitation with ammonium sulfate and control of pH, which had enabled John Northrop and Moses Kunitz (also at the Rockefeller Institute in Princeton, N.J.) to crystallize proteolytic enzymes. The virus had the solubility properties of a protein, and its apparent molecular weight as determined in the ultracentrifuge was found to be about 49 million. Stanley's report was quickly followed by that of Bawden and Pirie, who discovered that their preparation of TMV contained "0.5 per cent phosphorus and 2.5 per cent carbohydrate. The last two constituents can be isolated as nucleic acid of the ribose type from protein denatured by heating."[84] Although

Stanley then claimed that "it was possible to remove the nucleic acid and to obtain phosphorus-free protein possessing virus activity,"[85] several other plant viruses were also obtained in the form of crystalline or liquid-crystalline "nucleoproteins" having a ribonucleic acid component. The question whether the nucleic acid is essential for infectivity was debated for many years, but in 1956–1957 Gerhard Schramm and Heinz Fraenkel-Conrat (with their coworkers) independently showed the infectivity of TMV to be associated with its RNA component.[86] This shift in emphasis from the protein part of the virus to its nucleic acid paralleled the growing acceptance, after 1945, of the central role of nucleic acids in heredity, and that development will be considered later. It should be mentioned here, however, that before 1930 Francesco Sanfelice had called attention to the possibility that the fowl pox virus he had studied, as well as the chicken sarcoma virus discovered by Peyton Rous, were "nucleoproteids."[87]

In relation to the chemistry of ribonucleic acids, the discovery that some plant viruses are nucleoproteins was important in providing a source of RNA preparations in a relatively undegraded state. In their report on the physical properties of the RNA derived from TMV, Seymour Cohen and Wendell Stanley stated that the data for their preparation gave a value for its molecular weight of about 300,000.[88] Subsequent studies, with improved methods, gave even higher values, and there could be no doubt that, like thymus DNA, virus RNA is a macromolecule. Another important consequence of the studies on TMV and other plant viruses was the demonstration that in these cases one was dealing with natural assemblies of RNA and protein molecules rather than "protein nucleates" produced by the interaction of acidic nucleic acids with basic protein units during the isolation procedure. Such artificial combinations had been made in the past, and the question was raised repeatedly whether the term *nucleoprotein* was appropriate in all cases where a protein-nucleic acid complex was isolated.[89] Kossel and later investigators had provided analytical data showing that the protamines from the sperm of various fishes (salmon, herring, sturgeon, for example) differ in amino acid composition but are all characterized by a high content of arginine, along with lysine and histidine, and an acid-base type of interaction with chromosomal DNA was generally assumed. The histone Kossel found in the cell nuclei from several animal tissues was for a time thought to be a basic protein of a single type, but work during the 1950s showed such preparations to be composed of at least two different kinds of proteins, an "arginine-rich" type and a "lysine-rich" type. Subsequent studies, to be considered later, have begun to clarify some aspects of the interaction of the histones and other proteins with chromosomal DNA.

In addition to the conclusions drawn from the physical-chemical data acquired during the 1940s, the discussion about the structure of nucleic acids focused on the question of whether the presumed polynucleotide chains represent a random sequence of nucleotide units or a succession of tetranucleotide units. After a careful review of the conflicting chemical data for various nucleic acid preparations, Gulland stated: "There is at present no indisputable chemical evidence that the nucleotides are arranged in other than a random manner in the polynucleotides. It must be realized that the existence of tetranucleotide units, repeated throughout the molecule, would limit the potential number of isomers and hence diminish the possibilities of biological specificity. Thus the immense number of variations presented by a polynucleotide in which the nucleotides occur in a random sequence is reduced, in the case of an unbranched polynucleotide, to a single possible structure, varying only in length, if the same, uniform tetranucleotide occurs throughout."[90] The decisive evidence on this question came during 1946–1950 from the first reliable quantitative analyses of the base composition of nucleic acids, by Erwin Chargaff and his associates.

Although the introduction in 1944 of partition chromatography on paper strips was initially applied to the study of the amino acid composition of protein hydrolysates, the method was immediately adopted for the separation and identification of a large variety of biochemical substances, including the purines, pyrimidines, nucleosides, and nucleotides derived from nucleic acids. The work of the Chargaff group was also greatly aided by the general availability after 1945 of the photoelectric ultraviolet spectrophotometer. It had long been known from the work of Jacques Soret, Walter Hartley, and Charles Dhéré that purines and pyrimidines absorb ultraviolet light near 2600 Å, and Dhéré had described the absorption spectrum of yeast nucleic acid.[91] Before World War II, however, analytical spectrophotometers with quartz optics were highly specialized instruments and not generally available in biochemical laboratories. As we have seen, Otto Warburg had such an instrument, and during the 1930s Torbjörn Caspersson used one in his cytochemical study of nucleic acids. The method had been introduced in 1904 by August Köhler, who showed that cell nuclei absorb ultraviolet light; this was noted by William Bayliss, who added that "unfortunately, the method has not yet been made much use of, owing to the necessarily elaborate nature of the apparatus required."[92]

In 1950 Chargaff summarized the extensive analytical data collected in his laboratory for a number of carefully prepared DNA samples from various biological sources, including beef thymus, spleen, and liver; human sperm; yeast; tubercle bacilli. He stated that "we started in our work from the assumption that nucleic acids were complicated and intricate high polymers, comparable in this respect to the proteins, and that the determination of their structures and their structural differences would require the development of methods suitable for the precise analysis of all constituents of nucleic acids prepared from a large number of cell types."[93] The analytical data showed that the adenine, guanine, cytosine, and thymine were not present in equimolar proportion in the various DNA preparations, as had been claimed by Steudel in 1906 and reiterated by Levene in 1931. Moreover, the composition of the hydrolysates of DNA preparations from different biological sources differed widely, whereas that of the preparations from different tissues of the same biological species was the same. Chargaff concluded that "the desoxypentose nucleic acids extracted from different species thus appear to be different substances or mixtures of closely related substances of a composition constant for different organs of the same species and characteristic of the species. The results serve to disprove the tetranucleotide hypothesis." This statement was followed by another, which later assumed great importance in the historiography (and mythology) of molecular biology: "It is, however, noteworthy—whether this is more than accidental, cannot yet be said—that in all the desoxypentose nucleic acids examined thus far the molar ratios of total purines to total pyrimidines, and also of adenine to thymine and of guanine to cytosine, were not far off from 1."[94] A year later, Chargaff went a bit further: "As the number of examples of such regularity increases, the question will become pertinent whether it is merely accidental or whether it is an expression of certain structural principles that are shared by many desoxypentose nucleic acids, despite far reaching differences in their individual composition and the absence of a recognizable periodicity in their nucleotide sequence. It is believed that the time has not yet come to attempt an answer."[95]

As a reader of Chargaff's papers who had been a daily witness during the 1930s of Carl Niemann's efforts to produce analytical data in support of Max Bergmann's ill-fated periodicity theory of protein structure, I found to be wholly admirable these cautious statements of a theory of what later came to be called base-pairing in nucleic acids. The "hypnotic power of numerology" was later evident in the formulation of the genetic code, with four nucleotides and twenty amino acids providing the magic numbers.[96] We will consider that development later.

To sum up, the chemical achievements of the 1940s and 1950s established unequivocally the long-chain structure of RNA and DNA molecules in which the nucleotides are joined by 3′-5′ phosphodiester bonds, as well as the multiplicity and individuality of the nucleic acids in terms of their base composition. These achievements have been overshadowed in most historical accounts by the formulation in

1953 of the double-helical model of DNA by James Watson and Francis Crick.[97] Because of its immediate relevance to the problem of the chemical basis of heredity, that model attracted greater attention, especially after it received experimental support later in the 1950s. I will consider that development more fully shortly, and I only mention here two key chemical features of the model. One was the more precise definition of the concept of base-pairing than that implicit in Chargaff's conclusions from his analytical data, and the other was the idea of a two-strand complex in which one DNA strand runs in the $3' \rightarrow 5'$ direction, and the other in the "complementary" $5' \rightarrow 3'$ direction. These concepts were decisive in the successful approach during the 1960s to the still unsolved organic-chemical problem of the nucleotide sequences of nucleic acids.

The strategy of such sequence determinations resembled that of Sanger in his work on insulin. A large variety of purified nucleases were used to effect specific cleavages, and new chromatographic methods were introduced to separate the oligonucleotide fragments. To determine overlapping sequences, enzymatic or chemical procedures were developed involving the use of compounds labeled with ^{32}P or fluorescent substituents, and new electrophoretic methods were introduced to separate the labeled products and to identify the sequence of each fragment.[98] Among the first successes was the determination of the complete nucleotide sequences of relatively small (about 80 nucleotide units) RNA molecules by Robert Holley and his associates.[99] Two ingenious methods for the "sequencing" of DNA oligonucleotides were devised by Sanger and by Walter Gilbert, with their respective associates.[100] Sanger's procedure involves the use of a DNA polymerase (of which more later), a synthetic oligonucleotide primer, and the four $2',3'$-dideoxyribonucleoside triphosphates corresponding to the ^{32}P-labeled substrates (dATP, dGTP, dCTP, dTTP) in the polymerization reaction. The incorporation of the dideoxy compound terminates the process. In Gilbert's method, a DNA strand labeled with ^{32}P at one end is selectively cleaved by various chemical reagents at the two purines and the two pyrimidines. To these procedures was then added the one stemming from the exciting discovery by Werner Arber, Daniel Nathans, and Hamilton Smith of bacterial deoxyribonucleases ("restriction endonucleases") that interact specifically with interior nucleotides in both chains of a DNA duplex and cleave phosphodiester bonds in both chains to produce protruding oligonucleotide units.[101] I apologize for the inadequate description of these techniques, and only wish to emphasize that in its present form, the practice of "molecular biology" in the massive effort to "map the human genome" depends heavily on the use of purified enzymes, organic-chemical techniques, and new computerized instrumentation for the automatic sequencing of nucleic acids.

I return to the recognition during the 1930s that the plant viruses studied by Stanley and by Bawden and Pirie are large nucleoproteins. This came soon after the development of the electron microscope. After its invention in 1931 by Max Knoll and Ernst Ruska, several years passed before a resolving power of about 10 nm was attained.[102] Since physical-chemical studies on solutions of TMV had indicated it to consist of particles about 400 nm long and 12 nm wide, this instrument offered obvious opportunities for the direct measurement of the size and shape of viruses. Such studies were reported in 1941 by Thomas Anderson, and with the subsequent development of the "shadow-casting" technique by Robley Williams, electron microscopy became an important method for the examination of viruses, as well as DNA, proteins, and intracellular structures. For example, Cecil Hall showed that suitably prepared samples of salmon sperm DNA are very long fibrils about 20 nm wide.[103] There was also the remarkable finding by Salvador Luria and Anderson and by Helmuth Ruska that bacteriophages of the "T" group, which infect *Escherichia coli* B, are tadpolelike structures consisting of an oval body (about 65 × 80 nm) with a long tail (about 20 × 200 nm). This observation formed part of the background for the famous experiment of Alfred Hershey and Martha Chase, to be discussed later.[104]

William Astbury and Florence Bell, also working in the 1930s, interpreted their X-ray data on fibers prepared from calf thymus DNA to indicate that "the spacing of 3.3_4 Å along the fibre axis corresponds to a close succession of flat or flattish nucleotides standing out perpendicularly to the long axis of the molecule to form a relatively rigid structure."[105] Moreover, some ten years later Astbury concluded from their X-ray data that

> the pattern repeats itself along the axis of the molecule at a distance corresponding to the thickness of eight nucleotides or a multiple of eight nucleotides—most probably eight or sixteen nucleotides. The least possible value of the fibre period is 27 Å. . . . It is hardly likely . . . that the fact that the intramolecular pattern is found to be based on a multiple of four nucleotides is unrelated to the conclusion that has been drawn from chemical data that the molecule is composed of four different kinds of nucleotides in equal proportions. It seems improbable, too, to judge by the perfection of the X-ray fibre diagram, that these four different kinds of nucleotides are distributed simply at random.[106]

It is easy, in hindsight, to dismiss this statement, made at a time when Chargaff had begun to provide analytical data which led to the abandonment of the tetranucleotide hypothesis of DNA structure, and precise data were not available for the molecular dimensions (bond lengths and bond angles) for the constituents of nucleic acids. In the protein field, such data were collected during the 1940s by Robert Corey for amino acids and peptides and provided the basis for the demonstration by him and Linus Pauling that a significant contribution could be made to the understanding of protein structure by the construction of models in which careful account was taken of the interatomic distances and bond angles of hypothetical peptide chains, and by examination of the ways such models could be folded so as to permit maximum hydrogen bonding between -CO-NH- groups, either between parts of a single chain or between adjacent chains.[107] In particular, Pauling and Corey called attention to the stability of a helical structure ("α-helix") that, as has been mentioned, was subsequently found by John Kendrew to be a distinctive feature of the myoglobin molecule. Although speculations had been offered earlier regarding the helical coiling of long peptide chains, these proposals were either vague or incompatible with the molecular dimensions determined through X-ray studies on crystalline peptides.

In the nucleic acid field, by 1951 such studies had given the dimensions of adenine and guanine, whose rings were found by June Broomhead to be almost identical in size and shape, and Sven Furberg had determined the three-dimensional structure of cytidine. In this nucleoside the pyrimidine ring and the ribofuranose ring are approximately perpendicular, thus invalidating Astbury's conclusion that the nucleotide units of nucleic acids are flat structures. Furberg then described molecular models for a hypothetical polynucleotide in which most of the atoms constituting each nucleotide (including the phosphorus atom) lie in planes 3.4 Å apart, thus explaining the strong 3.3_4 reflection reported by Astbury and Bell. In one of Furberg's two models, the ribofuranose rings and the phosphorus atoms form a spiral that encloses a column with the purines and pyrimidines almost stacked on top of each other; in the other, the ribose phosphate chain forms a central column from which the purine and pyrimidines stand out perpendicularly.[108] A year later, shortly before the appearance of the first paper by Watson and Crick, there appeared one by Pauling and Corey, who suggested a structure for DNA in which each of three polynucleotide chains is coiled into a helix, and that the three helices are intertwined; in their model, the phosphoryl groups are closely packed about the axis of the column, and the nitrogenous bases project radially, as in Furberg's second model.[109] What needs emphasis at this point in our story is that neither Furberg nor Pauling and Corey took account of Chargaff's analytical data.

In their first joint paper, Watson and Crick pointed out the deficiencies of the model proposed by Pauling and Corey and

> put forward a radically different structure for the salt of deoxyribose nucleic acid. This structure has two helical chains each coiled round the same axis. . . . The two chains (but not their bases) are related by a dyad perpendicular to the fibre axis. Both chains follow right-handed helices, but owing to the dyad the sequences of the atoms in the two chains run in opposite directions. Each chain loosely resembles Furberg's model No. 1; that is, the bases are on the inside of the helix and the phosphates on the outside. The configuration of the sugar and the atoms near it is close to Furberg's "standard configuration," the sugar being roughly perpendicular to the attached base. There is a residue on each chain every 3.4 Å in the *z*-direction. . . . The novel feature of the structure is the manner in which the two chains are held together by the purine and pyrimidine bases. The planes of the bases are perpendicular to the fibre axis. They are joined together in pairs, a single base from one chain being hydrogen-bonded to a single base from the other chain, so that the two lie side by side with identical coordinates. One of the pair must be a purine and the other a pyrimidine for bonding to occur. . . . These pairs are adenine (purine) with thymine (pyrimidine), and guanine (purine) with cytosine (pyrimidine).[110]

This paper was immediately followed in *Nature* by two others, dealing with X-ray studies on DNA fibers. The first, by Maurice Wilkins, Alexander Stokes, and Herbert Wilson, reported that diffraction patterns similar to those given by calf thymus DNA had been obtained with nucleoproteins from sperm heads or T2 bacteriophage, indicating that such structures are present in biological systems and are not artifacts of preparation. The authors also discussed the available X-ray data in terms of a helical structure of DNA. In the second paper, Rosalind Franklin and Raymond Gosling (their colleagues at King's College, London), presented striking X-ray photographs of two forms (A and B) of calf thymus DNA and concluded that the diffraction pattern of the B form was compatible with the Watson-Crick model; a few months later they showed that the data for the more regularly ordered A form also were consistent with a double-helical structure.[111]

In their first joint paper, Watson and Crick stated that "it has not escaped our notice that the specific pairing we have postulated immediately suggests a possible copying mechanism for the genetic material,"[112] and this idea was developed in a paper that appeared a few weeks later: "The phosphate-sugar backbone of our model is completely regular, but any sequence of the pairs of bases can fit into the structure. It follows that in a long molecule many different permutations are possible, and it therefore seems likely that the precise sequence of bases is the code which carries the genetic information. If the actual order of the bases on one of the pairs of chains were given, one could write down the exact order of the bases on the other one, because of the specific pairing. Thus one chain is, as it were, the complement of the other, and it is this feature which suggests how the deoxyribonucleic acid might duplicate itself."[113] Five years later, striking experimental evidence in support of this theory of DNA replication was provided by Matthew Meselson and Franklin Stahl and by Arthur Kornberg and his associates. More will be said about these important contributions.

The fame rightly accorded to the Watson-Crick model of DNA has elicited a still-continuing discussion of the reception it received during the 1950s. The discourse was enlivened by the appearance in 1968 of Watson's book *The Double Helix*. Questions were raised regarding its accuracy in several matters—such as whether there was, in fact, a race with Linus Pauling—and there was outrage at Watson's cruel description of Rosalind Franklin, as well as at his comments about David Keilin. I will say no more about the book, except to express my opinion that it tells as much about the author as about the scientific development in which he played a part.[114]

Over the years there has been a series of celebrations of the appearance of the 1953 papers by Watson and Crick, the most recent ones in 1993. In his contribution on one of these occasions Crick stated:

> First and foremost, I should remind you of Rosalind Franklin, whose contributions have not been sufficiently acknowledged in these meetings on the fortieth anniversary of the discovery. It was Rosalind who clearly showed the existence of the two forms of DNA—the A and the B form. It was Rosalind who painstakingly determined the density, the exact cell dimensions, the symmetry of the A form, evidence that suggested very strongly that the structure had two chains (not just one), running in opposite directions. . . . On the chemical side it would not have been possible to build correct models without the general chemical formula for DNA, established largely by the work of Lord Todd and his colleagues. But perhaps the most vital information was provided by the careful work of Erwin Chargaff, which led him to his rule for the relative amounts of the four bases.[115]

On another such occasion, however, Francis had this to say about the reception of the double helix: "The reaction of many biochemists, such as Joseph Fruton, ranged from coolness to muted hostility. They had long considered the biochemistry of the gene to be based on proteins, not nucleic acids, and thought the problem far too difficult to tackle in the immediate future. It did not help that the structure had been put forward by two people who were obviously not card-carrying biochemists."[116] I cannot allow this assertion to go unchallenged, not only on personal grounds but also because it perpetuates a myth about the past views of "card-carrying" biochemists about the nature of the gene. As for my own view, before about 1940 I was more concerned with defending the basic concepts of protein chemistry in the face of transitory fashions such as the "cyclol" and "periodicity" hypotheses than about the chemical nature of the gene. When my friends Colin MacLeod and Maclyn McCarty at the Rockefeller Institute told me about the work they were doing with Oswald Avery on the pneumococcal transforming factor, I did not have to change my belief in accepting the idea that the gene might be DNA. As for my opinion of the Watson-Crick hypothesis, when Sofia Simmonds and I were completing the second edition of our *General Biochemistry* in 1957, we wrote:

> Although much still remains to be done to purify and characterize individual nucleic acids, stimulating hypotheses have been advanced about the manner in which the polynucleotide chains are arranged in space. The "pairing" of the nitrogenous bases in DNA preparations, together with X-ray diffraction data, have provided the basis for an ingenious speculation by Crick and Watson, who have proposed a helical structure for DNA. It is assumed that two polynucleotides are coiled in such a manner that an adenine of one chain is hydrogen-bonded to a thymine of the other, . . . and a guanine of one chain is bonded to a cytosine of the other. . . . The recognition of the role of RNA and DNA in the intracellular synthesis of specific proteins has led to stimulating speculations about the role of nucleic acids as "templates" in protein synthesis. In particular, the DNA model of Crick and Watson has been assumed as a basis for further hypotheses about the manner in which the specific structure of a DNA molecule might cause the specific alignment of activated amino acid units in the sequence present in the completed protein. However, much further experimental work is needed on the chemical structure of individual nucleic acids, and on the enzymic mechanisms of protein synthesis, before the status of such hypotheses can be properly assessed.[117]

There is cautious skepticism in these somewhat awkward sentences, written before the publication of the Meselson-Stahl paper, but no "coolness" or "muted hostility" was expressed or intended. I can only surmise that Francis, who showed me much kindness during my stay in Cambridge during 1962–1963, may have accepted too readily some idle gossip. For the record I must add that in a volume published in

1955, Erwin Chargaff wrote as follows about the Watson-Crick model: "This hypothesis . . . has much to recommend itself on aesthetical grounds; it makes good use of several experimentally established facts. Whether it does more than to describe the structure of that portion of the processed preparation for which diffraction patterns were obtained, remains, however, to be established."[118] Although neither Chargaff nor I were professional historians, and his familiarity with the nucleic acid field far exceeded mine, we were both "card-carrying" biochemists of a generation that considered some knowledge of the history of that field to be a requisite in our education and in our role as teachers. A part of that education was acquaintance with the many beautiful theories that stimulated productive research but were extensively modified in the process.

The other part of Francis's statement, about the adherence of biochemists to the protein theory of the gene, is an item of mythology that belongs with the canard that they considered cells to be "bags of enzymes." My reading of of the historical record, from Miescher to Watson and Crick, tells me that the chief adherents of that theory were members of the genetics community, not the biochemists, and this explains to me why George Beadle felt impelled to write: "I have said many times that I regard the working out of the detailed structure of DNA one of the great achievements of biology in the twentieth century, comparable in importance to the achievements of Darwin and Mendel in the nineteenth century. I say this because the Watson-Crick structure immediately suggested how it replicates or copies itself with each cell generation, how it is used in development and function, and how it undergoes the mutational changes that are the basis of organic evolution."[119] To see this judgment in historical perspective, it is necessary to sketch the development of genetics during the first half of the twentieth century, and to consider the speculations and experimental results relating to the chemical nature of the genetic material after the "rediscovery" of Gregor Mendel.

The Theory of the Gene

In his book *Intracelluläre Pangenesis*, published in 1889, the botanist Hugo de Vries wrote that "just as physics and chemistry go back to molecules and atoms, the biological sciences have to penetrate to these [hereditary] units in order to explain, by means of their combinations, the phenomena of the living world."[120] The hypothesis he put forward included the following points:

> Hereditary units are independent units, from the numerous and various groupings of which specific characters originate. Each of them can vary independently from the others; each can of itself become the object of ex-

> perimental treatment in our culture experiments. . . . According to the hypothesis concerning their nature, these units have been given different names. For the one adopted by me I have chosen the name, pangen. . . . The pangens are not chemical molecules, but morphological structures, each built up of numerous molecules. They are life-units, the characters of which can be explained in an historical way only. . . . At each cell division every kind of pangen present is, as a rule, transmitted to two daughter cells. . . . An altered numerical relation of the pangens already present, and the formation of new kinds of pangens must form the two main factors of variability.[121]

During the succeeding years de Vries conducted hybridization experiments on some twenty plant species belonging to different families to test his hypothesis, and in 1900 he reported that each case met a "law of segregation" *(Spaltungsgesetz, loi de ségrégation)* of the hereditary units. Although he adopted Darwin's term *pangenesis,* because de Vries considered the concept of independent segregation to offer an explanation of the mechanism of evolution, his theory had no place for Darwin's *gemmules,* and he later offered a "mutation" theory as an alternative to the theory of natural selection. In the 1900 paper de Vries also stated that his conclusions had been "put forward a long time ago by Mendel for a special case (peas)," and after citing Gregor Mendel's 1865 paper (published in 1866), he added the footnote: "This important report is cited so seldom that I only first learned of it after I had completed most of my experiments and had derived the conclusions communicated in the text."[122]

Immediately afterward, Carl Correns reported his hybridization experiments with peas and maize, observing that "in the sixties, the abbot Gregor Mendel in Brünn, through very extensive experiments lasting many years with peas, had not only obtained the same result as de Vries and I, but had given exactly the same explanation, so far as it was possible to do so in 1866."[123] A few weeks later there appeared another report on the hybridization of peas, by Erich Tschermak von Seysenegg, with the postscript: "The experiments published by Correns just now, also dealing with the artificial crossing of different varieties of *Pisum sativum* and observations of hybrids allowed to undergo self-fertilization for several generations, confirm as do mine the Mendelian theory. The simultaneous 'discovery' of Mendel by Correns, de Vries and myself is especially pleasing to me. Even during the second year of research, I thought that I had found something new."[124] With the emergence of genetics as a prominent biological discipline there has grown up a Mendel "industry" among historians of science, and only some of the highlights of their accounts and interpretations can be cited here.

Mendel found that when the pollen of a variety (for example, tall plants) of *Pisum sativa* was artificially placed on the stigma of a variety with a "differentiating character" *(differenzierendes Merkmal)* such as short plants, the resulting seeds produced hybrids (F_1 generation) that were tall (the dominant character). If the hybrids were allowed to undergo self-fertilization, the seeds gave rise to plants (F_2 generation) of the two original parental varieties in the ratio of 3 tall to 1 short. Upon self-fertilization, the short F_2 plants produced only seeds leading to the same variety, whereas the tall F_2 plants gave rise to both varieties in the ratio of 3 tall to 1 short. The latent character (in this case, short plants) was termed the recessive member of an antagonistic pair. Mendel investigated other pairs of carefully selected characters, such as round vs. wrinkled peas, green vs. yellow peas, and concluded that "it is now evident that the hybrids having each of the two differentiating characters form seeds of which one half again develops the hybrid form, whereas the other [half] gives the other plants, which remain constant, and receive in equal parts the dominant and recessive character."[125]

In addition to this "law of segregation," Mendel derived what came to be called his second "law of independent assortment." Thus if hybrids were obtained by the mating of two parents differing in two characters (for example, tall plants and round seeds vs. short plants and wrinkled seeds), the F_1 generation was tall and had round seeds, the latter being dominant over wrinkled seeds. Upon self-fertilization, the 3:1 ratio in the F_2 generation was maintained for each antagonistic pair of characters, and per 16 plants there were 9 tall-round, 3 tall-wrinkled, 3 short-round, and 1 short-wrinkled. Mendel considered his results to depend on "the material composition and arrangement of the elements in the [fertilized egg] cell in a viable union"[126] but offered no speculation about the chemical nature of these "elements."

Over the years there has been extensive discussion of the possible reasons for the neglect of Mendel's paper until 1900. Special emphasis has been placed on the negative reaction of the influential botanist Carl Nägeli, expressed in his correspondence with Mendel, and it is significant that Nägeli did not mention Mendel in his 1884 book on heredity.[127] There has also been debate over the question whether Mendel was primarily a hybridist and was not trying to discover new laws of inheritance, and about Ronald Fisher's claim that Mendel's data "were too good to be true."[128] It should be recalled that between 1870 and 1900, cytological studies had led to the recognition of the role of the cell nucleus and its chromatin-containing chromosomes as possible bearers of hereditary determinants. This new knowledge prepared botanists such as de Vries, Correns, and Tschermak, who were studying plant hybrids, to appreciate the importance of Mendel's work, once they learned of it.

Moreover, by the 1890s, Nägeli's concept of blending inheritance and his micellar theory of an albuminoid protoplasm had lost most of their earlier appeal after the acceptance of the ideas of August Weismann. To this I would add that Mendel's style of research, with careful study of one character at a time, his handling of the numerical data, and his choice of *Pisum sativa,* seems to have placed him somewhat above the hybridists—Joseph Koelreuter, Karl Friedrich Gaertner—whose tradition he continued, as well as his contemporary, Charles Naudin. It may never be known whether Mendel formulated his theory long before all the data had been gathered, and to what extent the collection of the data was influenced by his "preconception of what might be true," but these uncertainties do not alter the fact that after 1900 Mendel's laws became part of the foundation of a distinctive new discipline, which provided an experimental basis for the study of the mechanism of biological evolution.[129]

In 1902 William Bateson introduced "Mendelism" in Britain and proposed a terminology that was quickly adopted:

> By crossing two forms exhibiting antagonistic characters, cross-breeds were produced. The generative cells of these cross-breeds were shown to be of two kinds, each being pure in respect of *one* of the parental characters. This purity of the germ-cells, and their inability to transmit both of the antagonistic characters, is the central fact proved by Mendel's work. Such characters we propose to call *allelomorphs* [later shortened to *alleles*], and the zygote formed by the union of a pair of opposite allelomorphic gametes, we shall call a *heterozygote.* Similarly, the zygote formed by the union of gametes having similar allelomorphs, may be spoken of as a *homozygote.*[130]

Bateson also suggested the use of "F_1" and "F_2" to denote the first and second filial generations, and in 1902 he established the validity of Mendel's law of segregation in animals (poultry). Shortly afterward Lucien Cuénot and William Castle showed its applicability to the inheritance of hair color in mice.[131] During the course of this early work, many apparent exceptions were found, however, to the law of independent assortment; a convincing explanation later came from the work of Thomas Hunt Morgan and his associates.

Of special importance to our story is the 1902 report of the clinician Archibald Garrod on the inheritance of alcaptonuria. After collecting data on the incidence of the condition among members of individual families, he concluded that "a very large proportion are children of first cousins" and offered a Mendelian explanation by quoting Bateson: "We note that the mating of first cousins gives exactly the condition most likely to enable a rare, and usually recessive character to show itself.

If the bearer of such a gamete mate with individuals not bearing it the character will hardly ever be seen; but first cousins will frequently be the bearers of similar gametes, which may in such unions meet each other and thus lead to the manifestation of the peculiar recessive characters in the zygote."[132] In his paper Garrod suggested that two other "chemical abnormalities"—albinism and cystinuria—might belong to the group of congenital conditions represented by alcaptonuria, and in lectures he gave in 1908 he added pentosuria. Thus not only did Garrod provide the first evidence of the applicability of Mendelian genetics to humans, but, as we noted earlier, he perceived the relationship of hereditary factors to enzymes. That this meeting of two newly emergent scientific disciplines—biochemistry and genetics—was fathered by a physician serves as a reminder of the role that medical (and agricultural) practice has played in the development of fundamental biological knowledge.

In this connection, it may be appropriate to recall that in 1896 Edmund Wilson stated:

> In its physiological aspect . . . inheritance, in successive generations, is of like forms of metabolism; and this is effected through the transmission from generation to generation of a specific substance or idioplasm which we have reason to identify with chromatin. This remains true however we may conceive the morphological nature of the idioplasm—whether as a microcosm of invisible germs or pangens, as conceived by de Vries, Weismann, and Hertwig, as a storehouse of specific ferments as Driesch suggests, or as a complex molecular substance grouped in micellae as in Nägeli's hypothesis. It is true, as Verworn insists, that the cytoplasm is essential to inheritance; for without a specifically organized cytoplasm the nucleus is unable to set up specific forms of synthesis.[133]

In 1906 Bateson proposed that the area of science dealing with heredity and variation be named genetics. It was recognized that Mendel's approach required the use of large numbers of carefully selected organisms if statistically significant quantitative results were to be obtained. This attitude was emphasized by Wilhelm Johannsen, who stated, however, that "we pursue the study of heredity *with* mathematics, not *as* mathematics."[134] To denote more precisely the concepts emerging from the new science, he proposed that "one may designate the statistically appearing type as the appearance type or briefly and clearly, as the phenotype. Such phenotypes are in themselves measurable realities; just what can be typically observed; hence for a series of variations the centers about which the variants group themselves. In using the word phenotype the necessary reservation is only made that no further conclusion may be drawn from the appearance itself."[135] To denote the biological unit in the egg and sperm cells that determine the character of the progeny, Johanssen used the word *Gen* (English, *gene*) by dropping the first syllable of the *pangen* of Darwin and de Vries, and he emphasized the distinction between the concept of the genotype (the genetic constitution of the individual organism) and that of the phenotype. Although Johannsen noted that "no definite idea about the nature of the 'genes' is at present sufficiently grounded," he suggested that "they may be tentatively considered to be chemical factors of various kinds. . . . One may expect principally from general physical chemistry the viewpoints for theories regarding the action of chemical hereditary factors."[136]

I return to the years just before 1900 to recall the cytological discoveries that prepared the ground for the recognition of the importance of Mendel's laws. It had been demonstrated for several species that the number of chromosomes per nucleus is constant and usually even, with equal numbers coming from the sperm and the egg; at cell division, during mitosis, each chromosome appears to divide longitudinally, followed by nuclear division to give two nuclei each having a complete (diploid) set of chromosomes; in the special case of the germ cells, there occurs during meiosis a reduction in chromosome number during the last two divisions before the production of mature gametes, so that each egg or sperm has one-half (haploid) the full set of chromosomes; the maternal and paternal members of homologous sets of chromosomes form pairs (synapsis). After 1900 several investigators, notably Theodor Boveri and Walter Sutton, sought to define more clearly the relation between chromosomes and hereditary factors, and Sutton wrote:

> We have seen reason . . . to believe that there is a definite relation between chromosomes and allelomorphs or unit characters but have not before inquired whether an entire chromosome or only part of one is to be regarded as the basis of a single allelomorph. The answer must unquestionably be in favor of the latter possibility, for otherwise the number of distinct characters possessed by an individual could not exceed the number of chromosomes in the germ-products; which is undoubtedly contrary to fact. We must, therefore, assume that some chromosomes at least are related to a number of different allelomorphs. If then, the chromosomes retain their individuality, it follows that all the allelomorphs represented by any one chromosome must be inherited together.[137]

The suggestion that an individual character is associated with a particular chromosome had been put forward in 1901 by Sutton's teacher Clarence McClung, who proposed that an unpaired accessory chromosome, reported by Hermann Henking in 1891, carries the factor for maleness. Because the status of the body observed by

Henking was uncertain, it was denoted X. Soon afterward, Nettie Stevens and Edmund Wilson independently found that in addition to an X chromosome, some organisms also have an unlike smaller partner (named Y), while in other organisms the X chromosome is unpaired.[138] Their finding was quickly confirmed by other investigators. These sex chromosomes were to play a large role in the study of the mechanism of Mendelian inheritance, and the ideas expressed by Sutton received decisive experimental support in the work of Thomas Hunt Morgan and his associates on the transmission of sex-linked characteristics.

Much has been written by geneticists and historians of science about the achievements of the Morgan group at Columbia University, which included Alfred Sturtevant, Calvin Bridges, and Hermann Muller, and I will refer to some of these writings later. The remarkable series of publications from Morgan's laboratory on the genetics of the vinegar fly *Drosophila melanogaster* opened with a brief paper that began as follows:

> In a pedigree culture of Drosophila which had been running for nearly a year through a considerable number of generations, a male appeared with white eyes. The normal flies have red eyes. The white-eyed male, bred to his red-eyed sisters, produced 1,288 red-eyed offspring (F_1), and 3 white-eyed males. The occurrence of these three white-eyed males (F_1) (due evidently to further sporting) will in the present communication be ignored. The F_1 hybrids, inbred, produced 2,459 red-eyed females, 1,001 red-eyed males, 782 white-eyed males. *No white-eyed females appeared.* The new character showed itself therefore to be sex limited in the sense that it was transmitted only to the grandsons.[139]

Morgan concluded that the factor for red eyes (R) is closely associated with the X-chromosome and that the white-eyed male had arisen through the fertilization, by male-producing sperm, of an egg that had undergone mutation with the loss of R. Shortly afterward, another sex-linked mutant type ("miniature wing") was found, and a female fly having the white-eye mutation in one X-chromosome and the miniature wing mutation in its homolog was found to produce male offspring that included a double mutant with white eyes and miniature wings. To explain this result, Morgan drew on Frans Janssens's finding in 1909 that, during meiosis, there are exchanges between homologous chromosomes, and Morgan suggested that the recombination of genes leading to the double mutant was a consequence of "crossing-over." Other sex-linked mutations were found to exhibit gene linkage, and this suggested that the theory of a linear order of genes in the chromosome might be tested by determining the frequency of crossing-over for pairs of sex-linked factors.

In 1913 Alfred Sturtevant could write: "It has been possible to arrange six sex-linked factors in Drosophila in a linear series, using the number of cross-overs per 100 cases as an index of the distance between any two factors. This scheme gives consistent results, in the main. . . . These results are explained on the basis of Morgan's application of Janssens' chiasmatype hypothesis to associative inheritance. They form a new argument in favor of the chromosome view of inheritance, since they strongly indicate that the factors investigated are arranged in a linear series, at least mathematically."[140] Sturtevant thus produced the first chromosome "map," forerunner of the ever more detailed genetic maps up to those based on DNA sequences.[141]

In later work by the Morgan group, not only were apparent exceptions to the linear order shown to be consistent with the theory, but evidence was offered to strengthen it. For example, Calvin Bridges explained the appearance of the white-eyed males in the F_1 generation by showing that the females giving rise to such offspring were XXY in composition, as a consequence of the failure of chromosomes paired at meiosis to separate ("nondisjunction") and of the passage of both members of the pair to the same gamete.[142]

In 1915 Morgan and his three students (they received their Ph.D.'s at Columbia between 1912 and 1916) summarized their findings in a book,[143] and Morgan later formulated the "theory of the gene" as follows:

> *The theory states that the characters of the individual are referable to paired elements (genes) in the germinal material that are held together in a definite number of linkage groups; it states that the members of each pair of genes separate when the germ-cells mature in accordance with Mendel's first law, and in consequence each germ-cell comes to contain one set only; it states that the members belonging to different linkage groups assort independently in accordance with Mendel's second law; it states that an orderly interchange—crossing-over—also takes place, at times, between elements in corresponding linkage groups; and it states that the frequency of crossing-over furnishes evidence of the linear order of the elements in each linkage group and of the relative position of the elements with respect to each other.* These principles, which, taken together, I have ventured to call the theory of the gene, enable us to handle problems of genetics on a strictly numerical basis, and allow us to predict, with a great deal of precision, what will occur in any given situation. In these respects the theory fulfills the requirements of a scientific theory in the fullest sense.[144]

In this formulation, no statement is made about the nature of genes although cytological studies (performed principally by Bridges) on chromosomes played a decisive role in the development of the theory. No doubt such studies were encouraged and aided by Edmund Wilson, Morgan's colleague at Columbia.

In 1928 Morgan, Sturtevant, and Bridges moved from Columbia to the California Institute of Technology. There, Bridges gave further evidence of his prowess as a cytologist. The large strands of basophilic material in the salivary glands of dipteran larvae had been observed in 1881, but it was not until 1933 that Emil Heitz and Hans Bauer showed these strands to be chromosomes. Later in the same year, Theophilus Painter reported that the succession of stained bands in the salivary X chromosome of Drosophila corresponded to the linear sequence of gene loci as determined in crossing-over experiments, and by 1935 Bridges had performed such an analysis with remarkable detail for all four chromosomes. He recognized 725 bands for the X chromosome, 1,320 for the second, 1,450 for the third, and 45 for the fourth.[145] Apart from the increased precision with which Bridges could identify gene loci of linkage maps with only a few chromatin bands, he also demonstrated the recurrence of certain sequences of bands, and interpreted such "duplication" as a means "for evolutionary increase in the lengths of chromosomes with identical genes which could mutate separately and diversify their effects."[146]

Moreover, during the 1930s Bridges made an important cytological contribution to the interpretation of changes observed by Sturtevant in the so-called Bar mutation—a sex-linked dominant that reduces the size of the Drosophila eye—such as reversion to wild type or the appearance of a more extreme type ("double-Bar"). The more pronounced phenotypic effect of double-Bar was attributed to a "position effect," thus introducing into genetic theory the idea that genes do not act as independent units and that their activity may be influenced by their location in the chromosome. Bridges showed that the Bar mutation itself is associated with a repeat of a chromosomal section corresponding to about seven bands and that in double-Bar this section is present in triplicate.[147] The discovery of the position effect led Theodosius Dobzhansky (who joined Morgan in 1927) to conclude that the genetic system of the chromosome "is a continuum of a higher order, since the independence of the units is incomplete—they are changed if their position in the system is altered. A chromosome is not merely a mechanical aggregation of genes, but a unit of a higher order." Another noted geneticist, Richard Goldschmidt, went further and stated that "gene mutation and position effect are one and the same thing. This means that no genes are existing but only points, loci, in a chromosome which have to be arranged in proper order or pattern to control normal development. Any change in this order may change some detail of development, and this is what we call a mutation. . . . Better, then, give up the conception of the gene except for simple descriptive purposes."[148] This dissent from Morgan's formulation of the theory of the gene echoed the opinions expressed earlier by William Bateson and William Castle.[149]

These views were not shared by the geneticist Milislav Demerec, who wrote in 1935 that "a gene is a minute organic particle, probably a single large molecule, possessing the power of reproduction, which power is one of the main characteristics of living matter. Changes in genes (mutations) are visualised as changes or rearrangements within molecular groups of a gene molecule."[150] During 1920–1940, knowledge about the nature of such "organic particles" was sought through the artificial induction of mutation and cytochemical studies on chromosomes. As Lewis Stadler later emphasized, however, the concept of "the gene" in the observation of the effects of mutagenic agents (the "operational" gene) was not the same as the concept of a unit molecule (the "hypothetical" gene), thus contributing to the controversy of the 1930s about the nature of the gene. In particular, he called attention to the studies of Barbara McClintock on mutational behavior in maize, and her evidence for the transposition of chromosomal elements.[151]

Before 1900 there had been much discussion of "discontinuous variation" as a factor in the appearance of new species, but it was Hugo de Vries who explicitly linked evolution to the new genetics by attributing mutation to changes in the number and kind of "pangens." Although later work, notably by Otto Renner, showed that his conclusions about mutation in the evening primrose *Oenothera Lamarckiana* required revision, de Vries's ideas focused the attention of geneticists on the experimental study of mutation.[152] However, the frequency of spontaneous mutation in higher plants or animals such as flies or mice was too low to permit systematic study of the mutation process. It was Hermann Muller, the most boldly imaginative member of Morgan's group, who sought ways to increase the frequency in Drosophila by artificial means, and who reported in 1927 that intense irradiation of Drosophila sperm with X-rays "induces the occurrence of true 'gene mutations' in a high proportion of the treated germ cells. Several hundred mutants have been obtained in this way in a short time and considerably more than a hundred of the mutant genes have been followed through three, four, or more generations. They are (nearly all of them, at any rate) stable in their inheritance, and most of them behave in the manner typical of the Mendelian chromosomal mutant genes found in organisms generally."[153] The mutagenic effect of X-rays (or radium emanation) was discovered independently by Lewis Stadler, who used barley seeds and who showed that the mutation rate is proportional to the radiation dosage. Similar effects were observed with maize and other plants treated with X-rays or ultraviolet light. Also,

Painter and Muller described the breakage and translocation of Drosophila chromosomes after their exposure to X-rays.[154]

As has been often recounted by historians of the "origins of molecular biology," during the 1930s the geneticist Nikolai Timoféev-Ressovsky and the physicist Karl Zimmer applied to the process of X-ray induced mutation the "hit" and "target" theory previously introduced into radiobiology. Together with the theoretical physicist Max Delbrück, they published in 1935 a paper in which Delbrück discussed the problem of the "gene molecule" from a quantum-mechanical point of view.[155] This paper was later given prominence by Erwin Schrödinger, one of the founders of quantum mechanics:

> Delbrück's model, in its complete generality, seems to contain no hint as to how the hereditary substance works. Indeed, I do not expect that any detailed information on this question is likely to come from physics in the near future. The advance is proceeding and will, I am sure, continue to do so, from biochemistry under the guidance of physiology and genetics. . . . From Delbrück's general picture of the hereditary substance it emerges that living matter, while not eluding the "laws of physics" as established up to date, is likely to involve "other laws of physics" hitherto unknown, which, however, once they have been revealed, will form just as integral a part of this science as the former.[156]

Although the scientific concepts developed in Schrödinger's book do not appear to have had significant direct influence on research in genetics, there can be no doubt that the book itself—and the possibility of finding "other laws of physics"—appealed to some young physicists "suffering from a general professional malaise in the immediate post-war period."[157]

In addition to the early use of X-rays or ultraviolet light to induce artificial mutations, organic chemical agents of the type introduced into warfare during 1914–1918 and similar ones developed later have been found to be effective mutagens. Studies during the 1930s indicated that "mustard gas" (2,2′-dichloroethyl sulfide) can promote the regression of tumors in animals and human subjects, but its toxicity clearly limited its use as a therapeutic agent.[158] Between the two world wars, and as part of the research effort during World War II, several related compounds were made; notably the "nitrogen mustards," such as tris(2-chloroethyl)amine, were prepared, and studies on the reactions of both kinds of war gases with protein and nucleic acid constituents showed these agents to effect a variety of alkylation reactions involving sulfonium or ethylene imonium intermediates.[159] More immediately relevant to genetics was the discovery by Charlotte Auerbach and John Michael Robson that mustard gas is as effective as X-rays in producing mutations and chromosome rearrangements.[160] The more recent development of chemical mutagenesis will be mentioned later.

I return to the 1930s, and the use of cytochemical methods in the study of chromosomes and the chemical nature of genes. Of special importance was the work of Torbjörn Caspersson, who showed, by means of ingenious ultraviolet microspectroscopy of salivary chromosomes, that the "euchromatin" bands (those that stain intensely with the Feulgen nucleal reagent but which disappear during the interphase and prophase stages of mitosis) are also regions of high absorbance at 2600 Å, where nucleic acids absorb strongly.[161] Among his conclusions were the following:

> Intimately linked with the nucleic acid in the transverse bands there is also a protein component. This is dissolved out after prolonged digestion. If the genes are considered to be chemical substances, there can be only one known class of substances to which they may be reckoned to belong, namely the proteins in the broader sense, because of the inexhaustible possibilities of variation which they offer. In the localization of genes in chromosomes it seems more likely the carriers of the genes are just those highly-organized bands built on a nucleic acid skeleton than the less highly differentiated non-absorptive intermediate segments. . . Such being the case, the role of nucleic acid appears to be largely that of a structure-determining support substance.[162]

Subsequent work showed, however, that the action of pepsin, while reducing the chromosomal volume, does not destroy the chromatin strand; this was taken to indicate that the chromosome has a continuous fibrous histonelike structure. Furthermore, treatment with crude nuclease preparations destroyed the Feulgen-reactive material without disrupting chromosomes, suggesting that they "consist of polypeptide fibers to which various elements are permanently attached."[163] There was considerable uncertainty about this picture, as well as about the role of "heterochromatin" (which is stained during the interphase and prophase stages of mitosis), with which the chromosomal histone was thought to be associated.[164]

These opinions of the 1930s and 1940s about the chemical nature of the gene, at a time when the linear polypeptide structure of proteins was being questioned and no reliable chemical evidence was at hand to disprove the tetranucleotide theory of nucleic acid structure, echoed the views expressed by biologists who were more inclined to chemical speculation than Morgan, Sturtevant, Bridges, Stadler, and the many other investigators of lesser renown whose work often goes unmentioned by historians of the biological sciences. I offer some further examples of the acceptance of the view that genes are likely to be proteins. In 1935 Hermann Muller stated that according to his estimate, "the gene length would be between 6 and 30 times as great as its diameter. . . . This is in agreement with the fact that proteins and other chain-like molecules in general are chain-like, being much larger in one dimension than in the other two."[165] In 1942 J. B. S. Haldane stated that "The size of a gene is roughly that of a protein molecule, and it is very probable that the genes are proteins."[166] In 1945 George Beadle wrote: "The gene is made up of protein or nucleoprotein. It may correspond to a single giant molecule, or it may be a discrete unit of higher order made up of a group of protein or nucleoprotein molecules, with or without the addition of other substances."[167] In 1947 Muller went further than before: "It seems likely on general considerations that there is a limited number of possible types of building blocks in the gene, and that genes differ only in the arrangements and number of these. Under differences in arrangement may here be included not only changes in the linear sequence of amino acids in the polypeptide chains but also changes in shape, involving folding of chains, and attachments between R groups, whereby their active surfaces would acquire very different properties."[168] This theory of the gene as a protein or nucleoprotein reminds one of the discourse during 1870–1900 about the chemical nature of "gemmules," "idioplasm," or "pangens," when much less was known about the chemistry of proteins or nucleic acids. Also, as we saw earlier, in 1902 Archibald Garrod raised the matter of the relation of hereditary factors to enzymes, and for several decades afterward there was much discussion of the relationship of genes to enzymes. For example, in writing about inherited metabolic disorders, William Bateson stated:

> If . . . a disease descends through the affected persons as a dominant, we may feel every confidence that . . . there is something *present*, probably a definite chemical substance, which has the power of producing the affection. . . . On the contrary, when the disease is recessive we recognize that its appearance is due to the *absence* of some ingredient which is present in the normal body. So, for example, . . . as Garrod has shown, alkaptonuria must be regarded as due to the absence of a certain ferment which has the power of decomposing the substance alkapton. . . . We may draw from Mendelian observations the conclusion that in at least a large group of cases the heredity of characters consists in the power to produce something with properties resembling those of ferments.[169]

This idea encouraged Arthur Moore, a student of Jacques Loeb's, to offer a biochemical explanation of partial dominance in terms of a reduction of the amount of enzyme in hybrids from a cross between a dominant and a recessive factor.[170] Later, however, Bateson wrote: "We must not lose sight of the fact that though the factors operate by the production of enzymes, of bodies on which the enzymes act . . . yet these bodies themselves can scarcely be themselves genetic factors, but consequences of their existence."[171] As Richard Goldschmidt put it, "The hereditary factor is a determiner for a given mass of ferments; and we can demonstrate it by the fact that a quantitative difference in the potency of hereditary factors causes a parallel, quantitatively different enzyme formation."[172]

Among the first such biochemical studies were those of Huia Onslow, who reported in 1915 that recessive whiteness in rabbits is due to the absence of tyrosinase, an enzyme shown a few years earlier to catalyze the conversion of tyrosine into colored products.[173] Also, Muriel Wheldale (later Mrs. Onslow) collaborated with Rose Scott-Moncrieff and Robert Robinson in studies on the genetics of the formation of flower pigments.[174] It is noteworthy that the Onslows were both working in the Cambridge Biochemical Laboratory, where J. B. S. Haldane initiated his researches (with Scott-Moncrieff) on the genetics of flower pigmentation in the hope of correlating differences in the chemical structure of the pigments with differences in the enzymic constitution of various strains.[175] Indeed, Haldane's important theoretical contributions to enzyme chemistry were in part a consequence of his interest in the relation of genes to enzymes. Onslow's studies on hair color in mammals were extended by Sewall Wright, who wrote: "The very fact that it has been relatively easy to isolate unit factors in work on color inheritance suggests that in this case the chain of processes between germ cell and adult may be relatively simple. Observations which indicate that melanin pigment is formed in the cytoplasm of cells by the secretion of oxidizing enzymes from the nucleus suggest that the chain may be very short indeed when it is remembered that genetic factors are probably characters of the chromosomes and that these seem to be distributed unchanged from the germ cell to all other cells."[176] During 1910–1920, several biologists suggested that the gene for a given character is the specific enzyme involved in the chemical reaction that produces that character, and that the replication of hereditary factors is an autocatalytic process. For example, Arend Hagedoorn postulated that "each of the several transmittable genetic factors for the development of an organism is a definite chemical substance which has the property of being a ferment for its own formation (an autokatalyzer)."[177]

This concept was developed by Leonard Troland, who (as a Harvard undergraduate) offered the hypothesis that the "hereditary determinants" in the cell nucleus are enzymes,[178] and later (after receiving a Ph.D. in psychology) he presented his theory of "heterocatalysis": "It is obvious that exact similarity of the force patterns of the catalyzing and catalyzed systems is not essential. Indeed, the catalytic effect which is based upon direct similarity of structure between two systems should be much weaker than that which accompanies certain types of structural *correspondence,* such as that existing between a body and its mirror image, or between a lock and a key."[179] Troland's concept of the gene as an autocatalyst attracted Hermann Muller, who, while rejecting the idea that genes are enzymes, in 1922 expressed the view that in the self-propagation of the gene,

> it reacts in such a way as to convert some of the common surrounding material into an end-product identical in kind with the original gene itself. This action fulfills the chemist's definition of "autocatalysis." . . . But the most remarkable feature of the situation is not this oft-noted autocatalytic action in itself—it is the fact that, when the structure of the gene becomes changed, through some "chance variation," the catalytic property of the gene may be correspondingly changed, in such a way as to leave it still *autocatalytic.* . . . The question as to what the general principle of gene construction is, that permit this phenomenon of mutable autocatalysis, is the most fundamental question of genetics.[180]

Whereas before 1930 Muller stated that "At present any attempt to tell the chemical composition of genes is only guesswork," with the subsequent acceptance of the view that enzymes are proteins and the identification of some viruses as nucleoproteins, he and other biologists agreed with the statement that the material of the genes "must be protein, and may very likely be nucleo-protein."[181] For example, J. B. S. Haldane surmised that "the gene is within the range of size of protein molecules, and may be a nucleo-protein molecule like a virus. If so, the chemists will say, we must conceive reproduction as follows. The gene is spread out in a flat layer, and acts as model, another gene forming on top of it from pre-existing material such as amino acids."[182] As we shall see, such "template" hypotheses came into favor after 1945.[183] Before then, the idea that genes are autocatalytic proteins led to speculations such as that of Max Delbrück, who suggested in 1941 that such proteins are made not from amino acids but from amino aldehydes by a "catalytic mechanism that would be highly specifically *autocatalytic.*"[184]

Along with these and related speculations,[185] during the 1940s there were important experimental studies on the relation of genes to enzymes, and further advances in the definition of the chemical composition of chromosomes. As regards the first, in 1936 George Beadle and Boris Ephrussi reported their work on the genetic control of eye color in Drosophila.[186] They showed that a larval eye disk taken from a mutant ("vermilion," v^+), implanted in the abdomen of a wild-type larva, developed into an adult structure having the dull-red eye color characteristic of the wild-type fly. Beadle and Ephrussi concluded that a hormonelike substance made by the wild-type organism is a precursor of the normal red pigment. As has been mentioned, later work by Adolf Butenandt showed this substance to be kynurenine, a known intermediate in the metabolic breakdown of tryptophan in mammals. Another of the eye color mutants of Drosophila ("cinnabar," cn^+) turned out to be unable to convert kynurenine into the red eye pigment, and it was clear that each of the two steps in the metabolic sequence tryptophan → kynurenine → pigment is under the control of a separate gene. Although as Beadle stated in 1939, there was no "direct evidence for [the] intervention of enzymes in the system, they are assumed throughout for the reason that for the moment they appear to provide a simple mechanism by which genes control reactions."[187]

In 1937 Beadle moved to Stanford, where he was joined by the biochemist Edward Tatum, who undertook to isolate the metabolic precursors of the Drosophila eye pigments. He identified the "v^+-hormone" as kynurenine, but that success was anticipated by Butenandt. It seems that this experience led Beadle and Tatum to turn to another organism for the biochemical study of gene action. As Beadle described it many years later, they "hit upon the idea of reversing the procedure we had been using to identify specific genes with particular chemical reactions. We reasoned that, if the one primary function of a gene is to control a particular chemical reaction, why not begin with known chemical reactions and then look for the genes that control them? In this way, we could stick to our specialty, genetics, and build on the work chemists had already done. The obvious approach was first to find an organism whose chemical reactions were well-known and then induce mutations in it that would block identifiable reactions."[188] The organism they selected was the fungus (ascomycete) *Neurospora crassa* because the previous work of Bernard Dodge and Carl Lindegren had shown that its relatively short life cycle, its mode of reproduction, and the segregation of its genes in a string of ascospores made it especially suitable for genetic analysis. Moreover, Tatum's friend Nils Fries had found that the nutritional requirements of Neurospora are quite simple, as it can grow in pure culture on a chemically defined medium of sucrose, inorganic salts (phosphates, nitrates, sulfates, and others), and biotin.[189] From these materials the organism makes all the amino acids for its proteins, the purines and pyrimidines for its nucleic acids, as well as its many other cell constituents, including compounds such as pyridoxine and thiamine. In their first report on Neurospora, Beadle and Tatum described studies with a mutant strain (obtained by irradiation with X-rays) which had lost the ability to make pyridoxine (a member of the vitamin B_6 group), and they concluded that "this inability to synthesize vitamin B_6 is transmitted as it should be if it were differentiated from normal by a single gene."[190]

There followed a series of studies in which Neurospora mutants were used to determine the sequence of chemical reactions in the biosynthesis of various metabolites, especially several amino acids (arginine, tryptophan, isoleucine, valine). In a progress report Beadle discussed the bearing of the results on the problem of the relationship of genes to enzymes, and suggested that

> the gene can be visualized as directing the final configuration of a protein molecule and thus determining its specificity. A given protein molecule, patterned after a particular gene, might become a component of a new gene like the one from which it was copied or it might become an antigenically active protein, an enzyme protein, or a storage protein. . . . If genes in some way direct the configuration of protein molecules during their elaboration, it is not necessary to assume that they function in any other way. . . . The protein components of enzymes . . . would have their specificities imposed fairly directly by genes and the one-to-one relation observed to exist between genes and chemical reactions should be a consequence. It should follow, indeed, that every enzymatically catalyzed reaction that goes on in an organism should depend directly on the gene responsible for the specificity of the enzyme concerned. Furthermore, for reasons of economy in the evolutionary process, one might expect that with few exceptions the final specificity of a particular enzyme would be imposed by only one gene.[191]

In this statement, Beadle inclined to the then-current view of genes as proteins or nucleoproteins, but the idea itself, termed by Norman Horowitz the "one gene one-enzyme hypothesis,"[192] had considerable fruitful impact on biological thought.

As regards the advances during the 1940s in the study of the chemical composition of chromosomes, André Boivin, Alfred Mirsky, and Arthur Pollister found that the DNA content of haploid and diploid nuclei is roughly in the ratio 1:2, in agreement with Boveri's conclusion that the chromatin doubles with each cell division,[193] and to the knowledge about the chemical constitution of DNA, discussed earlier, Gerard Wyatt added the finding that some DNA preparations contain 5-methylcytosine.[194] There was renewed interest in the chemistry of chromatin after the preparation from nuclei of dispersed DNA-protein particles was described in 1959 by Geoffrey Zubay and Paul Doty.[195] After enzymatic digestion of the DNA component, ion-exchange chromatography of such extracts was shown by James Bonner's group to yield several histone fractions, whose amino acid composition and sequence were determined by Emil Smith and Robert DeLange.[196] As had been shown by Albrecht Kossel, like the protamines of fish sperm, the histones of calf thymus are basic proteins but differ in the ratio of their lysine/arginine content. Work by other investigators gave support to the suggestion by Edgar Stedman that one of the functions of the histones is to inhibit ("repress") DNA replication[197] and that they participate in the coiling of DNA strands in the chromosome. In particular, Roger Kornberg offered an attractive model of chromatin as a chain of repeating "nucleosome" units, each of which is composed of an assembly of histones around which there is wrapped a segment of a "supercoiled" DNA pair.[198] In this connection it should be mentioned that over the years since the appearance of the Watson-Crick model of DNA as a right-handed double helix, the question has been raised as to the extent to which that conformation corresponds to that of intracellular DNA, and it has been stated recently that "despite years of experimental study, it seems to us that there is very little reason to believe that the condensed chromatin fiber contains *substantial* amounts of any *regular* helical structure."[199] It was recognized that the features of the model that had been confirmed experimentally by Meselson and Stahl and by Arthur Kornberg (about whom more later) and his group by 1960[200] do not require the elegant symmetry of the acclaimed double helix. Alexander Rich and his associates have found that for some synthetic oligo-deoxyribonucleotides, a left-handed zigzag (Z-DNA) conformation is preferred, and single crystal X-ray diffraction studies by Olga Kennard of such compounds indicate considerable conformational flexibility, made evident in their combination with other substances.[201]

In this section I have attempted to trace the development of the "theory of the gene" formulated by Johanssen and Morgan. It is now customary to refer to the "classical" theory, with genes as indivisible units of biological transmission, recombination, mutation, and function. Morgan, in his 1926 statement, refers to Mendel's concept of the transfer of hereditary units from parents to progeny, and the reassortment of the linear order of genes by "crossing-over," whose physical reality was established only in 1931 by Barbara McClintock for *Zea mays* and by Curt Stern for Drosophila.[202] In that statement, Morgan mentions mutation only in connection with de Vries's theory (not Muller's theory of 1922), and, as regards "function," Morgan does not cite Johanssen's distinction between the genotype and phenotype, nor does he refer to Bateson's interpretation of Garrod's findings. Like Johanssen, Morgan then considered genes to be solely units of numerical calculation, and he eschewed speculation about their physical or chemical nature. Morgan's mathematical needs did not include the kind of statistical analysis introduced by the anti-Mendelian biometricians Karl Pearson and Walter Weldon and later applied by Ronald Fisher, J. B. S. Haldane, Sewall Wright, Sergei Chetverikov, Nikolai Dubinin, and Morgan's pupil Theodosius Dobzhansky to the study of the evolutionary process in populations of wild-type Drosophila and other organisms.[203] According to Alexander Weinstein, another of Morgan's students, Morgan was skeptical about Darwin's theory of natural selection because of its teleological flavor.[204] Also, although he had been an embryologist before turning to Drosophila, Morgan stated that the "cytoplasm may be ignored genetically,"[205] at a time when the concept of cytoplasmic inheritance was being defended by leading embryologists. For example, Ross Harrison wrote: "The prestige of success enjoyed by the gene theory might easily become a hindrance to the understanding of development by directing our attention solely to the genome, whereas cell movement, differentiation and in fact all developmental processes are actually effected by the cytoplasm."[206] Before 1940 the dominance of Morgan's formulation of the theory of the gene was more pronounced in the United States than in France and Germany, where it met stronger resistance from experimental embryologists and proponents of modernized versions of Lamarck's theory of the evolutionary role of acquired characters.[207] To these aspects of the "classical" era, which drew its conceptual structure primarily from studies on higher plants and animals, it should added that before 1940 Øjvind Winge had demonstrated the sexual reproduction of brewer's yeast, with the same kind of mitosis and meiosis as that observed in higher plants.[208] The recognition after 1940 of the applicability of genetic techniques to the study of bacteria and viruses, and the identification of the gene as DNA, vastly enlarged the scope of biochemical genetics. The early stages of that development will be considered in the next section.

To conclude this section, I offer a few comments about some recent writings about the development of the theory of the gene, with special attention to the role of Morgan and his associates. Apart from Garland Allen's full-length biography, the most extensive account of the research of the Drosophila group has been Robert Kohler's book, with a title resembling that of William Golding's famous novel.[209] Whereas in his earlier treatment of the institutional development of biochemistry, Kohler adopted the sociological approach of Joseph Ben-David, in this book Kohler uses the anthropological approach of Bruno Latour to present a "case study" of "laboratory life" in terms of its "moral economy" and "material culture." He states that "readers familiar with the history of genetics will note the absence here of any systematic discussion of the major concepts and discoveries of classical *Drosophila* genetics. I go into considerable detail about the production process—instruments, procedures, strategies—but not about the products of research."[210] There is much valuable information in Kohler's book, and also evidence of his penchant for facile generalization, which makes for easy reading but also for lapses from sound historical scholarship. If Bridges and Sturtevant are rightly given pride of place, surely Morgan played a larger role than that assigned to him by Kohler, and Muller also deserved more attention.[211] Kohler's exaltation of Drosophila to the status of leading player is an effective literary device, and reminiscent of Latour's view of the bacterium in the work of Pasteur, but it smacks of fiction. A similar account might perhaps be written about the role of maize, with Rollins Emerson and Barbara McClintock among the human "actants." I welcome the interest of former rigidly externalist historians and former members of the "social constructionist" group in the instruments, techniques, and organisms used in research on biological problems, but I deplore their view that it is possible to understand the "nature of experiment" without close consideration of the changes in the ideas (hypotheses, preconceptions, theories) that animated the use of such items of the "material culture," and of the interplay of thought and action during the conduct of a series of experiments. In the case of Morgan's Drosophila group, this interplay involved, as was not uncommon even before the 1940s, a team effort and, according to Jack Schultz, a "combination of Morgan's skepticism and Muller's system-building," together with "Sturtevant's extraordinary analytical powers, and the brilliance of Bridges' experimental talents."[212] Also, as rightly noted by Nils Roll-Hansen, Edmund Wilson played a significant role as an adviser.[213] Other recent writings of a sociological or philosophical nature about the development of the theory of the gene are included in the bibliography.[214]

The Biochemical Genetics of Bacteria and Viruses

The close connection between the development of enzyme chemistry and bacterial physiology since the middle of the nineteenth century, evident in the work of Louis Pasteur on fermentation and the later studies of Emile Duclaux, Sergei Winogradsky, and Martinus Beijerinck, was not reflected in a comparable interest during 1900–1940 in the possibility that the ability of bacteria to "adapt" to changes in the chemical composition of culture media might be amenable to genetic analysis. In a perceptive discussion of this historical problem, William Summers noted that in 1900 Beijerinck wrote: "Though the culture of microbes, compared to that of higher plants and animals is subject to many difficulties, it cannot be denied that, these once mastered, microbes are extremely useful material for the investigation of the laws of heredity and variability."[215] In part, the failure to take up this suggestion may have been due to the belief of the founders of the germ theory of disease in the fixity of bacterial species, although Rudolf Massini reported in 1907 the finding of a variant of what Theodor Escherich had named *Bacillus coli commune,* and Summers cited a succession of medical investigators, among them Paul de Kruif, who described other such variants before 1940.[216] For biochemists, and for such biochemically minded medical investigators as Paul Fildes, bacteria were valuable organisms for the study of metabolic pathways and for the discovery of new growth factors.[217] Of particular historical importance were the report of Frédéric Diénert in 1900 on the adaptation of yeast to the fermentation of galactose, and Jacques Monod's doctoral thesis (1942) on this subject (of which more later). In the intervening years, the problem of the enzymatic adaptation of microorganisms was actively studied by Henning Karström and John Yudkin, among others.[218]

In parallel with the work of Beadle and Tatum on Neurospora, that of Oswald Avery and his associates at the Rockefeller Institute for Medical Research on the pneumococcus *(Diplococcus pneumoniae)* marks the emergence of biochemical genetics (or molecular biology) as a distinctive research specialty. Avery's group had been studying the chemical basis of the differences among strains of this organism for many years. In 1926 Avery and Heidelberger showed that the serological difference between the two types denoted II and III resides in the slimy polysaccharide capsule that surrounds the virulent form, which grows in glistening smoothly rounded (S) colonies. For each type, there was known a corresponding variant of markedly attenuated virulence, which gives rough (R) colonies. During the 1930s Avery's associate Walther Goebel showed that one of the genetic characters of each of the virulent types was the ability to make a structurally distinctive polysaccharide composed of particular sugar units.[219]

In 1928 Fred Griffith reported that when mice were given living Type II (R) pneumococci together with heat-inactivated Type III (S) cells, many of the animals died, and their blood contained living Type III organisms. This discovery indicated that the living avirulent R bacteria had acquired something from the dead Type III (S) cells that caused the transformation of the Type II (R) cells into virulent organisms having the polysaccharide of the Type III (S) cultures. He did not interpret his remarkable finding in terms of the transfer of hereditary material but suggested that "when the R form of either type is furnished under suitable conditions with a mass of the S form of the other type, it appears to use that antigen as a pabulum from which to build up a similar antigen thus to develop into an S strain of that type."[220]

Griffith's observation was quickly confirmed and extended by Avery's associates Martin Dawson and James Alloway, and by 1933 aqueous solutions of the "transforming principle" were available for further study.[221] The effort to purify it was begun by Colin MacLeod during 1934–1937, resumed in 1940, and completed by Maclyn McCarty.[222] In 1944 there appeared a paper that began as follows: "Biologists have long attempted by chemical means to induce in higher organisms predictable and specific changes which thereafter could be transmitted in series as hereditary characters. Among microorganisms the most striking example of inheritable and specific alterations in cell structure and function that can be experimentally induced and are reproducible under well defined and adequately controlled conditions is the transformation of specific types of Pneumococcus."[223] There followed the description, in meticulous detail, of the isolation of the active transforming principle from Type III pneumococci and its identification as "a highly polymerized and viscous form of sodium desoxyribonucleate" that gave negative reactions when tested for the presence of protein.[224] Apart from the evidence from chemical analysis, the strongest arguments for the identification of the transforming factor as DNA were its inactivation by a crude preparation of "desoxyribonucleodepolymerase" and its resistance to the action of ribonuclease and of the proteinases trypsin and chymotrypsin. Because of the importance of this enzymic inactivation of the transforming factor, McCarty undertook the purification of the DNA "depolymerase" from beef pancreas, and in 1948 Moses Kunitz crystallized this enzyme (now termed deoxyribonuclease I); these preparations were highly effective.[225]

After noting that others had likened the transforming factor to genes and viruses, Avery, MacLeod, and McCarty stated:

> It is, of course, possible that the biological activity of the substance described is not an intrinsic property of the nucleic acid but is due to minute amounts of some other substance adsorbed to it or so intimately associated with it as to escape detection. If, however, the biologically active substance isolated in highly purified form as the sodium salt of desoxyribonucleic acid actually proves to be the transforming principle, as the available evidence strongly suggests, then nucleic acids of this type must be regarded not merely as structurally important but as functionally active in determining the biochemical activities and specific characteristics of pneumococcal cells.[226]

Not only were the results confirmed by other investigators, but also within a few years transformations induced by DNA preparations from other bacteria *(Escherichia coli, Hemophilus influenze, Bacillus subtilis)* were described. In 1947 André Boivin (who worked with *E. coli*) wrote: "In bacteria—and, in all likelihood, in higher organisms as well—each gene has as its specific constituent not a protein but a particular desoxyribonucleic acid which, at least under certain conditions (directed mutations of bacteria), is capable of functioning *alone* as the carrier of hereditary character; therefore, in the last analysis, each gene can be traced back to a macromolecule of a special desoxyribonucleic acid."[227] Much has been written about

Avery's 1944 paper, published when he was sixty-six years old and planning his retirement, as well as about the fact that he was not chosen for a Nobel Prize before his death in 1955. Of special interest is the publication of a passage from his letter of 13 May 1943 to his brother: "Sounds like a virus—may be a gene. But with the mechanisms I am not now concerned—one step at a time and the first step is, what is the chemical nature of the transforming principle? Someone else can work out the rest."[228] Years later, during the 1970s, there was a flurry of discussion about Gunther Stent's omission of Avery's work from his account of "that was the molecular biology that was" and about his subsequent statement that it had been "premature" in the sense that "geneticists did not seem to be able to do much with it or build upon it."[229] As I read the historical record, that inability, at least in the case of Delbrück and Luria, came from their low opinion of DNA as the bearer of genetic factors. In 1952, however, Alfred Hershey, a member of their tightly-knit "phage group," and Martha Chase reported that

> when a particle of bacteriophage T2 attaches to a bacterial cell, most of the phage DNA enters the cell, and a residue containing at least 80 percent of the sulfur-containing protein of the phage remains at the cell surface. This residue consists of the material forming the protective membrane of the resting phage particle, and it plays no further role in infection after the attachment of phage to bacterium. These facts leave in question the possible function of the 20 percent of sulfur-containing protein that may or may not enter the cell. . . . We infer that the sulfur-containing protein has no function in phage multiplication, and that DNA has some function. Our experiments show clearly that a physical separation of the phage T2 into genetic and non-genetic parts is possible. . . . The chemical identification of the genetic part must wait, however, until some of the above questions asked above have been answered.[230]

This statement may be compared with that quoted above from the paper by Avery, MacLeod, and McCarty. In their experiments, Hershey and Chase used ^{32}P to label the DNA of the phage and ^{35}S the phage proteins, and because earlier electron microscopy had shown that T2 phage attaches itself to the host cell by means of its tail, violent agitation was used to break off the DNA-containing headpiece. Although control experiments were performed to eliminate the factors of uncertainty inherent in this procedure, the results did not constitute proof of the DNA nature of the infective material.[231] Indeed, the chemical evidence for the cautiously worded conclusion that the genetic material of the phage is associated with its DNA and not its protein was subject to greater doubt than that expressed in regard to the pneumococcal transforming factor. In the case of the latter, Avery's cautious reservation that the activity might be "due to minute amounts of some other substance" was emphasized by other investigators. For example, Alfred Mirsky wrote in 1951 that "it is difficult to eliminate the possibility that the minute amounts of protein that probably remain attached to DNA, though undetectable by the tests applied are necessary for activity. . . . There is accordingly some doubt whether DNA is itself the transforming agent, although it can be regarded as established that DNA is at least part of the active principle."[232] Such doubts were reiterated by others and are reminiscent of the earlier debates about the protein nature of enzymes and of the hormone insulin. For this reason, Rollin Hotchkiss and Stephen Zamenhof expended much effort to purify transforming factors as much as possible, and by 1952 active preparations having no more than about 0.02 percent protein were available.[233]

Much has been written about the emergence after World War II of the "phage group," led by Max Delbrück, especially in relation to the "origins of molecular biology."[234] In setting themselves the task of studying bacterial viruses (in particular, the T phages of *E. coli* B) in a quantitative manner, they continued a line of research that had been actively pursued since the discovery of the bacteriophages by Frederick Twort and Félix d'Hérelle during World War I.[235] During the 1920s it was hoped that these agents might be useful in the treatment of infectious diseases, and although this expectation was not fulfilled, by 1926 it had been learned that the adsorption of the phage particle to the host cell, followed by the penetration of the particle into the bacterium, leads to the intracellular multiplication of the virus and lysis of the host cell, with the release of the phage progeny. Significant advances were subsequently made through the work of André Gratia, Frank Burnet, Albert Krueger, Eugène and Elisabeth Wollman, and especially Max Schlesinger (among others), in the development of methods for studying the infective cycle.[236] Delbrück's entry into the bacteriophage field was marked by the appearance in 1939 of a joint paper with Emory Ellis, who had been working on the "step curves" in the multiplication of virus particles. The technique they described permitted the determination of the effect of changes in the physical and chemical environment on the time of the cycle and on the number of virus particles produced per host cell. The paper began with the statement that "certain large protein molecules (viruses) possess the property of multiplying within living organisms. This process, which is at once so foreign to chemistry and so fundamental to biology, is exemplified in the multiplication of bacteriophage in the presence of susceptible bacteria. . . . Bacteriophage . . . can be concentrated, purified, and generally handled like nucleoprotein, to which class it apparently belongs."[237]

That this process was not as "foreign to chemistry" as Delbrück then supposed was shown during the 1940s by Seymour Cohen, who reported that the nucleic acid of T2 phage is exclusively of the DNA type, and that after a delay (7–10 minutes) following infection of *E. coli* cells, the metabolic activity of the host cells is channeled into the production of virus DNA in amounts much greater than the DNA of normal bacteria. In this work Cohen used ^{32}P-labeled phosphate to show that the virus DNA is made by the enzymic apparatus of the host cell and that the phosphorus of this DNA is largely derived from inorganic phosphate in the culture medium. In other experiments he also demonstrated that although the rate of protein formation after phage infection is essentially the same as before infection, most of the newly formed protein is of the type associated with the membrane of the phage head.[238] Also, Robert Sinsheimer reported important chemical studies on the DNA of the phage ϕX174.[239] Indeed, in 1951, before the publication of the Hershey-Chase paper, John Northrop, who had embarked on phage research during the 1930s in the belief that viruses are solely protein in nature, concluded from his studies on the formation of phage in *Bacillus megatherium* that "the nucleic acid may be the essential, autocatalytic part of the molecule, as in the case of the transforming principle of pneumococcus, and the protein may be necessary only to allow entrance to the host cell."[240]

As was mentioned earlier, de Vries's theory of mutation attracted the attention of microbiologists interested in the causes of bacterial variation such as the conversion of a form unable to ferment lactose (now denoted *lac*$^-$) into one able to do so (*lac*$^+$). Also, after the discovery of the bacteriophages, variants were found that were resistant to lysis by these bacterial viruses. During the 1920s and 1930s there was much discussion of the question whether such bacterial variation occurs spontaneously in the absence of lactose or phage, and the "mutants" then continue to multiply under the particular composition of the growth medium, or whether the variation is caused by the direct action of lactose or phage in a kind of adaptation to the change in the growth medium. Among the leading participants in the discussion were Philip Hadley, Joseph Arkwright, and Isaac Lewis, as well as the virologists Félix d'Hérelle, André Gratia, and Frank Burnet.[241] By the mid-1930s, largely owing to the work of Lewis, microbiologists inclined toward the first of these alternatives.

This problem, as it related to the appearance of phage-resistant *E. coli,* was taken up by Max Delbrück with Salvador Luria, who had previously worked on the effect of radiation on *E. coli.*[242] In their joint 1943 paper, based on the experimental work of Luria, they described a "fluctuation test," which involves the plating of a culture of sensitive bacteria in the presence of excess phage, and they established in elegant fashion the applicability of Poisson's "law of large numbers" to the statistical analysis of the behavior of bacterial populations. Their data on the distribution of the variant clones (populations of genetically identical cells, derived from the multiplication of a single cell) were more consistent with the view that phage-resistance in *E. coli* is a consequence of spontaneous mutation than with the view that it is a mechanism involving the direct action of phage.[243] This fluctuation test was subsequently applied to the study of variation in several microorganisms with respect to other genetic characters, such as resistance to antibiotics, with a similar result. A further important contribution to the study of bacterial populations was the technique of "replica plating," introduced by Joshua Lederberg and Esther Lederberg in 1952. This method, which involved placing a cloth fabric over the plate of bacterial colonies and transferring individual impressions to sterile plates containing different culture media, permitted the isolation of mutants in the absence of the inducing agent.[244] Although the results of the application of such techniques strongly supported the conclusion drawn by Luria and Delbrück, its generality has recently been questioned, and evidence has been offered for the view that, in some cases, nonrandom variation can occur.[245] To this may be added the fact that the rejection by geneticists of the contribution of the noted physical chemist Cyril Hinshelwood—who presented during the 1950s a kinetic explanation of bacterial adaptation—has been somewhat modified in the light of recent research.[246]

In addition to the discoveries regarding bacterial transformation and mutation, before 1950 it had been shown that bacteria can undergo the kind of reassortment ("recombination") of genetic factors that are characteristic of sexual reproduction and that formed the basis of Morgan's theory of the gene. This development was an outgrowth of the work of Beadle and Tatum on nutritional mutants of Neurospora, with Tatum's preparation in 1945 of similar "auxotrophs" of *E. coli*, strain K-12, whose wild-type "prototrophs" can grow on a medium composed solely of glucose and inorganic salts. Shortly afterward he was joined at Yale by Joshua Lederberg, who demonstrated that upon mixing two triple *E. coli* mutants—one requiring threonine, leucine, and thiamine for growth, the other requiring biotin, phenylalanine, and cystine—there were produced organisms of several types, including the wild-type form and double mutants no longer exhibiting a requirement for one of the above substances. In their brief paper announcing this discovery, Lederberg and Tatum also noted that "using the triple mutants mentioned, except that one was resistant to the coli phage T1 (obtained by the procedure of Luria and Delbrück), nutritionally wild-type strains were found in both sensitive and resistant categories," and they concluded that "these experiments imply the occurrence of a sexual process in the bacterium *Escherichia coli*."[247]

Much of the cytological evidence for the conjugation of the mature haploid sex cells of animals, higher plants, and sexual fungi, followed by the fusion of the nuclei and mitosis, was available before the formulation of the theory of the gene. In the case of bacteria, in contrast, the genetic evidence provided by Lederberg and Tatum preceded the cytological studies, largely by means of electron microscopy, which provided the physical basis for the interpretation of the mechanism of genetic recombination. During the 1950s a series of discoveries brought bacteria to the forefront, and *Escherichia coli* K-12 became for a time the most important organism in genetic research. Among the high points of that development were the discovery that the low frequency of recombination (about 1 in 10^5 of the parental population) found in the experiments of Lederberg and Tatum was due to the rare occurrence of a mutant (Hfr, high frequency of recombination) and that upon isolation and use of this mutant the frequency was increased to nearly 100 percent. It was concluded that the Hfr mutant carries an autonomous transmissible sex factor (F, for fertility) and that there are two mating types of *E. coli,* F^+ and F^-, corresponding to the male and female forms of higher organisms. There then followed the outstanding work of William Hayes and of Elie Wollman and François Jacob, who showed, by means of a technique of interrupted mating, that conjugation and gene transfer are followed by a progressive integration of the genetic elements of the zygote. They also concluded that the bacterial chromosome existed in the form of a continuous ("circular") double strand.[248] The technique was used by Austin Taylor to provide a genetic map of *E. coli* K-12.[249]

For a time, there was some discussion whether the term *chromosome* was appropriate in connection with the DNA of bacteria and phages. During the 1940s Carl Robinow demonstrated in bacilli such as *E. coli,* after fixation with osmic acid, the presence of sets of small bodies that stained with basic dyes and therefore were considered to be composed of chromatin. Later, electron microscopic observation of DNA released by osmotic shock, or by autoradiography of cells treated with ^3H-labeled thymidine, showed the presence of very long strands, from which the "chromatin bodies" had arisen by breakage.[250]

Also, during the 1950s Esther Lederberg discovered that irradiation with ultraviolet light causes the lysis of *E. coli* K-12, indicating that this strain carries a phage; it was named *lambda* by analogy to the killer factor *(kappa)* identified by Tracy Sonneborn in *Paramecium.*[251] The finding of *lambda* brought the Lederberg group into the field of lysogeny, whose long history began during the 1920s with the work of Félix d'Hérelle, Jules Bordet, and Frank Burnet, was developed by Eugène and Elisabeth Wollman during the 1930s, and was brought to fruition by André Lwoff after World War II.[252] As we shall see, the subsequent investigation of *lambda* phage by Seymour Benzer marked a new stage in the mapping of chromosomes.

In parallel with the work in Lederberg's laboratory on *E. coli,* his student Norton Zinder studied the genetics of auxotrophs of *Salmonella typhomurium,* most of whose strains were known to be lysogenic. He found that the genetic exchange in this organism involves the transfer from the prototroph of a filterable material (FA) that can carry with it factors for a large variety of genetic characters, in a process that Zinder and Lederberg termed transduction.[253] Lederberg introduced the word *plasmid* to denote autonomous genetic elements, such as F factor or *lambda,* which can multiply independently of the bacterial chromosome. Subsequent work, largely by Haruo Ozeki and by Werner Arber and his associates, has shown that a distinction needs to be made between plasmids that are only autonomous and those that can also be integrated with the bacterial chromosome.[254] The latter have also been termed episomes.

In one of the many brilliant investigations in this field during the 1950s, Seymour Benzer applied the genetic exchange by recombination to map a region of the T4 bacteriophage, which can lyse both the K and B strains of *E. coli,* by using mutants of the phage unable to form plaques (circular clear areas) with K. An important outcome of his work was a clearer operational definition of the gene. What had once been considered to be an indivisible unit of genetic recombination, mutation, and function was redefined in terms of a distinct unit of recombination ("recon"), a unit of mutation ("muton"), and of function. Because the last was expressed in various physiological or biochemical ways, it was named cistron, in line with the "cis-trans" genetic test for the complementation of two chromosome strands devised by Edward Lewis.[255] To this nomenclature, Jacques Monod and François Jacob added *operon,* in connection with their studies on the genetic control of the formation of enzyme proteins. I will discuss this work later in the context of the parallel development of research on the biochemical mechanism of protein synthesis.

I have thus far in this section sketched only a few of the major contributions to the exciting development of bacterial and phage genetics during about 1940–1960. It seemed to many at the time that these achievements overshadowed the equally exciting development of knowledge about the chemistry and metabolism of proteins and carbohydrates, about enzymes, and about the energetics of biochemical processes. At that time also, numerous biochemists were engaged in the study of what has been called genetic chemistry. In 1957 Robert Sinsheimer stated that the goals of that effort are "to achieve an understanding of the physical nature of the hereditary units and of the variations these units undergo during the various phases of the life of the cell, to learn how these units are reproduced from generation to generation, to learn how they are modified by extrinsic or intrinsic factors, and to learn in precise chemical terms how they influence the manifold activities of the cells in which they reside."[256] In the succeeding sections of this chapter, I offer a necessarily brief account of efforts during 1960–1990 to attain these goals. Before doing so, I deem it necessary to consider the impact of the new knowledge gained from the study of bacterial and viral genetics on the long-continued investigation of the genetics of the somatic cells of animals and higher plants.[257]

That story begins, in modern times, with the development of a method for the cultivation of animal tissues by the embryologist Ross Harrison, who was engaged around 1910 in the study of the growth of nerve fibers. His contemporary, Montrose Thomas Burrows, an associate of Alexis Carrel (who received the public accolades) also contributed to that effort. They were followed by Albert Fischer, but it was not until the 1940s and 1950s that the *in vitro* culture of animal cells was achieved, through the work of Wilton Earle, the biochemist Harry Eagle, and most notably Theodore Puck and Philip Marcus in devising a method for producing colonies of such cells.[258] Henry Harris has suggested that the main reason why this had not been achieved earlier "is that those who worked in the field of tissue culture at the time were simply not sufficiently impressed with the importance of applying quantitative microbiological techniques to somatic cells. The decisive impetus eventually came from the remarkable flow of information generated by experiments with bacteriophages."[259]

Although numerous metabolic variants of cultured animal cells have been identified, it does not appear to be clear which of them (if any) are the result of a genetic mutation.[260] On the other hand, the work of Renato Dulbecco, who applied the plaque-counting technique of phage genetics, has shown that the DNA of the tumor-producing polyoma virus can act as a transforming factor in cultured animal cells. This achievement reminded geneticists of the studies of Peyton Rous in 1912 on the soluble fowl sarcoma virus.[261] After the work of Dulbecco, other animal tumor viruses were found to be transforming agents, among them a simian virus (SV40) that became an object of close biochemical study. Moreover, as mentioned earlier in connection with the role of proteins in immune reactions, the invention of a method for the production of specific monoclonal antibodies depended on the prior development of procedures for the hybridization of cultured somatic cells and the discovery by Henry Harris and John Watkins of the virus-induced fusion of such cells.[262]

DNA Replication and Recombination

In their famous second 1953 paper in *Nature*, Watson and Crick wrote:

> Previous discussions of self-duplication have usually involved the concept of a template, or mould. Either the template was supposed to copy itself directly or it was to produce a "negative," which in its turn was to act as a template and produce the original "positive" once again. In no case has it been explained how it would do this in terms of atoms and molecules. Now our model for deoxyribonucleic acid is, in effect, a *pair* of templates, each of which is complementary to the other. We imagine that prior to duplication the hydrogen bonds are broken, and the two chains unwind and separate. Each chain then acts as a template for the formation onto itself of a new companion chain, so that eventually we shall have *two* pairs of chains, where we had only one before. Moreover, the sequence of the pairs of bases will have been duplicated exactly. A study of our model suggests that this duplication could be done most simply if the single chain (or the relevant portion of it) takes up the helical configuration. We imagine that at this stage in the life of the cell, free nucleotides, strictly polynucleotide precursors, are available in quantity. From time to time the base of a free nucleotide will join up by hydrogen bonds to one of the bases on the chain already formed. We now postulate that the polymerization of these monomers to form a new chain is only possible if the resulting chain can form the proposed structure. . . . Whether a special enzyme is required to carry out the polymerization, or whether the single helical chain already formed acts effectively as an enzyme, remains to be seen. . . . For the moment, the general scheme we have proposed for the reproduction of deoxyribonucleic acid must be regarded as speculative. Even if it is correct, it is clear from what we have said that much remains to be discovered before the picture of genetic duplication can be described in detail. What are the polynucleotide precursors? What makes the pair of chains unwind and separate? What is the precise role of the protein? Is the chromosome one long pair of deoxyribonucleic acid chains, or does it consist of patches of the acid joined together by protein? Despite these uncertainties we feel that our proposed structure for deoxyribonucleic acid may help to solve one of the fundamental biological problems—the molecular basis of the template needed for genetic replication.[263]

The status of this "speculative" scheme remained uncertain for several years, as is evident from the discussion of the paper by Max Delbrück and Gunther Stent (presented by Stent) at the June 1956 symposium on the "chemical basis of heredity" at the McCollum-Pratt Institute.[264] At that meeting, Joshua Lederberg remarked that "an optimistic report of this symposium might indicate a new era in genetic study, where factorial descriptions are (about to be) replaced by chemical ones. The details of the speculations which have been put forward here may all prove to be wrong, but we are at a stage in the development of chemical genetics where we need stepping stones as well as foundation blocks."[265]

As we have seen, the first decisive experimental evidence in favor of the Watson-Crick speculation regarding the replication of DNA came in 1958 from the work of Matthew Meselson and Franklin Stahl, and of Arthur Kornberg and his associates. Meselson and Stahl took advantage of the increase in the buoyant density of DNA by the replacement of its ^{14}N by ^{15}N. A culture of *E. coli* that had been fully labeled with ^{15}N was allowed to multiply in a ^{14}N medium and was subjected at intervals to equilibrium ultracentifugation in a cesium chloride gradient. Meselson and Stahl found that when the cell population had doubled, the DNA in the culture was half-labeled. They concluded "that the nitrogen of a DNA molecule is divided equally between two physically continuous subunits; that, following duplication, each daughter molecule receives one of these; and that the subunits are conserved through many duplications."[266] This mode of replication is commonly termed semiconservative.

The historical background of the important contributions of Arthur Kornberg's group to the study of DNA replication lies in tortuous development of knowledge about the intermediate metabolism of nitrogen compounds in animal organisms. Some features of that development were described in the preceding chapter. Much was said about the formation of urea as a breakdown product of the metabolism of proteins in man and other "ureotelic" terrestrial mammals, but more needs to be said about the formation of uric acid, the principal nitrogenous excretory product in the "uricotelic" birds and snakes. The finding during the first decades of the nineteenth century that the appearance of uric acid in human tissues is associated with the affliction known as gout, especially among the propertied classes, spurred speculation about the ways in which urea might be a metabolic precursor of uric acid.[267] Although work during the 1930s in Hans Krebs's laboratory on the biosynthesis of uric acid in pigeon liver slices disproved these ideas and provided fruitful clues for later investigators, the solution of the problem began to appear only with the introduction of the isotope technique into metabolic research. During the 1950s the work of the groups associated with John Buchanan and Robert Greenberg showed that the purine ring is assembled in a series of reactions in which ribose-5-phosphate is converted to inosine-5′-phosphate (IMP) by the addition of the two carbons and nitrogen of glycine, the amido-nitrogen of glutamine, the amino-nitrogen of aspartate, two carbons from formyl tetrahydrofolate, and one carbon from CO_2.[268] I will not recount the sequence of the eleven known enzyme-catalyzed reactions in this process and will note only that the enzymes, and the mechanisms of their action, have been studied to good purpose by biochemists and organic chemists during the succeeding three decades. The purine nucleotides of RNA—AMP and GMP—arise by the enzyme-catalyzed ATP-dependent amination of UMP, with glutamine as the donor of its amide nitrogen, and the corresponding di- and triphosphates are formed in enzyme-catalyzed reactions with ATP.[269] The relatively rapid elucidation of the pathway in the biosynthesis of these purine nucleotides was a consequence of the ready availability of uric acid, a compound of known structure, and the fact that it is formed by the oxidation of hypoxanthine, via xanthine, a reaction catalyzed by the long-known xanthine oxidase.

The parallel study during the 1950s of the metabolic pathway in the biosynthesis of the pyrimidine nucleotides of RNA found a fruitful starting point in orotic acid (uracil-6-carboxylic acid), which had been isolated from whey and synthesized at the turn of the century. After the finding that orotic acid can meet the pyrimidine requirement for the growth of some bacteria, Arthur Kornberg's group showed that it is formed by the enzyme-catalyzed cyclization of N-carbamoylaspartate to dihydroorotate and dehydrogenation of the latter in an NAD-dependent reaction. The condensation of orotic acid with 5-phosphoribosyl-1-pyrophosphate yields orotidine monophosphate, which is decarboxylated to form uridine-5′-phosphate (UMP), and the reaction of UMP with ATP yields UTP, whose ATP-dependent amination by glutamine yields CTP.[270]

If the biosynthesis of the purine and pyrimidine ribonucleotides presented relatively few major biochemical surprises at a time when analogous enzyme-catalyzed reactions were being encountered in the study of other metabolic pathways, the study of the formation of the corresponding deoxyribonucleotides revealed the operation of unexpected enzyme mechanisms. Peter Reichard and his associates isolated from *E. coli* an oligomeric "reductase" that effects the replacement of the hydroxyl group at the 2′-position of ribonucleoside diphosphates by a hydrogen atom. The catalytic action of this enzyme involves the participation of a tyrosine unit in the form of a free radical, an oxygen-linked iron center, a small dithiol protein as the ultimate reductant, as well as superoxide dismutase. Another ribonucleotide reductase, from *Lactobacillus leishmannii*, contains cobalamin (vitamin B_{12}) in place of the tyrosine radical and iron center, and JoAnne Stubbe has proposed that the mechanisms of the catalytic action of the two enzymes have in common the formation of a "thiyl" ($RS^{\top}$) radical. *E. coli* also has a a ribonucleotide reductase with an iron-sulfur center and glycyl radical; this enzyme is operative during anaerobic growth, and formate is the hydrogen donor.[271]

Before returning to the 1950s, and the work of Arthur Kornberg's group on the replication of DNA, it should be mentioned that deoxythymidylate (dTMP) has been shown to be derived from dUMP, with N^5,N^{10}-methylenetetrafolate as the source of the methyl group, and that the 5-methyl cytosine found in some DNAs, and the 5-hydroxymethyl cytosine of T4 phage, appear to be formed by enzyme-catalyzed modification of cytosine units of preformed DNA.[272]

In 1956 Kornberg summarized the work done in his laboratory on the biochemistry of nucleotides and also described his initial studies on the enzymatic synthesis of DNA in *E. coli* extracts. In these experiments ^{32}P-labeled thymidine-5′-$\overset{*}{P}$PP (the asterisk denotes the position of the label) was used, and it was found that for optimum isotope incorporation into acid-insoluble material (containing DNA) there were required, in addition to the triphosphates of the three other DNA deoxyribonucleotides, ATP, Mg^{2+}, two soluble fractions and a "primer" fraction obtained by heating a mixture of DNA and the soluble fractions.[273] So began a brilliant series of investigations that marked a new stage in the development of genetic chemistry and that also revealed the complexity of the biochemical systems for the replication of DNA.

In a remarkable series of research papers during 1958–1969, Kornberg and his associates described the purification of the *E. coli* DNA polymerase (now usually termed pol I) and its properties as a protein and as an enzyme.[274] They showed that the polymerization reaction involves the successive transfer of the nucleoside-5′-phosphoryl group of a nucleoside triphosphate to the 3′-hydroxyl of the terminal nucleotide unit of a "primer" chain, with the release of pyrophosphate, whose hydrolysis by a pyrophosphatase makes the process very exergonic. The sequence of the addition of nucleotides to the growing primer chain was found to be determined by the constitution of a "template" DNA, and in the resulting double stranded DNA, the Chargaff "rules"—adenine equals thymine, and guanine equals cytosine—were obeyed. What attracted most attention was the finding, based on the nearest-neighbor frequencies of the bases, that the replication proceeded in an antiparallel complementary fashion, in accord with the Watson-Crick hypothesis. In addition, the highly purified pol I, a monomeric protein (molecular weight, 109,000), was able to effect the hydrolytic cleavage of terminal internucleotide bonds (an "exonuclease" action) and to catalyze, in the presence of pyrophosphate, the reversal of the polymerization. Also, pol I was subjected to limited proteolysis by subtilisin to yield two protein fractions, the larger of which (also known as the Klenow fragment) retained the polymerase and the 3′ to 5′ exonuclease activities, and the other the 5′ to 3′ exonuclease. Later work indicated that pol I has separate active sites for the polymerase and the 3′ to 5′ exonuclease reactions.[275]

The initial studies of the Kornberg group on pol I opened a large field of investigation, with important discoveries about the chemical events in DNA replication in various biological systems, and about the relevance of this chemical knowledge to that gained by the methods of genetics. For example, during the early 1960s Julius Marmur, Paul Doty, and their associates reported that irradiation of double-stranded DNA by ultraviolet light weakens the linkages between the two strands, and Adolf Wacker showed that this treatment leads to the formation of a butane derivative linking two thymine units; Richard Setlow then found such dimerization for cytosine as well.[276] After the discovery of the DNA polymerases, and their intrinsic exonuclease activity, it was shown that this structural damage could be repaired by hydrolytic excision of the affected fragment, replacement of the excised unit by polymerase action, and junction of the terminal nucleotides by means of "DNA ligase."[277] In addition to this repair mechanism, bacteria can "recover" from the effect of ultraviolet irradiation by incubation in light. This photoreactivation, studied by Albert Kelner and Renato Dulbecco, has been explained by the discovery of a "DNA photolyase" that catalyzes the cleavage of the pyrimidine dimers by electron transfer from light-excited reduced flavin.[278]

The ability of pol I to effect the synthesis of a genetically active DNA was strikingly demonstrated in 1967 by the use of the single-stranded phage φX174 as the template.[279] The problem of the replication of both chains of a DNA duplex proved to be more elusive. As has been noted, in 1953 Watson and Crick raised the question of the mechanism of the "unwinding" of their double-stranded antiparallel model of DNA into separate strands for the synthesis of new complementary DNA chains. The important studies of Doty and Marmur on the conditions favorable for reversible strand separation and hybridization and the finding by Reiji Okazaki that the newly formed DNA appears first in relatively short (1000–2000 nucleotide long) segments provided support for the idea, also advanced by Kornberg, that the growth of the primer ("lagging") strand is discontinuous, and that the short chains are then joined by the action of DNA ligase, while the template ("leading") strand is being replicated in continuous fashion.[280] This mechanism involves the existence of a shifting "replication fork" at which the 5′ to 3′ synthesis of the two new chains is moving in opposite linear directions.

Whatever hopes may have been held during the 1960s that pol I is indeed the sole enzyme involved in the biosynthesis of *E. coli* DNA were dashed by the report of Paula DeLucia and John Cairns that they had isolated a viable mutant that contained almost no pol I but was more sensitive to ultraviolet radiation than the parent *E. coli* strain.[281] Two other DNA polymerases were then identified in *E. coli,* one of which (now termed pol III) proved to be more effective than pol I in DNA replication. Although the action of pol III was found to exhibit some features of the above mechanism, the catalytic entity turned out to be a complex multiprotein assembly, whose dissection and reconstitution was described in 1975 by Arthur Kornberg and his associates.[282] Among the components of the assembly is an ATPase-like enzyme ("helicase") that effects the unwinding of the DNA duplex, and many other helicases were found later.[283] Another class of enzymes, named topoisomerases, which catalyze the transient hydrolytic cleavage of internucleotide bonds, was discovered during the 1960s and has been invoked as agents in genetic recombination.[284] The formation of such "nicks" allows the passage of DNA strands to pass through each other, a process corresponding to the "crossing-over" mechanism invoked by Morgan, and was shown by Robin Holliday in 1964 to lead to the appearance of X-shaped "heteroduplex" structures.[285] The holoenzyme pol III also includes a "DNA primase" that promotes both the initiation of replication by adding ribonucleotides to the 5′ end of incipient Okazaki fragments, and the hybridization of such RNA units with complementary units in the DNA template strand.[286] Moreover, the role of the histonelike proteins associated with bacterial chromosomes remains to be elucidated.[287] In short, what seemed in 1958, with the discovery of pol I, to be a relatively simple chemical mechanism of DNA synthesis in prokaryotes had become only a dozen years later a vastly more complex process, thanks to the close interplay, involving many investigators, of outstanding genetic analysis and equally distinguished biochemical research.[288] The genetic analysis of recombination-deficient mutants of *E. coli* led to the discovery of the "restriction endonucleases," mentioned above in connection with the determination of the nucleotide sequence of DNA preparations. The remarkable specificity of these nucleases has made them valuable tools, along with the DNA polymerases, in constructing recombinant DNAs.[289] This technique, which has attracted public attention, involves the joining of a DNA segment to a carrier ("vector") plasmid or phage that can be replicated in a suitable host cell, such as *E. coli* K-12.[290] The segment to be cloned may be one obtained by digestion with a restriction endonuclease or by organic chemical synthesis. After the work of Gobind Khorana and his associates during the 1960s on the synthesis of oligonucleotides, the methods for such chemical synthesis have been greatly improved and, as in the case of polypeptides, polydeoxyribonucleotides can now be made on polymer supports in automatic "synthesizers" using various protecting groups and, most recently, phosphoramidate derivatives in the coupling reaction.[291] The emergence during the

1980s of the "biotechnology" industry, both as divisions of established pharmaceutical and chemical companies and as new commercial enterprises, was spurred by the development of DNA recombination as a means of making important therapeutic agents; one of the early successes was the synthesis of human proinsulin by *E. coli* in this manner.[292]

These achievements of the 1980s were in some measure overshadowed by the invention by Kary Mullis and his colleagues at the Cetus Corporation of the "polymerase chain reaction" (PCR) for the "amplification" of DNA.[293] The method combines the knowledge provided by Doty and Marmur on the separation by heat of the strands of a DNA duplex with that provided by Arthur Kornberg and his associates on the mechanism of the action of the *E. coli* pol I. In each cycle, after the strand separation by heat, suitable primers are annealed, and each strand is duplicated by DNA polymerase. The enzyme now used (Taq pol I) is derived from the thermophilic bacterium *Thermus aquaticus*, and the method has been automated, so that 30–50 cycles can be performed in a single vessel. It appears that PCR experiments had been conducted around 1970 by Kjell Kleppe and Ian Molineux in Khorana's laboratory but were not developed further.[294]

To the above sketch of the emergence of the recombinant DNA techniques and their impact on the transformation of much of molecular biology into genetic chemistry must be added the recollection that in 1975 the pioneers in that development demonstrated the capacity of the members of a research discipline to set up strict guidelines for the continuance of a controversial line of research, and these rules were then adopted by the National Institutes of Health and by comparable governmental agencies in other countries.[295]

The knowledge gained from the study of DNA replication in bacteria has guided research on the more complex biochemical systems in eukaryotes. After Frederick Bollum and Van Potter identified in 1958 a DNA polymerase in calf thymus, many enzymes of this class were found in mammalian tissues and successively denoted α, β, γ, and so on. Thus far, the best known of these enzymes has been DNA polymerase α, a multiprotein assembly that exhibits "primase" activity.[296] Studies on the replication of SV40 DNA have indicated the participation of an unwinding protein ("T antigen"), of DNA polymerase δ, which possesses a $3'$ to $5'$ exonuclease activity, of a $5'$ to $3'$ exonuclease associated with DNA polymerase δ, and DNA ligase.[297] Additional protein components have been identified in other eukaryotic DNA polymerase holoenzymes.

The complexity of the assembly of enzymes and other proteins involved in the replication of DNA makes possible the recognition and correction of "errors" such as mismatches of complementary nucleotide units and the repair of damaged sections of a DNA chain. In large part, the "fidelity" of the process is due to the specificity of the polymerases themselves, suggesting a discriminating interaction with both members of a complementary nucleotide pair at each step of the chain elongation.[298]

There are still considerable gaps between the cytological and the biochemical knowledge about chromosomal DNA replication and recombination in plants and animals during the processes of meiosis and mitosis, although much progress has been made since the 1950s, when Daniel Mazia reported the isolation of the mitotic apparatus of the sea urchin egg and found it to be largely protein in nature.[299] In seeking chemical explanations for the remarkable chromosomal movements in mitosis, he called attention to Hans Weber's suggestion that these movements are comparable to muscular contraction dependent on the cleavage of ATP.[300] During the 1960s the use of glutaraldehyde as a fixative by cytologists, notably Keith Porter, led to the recognition that what was once thought to be clear cytoplasm is a meshwork of thin filaments, which were named microtubules.[301] Subsequent work, by Shinya Inoué and other investigators, has shown that chromosome segregation in both meiosis and mitosis involves the participation of microtubules bearing proteins named kinesin and dynein, which, like actomyosin, couple the cleavage of ATP to molecular movement.[302] Other components of eukaryotic chromosomes, in addition to the histone-DNA "nucleosomes," mentioned above, are the chromatin-containing "centromeres," whose proteins appear to be involved in the regulation of the mitotic process, and the "telomeres," DNA-protein structures at the ends of chromosomes, with repetitive stretches of guanine nucleotides, which were shown by Elizabeth Blackburn to be synthesized by ribonucleoprotein enzymes named telomerases.[303] Among the many other features of DNA replication and recombination whose biochemical mechanisms are certain to receive continued attention are the chromosome rearrangements involving "transposons," DNA segments that move from one genetic location to another.[304] To these efforts must be added the ambitious and expensive governmental Human Genome Project, about whose advisability there has been considerable difference of opinion among investigators in this field, and which appears to have attracted the interest of commercial companies.[305]

The Role of RNA in Protein Synthesis

In 1941 Torbjörn Caspersson interpreted his cytochemical findings, mentioned earlier, as indicating that euchromatin controls the formation within the nucleus of specific proteins, while the heterochromatin is involved in the synthesis of histonelike proteins that move out of the nucleus and induce the formation of RNA, which controls the formation of cytoplasmic proteins.[306] This theory is reminiscent of the view, held by late-nineteenth-century cytologists, that the nucleus is the principal seat of intracellular synthesis.[307] Among those who questioned the validity of Caspersson's interpretation was the Belgian embryologist Jean Brachet, who had been using cytochemical staining techniques to study the role of nucleic acids in the development of sea urchin eggs. Brachet found large amounts of RNA in the cytoplasm, and during the 1940s advanced the theory that this RNA is involved in protein synthesis and is associated with the subcellular particles identified by Albert Claude as microsomes.[308]

After the studies by Charles Garnier during the 1890s on the basophilic components in the cytoplasm of glandular cells, his observations were extended to various animal tissues by other investigators.[309] It was not until the 1930s, however, that the availability of the ultracentrifuge for the separation of particulate subcellular elements, and of the electron microscope for their visualization, made possible the achievements of Albert Claude and later of Keith Porter and George Palade in defining the nature of what were variously called microsomes, or small particulate components. These particles were named ribosomes after chemical analysis during the 1950s by Philip Siekevitz and the research group associated with Paul Zamecnik showed them to be ribonucleoprotein particles.[310] Soon afterward, Zamecnik's group also showed that the incorporation by ribosomes of ^{14}C-labeled amino acids into material precipitable by trichloroacetic acid, and assumed to represent protein synthesis, required the presence of ATP, a "soluble RNA" ("sRNA," later termed transfer RNA or tRNA), and GTP. The process appeared to involve the intermediate formation, with the release of pyrophosphate (PP), of amino acyl adenylates (RCO-5′-AMP), termed activated amino acids.[311] Such compounds were similar in their chemical reactivity to the acyl chlorides, acyl anhydrides, and thiol esters (RCO-X) all used in the laboratory synthesis of peptides,[312] and in 1941 Fritz Lipmann had suggested that "the acyl phosphates of amino acids might likewise condense with the amino groups of other amino acids to form the proteins."[313] Four years later he reported his experiments on the enzyme-catalyzed acetylation of sulfanilamide to produce what he termed a "peptidic" bond. As has been mentioned, this work led to the discovery of coenzyme A.

In his studies on the possible role of acetyl phosphate as a metabolic acetylating agent, Lipmann used the known reaction of RCO-X compounds with hydroxylamine (NH_2OH) to form hydroxamic acids (RCO-NHOH), which give a red product with ferric salts, and this test was applied by Mahlon Hoagland, Elizabeth Keller, and Paul Zamecnik to identify in liver extracts the presence of "activating" enzymes, which catalyze the reaction of amino acids with ATP.[314] In subsequent work, notably by Paul Berg's group, the preferred enzyme assay was the measurement of the rate of the amino acid–dependent exchange of ^{32}P between ATP and ^{32}PP.[315]

Before recounting the further development of research on the ribosomal synthesis of proteins, I insert a passage from Lipmann's autobiography, published in 1971, and offer some comments about it. In that book, Lipmann referred to his early thoughts about protein synthesis: "At that time much effort was made to show that under special conditions, proteolytic enzymes could perform a peptide condensation (Figure 1). . . . I naively thought then that protein synthesis could be more or less solved if one understood the mechanism of amino acid activation. It took me a long time to realize that, in contrast to most other biosyntheses in the making of a protein, this was just a premise."[316] Because "much [of the] effort" mentioned by Lipmann was being expended in my laboratory, I was dismayed at his documentation of that statement, for "Figure 1" was a photograph of two test tubes labeled *A* and *B* and entitled "Proteinlike material from peptide [*sic*] digest of egg albumin. A. Without chymotrypsin. B. After 45-min incubation with chymotrypsin solid gel has formed. [Reprinted from the *Journal of the American Chemical Society,* vol. 73, 1288 (1951). Copyright 1951 by the American Chemical Society. Reprinted with permission of the copyright owner.]" The author of that paper was not named; it was Henry Tauber, and his article dealt with the formation of plasteinlike products obtained by the action of chymotrypsin on the products of the digestion of egg albumin by pepsin. The term *plastein* had been applied to an ill-defined product that precipitates when a concentrated peptic digest of a protein is incubated with pepsin at pH 4. It was described by Vasili Zavyalov (Sawyalow) in 1901 and later studied by several investigators, notably Hardolph Wasteneys and Henry Borsook, who considered the formation of plastein to represent a reversal of proteolysis.[317] In his egregious distortion of the historical record, Lipmann chose to ignore the difference between studies on plastein formation and those on the enzymatic synthesis of peptide bonds in clearly defined chemical reactions, conducted by my associates and others, notably the groups associated with Max Brenner and Theodor Wieland. We showed during the 1950s that the free energy values Lipmann used for the hydrolysis of peptide bonds were incorrect and that the amides and oxygen esters of acylamino acids or peptides could participate effectively in specific proteinase-catalyzed peptide chain elongation reactions which we termed transpeptidation reactions.[318] At that time, the relevance of our chemical studies to the problem of the intracellular mechanism of protein synthesis was uncertain. As defined in our work, the term *transpeptidation* referred to reactions catalyzed by proteinases of defined specificity of action at peptide bonds. The question then was: are there free peptide intermediates in the pathway of protein biosynthesis, and do they undergo transpeptidation reactions, or do proteins arise solely by the assembly of activated amino acids? In 1951 Christian Anfinsen and Daniel Steinberg isolated crystalline radioactive egg albumin from minced oviduct that had been incubated with $^{14}CO_2$, subjected the protein to partial enzymic hydrolysis, and found that the aspartic acid in the separated products was labeled in a nonuniform manner. This result was consistent with the hypothesis that the protein had been synthesized by way of peptide intermediates. In subsequent work the Anfinsen group obtained similar results for the synthesis of insulin and ribonuclease A in the appropriate tissues.[319] On the other hand, Peter Campbell and Thomas Work found uniform labeling in the biosynthesis of lactoglobulin and casein, in accord with the view that these proteins had arisen from free amino acids.[320] Later research established firmly the validity of the approach initiated by Paul Zamecnik and his associates, and the possible intermediate formation of free peptides and their participation in proteinase-catalyzed transpeptidation was discounted. Our work proved its worth, however, in the 1970s, when Michael Laskowski Jr. applied it in studies on peptide bond synthesis in a protein system and spurred the development of the use of proteinases as catalysts in the preparative synthesis of complex peptides.[321] Moreover, as will be seen shortly, peptides linked to RNA are now considered to be intermediates in the synthesis of proteins, and the term *transpeptidation* was adopted by Lipmann in 1969 to denote the peptide-forming step in the ribosome-mediated elongation of peptide chains.[322]

I return to the 1950s, when there was much speculation about "templates," which direct the order of the addition of amino acid units to a growing peptide chain, so as to produce the unique sequence in a protein. The general concept of a template, advanced in relation to the interaction of an enzyme and its substrate, or of an antibody with an antigen, had also been invoked earlier in connection with the replication of genes.[323] In 1952 Alexander Dounce suggested that a nucleic acid template is first phosphorylated to produce a polypyrophosphate which reacts with amino acids, followed by their polymerization, and in 1953 Fritz Lipmann offered a similar scheme, except that instead of specifying a nucleic acid template, he proposed that "amino-acid-specific activation spots are lined up on a structure in a demanded sequence," followed by a condensation that "forms the polypeptide chain laid down through the specific attraction of the active centers for particular amino acids."[324]

A decisive advance came with the finding, in Zamecnik's laboratory, of "soluble" RNA (sRNA, now tRNA) to which a ^{14}C-labeled amino acid was attached.[325]

This important discovery was reported in the same year that Francis Crick proposed that protein synthesis occurs on

> an RNA template in the cytoplasm. The obvious place to locate this is in the microsomal particles, because their uniformity of size suggests that they have a regular structure. It is also follows that the synthesis of at least some of the microsomal RNA must be under the control of the DNA of the nucleus. . . . Granted that the RNA of the microsomal particles, regularly arranged, is the template, how does it direct the amino acids unto the correct order? . . . It is a natural hypothesis that the amino acid is carried to the template by an "adaptor" molecule, and that the adaptor is the part which actually fits on to the RNA. In its simplest form one would require twenty adaptors, one for each amino acid. What sort of molecules such adaptors might be is anybody's guess. . . . There is one possibility which seems inherently more likely than any other—that they might contain nucleotides. This would enable them to join on to the RNA template by the same "pairing" of bases as is found in DNA.[326]

While Crick was musing about "adaptors," Paul Berg had demonstrated in 1956 that the ATP-dependent synthesis of acetyl-CoA by *E. coli* extracts involved the formation of an intermediate acyladenylate, with the release of PP, and two years later he reported the formation of such an intermediate in the RNA-dependent incorporation of labeled amino acids into the acid-insoluble material in these extracts.[327] Subsequent work in the laboratories of Berg and Zamecnik showed that the amino acyl group is linked to a hydroxyl group of the 3′-terminal adenylyl (A) unit of tRNA, and that this unit, together with two cytidylyl (C) units in the sequence RNA-C-C-A, form a structure common to the various amino acid specific tRNAs then identified.[328]

As mentioned earlier, the tRNAs turned out to be relatively small polyribonucleotides, and in 1965 Robert Holley and his associates reported the complete nucleotide sequence of the alanine-specific tRNA from yeast. Since then, the sequences of very many tRNAs from more than a hundred organisms have been determined and are usually represented in printed reports as variants of a "canonical" cloverleaf structure, based on hydrogen bonding of adenine-uracil (A-U) and guanine-cytosine (G-C) pairs, with many instances of the appearance of dihydrouracil or thymine units. Studies by means of X-ray crystallography or nuclear magnetic resonance spectroscopy have indicated considerable variability and flexibility in the three-dimensional structure of the tRNAs. Moreover, the suggestion in 1972 by Robert Loftfield[329] that the aminoacylation of tRNA is a concerted reaction, without the formation of an intermediate aminoacyl adenylate, has received additional experimental support, and the specificity of the "aminoacyl-tRNA synthetases" (as the "activating enzymes" are now known) is not as absolute as once thought.[330] We will return to some aspects of the role of RNA in protein synthesis later, in connection with the emergence of the "genetic code," but must first consider a parallel development, initiated by the studies of Jacques Monod on "enzyme induction."

From Enzyme Adaptation to the Operon

In 1947 Monod prefaced a report on his work by stating that "it is generally recognized that one of the main problems of modern biology is the understanding of the physical basis of specificity, and of the mechanisms by which specific molecular configurations (or multi-molecular patterns) are developed, maintained, and differentiated. The means, the experimental tools for this study, are found in those experiments which result in the formation, or suppressing the synthesis, or modifying the distribution of a specific substance or substances."[331] Monod began his studies on enzyme adaptation during the 1940s with the examination of the growth response of washed bacteria grown on a sugar such as maltose (α-glucosylglucose) or lactose (β-galactosylglucose) to a medium containing the other sugar. He extracted from such bacteria the enzyme "lactase" (a β-galactosidase), which hydrolyzes lactose, and the enzyme "amylomaltase," which catalyzes a transglycosylation reaction leading to the formation of a starchlike polysaccharide. Monod found that although either lactose or galactose could induce the formation of lactase, only maltose was able to induce the formation of amylomaltase. From the study of mutants with respect to the lactose and maltose characters, he concluded that "such observations could hardly be interpreted without assuming that the formation of each of these enzymes depends not only on an external specific stimulus, but in addition on the presence (let us rather say 'the functional presence') of at least one specific hereditary determinant."[332]

There can be little doubt that by 1950 Monod's approach to the problem of enzyme adaptation was influenced by Joshua Lederberg's work of the 1940s on the genetics of *E. coli* K-12. In that work Lederberg used *lac*⁻ mutants, some of which lacked β-galactosidase, found in *o*-nitrophenylgalactoside an excellent chromogenic substrate for this enzyme, and showed that while *lac*⁻ cells produced little of the enzyme when grown on lactose, they produced substantial amounts when grown on alkyl galactosides, which are much poorer substrates. This finding led Lederberg to state in 1950 that "since adaptation is presumably a physicochemical rather than an entelechist process, such deviations are not surprising but suggest the need for revising 'adaptive enzyme formation' in favor of a more general term connoting 'enzyme formation under environmental influence.' "[333] In 1949 Monod was joined at the Pasteur Institute by Melvin Cohn, and they embarked on an intensive investigation of the genetic control of the formation of *E. coli* β-galactosidase. A large number of synthetic galactosides were tested as substrates and inducers, and among them phenyl-β-thiogalactoside was exceptional in its high capacity as an inducer, though completely resistant to hydrolysis by the enzyme. In a 1952 report of their work they proposed "to abandon the term 'enzymatic adaptation' to adopt that of 'induced biosynthesis of enzymes' which is precise and sufficiently descriptive to be readily understood as relating to *the induction of the biosynthesis of enzymes under the influence of specific substances.*"[334]

During the course of Cohn's work at the Pasteur Institute, much was learned about the properties of β-galactosidase, a putative precursor (Pz) of the enzyme was identified but soon abandoned, experiments with labeled amino acids showed that the protein was newly synthesized, and it was concluded that the "dynamic state" of proteins in animal organisms did not apply to intracellular protein metabolism.[335] Monod's subsequent collaboration with François Jacob, who had worked with Elie Wollman on lysogeny, led to a genetic analysis of the β-galactosidase system of *E. coli* and to the conclusion that "in the synthesis of many proteins, there exists a dual genetic determinism, with the participation of two functionally distinct genes: one (the structural gene) responsible for the structure of the molecule, the other (the regulatory gene) governing the expression of the first by means of a repressor."[336] According to this model, the regulatory gene *(i)* forms an inhibitor ("repressor") of the action of the structural gene *(z)* and of one *(y)*, which controls the formation of a "permease" considered to catalyze the entry of a galactoside into the cell.[337] To explain the coordinated effect of the regulatory genes, Jacob and Monod then concluded that

> it seems necessary to invoke a new genetic entity, an "operator" which would be: (a) adjacent to a group of genes and controlling their activity; (b) sensitive to a repressor produced by a particular regulatory gene. . . . The hypothesis of the operator implies that between the classical gene, an independent unit of biochemical function, and the entire chromosome, there exists an intermediate genetic organization [that] comprises the *units of coordinated expression (operons)* constituted by an operator and the group of structural genes which it coordinates. By means of the operator, each operon would be subject to the action of a repressor whose synthesis would be determined by a regulatory gene.[338]

Some of the ideas embodied in the Jacob-Monod model had been put forward by others—for example, by Martin Pollock in the case of the "organizer" or by Henry Vogel and by Richard Yates and Arthur Pardee in the case of the "repressor."[339] Contrary to expectation, the *gal* repressor turned out to be an oligomeric protein, rather than a polynucleotide.[340] Moreover, later work by Ellis Englesberg and his associates demonstrated a case where, in the absence of repressor, the regulatory gene can "activate" the operon, and other investigators identified additional controlling elements in the *gal* operon.[341] It must be emphasized, however, that although the restrictive definition originally assigned to the operon by Jacob and Monod has been extensively modified,[342] that concept played a significant role in the subsequent study of the mechanisms of the "transcription" of the nucleotide sequence of chromosomal DNA into the amino acid sequence of the proteins made in the cytoplasm, and in the discovery of "messenger RNA." Before considering the development of knowledge about these mechanisms, it is necessary to recall some of the parallel events in the formulation of the "genetic code."

More will be said about the contributions of Monod and Jacob to the study of metabolic regulation. As regards the operon, I conclude this section with the following comment by Melvin Cohn: "Very often a desire for generalization with elegance plus parsimony is so compelling that other great discoveries are blotted out. Having revealed negative control by the *lac* repressor, truly a spectacular achievement, the Pasteur group was resistant to any challenge to its universality. Any hint of other mechanisms of control was shrugged off on intuitive grounds, sometimes correctly, sometimes incorrectly."[343]

The Genetic Code

The first formulations of the genetic code during the 1950s were further examples of the hypnotic power of numerology, like the periodicity hypotheses of protein structure from Albrecht Kossel to Max Bergmann and Carl Niemann. If these noted protein chemists extrapolated imaginatively from their analytical data, George Gamow used only two numbers (4 and 20) to devise arrangements of the four known nucleotides of DNA (abbreviated A, C, G, T), and chose the twenty most frequently found protein amino acids. Gamow, a respected theoretical physicist and proponent of the Big Bang theory of the origin of the universe, an engaging and bibulous playing-card juggler, author during the 1930s of the delightful Mr. Tompkins popular science books, and collaborator of Edward Teller in the design of the hydrogen bomb, found in the 1953 papers of Watson and Crick a challenge to his playful ingenuity.[344] Gamow devised a scheme, illustrated by means of playing cards, that involved sets of three adjacent nucleotides per amino acid unit ("triplet" code) in a sequence of overlapping triplets.[345] That proposal spurred Francis Crick and his colleagues to examine the coding problem more critically and to use knowledge gained from genetic experiments to test the possible validity of Gamow's scheme and its variants. By 1961 they had concluded that the nucleotides of each triplet (Sydney Brenner named it a "codon") did not belong to any other triplet ("nonoverlapping" code); that sets of triplets are arranged in continuous linear sequence starting at a fixed point on a polynucleotide chain, without breaks ("commaless" code), thus determining how a long sequence is to be read off as triplets; and that more than one triplet can code for a particular amino acid ("degenerate" code).[346]

In 1961, however, the search for the genetic code was transformed from an exercise in numerology, aided by genetic analysis, into an experimental biochemical investigation involving the intensive (and competitive) efforts of several laboratories. That research began with the dramatic announcement by Marshall Nirenberg that he and Heinrich Matthaei had found polyuridylic acid (polyU, prepared from UDP by the action of polynucleotide phosphorylase, of which more later) to promote specifically the incorporation of ^{14}C-labeled L-phenylalanine into an acid-insoluble "polyphenylalanine" by *E. coli* ribosomes in an ATP-dependent process that also required the sRNA fraction of an *E. coli* extract. They concluded that "polyuridylic acid contains the information for the synthesis of a protein having many of the characteristics of poly-L-phenylalanine. . . . One or more uridylic acid residues therefore appear to be the code for phenylalanine. Whether the code is of the singlet, triplet, etc., type has not yet been determined. Polyuridylic acid seemingly functions as a synthetic template or messenger RNA, and this stable, cell-free *E. coli* system may well synthesize any protein corresponding to meaningful information contained in added RNA."[347] This exciting paper was quickly followed by further reports from Nirenberg's laboratory and from that of Severo Ochoa, where the enzymatic synthesis of polyribonucleotides had previously been discovered. I will not list all the code assignments that emerged during 1961–1963 from this work but will only note the finding that polyadenylic acid (polyA) coded only for lysine, thus producing a soluble oligopeptide whose chemical nature could be defined more clearly than in the case of very insoluble oligophenylalanine peptides, and that mixed polyUA (5:1) promoted the incorporation of phenylalanine, tyrosine, leucine, and isoleucine.[348] Subsequently, Nirenberg and his associates showed that coding assignments could be made by determining the effect of trinucleotides (for example, UUU, UUC, UCC, CCC) on the binding of ^{14}C-labeled aminoacyl-sRNA to *E. coli* ribosomes.[349] Then came the elegant contributions of Gobind Khorana, who with his associates synthesized polydeoxyribonucleotides with repeating di-, tri-, and tetranucleotide units such as d[TC·GA], d[TTC·GAA], and d[TATC·GATA], used them to make (with DNA-dependent RNA polymerase and the appropriate ribonucleotide triphosphates) polyUC, polyUUC, and polyUAUC respectively for test of the amino acid code assignments. Thus polyUAUC, with codons in the sequence UUA-CUU-ACU-UAC, promoted the incorporation of amino acids in the sequence Tyr-Leu-Ser-Ile.[350] The genetic code that emerged from these studies is extremely "degenerate," with six codons for arginine, leucine, and serine; four for alanine, glycine, proline, threonine, and valine; and three for isoleucine. Except for methionine and tryptophan, with only one codon, the other amino acids have two codons. This degeneracy led Crick to propose the "wobble hypothesis," which assigned the "standard" position to the first two nucleotides in a codon and variation to the third nucleotide.[351] Three triplets (UAA, UAG, UGA) did not appear to code for any amino acid and were termed "nonsense" codons, but they were later shown to be involved in the termination step of polypeptide synthesis.[352] I will return to these features of the genetic code in connection with the parallel studies on the role of messenger RNA and the mechanism of ribosomal protein synthesis.

A striking test of the relevance of the results of these *in vitro* studies to the synthesis of proteins in biological systems was provided during the 1960s by Charles Yanofsky and his associates. That work was the capstone of a sustained program of research on the *E. coli* enzyme tryptophan synthetase, which they found to be composed of two subunits (A and B). In parallel with the purification of the A component and determination of its amino acid sequence, genetic analysis of numerous artificially induced *E. coli* K-12 mutants permitted a test of the concept that the sequence of nucleotides in DNA is "colinear" with that of amino acids in a protein made under its control. Moreover, the Yanofsky group showed that in two mutants, the glycine unit in a particular position in the protein had been replaced by an arginine or glutamic acid unit with loss of functional activity and that, upon mutational reversion, each of these mutants produced further mutants in which the amino acid in that position had been changed to threonine, serine, glycine, alanine, or valine, with restoration of partial or total activity. These changes could be correlated with the changes in the DNA triplets assigned to these amino acids in the formulation of the genetic code[353] and also provided an example of what came to called a "missense" mutation leading to the "suppression" of a previously determined genetic character.[354]

Since these studies of the 1960s, the triplet assignments for mitochondrial DNA has been shown to differ in some respects from those in the "standard" set of sixty-four codons, which appears to apply to the control of protein synthesis in nearly all organisms for which data are available. Among the exceptions are ciliated protozoa *(Paramecium, Tetrahymena)*.[355]

The Multiple Roles of RNA

As was mentioned above, the decipherment of the genetic code involved the use of polyribonucleotides prepared by the action of polynucleotide phosphorylase. This enzyme had been discovered in 1955 at the laboratory of Severo Ochoa by Marianne Grunberg-Manago in extracts of *Azotobacter vinelandii,* and had been shown to effect the polymerization, without the need of a template, of ribonucleoside-5′-diphosphates (ADP, CDP, UDP) in a transphosphorylation reaction analogous to that catalyzed by glycogen phosphorylase.[356] Similar enzymes were identified in *E. coli* and other microorganisms and, as in the case of pol I in relation to the biosynthesis of DNA, polynucleotide phosphorylase was at first thought to be the catalyst of intracellular RNA synthesis. That function was soon assigned to a different enzyme system, of which more shortly, but the finding that the enzymatically produced polymers from ADP and UDP (polyA and polyU) can associate ("anneal") to form a duplex marked the beginning of the experimental study of the "hybridization" of DNA and RNA chains. These studies, by Alexander Rich and by Paul Doty's group, and the demonstration by Benjamin Hall and Solomon Spiegelman of the association of T2 phage DNA and RNA strands, not only offered additional support for the Watson-Crick hypothesis but also provided the basis for the invention of valuable techniques, notably the blotting method devised for DNA fragments by Edwin Southern and later the so-called Northern blot method for RNA.[357]

In the "genetic operator model" proposed in 1961, Jacob and Monod used the term *messenger RNA* (or mRNA) to denote "a very short-lived intermediate both rapidly formed and rapidly destroyed during the process of information transfer." They assumed that it should be a polynucleotide whose base composition reflects that of DNA and that "it should, at least temporarily, or under certain conditions, be found associated with ribosomes, since there are good reasons to believe that ribosomes are the seat of protein synthesis."[358] In support of their assumptions, they cited the report of Eliot Volkin and Lazarus Astrachan in 1957 that T2 phage-infected *E. coli* cells produce a "minor species" of RNA whose base composition is similar to that of the phage DNA, and which undergoes rapid metabolic turnover.[359] At the time, this report attracted little attention, but a few years later its importance was established by work in several laboratories.[360] Also, enzymes that catalyze the DNA-dependent synthesis of polyribonucleotides were identified in many microbial, plant, and animal organisms and are now called RNA polymerases. The initial studies during the early 1960s, notably by the groups associated with Samuel Weiss, Jerard Hurwitz, and Paul Berg showed that in this process the triphosphates of all four nucleosides (ATP, GTP, CTP, UTP) were required, that the chain elongation proceeded in the 5′ to 3′ direction, and that the base composition of the resulting polyribonucleotides was complementary to that of the DNA strand that primed the synthesis.[361] As in the case of the DNA polymerases, the groundwork for the further study of these enzymes was provided by the investigation in many laboratories of the DNA-dependent *E. coli* RNA polymerase. By 1970 it was recognized that the action of this enzyme involves the operation of a multiprotein catalytic assembly (a "holoenzyme") composed of at least five subunits, one of which (denoted σ) is essential for the initiation of the polymerization, and some of them interact with the DNA template during the course of RNA synthesis.[362] In addition to the "sigma" unit, a "rho" protein was identified as a termination factor.[363] By 1970 it had also been shown by several investigators that upon infection of *E. coli* with a DNA phage such as T7, the production of phage protein is accompanied by the appearance of a phage-specific RNA polymerase, and after infection by an RNA phage such as Qβ the RNA directs the synthesis of an RNA-dependent enzyme, whose subunit structure is somewhat different from that of the DNA-dependent RNA polymerase.[364]

Three DNA-dependent RNA polymerases have been identified in eukaryotic cells; the one denoted pol II catalyzes the synthesis of mRNA, while pol I and III are considered to be responsible for the formation of ribosomal RNA and of tRNA, respectively. These "holoenzymes" are more complex assemblies than the *E. coli* RNA polymerase; pol II is composed of about ten subunits, and its action is regulated by many "transcription factors," whose number and variety have been increasing as more species of organisms are studied. Among these factors are DNA-binding proteins that interact specifically with nucleotide sequences (for example, TATA) that serve as a point of chain initiation, or proteins with a spatial arrangement of cysteine and histidine units that readily chelates zinc ions to form "zinc fingers" for interaction with DNA, or proteins that bind DNA by virtue of a hydrophobic repeating sequence of leucine units to form "leucine zippers," or proteins that can form "polar zippers," with repeating Glu-Glu-Lys-His sequences.[365] To these quaintly named transcription factors, and others to which bizarre acronyms have been assigned, must be added the "CAP" protein, which combines with cyclic AMP to form still another DNA-binding regulatory agent.[366] Moreover, because all the nucleotides in eukaryotic DNA do not appear in the functional "mature" mRNA, it was recognized that this RNA is produced from the primary RNA transcript in what were named splicing reactions (whose enzymology will be considered shortly), in which the regions complementary to the so-called exons of the DNA are retained, and the intervening regions, complementary to the DNA "introns," are excised.[367] Also, it was found that the mRNA product has a 3′-terminal polyadenylate tail.[368]

The most exciting biochemical developments in this field during the 1970s came from the study of the infection of animal cells by oncogenic RNA viruses. In adjacent papers published in 1970 Howard Temin reported the discovery of an RNA-dependent DNA synthetase in the Rous sarcoma virus, and David Baltimore announced the independent finding of a similar enzyme in the mouse leukemia virus.[369] These enzymes, which came to be called reverse transcriptases (the viruses were named retroviruses), were then isolated in considerably purified form, and the mechanism proposed for the RNA-directed synthesis of DNA involves a series of steps in which there occurs hybridization with a DNA primer, elongation of the DNA strand, and removal of the RNA by a ribonuclease.[370] Apart from requiring a revision of the "Central Dogma" of the unidirectional flow of genetic "information" from DNA to RNA to protein, the availability of the reverse transcriptases made possible the synthesis of "complementary (or copy) DNA" (now abbreviated cDNA) on an RNA template when provided with an oligodeoxyribonuucleotde primer such as oligo-dT, which can pair with the poly-A sequence at the 3′-end of most eukaryotic mRNA strands. With modern synthetic methods, any desired primer containing 15 to 20 nucleotides can readily be prepared.[371] The cDNA made with a "mature" mRNA may be said to represent the DNA that directed the formation of that mRNA, minus the noncoding "introns." It should be added that organic chemical synthesis has also provided "antisense" oligonucleotides, with sequences complementary to natural mRNAs, which have been tested as inhibitors of gene action.[372]

The addition to the known enzymic tools of nucleic acid chemistry (DNA polymerase, DNA ligase, and so on) of the RNA-dependent DNA synthetases and of the restriction endonucleases that produce DNA fragments with "sticky" ends spurred the insight and enterprise of the research groups associated with Paul Berg, and of Stanley Cohen and Herbert Boyer, in opening the age of DNA recombinant technology and the cloning of genes by joining DNA to a linear bacterial plasmid or phage DNA which served as a carrier ("vector") for the kind of transduction into *E. coli* K-12 discovered by Joshua Lederberg and Norton Zinder.[373] One of the early practical fruits of this dramatic development was the synthesis of human insulin by *E. coli*.[374]

Another source of excitement during the 1980s, though less intense than that provided by gene technology, was the demonstration by Sidney Altman and by

Thomas Cech, with their respective associates, that some RNA molecules catalyze the cleavage of internucleotide bonds, thus enlarging the view of the chemical nature of enzymes to include what have come to be called ribozymes. This development began with Altman's discovery of a precursor of tyrosine transfer RNA in *E. coli,* and led to the isolation of a ribonuclease (denoted RNase P) in the form of a nucleoprotein whose enzymic activity resides in its nucleic acid portion.[375] In the presence of Mg^{2+}, this endonuclease catalyzes the hydrolysis of the precursor at a U-G internucleotide bond, with the formation of the "mature" tRNA having a G-5′-P end and the C-C-A-3′-OH amino-acid acceptor end. Studies on the specificity and the chemical mechanism of the action of RNase P have indicated that only certain nucleotide regions (including a portion of the 3′-OH end) of the precursor are required for enzyme-substrate interaction. It was suggested that RNase P is a metalloribozyme and that the function of the protein component is to stabilize the catalytically active conformation of the enzyme.[376] Catalytic nucleoproteins similar to RNase P, with RNAs of about 400 nucleotide units but differing in nucleotide sequence, have been found in many organisms. Hypothetical secondary structures, with linear duplex stretches based on optimum traditional base pairing (plus some "noncanonical" G-U pairs), as in Hans Zachau's four-loop cloverleaf model of the tRNAs, have been proposed for the catalytic RNAs.[377]

We have seen that the formation of mRNA from the precursors generated by the action of DNA-dependent RNA polymerases involves "splicing," with the excision in the transcripts of "introns." This process was discovered by Phillip Sharp and his associates during the 1970s, and subsequent work showed that it is effected by small nuclear ribonucleoprotein particles ("snRNP," pronounced "snurp"!) associated in "spliceosomes."[378] Also, by 1980 genetic analysis had shown that the synthesis of some proteins—for example, globin—is controlled in eukaryotic organisms by segments of chromosomal DNA separated by noncoding nucleotide sequences.[379] There was therefore great interest in the 1981 report of Thomas Cech and his associates that a ribosomal RNA of the ciliated protozoan *Tetrahymena thermophilia* contains an "intervening sequence" (IVS) of about 400 nucleotides that is excised in a process initiated by GTP or, as was later found, by guanosine itself.[380] The Cech group showed that the 3′-OH of guanosine attacks an internucleotide bond in a transesterification reaction, with the attachment of guanosine to the 5′-phosphoryl end of the IVS, and the newly liberated 3′-OH of the "exon" attacks the far end of the "intron" in another transesterifcation reaction to produce a circular IVS that is further modified into a linear polyribonucleotide. Because no protein catalyst appeared to be involved in this process, at that time Cech considered it to be nonenzymic in nature. After the 1983 report from Altman's laboratory on the enzymic activity of the RNA of RNase P, however, Arthur Zaug and Cech showed that the linear IVS also has the properties of an enzyme.[381]

"Self-splicing" ribozymes of the kind found in *Tetrahymena* and denoted group I introns have also been identified by Marlene Belfort in bacteriophage.[382] In another type (group II introns), found in fungi and plants, the process is initiated by the 2′-OH of an internal adenosine to form a 2′-5′ phosphodiester. Still other catalytic RNAs, denoted hammerhead and hairpin ribozymes, produce products with 2′,3′-cyclic phosphate and 5′-OH ends.[383] Clearly, the ribozyme field has become a thriving subdiscipline. It has also spawned new theories of the origin of life and attracted the attention of some philosophers of science.[384]

The Elongation of Peptide Chains

During the 1950s studies in several laboratories, notably that of Paul Zamecnik, on the distribution of protein-bound ^{14}C among the subcellular fractions (nuclear, mitochondrial, microsomal, soluble) of mammalian liver, after the administration of ^{14}C-labeled amino acids to living animals, indicated that the microsomal fraction is the earliest to be labeled. Parallel electron microscopic study and chemical analysis, largely by George Palade and Philip Siekevitz, showed that the microsomes represent particles composed of RNA and protein, and attached to lipoprotein components of the cytoplasmic "endoplasmic reticulum."[385] The ribonucleoprotein particles obtained after removal of the lipids with detergents such as deoxycholate were termed "ribosomes," and much of the subsequent work on the intracellular mechanism of the elongation of polypeptide chains was focused on the chemical nature and function of these particles. Chemical analysis showed them to be composed of 50–60 percent RNA and 40–50 percent protein and also to contain polyamines such as spermidine or putrescine. The ribosomes studied most intensively were those of *E. coli,* and around 1960 several investigators, notably Charles Kurland, Masayasu Nomura, Mary Petermann, and Alfred Tissières, showed that they have a sedimentation coefficient of about 70S and are composed of two subunits (50S and 30S).[386] Subsequent work during the 1960s and 1970s by the groups associated with Nomura, Tissières, and Heinz Wittmann demonstrated that the larger subunit consists of two RNA molecules (23S and 5S) and about 34 proteins, while the smaller subunit is composed of one 16S RNA and about 20 proteins.[387] Indeed, in 1967 Nomura's group effected the reconstitution of the *E. coli* 30S subunit from the disassembled RNA and protein constituents, and several years later they reconstituted the 50S unit of *B. stearothermophilus.*[388] In eukaryotic cells, the ribosomes (80S) are composed of 60S and 40S subunits.

Parallel studies by Howard Dintzis on the incorporation of labeled amino acids into the hemoglobin of immature red blood cells (reticulocytes) indicated that the synthesis of this protein is a stepwise process starting with the amino-terminal amino acids and that the carboxyl-terminal amino acids are attached last.[389] With the discovery, at about the same time, of tRNA and mRNA, and with the formulation of the genetic code, during the 1960s the problem of the biochemical mechanism of protein synthesis became one of identifying the role of ribosomal components in their interactions with aminoacyl-tRNA's and mRNA in the initiation of the polymerization, the stepwise elongation of the peptide chain, and the termination of the process. By 1970 the main features of the process had been delineated, and shown to be much more complex than had been envisaged in the various template hypotheses offered during the 1950s. Several important aspects of the problem are still under active study.

During the 1960s it was found that the coat protein of some phages had an amino-terminal N-formylmethionyl (fMet) unit when synthesized in a cell-free system in the presence of phage RNA, although this unit was lacking in the coat protein isolated from the intact phage. This result indicated that the fMet unit might be a unique initiation component in protein synthesis, and that it is then removed by a hydrolytic process. Mario Capecchi found this inference to apply to the biosynthesis of *E. coli* proteins, and shortly afterward Nomura showed that such synthesis is initiated by the binding of fMet-tRNA to the 30S ribosomal subunit.[390] In experiments on the biosynthesis of hemoglobin, no such amino-terminal acylamino acid could be found, but in 1970 Dintzis reported the presence of an unsubstituted methionyl unit as an amino-terminal component during the early stages of chain elongation.[391] It seemed likely, therefore, that in prokaryotes the initiating unit is fMet-tRNA, while in eukaryotes it is Met-tRNA. This surmise was supported by later work, although it was also found that in mitochondria and chloroplasts protein synthesis is initiated by the formyl compound. It also turned out, however, that the initiation step in both bacterial and mammalian cells involved the participation of GTP and of several proteins which act as "initiation factors." Work in the laboratories of George Brawerman, Robert Thach, Alexander Rich, and Harvey Lodish showed that, for *E. coli* ribosomes, in addition to the 30S and 50S subunits, fMet-tRNA, mRNA, and GTP, three such factors (IF-1, IF-2, IF-3) are required.[392] The functions originally assigned to these factors were later revised. IF-1 is known to stimulate the activity of the two other factors, but its own role is unclear. IF-3 promotes the dissociation of ribosomes by interacting with the 30S subunit, which combines with IF-2, GTP, and mRNA to form a complex that accepts fMet-tRNA, with the binding of an AUG codon of mRNA to the "anticodon" of tRNA. In some bacterial or viral mRNAs the

AUG initiator triplet has been found to be located near a purine-rich sequence such as GGAGG, which pairs with a complementary sequence at the 3′-end of 16S ribosomal RNA. The hydrolysis of GTP then promotes the reassociation of the two ribosomal subunits, with the release of the three initiation units, and the resulting complex of the 70S ribosome with mRNA and fMet-tRNA is considered to represent the agent for the formation of the amino-terminus of a polypeptide chain.[393] In active protein synthesis, many such complexes are strung out along an mRNA strand, to form "polysomes." Although the initiation step of protein synthesis in eukaryotic cells resembles that in *E. coli* and other prokaryotes in many respects, in addition to the fact that the amino terminus is provided by Met-tRNA instead of its formyl derivative, the initiation process in the cytoplasm of eukaryotes has been found to involve a considerably larger number of protein factors (usually denoted "eIF"), the 5′-end of mRNA has been found to be "capped" by a substituent, and the activity of the initiation complex has been found to be promoted by the ATP-dependent phosphorylation of hydroxyamino acid units in eIF-2.[394] The growing knowledge of the biochemical complexity of only the first step in ribosomal protein synthesis offers an indication of the magnitude of the challenge to present-day investigators in this field.

The study of the elongation phase of ribosomal polypeptide synthesis (usually termed translation) also began during the 1960s. The model then proposed envisaged the existence of two sites on the 70S ribosome, a "P" (peptidyl) site occupied by the growing peptidyl-tRNA, linked to the appropriate codon in mRNA, and an open "A" (acceptor) site for the entry of an aminoacyl-tRNA with an anticodon triplet complementary to the next codon in mRNA. The reaction of the amino group of the aminoacyl-tRNA at the ester bond linking the peptidyl group to RNA thereby transferred the growing chain to the A site, with the ejection of that RNA from the P site. According to the model, there then occurs a translocation of the lengthened peptidyl-tRNA from the A site to the P site. The utility of this model in the subsequent study, by numerous investigators, of the mechanism of peptide chain elongation in bacterial systems was made evident by the discovery that the 30S subunit promotes the interaction of mRNA and tRNA and that the 50S subunit has the properties of a "peptidyl transferase" or "transpeptidase."[395] Also, Fritz Lipmann's group identified three protein elongation factors, originally denoted T_u, T_s, and G, now named EF-Tu, EF-Ts, and EF-G, and later X-ray diffraction studies on the complexes of crystalline preparations of these proteins with GTP, GDP, and aminoacyl-tRNA have provided much knowledge of the role of these factors in the elongation process.[396] The role of the ribosomal RNA was unclear, however, as was the nature of the ribosomal component that catalyzes the transpeptidation reaction.[397] Earlier work on the peptidyl transferase associated with the 50S subunit had shown that the antibiotic puromycin (an adenosyl phenylalanine derivative, found in 1959 to inhibit bacterial protein synthesis) can replace aminoacyl-tRNA in the ribosome-catalyzed reaction with fMet-tRNA, without a requirement for the 30S subunit, mRNA, or GTP.[398] After the discovery by Altman and Cech of RNA enzymes, the difficulty encountered in identifying a protein enzyme as the catalyst of the transpeptidation reaction led Harry Noller to propose that the 23S RNA component of the 50S subunit is the long-sought peptidyl transferase.[399] In eukaryotes, peptide chain elongation appears to involve a mechanism similar to that proposed for bacteria, with counterparts of EF-Tu (EF-1α, for binding aminoacyl-tRNA) and of EF-G (EF-2, for binding GTP and its hydrolysis in the translocation step).[400]

Peptide chain elongation ends in prokaryotes when one of the three mRNA "stop" codons (UAA, UAG, UGA) is reached and interacts with a "release" factor" (RF-1 or RF-2); that interaction is promoted by a third protein factor (RF-3). In eukaryotes, a complex of a protein factor (eRF) with GTP binds at a site occupied by a termination codon, and there follows cleavage of the peptidyl-tRNA ester bond, hydrolysis of GTP, and release of the polypeptide from the ribosome.[401]

In recent years, a problem in relation to the "universality" of the genetic code has been presented by Thressa Stadtman's discovery of selenocysteine, in which the sulfur atom of cysteine is replaced by selenium as a constituent of several enzyme proteins, and by the finding that the incorporation of this amino acid into such proteins is a consequence of the initial conversion of seryl-tRNA into selenocysteinyl-tRNA and the UGA-dependent insertion of selenocysteine in the elongation process, not by a "post-translational" modification of the seryl unit in the completed polypeptide chain.[402] As was mentioned previously, the UGA triplet was at first named a nonsense codon and later identified as a stop codon, but it was also found subsequently to code in some cases (as in mitochondrial protein synthesis) for tryptophan or glutamine. Such variations in the "standard" genetic code, of which there are now several instances, have been explained by the assumption that in its interaction with aminoacyl-tRNA, the ribosome, and accessory protein factors, mRNA carries more "messages" than those specified in the model of the chain elongation process outlined above, including the "recoding" of the standard mRNA triplets to account for such phenomena as "phaseshifts" or "suppression" observed upon genetic analysis.[403]

To the twenty-one kinds of amino acid units now considered to be incorporated into polypeptide chains by mRNA-directed ribosomal synthesis must be added the many other protein constituents which arise by "post-translational" modification.[404] Among them are hydroxyl derivatives of proline and lysine (in collagen), iodinated derivatives of tyrosine (in thyroglobulin), phosphoryl derivatives of serine (in casein), and a carboxylated derivative of glutamic acid (in prothrombin). The last example is of particular interest because it represents a striking example of the biological importance of such enzyme-catalyzed chemical modification. In 1971 Johan Stenflo discovered the presence of γ-carboxyglutamic acid in prothrombin, and John Suttie then showed that this amino acid is formed by the vitamin K–dependent carboxylation of a set of glutamyl units in the protein.[405] The $-CH(COOH)_2$ group is a powerful chelator of calcium ions, long known to be essential for blood coagulation. The clotting of blood is effected by the serine proteinase thrombin, which converts the soluble fibrinogen into insoluble fibrin and which is formed by the specific partial proteolysis of prothrombin, as in the case of the conversion of other "zymogens" to active proteinases, or of the formation of insulin from proinsulin.[406] Moreover, during the 1970s it was recognized that in eukaryotes the synthesis of some proteins involves the ribosomal production of a precursor bearing an amino-terminal peptide, about 15–25 units long and composed largely of hydrophobic amino acids. The work of numerous investigators, notably Günter Blobel, David Sabatini, Harvey Lodish, and Gottfried Schatz, has shown that such "signal" peptides facilitate the translocation of proteins into and across intracellular membranes and are removed from the amino terminus by selective proteolysis.[407]

It should be added that recent reports have raised the question whether it may be necessary to assign mRNA triplets to other unusual amino acids besides selenocysteine, including D-isomers of some of those listed in the standard code, or whether in all these cases only posttranslational modification is involved.[408] Also, it should be recalled that some microorganisms can synthesize peptides by a transpeptidation mechanism that does not involve the participation of aminoacyl-tRNAs. During the 1940s René Dubos and Rollin Hotchkiss, as well as Georgi Gause, isolated from strains of *Bacillus brevis* antibiotic peptides that were named tyrocidines and gramicidins.[409] The amino sequences of several of these peptides were determined by Richard Synge and by Bernhard Witkop. The one originally named gramicidin S turned out to be a cyclic decapeptide with two pentapeptide units (D-Phe-L-Pro-L-Val-L-Orn-L-Leu) joined head to tail, like the one named tyrocidine A $(\text{D-Phe-L-Pro-L-Phe-D-Phe-L-Asn})_2$, while gramicidin A was found to be a linear 15-unit peptide, with a formyl group at the amino end and an aminoethanol group $(-NHCH_2CH_2OH)$ at the carboxyl end.[410] Work in Edward Tatum's laboratory showed that the tyrocidines are made in a process different from the ribosomal synthesis of proteins, and that the amino acid composition of the culture medium determines the chemical nature of the product.[411] Several research groups, notably those of Søren Laland and of Kiyoshi Kurahashi, prepared from *B. brevis* active cell-free extracts and isolated two protein fractions which, together, effected the synthesis of gramicidin S.[412] Subsequent work by Horst Kleinkauf and Wieland Gevers, in Fritz Lipmann's laboratory, showed that this process involves the ATP-dependent formation of thiol esters ($NH_2CH(R)CO\text{-}SR'$), that one of two protein fractions activates L- or D-Phe while the other activates the other four amino acids, and that the sulfur compound is 4′-phosphopantetheine, found earlier to be the reactive component of coenzyme A. The mechanism that emerged from this outstanding work is that of a "thiotemplate" polymerization, in which a multiprotein complex, with each aminoacylthiol bound to a separate protein, directs the amino acid sequence of the polypeptide by virtue of the spatial arrangement of the protein subunits.[413] Recently, Stephen Kent and his associates have devised an ingenious chemical method of joining short peptide chains by the use of amino terminal cysteinyl units to form transient thiol esters that rearrange spontaneously into longer peptides.[414]

Site-Specific In Vitro *Mutagenesis*

It is appropriate, I believe, to conclude this chapter with some further words about the impact of the concepts and techniques of gene technology, as developed during the 1970s and 1980s, on the study of physiological processes in which proteins play a decisive role. After the discovery that mutations can be induced by the incorporation of plasmids into intracellular DNA, and the isolation of new enzymes (especially the polymerases, restriction endonucleases, and ligases) that act on DNA chains, as well as the invention of new or improved methods for the determination of the nucleotide sequence of DNA molecules and for the chemical synthesis of deoxyribonucleotides, this knowledge was combined to provide techniques for the production of proteins that differed from their natural counterparts by the replacement, deletion, transposition, or insertion of particular amino acid units. In such genetic engineering, whose technical details cannot be recounted here, the "standard" code is assumed to apply, and a synthetic deoxyribonucleotide of appropriate sequence is used as a primer for the formation of a mutant DNA whose expression yields the genetically altered protein. Such *in vitro* mutagenesis has proved to be valuable in the study of many biochemical problems—for example, in the work of Alan Fersht and his associates on the effect of changes in the structure of enzyme proteins on their catalytic activity.[415]

CHAPTER 9

The Whole and Its Parts

In the centuries-old philosophical discourse about materialism vs. vitalism, biological reductionism vs. organicism (holism), mechanism vs. purpose (teleology), or chance vs. necessity (determinism), the relation of the whole of a living entity to its chemical parts was considered by the disputants to require resort to the concept of "regulation."[1] Before about 1850 the regulation of such physiological processes as blood circulation, respiration, utilization of nutrients, nerve conduction, or sensory perception was thought to be comparable to the operation of man-made mechanical, chemical, or electrical devices with built-in controls, such as valves in mechanical pumps, thermostats in heating apparatus, or rheostats in electrical installations. Later, chemical substances were identified as "messengers" or "hormones" in the control of metabolic processes in animals and were considered to be involved in the maintenance of the "constancy of the internal environment" ("homeostasis"). Since about 1945, with the growth of knowledge about metabolic pathways in living organisms, from microbes to man, and about proteins and enzymes, the processes of biological regulation were increasingly studied in terms of the chemical interactions of known and newly discovered hormones with protein "receptors," and new analogies, such as "feedback inhibition" in electrical circuits, or new concepts, such as "allostery," were invoked. The knowledge so gained has been important not only in explaining the action of natural regulators and of well-known drugs but also in the design of new therapeutic agents.

Chemical Messengers

In 1902 William Bayliss and Ernest Starling reported that the injection of dilute hydrochloric acid into the duodenum of a dog whose spinal cord and neural connections to the small intestine had been severed elicited the secretion of pancreatic juice. They concluded that a substance, which they named secretin, had been formed in the intestine and transported by the blood to the pancreas. Three years later, at the suggestion of William Bate Hardy, Starling used the term *hormone* (Greek *hormao,* I arouse or excite) to denote what he called such "chemical messengers which, speeding from cell to cell along the blood stream, may coordinate the activities and growth of different parts of the body. . . . These chemical messengers . . . have to be carried from the organ where they are produced to the organ which they affect by means of the blood stream and the continually recurring physiological needs of the organism must determine their repeated production and circulation through the body."[2] In his famous experiments of the 1890s, Ivan Pavlov had studied the effect of acid in the intestine on the pancreatic secretion and had attributed it to a reflex through the central nervous system. Also, his student Leon Popielski and Emile Wertheimer had performed experiments similar to those of Bayliss and Starling but considered the effect to be due to an excitation of peripheral reflexes in the pancreas.[3]

For the experimental physiologists of the nineteenth century, whether inclined toward mechanism or toward vitalism, the nervous system constituted the regulatory apparatus of the animal organism. During 1810–1830, the studies of Charles Bell, François Magendie, Pierre Fluorens, and Johannes Müller laid the groundwork for the concept of the integrative action of the nervous system, as enunciated by Charles Sherrington,[4] and during the 1850s Claude Bernard emphasized the role of the neural control of the glycogenic function of the liver.[5] By the 1870s, however, there was a sizable body of evidence indicating that ductless or endocrine (Greek *endon,* within + *krinein,* to separate) glands, such as the adrenal and thyroid, secrete into the blood chemical substances which affect body functions.[6] Soon afterward, "internal secretions," the term Bernard used for the release of glucose and bile from the liver and which he applied to the release from glands such as the thyroid or adrenal medulla of other substances,[7] was seen by Charles Edouard Brown-Séquard to denote an additional mode of metabolic regulation: "We recognize that each tissue and, more generally, each cell of the organism secretes . . . special products or ferments into the blood and which through the medium of this liquid influence all the other cells thus integrated with each other by a mechanism other than the nervous system."[8] This view, which may be taken to mark the emergence of endocrinology as a distinctive specialty, also reflected the interest of pharmaceutical companies in the therapeutic use of organ extracts.[9] Indeed, after George Oliver and Edward Shäfer described in 1895 the effect of an extract of "suprarenal capsules" on the blood pressure, several investigators connected with such companies undertook the study of the active principle.[10] Among them was Franz Hofmeister's assistant

Otto von Fürth, who named it suprarenin. A German patent for "a method for the preparation of the blood-raising substance from adrenals" was issued in 1898 to Hofmeister and Fürth and assigned to the Höchst company. In 1900 Jokichi Takamine applied for a U.S. patent for a "glandular extractive product" he called adrenalin, and the right to use this name as a trademark was acquired by Parke, Davis and Company. John Jacob Abel and his assistant Thomas Aldrich began work in 1895 on the isolation of this substance, which they named epinephrin, and in 1901 Aldrich, who had moved to Parke, Davis in 1898, deduced a chemical formula whose validity was established when Friedrich Stolz, associated with the Höchst company, reported in 1904 the synthesis of the compound.[11] Among the other participants in that competition was Emil Abderhalden, in Emil Fischer's laboratory.

A striking example of the use of organotherapy had been provided in 1849 by Arnold Berthold, who concluded from his study of the effect of testicular transplants on the behavior of caponized cocks that there is a "testicular secretion" into the blood, but the importance of that work was not appreciated until much later, owing to the influence of Brown-Séquard. Also, by 1890 the discovery of the parathyroids by Ivar Sandström and by Eugène Gley completed the anatomical identification of the endocrine glands whose impairment or surgical removal had major physiological consequences.[12] During the 1890s a pattern was set for the interplay of physiological and chemical investigation that characterizes present-day research in endocrinology, and in which academic scientists collaborate (and compete) with their colleagues in commercial enterprises.

The purification of epinephrine, as it is now named, and the determination of its chemical structure as a derivative of catechol were achieved relatively rapidly. It was otherwise for other hormones identified before 1910. For example, in 1895 Eugen Baumann reported that he had obtained from an acid hydrolysate of thyroid protein an iodine-containing material that was active in the treatment of goiter, and shortly afterward Adolf Oswald isolated a protein (he named it thyreoglobulin) by extraction of the gland tissue with a salt solution. A sustained effort to determine the chemical structure of the active principle was not undertaken until 1910, when Edward Kendall began work on the problem while with Parke, Davis (he left soon afterward) and continued it at St. Luke's Hospital in New York and then at the Mayo Clinic, a famous center of thyroid surgery. Kendall repeated Baumann's work but used alkaline hydrolysis to isolate what he called thyroxin in crystalline form and suggested that it is a "diiodo-dihydroxyindol" related to tryptophan. Other biochemists, notably Henry Dakin, did not accept Kendall's chemical evidence. During the 1920s Charles Harington improved Kendall's isolation procedure and proved by synthesis that thyroxine is a tetraiodo derivative of an amino acid in which two phenyl groups are joined by an ether linkage.[13] Many years later, Harington's associates Rosalind Pitt-Rivers and Jack Gross found by means of chromatography that in addition to the tetraiodo compound (T_4), rat thyroid extracts and human plasma contain the corresponding triiodothyronine (T_3), and they also showed that T_3 is much more active physiologically than T_4.[14]

After the clinical demonstrations during the nineteenth century of the relation of the impairment of the thyroid to such diseases as cretinism or myxedema, and Adolf Magnus-Levy's finding that the administration of thyroid extract increases oxygen consumption, there were speculations about the role of the thyroid in promoting detoxification, but before 1950 the biochemical events in the physiological processes affected by thyroxine were unknown, except for the observation that it is one of the chemical agents that can uncouple oxidative phosphorylation. More recent work has shed much light on this problem, and I will return to it later. Mention should be made, however, of the important physiological discoveries in this field during 1910–1930 by Philip Smith, Herbert Evans, and especially Leo Loeb and Max Aron, showing that the thyroid secretion is under the hormonal control of the anterior pituitary, which produces a "thyroid stimulating hormone" (TSH, or "thyrotropin").[15] As has been noted, the secretion of TSH from the pituitary is in turn stimulated by a "thyrotropin releasing factor" (TRF) elaborated by the hypothalamus.

As for insulin, whose history was sketched earlier, and secretin, the determination of their chemical structure was achieved only after 1945, with the development of new methods for the purification of peptides, the determination of their amino acid sequence, and their unequivocal synthesis. These methods proved to be decisive in the isolation and characterization of secretin and other peptide hormones such as those from the pancreas (glucagon), the gastrointestinal tract (gastrin, cholecystokinin), the posterior pituitary (oxytocin, vasopressin), the anterior pituitary (prolactin, adrenocorticotropic hormone [ACTH]), melanocyte stimulating hormones [MSH], and the pineal gland (melatonin), as well as the parathyroid hormone and the several "releasing factors" and somatostatin of the hypothalamus. Moreover, as Brown-Séquard surmised a century earlier, organs that were not included among the endocrines also produced active principles transmitted by the blood. Among those that turned out to be peptide hormones were angiotensin I (kidney), the enkephalins (brain), and the atrial natriuretic factor (heart).[16] Some aspects of the physiological action of these peptide hormones will be considered later. The efflorescence of the art of peptide synthesis after 1945 was due in large part to the recognition of the physiological importance of such hormones, and to the interest of pharmaceutical companies in making related compounds of possible therapeutic value. Although peptide chemistry may be said to have begun with Emil Fischer's hope to synthesize a protein, it was the spur provided by research on peptide hormones (and peptide antibiotics) that led to the achievement of this objective some seventy years later. It should also be noted that the closer study of the peptide hormones known before 1970, and the later discovery of new ones, were greatly aided by the quantitative "radioimmunoassay" method described for insulin in 1972 by Solomon Berson and Rosalyn Yalow.[17]

During the 1930s—after the chemical synthesis of epinephrine and thyroxine, but before the post–World War II successes in the study of mammalian peptide hormones—the center of the endocrinological stage was occupied by the steroid hormones elaborated by the adrenal cortex and the gonads. The story of the adrenal cortical hormones, as it is usually told, properly begins with the 1855 report of Thomas Addison on the clinical aspects of the impairment of the "suprarenal capsules."[18] A year later, Brown-Séquard found that adrenalectomy led to the death of experimental animals, but subsequent work showed that addition of salt to their diet prolonged their life. Attempts to prepare active aqueous organ extracts were unsuccessful until 1930, when lipid solvents began to be employed for this purpose.[19] Immediately afterward, the research groups associated with Edward Kendall, Tadeusz Reichstein, and Oscar Wintersteiner succeeded in crystallizing several active compounds, one of which ("aldosterone") had a striking effect on electrolyte balance, while others ("cortisol," "cortisone," "corticosterone") were active in controlling carbohydrate metabolism.[20] As for the sex hormones, in 1923 Edgar Allen and Edward Doisy obtained active lipid-soluble extracts of ovaries, and in 1929 Selmar Aschheim and Bernhard Zondek found pregnancy urine to contain ovarian hormones. Soon afterward, Doisy, Adolf Butenandt, and Guy Marrian isolated from this source compounds later named estrone, estriol, and pregnanediol.[21] At the same time, George Corner isolated "progesterone" from the corpus luteum of the ovary. With the elucidation in 1932 of the structure of cholesterol and the bile acids, largely owing to the work of Heinrich Wieland, Adolph Windaus, and Otto Diels, and the chemical insight of Harold King and Otto Rosenheim, there quickly followed the determination of the structure of the adrenal steroids, the ovarian hormones, and the testicular hormone "testosterone." In the ensuing development of this field during the 1940s and 1950s, with an emphasis on the chemical synthesis of these hormones, whose isolation from natural sources was commercially impractical, the research in laboratories of American, Swiss, and German pharmaceutical companies outpaced the efforts of university chemists. Of particular historical interest is the emergence in

Mexico of the Syntex company, and the role of Russell Marker in devising a synthesis of progesterone from a readily available plant constituent (diosgenin), and (later) Carl Djerassi's synthesis of the oral contraceptive.[22]

Although plants had long been known to contain a "growth factor" or "auxin" (Greek *auxein,* to grow), which is carried by the sap from one organ to another, it was not until 1932 that Friedrich Went and Fritz Kögl called such substances phytohormones. Indeed, Kögl reported in 1933 the isolation from human urine of "auxin a," which he considered to be the natural plant hormone, and of a "heteroauxin" (indole-3-acetic acid) thought to be an artifact. Later work showed the "discovery" of auxin a to be false, and indoleacetic acid to be the natural plant growth substance. Also, plants and fungi were found to produce phytohormones named gibberellins.[23] The problem of the definition of the term *hormone* also arose in connection with the studies of Boris Ephrussi on the genetic determination of the eye color of Drosophila, and with those of Peter Karlson on the control of insect metamorphosis by the factors termed ecdysone and juvenile hormone.[24] If during the 1930s some biochemists thought it logical to group enzymes, hormones, and vitamins together as agents of metabolic change,[25] it was clear that Starling's definition in 1905 of the hormone concept had undergone considerable change. To agents such as thyroxine, insulin, or aldosterone, elaborated by glandular tissues, had been added those released by neuronal cells and synapses. These products of "neurosecretion" (a term introduced by Ernst and Berta Scharrer during the 1930s) include peptides such as oxytocin and the enkephalins, as well as "neurotransmitters" such as acetylcholine (discovered by Otto Loewi and Henry Dale during the 1920s), noradrenaline, glutamic acid, and γ-aminobutyric acid (GABA, a decarboxylation product of glutamic acid).[26] After 1950 the role of these and other hormonelike agents in the regulation of metabolic processes in animal organisms became an active area of biochemical investigation.

Homeostasis

In 1926, after citing statements by Charles Richet and Claude Bernard, Walter Cannon wrote:

> The steady states of the fluid matrix of the body are commonly preserved by physiological reactions, i.e., by more complicated processes than are involved in simple physical-chemical equilibria. Special designations, therefore, are appropriate—"homeostasis" to designate stability of the organism; "homeostatic conditions," to indicate details of the stability; and "homeostatic reactions," to signify means for maintaining stability. I suggest the

> following: 1. In an open system, such as our bodies represent, composed of unstable structure and subjected constantly to disturbance, constancy is in itself evidence that agencies are acting or are ready to act to maintain this constancy. . . . 2. If a homeostatic condition continues, it does so because any tendency toward change is automatically met by increased effectiveness of a factor or factors which lessen the change. . . . 3. A homeostatic agent does not act at the same point. . . . 4. Homeostatic agents, antagonistic in one region of the body, may be cooperative in another region. . . . 5. The regulating system which determines a homeostatic state may comprise a number of cooperating factors brought into action at the same time or successively. . . . 6. When a factor is known which can shift a homeostatic state in one direction, it is reasonable to look for automatic control of that factor or for a factor or factors having an opposing effect.[27]

This brief article marks the appearance of the word *homeostasis*, and a few years later Cannon developed the theme more fully in a review article and in a book for a general audience.[28]

In the background of Cannon's definition of homeostatis lies the concept of the "milieu de l'intérieur" presented in 1853 by Charles Robin and François Verdeil and, more prominently, of the "milieu intérieur" developed during 1857–1878 by Bernard in relation to the maintenance of the constancy of the chemical composition of the blood and its temperature.[29] In particular, during the first decades of the twentieth century, Bernard's statement that "la fixité du milieu intérieur est la condition de la vie libre, indépendante"[30] (often translated as "the constancy of the internal environment is the condition for free and independent life") attracted the attention of physiologists inclined toward "organismic" or "holistic" experimental approaches, notably John Scott Haldane and Joseph Barcroft in England, Ivan Pavlov in Russia, and Lawrence Henderson and Cannon in the United States.

Before 1926 Cannon had done extensive work on the autonomic nervous system, while the research of Haldane, Barcroft, and Henderson had dealt largely with respiratory physiology. Haldane (with John Priestley) showed in 1905 that the regulation of breathing depends more on the CO_2 content of the inspired air than on its O_2 content.[31] In 1901, with Haldane, Barcroft devised an improved apparatus for the study of the gas exchange in the blood, and in his later work he contributed much to the knowledge of the interaction of hemoglobin with O_2 and CO_2.[32] Henderson, more inclined to theory than to experiment, made his most important scientific contribution in 1908, with the derivation of an equation for the behavior of buffer solutions and its relevance to the acid-base balance in the blood.[33] In addition to the fact the Cannon drew his inspiration from the writings of Claude Bernard and Henderson drew his from the equilibrium theory of Josiah Willard Gibbs, it is noteworthy that the two Harvard professors extended their views on biological regulation to sociological problems, with opposite political orientations. At the time of the economic depression of the 1930s, Cannon backed the New Deal policies of the Roosevelt administration. Later he supported efforts to aid the Spanish government, headed by the physiologist Juan Negrin, in its resistance to the rebellion led by Francisco Franco, and Cannon's collegial tie to Ivan Pavlov led him during the 1940s to promote closer relations between the United States and the Soviet Union. On the other hand, Henderson opposed the policies of the New Deal and, in espousing the sociological theories of Vilfredo Pareto, advocated a political policy more in accord with the views and aspirations of American businessmen.[34]

After 1945 the language used in the interpretation of the phenomena in biological regulation was strongly influenced by that developed during World War II in connection with the invention of electronic control devices for military purposes, and especially by the appearance in 1948 of a book entitled *Cybernetics, or the Control and Communication in the Animal and the Machine* by the mathematician Norbert Wiener.[35] Of that book, the noted Dutch mathematician Hans Freudenthal wrote: "It has contributed to popularizing a way of thinking in communication theory terms, such as feedback, information, control, input, output, stability, homeostasis, prediction, and filtering. On the other hand, it has also contributed to spreading mistaken ideas of what mathematics really means. *Cybernetics* suggests that it means embellishing a non-mathematical text with terms and formulas from highbrow mathematics. This is a style that is too often imitated by those who have no idea of the mathematical words they use. Almost all so-called applications of information theory are of this kind."[36] In 1949 the endocrinologist Roy Hoskins used the phrase "servo (feed-back) mechanism" in connection with the reciprocal control of the pituitary by the thyroid, and the work of Carl Moore and Dorothy Price during the early 1930s on the control of hypophyseal activity by the gonads came to be cited as a striking example of "negative feedback."[37] In these applications of feedback theory to endocrinological problems, one dealt with the interconnection of "black boxes" arranged in "loops" and, as in the use of thermochemical data to study "energy metabolism," the events in the black boxes were unknown, the "input" and "output" being the only observable entities.[38]

During the 1950s, with the introduction of bacterial mutants into the study of the metabolic pathways in the biosynthesis of amino acids, purines, and pyrimidines, and with the purification of some of the enzymes involved in these pathways, several investigators found evidence of feedback inhibition at the cellular level. In 1950 Joseph Gots reported that the presence of exogenous purines in cultures of sulfonamide-inhibited *E. coli* inhibited the accumulation of one of the precursors of purine biosynthesis.[39] A few years later Richard Yates and Arthur Pardee showed that cytidine triphosphate (CTP) inhibits the first committed step in the pathway of its biosynthesis in *E. coli,* and Edwin Umbarger found that isoleucine specifically inhibits the first enzyme-catalyzed reaction in the sequence of steps leading to its synthesis from threonine.[40] Umbarger's paper began as follows:

> Recent developments in automation have led to the use in industry of machines capable of performing operations that have been compared with certain types of human activity. In the internally regulated machine, as in the living organism, processes are controlled by feedback loops that prevent any one phase of the process from being carried to a catastrophic extreme. The consequences of such feedback control can be observed at all levels of organization in a living animal—for example, proliferation of cells to form a definite structure, the maintenance of muscle tone, and such homeostatic mechanisms as temperature regulation and maintenance of a relatively constant blood sugar level.[41]

As has been mentioned, during the 1950s Henry Vogel and others showed that the product of a microbial biosynthetic pathway may cause the "repression" of the DNA-directed formation of enzyme proteins involved in that pathway.[42] This mechanism of biochemical regulation affects the *amount* of a critical enzyme, whereas end-product inhibition alters the *activity* of such an enzyme already present in the cell. This distinction was clearly defined by Arthur Pardee in 1958 at a Ciba Foundation symposium on the regulation of cell metabolism, organized by Hans Krebs.[43] At that meeting, Britton Chance presented a valuable discussion of the limitations in drawing an analogy between negative feedback in electronic circuits and the metabolic control of enzyme-catalyzed processes.[44]

In his studies on the inhibition by isoleucine of the enzyme that converts threonine to α-ketoglutarate ("threonine deaminase") in the pathway of isoleucine synthesis, Umbarger also showed that the activity of another threonine deaminase, associated with the metabolic breakdown of threonine in *E. coli,* is not inhibited by isoleucine. Moreover, he found that the kinetics of the biosynthetic enzyme did not conform to the Michaelis-Menten equation but gave a sigmoid curve in a plot of the initial velocity vs. substrate concentration. These findings were confirmed by Jean Pierre Changeux in Jacques Monod's laboratory.[45] Subsequent research on the problem presented by the work of Umbarger and of Yates and Pardee shifted from threonine deaminase (and other enzymes found to exhibit a similar response to end products) to the intensive study of the *E. coli* enzyme aspartate transcarbamoylase (ATCase) which catalyzes the reaction of aspartate and carbamoyl phosphate, the first committed step in the biosynthesis of pyrimidines. In 1961 John Gerhart and Arthur Pardee showed that the sensitivity of ATCase to CTP, with accentuation of the sigmoid curve for plots of initial velocity vs. aspartate concentration, can be abolished without loss of catalytic activity, indicating that CTP and aspartate are bound at separate sites in the enzyme protein.[46] Decisive evidence in support of this conclusion came from ultracentrifuge studies by Gerhart and Howard Schachman, with the physical separation of subfractions of ATCase, one containing the catalytic site and the other the binding site for CTP.[47] Later work showed that the enzyme is composed of twelve polypeptide chains, six of which contain the catalytic site and six the "regulatory" site, and that this assembly undergoes conformational change on conversion from the unliganded to the liganded state.[48]

Allostery

At the conclusion of the 1961 Cold Spring Harbor symposium, Jacques Monod and François Jacob stated: "As the reports here have shown, end product inhibition is extremely widespread in bacteria, insuring immediate and sensitive control of the rate of metabolite biosynthesis in most, if not all, pathways. From the point of view of mechanisms, the most remarkable feature of the Novick-Szilard-Umbarger effort is that the inhibitor *is not a steric analogue of the substrate.* We propose therefore to designate this mechanism as 'allosteric inhibition.'"[49] In a paper published two years later, Monod, Changeux, and Jacob noted that the dissociation curve of the tetrameric oxyhemoglobin is sigmoid (the "Bohr effect"), whereas that of the monomeric oxymyoglobin accords with what they termed the Michaelis-Henri relation, and stated that "in the case of haemoglobin, there is complete evidence that the regulatory effect, i.e., the cooperative binding of oxygen, is related to a reversible, discrete conformational alteration of the protein, i.e., in our nomenclature to an allosteric transition. Actually, thanks to the considerable work which has been devoted to it, the haemoglobin system provides the most valuable model from which to start in the further analysis and interpretation of allosteric effects in general."[50] Not mentioned in this paper is the fact that in 1951 Jeffries Wyman (with David Allen) had stated that "the reason why certain acid groups are affected by oxygenation is simply the alteration in their position and environment which results from the change in the configuration of the hemoglobin molecule as a whole accompanying oxygenation," and suggested that "if we are prepared to accept hemoglobin as an enzyme, its behavior might give us a hint as to the kind of process to be looked for in enzymes generally."[51]

Wyman and Monod became well acquainted during the 1950s, when Wyman served for several years as science attaché at the American embassy in Paris, and presumably had occasion to discuss Wyman's thermodynamic analysis of the Bohr effect. In 1965 there appeared the famous paper by Monod, Wyman, and Changeux in which a "plausible" model was presented for the nature of allosteric transition. After distinguishing between "homotropic" interactions between identical ligands and "heterotropic" interactions between different ligands, the properties of allosteric proteins were defined as follows:

> (1) Most allosteric proteins are polymers, or rather oligomers, involving several identical units. (2) Allosteric interactions frequently appear to be correlated with alterations of the *quaternary* structure of the proteins (i.e., alterations of the bonding between subunits). (3) While heterotropic effects may be either positive or negative (i.e., co-operative or antagonistic), homotropic effects appear to be always co-operative. (4) Few, if any, allosteric systems exhibiting *only* heterotropic effects are known. In other words, co-operative homotropic effects are almost invariably observed with at least one of the two (or more) ligands of the system. (5) Conditions, or treatments, or mutations, which alter the heterotropic interactions also simultaneously alter the homotropic interactions.[52]

It was assumed that oligomeric enzymes can exist in two conformational states (R and T), and the effect of the interaction of a ligand (or ligands) in regions of the protein distant from the active site was expressed in terms of an equilibrium constant (L) for the R-T transition and equilibrium constants (K_R and K_T) for the binding of a ligand at the same site of the protein in each of the two conformational states. It was further assumed that, as shown by Wyman for hemoglobin, such equilibria are "linked" so that a ligand that binds preferentially to the T state will promote the transition from R to T. Moreover, it was assumed that the monomeric subunits are arranged in a symmetrical fashion and that this symmetry is conserved during changes in protein conformation.

Soon afterward, several investigators, notably Daniel Koshland and Daniel Atkinson, offered somewhat different "sequential" models that also explained the sigmoid kinetics of some oligomeric enzyme proteins, and it was also noted that this kinetic behavior can be exhibited by monomeric enzymes.[53] In particular, they included the factor of flexibility in the region of the catalytic site, as presented in Koshland's "induced-fit" theory of enzyme-substrate interaction.[54] In considering the present status of the Monod-Wyman-Changeux model, there can be no question of its role in adding to the language of biochemistry new words that have been useful in the description and interpretation of the behavior of many oligomeric enzymes, most notably ATCase and glycogen phosphorylase.[55] Owing to Wyman's extensive prior studies on the hemoglobin system, the thermodynamic treatment offered in his paper with Monod and Changeux cannot be faulted although, as we have seen, equilibrium thermodynamics alone has not always been a sure guide in the interpretation of biochemical processes in steady-state biological systems. Wyman later wrote of "Jacques' quickness of response to an idea, which was one of his great qualities. The paper as it stands was written almost wholly by him and was presented to me more or less as a fait accompli for discussion and criticism."[56] John Edsall, Wyman's lifelong friend, has also stated: "It included an extensive discussion of symmetry relations in the assembly of oligomeric proteins, almost entirely due to Jacques. Jeffries, in spite of his long-standing interest in the symmetry (or asymmetry) of ligand binding curves, and their implications, has always remained skeptical concerning these symmetry arguments."[57] As Max Perutz showed in 1970, one of Monod's main assumptions—that only the interaction between subunits is altered by allosteric effectors—was not borne out in the case of hemoglobin.[58] Moreover, monomeric enzymes that act on polymeric substrates, such as lysozyme or pepsin, exhibit conformational flexibility at their extended sites upon interaction with their substrates and in response to the binding of ligands at other sites in the enzyme.[59]

Second Messenger

In 1964 Earl Sutherland introduced the term *second messenger* to denote the cyclic 3′,5′-adenosine phosphate (cyclic AMP, cAMP) which he and Theodore Rall had discovered in 1957, and whose chemical structure was established by David Lipkin and Roy Markham.[60] This compound was found during the course of Sutherland's studies on the effect of epinephrine or glucagon on the activity of glycogen phosphorylase in dog liver slices. As was mentioned earlier, during the 1920s Carl and Gerty Cori had shown that epinephrine promotes the conversion of liver glycogen into blood glucose.[61] In his Nobel Prize address, Sutherland gratefully acknowledged his debt to the Coris, in whose laboratory he began his scientific career and participated in the research on the effect of hormones on enzymes. The far-reaching consequences of Sutherland's discovery of cyclic AMP and the research of his group on its mode of action represent a transformation of the study not only of hormone function but also of other aspects of biochemical regulation, comparable to the transformation of genetics after the acceptance of the Watson-Crick model of DNA.

By 1970 work in Sutherland's laboratory had shown that the intracellular level of cAMP depends on the activity of a membrane-bound "adenyl cyclase" (now adenylyl cyclase), an enzyme that catalyzes the formation of cAMP from ATP, and of a phosphodiesterase that inactivates cAMP by catalyzing its hydrolysis to 5′-AMP.[62] It was also shown that cAMP mediates the hyperglycemic effect of epinephrine or glucagon by stimulating the conversion of the inactive "b" form of glycogen phosphorylase into its active "a" form. As we have seen, this process was also discovered by Carl and Gerty Cori. Another of their former students, Edwin Krebs, in association with Edmond Fischer, then showed that the conversion of muscle phosphorylase b (a two-subunit protein) to the tetrameric phosphorylase a involves the action of a calcium-dependent "phosphorylase b kinase," which catalyzes the phosphorylation, by ATP, of a specific serine unit in the b form. This kinase must, however, be activated by a "phosphorylase b kinase kinase," whose catalytic activity depends on its allosteric interaction with cAMP.[63] The reverse conversion of phosphorylase a to the b form is effected by a protein phosphatase ("phosphorylase a phosphatase"). As was shown by Joseph Larner (another former member of the Cori group) and his associates, cAMP affects the action of the glycogen synthase system in an analogous fashion, with the important difference that in this case it is the active form that undergoes the cAMP-dependent phosphorylation and is converted to the inactive form, thus inhibiting glycogen synthesis.[64] At this stage of the unraveling of mechanism of the regulatory action of cAMP in the metabolic breakdown and synthesis of glycogen, there had thus been brought into play "protein kinases" analogous in their chemical action to that of the known ATP-dependent enzymes in the glycolytic pathway (for example, phosphofructokinase), and to that of the known hydrolytic enzymes that catalyze the dephosphorylation of phosphoproteins such as casein or phosvitin. In relation to the problem of the hormonal regulation of metabolic processes, however, the center of the stage was occupied by the mammalian adenylyl cyclases, although such enzymes had also been found in insects, higher plants, and bacteria.[65]

After 1970 an early clue to the complexity of the biochemical mechanism of the conversion of ATP to cAMP, a reaction readily effected in the test tube by means of barium hydroxide, came from the work of Martin Rodbell and his associates, who showed in 1971 that, in addition to epinephrine, GTP is needed for the activation of adenylyl cyclase.[66] This important finding led to the demonstration, by Alfred Gilman, that plasma membrane proteins that bind guanosine nucleotides (GDP and GTP) and hence are named G proteins, are involved in that process, and that the GDP-bound form is inactive, but when the GDP is replaced by GTP, the activity of the GTP-protein is stimulated by its interaction with a specific membrane "receptor" of the hormone.[67] Of particular biochemical interest is the fact that the G proteins that activate adenylyl cyclase are composed of three different subunits—one denoted α, which has GTPase activity, and two other coupled smaller ones, denoted β and γ. The α-subunit separates from the other two upon binding GTP, and the trimeric structure is restored upon enzymatic hydrolysis of the GTP. The α-subunit is also substituted at its amino terminus by a fatty acid (myristic or palmitic acid) that serves as a means of attachment of the G protein to the plasma membrane.[68] In their catalytic properties, the G proteins are members of a large family of GTPases, which include the smaller "ras" proteins (so called because they were first identified as products generated by the rat sarcoma virus), and those mentioned earlier as participants in the elongation of polypeptide chains in *E. coli*.[69] In recent years, the G proteins have been (and continue to be) objects of intensive study, because of both their importance in relation to the activation of adenylyl cyclase and the possibility that several human diseases may arise from changes in their physiological function.[70]

The achievements initiated by the work of Earl Sutherland thus brought to the fore the role of protein kinases and membrane "receptors" in the regulation of metabolic pathways by hormones such as epinephrine, insulin, and glucagon (the "hyperglycemic-glycogenolytic factor," HGF). Before 1950 some biochemists tended to focus their attention on the *in vitro* effect of a hormone on the activity of a key enzyme in a known metabolic pathway.[71] For example, during the 1940s the Coris and their associates showed that the inhibition of hexokinase by extracts of the anterior pituitary could be counteracted by insulin.[72] In perceptive review articles, William Stadie criticized this approach, and Oscar Hechter wrote that "the basic mechanism of epinephrine action cannot be action on cell factors which act primarily to regulate 'phosphorylase' activity but must be a more generalized reaction applicable to cells generally."[73] Hechter also stressed the possible role of hormones in transfer mechanisms and noted that since the time of Paul Ehrlich and John Newport Langley, pharmacologists had invoked the concept of "receptors" in cell membranes as the sites of specific interaction with drugs.

Protein Kinases and Phosphoprotein Phosphatases

In a review article published in 1955, Gertrude Perlmann wrote: "The biological function of phosphoproteins is still unknown. Owing to their abundance in embryonic tissue and in foodstuffs it has often been suggested that one of the roles of these materials is that of supplying phosphorus and the essential amino acids to the developing organism."[74] While that article was still in press, Eugene Kennedy demonstrated the ATP-dependent phosphorylation by an extract of rat liver mitochondria of serine units in casein, and a few years later Fritz Lipmann showed a similar protein kinase action on the egg protein phosvitin.[75] In ensuing work, not only glycogen phosphorylase and glycogen synthase but other proteins were also found to be substrates of protein kinases that catalyze the phosphorylation of serine (or threonine) units, and other kinases were found to phosphorylate the hydroxyl group of tyrosine units.[76] The transformation of the field of phosphoprotein chemistry during the 1960s, as a consequence of the work of the groups associated with Earl Sutherland, Edwin Krebs, Edmond Fischer, and Joseph Larner, is evident in a splendid review article by George Taborsky, who could only sketch the "explosively expanding research" on the protein kinases.[77] Twenty years after the publication of that article, Edwin Krebs wrote: "A phenomenon that has occurred during the rapid spread of interest in protein phosphorylation has been been the recruitment of groups who suddenly became aware of the fact that protein phosphorylation applies to their field of research. . . . These include, for example, investigators interested in muscle contraction, oocyte maturation and the cell cycle, protein synthesis, virology, transcriptional regulation, lymphocyte activation, secretion, and ion channels, to name but a few."[78] In this development, new methods of protein chemistry, peptide synthesis, and gene technology have made possible the characterization of the structure, specificity, and function of two classes of protein kinases: those that catalyze the phosphorylation of serine or threonine units and those that act at tyrosine units.

Among the many serine/threonine protein kinases are, in addition to the oligomeric cAMP-dependent enzymes, the pyruvate dehydrogenase complex and some of the protein factors in gene transcription and in ribosomal polypeptide synthesis, mentioned earlier. Others identified in eukaryotic organisms are a "protein kinase C," which phosphorylates numerous cell proteins, as well as more specific enzymes such as those that phosphorylate a region of the rhodopsin complex, or the light chain of myosin.[79] Nearly all of these protein kinases are oligomeric enzymes, with a separation of the regulatory site from the catalytic site on different subunits. The activity of some of the protein kinases depends on an association with "calmodulin," a protein with an avidity for calcium ions, and others are activated by interaction with a diacylglycerol.[80] The protein kinases have extended active sites, and in those cases where the site of phosphorylation has been identified, the serine unit that is phosphorylated is flanked in the peptide chain by basic (Arg or Lys) amino acid units.[81]

The existence of protein kinases that specifically phosphorylate tyrosine units was initially inferred from the discovery in 1979 of O-phosphotyrosine as a protein constituent, and the first enzymes of this group to be detected are associated with viral transforming proteins.[82] The importance of such "protein-tyrosine kinases" in animal metabolism became evident upon the demonstration that they are involved in the mechanism of insulin action, about which more shortly. At this point I only note that in this mechanism, the "signal" received upon interaction of insulin with its membrane receptor has been shown to be "switched" from the protein-tyrosine kinase to a set of serine/threonine kinases (termed the MAPK cascade) by means of a "Raf kinase."[83] Protein-tyrosine kinases have also been found to participate in such processes as the activation of immunocompetent cells, mitogenesis, and the aggregation of blood platelets.[84]

To the protein kinases mentioned above may be added others that have been objects of intensive study—for example, those that depend upon activation by cyclic 3′,5′-guanosine phosphate (cGMP), formed by a "guanylyl cyclase," and those involved in the phosphorylation of the inositol portion of phosphatidyl inositol, which is then cleaved by a phospholipase to form the "second messengers" inositol triphosphate (which promotes the intracellular mobilization of calcium ions) and diacyl glycerol (which activates a widely distributed "protein kinase C").[85]

The action of the protein kinases in the posttranslational phosphorylation of proteins is reversed by the protein phosphatases. Such enzymes had been found in animal tissues, with casein as the substrate, and it was also shown that they hydrolyze phosphoamides.[86] More recently, it has been recognized that protein phosphatases are involved in the regulatory control of cellular processes as diverse as cell growth, the biosynthesis of proteins, fatty acids, and sterols, and in glycogen metabolism. One large group comprises enzymes that catalyze the hydrolysis of phosphoseryl and phosphothreonyl units, as in the case of glycogen synthase, glycogen phosphorylase kinase, or myosin, and are oligomeric proteins with catalytic subunits and regulatory subunits sensitive to cAMP and calcium ions.[87] Another group consists of protein phosphatases that act specifically at phosphotyrosyl units, although some members dephosphorylate serine and threonine units as well. Among these protein tyrosine phosphatases are intracellular enzymatic components of transmembrane receptors of signals received upon their interaction with hormones such as insulin.[88]

Receptors for Hormones and Neurotransmitters[89]

The emergence of pharmacology as a research specialty during the second half of the nineteenth century, and the efforts of organic chemists to synthesize compounds of therapeutic value, marked a new stage in the interplay of biological and chemical thought about the action of drugs on physiological processes in animal organisms. In the early 1870s Thomas Lauder Brunton and Thomas Fraser suggested that such effects involve the chemical interaction of drugs with cell constituents, and in 1901 Sigmund Fränkel summarized the experimental evidence in favor of this view.[90] A spur to the adoption of the "receptor theory," as formulated by Paul Ehrlich and John Newport Langley, was Emil Fischer's hypothesis of the specific interaction of an enzyme with its substrate, and during the 1920s Alfred Joseph Clark's development of the theory drew on contemporary work in the enzyme field.[91]

Ehrlich's version of the theory stemmed from his studies on the staining of tissues by dyes and was then elaborated in connection with his work on immunity and chemotherapy, while Langley's approach was from his research on the action of drugs such as nicotine and curare, and of adrenal extracts, on the autonomous nervous system. In 1905 Langley concluded that "in all cells two constituents at least must be distinguished, (1) substances concerned with carrying out the chief functions of the cells, such as contraction, secretion, the formation of special metabolic products and (2) receptive substances especially liable to change and capable of setting the chief substance in action," and he later stated that "my theory of the action is in general on the lines of Ehrlich's theory of immunity. I take it that the contractile molecule has a number of 'receptive' or side-chain radicles, and the nicotine by combining with one of these causes tonic contraction, and by combining with another, causes twitching."[92]

During 1910–1930, many biologists preferred the "surface adsorption" interpretation of the interaction of cells with chemical agents, based on the physical-chemical approach of the colloid chemists, to Ehrlich's idea of specific chemical combination, derived from the organic-chemical hypothesis of Emil Fischer. For example, the pharmacologist Walther Straub and the endocrinologist Eugène Gley both questioned the validity of Ehrlich's theory and, in an extended discussion of the matter, William Bayliss stated that "to be satisfied with the fact that the chemical composition of such substances as 'immune bodies' is so complex that one may quietly ascribe all their properties to it, is the antithesis of a view leading to progress."[93]

Several of the developments discussed in previous chapters greatly influenced the subsequent fate of the receptor theory of hormone action. During the 1930s, with the isolation of numerous enzymes in the form of crystalline proteins and the elucidation of their relationship to coenzymes, the groundwork was laid for the post–World War II study of the interaction of trace biologically active substances with proteins. As Gregory Pincus and Kenneth Thimann put in 1948, "The future of hormone research will thus be in large part bound up with advances in general biochemistry and physiology."[94] Among these advances was the formulation, by the physical chemist George Scatchard, of an equation for the binding of small molecules and ions by proteins, as well as the invention of improved methods for the quantitative measurement of such binding.[95] Also, electron microscopy had provided new insights into the problem of cell permeability, and during the 1950s Rachmiel Levine provided support for the view that hormones such as insulin promote the passage of metabolites across plasma membranes.[96] Sutherland's discovery of cAMP and his proposal that adenyl cyclase spans such membranes, with the binding site for the hormone outside the cell and the catalytic unit inside the cell, spurred research during the 1970s on the chemical nature and function of hormone receptors. In what has became one of the most active areas of biological investigation, the application of the techniques of protein chemistry and genetics, of radiochemistry, and of X-ray crystallography, has opened the way to a better understanding of the chemical basis of the role of receptors in biological regulation.

At the present writing, it appears that in the case of insulin, which modulates the metabolism not only of carbohydrates but of lipids and proteins as well, the receptor is a part of a membrane-bound assembly of protein subunits, two of which contain binding sites for the hormone and are linked to two others joined across the plasma membrane to two intracellular protein tyrosine kinases.[97] The receptors for the binding of peptide hormones such as ACTH appear to be transmembrane proteins linked inside the cell to G proteins that stimulate the enzymatic conversion of phospholipids to inositol phosphates and diacylglycerol.[98] It seems that in the case of triiodotyrosine and steroids the protein receptors are cytoplasmic units that promote the action of these hormones on gene expression.[99]

The concept of the chemical transmission of neural impulses was advanced by Thomas Elliott in 1904, in relation to the action of epinephrine, but that idea gained general acceptance only some years later, with Otto Loewi's discovery of what he called *Vagusstoff*, and its identification as acetylcholine by Henry Dale and Harold Dudley.[100] As was noted above, Langley's receptor theory came from his studies on the action of nicotine and curare on muscle, and later work showed that acetylcholine acts not only at the "nicotinic" receptors in autonomic ganglia and at neuromuscular junctions but also at receptors for the toadstool toxin muscarine at postanglionic parasympathetic nerve endings.[101] Extensive work by numerous investigators has shown that the nicotinic "cholinergic" receptor is a transmembrane glycoprotein assembly of six subunits, of which two contain the binding sites for acetylcholine or a toxin. In the postsynaptic membranes, such receptors act as ion channels, opened by interaction with acetylcholine, and reclosed upon its hydrolysis by a specific esterase. The binding of acetylcholine to transmembrane muscarinic receptors activates G proteins, with inactivation of adenylyl cyclase, activation of a phospholipase, and opening of an ion channel.

Among the known neurotransmitters are several amino acids. For example, glutamic and aspartic acids, like acetylcholine, stimulate receptors on postsynaptic membranes to transmit a neuronal impulse. On the other hand, γ-aminobutyric acid (formed by decarboxylation of glutamic acid) and glycine, whose receptors are also multisubunit proteins, are inhibitors of this process, and the glycine receptor has an affinity for strychnine.[102] In addition to the enkephalins, produced by specific partial proteolysis of longer peptide chains and discovered by Hans Kosterlitz during the 1950s, several of the longer-known peptide hormones (secretin, gastrin, oxytocin, and others) have been found to be neurotransmitters and to stimulate membrane-bound guanylyl cyclases.[103]

The so-called "catecholamine" neurotransmitters, all derived from tyrosine, are 3-hydroxytyrosine ("dopa"), its decarboxylation product ("dopamine"), and epinephrine $[C_6H_3(OH)_2\text{-}CH(OH)CH_2NHCH_3]$. They have distinctive transmembrane "adrenergic" receptors, some acting through G proteins and coupled to adenylyl and guanylyl cyclases, phospholipases, ion channels, and phosphodiesterases.

The range of the hormone concept, as defined by Starling, was further broadened during the 1950s, when the embryologist Rita Levi-Montalcini and her biochemical associate Stanley Cohen described a "nerve growth factor" in mammalian tissues and fluids that Cohen isolated and found to be a 44,000-molecular weight protein. A few years later, Cohen found an "epidermal growth factor," which he identified as an 49-amino acid peptide.[104]

In a remarkable development during the 1980s, there was added to the list of physiologically important small molecules nitric oxide (NO), an unstable and highly toxic radical. Known since the eighteenth century, its discovery has been attributed to Joseph Priestley in 1772.[105] Among the first to recognize the possible regulatory role of NO were Robert Furchgott and his associates, who discovered that the relaxation of blood vessels by acetylcholine involves the action of this agent on muscarinic receptors in endothelial cells with the release of a substance responsible for the vasodilatation.[106] At about the same time, other investigators, in particular Salvador Moncada, found arginine to be essential to the ability of macrophages in the immune system to kill tumor cells and microbes, and showed that this amino acid is decomposed to citrulline and NO.[107] It was also found that NO is involved in the control of the excitatory action of glutamic acid in the brain.[108] It is a measure, I believe, of the present state of the development of enzyme chemistry that within a few years the complex mechanism of the formation of NO from arginine was largely elucidated by several investigators, notably Michael Marletta and Solomon Snyder. The oxidation involves reduced NAD, FMN and FAD flavoproteins, a heme protein of the cytochrome P450 type, and a pterin cofactor. This assembly is termed nitric oxide synthase and exists in animal tissues in the form of different "isozymes."[109] Of particular interest was the finding that NO activates a soluble guanylyl cyclase by binding to a heme protein discovered to be a constituent of this enzyme, and the subsequent demonstration that carbon monoxide, long known to bind tightly to heme proteins, also acts as a regulatory agent.[110] The pharmacological implications of these discoveries are certain to be important; for example, it has been reported that a possible mechanism of the action of aspirin may involve its inhibition of the catalytic activity of nitric oxide synthase.[111] Moreover, work in the laboratory of Eric Kandel has implicated nitric oxide as a factor in the storage of memory in the brain.[112] It is clearly too early to assess the impact of these recent discoveries on the study of brain function, but there can be little doubt that neurochemistry, with its beginnings during the nineteenth century in the work of Fourcroy, Vauquelin, Couerbe, and especially Thudichum,[113] has become one of the most exciting areas of biochemical investigation. One may perhaps expect to learn more about the function of chemical entities denoted as sphingomyelin, cephalin, and cerebrosides, or the brain "proteolipid" described by Jordi Folch-Pi in 1951.[114]

To these currently active lines of research on the role of receptors may be added studies on the regulation of what has come to be called apoptosis (Greek *apo,* away from; *ptosis,* fall), or cell death. This process appears to involve the stimulation of membrane receptors that initiate the activation of intracellular proteinases.[115]

The variety and complexity of the regulatory mechanisms developed during the course of biological evolution, as revealed by the application of present-day biochemical and genetic techniques, raise anew the question whether the "whole" of a living organism will ever be fully understood in terms of the properties and interactions of its separate "parts." During the twentieth century, eminent biologists and philosophers—Edmund Wilson, Alfred Whitehead, Alex Novikoff, André Lwoff, Ernst Mayr, François Jacob, François Gros, to name only a few—have expressed their views about the "organization" or "integration" of the chemical constituents and processes in living things.[116] Philosophers of biology may perhaps continue to debate the merits of "antireductionism," but it is certain that biochemists, geneticists, cytologists, and physiologists will continue their efforts to reconstitute parts of the "whole" by experimental combination of well-known and newly discovered chemical entities or assemblies. Whether such efforts will lead to the planned or accidental creation of a living organism is now a matter suitable only for science fiction. More important, in my opinion, is that, as François Gros put it: "Scientists are well aware of the fact that only reading the genes will not explain everything, and that it is not really useful unless account is taken of the intra- or extracellular environment."[117]

Introduction

1 Bloch (1994), pp. 65–91.

2 Woodward (1956, 1989); Cornforth (1993); Hoffmann (1995), pp. 90–100.

3 See, for example, Racker (1983).

4 Thackray and Merton (1975).

5 Kohler (1982).

6 Grmek (1995).

7 Reingold (1981); Whitaker (1986); Forman (1991); Gillispie (1994); Brush (1995). See also Lurçat (1989).

8 Hull (1979).

9 Feyerabend (1975), p. 302. See McEvoy (1975); Feyerabend (1995), pp. 139–152.

10 Racker (1984), p. 2902.

11 Fruton (1976), p. 233.

12 Brush (1995), p. 231. See also Brock (1976); Reingold (1990).

13 Medawar (1963), p. 377. See also Hoffmann (1988).

14 For Bernard, see Wolff (1967); Robin (1979). For Pasteur, see Geison (1995); Perutz (1995); Latour (1995).

15 Wittgenstein (1929). I am indebted to Professor J. Mitchell Morse for the source of this quotation. See also Reichvarg and Jacques (1991); Cooter and Pumphrey (1994).

1. On the History of Scientific Disciplines

1 Kay (1993); Abir-Am (1995).

2 Morange (1994), pp. 5–6.

3 Morange (1992), p. 71. See also Fantini (1988).

4 Fruton (1992a), pp. 195–214; Morange (1994), p. 12.

5 The italicized words represent abstractions related to other sets of abstractions, such as *vocation* and *profession*. See Tondl (1979) and Pelikan (1984).

6 Lippmann (1934).

7 Hill (1956); Cranefield (1957, 1966); Coleman and Holmes (1988); Rasmussen (1997a).

8 Fruton (1976a), p. 330; references are cited in the article.

9 Bence-Jones (1867), p. 2; Foster (1877), p. 494.

10 Klein (1972a); Brush (1965, 1966); Tolman (1938); Kuhn (1978).
11 Hund (1977); Kober (1980); Berson (1996); Kikuchi (1997).
12 Bohr (1933). For Bohr, see Heilbron and Kuhn (1969) and Pais (1991); for Schrödinger, see Moore (1989). For their place in the history of molecular biology, see Fleming (1968); Stent (1968a, 1969); Olby (1974), pp. 227–247; Yoxen (1979); Abir-Am (1985); Symonds (1988); Fischer and Lipson (1988); Keller (1990); Perutz (1990a); Kay (1992, 1997); Morange (1994), pp. 94–103.
13 Shannon and Weaver (1949); Quastler (1953); Lwoff (1962), pp. 89–97; Johnson (1970); Denbigh and Denbigh (1985); Darrigol (1990); Kay (1995); Sarkar (1996). For Neumann, see Aspray (1990).
14 Ewald (1962); Law (1973); Bragg (1975); Glusker (1981); Hauptmann and Blessing (1987); Hildebrandt (1993).
15 Crowther (1974); Steitz (1986); Eisenberg (1994); Kim (1995).
16 Kendrew (1968); Olby (1974), pp. 225–265; (1985); Rossmann (1994).
17 Weaver (1970a), p. 582. See also Kohler (1977a, 1991); Abir-Am (1982, 1984); Fuerst (1984); Fruton (1992a), pp. 211–212; (1994), pp. 92–94; Kay (1993); Morange (1994). For Weaver, see Rees (1987).
18 Astbury (1947a), p. 326. See also Olby (1974), pp. 41–70.
19 Beadle (1951); McElroy and Glass (1957).
20 Burian (1993) has suggested that a distinction be made between molecular biology and molecular genetics because the latter has a more clearly defined set of research problems. See also Burian (1996). Recently, the trend to label biological and medical specialties as "molecular" has been resisted by investigators in the field of "chemical ecology;" see Eisner and Meinwald (1995).
21 For my personal involvement in the academic politics on that issue, see Fruton (1994).
22 Kohler (1982), p. 333. According to the entry on p. 388, the source of the quotation is a letter from the physiologist A. V. Hill to the endocrinologist C. N. H. Long, dated 2 June 1936, and reflecting an attitude prevalent during Hill's youth.
23 Ben-David (1960, 1968, 1971, 1991); Kohler (1982); Fruton (1992a), pp. 183–186. For Ben-David, see Schott (1993).
24 Broad (1980); Kohler (1980); Gillispie (1980); Reingold (1981); Holmes (1981); Whitaker (1984); Brush (1995).
25 Among these books were those of Kopp (1843–1847); Wurtz (1869); Hoefer (1872); Meyer (1889); Schorlemmer (1894); Freund (1904); Hjelt (1916); Walden (1941), and especially Partington (1961–1970). See Weyer (1971, 1973, 1974a). A reviewer of the fourth volume of Partington's treatise deplored the author's "obsessive concern with facts," found "little sign of a trained historian's discrimination," and considered the book to be "firmly imbedded in the old, and by now weary tradition of chemical history by chemists for chemists" [Thackray (1966), p. 126], and Reingold (1981), p. 278, wrote "I am indebted to and very respectful of Partington, but want a different history of chemistry."
26 For example, see Cohen (1956); Crombie (1963), pp. 1–11; Guerlac (1977); Williams (1965); Holmes (1992a). See also Ravetz (1971) and Ackermann (1985).
27 Holmes (1981), p. 61.
28 For some of the recent discussions of the academic profession, see Clark (1983, 1984, 1987); Bourdieu (1984); Becher (1989).
29 Merton (1968a, 1979, 1987). See also Coser (1975); Hollinger (1983); Zuckerman (1989); Mendelsohn (1989).
30 Bernal (1939), p. xiii.

31 Sheehan (1985), pp. 303–336. See also Werskey (1978); McGucken (1984); Olwell (1996). For the U.S. counterparts, see Kuznick (1987).
32 Merton (1957), pp. 552–553.
33 Barber (1952); Barber and Hirsch (1962); Hagstrom (1965); Storer (1966); Merton (1973); Coser (1975); Mulkay (1976); Zuckerman (1977b); Swazey et al. (1993); Etzkowitz (1996).
34 Chubin (1990); David et al. (1992, 1993); Alberts and Shine (1994); Wade (1994); Griffiths et al. (1995); Cohen et al. (1995).
35 Karlson (1986); Chargaff (1986), pp. 193–205; Fruton (1990a); (1992a), pp. 255–258; Kornberg (1995).
36 Edge (1994), p. 366.
37 Kuhn (1962), p. 8; Barnes (1982), p. x.
38 Kuhn (1977), p. 297; (1993), p. 315. See Hoyningen-Huene (1993).
39 Barnes and Dolby (1970); Krohn (1982), p. 327.
40 Kuhn (1977), pp. xxi–xxii; Merton (1979), pp. 71–109.
41 Bloor (1976). For criticisms, see Trigg (1978); Laudan (1981); Turner (1981); Mounce (1985); Roth (1987); Hadden (1988); Hesse (1988); and especially Bunge (1991, 1992) and Schmauss (1994), pp. 256–264.
42 Latour and Woolgar (1979, 1986); Knorr-Cetina (1981). For the citation analysis, see Hicks and Potter (1991).
43 Burgus and Guillemin (1970); see also Guillemin (1978) and Wade (1981).
44 Du Vigneaud (1952).
45 Hacking (1988), pp. 279–280.
46 Latour and Woolgar (1979), pp. 29–30, 36, 64, 88. For Latour, see Koch (1995).
47 Latour (1984), pp. 89, 165–166. See also Bowker and Latour (1987).
48 Salaman (1949); Kahane (1978); Carpenter (1994). Also recall van Gogh's *The Potato Eaters*.
49 Woolgar (1983), p. 468.
50 Among such reports were those of Collins and Pinch (1982); Zenzen and Restivo (1982); Mulkay and Gilbert (1984); Pickering (1984); Lynch (1985); Pinch (1986); Traweck (1988). Some have shifted to the social study of experimentation; see Gooding (1990); Clarke and Fujimura (1992); Pickering (1995).
51 See Tilley (1981); Freudenthal (1984); Boudon (1990), pp. 311–317; Bunge (1991, 1992, 1996); Brown (1991, 1994); Cole (1992, 1996); Sismondo (1993); Segerstrale (1993); Roth (1996); Roll-Hansen (1998).
52 This term was taken from the writings of E. P. Thompson (1968, 1991), who studied the peasants, artisans, and laborers in England during the period around 1800. For Thompson, see Saville (1993) and Daston (1995).
53 For McClintock, see Federoff and Botstein (1992).
54 Shapin (1989); Pumphrey (1995).
55 Morrell (1972); Geison (1978); Fruton (1990a); Geison and Holmes (1993).
56 Holden (1993); Aronowitz and DiFazio (1994), pp. 139–169.
57 Stone (1971); Shapin and Thackray (1974); Pyenson (1977); Kragh (1987), pp. 174–181. For bio-bibliographic sources in the chemical and biological sciences, see Fruton (1995a). For examples of valuable studies on women scientists, see Rossiter (1982, 1993, 1995) and Creese (1991).
58 Beveridge (1951); Bunge (1962); Taylor and Barron (1963); Medawar (1969); Krebs and Shelley (1975); Sasso (1980); Gruber (1981); Holmes (1981); Aris et al. (1983); Wallace and Gruber (1989);

Gholson et al. (1989); Weissberg (1993). The term *creativity* became popular among makers of U.S. science policy after the U.S.S.R. launched *Sputnik;* see Golovin (1963).

59 James (1880).

60 Crombie (1995), p. 230; Merton (1965, 1983).

61 Hankins (1979); Grmek (1980); Kragh (1987), pp. 169–181; Williams (1991); Fruton (1992a), pp. 215–228. Nevertheless, the *Dictionary of Scientific Biography* (1970–1990) is, for me, the single most important reference work in the history of science.

62 For example, Holmes (1974) for Claude Bernard; Jacques (1987) for Marcelin Berthelot; Geison (1995) for Louis Pasteur.

63 Bloch (1954); Geyl (1958); Tosh (1984); Kragh (1987); Marwick (1989).

64 Reingold (1990). For oral history, see Weiner (1988) and Pang (1989).

65 For Galton, see Forrest (1974); for the eugenics movement, see Ludmerer (1972) and Kühl (1994); for Haber, see Stern (1987); Harris (1992); Stoltzenberg (1994); Szöllosi-Janze (1996).

66 Pascal (1960); May (1979); Leibowitz (1989); Lejeune (1989).

67 Millar and Millar (1983, 1985, 1988); Russell (1988a); Fruton (1992a), pp. 221–226; (1994), pp. 1–7.

68 Crombie (1988, 1994, 1995). See also Hacking (1982); Vicedo (1995).

69 Crombie (1995), p. 237. For Crombie, see North (1996).

2. *The Institutional Settings*

1 Ziman (1994), p. 276.

2 For Boerhaave, see Lindeboom (1968); Underwood (1977). For Cullen and Black, see Kent (1950); Guerlac (1970); Donovan (1975); Simpson (1980).

3 For the Rouelle brothers and Macquer, see Guerlac (1959) and Rappaport (1960); for Chevreul, see Costa. See also Holmes (1989a).

4 Berman (1963). For Fourcroy, see Smeaton (1962) and Kersaint (1966); for Vauquelin, see Kersaint (1958) and Queruel (1994); for Braconnot, see Nicklès (1856); for Robiquet, see Berman (1975); for Pelletier, see Dillemann et al. (1989); for Caventou, see Delépine (1951). See also Crosland (1994).

5 Jadin (1933).

6 For Gay-Lussac, see Crosland (1978); for Thenard, see Thenard (1950); for Dumas, see Kapoor (1969) and Klosterman (1985); for Boussingault, see McCosh (1984) and Kahane (1988); for Pelouze, see Dumas (1885). See also Crosland (1967, 1992).

7 Kapoor (1973); Fisher (1973); Brooke (1975); Novitski (1992). For Gerhardt, see also Tiffeneau (1916, 1918).

8 For Berthelot, see Jungfleisch (1913) and Jacques (1987). For Wurtz, see Friedel (1885); Williamson (1885); Brooke (1976); Carneiro (1993). For Friedel, see Crafts (1900). For Pasteur, see Duclaux (1895); Dubos (1950); Dagognet (1967, 1994); Geison (1974, 1995).

9 Paul (1972a, 1972b, 1985, 1996); Fox and Weisz (1980); Weisz (1983); Hulin (1990); Telkes (1990).

10 For Grimaux, see Adam (1911); for Deville, see Oesper and Lemay (1950); for Lespieau, see Nye (1993a), pp. 142–162; for Le Châtelier, see Pascal (1937); for Urbain, see Job (1939) and Courrier (1974); for Haller, see Ramart (1926); for Raoult, see Kuslan (1975); for Grignard, see Courtot (1936) and Aubry (1984–1986); for Sabatier, see Nye (1977) and Wojtkowiak (1989); for Perrin, see Nye (1972) and Achinstein (1994). See also Guéron and Magat (1971); Nye (1986).

11 Charpentier-Morize (1989).

12 Hufbauer (1982).
13 For Stromeyer, see Zaunick (1942); for Gmelin, see Mani (1956) and Fluck (1989); for Mitscherlich, see Melhado (1980) and Schuett (1992).
14 The extensive biographical literature about Liebig includes Volhard (1909); Paolini (1968); Brock (1972, 1981, 1984, 1990, 1997); Morrell (1972); Turner (1982); Fruton (1988a); Holmes (1989b); Finlay (1991); Billig (1994).
15 For Wöhler, see Van Klooster (1944), Valentin (1949), and Keen (1976). For Bunsen, see Lockemann (1949), Meinel (1978), and Lotze (1986).
16 For Scheele, see Bokland (1975); for Bergman, see Smeaton (1970); for Berzelius, see Jorpes (1966), Melhado (1981), and Melhado and Frängsmyr (1992); for Mulder, see Snelders (1982, 1986).
17 Flemming (1967); Schmauderer (1969); Gustin (1975); Borscheid (1976); Huhle-Kreutzer (1989); Tuchman (1993).
18 For Williamson, see Harris and Brock (1976) and Paul (1978); for Odling, see Marsh (1921); Thornton and Wiles (1956); for Couper, see Anschütz (1909) and Duff (1987); for Frankland, see Russell (1986) and Hamlin (1990). For Kekulé, see Anschütz (1929), Gillis (1966), Rocke (1981, 1985), Goebel (1984), and Wotiz (1993). For Hofmann, see Perkin (1896), Lepsius (1918), Roberts (1976), Brock (1984), Oelsner (1989), Meinel (1992), and Meinel and Stolz (1992). For Perkin, see Meldola (1908).
19 For Kolbe, see Meyer (1884); Rocke (1993a,1993b, 1995), and Rocke and Heuser (1994). For Baeyer, see Perkin (1923); Rupe (1932); Prandtl (1949, 1952); Schmorl (1952), Huisgen (1986), and Fruton (1990a), pp. 118–162. For Fischer, see Hoesch (1921), Fischer (1922), Feldman (1973), Remane (1984), and Fruton (1990a), pp. 163–229.
20 For brief biographical sketches and references, see Fruton (1990a). For Wieland, see also Witkop (1992).
21 For Claus, see Costa (1971a); for Fittig, see Fittig (1870); Fichter (1911); for Wallach, see Ruzicka (1932) and Hückel (1961); for Auwers, see Meerwein (1939); for Windaus, see Butenandt (1961), Dimroth (1986) and Schuett (1988); for Hantzsch, see Hein (1941).
22 For a list of American and British doctoral students of chemistry at German universities between 1840 and 1914, see Jones (1983, 1991).
23 For the travail of the Privatdozent, see Busch (1959, 1963), and for the assistantship system, see Bock (1972). A valuable compilation of data on the composition and social origins of the faculties of German universities has been provided by Ferber (1956). More recent writings on these subjects include those of Schmauderer (1973a), Burchardt (1978, 1980), Jarausch (1982, 1983), Craig (1983), McClelland (1983), Johnson (1985a, 1985b, 1989), Schwabe (1988), and Fruton (1990a).
24 Prandtl (1952).
25 According to Jost (1966), p. 2, Ostwald was appointed "only after Landolt, Lothar Meyer, Cl. Winkler, and J. H. van't Hoff had declined." For Ostwald, see Donnan (1933), Hiebert and Körber (1978), Leegwater (1986), and Hakfoort (1992). For van't Hoff, see Cohen (1912), Snelders (1984), and Fischmann (1985). For Arrhenius, see Riesenfeld (1931) and Crawford (1996). See also Dolby (1976); Servos (1990), pp. 1–45; Barkan (1992); Laidler (1993).
26 Lenz (1910), pp. 306–310. For Nernst, see Mendelsohn (1973), Hiebert (1978), and Barkan (1994).
27 Harteck (1960).

28 For the emergence of the German dye industry, see Caro (1892); Perkin (1896); Haber (1958); Beer (1959); Hardie and Pratt (1966); Schmauderer (1969, 1973b); Stolz (1980, 1981); Travis (1993). For Caro, see Bernthsen (1912); for Graebe, see Duden and Decker (1928); for Liebermann, see Wallach and Jacobson (1914); for Duisberg, see Stock (1935) and Flechtner (1981); for Weinberg, see Ritter and Zerweck (1956); for Knorr, see Flemming (1965, 1967). For the early history of the Höchst firm, see Fischer (1958). According to my count, at least 81 of the 179 pupils of the Baeyer school whom I identified as having entered the chemical industry were employed by BASF, Bayer, and Höchst; see Fruton (1990a), pp. 158–161. See also Lundgren (1975).

29 Leprieur and Papon (1979); for Mulhouse, see *Société Industrielle de Mulhouse* (1902, 1994) and Fox (1984); for Geigy, see Bürgin (1958); for Zurich, see Eugster (1983); for Geneva, see Cherbuliez (1980); for Ruzicka, see Prelog and Jeger (1980). See also Perkin (1896) and Schmauderer (1971).

30 McClelland (1980).

31 Sachse (1928); Pfetsch (1970, 1974); Brocke (1991a, 1991b).

32 Craig (1984); Legée (1987–1988).

33 Ackerknecht (1988); Eckart (1991); Scholz (1991); Zirnstein (1991). For Koch, see Dolman (1973); Brock (1988); Weindling (1992).

34 Reinke (1925), pp. 176–177.

35 Humboldt (1970); Weber (1973).

36 The most valuable source is the book edited by Vierhaus and Brocke (1990). See also Pfetsch (1970); Burchardt (1975, 1977); Wendel (1975); Moy (1989); Johnson (1990).

37 Letter to Emil Abderhalden, 14 March 1912 (Bancroft Library).

38 For the Notgemeinschaft, see Zierold (1968) and Nipperdey and Schmugge (1970). The extensive literature on the emigration of German scientists and scholars is summarized in Kröner (1969) and Strauss et al. (1991). See also Pross (1955); Widmann (1973); Möller (1984); Born (1984); Hoch (1987); Fischer (1994); Halary (1994); Ash and Sollner (1996).

39 Deichmann (1992); Macrakis (1993a, 1993b).

40 Ironmonger (1957); Vernon (1957); Cardwell (1972); Berman (1978); Golinski (1992); MacLeod (1994). For Rumford, see Brown (1979). For Davy, see Knight (1992). For Faraday, see Williams (1965); Gooding and James (1985); Thomas (1981); Cantor (1991). For Tyndall, see Eve and Creasey (1945); Brock et al. (1981); for Dewar, see Findlay (1947) and Costa (1971b); for Bragg, see Caroe (1978); for Mond, see Cohen (1956). See also MacLeod (1971).

41 For Wollaston, see Goodman (1976); for Prout, see Brock (1965); for Dalton, see Cardwell (1968); for Joule, see Rosenfeld (1973). For Manchester, see Kargon (1977) and Bud and Roberts (1984); for Roscoe, see Roscoe (1906) and Thorpe (1916). For Schorlemmer, see Spiegel (1892); Henderson (1976); Heinig (1979), Benfey and Travis (1992) and Gelius (1996). For Perkin, see Greenaway et al. (1932); for Robinson, see Todd and Cornforth (1976); Robinson (1976) and Williams (1990). For Todd, see Todd (1983).

42 For Oxford, see Brewer (1957); Bowen (1965); Hartley (1965). For Cambridge, see Mann (1957); Berry and Moelwyn-Hughes (1963); Roberts (1980).

43 Burns (1993). For Turner, see Brock (1986); for Frankland, see Russell (1996); for Graham, see Clark (1991). For St. Bartholomew's, see Templeton (1982); for Guy's, see Coley (1988).

44 For Armstrong, see Rodd (1947); Eyre (1958). For Lapworth, see Robinson (1947); Schofield (1995). For Lowry, see Allsop and Waters (1947) and Saltzman (1997); for Ingold, see Green (1966), Roberts (1966), Shoppee (1972); Nye (1993), pp. 196–223; Schofield (1994).

45 Chick et al. (1971); Neuberger (1985).

46 Austoker and Bryder (1989). For Dale, see Feldberg (1970); for Harington, see Himsworth and Pitt-Rivers (1972); for Medawar, see Medawar (1990) and Mitchison (1990); for Fletcher, see Fletcher (1957); for Mellanby, see Dale (1955) and Platt (1956). For Hopkins, see Needham and Baldwin (1949); Needham (1962); Pirie (1983); Kamminga and Weatherall (1996); Weatherall and Kamminga (1992, 1996).

47 Holter and Møller (1976). This volume contains biographical articles for the persons named in the text; for Kjeldahl, see also McKenzie (1994). For an account of the Carlsberg Foundation, see Pedersen (1941).

48 Delaunay (1962); McElheny (1966); Morange (1991); Gaudillière (1993, 1996).

49 For Nencki, see Bickel (1972). For the Kazan school, see Vinogradov (1965); Lewis (1994); Brooks (1995).

50 For Boussingault, see Aulie (1970) and McCosh (1984); for Ritthausen, see Osborne (1913). For the development of agricultural chemistry, see Russell (1966); Rossiter (1975); Finlay (1988); Marcus (1988, 1990); Bud (1992, 1993); Brassley (1994).

51 For Schultze, see Rothschuh (1969); for Purkyně, see Kruta (1969); for Müller, see Haberling (1924) and Koller (1958); for Schwann, see Watermann (1960); for Helmholtz, see Königsberger (1902–1903), Cahan (1993), and Holmes (1994); for Brücke, see Brücke (1928) and Lesky (1970); for Virchow, see Ackerknecht (1963); for du Bois-Reymond, see Marseille (1968) and Rothschuh (1971). See also Engel (1989) and Cunningham and Williams (1992), pp. 14–72.

52 Rosen (1936); Schröer (1967); Fye (1986); Zupan (1987); Cranefield (1988); Lenoir (1988); Gerabek (1991). See also Bonner (1965). For the quotation, see du Bois-Reymond (1927), p. 54.

53 Eulner (1970).

54 Lenz (1910), pp. 154–164. For Boll, see Hubbard (1977); for Thierfelder, see Klenk (1931) and Lesch (1990); for Herrick, see Herrick (1949).

55 For Kühne, see Schalck (1940) and Fruton (1990a), pp. 72–117; for Voit, see Frank (1908) and Holmes (1984, 1988); for Abderhalden, see Hanson (1970) and Picco (1981), pp. 24–30.

56 Lenz (1910), pp. 165–176.

57 Büttner (1983). For Vienna, see Lesky (1976), pp. 106–117, 222–228; Schmidt (1991).

58 Kuschinsky (1968). For Buchheim, see Schmiedeberg (1912) and Bruppacher-Celler (1971). For Schmiedeberg, see Meyer (1922) and Koch-Weser and Schechter (1978); for Abel, see Parascandola (1992); for Cushny, see Abel (1926) and MacGillivray (1968); for Hofmeister, see Pohl and Spiro (1923) and Fruton (1990a), pp. 163–229. See Legée (1987).

59 For Sigwart and Schlossberger, see Simmer (1955) and Hesse (1976). For the establishment of the science faculty at Tübingen, see Engelhardt and Decker-Hauff (1963). For Hoppe-Seyler, see Baumann and Kossel (1895) and Fruton (1990a), pp. 72–117; for Baumann, see Spaude (1973) and Bäumer (1996); for Kossel, see Jones (1953) and Gerber and Sauer (1985); for Knoop, see Thomas (1948); for Embden, see Deuticke (1933); for Loewi, see Dale (1962) and Lembeck and Giere (1968). See also Wankmüller (1980).

60 For Hüfner, see Zeynek (1908); for Klenk, see Butenandt (1973); for Butenandt, see Karlson (1990).

61 For Schmidt, see Zaleski (1894); for Huppert, see Koerting (1967); for Nencki, see Bickel (1972); for Bunge, see Schmidt (1974); for Pregl, see Philippi (1962).

62 Hoppe-Seyler (1877a), p. I; (1884); Pflüger (1877), pp. 363, 365; du Bois-Reymond (1912), vol. 1, pp. 630–653; Verworn (1901); Kossel (1908).

63 Ganss (1937), p. 68. See also Maehle et al. (1990).

64 Lorenz (1984). For Thomas, see Thomas (1954); Ihde (1990).

65 For Feulgen, see Felix (1957) and Kasten (1964). For Michaelis, see Fruton (1990b). For Berlin, see Sauer et al. (1961); Krebs and Lipmann (1974); Fischer (1994). For appeals to liberate physiological chemistry, see Ackermann (1931a), Abderhalden (1931, 1932), and Schütte (1976).

66 Hopkins (1926), pp. 34–35.

67 E. Fischer to T. W. Engelmann, 24 February 1904 (Archiv der Humboldt Universität, Akte 1450, Bl. 172–173).

68 Trott zu Solz et al. (1912), pp. 70, 87. See also Fischer (1924).

69 Roger (1979); Caron (1988). Historians of biology now appear to prefer the term *life sciences*.

70 Engelhardt and Decker-Hauff (1963), p. 35.

71 For the history of nineteenth-century biology, see Müller (1902); Nordenskiøld (1928); Coleman (1971); Hünemörder and Scheele (1977); Jahn et al. (1985). For the history of botany, see Sachs (1890); Reed (1942). For Schleiden, see Möbius (1904); Studnicka (1933); Buchdahl (1973); Franke (1988). For Mohl, see Pelz (1987); for Hofmeister, see Goebel (1924); for Nägeli, see Wilkie (1960, 1961); for Sachs, see Pringsheim (1932) and Gimmler (1982); for de Bary, see Jost (1930) and Robinson (1971); for Strasburger, see Tischler (1913) and Robinson (1976); for Pfeffer, see Bünning (1975) and Sucker (1988); for Cohn, see Cohn (1901), Geison (1971), and Hoppe (1983); for Brefeld, see Falck (1925), Dolman (1970), and Höxtermann (1996).

72 Uschmann (1959), p. 41. For Haeckel, see also Uschmann (1961, 1979).

73 For O. Hertwig, see Weissenberg (1959), Churchill (1970), and Weindling (1991); for R. Hertwig, see J. Oppenheimer (1972); for Roux, see Koch (1929) and Mocek (1974); for Driesch, see Wenzl (1951), Churchill (1969), and Freyhofer (1982); for Waldeyer, see Sobotta (1923); for Flemming, see Peters (1968); for Weismann, see Churchill (1970, 1986); for Boveri, see Baltzer (1962) and Moritz (1993).

74 Querner (1972a), p. 187.

75 Freund and Berg (1963–1966); Bradbury (1967); Sandritter and Kasten (1964); Clark and Kasten (1983).

76 Coleman (1965); Robinson (1979).

77 Harwood (1993). For Correns, see Wettstein (1938); for Baur, see Schiemann (1934); for Goldschmidt, see Stern (1967), Allen (1974), and Caspari et al. (1980); for Wettstein, see Melchers (1987).

78 Beyerlein (1991).

79 For Bourquelot, see Bougault and Hérissey (1921); for Delépine, see Kauffman (1976).

80 Lesch (1984). For Magendie, see Olmsted (1944), Albury (1977), and Théodoridès et al. (1983). The literature on Bernard is enormous; the most valuable recent writings are those by Grmek (1967a, 1967b, 1970, 1973); Schiller (1967b); Holmes (1974); Michel et al. (1991). For Barreswil, see Peumery (1986); for Pelouze, see Dumas (1885) and Berman (1974). See the report by Bernard (1872, 1979).

81 For Bert, see Mani (1966) and Jacquemin (1988); for d'Arsonval, see Culotta (1970a); for Dastre, see Delezenne (1917); for Ranvier, see Jolly (1923).

82 For Raspail, see Weiner (1968); for Robin, see Pouchet (1886) and Grmek (1975).

83 Buican (1984); Burian et al. (1988).

84 For Chabry, see Fischer and Scott (1984); for Cuénot, see Courrier (1957a) and Buican (1982); for Delage, see Fischer (1979); for Caullery, see Fauré-Fremiet (1964) and Telkes (1993); for Guyénot,

see Fischer (1995); for Giard, see Gohau (1979); for Bonnier, see Jumelle (1924) and Davy de Virville (1970); for Maquenne, see André (1937); for Guillermond, see Emberger (1946); for Molliard, see Combes (1946); for Devaux, see Genevois (1956) and Kaplan (1956).

85 Fleury (1966). For Béchamp, see Guédon (1978); for Schützenberger, see Friedel (1898) and Davis (1929); for Duclaux, see Roux (1904) and Vermenouze (1992); for Bertrand, see Courrier (1972) and Gaudillière (1991).

86 For Brown-Séquard, see Borell (1976b) and Aminoff (1993); for Gley, see Grmek (1990) and Bange (1995).

87 Gilpin (1968); Ben-David (1970); Ben-David and Zloczower (1962); Paul (1972b); Day (1992).

88 For the Wurtz statement, see Paul (1972b), p. 8. For Duruy, see Horvath-Peterson (1984).

89 Fox (1973); Crosland (1975); Nye (1986); Charle (1990); Hulin (1990).

90 Burian et al. (1988); Burian and Gayon (1990); Gaudillière (1990, 1992, 1993); Debru et al. (1994). For Wollman, see Heim (1959); for Lwoff, see Lwoff (1971); for L'Héritier and Teissier, see Petit (1990); for Ephrussi, see Sapp (1987), pp. 123–167; for Monod, see Lwoff (1977) and Gaudillière (1992); for Fromageot, see Desnuelle (1958); for Desnuelle, see Desnuelle (1983), for Roche, see Polonovski (1993).

91 For Prout, see Brock (1965, 1985); for Marcet, see Coley (1967a). The quotation is from Coley (1967b), p. 180.

92 Johnson (1803), vol. 1, p. iv. I have been unable thus far to find biographical information about the author. The text bears some resemblance to parts of Fourcroy's *Système des Connaissances Chimiques* of 1801–1802.

93 Berzelius (1806, 1808, 1813).

94 See Pelling (1978), pp. 113–145 for the response of some British physicians to Liebig's suggestion that among the products of putrefaction may be "morbid poisons" that can affect the blood and tissues in a manner to produce disease.

95 For Simon, see Minding (1844); for Kingzett, see Morgan (1935); for Kletzinsky, see Hoswell (1882).

96 For Bence Jones, see Coley (1973) and Putnam (1993); for Thudichum, see Drabkin (1958) and Rafaelson (1982); for MacMunn, see Keilin (1966), pp. 86–116, 355–357. Thudichum's chief patron was Sir John Simon; for Simon, see Lambert (1963); Cunningham and McGowan (1992); Sourkes (1993b).

97 Geison (1972, 1978); Butler (1988). For Sharpey, see Taylor (1971).

98 Langley (1917), p. xxv.

99 Sharpey-Shafer (1927), p. 30. For more on Gamgee, see D'Arcy Thompson (1974); for Pavy, see Bywaters (1916); Tattersall (1996).

100 Weatherall and Kamminga (1992, 1996) and Kamminga and Weatherall (1996) have provided a valuable account of the history of the development of biochemistry at Cambridge during 1898–1949.

101 Hopkins (1926), pp. 35–36.

102 For Hardy, see Hopkins (1934) and Bate-Smith (1964); for Barcroft, see Franklin (1953); for Halliburton, see Morgan (1983); for Moore, see Hopkins (1927); for Garrod, see Bearn (1993); for Harden, see Smedley-Maclean (1941); for Schryver, see Harden (1930); for Plimmer, see Lowndes (1956).

103 Hartree (1981).

104 For Chibnall, see Chibnall (1987) and Synge and Williams (1990); for Young, see Randle (1990); for Todd, see Todd (1983); for Perutz, see Perutz (1980); for Bernal, see Hodgkin (1980).

105 For Bayliss, see Bayliss (1961); for Donnan, see Freeth (1957); for Rideal, see Eley (1976); for Adam, see Carrington et al. (1974).

106 Norrish (1969).

107 Weatherall and Kamminga (1992), p. 11; Morton (1972).

108 Morton (1969, 1972). For Sherrington, see Swazey (1975); for Peters, see Thompson and Ogston (1983); for Burn, see Bülbring and Walker (1984); for Blaschko, see Blaschko (1983); for Morton, see Glover et al. (1978); for Heilbron, see Cook (1960). See also Weindling (1996).

109 For Rosenheim, see King (1956); for Bayliss and Starling, see Verney (1956), Bayliss (1961), Wilson (1968), and Hill (1969); for Drummond, see Young (1954); for Haldane, see Pirie (1966) and Clark (1968); for Rimington, see Rimington (1995); Neuberger (1996).

110 For Astbury, see Bernal (1963); Olby (1974), pp. 41–70; Witkowski (1980a); for Cohen, see Raper (1935); for Dakin, see Clarke (1952) and Hartley (1952); for Hartley, see Dale (1957); for Raistrick, see Birkinshaw (1972).

111 For Hill, see Katz (1978); for Leathes, see Peters (1958). For Krebs, see Kornberg and Williamson (1984), Holmes (1991, 1993), and Weatherall and Kamminga (1992), p. 83.

112 For Walker, see Kendall (1935); for Sharpey-Shafer, see Hill (1935); for Barger, see Hill (1940).

113 Smith and Nicolson (1989). For Davidson, see Neuberger (1973); for Paton, see Cathcart (1929); for Cathcart, see Wishart (1954).

114 Tansey (1989); Liebenau (1990); James (1994).

115 Hastings (1971); Morgan (1980).

116 For Fredericq, see Florkin (1979); for Beneden, see Brachet (1923); for A. Brachet, see Winiwarter (1933); for Gratia, see Welsh (1967); for Bacq, see Lecomte (1985); for Florkin, see Bacq and Brachet (1981); for Errera, see Fredericq and Massart (1908); for J. Brachet, see Pirie (1990) and Chantrenne (1990).

117 Gillis (1966). For Spring, see Ross (1975).

118 Jorissen and Reicher (1912); Hoytinck (1970). For Cohen, see Donnan (1948); for van der Waals, see Prins (1976); for Bijvoet, see Groenwege and Peerdemann (1983); for Kruyt, see Overbeek (1960); for Holleman, see Wibaut (1955); for Jaeger, see Klooster (1973); for Kögl, see Lynen (1959) and Karlson (1986). See also Willink (1991).

119 Davies (1970); Williams (1975).

120 For Pekelharing, see Erdmann (1964); for Hamburger, see Cohen (1908); for Einthoven, see Hoogewerf (1971) and Snellen (1995); for Eijkman, see Lindeboom (1971); for de Vries, see Allen (1969) and van der Pas (1976); for Beijerinck, see Iterson et al. (1940); for Kluyver, see van Niel (1949), Woods (1957), and Kamp et al. (1959).

121 Veibel (1939, 1943, 1968); Bak (1974). For Thomsen, see Bjerrum (1909) and Kragh (1982); for Brønsted, see Christiansen et al. (1949) and Bell (1950); for Bjerrum, see Guggenheim (1960) and Kauffman (1980); for Jørgensen, see Kauffman (1960); for Biilmann, see Stock (1989); for Bohr, see Tigerstedt (1911) and Edsall (1972); for Krogh, see Rehberg (1951), Jørgensen (1976), and Schmidt-Nielsen (1995); for Lundsgaard, see Kruhoffer and Crone (1972); for Johannsen, see Churchill (1974) and Roll-Hansen (1978).

122 For O. Hammarsten, see Thunberg (1933); for Thunberg, see Kahlson (1953); for Euler, see Lynen (1965); for E. Hammarsten, see Jorpes (1969); for Jorpes, see Roden and Feingold (1985); for Theo-

rell, see Dalziel (1983); for Caspersson, see Mayall (1984); for Svedberg and Tiselius, see Claesson and Pedersen (1972), Kekwick and Pedersen (1974), Kerker (1976), Pedersen (1983), and Kay (1988); for Blix, see Tigerstedt (1904); for Granit, see Enroth-Cugell (1994); for Liljestrand, see Euler (1968).

123 Babkin (1949); Sechenov (1965); Grigorian (1974); Gray (1979). For an interesting account of the institutional features of the universities in Czarist Russia, see Petrunkievich (1920).

124 For Botkin and Danilevski, see Bezkorovainy (1980); for Bakh, see Shamin (1970); for Chodat, see Lendner (1934); for Battelli, see Morsier and Monnier (1977).

125 For Khodnev, see Figurovski and Soloviev (1954); for Zinin, see Brooks (1995); for Borodin, see Soloviev (1970) and Davies (1995); for Bidder, see Achard (1969) and Bing (1973); for Zinin, see Brooks (1995); for Tammann, see Tammann (1976); for Bunge, see Schmidt (1974) and Portmann (1974); for Lunin, see Bezkorovainy (1973a). For Tartu University, see Siilivask (1985).

126 For A. Schmidt, see Jorpes (1951); for Sakharin, see Gukasyan (1948); for Gulevich, see Bezkorovainy (1974); for Pashutin, see Veselkin (1950).

127 For Braunstein, see Braunstein (1982); for Oparin, see Adams (1990d); for Engelhardt, see Engelhardt (1982) and Kisselev (1990); for Belozerski, see Shamin (1990a); for Severin, see Severin (1983); for Shemyakin, see Shamin (1990b).

128 For Famintzin, see Senchenkova (1960); for Navashin, see Lewitsky (1931); for Timiryazev, see Gaissinovich (1985) and Berg (1990); for Prianishnikov, see Joffe (1948) and Senchenkova (1975); for Palladin, see Senchenkova (1974); for Chetverikov, see Adams (1968, 1990b); for Filipchenko, see Adams (1990c); for Vavilov, see Adams (1978); for Astaurov, see Berg (1979) and Adams (1990a); for Lysenko, see Soyfer (1994).

129 For Mechnikov, see Hutchinson (1985) and Tauber and Chernyak (1991); for Vinogradsky, see Waksman (1953), Thornton (1953), and Courrier (1957b); for Ivanovski, see Lechevalier (1972).

130 For Hjelt, see Enkvist (1972), pp. 66–83; for Aschan, see Hückel (1941); for Marchlewski, see Ostrowski (1966); for Tsvet, see Robinson (1960); for Parnas, see Mochnacka (1956), Lutwak-Mann and Mann (1981), and Zielynska (1987).

131 For Zemplén, see Schmidt (1959); for Bruckner, see Medzihradszky (1981); for Szent-Györgyi, see Straub and Cohen (1987) and Moss (1988); for Straub, see Straub (1981); for Zervas, see Katsoyannis (1972), pp. 1–20; for the Weizmann Institute, see Sela (1987), Eisenberg (1990), and Katchalski-Katzir (1993).

132 For Volta, see Heilbron (1976); for Avogadro, see Crosland (1970a), Frické (1976) and Brooke (1981); for Cannizzaro, see Tilden (1912); for Ciamician, see Nasini (1926); for Moleschott, see Moser (1967) and Hagelhans (1985); for Mosso, see Aducco (1911); for Bottazzi, see Quagliariello (1944); for Rondoni, see Giordano (1958). For the Naples Zoological Station, see Müller (1996).

133 Trikojus (1978); Young (1976).

134 For Houssay, see Foglia (1980) and Cueto (1994); for Leloir, see Ochoa (1990); for Sols, see Sols (1990). For Chile, see Schmitt-Fiebig (1988).

135 Needham (1954–1974). For a searching discussion of this question, see Cohen (1994), pp. 378–488.

136 For PUMC, see Bowers (1972). For Wu, see Wu (1959) and Edsall (1995a); for Chen, see Chen (1981). For the post–World War II developments in China, see Cheng and Doi (1968); Needham (1978); Tang and Lee (1980). For Tsou, see Tsou (1990, 1995).

137 Tsuzuki and Yamashita (1968); Doke (1969); Sakaguchi and Ikeda (1970); Mizushima (1972); Tamiya (1973); Bartholomew (1989). For Takamine, see Davenport (1982b); for Akabori, see Edsall (1995b); for Umezawa, see Tsuchiya et al. (1990).

138 Bud (1993), pp. 197–199; Morris-Suzuki (1994).

139 Newell (1976). For Rush, see Miles (1953), Shryock (1971), and Carlson (1975); for Mitchill, see Hall (1934); for Woodhouse, see Smith (1918); for Hare, see Smith (1917); for Priestley, see Gibbs (1965), Schofield (1967), McEvoy (1978), and Graham (1995); for Cooper, see Malone (1926) and Cohen (1982); for Maclean, see Miles (1994).

140 For Silliman, see Fulton and Thomson (1947) and Brown (1989); for Silliman Jr., see Thomson (1975); for Norton, see Rossiter (1971).

141 Chittenden (1930). For Johnson, see Osborne (1911); for Atwater, see Rosenberg (1970); for Osborne, see Vickery (1931) and Fruton (1990c).

142 For Horsford, see Resneck (1970) and Rossiter (1975), pp. 49–88; for Gibbs, see Clarke (1910); for Genth, see Barker (1902); for Smith, see Silliman (1886); for Chandler, see Bogert (1931).

143 Cowen (1983); Wankmüller (1985).

144 For Cooke, see Forbes (1971); for Richards, see Hartley (1930), Ihde (1969), Conant (1970, 1974), and Kopperl (1976); for Crafts, see Cross (1919) and Ashdown (1928); for Remsen, see Noyes and Norris (1932) and Hannaway (1976).

145 For Michael, see Costa (1971c), Fieser (1975), and Ramberg (1995). For Nef, see Wolfrom (1960); for Noyes, see Hopkins (1944) and Adams (1952); for Gomberg, see Bailar (1970), McBride (1974), and Walling (1977).

146 Servos (1990); Laidler (1993). For Gibbs, see Wheeler (1952), Klein (1972b), and Daub (1976); for Noyes, see Pauling (1958) and Servos (1976, 1980); for Lewis, see Servos (1984), Calvin and Seaborg (1984), Calvin (1984) and Saltzman (1986); for Kraus, see Fuoss (1971); for Onsager, see Longuet-Higgins and Fischer (1978); for Harned, see Sturtevant (1980); for Scatchard, see Wilson and Ross (1973) and Edsall and Stockmayer (1980); for MacInnes, see Longsworth and Shedlovsky (1970); for Kirkwood, see Scatchard (1960).

147 Servos (1982, 1994).

148 For Garvan, see Gould (1994); for Adams, see Tarbell and Tarbell (1981); for Carothers, see Adams (1939) and Hill (1979); for Stieglitz, see Noyes (1939).

149 Swann (1988); Liebenau (1987); Liebenau et al. (1990); Bowie (1994); Galambos and Sturchio (1996); Cockburn and Henderson (1996).

150 For Agassiz, see Lurie (1960) and Winsor (1979); for Cope, see Osborn (1931); for Marsh, see Schuchert and LeVine (1940).

151 For Beaumont, see Rosen (1970) and Bylebyl (1970).

152 Geison (1987); Frank (1987); Pauly (1987b). For Loeb, see Osterhout (1930), Rothberg (1965), and Pauly (1987a).

153 Pauly (1984); Maienschein (1991). For Bowditch, see Cannon (1922) and Fye (1987); for Martin, see Fye (1985); for Brooks, see Conklin (1913), McCullough (1969), and Benson (1985); for Wilson, see Morgan (1940), Muller (1943), and Baxter (1979); for Morgan, see Sturtevant (1959) and Allen (1978); for Conklin, see Harvey (1958) and Atkinson (1985); for Harrison, see Abercrombie (1961), Oppenheimer (1966), Hamburger (1980), and Witkowski (1980a).

154 Kevles (1980); Rainger et al. (1988); Benson et al. (1991). For Whitman, see Davenport (1917); for Wheeler, see Evans and Evans (1970); for McClung, see Allen (1973); for Jennings, see Sonneborn (1975); for East, see Jones (1944); for Castle, see Dunn (1965); for Howell, see Erlanger (1951); for Lombard, see Davenport (1982a), pp. 51–76; for Carlson, see Ingle (1979); for

Cannon, see Cannon (1945) and Benison et al. (1987); for Erlanger, see Davis (1970) and Marshall (1983).

155 Lillie (1944); Barlow et al. (1993).

156 Shryock (1948); Daniels (1971); Reingold (1972, 1991).

157 Chittenden (1930), p. 321. For Chittenden, see Vickery (1945).

158 Kohler (1982), p. 99; Fruton (1990a), pp. 110–114. For Gooch, see Van Name (1931); for Johnson, see Vickery (1952); for Anderson, see Vickery (1962).

159 Pierson (1952, 1955).

160 Holmes (1984, 1987, 1988).

161 For Mendel, see Chittenden (1937), Rose (1969), and Rossiter (1994). For McCollum, see Day (1974). McCollum was associated with Osborne and Mendel for about a year immediately after receiving his Ph.D. (1908) in chemistry at Yale.

162 For Lewis, see Rose and Coon (1974); for Rose, see Roe (1981).

163 Chittenden (1930), p. 322.

164 Fruton (1951, 1982a, 1994). For Long, see Smith and Hardy (1975).

165 For Atwater, see Maynard (1962); for Armsby, see Jones (1994); for Babcock, see Ihde (1971); for Hart, see Elvehjem (1954); for Steenbock, see Apple (1989); for Wilson, see Burris (1992); for Elvehjem, see Burris et al. (1990); for Benedict, see Dubois and Riddle (1958); for Alsberg, see Davis (1948); for Lusk, see DuBois (1940); for Murlin, see Fenn (1961); for Cowgill, see Orten (1976). For Wisconsin, see Nelson and Soltvedt (1989); Ihde (1990).

166 Meites (1989).

167 For Van Slyke, see Corner (1965), pp. 274–280, Hastings (1976), and Edsall (1985a); for Peters, see Lavieties (1956) and Paul and Long (1958), for Henderson, see Cannon (1943) and Parascandola (1971a, 1971b).

168 Kohler (1982), p. 269.

169 Corner (1965), pp. 152–153, 559. Flexner's note was dated 6 February 1919. For Levene, see Van Slyke and Jacobs (1944) and Tipson (1957).

170 R. S. Tipson to J. S. Fruton, 27 October 1988; italics in original.

171 Hartley (1952); Clarke (1952); Hawthorne (1983).

172 Kohler (1982), p. 270.

173 Voegtlin (1939); MacNider (1946); Parascandola (1982, 1992).

174 Chittenden (1945); Edsall (1980).

175 Clark (1938); Harvey (1976), pp. 258–263.

176 Clark (1962); Vickery (1967); Harvey (1976), pp. 263–267.

177 For Lehninger, see Harvey (1976), pp. 267–273 and Fruton (1986).

178 Kohler (1982), p. 279.

179 For Gies, see Clarke (1956); for Clarke, see Shemin (1974), Vickery (1975), and Fruton (1994), pp. 22–32.

180 For Hecht, see Wald (1948); for Nelson, see Herriott (1955).

181 R. W. Berliner to J. S. Fruton, 29 July 1994.

182 For Schoenheimer, see Peyer (1972), Kohler (1977b), and Stetten (1982); for Chargaff, see Chargaff (1978) and Abir-Am (1980).

183 Reingold (1991), p. 17.

184 Zucker (1950); Sutherland et al. (1963); Edsall (1977).
185 Kalckar (1983); Cohn (1992); Larner (1992).
186 For Smith, see Agate (1975); for Evans, see Amoroso and Corner (1972) and Raacke (1983); for Corner, see Corner (1958, 1981) and Ramsey (1994); for Kendall, see Kendall (1971) and Ingle (1975). See also Greenberg (1972); Greagar (1996).
187 Bliss (1982, 1984). For Collip, see Barr and Rossiter (1973) and Li (1992). See also Murray (1971) and McRae (1987).
188 Young (1976).
189 Strickland (1972).
190 Ziman (1994); Braun (1993); Guston and Keniston (1994).
191 Lemaine et al. (1976); Oleson and Voss (1979); Geison (1981).
192 Crosland (1994); Hafner (1981).
193 Remane and Wiese (1992).
194 Hupke (1990).
195 Karlson (1977).
196 For France, see Courtois and Malangeau (1952); for Belgium, see Liébecq (1977); for Japan, see Shimazono (1991).
197 Edsall (1980).
198 Wilhelmi (1988); Bachelard (1988).
199 Crawford (1990); vom Brocke (1991b), pp. 185–242.
200 Liébecq (1992).

3. Philosophy, Chemistry, and Biology

1 Hall (1969).
2 Beckner (1959); Coleman (1971); Hull (1974); Goodfield (1975); Roll-Hansen (1979); Rosenberg (1985, 1989); Canguilhem (1988).
3 Wilson (1923), pp. 284–286.
4 Caldin (1961); Russell (1971); Theobald (1976); Woolley (1978); Kober (1982); Primas (1983); Weininger (1984); Klein and Trinajstic (1990); Rouvray (1992); Scerri (1993); Nye (1993a, 1993b); Christie (1994).
5 Althusser (1974); Hiebert (1987).
6 Bernard (1927), pp. v–vi. See also Medawar (1969), pp. 1–2.
7 Grmek (1968a, 1973); Holmes (1974). See also Robin (1979) and Michel (1991).
8 Naess (1972), pp. 76–77. For Naesé see Gullväg and Wetlesen (1982).
9 For the popularization of science, see Reichvarg and Jacques (1991); Curtis (1992); Cooter and Pumphrey (1994).
10 Mannheim (1952), p. 61. See Simonds (1978), pp. 160–200.
11 See, for example, Mach (1896); Cannon (1940); Grmek (1976).
12 Among the many recent books and journal articles (in addition to those cited above) in which these terms are defined and discussed are: Hanson (1958, 1969, 1971a, 1971b); Popper (1959, 1972); Quine (1960, 1969, 1976); Nagel (1961); Nash (1963); Hempel (1965, 1988); Bunge (1967); Lakatos and Musgrave (1970); Harré (1972); Toulmin (1972); Hollis and Lukes (1982); Hacking (1983); Shapere (1984, 1989); Ackermann (1985); Gellner (1985); Nersessian (1987); Amsterdamski (1992);

Suppes (1993); Margolis (1993); Musgrave (1993). For defenses of realism, see Leplin (1984); Trigg (1989); Bhaskar (1989); DeVitt (1991); Trout (1994); Nola (1995); Marsonnet (1995). For defenses of rationalism, see Brown (1984, 1989); Trigg (1993). For inductivism, see Cohen (1989). For pragmatism, see Appleby et al. (1994); Bunzl et al. (1995); Solomon (1995); Hollinger and Depew (1995). For statistical analysis, see Hacking (1990). For simplicity and beauty in scientific theories, see Goodman (1958); Sober (1975); McAllister (1989, 1991). For a valuable discussion of the development of the relationship between the history of science and the philosophy of science, see Nickles (1995).

13 For Bacon, see Pérez-Ramos (1988); for Condorcet, see Granger (1971); for Whewell, see Fisch (1991); for Comte, see Laudan (1971).

14 McGuiness (1987). Thomas Kuhn's famous 1962 book first appeared in this publication.

15 For Mach, see Hiebert (1973).

16 Meyerson (1926), p. 447; see also Meyerson (1921). For Meyerson, see Paul (1978).

17 Williams (1985), pp. 200–201. See, for example, Kohler (1975), p. 288 and Creagar (1995), p. 175. See also McEvoy (1997).

18 Simpson (1964), pp. 106–107; Mayr (1988), p. 21; (1996). For comments, see Shapere (1969); Fodor (1974); Fruton (1992a), pp. 161–166. See also Smocovitis (1996).

19 Dupré (1993); Rosenberg (1985, 1989, 1994); Waldrop (1992); Galison and Stump (1996).

20 Kant (1906), p. 26.

21 Friedman (1992), pp. 264–341.

22 Kragh (1989).

23 Engelhardt (1976); Esposito (1977); Caneva (1997).

24 Trommsdorff (1800–1807), vol. 1, p. 2. For the practical value and risks of analogical reasoning in chemistry, see Snelders (1994).

25 See, for example, Conant (1950); Multhauf (1966); Siegfried and Dobbs (1968); Perrin (1987); Holmes (1989a, 1995a, 1997a); Melhado (1989); Bensaude-Vincent (1990, 1993); Beretta (1993). See also Berry (1960); Jungnickel and McCormmach (1996).

26 In addition to the references cited above, see Hooykaas (1958); Paneth (1962); Ströker (1968); Hannaway (1975); Duhem (1985); Meinel (1988); Kim (1992); Klein (1994a). For Paracelsus, see Pagel (1982a) and Engelhardt (1994); for Boyle, see Clericuzio (1990, 1993) and Hunter (1994).

27 Nash (1950); Bradley (1955, 1992); Kargon (1966); Brock (1967); Knight (1967, 1968); Thackray (1970); Rocke (1984).

28 Smeaton (1963); Levere (1971); Goupil (1991); Fruton (1992a), pp. 246–249; Klein (1994b, 1995).

29 Bachelard (1932, 1953). For Bachelard, see Lecourt (1974) and Tiles (1984).

30 Crosland (1962); Dagognet (1969); Groult (1988); Locke (1992); Fruton (1992a), pp. 233–237.

31 Luisi and Thomas (1990); Hoffmann and Laszlo (1991); Laszlo (1993). See Latour (1990) for a restatement of his views about "inscription" in scientific practice.

32 For Newton, see Dobbs (1975, 1991); for Goethe, see Gray (1952); for Jung, see Jung (1968).

33 Medawar and Medawar (1983), pp. 26–27.

34 Lloyd (1968); Gotthelf and Lennox (1987); Freudenthal (1995).

35 Grmek (1962, 1972); Roger (1963, 1979); Sloan (1977); Klein (1980), pp. 115–121; Larson (1994); Wilson (1995); Barsanti (1995). For Ray, see Raven (1950); for Linneaus, see Lindroth (1973).

36 For Spallanzani, see Dolman (1975); for the Pasteur-Pouchet debate, see Farley and Geison (1974) and Geison (1995), pp. 110–142.

37 Oparin (1938, 1968); Haldane et al. (1954); Allen (1957); Fox (1956); Fox and Dose (1972); Miller and Orgel (1974); Kamminga (1993); Orgel and Crick (1993); Doolittle and Brown (1995); Podolsky (1996).
38 Kant (1902), pp. 303–311. See Williams (1973); Roll-Hansen (1976); Lenoir (1980); McLaughlin (1982, 1989, 1994); Zumbach (1984); Shanahan (1989).
39 Pittendrigh (1958); Lenoir (1982); Mayr (1988, 1992).
40 For Kielmeyer, see Kanz (1994). For Schelling and German romanticism, see Cunningham and Jardine (1990); Poggi and Rossi (1994).
41 Simmer (1958); Teich (1965).
42 Tiedemann and Gmelin (1826), vol. 1, p. 336.
43 Hunter (1837), pp. 216–217.
44 Reil (1799), p. 424. See Hüfner (1873).
45 Berzelius (1813), p. 4. See Jorgensen (1965).
46 Walden (1928); McKie (1944); Lipman (1964); Schiller (1967a); Brooke (1968).
47 Müller (1843), p. 15.
48 Mialhe (1856), p. 8. See also Hüfner (1873).
49 Beckner (1971); see also Beckner (1959). See Hopkins (1949).
50 Bergson (1936); for Bergson, see Pilkington (1976). Carrel (1935); Lecomte du Nouy (1964). Teilhard de Chardin (1955); see Ayala (1972).

4. *From Ferments to Enzymes*

1 Buchner (1897), pp. 119–120. For Buchner, see Harries (1917), Buchner (1963); Kohler (1971, 1972); Teich and Needham (1992), pp. 43–47.
2 Green (1898); MacFadyen et al. (1900).
3 Ahrens (1902), p. 494.
4 Forbes (1954); Smith (1983).
5 Sambursky (1959); Osler (1991).
6 Taylor (1953).
7 Forbes (1948); Liebmann (1956).
8 Beguin (1624), p. 414. For Beguin, see Patterson (1937) and Rattansi (1970).
9 Libavius (1964), pp. 103–104. For Libavius, see Hubicki (1973) and Hannaway (1975).
10 Hirsch (1950).
11 Lippmann (1919, 1931, 1954); Taylor (1949); Holmyard (1957); Yates (1964); Debus (1965); Multhauf (1966); Patai (1994).
12 Waite (1888), p. 139. For Paracelsus, see Pagel (1982a, 1984); Meier (1993).
13 Pagel (1944, 1955, 1956, 1962, 1972, 1982b); Heinecke (1995).
14 Boyle (1680), p. 122. See Clericuzio (1993).
15 Van Helmont (1662), p. 106.
16 Willis (1684), p. 9. In the same work, he wrote: "We are not only born and nourished by means of Ferments, but we also Dye; every Disease acts its Tragedies by the strength of some Ferment" (p. 14). For Willis, see Isler (1965) and Frank (1979).
17 Patin (1846), vol. 3, p. 795.

18 Metzger (1930); White (1932); Partington and McKie (1937–1938); Rappaport (1961); Jevons (1962); Lindeboom (1968); King (1975); Underwood (1977); Laupheimer (1992); Smith (1994).

19 Boerhaave (1735), vol 2, pp. 115–116.

20 Macquer (1777).

21 Guerlac (1957). See also Proctor (1995).

22 Partington (1962). For Cavendish, see Berry (1960); Jungnickel and McCormmach (1996).

23 Morris (1972); Crosland (1973); Musgrave (1976).

24 Lavoisier (1799), pp. 188–189. For a meticulous account and critical discussion of these experiments, see Holmes (1985), pp. 319–384; also see Golinski (1994); Crosland (1994). Although elsewhere Lavoisier denoted alcoholic fermentation by means of the equation "mout de raisin = acid carbonique + alcool," it is probable that the "sugar" he used was cane sugar or beet sugar (later termed saccharose or sucrose); beet sugar had been crystallized by Andreas Sigismund Marggraf around 1749. For Lavoisier's system of weights, see Holmes (1974), p. xxiv.

25 Gay-Lussac (1810).

26 For Dubrunfaut, see Fletcher (1940).

27 Bitting (1937).

28 Gay-Lussac (1810), p. 255.

29 Fabbroni (1799), pp. 301–302; Thenard (1803).

30 Astier (1813), p. 274; Erxleben (1813); Desmazières (1827); Colin (1825).

31 Farley (1972, 1974).

32 Merton (1961); Baumann (1972).

33 Clay and Court (1932); Bradbury (1967); Turner (1980); Wilson (1995); Ruestow (1996); Fournier (1996).

34 Cagniard-Latour (1838), p. 221.

35 Schwann (1839), p. 235.

36 Schwann (1837), p. 192.

37 Kützing (1837), pp. 386, 392. See Kützing (1960).

38 Berzelius (1836), p. 240.

39 Berzelius (1839), pp. 400–403.

40 Mitscherlich (1834), p. 281.

41 Liebig (1839), p. 262.

42 For example, see Mulder (1844–1851), pp. 49–51, and Gerhardt (1853–1856), vol. 4, p. 538.

43 Turpin (1838), p. 402.

44 [Liebig and Wöhler] (1839).

45 Delbrück and Schrohe (1904); Mathias (1959); Teich (1965, 1983).

46 Berthelot (1860a), vol. 2, pp. 655–656. See Jacques (1950).

47 For Laurent, see Kapoor (1973); for Delafosse, see Friedel (1878) and Taylor (1978); for Biot, see Crosland (1970b).

48 Metzger (1918); Hooykaas (1953); Burke (1966); Multhauf (1976). For Romé de Lisle, see Hooykaas (1975); for Haüy, see Hooykaas (1972).

49 For Mitscherlich, see Schuett (1984, 1992).

50 Herschel (1822).

51 Bates (1942); Lyle and Lyle (1964).

52 Kottler (1978); Geison (1995), pp. 56–85. See also Partington (1964), pp. 751–755.
53 Pasteur (1861a), p. 33; (1886).
54 Weyer (1974b); Mason (1976, 1987, 1989, 1991); Bentley (1995).
55 Pasteur (1857), p. 914. See Latour (1992).
56 Pasteur (1858a), p. 414. See Paul (1996).
57 Pasteur (1858b), p. 1012.
58 Pasteur (1860a), pp. 359–360.
59 Pasteur (1861), pp. 1263–1264.
60 Grünhut (1896); Höxtermann (1996).
61 Liebig (1870); Pasteur (1871); Volhard (1909), vol. 2, pp. 88–103.
62 Berthelot (1860b); Pasteur (1860c).
63 Gautier (1869); Ingencamp (1886).
64 Bernard (1878–1879), vol. 1, pp. 39–40.
65 Pasteur (1878), p. 1057.
66 Pasteur (1879), p. 54. For a somewhat different account of this episode, see Geison (1995), pp. 18–20.
67 Traube (1899), p. 74.
68 Manassein (1897); Buchner and Rapp (1898); Cochin (1880); Roux (1898). For Manassein, see Kästner (1996).
69 Pasteur (1862), p. 52.
70 Pasteur (1876a), pp. 7–8.
71 Pasteur (1876b).
72 Medawar (1979), p. 84.
73 Traube (1878), p. 1984. For Traube, see Bodländer (1895); Sourkes (1955), and Müller (1970).
74 Nägeli (1879), pp. 86–87. For Nägeli's speculations about the nature of protoplasm, see Robinson (1979), pp. 109–130.
75 Payen and Persoz (1933); for an English translation, see Friedman (1981), pp. 119–122.
76 Saussure (1819), p. 407.
77 Berzelius (1836), pp. 243, 245.
78 Tiedemann and Gmelin (1826–1827); Beaumont (1833).
79 Schwann (1836); Schwerin (1867). For Prout, see Brock (1965, 1985) and Baron (1973). For early medical use of pepsin, see Basslinger (1858). See also Holmes (1974), pp. 160–178; Hickel (1975); Sernka (1979); Davenport (1992).
80 Holmes (1974), pp. 197–246.
81 Mialhe (1856), pp. 35–36. See also Hüfner (1872).
82 Bourquelot (1896); Green (1899); Oppenheimer (1900).
83 Kühne (1876), p. 190; (1878a), p. 293. See Fruton (1990a), pp. 86–90.
84 Boyde (1977); Hofmann-Ostenhof (1978), p. 187; Fruton (1978).
85 Hoppe-Seyler (1870), pp. 564–565; (1876), pp. 1–17.
86 Plantefol (1968).
87 See Teich (1981).
88 Kohler (1973a); Crick (1988), p. 32. Among the other nominees for the 1907 Nobel Prize in Chemistry were Stanislao Cannizzaro, Theodor Curtius, Walther Nernst, and Wilhelm Ostwald; see Crawford et al. (1987), p. 172.
89 Buchner et al. (1903); Buchner (1966); Harries (1917); Kohler (1971).

90 Rubner (1913), p. 55; Kohler (1972).
91 Neumeister (1897b); Buchner et al. (1903), p. 38.
92 Willstätter (1949), p. 63.
93 Loeb (1906), p. 22.
94 Fischer (1898), p. 62. Because the lowercase symbols also came to be used to indicate the direction of optical rotation of many compounds, the Fischer convention was later modified by the introduction of small-capital D- and L- to denote the configuration; see Hudson (1948); Fruton and Simmonds (1958), pp. 74–84; Lichtenthaler (1992).
95 Fischer and Thierfelder (1894), p. 2037.
96 Fischer (1894), p. 1992.
97 Bourquelot (1896), pp. 133–134.
98 Pigman (1944).
99 Bourquelot (1896), p. 77.
100 Brücke (1861).
101 Willstätter (1922, 1927).
102 Jager (1890); Arthus (1896); Barendrecht (1904); Bayliss (1908). See also Rothen (1948).
103 Segal (1959); Segel (1975); Cleland (1963); Fromm (1975); Boyde (1980).
104 Laidler (1993), pp. 114–123, 232–249.
105 Duclaux (1899), pp. 129–170; Oppenheimer (1900), p. 14.
106 Van't Hoff (1898); Hill (1898); Oppenheimer (1900), p. 23.
107 Henri (1903); for Henri, see Debru (1990). See also Brown (1902).
108 Sørensen (1909); Michaelis and Menten (1913); Briggs and Haldane (1925); Haldane (1930); Bates (1973). For Haldane, see Sarkar (1992).
109 See, for example, Bender and Kezdy (1965).
110 Northrop et al. (1948).
111 E. Schütz (1885); J. Schütz (1900); Northrop (1924).
112 Lieben (1935), p. 365. See also Fruton (1979), pp. 8–9; for comments, see Mazumdar (1989), p. 14 and Morgan (1990), p. 141.
113 For the history of clinical enzymology, see Habricht (1978).
114 Fantini (1984), p. 16; Kohler (1971), p. 61; Florkin (1975), p. 26.
115 Hofmeister (1901), pp. 14, 26–27.
116 Hofmeister (1914), pp. 717, 722.
117 Florkin (1975), p. 35.
118 Florkin (1972), pp. 279–283.
119 Bertrand (1897a); Harden and Young (1906); Harden (1923). For Bertrand, see Gaudillière (1991).
120 Warburg (1929), p. 2.
121 Bayliss (1924), p. 326a.
122 Hopkins (1913), p. 214.

5. The Nature and Function of Proteins

1 For general accounts of the early developments, see Delage (1895); Wilson (1896); Mann (1906); Plimmer (1908); Schryver (1909); Cohnheim (1911); Kestner (1925); Lloyd and Shore (1938).
2 Fourcroy (1801), vol. 5, p. 117.

3 For Beccari, see Beach (1961); for Parmentier, see Kahane (1978). See also Carpenter (1994).
4 Heinemann (1939).
5 Fourcroy (1789a), p. 46.
6 Fourcroy (1789b).
7 Rudé (1966); Thompson (1968).
8 Chevreul (1847), p. 577.
9 Chevreul (1824), pp. 22–23.
10 André (1932); Costa (1962).
11 Chevreul (1824), p. 159.
12 Holmes (1963a); Hiebert (1964); Hickel (1979); Stephen (1984).
13 Dennstedt (1899); Vickery (1946).
14 Goodman (1972a).
15 Baker (1948–1955); Zanobio (1971); Pickstone (1973); Hacking (1985); Wilson (1995).
16 For Dutrochet, see Wilson (1947); Pickstone (1976); Schiller (1971), (1980), pp. 131–156.
17 Dujardin (1835). For Dujardin, see Fauré-Fremiet (1948).
18 Schwann (1839); Schleiden (1842); Virchow (1855), p. 23. See also Klein (1936) and Wilson (1944).
19 Geison (1969); Güttler (1972).
20 Huxley (1869), pp. 135–136. See Geison (1969), pp. 279–285 and Welch (1995). For Huxley, see Desmond (1994).
21 Helmholtz (1845), p. 47. See Olesko and Holmes (1993).
22 Graham (1861), pp. 184–185; reprinted in Graham (1876), pp. 553–554.
23 Pflüger (1875).
24 Liebig (1870); Kühne (1864); Hermann (1867).
25 Pflüger (1875), p. 309; Clausius (1865).
26 Pflüger (1875), pp. 334, 343. See also Pflüger (1872). For a structure based on Pflüger's idea of a living protein, see Latham (1887).
27 Kekulé (1878), p. 212.
28 Chittenden (1894), pp. 115–116.
29 Loew (1880, 1883, 1896); Loew and Bokorny (1881, 1882). For an account of Loew's career, which included stays in Japan and the United States, see Klinkowski (1941).
30 Baumann (1882); Nencki (1885), p. 343.
31 Bokorny (1922); Delbrück (1941).
32 Ehrlich (1885), p. 6; Verworn (1897, 1903). For Verworn, see Rothschuh (1976).
33 Baumann (1878); Bernard (1879), vol. 2, p. 517; Bunge (1889), p. 5; Neumeister (1903).
34 Hopkins (1913), p. 220.
35 Mulder (1844–1851), pp. 300–301.
36 Berzelius (1916), pp. 104–109. See Vickery (1950); Glas (1975).
37 Mulder (1839), p. 140.
38 Liebig (1841); Hofmann (1888), vol. 1, p. 185.
39 Dumas (1844), p. 41. See also Boussingault (1839) and Dumas (1841).
40 Gerhardt (1856), vol. 4, p. 432.
41 Berzelius (1843), pp. 535–536.
42 Hofmann (1888), vol. 1, p. 263; Laskowski (1846), p. 165.

43 Liebig (1846a), p. 133; Mulder (1846).
44 Lehmann (1855), vol. 1, p. 290.
45 Bopp (1849); Guckelberger (1848).
46 Liebig (1847a), p. 27.
47 Kapoor (1969); Fisher (1973).
48 Gay (1976).
49 Kekulé (1861), vol. 1, p. 758.
50 Bykov (1962); Larder (1967); Brooke (1987); Fisher (1996).
51 Couper (1858); Kekulé (1858). For Couper, see Dobbin (1934) and Duff (1987). See also Rouvray (1977).
52 Kekulé (1861), vol. 1, p. 157.
53 Kolbe (1877), p. 473. See LeBel (1874); Hoff (1875, 1960); Weyer (1977).
54 Kolbe (1845, 1858); Strecker (1850); Wurtz (1857); Béchamp (1854). See Weyer (1974b, 1977). For Wislicenus, see Beckmann (1904); Perkin (1905); Ramberg (1994).
55 Surrey (1954).
56 Brooke (1971); Russell (1987).
57 Jacques (1950, 1987).
58 Naquet (1867), p. 612. See also Kekulé (1861), vol. 1, p. 11.
59 Hoppe-Seyler (1877a), p. I.
60 Kingzett (1878), pp. 27–28. For Kingzett, see Morgan (1935).
61 For details, see Vickery and Schmidt (1931) and Vickery (1972).
62 Ritthausen (1866).
63 Hlasewitz and Habermann (1871, 1873).
64 Chibnall (1939).
65 Drechsel (1889), p. 426.
66 Kossel and Dakin (1904).
67 Mueller (1923); Barger and Coyne (1928); Toennies (1937).
68 Womack and Rose (1935); McCoy et al. (1935); Fischer (1901); Ehrlich (1904). For threonine, see also Carter (1979).
69 Hopkins and Cole (1902).
70 See, for example, Stryer (1988), p. 16; Berg and Singer (1992), p. 53.
71 Fowden (1958); Wagner and Musso (1983).
72 Corrigan (1969); Kögl and Erxleben (1939). For Kögl, see Karlson (1986). For the disproof of Kögl's claim, see Chibnall et al. (1940).
73 Lehmann (1855), vol. 1, pp. 449–450.
74 Kühne (1866), p. 48.
75 Moore (1898), p. 441. See also Neumeister (1897a), p. 312.
76 Cohnheim (1901); Loewi (1902). For Cohnheim, see Matthews (1978).
77 Fürth (1912), vol. 1, p. 77.
78 Loew (1896), p. 22.
79 Fischer (1906a), pp. 605–606.
80 Hofmeister (1908); Siegfried (1916); Synge (1943).
81 Fischer and Fourneau (1901), p. 2868.

82 Fischer (1902), p. 939.
83 Hofmeister (1902a, 1902b).
84 Fischer (1902), p. 940.
85 Fischer to Baeyer, 5 December 1905 (Bancroft Library).
86 Fischer (1907a), pp. 1761–1762.
87 Fischer (1923), p. 13.
88 Fischer (1907b), p. 1757. See Carpino et al. (1996).
89 Hofmeister (1908).
90 Curtius (1904). For azides, see Scriven and Turnbull (1988).
91 Fruton (1949, 1987a); Wieland and Bodanszky (1991); Wieland (1995a).
92 Other amino acids added to the list of protein constituents, but later withdrawn, included α-aminobutyric acid, α-aminocaproic acid (norleucine), "prolysine," "oxytryptophan," and β-hydroxyglutamic acid.
93 Fischer to T. W. Engelmann, 24 February 1904.
94 Osborne and Jones (1910); Osborne (1910). See also Fruton (1990c).
95 Fischer (1899b), p. 2451.
96 Denis (1856); Weyl (1877).
97 For Cohn, see Edsall (1961); Scatchard (1969); Diamond (1971).
98 For Méhu, see McCollum (1956).
99 Ritthausen (1872), p. 236. See also Schimper (1878).
100 Hoppe-Seyler (1877c), vol. 1, p. 76.
101 Osborne (1892), p. 663; Osborne and Jones (1910). For Osborne, see Vickery (1931) and Fruton (1990c, 1995b).
102 Wells and Osborne (1911). For Wells, see Long (1949).
103 Hünefeld (1840), p. 160.
104 Preyer (1871).
105 Stahnke (1979); Holmes (1995b).
106 McGucken (1969); James (1983, 1985).
107 Reichert and Brown (1909), pp. 326–327.
108 Landsteiner and Heidelberger (1923); Landsteiner (1945).
109 The latter term was subsequently used for all proteins, thus contributing to already considerable confusion in the nomenclature.
110 Stahnke (1981).
111 For Küster, see Brigl (1931); for Hans Fischer, see Treibs (1971). For a history of porphyrin chemistry, see Drabkin (1978); for vitamin B_{12}, see Eschenmoser and Wintner (1977).
112 Pasteur (1886), p. 36.
113 McPherson (1991).
114 Wichmann (1899), p. 584; Fischer (1913), p. 3288.
115 Tutton (1924), p. 5; Lorch (1974).
116 Metzger (1918); Burke (1966); Goodman (1969). For Metzger, see Freudenthal (1990).
117 Ewald (1962), p. 347. See Cruikshank et al. (1992).
118 Armstrong (1927), p. 478.
119 For Dickinson, see Pauling (1971); for Lonsdale, see Hodgkin (1976); for Hutchinson, see Mills (1939) and Smith (1939); for Bernal, see Hodgkin (1980); for Hodgkin, see Glusker (1994).

120 Lipson (1990); Bernal and Crowfoot (1934); Hodgkin (1979); Robertson (1972). For Robertson, see Arnott (1994).
121 Hamilton (1970); Rhodes (1993).
122 For the crystallization of proteins, see Weber (1991); McRee (1993); Blow et al. (1994). For X-ray diffraction analysis, see Jones (1984); Branden and Jones (1990); Janin (1990); Kleywegt and Jones (1995).
123 Graham (1876), p. 553. See Mokrushin (1962) and Ede (1993).
124 For Dutrochet, see Schiller (1971); for Nollet, see Torlais (1954).
125 Tyrrell (1964); Wheatley and Agutter (1996).
126 Picton and Linder (1892), p. 159.
127 Layton (1965); Teske (1969); Goodman (1972).
128 Zsigmondy (1909).
129 Schulz (1903), p. 101. See Edsall (1962).
130 Fischer (1923), p. 8.
131 Ibid., p. 40.
132 For Sørensen, see Linderstrøm-Lang (1939); for Adair, see Johnson and Perutz (1981).
133 Svedberg and Pedersen (1940); Pickels (1942); Williams (1979).
134 Svedberg (1937), p. 1961. See Ede (1996).
135 Graham (1876), p. 597; Roche (1935), p. 740.
136 Staudinger (1920, 1961); Morawetz (1985); Zanvoort (1988).
137 Fischer (1906a), pp. 607–608.
138 Abderhalden and Komm (1924); Vickery and Osborne (1928), p. 426; Gortner (1929), p. 344.
139 Ewald (1962), p. 631. See Polanyi (1921); Mark (1980).
140 Grimaux (1882); Carothers (1931).
141 For Natta, see Cerutti (1990).
142 Wilkie (1960, 1961).
143 For Hardy, see Bate-Smith (1964) and Morgan (1990).
144 Hardy (1900), p. 110.
145 Hardy (1903), p. xxix; (1906), p. 198.
146 F. G. H[opkins]., (1934), p. 1150.
147 Michaelis (1914); Clark (1928). See Szabadvary (1964).
148 Loeb (1912, 1922). For Loeb, see Osterhout (1930) and Pauly (1987a).
149 Cohn (1925); Cohn and Edsall (1943).
150 Tiselius (1937); Kay (1988); Vesterberg (1989).
151 Chick and Martin (1912).
152 Sørensen and Sørensen (1925), p. 26.
153 Hopkins (1930a); Anson and Mirsky (1931).
154 Wu (1931); Mirsky and Pauling (1936), pp. 442–443.
155 Huggins (1971).
156 Tanford (1980, 1989).
157 Vickery and Osborne (1928), p. 417.
158 Gortner (1929), p. 319; Haldane (1930), p. 115. See also Holter (1931).
159 Bergmann and Zervas (1932); Fruton and Bergmann (1939); Fruton (1970, 1987b); Merrifield (1993, 1995); Hilvert (1994).

160 Bergmann and Fruton (1937); Bergmann, Fruton, and Pollok (1939). For Kunitz, see Herriott (1989) and Fruton (1992a), pp. 29–31. For the subsequent use of specific proteinases in the study of protein structure, see Mihalyi (1978).

161 Linderstrøm-Lang, Hotchkiss, and Johansen (1938), p. 996.

162 Ibid., p. 996.

163 Frank (1936); Wrinch (1938). For Wrinch, see Julian (1984) and Abur-Am (1987).

164 Langmuir (1939), p. 611. For Langmuir, see Rosenfeld (1862, 1966); Süsskind (1976).

165 Pauling and Niemann (1939), p. 1860.

166 Ibid., p. 1867.

167 Bergmann (1938).

168 Fruton (1992a), pp. 50–52.

169 Goldschmidt (1938).

170 Chibnall (1942), p. 138. See also Witkowski (1985).

171 Gordon, Martin, and Synge (1941), p. 1385; Martin and Synge (1941). For Synge, see Gordon (1996).

172 Tswett (1906), p. 322. For Tsvet, see Williams and Weil (1953); Robinson (1960); Sakodynskii (1972); Senchenkova (1976); Berezkin (1989).

173 Schönbein (1861); Fischer and Schmidmer (1893); Goppelsroeder (1901), Ettre (1995).

174 Brand et al. (1945).

175 Consden, Gordon, and Martin (1944), p. 225.

176 Moore and Stein (1951); Dowmont and Fruton (1952).

177 Sanger (1952), p. 3.

178 Sanger (1945), p. 514. See also Fox (1945) and Fruton (1992a), pp. 35–45. The amino-terminal phenylalanine unit in insulin had been identified qualitatively in 1935 by Hans Jensen and Earl Evans.

179 Edman (1950). For Edman, see Partridge and Blombäck (1979) and Fruton (1992b).

180 Ryle, Sanger, Smith, and Kitai (1955). See Chadarevian (1996).

181 Moore and Stein (1973). See Ettre and Zlatkis (1979).

182 Ingram (1957).

183 Bernal and Crowfoot (1934), p. 794. See Hodgkin and Riley (1968) and Hodgkin (1979).

184 Bernal, Fankuchen, and Perutz (1938).

185 Kendrew et al. (1960).

186 Kendrew et al. (1958), p. 665.

187 Bragg, Kendrew, and Perutz (1950), p. 356.

188 Pauling and Corey (1950, 1951); see also Chan and Dill (1990); Pauling (1996). For Pauling, see Hager (1995); Goertzel and Goertzel (1995); Krishnamurthy (1996).

189 Hodgkin (1979), p. 142.

190 Wüthrich (1990); Roberts (1993).

191 Anfinsen (1973).

192 Schlesinger (1990); Creighton (1992); Ptitsyn (1995); Shortle et al. (1996).

193 Krishna and Wold (1993).

194 Pauli (1934), p. 111. See also Oppenheimer (1933).

195 Allen and Murlin (1925), p. 138.

196 Chance (1951).

197 Herriott (1947); Lundblad and Noyes (1984); Kaiser, Lawrence, and Rokita (1985); Leatherbarrow and Fersht (1987).
198 Bell (1941); Laidler and King (1983); Hammett (1940). For Hammett, see Shorter (1990, 1996). For the development of physical-organic chemistry, see Saltzman (1986); Thomas (1994); Roberts (1996). For enzyme studies, see Bender (1971); Jencks (1969); Koshland (1960); Walsh (1979).
199 Bender and Kezdy (1965); Polgar (1987). For Bender, see Westheimer (1995).
200 Brocklehurst, Willenbrock, and Salih (1987); Durell and Fruton (1954). The formation of a thiolester intermediate in the action of papain had been proposed by Weiss (1937).
201 Fruton (1970, 1976b, 1987b); Tang (1977); Takahashi (1995). The experimental work in my laboratory during 1960–1989 on the specificity and mechanism of pepsin action was performed largely by Georges Delpierre, Ken Inouye, Goverdhan Sachdev, and Irene Voynick.
202 Vallee and Geddes (1984); Auld and Vallee (1987); Morgan and Fruton (1978).
203 Zerner (1991). Note that this reaction is not a reversal of Wöhler's famous synthesis of urea from ammonium cyanate; see Shorter (1978).
204 For ribonuclease, see Walsh (1979), pp. 199–207; Breslow and Chapman (1996). For lysozyme, see Walsh (1979), pp. 299–307.
205 James et al. (1982); Umezawa (1972); Wolfenden (1972, 1976); Aoyagi and Umezawa (1975); Mattis, Henes, and Fruton (1977); Lolis and Petsko (1990). See also Koshland (1960); Huber (1988).
206 Jencks (1975, 1980, 1994); Page (1979). See also Weber (1975).
207 Northrop, Kunitz, and Herriott (1948).
208 Neurath (1957, 1975).
209 Davie and Fujikawa (1975); Scully (1992).
210 Steiner et al. (1975); Steiner (1992); Mains et al. (1977); Fisher and Scheller (1988); Peng Loh (1992).
211 Bainton (1981); Bohley (1987); Bond and Butler (1987); Hershko (1988); Mayer and Doherty (1992); Rechsteiner, Hoffman, and Dubiel (1993); Mason and Rivett (1994); Hilt and Wolf (1996); Coux et al. (1996).
212 Silverstein (1989); Mazumdar (1989, 1995); Moulin (1991); Gallagher et al. (1995).
213 Metchnikoff (1893, 1901). For Mechnikov, see Brieger (1974).
214 Ehrlich (1900), pp. 433–434.
215 Ibid., p. 437. See also Ehrlich (1910).
216 Ehrlich (1897), p. 306. For Ehrlich, see Bauer (1954); Witebsky (1954); Witkop (1981); Travis (1989); Cambrosio et al. (1993).
217 Pauli (1907); Biltz (1904). See Mazumdar (1974).
218 Arrhenius (1904, 1907); Nernst (1904). See Rubin (1980) and Luttenberger (1992).
219 For Michaelis, see Fruton (1990b); for his immunological views, see Mazumdar (1995), pp. 235–236.
220 Obermayer and Pick (1906).
221 Landsteiner (1945); Marrack (1938); Keating and Ousman (1991). See Mazumdar (1995), pp. 136–151, 214–278, 305–336.
222 Mazumdar (1995), p. 312; see also Goebel (1975). During my stay at the Rockefeller Institute (1934–1945) as a junior member of Max Bergmann's chemical research group, I had frequent occasion to sit next to Landsteiner in the lunch room and derived inspiration and encouragement from our conversations. Perhaps in his final years (he died in 1941) Landsteiner was more kindly disposed toward others at the institute. Perhaps insufficient account has been taken of Flexner's own authoritarian style as an administrator, in contrast to that of Herbert Gasser, his successor in

1935 as director. In any case, other senior members were far more unapproachable, notably the famous Alexis Carrel.

223 Wells (1928), p. 574.

224 Heidelberger and Avery (1923).

225 For Avery, see Goebel (1975), Dubos (1976), McCarty (1985), and Amsterdamska (1993); more will be said about Avery later.

226 Heidelberger (1977, 1979). See Kabat (1943, 1992); Kabat and Mayer (1961).

227 Breinl and Haurowitz (1930); Alexander (1931); Mudd (1932); Haurowitz (1937). See Mazumdar (1989), pp. 13–32. For Haurowitz, see Putnam (1994).

228 Pauling (1940). See Kay (1993), pp. 164–193 and Hager (1995), pp. 261–265.

229 Burnet (1941); Bergmann (1938). For Burnet, see Tauber and Podolsky (1993); Söderquist (1994).

230 Burnet and Fenner (1949); Monod (1950).

231 Billingham et al. (1954); Medawar (1958).

232 Tiselius and Kabat (1939).

233 Porter (1973); Edelman (1973).

234 Davies, Padlan, and Sheriff (1990).

235 Kunkel (1964); Putnam (1969). For Bence Jones, see Coley (1973) and Putnam (1993). For a nineteenth-century description of the Bence Jones protein, see Neumeister (1897a), pp. 804–807.

236 Jerne (1955, 1985); Burnet (1957, 1959); Talmage (1957, 1959); Lederberg (1959, 1988).

237 Nossal and Lederberg (1958); Köhler and Milstein (1975); Edwards (1981); Köhler (1985); Milstein (1985); Nossal (1986, 1995); Winter and Milstein (1991); Cambrosio and Keating (1995). For the prior development of techniques for the controlled fusion of cells, see Harris (1970, 1995).

238 Tramontano, Janda, and Lerner (1986); Pollack, Jacobs, and Schultz (1986); Lerner and Benkovic (1988); Lerner, Benkovic, and Schultz (1991); Charbonnier et al. (1995).

239 See Hamblin (1988)

240 Löwy (1992), p. 371.

241 Prusiner (1982, 1991); Baldwin et al. (1995).

242 Fermi et al. (1984).

243 Laskowski and Kato (1980); Birk (1987); Bode and Huber (1992); McGrath et al. (1995).

244 Davies and Cohen (1996); Zhao et al. (1996).

245 Frieden and Nichol (1981); Janin (1995); Jones and Thornton (1996).

6. Chemical Energy of Biological Systems

1 Lavoisier (1862), vol. 2, p. 691.

2 Ibid., p. 703. See Holmes (1985), pp. 440–468; McKie (1952); Fayet (1960); Poirier (1993); Donovan (1994); Bensaude-Vincent (1995); Viel (1996).

3 Lavoisier (1862), vol. 2, p. 332. See Guerlac (1976); Perrin (1989); Kremer (1990), pp. 40–83; Holmes (1997b). For the development of calorimetry, see Armstrong (1964).

4 McKie and Heathcote (1935); Mendelsohn (1964); Morris (1972); Holmes (1985), pp. 63–129, 151–223.

5 Hassenfratz (1791), p. 266. See Hall (1971).

6 Vauquelin (1792); Spallanzani (1807). For Spallanzani, see Dolman (1975); for Senebier, see Marx (1974); Pilet (1962); Legée (1991); for Saussure, see Pilet (1975). Regnault and Reiset (1849); for Regnault, see Fox (1975).

7 Magnus (1837), pp. 602–604. For Magnus, see Hofmann (1870).
8 Hünefeld (1840), p. 103. See Culotta (1970b, 1970c).
9 Bichat (1801), vol. 1, pp. 534–535. For Bichat, see Canguilhem (1970); Haigh (1984); Sutton (1984).
10 Du Bois-Reymond (1927), p. 19.
11 Allen and Pepys (1808); Brodie (1812); Legallois (1817); Chossat (1820); Despretz (1824); Dulong (1841). See Kremer (1990), pp. 84–163.
12 Holmes (1963a, 1964).
13 Kahlbaum (1904), pp. 67, 72.
14 Dumas (1841, 1844).
15 Dumas (1844), p. 139.
16 Liebig (1846a), p. 62; (1846b).
17 Liebig (1842), p. 110.
18 Lipman (1966, 1967); Benton (1974); Gregory (1977); Hall (1980); Lenoir (1982).
19 Mayer (1893), pp. 14, 101–103. For Mayer, see Caneva (1993).
20 Liebig (1842), pp. 273–274.
21 Lehmann (1855), vol. 1, pp. 269, 553–560; Culotta (1970c).
22 Bernard (1859), vol. 1, p. 396.
23 Stokes (1864). For Stokes, see Breathnach (1966) and Parkinson (1976). For indigo, see Epstein et al. (1967); McGovern and Michel (1990); Ghiretti (1994).
24 Friedländer (1915).
25 Bohr (1904); Hill (1913); Barcroft (1914); Haldane (1917); Henderson (1928); Anson and Mirsky (1930); Roughton (1970); Edsall (1972); Breathnach (1972); West (1996).
26 Helmholtz (1845), p. 72.
27 Ibid., p. 82.
28 Helmholtz (1848); Liebig (1850); Valentin (1855); Matteucci (1856). See Needham (1971); Kremer (1990); Olesko and Holmes (1993), pp. 55–82; Krüger (1994).
29 Traube (1861), (1899), p. 168.
30 Ibid., pp. 178–179. For Traube, see Sourkes (1955) and Müller (1970).
31 Bernard (1859), vol. 1, p. 342; Hermann (1867); Bert (1870), p. 118.
32 Pflüger (1872), p. 52.
33 Oertmann (1877), p. 395.
34 Pflüger (1875), p. 343.
35 Bernard (1878–1879), vol. 1, pp. 166, 172–173.
36 Ibid., pp. 167–171.
37 For a critical analysis of Helmholtz's 1847 treatise, see Bevilacqua (1993). For Mayer, see Heimann (1976) and Caneva (1993); for Colding, see Dahl (1978); for Joule, see Cardwell (1989). See also Kuhn (1959) and Fox (1971).
38 Helmholtz (1882), p. 73.
39 Ibid., p. 66. See Rosen (1959).
40 Hess (1840), p. 392; Helmholtz (1882), p. 9.
41 Berthelot (1875), p. 6. For Thomsen, see Kragh (1982, 1984). See Armstrong (1964); Schelar (1966); Dolby (1984); Médard and Tachoire (1994).
42 Rubner (1894). See also Kleiber (1961); Ihde and Janssen (1974).

43 Cardwell (1971); Kober (1980); Laidler (1993), pp. 83–130. For Carnot, see Challey (1971); for Clausius, see Wolff (1995).

44 Thomson (1852), pp. 141–142. For Kelvin, see Buchwald (1976) and Smith and Wise (1989). In 1849 Kelvin introduced the word *thermodynamic*, and a few years later, he and Joule proposed a temperature scale, with an "absolute zero" now set at –273.15°C.

45 Dolby (1984). For Duhem, see Miller (1971) and Jaki (1984).

46 For Horstmann, see Trautz (1930). For Gibbs, see Wheeler (1952); Klein (1972b); Deltete and Thorsell (1996). See Gibbs (1906) for his collected papers. See also Moutier (1885); Le Châtelier (1888).

47 Kragh (1993); Kragh and Weininger (1996).

48 Hoff (1884); Lewis and Randall (1923); Laidler (1993), pp. 114–126. For Guldberg and Waage, see Lund (1965).

49 Warburg (1914a), p. 256.

50 Nernst (1906); Baron and Polanyi (1913). See McGlashen (1966).

51 Hill (1912), p. 512.

52 Dubpernel and Westbrook (1978); Stock and Orna (1989).

53 Faraday (1839), pp. 249–250. For Faraday, see Williams (1965); Cantor (1991); Davenport (1991).

54 Helmholtz (1881), p. 290. See Clark (1976).

55 $E_o = E_h$ when the activities of the oxidized and reduced forms are equal. In biochemical slang, the oxidation-reduction potential is termed the redox potential or O-R potential.

56 Peters (1898); Haber (1901).

57 Palmer (1965); Stranges (1984).

58 Clark (1925, 1960); Michaelis (1929); Krebs and Kornberg (1957).

59 Ostwald (1900), p. 250.

60 Schrödinger (1944); Brillouin (1962); Riedl (1978); Prigogine and Stengers (1979); Bricmont (1996); Sarkar (1996).

61 Bence Jones (1867), p. 75.

62 Lehmann (1955), vol. 2, p. 356; Mialhe (1856), pp. 75–77.

63 Wöhler (1842); Wöhler and Frerichs (1848).

64 For Schönbein, see Snelders (1975).

65 Hadelich (1862).

66 Schönbein (1848).

67 Schönbein (1861). For Schönbein's correspondence, see Kahlbaum and Darbyshire (1899); Kahlbaum (1900); Kahlbaum and Thun (1900).

68 Schönbein (1863).

69 Kühne (1866), p. 214.

70 Pflüger (1875), pp. 300–301.

71 Hermann (1867); Nägeli (1879), p. 117; Pfeffer (1881), vol. 1, p. 374; Nencki and Sieber (1882); Loew (1896); Verworn (1903).

72 Traube (1858a), pp. 332–333.

73 Traube (1874), p. 882.

74 Ibid., p. 884.

75 Hoppe-Seyler (1876), pp. 15–16.

76 Bodländer (1899); Baeyer and Villiger (1900); Milas (1932).
77 Ehrlich (1885), p. 10.
78 Reinke (1883).
79 Bertrand (1894, 1897b). For Bertrand, see Courrier (1972).
80 Jaquet (1892). See also Pohl (1893, 1896).
81 Hofmeister (1901), p. 26. See Scheer (1939).
82 Battelli and Stern (1912).
83 Chodat and Bach (1903); Moore and Whitley (1909); Wheldale (1911).
84 Battelli and Stern (1910); Wieland (1922a), p. 3647.
85 Fenton (1894); Manchot (1902); Dakin (1912).
86 Taube (1965); Fridovich (1974, 1986, 1995); Malmström (1982).
87 Spitzer (1897); Burian (1906); Battelli and Stern (1909, 1911, 1914); Vernon (1911). For Battelli, see Monnier (1942).
88 For Warburg, see Krebs (1972a, 1979); Höxtermann (1984); Höxtermann and Sucker (1989); Werner and Renneberg (1991).
89 Barcroft and Haldane (1902); Brodie (1910); Oesper (1964); Holmes (1995c).
90 Warburg (1908); (1911), p. 416; (1914a), pp. 314, 335.
91 Warburg (1914b), pp. 253–254; Thunberg (1911); Mathews and Walker (1909).
92 Warburg (1924), pp. 479, 483, 485–486, 488. This paper, in a biochemical journal, was followed by Warburg (1925a), with the same title and message, but intended for the community of German organic chemists.
93 Haldane (1895), p. 213; Haldane and Smith (1896); Warburg and Negelein (1929).
94 Dixon and Thurlow (1925); Warburg (1925b), p. 252.
95 Warburg (1928), p. 1.
96 Willstätter (1922), p. 3623.
97 Hopkins (1926), p. 56.
98 Willstätter (1926).
99 Theorell (1947).
100 Keilin (1925), p. 312. See Keilin (1966) for his later historical account. For Keilin, see Mann (1964); Tate (1965); Slater (1977); King et al. (1988), pp. 3–90; Perutz (1986).
101 For Lankester, see Lester (1995).
102 MacMunn (1886), pp. 279–280.
103 MacMunn (1887); Levy (1889); MacMunn (1890), p. 329.
104 MacMunn (1914), pp. 72–73.
105 Keilin (1966), pp. 103–105.
106 Conant and Fieser (1925); Anson and Mirsky (1930); Drabkin (1978).
107 Hill (1926).
108 For Werner, see Kauffman (1966, 1994, 1997); Schwarzenbach (1966); Venanzi (1994).
109 Keilin (1925), p. 323.
110 Keilin (1927), p. 671.
111 Keilin (1929, 1930). See Dixon (1929), p. 385n.
112 Warburg (1932), p. 2; Keilin (1966), pp. 197–203.
113 Anson and Mirsky (1930), pp. 541–542.

114 Keilin and Hartree (1939). The term a_3 was used because two other cytochromes with bands near 600 nm had been identified in bacteria; see Lemberg (1961). As regards Warburg's treatment of this development, see Warburg (1949), and the review by Keilin (1950) of that book.

115 Chance (1953). In this work, Chance used a dual-beam spectrophotometer of his own design; see Chance (1991), pp. 9–10.

116 Altmann (1890); Bensley and Hoerr (1934). For Bensley, see Hoerr et al. (1957). Claude (1946); Hogeboom, Claude, and Hotchkiss (1946); Hogeboom, Schneider, and Palade (1948). For Claude, see Brachet (1988).

117 Kennedy and Lehninger (1949); Lehninger (1951).

118 Wikström, Krab, and Saraste (1981); Malmström (1982); Kadenbach (1983); Brunori and Chance (1988); Capaldi (1990); Howard and Rees (1991); Musser, Stowell, and Chan (1995); Iwata et al. (1995).

119 Ramirez et al. (1995); Larsson et al. (1995). A similar membrane-bound multiprotein "respiratory burst oxidase" has been found to be present in blood cells of the immune system; see Babior (1992).

120 Theorell (1947); Margoliash and Schejter (1966, 1984). The crystallization of cytochrome c was first reported by Gerhard Bodo (1955).

121 Lankester (1871), pp. 318–319.

122 Keilin (1966), pp. 223–224. For hemocyanin, see Redfield (1934), Dawson and Mallette (1945), and Van Holde and Miller (1995); for hemocuprein, see Mann and Keilin (1938); for polyphenol oxidase, see Kubowitz (1937); for the role of copper in iron metabolism, see Elvehjem (1935) and Lardy (1978). See also Gray and Solomon (1981) and Adman (1991).

123 Glusker (1991); Regan (1995).

124 Keilin (1966), pp. 289–315; Slater (1983, 1986); Trumpower (1990).

125 Omura and Sato (1964); Gunsalus and Sligar (1978); Walsh (1979), pp. 468–478; White and Coon (1980); Guengerich (1991); Porter and Coon (1991).

126 Wieland (1912a, 1912b). For Wieland, see Dane et al. (1942) and Witkop (1992).

127 Wieland (1913), p. 3328.

128 Hoppe-Seyler (1876), p. 5.

129 Osann (1853).

130 Graham (1876), p. 299.

131 Rey-Pailhade (1888), p. 44.

132 Abelous and Gérard (1899), p. 1025.

133 Heffter (1908).

134 Schardinger (1902); Trommsdorff (1909); Bredig and Sommer (1909); Bredig and Fiske (1912). For Bredig, see Kuhn (1962).

135 Bach (1911).

136 Battelli and Stern (1911); Einbeck (1914); Parnas (1910).

137 Wieland (1913), p. 3339.

138 Ibid., p. 3341.

139 Wieland (1922b), p. 502; Warburg (1923).

140 Gillespie and Liu (1931).

141 Thunberg (1920).

142 Fleisch (1924), p. 310; Szent-Györgyi (1924), p. 196.

143 Hopkins (1926), p. 52.
144 Palladin (1909).
145 Clark (1925), p. 171.
146 Hopkins (1926), p. 49.
147 Quastel and Whetham (1924); Thunberg (1926); Lehmann (1930); Borsook and Schott (1931).
148 Michaelis (1935). For Michaelis, see MacInnes and Granick (1958).
149 Hopkins (1921a), p. 297.
150 Ibid., p. 303.
151 Hopkins and Dixon (1922), p. 559.
152 Hopkins (1929).
153 Crook (1959); Meister and Tate (1976); Douglas (1987); Dolphin (1989).
154 Szent-Györgyi (1928), p. 1391.
155 Ibid., p. 1401.
156 Szent-Györgyi (1937), p. 73. See Staudinger (1978); King (1979).
157 For Haworth, see Hirst (1951); Stacey (1973); Isbell (1974).
158 Cardinale and Udenfriend (1974).
159 Hopkins (1906), pp. 395–396; see also Hopkins (1912).
160 Lunin (1881), p. 37. For Eijkman, see Carpenter and Sutherland (1995).
161 Hopkins (1930b); McCollum (1957).
162 Barron and Harrop (1928), p. 85. It seems likely that Warburg learned of this work during his visit to the United States in 1929.
163 Warburg, Kubowitz, and Christian (1930a), p. 496; (1930b), pp. 270–271.
164 Warburg and Christian (1931), p. 215; my emphasis.
165 Warburg and Christian (1932); (1933), p. 377.
166 Kühling (1899); Ellinger and Koschara (1934).
167 Stern and Holiday (1934); Kuhn (1935). For Kuhn, see Westphal (1968).
168 Theorell (1935); Kuhn, Rudy, and Weygand (1936).
169 Walsh (1980); Massey (1994).
170 Warburg and Christian (1933), p. 394.
171 Euler and Myrbäck (1923); Euler (1936).
172 Warburg and Christian (1935, 1936).
173 Warburg, Christian, and Griese (1935).
174 Schlenk (1942); Kornberg and Pricer (1950).
175 Vennesland and Westheimer (1954). See also Vennesland (1981) and Zerner (1992).
176 Thunberg (1937).
177 Kuhn (1935), p. 921. See also Quastel and Wooldridge (1927) and Quastel (1974).
178 Negelein and Wulff (1937); Warburg and Christian (1939), p. 40. According to Negelein and Haas (1935), p. 207, "The *Zwischenferment* is no more a dehydrogenase than the globin of hemoglobin is an oxygen transfer agent."
179 Krebs (1935a); Haas (1938); Warburg and Christian (1938); Negelein and Brömel (1939a). To my knowledge, the physiological role of D-amino acid oxidase in animal tissues still remains to be established; see Yagi (1971).
180 Negelein and Brömel (1939b); Warburg and Christian (1939); Bücher (1947, 1989).
181 Green, Dewan, and Leloir (1937), p. 948.

182 Dixon and Zerfas (1940); Parnas (1943); Racker (1955).
183 Slater (1983), p. 239. See also Ball (1974); for Ball, see Buchanan and Hastings (1989).
184 Duine and Jongejan (1989); Klinman and Mu (1994).
185 Stephenson and Strickland (1931); Elsden (1981); Collman (1996). For Stephenson, see Robertson (1949) and Mason (1992).
186 Baeyer (1870).
187 Bernard (1877a), p. 328.
188 Liebig (1847b), p. 330.
189 Buchner and Meisenheimer (1904), pp. 419–421.
190 Harden and Maclean (1911), p. 66.
191 Wohl (1907). See Engel (1996).
192 Harden (1903), p. 716.
193 Harden and Young (1906), p. 410.
194 Wróblewski (1901).
195 Harden and Young (1906), pp. 415–416, 418.
196 Levene and Raymond (1928).
197 Robison (1922); Robison and King (1931).
198 Embden and Laquer (1914, 1921).
199 Meyerhof (1918, 1926a).
200 Kluyver and Donker (1925, 1926).
201 Harden (1923), p. 109; Lebedev (1912). For Lebedev, see Bezkorovainy (1973b).
202 Neubauer and Fromherz (1911); Neuberg and Kerb (1913); Schoen (1928).
203 Dakin and Dudley (1913). For glutathione as a cofactor in glyoxylase action, see Knox (1960), pp. 271–282 and Douglas (1987), pp. 106–127.
204 Neuberg and Reinfurth (1919), p. 1681. See Engel (1996).
205 Connstein and Lüdecke (1919), p. 1385.
206 Dumas (1874), p. 104.
207 Parnas (1910), pp. 286–287.
208 Meyerhof (1926a, 1930). For Meyerhof, see Muralt (1952) and Schweiger (1986). For the determination of lactic acid, Meyerhof used a newly improved version of the Warburg manometers.
209 A similar enzyme ("phosphatese") had been postulated by Euler and Kullberg (1911) for the initiation of the fermentation of glucose. For hexokinases, see Purich et al. (1973).
210 Embden (1924).
211 Ohle (1931).
212 Fischer and Baer (1932); Smythe and Gerischer (1933); Fischer (1960). For Fischer, see Sowden (1962); Stanley and Hassid (1969). The synthetic method used by Fischer and Baer was based on the prior work of Rosenmund and Zetzsche (1921); it was introduced into peptide chemistry by Bergmann and Zervas (1932).
213 Embden, Deuticke, and Kraft (1933); an English translation is given in Kalckar (1969), pp. 67–72. For Embden, see Deuticke (1933) and Cori (1983).
214 Meyerhof and Lohmann (1934). For later work on aldolases, see Horecker, Tsolas, and Lai (1972).
215 Rose (1995); Knowles and Albery (1977). See also Morse and Horecker (1968).
216 Shaffer and Ronzoni (1932), p. 258.

217 Meyerhof and Kiessling (1935).
218 Elvehjem et al. (1937).
219 Stephenson (1949); Smith (1955).
220 Auhagen (1932); Peters (1950); Williams and Spies (1938); Lohmann and Schuster (1937); Haake (1987).
221 Needham (1971); Kremer (1990), pp. 308–453; Harold (1986), pp. 389–426.
222 Fick and Wislicenus (1865). For Fick, see Schenck (1902) and Bezel (1979); for Wislicenus, see Beckmann (1904); Perkin (1905); Ramberg (1994).
223 Pettenkofer and Voit (1866), p. 572.
224 Liebig (1870); Hermann (1867); Nasse (1869); Pflüger (1875); Fürth (1903). For Liebig's famous meat extract, see Finlay (1995).
225 Fick (1882), (1903–1905), vol. 4, pp. 41–63.
226 Hill (1912), p. 507. See also Hill (1950, 1959).
227 Fletcher and Hopkins (1907).
228 Hopkins (1921b), p. 361.
229 Hill (1912), p. 506.
230 Meyerhof (1925), p. 995.
231 Hill (1932).
232 Fiske and SubbaRow (1927, 1929); Eggleton and Eggleton (1927, 1928); Lipmann (1977). For SubbaRow, see Gupta et al. (1987).
233 Fiske and SubbaRow (1929); Lohmann (1928, 1935); Davenport and Sacks (1929); Makino (1931a); Lythgoe and Todd (1945). See also Schlenk (1987) and Maruyama (1991).
234 Lundsgaard (1930), p. 177. See also Kalckar (1969), pp. 344–353.
235 Lehnartz (1933), p. 966; Schmidt (1928).
236 Lohmann (1934).
237 Meyerhof and Lohmann (1931), p. 576.
238 Lohmann (1934), p. 276.
239 Engelhardt and Lyubimova (1939); Engelhardt (1942, 1982). See also Kisselev (1990).
240 Fürth (1919).
241 Muralt and Edsall (1930).
242 Weber and Portzehl (1952).
243 Szent-Györgyi (1951); Straub (1981). See also Perry (1989).
244 Bailey (1954); Szent-Györgyi (1953); Mannitzer and Goody (1976).
245 Huxley, A. F. (1953, 1957, 1980); Hanson and Huxley (1955); Huxley, H. E. (1969, 1971, 1990). For a fuller account of the early development of the "sliding mechanism" theory, see Needham (1971), pp. 237–307, and for later developments, Hibberd and Trentham (1986) and Pollard and Cooper (1986). See also Morales (1995).
246 Needham (1971), pp. 514–577; Harold (1986), pp. 405–422; Carlier (1991); Mannherz (1992). An actin has also been found in yeast cells; see Buzan and Frieden (1996).
247 Parnas, Ostern, and Mann (1934), p. 68.
248 Meyerhof (1930); Hahn (1931).
249 Meyerhof and Kiessling (1935); Warburg and Christian (1942). See also Wold (1971).
250 Sutherland, Posternak, and Cori (1949); Greenwald (1925). See also Ray and Peck (1972).

251 Meyerhof and Kiessling (1935); Lohmann (1935); Kalckar (1942).
252 Needham and Pillai (1937); Green, Needham, and Dewan (1937); Dische (1937); Meyerhof, Ohlmeyer, and Möhle (1938).
253 Rapkine (1938); Racker and Krimsky (1952); Racker (1954); Harris and Waters (1976). For Rapkine, see Karp and Karp (1988).
254 Negelein and Brömel (1939b); Bücher (1947, 1989); Racker (1980). See also Warburg et al. (1954).
255 Lipmann (1939a, 1939b).
256 Kalckar (1942). See also Kalckar (1990).
257 Kalckar (1941); Lipmann (1941). See also Meyerhof (1944).
258 The terms *exergonic* and *endergonic* were used by Charles Coryell (1940) as counterparts of Berthelot's "exothermic" and "endothermic" to emphasize the distinction between free-energy changes and changes in the heat content (enthalpy) in chemical reactions.
259 Lipmann (1941), pp. 153–154. For Lipmann, see Kaplan and Kennedy (1966); Lipmann (1971, 1984); Chapeville and Haenni (1980); Kleinkauf et al. (1988).
260 Atkinson (1977), p. 265.
261 Fruton (1988), p. 169.
262 Parnas and Baranowski (1935); Cori and Cori (1936); Cori, Cori, and Schmidt (1939); Green and Cori (1943). See also Kalckar (1963).
263 Pasteur (1876c), pp. 251–252. His observation of this phenomenon was reported in Pasteur (1861).
264 Pasteur (1876c), p. 246.
265 Pfeffer (1881), vol. 1, p. 370; Wortmann (1880).
266 Meyerhof (1918), p. 174; (1926b), p. 257.
267 Warburg (1926), p. 241.
268 Warburg (1956, 1962); Weinhouse et al. (1956).
269 Lipmann (1933); Stern and Melnick (1941); Engelhardt (1974).
270 Burk (1939); Krebs (1972b); Racker (1974, 1988).
271 Thunberg (1920).
272 Toenniessen and Brinkmann (1930).
273 Szent-Györgyi (1937); Stare and Baumann (1936).
274 Krebs and Johnson (1937). See also Krebs (1970, 1981) and Holmes (1993), pp. 396–400.
275 Knoop and Martius (1936); Martius and Knoop (1937); Martius (1937, 1982). See also Wagner-Jauregg and Rauen (1935).
276 Krebs and Eggleston (1940), p. 442.
277 Krebs (1943), p. 229; Ogston (1948); Potter and Heidelberger (1949). For the "three point attachment" concept, see Easson and Stedman (1933); Bergmann et al. (1935); Cornforth (1976); Bentley (1983).
278 Klein and Harris (1938); Lipmann (1945); Nachmansohn and Berman (1946).
279 Williams (1943), p. 267.
280 Baddiley et al. (1953); Baddiley (1955); Moffatt and Khorana (1961).
281 Martius and Lynen (1950); Lynen et al. (1951, 1959); Stern, Ochoa, and Lynen (1952); Ochoa (1954); Lynen (1954). For Lynen, see Wood (1978); Krebs and Decker (1982).
282 Reed et al. (1951); Reed (1957); Hucho (1975); Stryer (1988), pp. 279–283. A similar multienzyme system was found to be operative in the conversion of α-ketoglutarate to succinyl-CoA. See Welch (1985); Srere (1987).

283 Kornberg and Elsden (1961), pp. 404–420; Kornberg (1989).
284 Ball (1942), p. 22.
285 Chance (1956); Chance et al. (1962).
286 Engelhardt (1932); Runnström and Michaelis (1935); Kalckar (1937); Belitser and Tsibakova (1939).
287 Ochoa (1941, 1943). For Ochoa, see Gomez-Sanchez (1989).
288 Ogston and Smithies (1948); Robbins and Boyer (1957); Benzinger et al. (1959).
289 Kennedy and Lehninger (1949), p. 970.
290 Béchamp (1883). For a partisan account of his dispute with Pasteur, see Hume (1923).
291 Altmann (1890); Hardy (1899), pp. 160–161.
292 Michaelis (1900).
293 Bensley and Hoerr (1934); Claude (1946, 1950); de Duve and Beaufray (1981).
294 Palade (1952a, 1952b); Sjöstrand (1943, 1953). See Rasmussen (1995, 1997b) for valuable accounts of this work.
295 For the early history of the electron microscope, see Freundlich (1963). For a discussion of the limitations and possible errors in the interpretation of electron micrographs, see Rasmussen (1993) and Penman (1995).
296 Lehninger (1951, 1954, 1955).
297 Green (1959); Crane et al. (1959). For Green, see Beinert and Stumpf (1983).
298 Chance and Williams (1956a, 1956b).
299 Cohn and Drysdale (1955); Boyer, Falcone, and Harrison (1954).
300 Pullman et al. (1960); Racker and Horstmann (1967). See also Racker (1965, 1967, 1979, 1988); Allchin (1996). For Racker, Schatz (1996).
301 Bergmann and Fruton (1944); Fruton (1988b).
302 Meyerhof (1945).
303 Kagawa (1984); Futai et al. (1989); Penefsky and Cross (1991); Pedersen and Amzel (1993); Abrahams et al. (1994); Boyer (1997).
304 Parascandola (1974); Lardy and Phillips (1943); Loomis and Lipmann (1948). Thyroxine is another of the known uncoupling agents; see Hoch and Lipmann (1954).
305 Racker (1967, 1983, 1985). See also Racker and Racker (1981).
306 Slater (1953); Mitchell (1961). For Mitchell, see Mitchell (1981); Weber (1991); Slater (1994).
307 Mitchell (1979a), p. 1; (1991), pp. 297–298.
308 Williams (1961, 1962, 1979); Boyer (1977, 1981, 1993); Robinson (1984); Robertson (1995). Boyer, Chance, Ernster, Mitchell, Racker, and Slater (1977); I deplore the absence of Lehninger's name in that list.
309 Harold (1978); Crofts (1979).
310 For general historical accounts, see Ernster and Schatz (1981); Ernster (1984, 1993); Rothstein (1984); Harold (1986), pp. 58–90; Senior (1988); Robinson (1997). For a sociological analysis of the reception of the chemiosmotic hypothesis, see Gilbert and Mulkay (1984).
311 Schiller (1980), pp. 131–156.
312 Traube (1867).
313 Pfeffer (1877); van't Hoff (1887).
314 Overton (1899, 1901); for Overton, see Collander (1962–1963). For later versions of the bilayer lipoid model, see Gorter and Grendel (1925); Danielli (1975); Singer and Nicolson (1972); Jacobson et al. (1995). For Gorter, see Seeder (1954).

315 For valuable summaries, see Brooks and Brooks (1941); Davson and Danielli (1943); Brown and Danielli (1954). For Osterhout, see Blinks (1974) and Slayman (1993). For Danielli, see Stein (1986).

316 Onsager (1969); for Onsager, see Lyons (1981). See also Rosenberg (1954).

317 Skou (1965, 1989); Skulachev (1991); Lingrel and Kuntzweiler (1994).

318 Racker (1972); Pickart and Jencks (1984); Carafoli (1987).

319 Crane (1965, 1983); Semenza et al. (1984); Wright et al. (1986).

320 Pressman (1965); Ovchinnikov (1979); Anderson (1984).

321 Priestley (1774), pp. 89–92; Ingen-Housz (1780), p. 17; Senebier (1788), pp. 410–429. For Ingen-Housz, see Reed (1949); for Senebier, see Legée (1991). See also Hill (1970) and Gest (1991).

322 Saussure (1804).

323 Pelletier and Caventou (1817). See Höxtermann (1980).

324 Sachs (1862); Engelmann (1881, 1882). For Engelmann, see Kingreen (1972) and Kamen (1986a).

325 Arnon (1955).

326 For Winogradsky, see Waksman (1953) and Courrier (1957b).

327 Wurmser (1925); van Niel (1931, 1935, 1941).

328 Hill (1939, 1965). For Hill, see Myers (1974); Bendall (1994).

329 Höxtermann (1995); Scheer (1991).

330 Warburg (1919, 1920); Warburg and Negelein (1922, 1923).

331 Warburg (1958, 1962); Emerson and Arnold (1933); Emerson and Lewis (1943). See also Zallen (1993b) and Kamen (1995).

332 Baeyer (1870); Bayliss (1924), pp. 564–568.

333 Ruben, Kamen, and Hassid (1940). For Blackman, see Briggs (1948).

334 Ruben (1943).

335 Kamen (1963, 1986b, 1994).

336 Calvin (1953, 1956).

337 Horecker (1962, 1982).

338 Arnon (1959, 1961, 1967).

339 Arnon (1965, 1984).

340 Barber (1992); Prince (1996).

341 Jagendorf and Uribe (1966); Jagendorf (1967).

342 Glazer and Melis (1987); Henderson et al. (1990); Stoeckenius (1994); Diesenhofer et al. (1995). See also Harold (1986), pp. 92–126.

343 Racker and Stoeckenius (1974).

344 Zallen (1993a). To the scientists mentioned in this valuable paper as having had an influence on many productive investigators in this field may be added Roger Stanier; see Stanier (1980) and Clarke (1986).

345 Moser (1842); Müller (1856).

346 Boll (1876, 1877). For Boll, see Gamgee (1877), Haltenhoff (1880); Hubbard (1976, 1977); Baumann (1977).

347 Kühne (1878b). See also Thudichum (1881a).

348 For Helmholtz's letter to Boll, see Belloni (1982), p. 133.

349 Wald (1968); Applebury (1993).

350 Hecht and Williams (1922); Hecht, Shlaer, and Pirenne (1942). See also Hecht (1937, 1938). For Hecht, see Wald (1948).

351 Wald (1936). See Karlson (1978).
352 Hubbard and Wald (1952); Wald (1968). For parallel work in the laboratory of Richard Alan Morton, see Collins (1954) and Glover et al. (1978).
353 Wald (1964).
354 For recent accounts of this rapidly developing area of research, see Stryer (1986, 1991); Khorana (1988, 1992); Applebury and Hargrave (1986); Nathans (1987); Stryer et al. (1996); Liu et al. (1996); Chen et al. (1996).
355 Lillie (1924), p. 171.
356 Kluyver and Donker (1925), p. 618.
357 Kostychev (1926); Quastel (1926).
358 Euler (1938); Stern (1939), p. 318.

7. Pathways of Biochemical Change

1 Spallanzani (1789), vol. 1, p. 389. This English translation of Spallanzani's book contains a translation of Stevens's 1777 dissertation. For Stevens, see Day (1976).
2 Hallé (1791).
3 Fourcroy (1801), vol. 5, pp. 660–661.
4 Holmes (1963a).
5 Cuvier (1828), p. 117.
6 Beaumont (1833), pp. 36–37.
7 For Magendie, see Olmsted (1944); Grmek (1974); Albury (1977); Théodoridès et. al. (1983). For Tiedemann, see Mani (1956) and Kruta (1976).
8 Magendie (1816–1817), vol. 2, p. 395.
9 Ibid., vol. 1, pp. 19–20.
10 Gottschalk (1921); Bing (1971); Mani (1976). The word *métabolisme* does not appear to have been adopted by French physiologists until the beginning of the twentieth century but was used by some French chemists around 1858. It was not used by Claude Bernard, who wrote of "circulation chimique"; see Bernard (1965), p. 262.
11 Hardy (1936), p. 898.
12 Hering (1888), p. 35.
13 Tiedemann and Gmelin (1826–1827); Schwann (1836); Bouchardat and Sandras (1842). For Gmelin, see Pietsch and Beyer (1939).
14 Thomson (1843), pp. 651, 656. For Thomson, see Morrell (1969, 1972).
15 Schwann (1839), p. 257.
16 Mendelsohn (1963); Maulitz (1971).
17 Magendie (1841).
18 Boussingault (1839). See Aulie (1970); McCosh (1984); Holmes (1987).
19 Liebig (1846b), p. 228.
20 Milne Edwards (1862), vol. 7, p. 542.
21 Bernard (1853), p. 7.
22 Bernard (1848). For valuable accounts of the development of Bernard's ideas, as recorded in his research notes, see Grmek (1968a, 1968b) and Holmes (1974).
23 Bernard (1853), p. 9.

24 Liebig (1842), p. 86.
25 Bernard (1855a), vol. 1, p. 85. What Bernard termed glycogenesis was later defined more accurately as gluconeogenesis.
26 Ibid., p. 97.
27 See Grmek (1986b), pp. 220–224.
28 Bernard (1855b), p. 467.
29 Bernard (1857b, 1857c); Sanson (1857).
30 Young (1937); Mani (1964).
31 Bernard (1877a), p. 430.
32 Bernard (1857b), p. 583.
33 Bernard (1877b), pp. 524–525.
34 Bernard (1878–1879), vol. 2, pp. 40–41.
35 Bernard (1848), p. 317.
36 Bernard (1878–1879), vol. 1, p. 227.
37 Cited in Grmek (1968b), p. 231 and Bernard (1979), p. 56.
38 Baeyer (1870); Nencki (1872).
39 Hoppe-Seyler (1884), pp. 17, 26.
40 Voit (1891), pp. 245–246.
41 Pflüger (1903), p. 168.
42 Lüthje (1904); Pflüger (1910), p. 302.
43 Mering and Minkowski (1890). See Houssay (1952); Zeller and Bliss (1990).
44 Pratt (1954); Bliss (1982). It has been asserted that Nicolas Paulescu anticipated the work of Banting and Best; see Teichman and Aldea (1985); Lestradet (1993).
45 Lusk (1910).
46 Haworth and Percival (1932); Haworth, Hirst, and Isherwood (1937). For Haworth, see Hirst (1951).
47 Meyer (1943); Rundle and French (1943).
48 Cori and Cori (1929), pp. 401–402. See Cori (1931).
49 Houssay (1936). For Houssay, see Young and Foglia (1974); for Long, see Smith and Hardy (1975).
50 For Carl Cori, see Cori (1969), Randle (1986), and Cohn (1992, 1996). For Gerty Cori, see Larner (1992). See also Kalckar (1983) and Ochoa (1985).
51 Cori and Cori (1936); Cori, Cori, and Schmidt (1939); Cori and Cori (1940). Parnas (1937) had observed the disappearance of inorganic phosphate during glycogen breakdown in muscle extracts. See Willstätter and Rohdewald (1940).
52 Hanes (1940).
53 Green and Cori (1943).
54 Krebs and Fischer (1962).
55 Klein and Helmreich (1985); Helmreich (1995).
56 Leloir (1964, 1971); Leloir and Cardini (1952); Leloir et al. (1959); Leloir and Paladini (1983).
57 Hassid (1969); Strominger (1970).
58 Ryman and Whelan (1971).
59 Cori and Cori (1952); Cori (1954). Rudolf Schoenheimer (1929) had investigated the chemical aspects of von Gierke's disease and concluded that it involves liver enzymes rather than a defect in glycogen; also see Huijing (1979).

60 Stanbury et al. (1983).
61 Larner (1990).
62 Sutherland and Rall (1958, 1960); Lipkin, Cook, Markham (1959); Sutherland (1972). See Woolf (1975) and Sinding (1996) for the sociological aspects of the discovery of cAMP. For Sutherland, see Cori (1978).
63 Kurzer and Sanderson (1956); ten Hoor (1996).
64 Bérard (1817), p. 296.
65 Prévost and Dumas (1823), pp. 100–101.
66 Wallach (1901), vol. 1, pp. 206–208; see also Moore (1978).
67 Liebig (1843), p. 132.
68 Bidder and Schmidt (1852); see Holmes (1987).
69 Bischoff (1853), p. 74.
70 Bischoff and Voit (1860), p. 268; see Holmes (1987).
71 Voit (1870), p. 399.
72 Hoppe-Seyler (1873), pp. 404, 412; (1866), p. 138.
73 Pflüger (1893).
74 Folin (1905), pp. 122–123. For Folin, see Meites (1989).
75 Bernard (1865), p. 228.
76 Schoenheimer and Rittenberg (1938), p. 222.
77 Dumas (1856), p. 549.
78 Schultzen and Nencki (1869), pp. 566–567.
79 Knieriem (1874).
80 Schroeder (1882).
81 Hoppe-Seyler (1873), p. 415.
82 Hofmeister (1882).
83 Bunge (1902), p. 168.
84 Cohnheim (1901), p. 465. For Cohnheim, see Matthews (1978).
85 Loewi (1902), p. 316. For Loewi, see Dale (1962); Lembeck and Giere (1968). See also Abderhalden and Rona (1905); Henriques and Hansen (1905); Wolf (1996).
86 Willcock and Hopkins (1907). See Cathcart (1912).
87 Van Slyke and Meyer (1913), p. 206.
88 Salkowski (1877); Schmiedeberg (1877); Drechsel (1880).
89 Drechsel (1890); Schulze and Steiger (1887); Schulze and Winterstein (1897). For Schulze, see Winterstein (1914).
90 Jaffé (1877, 1878a, 1878b). For Jaffé, see Ellinger (1913). See also Akers and Dromgoodle (1982).
91 Kossel and Dakin (1904).
92 Clementi (1915). See Baldwin (1940).
93 Maynard (1958). See also Van Slyke (1929).
94 Werner (1923); Loeffler (1917); (1920), p. 177; Kase (1931).
95 Krebs and Henseleit (1932). See Holmes (1991), pp. 277–341.
96 Krebs (1947).
97 Krebs (1970), p. 161.
98 For a detailed account of the "ornithine effect," see Holmes (1991), pp. 283–306.
99 Wada (1930). See also Ackermann (1931b).

100 Krebs and Henseleit (1932), p. 53.
101 Hofmeister (1894), p. 214; Loewi (1898), p. 521. See Holmes (1991), pp. 356–358.
102 Cohen and Hayano (1946).
103 Ratner (1954, 1977).
104 Jones, Spector, and Lipmann (1955).
105 Krebs (1942), pp. 764–765. See also Collatz (1976).
106 Piria (1844).
107 Boussingault (1868); Pfeffer (1872).
108 Schulze and Barbieri (1880). See Chibnall (1939).
109 Prianishnikov (1904), p. 42. For Prianishnikov, see Senchenkova (1975).
110 Prianishnikov (1924), pp. 421–423.
111 Neubauer (1909); Embden and Schmitz (1910).
112 Knoop (1910).
113 Knoop and Oesterlin (1925).
114 Krebs (1935a, 1935b). See Holmes (1991).
115 Schulze and Bosshard (1883).
116 Meister (1956); Ginsburg (1972); Zalkin (1993).
117 Braunstein and Kritzmann (1937); Kritzmann (1938); Braunstein (1939, 1982); Torchinsky (1987). The biological occurrence of transamination reactions had been suggested by the earlier biochemical studies of Dorothy Needham (1930) and the chemical work of Robert Herbst and Lewis Engel (1934); see also Herbst (1944).
118 Schlenk and Snell (1945); Meister (1955); Snell (1979, 1993). For the mechanism of enzyme-catalyzed transamination reactions, see Emery and Akhtar (1987).
119 Dakin (1912), p. 2.
120 Hopkins (1913), p. 214.
121 Fischer (1907a), p. 1753.
122 Dakin (1912), p. 2.
123 Hofmeister (1900), p. 3877.
124 Liebig (1853); Jaffé (1874); Hunter (1912).
125 Miller and Waelsch (1957).
126 Mendel and Jackson (1898), p. 28.
127 Ellinger (1904); Kotake (1935); Butenandt et al. (1943).
128 Wolkow and Baumann (1891), pp. 237, 277. For Baumann, see Tiemann (1896) and Orten (1956)
129 Neubauer and Falta (1904), p. 91.
130 Garrod and Hele (1905), p. 205.
131 Kaufman (1971). For the early history of folic acid, see Stokstad (1979) and Jukes (1980a).
132 Wellner and Meister (1981); Hayaishi (1974).
133 Bence-Jones (1867), p. 40.
134 Jaksch (1883); Minkowski (1884).
135 Neumeister (1897), p. 773. See also Rosenfeld (1895).
136 Magnus-Levy (1899). For Magnus-Levy, see Goldner (1955).
137 André (1932); de Romo (1989).
138 Knoop (1904). For Knoop, see Thomas (1948) and Ohlmeyer (1948).
139 Embden and Michaud (1908). For Embden, see Deuticke (1933).

140 Dakin (1909, 1912).
141 Friedmann (1913), p. 442. For Friedmann, see Mitchell (1956).
142 Jowett and Quastel (1935); MacKay et al. (1940). See Stadie (1945).
143 Edson and Leloir (1936); Leloir and Munoz (1939, 1944).
144 Lehninger (1945); Stadtman et al. (1951); Drysdale and Lardy (1953); Green (1954); Lynen (1955); Miller and Waelsch (1957).
145 Fraenkel and Friedman (1957); Bremer (1977).
146 Lawes and Gilbert (1866).
147 Collie (1893, 1907); Raper (1907), p. 1831.
148 Wakil (1962); Vagelos (1964); Wakil et al. (1983).
149 Smith (1994).
150 Schoenheimer and Rittenberg (1935), p. 156; Schoenheimer (1937, 1941). For Schoenheimer, see Peyer (1972); Kohler (1977b); Stetten (1982). Schoenheimer's interest in sterol metabolism began during his medical studies (M.D. Berlin 1922). After a year as a pathologist at the Moabit hospital, he enrolled in Karl Thomas's postgraduate program at Leipzig for young physicians who wished to improve their knowledge of chemistry; see Thomas (1954). (According to Kohler [1982], p. 37, none of the students in that program became biochemists.) While in Leipzig, Schoenheimer developed a new method of peptide synthesis; see Schoenheimer (1926). In 1926 he went to Ludwig Aschoff's institute of pathological anatomy in Freiburg and resumed work on sterol metabolism.
151 Hevesy (1923). For Hevesy, see Cockcroft (1967).
152 Soddy (1913), p. 400.
153 Conant (1970).
154 Aston (1933). For Aston, see Brock (1970).
155 Nier (1955).
156 Hevesy (1948), p. 130. There does not appear to have been any publication from Hevesy's laboratory with Schoenheimer as one of the authors. In his writings Schoenheimer referred to Hevesy's work but did not mention the association indicated by Hevesy in this quotation.
157 Urey et al. (1932). For Urey, see Garratt (1962) and Cohen et al. (1983).
158 Urey (1934); van Heyningen (1939).
159 Some contributions were of doubtful merit and invited the question whether, in these cases, it was a matter of "applying intermediate metabolism to the study of isotopes." For the early history of the scintillation counter, see Krebs (1955).
160 Schoenheimer (1937).
161 Weinhouse et al. (1944).
162 Crandall and Gurin (1949).
163 Schoenheimer, Ratner, and Rittenberg (1939), pp. 729–730. The entire passage was italicized in the original.
164 Borsook and Keighley (1935).
165 Madden and Whipple (1940).
166 Hogness, Cohn, and Monod (1955); Mandelstam (1958).
167 Hunter (1928).
168 Bloch and Schoenheimer (1941). See Bloch (1987).
169 Bergmann and Zervas (1927).
170 Mueller (1923); Barger and Coyne (1928).

171 Jukes (1980b).
172 Borsook and Dubnoff (1940, 1947).
173 Du Vigneaud (1952); Simmonds et al. (1943); Simmonds and du Vigneaud (1945). See also Cohn (1995).
174 His (1887); Hofmeister (1894). See Challenger (1951) and Schlenk (1984).
175 Cantoni (1953).
176 Walsh (1979), pp. 828–866; Walker (1979); Tabor and Tabor (1984); Stubbe (1994).
177 Wood and Werkman (1936); Werkman and Wood (1942). See Singleton (1997a, 1997b).
178 Evans (1940); Evans and Slotin (1941). For a valuable exchange on the history of this development, see Krebs (1974) and the accompanying letters from several investigators, especially Wood, pp. 88–90, 91–94. See also Buchanan and Hastings (1946); Hastings (1970); Vennesland (1991).
179 Stephenson (1949); Van Niel (1949); Woods (1953).
180 Wood (1972, 1982, 1985). See also Williams (1980).
181 Utter and Wood (1945, 1951); Utter and Keech (1960).
182 Hofmann (1943).
183 Moss and Lane (1971); Utter et al. (1975); Wood and Barden (1977).
184 Lynen et al. (1959); Obermayer and Lynen (1976). For Lynen, see Wood (1979); Krebs and Decker (1982).
185 Robinson (1917), p. 877. See also Schöpf (1937).
186 Robinson (1955), p. 1.
187 Dawson (1948), p. 204. See Battersby (1963); Leete (1965, 1967).
188 Bondzynski and Humnicki (1896); Windaus and Neukirchen (1919). For Windaus, see Butenandt (1961). During the nineteenth century, some biologists (for example Ernst Haeckel) became interested in the discovery that cholesteryl benzoate forms "liquid crystals"; see Kelker (1986); Janko et al. (1989).
189 Bernal (1932); Rosenheim and King (1932, 1934); Wieland, Dane, and Scholz (1932). For Diels, see Olsen (1962); for Wieland, see Witkop (1992). See also Bloch (1992).
190 Channon (1926).
191 Heilbron, Kamm, and Owens (1926).
192 Ruzicka (1959, 1973). See Pickles (1910).
193 Robinson (1934).
194 Schoenheimer and Breusch (1933).
195 Schoenheimer (1937); Sonderhoff and Thomas (1937), p. 203.
196 Popjak and Cornforth (1960); Bloch (1965).
197 Bonner and Arreguin (1949).
198 Langdon and Bloch (1953).
199 Wagner and Folkers (1961).
200 Beytia and Porter (1976); Bloch (1982, 1992); Schoepfer (1981, 1982).
201 Karrer and Wehrli (1933); Goodwin (1954).
202 Woodward and Bloch (1953). For Woodward, see Todd and Cornforth (1981).
203 Goldstein and Brown (1977). See also Hampton et al. (1996).
204 Fischer (1934–1940). For Fischer, see Watson (1965); Treibs (1971). Fischer was also chiefly responsible for the determination of the structure of the bile pigments (bilirubin, biliverdin, stercobilin), linear tetra-pyrroles that are the metabolic breakdown products of the porphyrins; see Gray (1983).

205 Shemin and Rittenberg (1945).
206 Shemin and Wittenberg (1950), Muir and Neuberger (1950). See also Neuberger (1990).
207 Shemin and Wittenberg (1951).
208 Ibid., pp. 326–328. See Shemin (1956).
209 Maitland (1950).
210 Macalpine, Hunter, and Rimington (1968); Warren et al. (1996).
211 Waldenström (1937).
212 Cookson and Rimington (1954). See Rimington (1995). For Rimington, see Neuberger (1996).
213 Goodwin (1968); Battersby and McDonald (1979).
214 Granick and Beale (1978); Porra and Meisch (1984); Scott (1994); Battersby (1994).
215 Jencks (1969); Klinman (1978); Northrop (1981); Cleland (1982); O'Leary (1989).
216 Garrod (1902), p. 1616. See Knox (1958); Childs (1970); Bearn (1993); Voswinckel (1993).
217 Garrod (1909), p. 80.
218 Hopkins (1913), p. 218.
219 Haldane (1937), p. 2.
220 See Williams (1950).
221 Beadle and Ephrussi (1936). See Burian et al. (1988).
222 Beadle and Tatum (1941). For the unexpected utility of unusual organisms, see Schmidt-Nielsen (1967).
223 Fildes (1941); Snell (1943).
224 Tatum, Bonner, and Beadle (1944). For Tatum, see Lederberg (1990).
225 Bonner (1946); Haskins and Mitchell (1949); Jakoby (1955).
226 Yanofsky and Crawford (1972); Snell (1975); Soda and Tanizawa (1979); Miles (1979); Yanofsky (1987, 1989); Botting (1995).
227 Butenandt et al. (1956).
228 Srb and Horowitz (1944).
229 Tatum (1945); Davis (1948); Lederberg (1950, 1956a).
230 Simmonds (1950); Davis (1955); Sprinson (1960); Cotton and Gibson (1967). See also Knox (1955); Wagner and Mitchell (1964); Umbarger (1978).
231 Woods (1940), p. 88. For Fildes, see Glasstone et al. (1973); for Woods, see Gale and Fildes (1965).
232 Woolley (1952).
233 Strominger (1970); Waxman and Strominger (1983). For earlier work on the effect of penicillin on the bacterial utilization of peptides, see Simmonds and Fruton (1951).
234 Boussingault (1838), p. 14. See Aulie (1970), pp. 446–447.
235 Way (1856); Lawes and Gilbert (1863). For Lawes, see van Dyke (1993).
236 Ville (1854); Boussingault (1854); Lawes, Gilbert, and Pugh (1861). For Ville, see Chevreul et al. (1955) and McCosh (1975).
237 Schloesing and Müntz (1877); Schloesing and Laurent (1890).
238 For Hellriegel, see Schadewaldt (1972).
239 For Beijerinck, see Iterson et al. (1940); for Winogradsky, see Waksman (1953).
240 Russell (1921); Fred, Baldwin, and McCoy (1932); Wilson (1957).
241 Zelitch (1989).
242 Burris and Wilson (1945); Wilson and Burris (1947); Virtanen (1938).
243 Carnahan et al. (1960); Mortenson et al. (1967).

244 Burns and Hardy (1975); Nutman (1976); Burris (1991, 1994); Rajagopalan and Johnson (1992). For a crystallographic study of nitrogenase, see Kim, Woo, and Rees (1993).
245 Gayon and Dupetit (1886). See Delwiche (1956).
246 Nason (1956); Najjar and Chung (1956).
247 Caetano-Anollés and Gresshoff (1991).
248 Schwendener (1868); Crombie (1886). For Schwendener, see Hoppe (1987).
249 Reinke (1873); Stahl (1877); Frank (1877); de Bary (1879). See Ainsworth (1976), pp. 96–100.
250 Bonnier (1886).
251 Ahmadjian (1993); Turner (1971). For Raistrick, see Birkinshaw (1972).
252 Sapp (1994).

8. The Chemical Basis of Heredity

1 Morgan (1919), p. 15.
2 Weaver (1970b). p. 67.
3 Holliday (1990), p. 136.
4 Schrader (1928); Baker (1955); Hughes (1959); Coleman (1965); Robinson (1979).
5 Ehrlich (1877). See Clark and Karsten (1983).
6 Hertwig (1876, 1885); Fol (1879).
7 Flemming (1879), (1882), p. 129. For Flemming, see Spee (1906) and Peters (1968). I will discuss the "nuclein bodies" and the work of Zacharias in the next section of this chapter and continue with the cytological studies of other biologists.
8 Strasburger (1880); Beneden (1883). For Beneden, see Brachet (1923) and Hamoir (1994).
9 Strasburger (1884); Boveri (1904); Waldeyer (1888). See Sirks (1952).
10 Roux (1883); Weismann (1883), (1891), vol. 1, pp. 175–176.
11 Delage (1895).
12 Weismann (1893), p. 453; italics in original. See Churchill (1968); Sander (1985).
13 Sourkes (1995).
14 Drabkin (1958); Sourkes (1993a).
15 Miescher (1897), vol. 1, p. 344.
16 Ibid., pp. 34–35.
17 Kühne (1866), p. 50.
18 Miescher (1871a). In a contemporary annual report, there is a note about "a peculiar phosphorus-rich albuminoid material which, according to Miescher, is the principal constituent of the nuclei of pus cells." See Maly (1873), 14. For Miescher, see Meuron-Landot (1965).
19 Hoppe-Seyler (1871), p. 501.
20 Miescher (1871b), p. 507.
21 Miescher (1897), vol. 2, p. 66.
22 Ibid., pp. 101–102.
23 Worm-Müller (1874); Wurtz (1880), p. 148; Kingzett (1878), p. 360; Thudichum (1881b), p. 186.
24 Miescher (1897), vol. 1, p. 36.
25 Ibid., p. 77.
26 His (1874), p. 152. The unnamed person referred to in the subtitle of this book was his nephew Friedrich Miescher.

27 Miescher (1897), vol. 2, pp. 94–98.
28 Zacharias (1881).
29 Hertwig (1885), pp. 290–291. A similar opinion was expressed by Albert Kölliker (1885), p. 42.
30 Altmann (1889); Wilson (1895), p. 4.
31 Miescher (1897), vol. 1, pp. 110–122. See Olby and Posner (1967) for a fanciful interpretation of this statement as an anticipation of the genetic code.
32 Weismann (1893), p. 50.
33 Heidenhain (1894), p. 548; Malfatti (1891).
34 Wilson (1896), p. 246.
35 Strasburger (1901), p. 359; (1909), p. 108.
36 Loeb (1906), p. 180; Burian (1906).
37 Wilson (1925), pp. 351, 652; italics in original. See Maienschein (1996). See also Mathews (1924), p. 89.
38 Levene and Bass (1931), p. 280. The other two monographs mentioned in the text were by Jones (1920a) and Feulgen (1923). See also Hammarsten (1924).
39 Nägeli and Loew (1878), pp. 339–340.
40 Kossel (1879, 1881).
41 Miescher (1874); Piccard (1874).
42 Kossel (1881).
43 Fischer (1899a, 1907c). For uric acid, see Kelley and Weiner (1978).
44 Kossel (1884, 1928).
45 Altmann (1892), p. 230.
46 Spitzer (1897); Pekelharing (1902); Mann (1906), p. 454. For a skeptical reaction to such claims, see Loeb (1906), p. 28.
47 Kossel and Neumann (1893); Miescher (1896), pp. 124–125.
48 Steudel (1901); Fischer and Roeder (1901); Kossel and Neumann (1894); Wheeler and Johnson (1903).
49 Levene (1903), p. 211.
50 Liebig (1847b); Haiser (1895); Bauer (1907); Neuberg and Brahm (1907).
51 Levene and Jacobs (1908, 1909).
52 Hammarsten (1894); Steudel (1907); Levene and Jacobs (1909).
53 Osborne and Harris (1902).
54 Steudel (1906); Levene (1909).
55 Levene and Jacobs (1912); Thannhauser and Dorfmüller (1917); Thannhauser and Ottenstein (1921); Levene and Bass (1931); Tipson (1945). For Thannhauser, see Martini and Schmidt (1955); Hohmann (1963); Seidler (1991).
56 Neumann (1898), p. 375. See Feulgen (1923), pp. 268–276. See also Olby (1974), p. 81 for a wishful interpretion of an ambiguous chemical finding.
57 Araki (1903); Iwanow [Ivanov] (1903); Sachs (1905).
58 Feulgen (1914).
59 Fischer, Bergmann, and Schotte (1920).
60 Bergmann, Schotte, and Leschinsky (1923).
61 Levene and Mori (1929); Levene et al. (1930); Fruton (1979b). Robert Robinson (1927) suggested xylose instead. See Dische (1955).

62 Jones (1920a), pp. 9, 11.
63 Lumb (1950).
64 Levene and Tipson (1935), p. 625; italics in original.
65 Gulland (1938).
66 Jones (1920b), p. 204.
67 Dubos and Thompson (1938); Schmidt and Levene (1938), p. 425.
68 Fruton and Bergmann (1939).
69 Kunitz (1940).
70 Todd (1983).
71 Brown and Todd (1952, 1955).
72 Carter (1950); Cohn (1950).
73 Markham and Smith (1951); Brown and Todd (1952). For Markham, see Elsden (1982). For later work on the mechanism of ribonuclease action, see Walsh (1979), pp. 199–207; Breslow and Chapman (1996).
74 Cohn and Volkin (1953).
75 Kunitz (1950); Laskowski (1967, 1982); Moore (1981).
76 Levene and Bass (1931), p. 289. See also Tipson (1945).
77 Makino (1931b).
78 Hammarsten (1924); Signer, Caspersson, and Hammersten (1938).
79 See Cecil and Ogston (1948) and Jungner et al. (1949).
80 Gulland (1947b); Gulland et al. (1947). For Gulland, see Haworth (1948); Manchester (1995); Booth and Hey (1996).
81 Stanley (1935, 1936). For Stanley, see Cohen (1986, 1990b). See also Wilkinson (1976); Markham (1977); Pirie (1984); Kay (1986); Helvoort (1991).
82 Ivanovski (1899); Beijerinck (1899). See Bos (1995).
83 Waterston and Wilkinson (1978). See also Palladino (1994).
84 Bawden et al. (1936); Pirie (1970, 1984). For Bawden, see Cohen (1990a); for Pirie, see Pirie (1986).
85 Stanley (1937), p. 329.
86 Gierer and Schramm (1956); Fraenkel-Conrat et al. (1957).
87 Sanfelice (1928). For Rous, see Andrewes (1971).
88 Cohen and Stanley (1942).
89 Greenstein (1944); Chargaff and Vischer (1948); Markham and Smith (1954); Butler and Davison (1957).
90 Gulland (1947a), p. 12.
91 See Holiday (1930).
92 Bayliss (1924), p. 9.
93 Chargaff (1950), pp. 202–203. For Chargaff, see Chargaff (1971, 1974, 1975, 1978); Abir-Am (1980).
94 Chargaff (1950), p. 206. See also Wyatt and Cohen (1953).
95 Chargaff et al. (1951), p. 229; Chargaff (1968a), p. 306.
96 Witkowski (1985).
97 Watson and Crick (1953a, 1953b); Crick and Watson (1954).
98 RajBhandary and Stuart (1966); Mandeles (1972); Deutscher (1993).
99 Holley et al. (1965); Holley (1968).

100 Sanger et al. (1973); Gilbert and Maxam (1973); Sanger (1981); Gilbert (1981). See also Smith et al. (1986); Wu (1993).

101 Arber (1978); Smith and Wilcox (1970); Nathans and Smith (1975); Roberts (1976); Nathans (1979); Kessler and Holtke (1986); Bird (1986).

102 Freundlich (1963); Marton (1968).

103 Hall (1956).

104 Luria and Anderson (1942); Ruska (1941); Hershey and Chase (1952); Anderson (1966).

105 Astbury and Bell (1938), p. 747.

106 Astbury (1947), p. 67.

107 Corey (1948); Pauling and Corey (1950, 1951).

108 Broomhead (1951); Furberg (1952).

109 Pauling and Corey (1953).

110 Watson and Crick (1953a), p. 737.

111 Wilkins, Stokes, and Wilson (1953); Franklin and Gosling (1953). For Franklin, see Klug (1968, 1974); Sayre (1975); Donohue (1976). See also Hamilton (1968); Olby (1974), pp. 385–423; Judson (1979), pp. 70–194; Wilson (1988). See also Moran et al. (1997).

112 Watson and Crick (1953a), p. 737.

113 Watson and Crick (1953b), p. 965.

114 Watson (1968, 1980). For reviews of Watson's book, see Bernal (1968); Chargaff (1968b); Lwoff (1968); Medawar (1968); Merton (1968b); Stent (1968b). See also Perutz, Wilkins, and Watson (1969).

115 Crick (1995), pp. 198–199.

116 Crick (1993), p. 17.

117 Fruton and Simmonds (1958), pp. 200–201, 748.

118 Chargaff (1955), p. 371; (1974).

119 Beadle (1969), p. 2.

120 Vries (1910), p. 13.

121 Ibid., pp. 69–74. For de Vries, see Heimans (1962); Allen (1969); van der Pas (1976); Campbell (1980); Meijer (1986); Lenay (1995).

122 Vries (1900), p. 85. In a French abstract, de Vries wrote: "This memoir, exceedingly beautiful for its time, was misunderstood and forgotten. It is very rarely cited, and then only for incidental observations." This translation is by Weinstein (1977), p. 361.

123 Correns (1900), p. 138. See Dunn (1973); Rheinberger (1995b).

124 Tschermak (1900), p. 239. The papers by de Vries, Correns, and Tschermak, with the reference to Mendel, as well as other papers dealing with the emergence of Mendelian genetics, were reprinted in Krizenecky (1965). See also Roberts (1929); Stern and Sherwood (1966); Stubbe (1963).

125 Mendel (1865), p. 17. For Mendel, see Iltis (1932); Orel (1995).

126 Mendel (1865), p. 41.

127 Nägeli (1884). For Nägeli's work, see Wilkie (1960, 1961), and for his attitude to Mendel, see Coleman (1965), pp. 142–145; Sandler and Sandler (1985), pp. 33–38.

128 Olby (1979); Orel and Hartl (1994). For the validity of Mendel's data, see Fisher (1936); Sturtevant (1965), pp. 12–16; Orel (1968); Sapp (1990). See also Kalmus (1983); Nissani (1994).

129 Sturtevant (1965); Dunn (1965, 1969).

130 Bateson and Saunders (1902), p. 126.

131 Cuénot (1902); Castle (1903). For Cuénot, see Buican (1984), pp. 73–135. For Castle, see Dunn (1965); Vicedo (1991).

132 Garrod (1902), p. 1617; Bateson and Saunders (1902), p. 133. See Knox (1958).

133 Wilson (1896), p. 326.

134 Johannsen (1909), p. 2. For Johannsen, see Churchill (1974); Roll-Hansen (1978a); Balen (1986).

135 Johannsen (1909), p. 123.

136 Ibid., pp. 124–125, 485.

137 Sutton (1903), p. 240.

138 McClung (1902); Henking (1891); Stevens (1905); Wilson (1905).

139 Morgan (1910), p. 120; italics in original.

140 Sturtevant (1913), p. 59. See also Sturtevant (1948).

141 Lederberg (1993).

142 Bridges (1916).

143 Morgan, Sturtevant, Muller, and Bridges (1915).

144 Morgan (1926a), p. 25; italics in original. See Wilson and Morgan (1920).

145 Heitz and Bauer (1933); Painter (1933); Bridges (1935).

146 Bridges (1935), p. 64.

147 Sturtevant (1925); Bridges (1936).

148 Dobzhansky (1936), p. 382; Goldschmidt (1938), p. 271. For Dobzhansky, see Levine (1995).

149 Castle (1919); Bateson (1926). See Coleman (1970).

150 Demerec (1935), p. 133.

151 Stadler (1954); McClintock (1941, 1950). See also Rhoades (1954); Comfort (1995).

152 Vries (1901–1902); Renner (1919); Cleland (1923).

153 Muller (1927), p. 84.

154 Stadler (1928); Painter and Muller (1929).

155 Timoféeff-Ressovsky, Zimmer, and Delbrück (1935). See Fano (1942); Zimmer (1966). For Timoféeff-Ressovsky, see Glass (1990); Berg (1990).

156 Schrödinger (1944), pp. 68–69.

157 Stent (1968a), p. 392.

158 Berenblum (1935).

159 Fruton, Stein, and Bergmann (1946). See Williams (1948); Brookes (1990).

160 Auerbach et al. (1947); Auerbach (1949, 1967); Boyland (1954). For Auerbach, see Beale (1995).

161 Caspersson (1936); Caspersson and Schultz (1951).

162 Ibid., p. 138.

163 Darlington (1947), p. 252. See also Darlington (1980); Lewis (1983); Olby (1990).

164 Schultz (1947); Brown (1966).

165 Muller (1935), pp. 410–411.

166 Haldane (1942), p. 22. See also Haldane (1938).

167 Beadle (1945), p. 18.

168 Muller (1947), p. 26.

169 Bateson (1909), pp. 232–233, 268.

170 Moore (1910, 1912).

171 Bateson (1913), p. 86.

172 Goldschmidt (1916), p. 100. For Goldschmidt, see Dietrich (1995).
173 Onslow (1915). See also Riddle (1909).
174 Scott-Moncrieff (1936).
175 Haldane (1966); Pirie (1966); Clark (1968).
176 Wright (1917), p. 224; (1941, 1953). For Wright, see Russell (1989).
177 Hagedoorn (1911), p. 33. See also Beijerinck (1917).
178 Troland (1914).
179 Troland (1917), p. 333. See Ravin (1977), pp. 19–31.
180 Muller (1922), pp. 33–34.
181 Gulick (1939), p. 241. See also Gulick (1938).
182 Haldane (1942), p. 44.
183 Stebbins (1972).
184 Delbrück (1941).
185 Waddington (1939, 1969); Schultz (1941); Goldschmidt and Kodani (1942).
186 Beadle and Ephrussi (1936); Ephrussi (1953). For Ephrussi, see Sapp (1987), pp. 123–162.
187 Beadle (1939), p. 61. For Beadle, see Kay (1989); Horowitz (1995).
188 Beadle (1963), p. 19.
189 Dodge (1927, 1939); Lindegren (1932, 1936); Fries (1938).
190 Beadle and Tatum (1941), p. 505. See Perkins (1992).
191 Beadle (1945), pp. 191–192.
192 Horowitz (1948); Horowitz and Leupold (1951).
193 Boivin et al. (1948); Ris and Mirsky (1949); Pollister and Leuchtenberger (1949); Mirsky (1951).
194 Wyatt (1950).
195 Zubay and Doty (1959).
196 Smith, DeLange, and Bonner (1970); DeLange and Smith (1971); Smith (1982).
197 Stedman and Stedman (1950); Simpson (1973).
198 Kornberg (1974); van Holde (1988); Wolffe (1995); Zlatanova and van Holde (1996); Gruss and Knippers (1996).
199 Crick et al. (1979); Stokes (1982); van Holde and Zlatanova (1995), p. 8376.
200 Meselson and Stahl (1958); Kornberg (1960). See Holmes (in press).
201 Record et al. (1981); Rich et al. (1984); Kennard and Hunter (1991); Kennard and Salisbury (1983).
202 Creighton and McClintock (1931); Stern (1931).
203 For Pearson, see Eisenhart (1974); for Weldon, see Pearson (1906); for Chetverikov and Dubinin, see Adams (1968); for Dobzhansky, see Adams (1994).
204 Weinstein (1980).
205 Morgan (1926b), p. 491.
206 Harrison (1937), p. 372. See also Conklin (1905), p. 220; Fruton (1992a), pp. 85–89. For Conklin, see Harvey (1958).
207 Sapp (1983, 1987); Buican (1984); Burian, Gayon, and Zallen (1988); Harwood (1992).
208 Winge (1935). See Hall and Patrick (1993).
209 Allen (1978); Kohler (1994).
210 Kohler (1994), p. 15.
211 For Muller, see Carlson (1966, 1971, 1981); Schultz (1967).
212 Schultz (1967), p. 297.

213 Roll-Hansen (1978b), pp. 165–177.

214 Kitcher (1983); Bowler (1989); Darden (1991); Kim (1994); Burian, Richard, and Van Steen (1996).

215 Beijerinck (1921), p. 37. See Brock (1990); Summers (1991); Theunissen (1996).

216 Massini (1907). See Dunn (1969), p. 71; Amsterdamska (1987).

217 Stephenson (1949); Gunsalus, Horecker, and Wood (1955); Kohler (1985a, 1985b).

218 Karström (1930, 1937); Yudkin (1938); Diénert (1900); Monod (1942, 1947); Spiegelman (1945).

219 Dubos (1976); Amsterdamska (1993).

220 Griffith (1928), p. 153. See Hayes (1966); Pollock (1970); Tooze (1978).

221 Dawson (1930); Dawson and Sia (1931); Alloway (1932).

222 McCarty (1985).

223 Avery, MacLeod, and McCarty (1944), p. 137.

224 Ibid., p. 152.

225 McCarty (1946); Kunitz (1950).

226 Avery, MacLeod, and McCarty (1944), p. 155.

227 Boivin (1947), p. 12. See also Hotchkiss (1979).

228 The full excerpt is given in Hotchkiss (1966), pp. 185–187, and Dunn (1969), pp. 61–63.

229 Stent (1968a); (1972), p. 84. For some of the other contributions to the discussion, see Wyatt (1972, 1975); Olby (1972); Cohen and Portugal (1975); Russell (1988b). See also Lederberg (1947); Fox and Allen (1964).

230 Hershey and Chase (1952), p. 54.

231 Wyatt (1974).

232 Mirsky (1951), pp. 132–133.

233 Hotchkiss (1952); Zamenhof (1957).

234 Cairns, Stent, and Watson (1966); Summers (1993); Kellenberger (1995).

235 Duckworth (1976); van Helvoort (1992).

236 Gratia (1921, 1931); Burnet (1929a, 1929b); Krueger (1931); Wollman and Wollman (1936, 1938); Schlesinger (1932, 1934a, 1934b).

237 Ellis and Delbrück (1939), p. 365. See Mullins (1972); Kay (1985); Summers (1993).

238 Cohen (1947, 1948). See also Cohen (1956, 1968, 1984, 1986, 1994).

239 See Sinsheimer (1960, 1968).

240 Northrop (1951), p. 732.

241 Arkwright (1921); Gratia (1921); d'Hérelle (1926); Burnet (1929b); Lewis (1934).

242 Ellis and Delbrück (1939); Luria (1939, 1984).

243 Luria and Delbrück (1943).

244 Lederberg and Lederberg (1952); Lederberg (1989).

245 Cairns, Overbaugh, and Miller (1988); Davis (1989); Lederberg (1989).

246 Hinshelwood (1946); Dean and Hinshelwood (1953, 1966). For Hinshelwood, see Thompson (1973).

247 Tatum (1945); Lederberg and Tatum (1946), pp. 558; Tatum and Lederberg (1947); Lederberg (1951, 1956a); Zuckerman and Lederberg (1986).

248 Hayes (1953); Jacob and Wollman (1958); Wollman, Jacob, and Hayes (1962). See Hayes (1968), pp. 650–699.

249 Taylor and Thoman (1964).

250 Robinow (1956); Kleinschmidt et al. (1961, 1962); Cairns (1963); Kellenberger (1966); Robinow and Kellenberger (1994).

251 E. M. Lederberg (1951); Lederberg and Lederberg (1953). See Hayes (1980). For Sonneborn, see Sapp (1987), pp. 87–122.
252 Bordet (1925); Burnet and McKie (1929); Wollman and Wollman (1938); Lwoff and Gutmann (1950); Lwoff (1953).
253 Zinder and Lederberg (1952); Lederberg (1956b); Morse, Lederberg, and Lederberg (1956); Hartman (1957).
254 Ozeki (1956); Arber et al. (1957); Dessoix and Arber (1962); Arber (1979).
255 Benzer (1955, 1957, 1959, 1961a, 1961b). For cis-trans test, see Lewis (1951). See also Pontecorvo (1952); Bonner (1956); Portin (1993).
256 Sinsheimer (1957), p. 1123.
257 The book by Harris (1995) is not likely to be soon surpassed.
258 Harrison (1910); Burrows (1911); Fischer (1946); Earle et al. (1956); Eagle (1959, 1960); Puck and Marcus (1955); Witkowski (1990).
259 Harris (1995), p. 50.
260 Siminovich (1976); Goss (1993).
261 Dulbecco (1969, 1976a). For Dulbecco, see Kevles (1993); for Rous, see Dulbecco (1976b). For later work on genetic transformation, see Smith et al. (1981).
262 Harris and Watkins (1965); Harris (1995), pp. 121–152.
263 Watson and Crick (1953b), pp. 966–967.
264 Delbrück and Stent (1957).
265 Ibid., p. 752.
266 Meselson, Stahl, and Vinograd (1957); Meselson and Stahl (1958), p. 588.
267 Brugsch and Schittenhelm (1910); Copeman (1969); Kelley and Wyngaarden (1974); Kelley and Weiner (1978).
268 Wolstenholme and O'Connor (1957); Buchanan and Hartman (1959); Goldthwait, Peabody, and Greenberg (1955). See also Buchanan (1986, 1994).
269 See Walsh (1979); Henikoff (1987).
270 Kornberg et al. (1955); Lieberman et al. (1955); Reichard (1959); Koshland and Levitski (1974); Zalkin (1993).
271 Thelander and Reichard (1979); Stubbe (1990); Reichard (1993); Licht, Gerfen, and Stubbe (1996).
272 Pagliotti and Santi (1977).
273 Kornberg (1957). See also Kornberg and Baker (1992), pp. 146–169.
274 For summaries, see Kornberg (1960, 1969).
275 Mizrahi and Benkovic (1988).
276 Kelner (1949); Dulbecco (1950); Setlow (1968).
277 For DNA ligase, see Gellert et al. (1968); Lehman (1974).
278 Sancar and Sancar (1988); Heelis et al. (1995); Friedberg et al. (1995).
279 Goulian, Kornberg, and Sinsheimer (1967).
280 Doty et al. (1960); Okazaki et al. (1968); Kornberg (1969). See Goulian (1971).
281 DeLucia and Cairns (1969); Okazaki et al. (1970); Gefter et al. (1971).
282 Wickner and Kornberg (1974); Schekman et al. (1974, 1975).
283 Geider and Hoffmann-Berling (1981); Lohman (1993); Lohman and Bjornson (1996).
284 Cairns (1963); Gellert (1981); Wang (1991, 1996).
285 Holliday (1964, 1990).

286 Brutlag et al. (1971); Giocomoni (1993).
287 Pettijohn (1988).
288 Bramhill and Kornberg (1988); Kornberg (1988); McHenry (1991); Kornberg and Baker (1992); Harrington and Winicov (1994).
289 Clark and Margulies (1965); Mertz and Davis (1972); Roberts (1976); Arber (1978); Radding (1982, 1991); Cox and Lehman (1987).
290 See Singer and Berg (1991); Berg and Singer (1992).
291 Davies and Gassen (1983); Ferretti et al. (1986).
292 Johnson (1983); Vane and Cuatrecasas (1984).
293 Mullis (1990); Bloch (1991); Templeton (1992); Mullis et al. (1994); Rabinow (1996).
294 Templeton (1992), pp. 58–60.
295 Berg and Singer (1995); Wright (1986, 1994).
296 Weissbach (1977); Lehman and Kaguni (1989).
297 Kelly (1988); Ishimi et al. (1988); Campbell (1993); Barbara and Huang (1995).
298 Loeb and Kunkel (1982); Modrich (1987); Radman and Wagner (1988); Kunkel (1992).
299 Mazia and Dan (1952); Mazia (1956, 1987).
300 Weber (1955).
301 Roberts and Hyams (1979).
302 Inoué (1969, 1981); Walker and Sheetz (1993); Barton and Goldstein (1996); Kleckner (1996).
303 Blackburn (1990); Blackburn and Greider (1995); Zakian (1995).
304 Grindley and Reed (1985); Mizuuchi (1992).
305 Davis (1990); Kevles and Hood (1992); Anderson (1993).
306 Caspersson (1941, 1947).
307 Wilson (1896), pp. 247–248.
308 Brachet (1941a, 1941b, 1942, 1947). See also Gale and Folkes (1953, 1955). For Brachet, see Pirie (1990); Chantrenne (1990); Burian (1994); Thieffry and Burian (1996).
309 Hagenau (1958).
310 Siekevitz and Zamecnik (1981); Rheinberger (1995, 1996).
311 Campbell (1960); Hoagland (1960); Fruton (1963); Rheinberger (1993).
312 Fruton (1949); Goodman and Kenner (1957); Gutte (1995).
313 Lipmann (1941), p. 154.
314 Hoagland, Keller, and Zamecnik (1956). See Hoagland (1990).
315 Berg et al. (1961).
316 Lipmann (1971), pp. 80–81.
317 Sawyalow (1901); Wasteneys and Borsook (1930).
318 Fruton (1957, 1963, 1988b); Brenner and Pfister (1951); Dobry et al. (1952); Jones et al. (1952); Wieland et al. (1960); Determann et al. (1963).
319 Anfinsen and Steinberg (1951); Steinberg and Anfinsen (1952); Steinberg, Vaughan, and Anfinsen (1956); Vaughan and Steinberg (1959).
320 Campbell and Work (1953); Campbell (1960, 1995).
321 Homandberg, Mattis, and Laskowski (1978); Fruton (1982b); Jakubke et al. (1985); Kullmann (1987).
322 Lipmann (1969).
323 Emerson (1945); Stebbins (1972).
324 Dounce (1952); Lipmann (1954), pp. 602–603. See also Dalgliesh (1953).

325 Hoagland, Zamecnik, and Stephenson (1957); Zamecnik (1950, 1979). See also Ogata et al. (1957).
326 Crick (1958), p. 155. See Mazia (1956b); Hoagland (1996).
327 Berg and Ofengand (1958).
328 Hecht et al. (1958); Berg (1961); Berg et al. (1961).
329 Loftfield (1972).
330 Schimmel and Söll (1979); Schimmel (1987); Giegé et al. (1993); Cavarelli and Moras (1993); Kisselev and Wolfson (1994).
331 Monod (1947), p. 224. For comments on the concept of specificity see Judson (1979), p. 608, and Stent (1979), pp. 426–427.
332 Monod (1950), p. 56. See Lwoff and Ullman (1978).
333 Lederberg (1951), p. 280.
334 Monod and Cohn (1952), p. 68.
335 Hogness, Cohn, and Monod (1955); Monod (1956); Cohn (1957). See also Mandelstam (1958).
336 Pardee, Jacob, and Monod (1959); Jacob and Monod (1959), p. 1283. The term *repression* was introduced by Vogel (1957) to distinguish it from *feedback inhibition.*
337 Cohen and Monod (1957).
338 Jacob et al. (1960), pp. 1727, 1729; Jacob and Monod (1961).
339 Pollock (1950, 1951, 1953); Vogel (1957); Yates and Pardee (1957). See Hayes (1968), pp. 712–735.
340 Gilbert and Müller-Hill (1966).
341 Englesberg et al. (1965); K. Ippen et al. (1968); Zubay et al. (1970); Riggs et al. (1971).
342 Yanofsky and Kolter (1982); Beckwith (1987).
343 Cohn (1989), p. 112.
344 Greenstein (1990).
345 Gamow (1954, 1955); Gamow and Ycas (1955). See Kragh (1996).
346 Crick (1958, 1959); Crick, Griffith, and Orgel (1957); Crick et al. (1961). See also Lanni (1964); Woese (1967).
347 Nirenberg and Matthaei (1961), p. 1601. Soon afterward, the specificity of UUU for phenylalanine was questioned; see Bretscher and Grunberg-Manago (1962), who reported the sizable incorporation of leucine.
348 Lengyel, Speyer, and Ochoa (1961); Nirenberg, Matthaei, and Jones (1962); Speyer et al. (1962); Matthaei et al. (1962). See Crick (1962); Judson (1979), pp. 470–489.
349 Nirenberg and Leder (1964); Bernfield and Nirenberg (1965).
350 Jones et al. (1966); Khorana (1968); Agarwal et al. (1972); Itakura et al. (1984). See also Khorana (1989).
351 Crick (1962).
352 Garen (1968).
353 Yanofsky (1963, 1987); Helinski and Yanofsky (1963); Yanofsky et al. (1964).
354 Champe and Benzer (1962); Brody and Yanofsky (1963); Sarabhai et al. (1964); Davies et al. (1964); Casselton (1971).
355 Fox (1987).
356 Grunberg-Manago, Ortiz, and Ochoa (1956); Singer (1958); Heppel et al. (1959). See also Grunberg-Manago (1962, 1989).
357 Rich (1960, 1989); Fresco et al. (1960); Hall and Spiegelman (1961); Nygaard and Hall (1964); Gillespie and Spiegelman (1965); McCarthy and Church (1970); Southern (1975).

358 Jacob and Monod (1961), pp. 346, 349.
359 Volkin and Astrachan (1957).
360 Nomura, Hall, and Spiegelman (1960); Brenner, Jacob, and Meselson (1961); Gros et al. (1961).
361 Weiss (1960); Hurwitz et al. (1960); Furth et al. (1962); Chamberlin and Berg (1962, 1964); Grunberg-Manago (1962).
362 Burgess (1971); Rüger (1972); von Hippel et al. (1984); McClure (1985); Russo and Silhavy (1992); Dombrowski et al. (1996).
363 Roberts (1969).
364 See, for example, Chamberlin et al. (1970); Kondo, Gallerini, and Weissmann (1970).
365 Pabo and Sauer (1984, 1992); Klug and Rhodes (1987); Berg (1990, 1994); Kim et al. (1993); Perutz (1994); Tang et al. (1996).
366 Roesler et al. (1988); Conaway and Conaway (1993); De Pamphilis (1993); Lalli and Sassone-Corsi (1994); Latchman (1995).
367 Padgett et al. (1986); Kolesko and Young (1995); Aso et al. (1995).
368 Sachel and Wahle (1993); Manley (1995).
369 Temin and Mizutani (1970); Baltimore (1970, 1976; 1995); Temin and Baltimore (1972); Temin (1976). For Temin, see Dulbecco (1995).
370 Gilboa et al. (1979).
371 For Central Dogma, see Crick (1966), p. 41; (1970); Watson (1970), pp. 330–331. Davies and Gassen (1983); Caruthers (1985); Wu and Grossman (1987a).
372 Izant and Weintraub (1984); Inouye (1988); Takayama and Inouye (1990); Nielsen et al. (1991); Leonetti et al. (1993).
373 Jackson, Symons, and Berg (1972); Cohen et al. (1973); Lobban and Kaiser (1973); Morrow et al. (1974); Berg (1981). See Singer and Berg (1991), pp. 223–431; Glover and Hames (1995).
374 Johnson (1983); Wengenmeyer (1983).
375 Altman (1971, 1989); Guerrier-Takada et al. (1983).
376 McClain, Guerrier-Takada, and Altman (1987); Pace and Smith (1990); Guerrier-Takada and Altman (1993); Altman et al. (1993); Yarus (1993); Gold et al. (1995); Warnecke et al. (1996).
377 Darr, Brown, and Pace (1992); Zachau et al. (1966). See Burbaum and Schimmel (1991); Cruse et al. (1994).
378 Berget, Moore, and Sharp (1977); Sharp (1987); Berget (1995); Krämer (1996).
379 Singer and Berg (1991), pp. 641–659.
380 Cech et al. (1981); Zaug, Grabowsky, and Cech (1983).
381 Zaug and Cech (1986); Cech (1987, 1990). See Doudna and Szostak (1989); Cate et al. (1996).
382 Belfort et al. (1985); Belfort and Perlman (1995).
383 Saldanha et al. (1993); Ahsen and Schroeder (1993); Leuhrsen et al. (1994); Lorsch and Szostak (1996).
384 Gilbert (1986); Lamond and Gibson (1990); Orgel and Crick (1993); Böhler et al. (1995); Culp and Kitcher (1989).
385 Fruton (1963); Palade (1975); Siekevitz and Zamecnik (1981); Rheinberger (1995).
386 Tissières et al. (1959, 1960); Kurland (1960); Nomura et al. (1960); Cohen and Lichtenstein (1960); Petermann (1964); Kohler et al. (1968); Cohen (1998).
387 Nomura et al. (1969); Traut et al. (1969); Wittmann (1982); Gassen (1982); Giri et al. (1984); Wittmann and Yonath (1988).

388 Nomura (1973).
389 Dintzis (1961).
390 Capecchi (1966); Nomura and Lowry (1967).
391 Wilson and Dintzis (1970).
392 Brawerman (1969); Thach et al. (1969); Kuechler and Rich (1970); Lodish (1970).
393 Nomura (1988); Noller (1991). See also Shine and Dalgarno (1974).
394 Nomura et al. (1984); Hershey (1989); Kozak (1991); Rhoads (1993); Samuel (1993); Dahmus (1994).
395 Lipmann (1969); Lucas-Lenard and Lipmann (1971). See also Nierhaus et al. (1997).
396 Moore (1995); Czworkowski and Moore (1996).
397 Noller (1984, 1991, 1993); Lieberman and Dahlberg (1995).
398 Yarmolinsky and de la Haba (1959); Munro and Marcker (1967); Krayevsky and Kukhanova (1979).
399 Noller, Hoffarth, and Zimniak (1992); Porse and Garrett (1995); Green and Noller (1997).
400 Moldave (1985).
401 Capecchi and Klein (1969); Caskey (1980).
402 Stadtman (1979, 1991); Low and Berry (1996).
403 Fox (1987); Atkins et al. (1992); Herschlag (1995); Gesteland and Atkins (1996).
404 Wold (1981); Yan et al. (1989); Singer and Berg (1997). For the posttranslational methylation or acetylation of histones, see Razin and Kafri (1994); Sternglanz (1996).
405 Stenflo et al. (1974); Swanson and Suttie (1985). The homologous aminomalonic acid (HOOC-CH(NH_2)-COOH) was found by Buskirk et al. (1984) in *E. coli* and atherosclerotic plaque. For the discovery of vitamin K, see Olson (1978); Almquist (1979); Jukes (1980c).
406 Neurath (1975); Steiner et al. (1975).
407 Sabatini and Blobel (1970); Sabatini et al. (1982); Walter, Gilmore, and Blobel (1984); Wickner and Lodish (1985); Lodish (1988); Hartl and Neupert (1990); Lithgow, Glick, and Schatz (1995).
408 Kreil (1994).
409 Hotchkiss and Dubos (1941); Gause and Brezhnikova (1944).
410 Consden, Martin, Gordon, and Synge (1947); Sarges and Witkop (1964).
411 Mach, Reich, and Tatum (1963).
412 Laland and Zimmer (1973).
413 Kleinkauf and Gevers (1969); Lipmann (1973); Aarstad et al. (1980); Döhren and Kleinkauf (1988).
414 Dawson et al. (1994). See also Kent (1988).
415 Fersht et al. (1984); Leatherbarrow and Fersht (1987); Smith (1985); Shortle and Botstein (1985); Wu and Grossman (1987).

9. The Whole and Its Parts

1 Bayliss (1906); Adolph (1961).
2 Bayliss and Starling (1902); Starling (1905), p. 340. See Bayliss (1924), pp. 712–713; Simmer (1978); Wright (1978). For Bayliss and Starling, see Bayliss (1961) and Hill (1969). See also Jorpes and Mutt (1973).
3 Pavlov (1902); Popielski (1901); Wertheimer and Lepage (1901). A French translation of Pavlov's book (Russian edition, 1897) appeared in 1901.
4 Sherrington (1906). See Brazier (1987); Clarke and Jacyna (1987); Finger (1994). For Sherrington, see Granit (1966).

5 Bernard (1853), p. 92.
6 Rolleston (1936); Medvei (1982); Wilson (1984); Wellbourn (1990).
7 Bernard (1859), vol. 2, p. 412. See Bange (1991).
8 Brown-Séquard and d'Arsonval (1891), p. 266. For Brown-Séquard, see Berthelot (1898); Deloume (1939); Olmsted (1946).
9 Borell (1976a, 1976b, 1976c); Tattersall (1995).
10 Oliver and Schäfer (1895); Davenport (1982b); Borell (1978); Weisser (1984).
11 Stolz (1904).
12 For Berthold, see Klein (1970); for Sandström, see Breimer and Sourander (1981); for Gley, see Grmek (1990).
13 Baumann (1895); Oswald (1899); Kendall (1919); Harington and Barger (1927). For Harington, see Himsworth and Pitt-Rivers (1972); Quastel (1977).
14 Pitt-Rivers (1978). For Pitt-Rivers, see Tata (1990, 1994).
15 Sawin (1988). See also Bornhauer (1951).
16 Li (1956, 1957); McCann (1988); Sowers (1980); Reichlin (1986); Wieland and Bodanszky (1991), pp. 136–190.
17 Yalow (1978). For Berson, see Rall (1990).
18 Addison (1855). See Thorn (1968).
19 Britton (1930); Gaunt and Eversole (1949).
20 Kendall (1971); Tait and Tait (1988).
21 Butenandt (1979).
22 See Jones et al. (1992) for a valuable set of articles dealing with the history of steroid chemistry during 1930–1960 and the role of pharmaceutical companies in that development. See also Oudshoorn (1990); Djerassi (1992); Perone (1993).
23 Linser (1966); Karlson (1963, 1982); Höxtermann (1994).
24 Gayon (1994); Karlson (1996).
25 See, for example, Ammon and Dirscherl (1938).
26 Scharrer and Scharrer (1937); Bargmann and Scharrer (1951). For Ernst Scharrer, see Bargmann (1966) and Scharrer (1975). See also Meites et al. (1975, 1978).
27 Cannon (1926). This article is reprinted in full in Langley (1973), pp. 246–249.
28 Cannon (1929, 1932). See Perlman (1977).
29 Robin and Verdeil (1853), vol. 1, p. 15; Bernard (1859), vol. 1, p. 42; (1872), pp. 181–182. See Holmes (1963b, 1963c); Grmek (1967b), pp. 117–150; Schiller (1967b), pp. 172–200; Robin (1979).
30 Bernard (1878–1879), vol. 1, p. 113.
31 Haldane and Priestley (1905); Haldane (1917, 1935). For Haldane, see Douglas (1936).
32 Barcroft and Haldane (1902); Barcroft (1908, 1914, 1934). For Barcroft, see Roughton and Kendrew (1949); Franklin (1953); Holmes (1969).
33 Henderson (1909, 1913, 1917, 1928). See Parascandola (1971a, 1971b); Edsall (1985a); Fry (1996).
34 Henderson (1935); Russett (1966); Cross and Albury (1987).
35 Wiener (1948, 1953, 1955).
36 Freudenthal (1976), p. 376. See Sarkar (1996).
37 Hoskins (1949); Moore and Price (1932). See Sawin (1988), pp. 190–193. For Moore, see Price (1974); Foreman (1992).
38 Machin (1964).

39 Gots (1950, 1957).
40 Yates and Pardee (1956); Umbarger (1956, 1961, 1992).
41 Umbarger (1956), p. 848.
42 See Brock (1990), pp. 190–194.
43 Pardee (1959); Krebs (1956, 1957).
44 Wolstenholme and O'Connor (1959), pp. 354–364.
45 Changeux (1961, 1963).
46 Gerhart and Pardee (1962, 1963).
47 Gerhart and Schachman (1965, 1968).
48 Lahue and Schachman (1986); Kantrowitz and Lipscomb (1988).
49 Monod and Jacob (1961), p. 391; italics in original. In 1954 Aaron Novick and Leo Szilard reported that tryptophan inhibits the formation of an unknown precursor in the synthesis of that amino acid by *E. coli*. In my opinion, the implication that their contribution was comparable in importance to that of Umbarger is deplorable. For a valuable account, see Creagar and Gaudillière (1996).
50 Monod, Changeux, and Jacob (1963), pp. 320–321. See Monod (1966).
51 Wyman and Allen (1951), p. 515. See also Wyman (1949, 1963, 1964, 1972); Wyman and Gill (1990). For Wyman, see Edsall (1980b, 1985b, 1990).
52 Monod, Wyman, and Changeux (1965), p. 89. For a reconsideration of the allosteric model, see Galzi et al. (1996).
53 Atkinson (1965); Atkinson et al. (1965, 1967); Koshland, Nemethy, and Filmer (1966); Koshland (1970); Hammes and Wu (1974).
54 Koshland (1963); Koshland and Neet (1968).
55 Stadtman (1966); Schachman (1988); Barford and Johnson (1989); Johnson (1992); Johnson and Barford (1994).
56 Wyman (1979), p. 223.
57 Edsall (1985b), pp. 160–161; (1990).
58 Perutz (1970).
59 Phillips (1972); Fruton (1974); James et al. (1982). For an extended discussion of the questions raised by the Monod-Wyman-Changeux model, see Debru (1987), pp. 281–328. See also Rosenberg (1985), p. 81; Fruton (1992a), pp. 120–125.
60 Sutherland (1972); Sutherland and Rall (1958, 1960); Lipkin, Cook, and Markham (1959); Sutherland et al. (1965); Robison et al. (1971); Jost and Rickenberg (1971). For Sutherland, see Cori (1978). See also Woolf (1975); Sinding (1996).
61 Cori (1931).
62 Appelman et al. (1973).
63 Krebs and Fischer (1962); Walsh, Perkins, and Krebs (1968).
64 Villar-Palasi and Larner (1970).
65 Perkins (1973); Peterkofsky et al. (1993).
66 Rodbell et al. (1971).
67 Gilman (1987); Casey and Gilman (1988); Taussig and Gilman (1995).
68 Wedegaertner et al. (1995); Clapham (1996).
69 Hall (1990); Bourne et al. (1991); Kaziro et al. (1991); Coleman and Sprang (1996).
70 Emala et al. (1994).

71 Green (1951).
72 Price et al. (1945); Cori (1946); Colowick et al. (1947).
73 Stadie (1954); Hechter (1955), p. 331.
74 Perlmann (1955), p. 25.
75 Burnett and Kennedy (1954); Rabinowitz and Lipmann (1960). Lipmann's interest in phosphoproteins stemmed from his association with P. A. Levene; in 1932, they reported the isolation of O-phosphoryl serine from a hydrolysate of casein.
76 Langan (1973).
77 Taborsky (1974), pp. 174–186.
78 Krebs (1994). See, for example, Hershey (1989); Taylor (1989); Johnson and Barford (1994).
79 Edelman et al. (1987); Wek (1994); Newton (1995); Finkelman et al. (1996). For calmodulin, see Cheung (1982); Kretsinger et al. (1986); O'Neil and DeGrado (1990).
80 Heizmann (1991); Bell and Burns (1991).
81 Kennelly and Krebs (1991); Kemp et al. (1994).
82 Hunter and Cooper (1985).
83 Daum et al. (1994); Graves et al. (1995); Huang and Ferrell (1996).
84 Cadena and Gill (1992); Schaller et al. (1993); Clark et al. (1994).
85 Goldberg et al. (1973); Greengard (1978); Garbers and Lowe (1994). Hokin (1984); Berridge (1987, 1995); Carafoli (1987); Majerus et al. (1988). See also Hannun (1994); Moolenaar (1995).
86 Ichihara (1933); Harris (1946); Norberg (1950); Singer and Fruton (1957); Meyer and Weinmann (1957); Glomset (1959).
87 Krebs and Beavo (1979); Cohen (1989); Cohen and Cohen (1989).
88 Fischer et al. (1991); Stone and Dixon (1994); Dixon (1995).
89 Lübke et al. (1976); Parascandola (1986); Roth (1973, 1988).
90 Brunton (1871); Fraser (1872); Fränkel (1901). See Bynum (1970).
91 Ehrlich and Morgenroth (1900); Ehrlich (1900); Langley (1905, 1906); Clark (1933, 1938).
92 Langley (1905), pp. 400–401; (1908), p. 297.
93 Straub (1912); Gley (1929); Bayliss (1924), p. 733. See also Barger and Dale (1910); Barger (1914).
94 Pincus and Thimann (1948), p. 4. See Jensen (1992).
95 Scatchard (1949); Hummel and Dreyer (1962); Fairclough and Fruton (1966).
96 Levine and Goldstein (1955).
97 Denton (1986); White and Kahn (1994); Hubbard et al. (1994).
98 Schwyzer (1975, 1995).
99 Cato et al. (1992).
100 Elliott (1905); Dale (1914); Loewi (1921); Dale and Dudley (1929). For Elliott, see Dale (1961); for Loewi, see Dale (1962). For Dale, see Feldberg (1970); Smith (1995); Tansey (1995). For Dudley, see Dale (1935).
101 This classification was proposed by Dale (1914). See Durell and Garland (1969); Changeux (1981); Karlin et al. (1983); Waser (1986); Hulme et al. (1990).
102 Monaghan et al. (1989); DeLorey and Olsen (1992).
103 Garbers (1989); Garbers and Lowe (1994).
104 Levi-Montalcini (1987, 1988); Cohen and Carpenter (1975); Carpenter and Cohen (1990).
105 Ainscough and Brodie (1995). See Partington (1962), p. 253.
106 Furchgott and Zawadski (1980); Palmer et al. (1987); Furchgott (1996).

107 Moncada et al. (1991).

108 Garthwaite (1991); Bredt and Snyder (1994).

109 Stuehr and Griffith (1992); Marletta (1993); Nathan and Xie (1994); Stone and Marletta (1994). For isozymes, see Markert and Whitt (1968); Markert et al. (1976).

110 Prabhakar et al. (1995).

111 Vane (1990); Amin et al. (1995).

112 O'Dell et al. (1991); Kandel and Abel (1995); Bailey et al. (1996).

113 Tower (1994). For Thudichum, see Sudhoff (1932); Drabkin (1958); Lambert (1963); Schulte (1984); McIlwain (1990); Dupré (1993); Sourkes (1993a, 1995).

114 Schlesinger (1981).

115 Nagata (1994); Vaux and Strasser (1996); Grimm et al. (1996).

116 Wilson (1923), pp. 284–286; Whitehead (1925), pp. 214–215; Hardy (1936), p. 899; Novikoff (1945), pp. 214–215; Lwoff (1962), pp. 99–100; Jacob (1970), p. 323; Mayr (1988), p. 14; Gros (1993), pp. 293–294. See Fruton (1992a), pp. 63–64, 107–110. See also Schaffner (1967); Goodfield (1975); Roll-Hansen (1979); Rosenberg (1989).

117 Gros (1993), p. 294.

Acc. Chem. Res.	*Accounts of Chemical Research*
Acta Chem. Scand.	*Acta Chemica Scandinavica*
Acta Cryst.	*Acta Crystallographica*
Acta Hist. Sci. Nat. Med.	*Acta Historica Scientarum Naturalium et Medicinalium*
Adv. Carb. Chem.	*Advances in Carbohydrate Chemistry and Biochemistry*
Adv. Cycl. Nucl. Res.	*Advances in Cyclic Nucleotide Research*
Adv. Enzymol.	*Advances in Enzymology*
Adv. Immunol.	*Advances in Immunology*
Adv. Protein Chem.	*Advances in Protein Chemistry*
Agr. Hist.	*Agricultural History*
Am. Chem. J.	*American Chemical Journal*
Am. J. Bot.	*American Journal of Botany*
Am. J. Physiol.	*American Journal of Physiology*
Am. J. Sociol.	*American Journal of Sociology*
Am. Nat.	*American Naturalist*
Am. Phil. Soc. Year Book	*American Philosophical Society Year Book*
Am. Sci.	*American Scientist*
Am. Soc. Rev.	*American Sociological Review*
Anal. Proc.	*Analytical Proceedings*
Anat. Anz.	*Anatomischer Anzeiger*
Anat. Rec.	*Anatomical Record*
Ang. Chem. (Int. Ed.)	*Angewandte Chemie (International Edition)*
Ann.	*Annalen der Pharmacie* (1832–39); *Annalen der Chemie und Pharmacie* (1840–73); *Justus Liebigs Annalen der Chemie* (1873–)
Ann. Acad. Roy. Belg.	*Annuaire de l'Académie Royale . . . de Belgique*
Ann. Chim.	*Annales de Chimie* (1789–1815); *Annales de Chimie et de Physique* (1816–1914)
Ann. Inst. Pasteur	*Annales de l'Institut Pasteur*
Ann. Missouri Bot. Garden	*Annals of the Missouri Botanical Garden*
Ann. N.Y. Acad. Sci.	*Annals of the New York Academy of Sciences*
Ann. Phys.	(Poggendorf's) *Annalen der Physik und Chemie*
Ann. Rev. Biochem.	*Annual Review of Biochemistry*
Ann. Rev. Biophys.	*Annual Review of Biophysics and Bioengineering*
Ann. Rev. Genetics	*Annual Review of Genetics*
Ann. Rev. Immunol.	*Annual Review of Immunology*
Ann. Rev. Med.	*Annual Review of Medicine*
Ann. Rev. Microbiol.	*Annual Review of Microbiology*
Ann. Rev. Neurosci.	*Annual Review of Neuroscience*
Ann. Rev. Pharmacol.	*Annual Review of Pharmacology and Toxicology*
Ann. Rev. Phys. Chem.	*Annual Review of Physical Chemistry*

Ann. Rev. Physiol.	*Annual Review of Physiology*
Ann. Rev. Plant Physiol.	*Annual Review of Plant Physiology*
Ann. Sci.	*Annals of Science*
Ann. Sci. Nat.	*Annales des Sciences Naturelles*
Arch. Anat. Micr.	*Archives d'Anatomie Microscopique*
Arch. Anat. Physiol.	(Müllers) *Archiv für Anatomie, Physiologie, und wissenschaftliche Medizin*
Arch. Biochem. Biophys.	*Archives of Biochemistry and Biophysics*
Arch. Biol.	*Archives de Biologie*
Arch. exp. Path. Pharm.	*Archiv für experimentelle Pathologie und Pharmakologie*
Arch. Fisiol.	*Archivio di Fisiologia*
Arch. Int. Hist. Sci.	*Archives Internationales d'Histoire des Sciences*
Arch. mikr. Anat.	*Archiv für mikroskopische Anatomie*
Arch. Zool. Exp. Gen.	*Archives de Zoologie Expérimentale et Générale*
Austral. J. Chem.	*Australian Journal of Chemistry*
Austral. J. Exp. Biol.	*Australian Journal of Experimental Biology*
Austral. J. Sci	*Australian Journal of Science*
Bact. Revs.	*Bacteriological Reviews*
Bayer. Akad. Wiss. Jahrbuch	*Bayerische Akademie der Wissenschaften Jahrbuch*
Beitr. chem. Physiol.	*Beiträge zur chemischen Physiologie und Pathologie; Zeitschrift für die gesamte Biochemie*
Ber. bot. Ges.	*Berichte der deutschen botanischen Gesellschaft*
Ber. chem. Ges.	*Berichte der deutschen chemischen Gesellschaft*
Ber. Wissen.	*Berichte zur Wissenschaftsgeschichte*
Biochem. J.	*Biochemical Journal*
Biochem. Soc. Bull.	*Biochemical Society Bulletin*
Biochem. Z.	*Biochemische Zeitschrift*
Biochim. Biophys. Acta	*Biochimica et Biophysica Acta*
Biog. Mem. FRS	*Biographical Memoirs of Fellows of the Royal Society of London*
Biog. Mem. NAS	*Biographical Memoirs, National Academy of Sciences of the United States*
Biol. Bull.	*Biological Bulletin*
Biol. Revs.	*Biological Reviews*
Biol. Z.	*Biologisches Zentralblatt*
Biophys. Chem.	*Biophysical Chemistry*
Bot. Z.	*Botanische Zeitung*
Brit. J. Exp. Pathol.	*British Journal of Experimental Pathology*
Brit. J. Hist. Sci.	*British Journal for the History of Science*
Brit. J. Phil. Sci.	*British Journal for the Philosophy of Science*
Brit. Med. J.	*British Medical Journal*
Bull. Acad. Med.	*Bulletin de l'Académie Nationale de Médecine*
Bull. Hist. Chem.	*Bulletin for the History of Chemistry*
Bull. Hist. Med.	*Bulletin of the History of Medicine*

Bull. Soc. Chim.	*Bulletin de la Société Chimique de France*
Bull. Soc. Chim. Biol.	*Bulletin de la Société Chimie Biologique*
Can. J. Biochem.	*Canadian Journal of Biochemistry*
Carlsberg Lab. Comm.	*Carlsberg Laboratory Communications*
Chem. Ber.	*Chemische Berichte*
Chem. Biol.	*Chemistry and Biology*
Chem. Brit.	*Chemistry in Britain*
Chem. Eng. News	*Chemical and Engineering News*
Chem. Ind.	*Chemistry and Industry*
Chem. Intell.	*Chemical Intelligencer*
Chem. Revs.	*Chemical Reviews*
Chem. Soc. Revs.	*Chemical Society Reviews*
Chem. Z.	*Chemiker Zeitung*
Cold Spring Harbor Symp.	*Cold Spring Harbor Symposia on Quantitative Biology*
Compt. Rend.	*Comptes Rendus Hebdomadaires des Séances de l'Académie des Sciences, Paris*
Compt. Rend. Carlsberg	*Comptes Rendus des Travaux du Laboratoire Carlsberg*
Compt. Rend. Soc. Biol.	*Comptes Rendus . . . de la Société de Biologie*
Crit. Revs. Biochem.	*Critical Reviews in Biochemistry*
Curr. Top. Cell Reg.	*Current Topics in Cell Regulation*
Deutsche med. Wchschr.	*Deutsche medizinische Wochenschrift*
DSB	*Dictionary of Scientific Biography*
Erg. Enzymforsch.	*Ergebnisse der Enzymforschung*
Erg. Physiol.	*Ergebnisse der Physiologie*
Eur. J. Biochem.	*European Journal of Biochemistry*
FASEB J.	*FASEB Journal*
FEBS J.	*FEBS Journal*
FEBS Lett.	*FEBS Letters*
Fed. Proc.	*Federation Proceedings*
Helv. Chim. Acta	*Helvetica Chimica Acta*
Hist. Phil. Life Sci.	*History and Philosophy of the Life Sciences*
Hist. Sci.	*History of Science*
Hist. Stud. Phys. Biol. Sci.	*Historical Studies in the Physical and Biological Sciences*
Int. J. Dev. Biol.	*International Journal of Developmental Biology*
Int. J. Peptide Res.	*International Journal of Peptide and Protein Research*
Irish J. Med. Sci.	*Irish Journal of Medical Science*
J. Agr. Res.	*Journal of Agricultural Research*
Jahrb. wiss. Bot.	*Jahrbücher für wissenschaftliche Botanik*
Jahres-Ber.	*Jahres-Bericht über die Fortschritte der Chemie*
J. Am. Chem. Soc.	*Journal of the American Chemical Society*
J. Am. Med. Assn.	*Journal of the American Medical Association*
Jap. Stud. Hist. Sci.	*Japanese Studies in the History of Science*
J. Bact.	*Journal of Bacteriology*
J. Biochem.	*Journal of Biochemistry* (Tokyo)

J. Biol. Chem.	*Journal of Biological Chemistry*
J. Botany	*Journal of Botany*
J. Cell Biol.	*Journal of Cell Biology*
J. Cell Sci.	*Journal of Cell Science*
J. Chem. Ed.	*Journal of Chemical Education*
J. Chem. Soc.	*Journal of the Chemical Society*
J. Chromat.	*Journal of Chromatography*
J. Exp. Med.	*Journal of Experimental Medicine*
J. Exp. Zool.	*Journal of Experimental Zoology*
J. Gastroenter.	*Journal of Gastroenterology and Hepatology*
J. Genetics	*Journal of Genetics*
J. Heredity	*Journal of Heredity*
J. Hist. Biol.	*Journal of the History of Biology*
J. Hist. Med.	*Journal of the History of Medicine and Allied Sciences*
J. Hist. Neurosci.	*Journal of the History of the Neurosciences*
J. Hygiene	*Journal of Hygiene*
J. Immunol.	*Journal of Immunology*
J. Lab. Clin. Med.	*Journal of Laboratory and Clinical Medicine*
J. Med. Biog.	*Journal of Medical Biography*
J. Membrane Biol.	*Journal of Membrane Biology*
J. Neurochem.	*Journal of Neurochemistry*
J. Nutrition	*Journal of Nutrition*
Johns Hopkins Med. J.	*Johns Hopkins Medical Journal*
J. Org. Chem.	*Journal of Organic Chemistry*
J. Path. Bact.	*Journal of Pathology and Bacteriology*
J. Pharm. Exp. Ther.	*Journal of Pharmacology and Experimental Therapeutics*
J. Phys. Chem. Hist. Nat.	*Journal de Physique, Chimie, Histoire Naturelle et des Arts*
J. Physiol.	*Journal of Physiology*
J. prakt. Chem.	*Journal für praktische Chemie*
J. Roy. Soc. Med.	*Journal of the Royal Society of Medicine*
J. Soc. Hist.	*Journal of Social History*
J. Theor. Biol.	*Journal of Theoretical Biology*
Klin. Wchschr.	*Klinische Wochenschrift*
Koll. Z.	*Kolloid Zeitschrift*
Med. chem. Unt.	*Medicinisch-chemische Untersuchungen* (Tübingen)
Med. Hist.	*Medical History*
Med. Hist. J.	*Medizin-Historisches Journal*
Med. Welt	*Medizinische Welt*
Mem. Acad. Stanislas	*Mémoires de l'Académie de Stanislas*
Microbiol. Revs.	*Microbiological Reviews*
Microchem. J.	*Microchemical Journal*
Mol. Cell. Biochem.	*Molecular and Cellular Biochemistry*
Mon. Chem.	*Monatshefte für Chemie*
Münch. med. Wchschr.	*Münchener medizinische Wochenschrift*

Naturwiss.	*Naturwissenschaften*
Naturw. Rund.	*Naturwissenschaftliche Rundschau*
New Eng. J. Med.	*New England Journal of Medicine*
Not. Acad. Sci.	*Notes et Discours, Académie des Sciences Paris*
Notes Roy. Soc.	*Notes and Records of the Royal Society London*
NTM	*Schriftenreihe für Geschichte der Naturwissenschaften, Technik und Medizin*
Obit. Not. FRS	*Obituary Notices of Fellows of the Royal Society of London*
Persp. Biol. Med.	*Perspectives in Biology and Medicine*
Persp. Sci.	*Perspectives on Science*
Pflügers Arch.	*Pflügers Archiv für die gesamte Physiologie des Menschen und der Tiere*
Pharmacol. Revs.	*Pharmacological Reviews*
Pharm. Hist.	*Pharmacy in History*
Phil. Mag.	*Philosophical Magazine*
Phil. Sci.	*Philosophy of Science*
Phil. Soc. Sci.	*Philosophy of the Social Sciences*
Phil. Trans.	*Philosophical Transactions of the Royal Society of London*
Physiol. Revs.	*Physiological Reviews*
PNAS	*Proceedings of the National Academy of Sciences, U.S.A.*
Proc. Am. Acad. Arts Sci.	*Proceedings of the American Academy of Arts and Sciences*
Proc. Am. Phil. Soc.	*Proceedings of American Philosophical Society*
Proc. Chem. Soc.	*Proceedings of the Chemical Society*
Proc. Roy. Inst.	*Proceedings of the Royal Institution*
Proc. Roy. Soc.	*Proceedings of the Royal Society of London*
Proc. Roy. Soc. Edin.	*Proceedings of the Royal Society of Edinburgh*
Proc. Soc. Exp. Biol. Med.	*Proceedings of the Society for Experimental Biology and Medicine*
Proc. Welch Conf.	*Proceedings of the Robert A. Welch Conferences*
Prog. Biophys.	*Progress in Biophysics and Molecular Biology*
Prog. Nucleic Acid Res.	*Progress in Nucleic Acid Research and Molecular Biology*
Quart. Rev. Biol.	*Quarterly Review of Biology*
Quart. Revs. Chem. Soc.	*Quarterly Reviews of the Chemical Society*
Schw. med. Wchschr.	*Schweizerische medizinische Wochenschrift*
Schw. Z. allgem. Path. Bakt.	*Schweizerische Zeitschrift für die allgemeine Pathologie und Bakteriologie*
Sci. Am.	*Scientific American*
Sci. Mon.	*Scientific Monthly*
Sci. Prog.	*Science Progress*
Skand. Arch. Physiol.	*Skandinavisches Archiv für Physiologie*
Soc. Stud. Sci.	*Social Studies of Science*
Stud. Hist. Biol.	*Studies in History of Biology*
Stud. Hist. Phil. Sci.	*Studies in History and Philosophy of Science*
Sudhoffs Arch.	*Sudhoffs Archiv für Geschichte der Medizin und Naturwissenschaften*
Symp. Soc. Exp. Biol.	*Symposia of the Society for Experimental Biology*
Symp. Soc. Gen. Microbiol.	*Symposia of the Society for General Microbiology*

Tech. Cult.	*Technology and Culture*
TIBS	*Trends in Biochemical Sciences*
Trans. Conn. Acad. Arts Sci.	*Transactions of the Connecticut Academy of Arts and Sciences*
Trans. Faraday Soc.	*Transactions of the Faraday Society*
Trans. N.Y. Acad. Sci.	*Transactions of the New York Academy of Sciences*
Trudy Inst. Ist. Est.	*Trudy Instituta Istorii Estestvozania i Tekhniki*
Verhandl. Nat. Heidelberg	*Verhandlungen des Naturhistorisch-Medizinischen Vereins zu Heidelberg*
Verhandl. path. Ges.	*Verhandlungen der deutschen pathologischen Gesellschaft*
Verhandl. Schw. Nat. Ges.	*Verhandlungen der Schweizerischen Naturforschenden Gesellschaft*
Viert. Nat. Ges. Zurich	*Vierteljahrschrift des Naturforschenden Gesellschaft in Zürich*
Virchows Arch.	*Archiv für pathologische Anatomie und Physiologie und klinische Medizin*
Vit. Horm.	*Vitamins and Hormones*
Wiener klin. Wchschr.	*Wiener klinische Wochenschrift*
Yale J. Biol. Med.	*Yale Journal of Biology and Medicine*
Z. allgem. Physiol.	*Zeitschrift für allgemeine Physiologie*
Z. ang. Chem.	*Zeitschrift für angewandte Chemie* (after 1947, *Angewandte Chemie*)
Z. anorg. Chem.	*Zeitschrift für anorganische und allgemeine Chemie*
Z. Bakt. Parasit.	*Zentralblatt für Bakteriologie, Parasitenkunde . . .*
Z. Biol.	*Zeitschrift für Biologie*
Z. Bot.	*Zeitschrift für Botanik*
Z. Elektrochem.	*Zeitschrift für Elektrochemie*
Z. ges. inn. Med.	*Zeitschrift für die gesamte innere Medizin*
Z. Immun.	*Zeitschrift für Immunitätsforschung und experimentelle Therapie*
Z. klin. Med.	*Zeitschrift für klinische Medizin*
Z. Naturforsch.	*Zeitschrift für Naturforschung*
Z. physik. Chem.	*Zeitschrift für physikalische Chemie*
Z. Physiol.	*Zentralblatt für Physiologie*
Z. physiol. Chem.	(Hoppe-Seyler) *Zeitschrift für physiologiche Chemie* (now *Biological Chemistry Hoppe-Seyler*)
Z. Unt. Nahrung.	*Zeitschrift für Untersuchung der Nahrungsund Genussmittel*
Z. wiss Zool.	*Zeitschrift für wissenschaftliche Zoologie*

Aarstad, K., Zimmer, T. L., and Laland, S. G. (1980). The fidelity of gramicidin S synthetase. *Eur. J. Biochem.* 112:335–338.

Abderhalden, E. (1931, 1932). Physiologische Chemie als selbstständiges Fach. *Der Biologe* 1:159–162; 2:166–167.

Abderhalden, E., and Komm, E. (1924). Ueber die Anhydridstruktur der Proteine. *Z. physiol. Chem.* 139:181–204.

Abderhalden, E., and Rona, P. (1905). Ueber die Verwertung der Abbauprodukte des Caseins im tierischen Organismus. *Z. physiol. Chem.* 44:198–205.

Abel, J. J. (1926). Arthur Robertson Cushny and pharmacology. *J. Pharm. Exp. Ther.* 27:265–286.

Abelous, E., and Gérard, E. (1899). Sur la coexistence d'une diastase réductrice et d'une diastase oxydante dans les organes animaux. *Compt. Rend,* 129:1023–1025.

Abercrombie, M. (1961). Ross Granville Harrison. *Biog. Mem. FRS* 7:111–126.

Abir-Am, P. (1980). From biochemistry to molecular biology: DNA and the acculturated critic of science Erwin Chargaff. *Hist. Phil. Life Sci.* 2:3–60.

——— (1982). The discourse of physical power and biological knowledge in the 1930s: A reappraisal of the Rockefeller Foundation's "policy" in molecular biology. *Soc. Stud. Sci.* 12:341–382.

——— (1984). Beyond deterministic sociology and apologetic history: Reassessing the impact of research policy upon new scientific disciplines. *Soc. Stud. Sci.* 14:252–263.

——— (1985). Themes, genres, and orders of legitimation in the consolidation of new scientific disciplines: Deconstructing the historiography of molecular biology. *Hist. Sci.* 23:73–117.

——— (1987). Synergy or clash: Disciplinary and marital strategies in the careers of the mathematical biologist Dorothy Wrinch. In *Uneasy Careers and Intimate Lives* (P. G. Abir-Am and D. Outram, eds.), pp. 239–280. New Brunswick, N.J.: Rutgers University Press.

——— (1995). "New" trends in the history of molecular biology. *Hist. Stud. Phys. Biol. Sci.* 26:167–196.

Abrahams, J. P., Leslie, G. W., Lutter, R., and Walker, J. E. (1994). Structure at 2.8 Å resolution of F_1-ATPase from bovine heart mitochondria. *Nature* 370:621–628.

Achard, T. (1969). *Der Physiologe Friedrich Bidder.* Zurich: Juris.

Achinstein, P. (1994). Jean Perrin and molecular reality. *Persp. Sci.* 2:396–427.

Ackerknecht, E. H. (1953). *R. Virchow: Doctor, Statesman, Anthropologist.* Madison: University of Wisconsin Press.

——— (1988). Friedrich Theodor Althoff und the deutschen Universitäten von 1900. *Schw. med. Wchschr.* 118:812–813.

Ackermann, D. (1931a). Physiologische Chemie als selbstständiges Fach: Neugestaltung des physiologisch-chemischen Unterwelts. *Klin. Wchschr.* 10:175–176, 1135–1136.
——— (1931b). Ueber den biologischen Abbau des Arginin zu Citrullin. *Z. physiol. Chem.* 203:66–69.
Ackermann, R. J. (1985). *Data, Instruments, and Theory.* Princeton: Princeton University Press.
Adam, P. (1911). Louis Edouard Grimaux. *Bull. Soc. Chim.* [4]9:i–xxxvi.
Adams, M. B. (1968). The founding of population genetics: Contributions of the Chetverikov school. *J. Hist. Biol.* 1:23–39.
——— (1978). Nikolai Ivanovich Vavilov. *DSB* 15:505–513.
——— (1990a). Boris Lvovich Astaurov. *DSB* 17:35–39.
——— (1990b). Sergei Sergeevich Chetverikov. *DSB* 17:155–165.
——— (1990c). Iurii Aleksandrovich Filipchenko. *DSB* 17:297–303.
——— (1990d). Aleksandr Ivanovich Oparin. *DSB* 18:695–700.
——— (ed.) (1994). *The Evolution of Theodosius Dobzhansky.* Princeton: Princeton University Press.
Adams, R. (1939). Wallace Hume Carothers. *Biog. Mem. NAS* 20:293–309.
——— (1952). William Albert Noyes. *Biog. Mem. NAS* 27:179–208.
Addison, T. (1855). *On the Constitutional and Local Effects of Disease of the Suprarenal Capsules.* London: Highley.
Adman, E. T. (1991). Copper protein structures. *Adv. Prot. Chem.* 42:145–195.
Adolph, E. F. (1961). Early concepts of physiological regulation. *Physiol. Revs.* 41:737–770.
Aducco, V. (1911). Angelo Mosso. *Rend. Accad. Lin.* [5]20:841–878.
Agarwal, K. L., Yamazaki, A., Cashion, P. J., and Khorana, H. G. (1972). Chemical synthesis of polynucleotides. *Ang. Chem. (Int. Ed.)* 11:451–459.
Agate, F. J. (1975). Philip Edward Smith. *DSB* 12:472–477.
Ahmadjian, V. (1993). *The Lichen Symbiosis*, 2d ed. New York: Wiley.
Ahrens, F. B. (1902). Das Gährungsproblem. *Sammlung chemischer und chemischtechnischer Vorträge* 7:445–495.
Ahsen, U. V., and Schroeder, R. (1993). RNA as a catalyst: Natural and designed ribozymes. *BioEssays* 15:299–307.
Ainscough, E. W., and Brodie, A. M. (1995). Nitric oxide: Some old and new perspectives. *J. Chem. Ed.* 72:686–692.
Ainsworth, C. (1976). *Introduction to the History of Mycology.* Cambridge: Cambridge University Press.
Akers, H. A., and Dromgoodle, E. V. (1982). Ornithine isn't just for the birds anymore! *TIBS* 7:156–157.
Alberts, B., and Shine, K. (1994). Scientists and the integrity of research. *Science* 266:1660–1661.
Albury, W. R. (1977). Experiment and explanation in the physiology of Bichat and Magendie. *Stud. Hist. Biol.* 1:47–131.
Alexander, J. (1931). Some intercellular aspects of life and disease. *Protoplasma* 14:296–306.
Allchin, D. (1996). Cellular and theoretical chimeras: Piecing together how cells process energy. *Stud. Hist. Phil. Sci.* 27:31–41.
Allen, G. (1957). Reflexive catalysis, a possible mechanism of molecular duplication in prebiological evolution. *Am. Nat.* 91:65–78.
Allen, G. E. (1969). Hugo de Vries and the reception of the "mutation theory." *J. Hist. Biol.* 2:55–87.
——— (1973). Clarence Erwin McClung. *DSB* 8:586–590.
——— (1974). Opposition to the Mendelian chromosome theory: The physiological and developmental genetics of Richard Goldschmidt. *J. Hist. Biol.* 7:49–92.

——— (1978). *Thomas Hunt Morgan: The Man and His Science.* Princeton: Princeton University Press.

Allen, R. S., and Murlin, J. R. (1925). Biuret-free insulin. *Am. J. Physiol.* 75:131–139.

Allen, W., and Pepys, W. H. (1808). On the changes produced in atmospheric air, and oxygen gas, by respiration. *Phil. Trans.* 98:249–281.

Alloway, J. L. (1932). The transformation in vitro of R pneumococci into S forms of different specific types by the use of filtered pneumococcus extracts. *J. Exp. Med.* 55:91–99.

Allsop, C. B., and Waters, W. A. (1947). Thomas Martin Lowry. In *British Chemists* (A. Findlay and W. H. Mills, eds.), pp. 402–418. London: Chemical Society.

Almquist, H. J. (1979). Vitamin K: Discovery, identification, synthesis, functions. *Fed. Proc.* 38:2687–2689.

Althusser, L. (1974). *Philosophie et philosophie spontanée des savants.* Paris: François Maspero.

Altman, S. (1971). Isolation of tyrosine tRNA precursor molecules. *Nature New Biology* 229:19–21.

——— (1989). Ribonuclease P, an enzyme with a catalytic RNA subunit. *Adv. Enzymol.* 62:1–36.

Altman, S., Kirsebom, L., and Talbot, S. (1993). Recent studies of ribonuclease P. *FEBS J.* 7:7–14.

Altmann, R. (1889). Ueber Nucleinsäuren. *Arch. anat. Physiol.*, pp. 524–536.

——— (1890). *Die Elementarorganismen und ihre Beziehung zu den Zellen.* Leipzig: Voit.

——— (1892). Ueber Kernstruktur und Netzstruktur. *Arch. anat. Physiol.*, pp. 223–230.

Ames, B. N., and Hartman, P. E. (1963). The histidine operon. *Cold Spring Harbor Symp.* 28:349–356.

Amin, R. A., Vyas, P., Attur, M., Leszczynska-Piziak, J., Patel, I. R., Weissmann, G., and Abramson, S. B. (1995). The mode of action of aspirin-like drugs: Effect on inducible nitric oxide synthase. *PNAS* 92:7926–7930.

Aminoff, M. J. (1993). *Brown-Séquard: A Visionary of Science.* New York: Raven.

Ammon, R., and Dirscherl, W. (1938). *Fermente, Hormone, und Vitamine.* Leipzig: Thieme.

Amoroso, E. C., and Corner, G. W. (1972). Herbert McLean Evans. *Biog. Mem. FRS* 18:83–186.

Amsterdamska, O. (1987). Medical and bacteriological constraints: Early research on variation in bacteriology. *Soc. Stud. Sci.* 17:657–687.

——— (1993). From pneumonia to DNA: The research career of Oswald T. Avery. *Hist. Stud. Phys. Biol. Sci.* 24:1–40.

Amsterdamski, S. (1992). *Between History and Method: Disputes about the Rationality of Science.* Dordrecht: Kluwer.

Anderson, C. (1993). Genome project goes commercial. *Science* 259:300–302.

Anderson, O. S. (1984). Gramicidin channels. *Ann. Rev. Biochem.* 46:531–548.

Anderson, T. F. (1966). Electron microscopy of phages. In *Phage and the Origins of Molecular Biology* (J. Cairns et al., eds.), pp. 63–78. Plainview, N.Y.: Cold Spring Harbor Laboratory.

André, E. (1932). Histoire du développement de la chimie des corps gras. *Bull. Soc. Chim.* [4]51:1–28, 145–170.

André, G. (1937). Léon Maquenne. *Not. Acad. Sci.* 1:67–90.

Andrewes, C. (1971). Francis Peyton Rous. *Biog. Mem. FRS* 17:643–662.

Anfinsen, C. B. (1973). Principles that govern the folding of peptide chains. *Science* 181:223–230.

Anfinsen, C. B., and Steinberg, D. (1951). Studies on the biosynthesis of ovalbumin. *J. Biol. Chem.* 189:739–744.

Anschütz, R. (1909). Life and work of Archibald Scott Couper. *Proc. Roy. Soc. Edin.* 29:193–273.

——— (1929). *August Kekulé.* Berlin: Verlag Chemie.

Anson, M. L., and Mirsky, A. E. (1930). Hemoglobin, the heme pigments, and cellular respiration. *Physiol. Revs.* 10:506–546.

——— (1931). Protein coagulation and its reversal. *J. Gen. Physiol.* 14:725–732.

Aoyagi, T., and Umezawa, H. (1975). Structures and activities of protease inhibitors of microbial origin. In *Proteases and Biological Control* (E. Reich, D. B. Rifkin, and E. Shaw, eds.), pp. 429–454. Plainview, N.Y.: Cold Spring Harbor Laboratory.

Appelman, M. M., Thompson, W. J., and Russell, T. R. (1973). Cyclic nucleotide phosphodiesterase. *Adv. Cyclic Nucl. Res.* 3:65–98.

Apple, R. D. (1989). Patenting university research: Harry Steenbock and the Wisconsin Alumni Research Foundation. *Isis* 80:375–394.

Applebury, M. L. (1993). Establishing the molecular basis of vision: Hecht and Wald. In *The Biological Century* (R. B. Barlow et al., eds.), pp. 178–201. Woods Hole, Mass.: Marine Biological Laboratory.

Applebury, M. L., and Hargrave, P. A. (1986). Molecular biology of the visual pigments. *Vision Research* 26:1881–1895.

Appleby, J., Hunt, L., and Jacob, M. (1994). *Telling the Truth about History.* New York: Norton.

Araki, T. (1903). Ueber enzymatische Zersetzung der Nukleinsäure. *Z. physiol. Chem,* 38:84–97.

Arber, W. (1978). Restriction endonucleases. *Ang. Chem. (Int. Ed.)* 17:73–79.

——— (1979). Promotion and limitation of genetic exchange. *Science* 205:361–365.

Arber, W., Kellenberger, E., and Weigle, J. (1957). La défectuosité du phage lambda tranducteur. *Schw. Z. allgem. Path. Bakt.* 20:659–665.

Aris, R., Davis, H. T., and Steuwer, R. H. (eds.) (1983). *Springs of Scientific Creativity.* Minneapolis: University of Minnesota Press.

Arkwright, J. A. (1921). Variation in bacterial agglutination both by salts and by specific serum. *J. Path. Bact.* 24:36–60.

Armstrong, G. T. (1964). The calorimeter and its influence on the development of chemistry. *J. Chem. Ed.* 41:297–307.

Armstrong, H. E. (1927). Poor common salt! *Nature* 120:478.

Arnon, D. I. (1955). The chloroplast as a complete photosynthetic unit. *Science* 122:9–16.

——— (1959). Conversion of light into chemical energy in photosynthesis. *Nature* 184:10–21.

——— (1961). Cell-free photosynthesis and the energy conversion process. In *Light and Life* (W. D. McElroy and B. Glass, eds.), pp. 489–569. Baltimore: Johns Hopkins University Press.

——— (1965). Ferredoxin and photosynthesis. *Science* 149:1460–1470.

——— (1967). Photosynthetic activity of isolated chloroplasts. *Physiol. Revs.* 47:317–358.

——— (1984). The discovery of photosynthetic phosphorylation. *TIBS* 9:258–262.

Arnott, S. (1994). John Monteath Robertson. *Biog. Mem. FRS* 39:350–362.

Aronowitz, S., and DiFazio, W. (1994). *The Jobless Future.* Minneapolis: University of Minnesota Press.

Arrhenius, S. (1904). Die Serumtherapie von physikalisch-chemischen Standpunkte. *Z. Elektrochem.* 10:661–664.

——— (1907). *Immunochemistry: The Application of the Principles of Physical Chemistry to the Study of the Biological Antibodies.* New York: Macmillan.

Arthus, M. (1896). *Nature des enzymes.* Paris: Jouve.

Ash, M. G., and Sollner, A. (eds.) (1996). *Forced Migration and Scientific Change: Emigre German-speaking Scientists and Scholars after 1933.* Cambridge: Cambridge University Press.

Ashdown, A. (1928). James Mason Crofts. *J. Chem. Ed.* 5:911–921.

Aso, T., Conaway, J. W., and Conaway, R. C. (1995). The RNA polymerase II elongation complex. *FEBS J.* 9:1419–1428.

Aspray, W. (1990). *John von Neumann and the Origin of Modern Computing.* Cambridge: MIT Press.

Astbury, W. T. (1947a). On the structure of biological fibres and the problem of muscle. *Proc. Roy. Soc.* B134:303–328.

——— (1947b). X-ray studies of nucleic acids. *Symp. Soc. Exp. Biol.* 1:66–76.

Astbury, W. T., and Bell, F. O. (1938). X-ray studies of thymonucleic acid. *Nature* 141:747–748.

Astier, C. (1813). Expériences faites sur le sirop de raisin. *Ann. Chim.* 87:271–285.

Aston, F. W. (1933). *Mass-spectra and Isotopes.* London: Arnold.

Atkins, J. F., Weiss, R. B., Thompson, S., and Gesteland, R. F. (1991). Towards a genetic dissection of the basis of triplet coding, and its natural subversion: programmed reading frame shifts and hops. *Ann. Rev. Gen.* 25:201–228.

Atkinson, D. E. (1965). Biological feedback control at the molecular level. *Science* 150:851–857.

——— (1977). *Cellular Energy Metabolism and Its Regulation.* New York: Academic.

Atkinson, D. E., et al. (1965). Kinetics of regulatory enzymes. Kinetic order of the yeast diphosphopyridine nucleotide isocitrate dehydrogenase reaction and a model for the reaction. *J. Biol. Chem.* 240:2682–2690.

——— (1987). *Dynamic Models in Biochemistry.* Menlo Park, Colo.: Benjamin.

Atkinson, J. W. (1985). E. G. Conklin on evolution: The popular writings of an embryologist. *J. Hist. Biol.* 18:31–50.

Aubry, J. (1984–1986). Victor Grignard. *Mem. Acad. Stanislas* [7]13/14:247–261.

Auerbach, C. (1949). Chemical mutagenesis. *Biol. Revs.* 24:355–391.

——— (1967). The chemical production of mutations. *Science* 168:1141–1147.

Auerbach, C., Robson, J. M., and Carr, J. G. (1947). The chemical production of mutations. *Science* 105:243–247.

Auhagen, E. (1932). Co-Carboxylase, ein neues Co-Enzym der alkoholischen Gärung. *Z. physiol. Chem.* 204:149–167.

Auld, D. S., and Vallee, B. L. (1987). Carboxypeptidase A. In *Hydrolytic Enzymes* (A. Neuberger and K. Brocklehurst, eds.), pp. 201–255. Amsterdam: Elsevier.

Aulie, R. P. (1970). Boussingault and the nitrogen cycle. *Proc. Am. Phil. Soc.* 114:435–479.

Austoker, J., and Bryder, L. (1989). *Historical Perspectives on the Role of the MRC.* Oxford: Oxford University Press.

Avery, O. T., MacLeod, C. M., and McCarty, M. (1944). Studies on the nature of the substance inducing transformation of pneumococcal types. Induction of transformation by a desoxyribonucleic acid fraction isolated from Pneumococcus Type III. *J. Exp. Med.* 79:137–158.

Ayala, F. J. (1972). The evolutionary thought of Teilhard de Chardin. In *Biology, History, and Natural Philosophy* (A. D. Brock and W. Yourgrau, eds.), pp. 207–216. New York: Plenum.

Babior, B. M. (1992). The respiratory burst oxidase. *Adv. Enzymol.* 65:49–65.

Babkin, B. P. (1949). *Pavlov.* Chicago: University of Chicago Press.

Bach, A. (1911). Zur Kenntnis der Reduktionsfermente. I. Mitteilung. Ueber das Schardinger-enzym (Perhydridase). *Biochem. Z.* 31:443–449.

Bachelard, G. (1932). *Le pluralisme cohérént de la chimie moderne.* Paris: Vrin.

——— (1953). *Le matérialisme rationel.* Paris: PUF.

Bachelard, H. S. (1988). A brief history of neurochemistry in Britain and of the neurochemical group of the British Biochemical Society. *J. Neurochem* 50:992–995.

Bacq, Z. M., and Brachet, J. (1981). Marcel Florkin. *Ann. Acad. Roy. Belg.* 147:41–98.

Baddiley, J. (1955). The structure of coenzyme A. *Adv. Enzymol.* 16:1–21.

Baddiley, J., Thain, E. M., Novelli, D., and Lipmann, F. (1953). The structure of coenzyme A. *Nature* 171:76.

Baeyer, A. (1870). Ueber die Wasserentziehung und ihre Bedeutung für das Pflanzenleben und die Gärung. *Ber. chem. Ges.* 3:63–75.

Baeyer, A., and Villiger, V. (1900). Benzoylwasserstoffsuperoxide und die Oxydation des Benzaldehyds in der Luft. *Ber. chem. Ges.* 33:1569–1585.

Bailar, J. C., Jr. (1970). Moses Gomberg. *Biog. Mem. NAS* 41:141–173.

Bailey, C. H., Bartsch, D., and Kandel, E. (1996). Toward a molecular definition of long-term memory storage. *PNAS* 93:13445–13452.

Bailey, K. (1954). Structure proteins II: Muscle. In *The Proteins* (H. Neurath and K. Bailey, eds,), pp. 951–1055. New York: Academic.

Bainton, D. F. (1981). The discovery of lysosomes. *J. Cell Biol.* 91:66s–76s.

Bak, T. A. (1974). The history of physical chemistry in Denmark. *Ann. Rev. Phys. Chem.* 25:1–10.

Baker, J. R. (1948–1955). The cell theory: A restatement, history, and critique. *Quarterly Journal of Microscopical Science* 89:103–125; 90:87–108; 93:157–190; 94:407–440; 96:449–481.

Baldwin, E. (1940). *Introduction to Comparative Biochemistry,* 2d ed. Cambridge: Cambridge University Press.

Baldwin, M. A., Cohen, F. E., and Prusiner, S. B. (1995). Prion protein isoforms, a convergence of biological and structural investigations. *J. Biol. Chem.* 270:19197–19200.

Baldwin, R. S. (1975). Robert R. Williams. *J. Nutrition* 105:1–14.

Balen, C. M. van (1986). The influence of Johannsen's discoveries on the constraint structure of the Mendelian research program. *Stud. Hist. Phil. Sci.* 17:175–204.

Ball, E. G. (1942). Oxidative mechanisms in animal tissues. In *A Symposium on Respiratory Enzymes,* pp. 16–32. Madison: University of Wisconsin Press.

——— (1974). The development of our current concepts of biological oxidations. *Mol. Cell. Biochem.* 5:35–46.

Baltimore, D. (1970). Viral RNA–dependent DNA polymerase. *Nature* 226:1209–1211.

——— (1976). Viruses, polymerases, and cancer. *Science* 192:632–636.

——— (1995). Discovery of the reverse transcriptase. *FASEB J.* 9:1660–1665.

Baltzer, F. (1962). *Theodor Boveri.* Stuttgart: Wissenschaftliche Verlagsgesellschaft. English trans. 1967, Berkeley: University of California Press.

Bange, C. (1991). Les glandes a sécretion interne d'après Claude Bernard: Naissance, diffusion et postérité d'un concept. In *La nécessité de Claude Bernard* (J. Michel, ed.), pp. 83–108. Paris: Méridien Klincksieck.

——— (1995). Le role des faits expérimentaux et les concepts dans l'élaboration de la connaissance scientifique, selon le physiologiste Eugène Gley. In *Les Savants et l'epistémologie vers la fin du XIXe siècle* (M. Panza and J. C. Pont, eds.), pp. 245–262. Paris: Blanchard.

Barbara, R. A., and Huang, L. (1995). Reconstitution of mammalian DNA replication. *Prog. Nucleic Acid Res.* 51:93–122.

Barber, A. W. (ed.) (1992). *The Photosystems.* Amsterdam: Elsevier.

Barber, B. (1952). *Science and the Social Order.* New York: Free Press.

Barber, B., and Hirsch, W. (eds.) (1962). *The Sociology of Science.* New York: Free Press.

Barcroft, J. (1908). Zur Lehre vom Blutgaswechsel in den verschiedenen Organen. *Erg. Physiol.* 7:699–794.

——— (1914). *The Respiratory Function of the Blood.* Cambridge: Cambridge University Press.

——— (1934). *Features in the Architecture of Physiological Function.* Oxford: Oxford University Press.

Barcroft, J., and Haldane, J. S. (1902). A method of estimating the oxygen and carbonic acid in small amounts of blood. *J. Physiol.* 28:232–240.

Barendrecht, H. P. (1904). Enzymwirkung. *Z. physiol. Chem.* 49:456–482.

Barford, D., and Johnson, L. N. (1989). The allosteric transition of glycogen phosphorylase. *Nature* 340:609–616.

Barger, G. (1914). *The Simpler Natural Bases.* London: Longmans, Green.

Barger, G., and Coyne, F. P. (1928). The amino-acid methionine; constitution and synthesis. *Biochem. J.* 22:1417–1425.

Barger, G., and Dale, H. H. (1910). Chemical structure and sympathomimetic action of amines. *J. Physiol.* 41:19–59.

Bargmann, W. (1966). Ernst Scharrer. *Anat. Anz.* 119:119–127.

Bargmann, W., and Scharrer, E. (1951). The site of origin of hormones of the posterior pituitary. *Am. Sci.* 39:255–259.

Barkan, D. K. (1992). A usable past: Creating disciplinary space for physical chemistry. In *The Invention of Physical Science* (M. J. Nye et al., eds.), pp. 175–202. Dordrecht: Kluwer.

——— (1994). Simply a matter of chemistry? The Nobel Prize for 1920. *Persp. Sci.* 2:357–395.

Barker, G. F. (1902). Frederick Augustus Genth. *Biog. Mem. NAS* 4:201–231.

Barlow, R. B., et al. (eds.) (1993). *The Biological Century.* Woods Hole, Mass.: Marine Biological Laboratory.

Barnes, B. (1982). *T. S. Kuhn and Social Science.* London: Macmillan.

Barnes, S. B., and Dolby, R. G. A. (1970). The scientific ethos: A deviant viewpoint. *Archives Européennes de Sociologie* 11:3–25.

Baron, J., and Polanyi, M. (1913). Ueber die Anwendung des zweiten Hauptsatzes der Thermodynamik auf Vorgänge im tierischen Organismus. *Biochem. Z.* 53:1–20.

Baron, J. H. (1973). One-hundred-and-fifty years of measurements of hydrochloric acid in gastric juice. *Brit. Med. J.* (4):600–601.

Barr, M. L., and Rossiter, R. J. (1973). James Bartram Collip. *Biog. Mem. FRS* 19:235–267.

Barron, E. S. G., and Harrop, G. (1928). Studies on blood metabolism, 2: The effect of methylene blue and other dyes upon the glycolysis and lactic acid formation of mammalian and avian erythrocytes. *J. Biol. Chem.* 74:65–87.

Barsanti, G. (1995). La naissance de la biologie: Observations, théories, métaphysiques en France, 1740–1810. In *Nature, Histoire, Société* (C. Blanckaert et al., eds.), pp. 197–228. Paris: Klincksieck.

Bartholomew, J. R. (1989). *The Formation of Science in Japan.* New Haven: Yale University Press.

Barton, N. R., and Goldstein, L. S. B. (1996). Going mobile: Microtubule motors and chromosome segregation. *PNAS* 93:1735–1742.

Bary, A. de (1879). *Die Erscheinung der Symbiose.* Strassburg: Trübner.

Basslinger, J. (1858). *Pepsin, seine physiologischen Erscheinungen und therapeutische Wirkungen gegen Verdauungsschwäche.* Vienna: Zimarski.

Bates, F. J. (1942). *Polarimetry, Saccharimetry, and the Sugars.* Washington, D. C.: U. S. Government Printing Office.

——— (1973). *Determination of pH. Theory and Practice.* 2d Ed. New York: Wiley.

Bate-Smith, E. C. (ed.) (1964). *Sir William Bate Hardy: Biologist, Physicist, and Food Scientist.* Cambridge: Cambridge University Press.

Bateson, W. (1909). *Mendel's Principles of Heredity.* Cambridge: Cambridge Cambridge University Press.

——— (1913). *Problems of Genetics.* New Haven: Yale University Press.

——— (1926). The mechanism of Mendelian heredity. *Science* 44:536–543.

Bateson, W., and Saunders, E. R. (1902). The facts of heredity in the light of Mendel's Discovery. *Reports to the Evolution Committee of the Royal Society of London* 1:125–160.

Battelli, F., and Stern, L. (1909). Die akzessorische Atmung in den Tiergeweben. *Biochem. Z.* 21:488–509.

——— (1910). Die Katalase. *Erg. Physiol.* 10:531–597.

——— (1911). Die Oxydation der Bernsteinsäure durch Tiergewebe. *Biochem. Z.* 30:172–194.

——— (1912). Die Oxydationsfermente. *Erg. Physiol.* 12:96–268.

——— (1914). Einfluss der mechanischen Zerstörung der Zellstruktur auf die verschiedenen Oxydationsprozesse der Tiergewebe. *Biochem. Z.* 67:443–471.

Battersby, A. R. (1963). The biosynthesis of alkaloids. *Proc. Chem. Soc.*, pp. 189–200.

——— (1994). How nature builds the pigments of life: The conquest of vitamin B_{12}. *Science* 264:1551–1557.

Battersby, A. R., and McDonald, E. (1979). Origins of the pigments of life: The type-III problem in porphyrin biosynthesis. *Acc. Chem. Res.* 12:14–22.

Bauer, F. (1907). Ueber die Konstitution der Inosinsäure und die Muskelpentose. *Beitr. chem. Physiol.* 10:345–357.

Bauer, H. (1954). Paul Ehrlich's influence on chemistry and biochemistry. *Ann. N.Y. Acad. Sci.* 59:150–167.

Baumann, C. (1977). Franz Boll. *Vision Research* 17:1267–1268.

Baumann, E. (1878). *Ueber die synthetische Prozesse im Tierkörper.* Berlin: Hirschwald.

——— (1882). Ueber das von O. Loew und T. Bokorny erbrachten Nachweis der chemischen Ursache des Lebens. *Pflügers Arch.* 29:400–421.

——— (1895). Ueber das Vorkommen von Iod im Thierkörper. *Z. physiol. Chem.* 21:319–330.

Baumann, E., and Kossel, A. (1895). Zur Erinnerung an Felix Hoppe-Seyler. *Z. physiol. Chem.* 21:i–lxi.

Baumann, H. U. (1972). *Ueber Mehrfach Entdeckungen.* Münster: n.p.

Bäumer, B. (1996). *Von der physiologischen Chemie zur frühen biochemischen Arzneimittelforschung.* Stuttgart: Deutscher Apotheker Verlag.

Bawden, F. C., Pirie, N. W., Bernal, J. D., and Fankuchen, I. (1936). Liquid crystalline substances from virus-infected plants. *Nature* 138:1051–1054.

Baxter, A. L. (1976). Edmund B. Wilson as a preformatist: Some reasons for his acceptance of the chromosome theory. *J. Hist. Biol.* 9:29–57.

Bayliss, L. E. (1961). William Maddock Bayliss, 1860–1924: Life and scientific work. *Persp. Biol. Med.* 4:460–479.

Bayliss, W. M. (1906). Die chemische Koordination der Funktion des Körpers. *Erg. Physiol.* 5:664–697.

——— (1908). *The Nature of Enzyme Action.* London: Longmans, Green.

——— (1924). *Principles of General Physiology.* 4th ed. London: Longmans, Green.

Bayliss, W. M., and Starling, E. (1902). The mechanism of pancreatic secretion. *J. Physiol.* 28:325–353.

Beach, E. F. (1961). Beccari of Bologna: The discoverer of vegetable protein. *J. Hist. Med.* 16:354–373.

Beadle, G. W. (1939). Physiological aspects of genetics. *Ann. Rev. Physiol.* 1:41–62.

——— (1945). Biochemical genetics. *Chem. Revs.* 37:15–96.

——— (1951). Chemical genetics. In *Genetics in the 20th Century* (L. C. Dunn, ed.), pp. 221–239. New York: Macmillan.

——— (1963). *Genetics and Modern Society.* Philadelphia: American Philosophical Society.

——— (1969). Genes, chemistry, and the nature of man. In *Biology and the Physical Sciences* (S. Devons, ed.), pp. 1–13. New York: Columbia University Press.

Beadle, G. W., and Ephrussi, B. (1936). The differentiation of eye pigments in Drosophila as studied by transplantation. *PNAS* 21:225–247.

Beadle, G. W., and Tatum, E. L. (1941). Genetic control of biochemical reactions in Neurospora. *PNAS* 27:499–506.

Beale, G. H. (1995). Charlotte Auerbach. *Biog. Mem. FRS* 41:20–42.

Bearn, A. G. (1993). *Archibald Garrod and the Individuality of Man.* Oxford: Clarendon.

Beaumont, W. (1833). *Experiments and Observations on the Gastric Juice and the Physiology of Digestion.* Plattsburg, N.Y.: Allen.

Béchamp, A. J. (1854). De l'action des protosels de fer sur la nitronaphthalene et la nitrobenzine: Nouvelle méthode de formation des bases de Zinin. *Ann. Chim.* [3]42:186–196.

——— (1883). *Les Microzymas.* Paris: Baillière.

Becher, T. (1989). *Academic Tribes and Territories: Intellectual Enquiry and the Cultures of Disciplines.* Milton Keynes, England: Open University Press.

Beckmann, E. (1904). Johannes Wislicenus. *Ber. chem. Ges.* 37:4861–4946.

Beckner, M. (1959). *The Biological Way of Thought.* New York: Columbia University Press.

——— (1971). Organismic biology. In *Man and His Nature* (R. Munson, ed.), pp. 54–61. New York: Dell.

Beckwith, J. (1987). The operon: An historical account. In *Escherichia coli and Salmonella typhimurium* (F. C. Neidhardt, ed.). Washington, D. C.: American Society for Microbiology.

Beer, J. J. (1959). *The Emergence of the German Dye Industry.* Urbana: University of Illinois Press.

Beguin, J. (1624). *Les Elemens de chymie.* 3d ed. Paris: Le Maistre.

Beijerinck, M. W. (1899). Ueber ein Contagium vivum fluidum als Ursache der Fleckenkrankheit der Tabakblätter. *Z. Bakt. Parasit.* [2]5:27–33.

——— (1917). The enzyme theory of heredity. *Proceedings of the Royal Academy of Sciences, Amsterdam* 19:1275–1289.

——— (1921). On different forms of hereditary variation of microbes [1900]. In *Verzamelde Geschriften van M. W. Beijerinck* 4:37–48. Delft: Brinckmann and Obernetter.

Beinert, H., and Stumpf, P. K. (1983). David Ezra Green. *TIBS* 8:434–436.

Belfort, M., Pedersen-Lane, J., Wegt, D., Ehrenman, K., Maley, G., Chu, F., and Maley, F. (1985). Processing of the intron-containing thymidylate synthase (td) gene of phage T4 is at the RNA level. *Cell* 41:375–382.

Belfort, M., and Perlman, P. S. (1995). Mechanisms of intron mobility. *J. Biol. Chem.* 270:30237–30240.

Belitser, V. A., and Tsibakova, E. T. (1939). [The mechanism of phosphorylation associated with respiration]. *Biokhimiya* 4:516–535.

Bell, R., and Burns, D. (1991). Lipid activation of protein kinase C. *J. Biol. Chem.* 266:4661–4664.

Bell, R. P. (1941). *Acid-Base Catalysis.* Oxford: Clarendon Press.

——— (1950). The Brønsted memorial lecture. *J. Chem. Soc.*, pp. 409–419.

Belloni, L. (1982). Hermann Helmholtz and Franz Boll. *Med. Hist. J.* 17:129–137.

Bence-Jones, H. (1867). *Lectures on Some Applications of Chemistry and Mechanics to Pathology and Therapeutics.* London: Churchill.

Bendall, D. S. (1994). Robert Hill. *Biog. Mem. FRS* 40:143–170.

Ben-David, J. (1960). Scientific productivity and academic organization in nineteenth-century Germany. *Am. Soc. Rev.* 25:827–843.

——— (1968). The universities and the growth of science in Germany and the United States. *Minerva* 7:1–35.

——— (1970). The rise and decline of France as a scientific centre. *Minerva* 8:160–179.

——— (1971). *The Scientist's Role in Society: A Comparative Study.* Englewood Cliffs, N.J.: Prentice-Hall.

——— (1991). *Scientific Growth: Essays on the Social Organization and Ethos of Science* (G. Freudenthal, ed.). Berkeley: University of California Press.

Ben-David, J., and Zloczower, A. (1962). Universities and academic systems in modern societies. *European Journal of Sociology* 3:45–84.

Bender, M. L. (1971). *Mechanisms of Homogeneous Catalysis from Protons to Proteins.* New York: Wiley-Interscience.

Bender, M. L., and Kezdy, F. J. (1965). Mechanism of the action of proteolytic enzymes. *Ann. Rev. Biochem.* 34:49–76.

Beneden, E. van (1883). Recherches sur la maturation de l'oeuf et la fécondation. *Arch. Biol.* 4:265–640.

Benfey, T., and Travis, T. (1992). Carl Schorlemmer. The red chemist. *Chem. Ind.*, pp. 441–444.

Benison, S., Barger, A. C., and Wolfe, E. L. (1987). *Walter B. Cannon: The Life and Times of a Young Scientist.* Cambridge: Harvard University Press.

Bensaude-Vincent, B. (1983). Une méthodologie révolutionnaire dans la chimie française. *Ann. Sci.* 30:189–196.

——— (1990). A view of the chemical revolution through contemporary textbooks: Lavoisier, Fourcroy, and Chaptal. *Brit. J. Hist. Sci.* 23:435–460.

——— (1993). *Lavoisier: Mémoires d'une revolution.* Paris: Flammarion.

Bensley, R. R., and Hoerr, N. L. (1934). Studies on cell structure by the freeze-drying method, 5: The chemical basis of the organization of the cell; 6: The preparation and properties of mitochondria. *Anat. Rec.* 60:251–266; 449–455.

Benson, K. R. (1985). American morphology in the late nineteenth century: The biology department at Johns Hopkins University. *J. Hist. Med.* 18:163–205.

Benson, K. R., Maienschein, J., and Rainger, R. (eds.) (1991). *The Expansion of American Biology.* New Brunswick, N.J.: Rutgers University Press.

Bentley, R. (1983). Three-point attachment: Past, present, but no future. *Trans. N. Y. Acad. Sci.* [2]41:5–24.

——— (1995). From optical activity in quartz to chiral drugs: Molecular handedness in biology and medicine. *Persp. Biol. Med.* 38:188–229.

Benton, E. (1974). Vitalism in nineteenth-century scientific thought: A typology and reassessment. *Stud. Hist. Phil. Sci.* 5:17–48.

Benzer, S. (1955). Fine structure of a genetic region in bacteriophage. *PNAS* 41:344–354.

——— (1957). The elementary units of heredity. In *The Chemical Basis of Heredity* (W. D. McElroy and B. Glass, eds.), pp. 70–93. Baltimore: Johns Hopkins University Press.

——— (1959). On the typology of the genetic fine structure. *PNAS* 45:1607–1620.

——— (1961a). Genetic fine structure. *Harvey Lectures* 56:1–21.

——— (1961b). On the typography of the genetic fine structure. *PNAS* 47:403–415.

Benzinger, T., Kitzinger, C., Hems, R., and Burton, K. (1959). Free-energy changes of the glutaminase reaction and the hydrolysis of the terminal pyrophosphate bond of adenosine triphosphate. *Biochem. J.* 71:400–407.

Bérard, J. F. (1817). Essai sur l'analyse des substances animales. *Ann. Chim.* [2]5:290–298.

Berenblum, I. (1935). Experimental inhibition of tumour induction by mustard gas and other compounds. *J. Path. Bact.* 40:549–558.

Beretta, M. (1993). *The Enlightenment of Matter: The Definition of Chemistry from Agricola to Lavoisier.* Canton, Mass.: Science History Publications.

Berezkin, V. G. (1989). Biography of Mikhail Semenovich Tswett and translation of Tswett's preliminary communication on a new category of adsorption phenomena. *Chem. Revs.* 89:279–285.

Berg, J. M. (1990). Zinc fingers and other metal-binding domains. *J. Biol. Chem.* 265:6513–6516.

——— (1994). Zinc finger domains: From prediction to design. *Acc. Chem. Res.* 28:14–19.

Berg, P. (1961). Specificity in protein synthesis. *Ann. Rev. Biochem.* 30:293–324.

——— (1981). Dissections and reconstructions of genes and chromosomes. *Science* 213:296–303.

Berg, P., Bergmann, F. H., Ofengand, E. J., and Dieckmann, M. (1961). The enzymic synthesis of amino acyl derivatives of ribonucleic acid, 1: The mechanism of leucyl-, valyl-, isoleucyl-, and methionyl-ribonucleic acid formation. *J. Biol. Chem.* 236:1726–1734.

Berg, P., and Ofengand, E. J. (1958). An enzymatic mechanism for linking amino acids to RNA. *PNAS* 44:78–86.

Berg, P., and Singer, M. F. (1992). *Dealing with Genes: The Language of Heredity.* Mill Valley, Calif.: University Science Books.

——— (1995). The recombinant DNA controversy: Twenty years later. *PNAS* 92:9011–9013.

Berg, R. L. (1979). The life and work of Boris L. Astaurov. *Quart. Rev. Biol.* 54:397–416.

——— (1990). In defense of Timoféeff-Ressovsky. *Quart. Rev. Biol.* 65:457–479.

Berget, S. M. (1995). Exon recognition in vertebrate splicing. *J. Biol. Chem.* 270:2411–2414.

Berget, S. M., Moore, C., and Sharp, P. A. (1977). Spliced RNA segments at the 5′-terminus of late adenovirus 2 mRNA. *PNAS* 74:3171–3175.

Bergmann, M. (1938). The structure of proteins in relation to biological problems. *Chem. Revs.* 22:423–435.

Bergmann, M., and Fruton, J. S. (1937). On proteolytic enzymes, 13: Synthetic substrates for chymotrypsin. *J. Biol. Chem.* 118:405–415.

——— (1944). The significance of coupled reactions for the enzymatic hydrolysis of proteins. *Ann. N.Y. Acad. Sci.* 45:409–423.

Bergmann, M., Fruton, J. S., and Pollok, H. (1939). The specificity of trypsin. *J. Biol. Chem.* 127:643–648.

Bergmann, M., Schotte, H., and Leschinsky, W. (1923). Ueber die ungesättigte Reduktionsprodukte der Zuckerarten und ihre Umwandlungen. V. Ueber 2-Desoxy-glucose (Glucodesose). *Ber. chem. Ges.* 56:1052–1059.

Bergmann, M., and Zervas, L. (1927). Synthese von Glykocyamin aus Arginin und Glykokoll. *Z. physiol. Chem.* 172:277–288.

——— (1932). Ueber ein allgemeines Verfahren der Peptidsynthese. *Ber. chem. Ges.* 65:1192–1201.

Bergmann, M., Zervas, L., Fruton, J. S., Schneider, F., and Schleich, H. (1935). On proteolytic enzymes, 5: On the specificity of dipeptidase. *J. Biol. Chem.* 109:325–346.

Bergson, H. (1946). *L'Energie spirituelle.* Paris: Alcan.

Berman, A. (1963). Conflict and anomaly in the scientific orientation of French pharmacy, 1800–1873. *Bull. Hist. Med.* 37:440–462.

——— (1974). Théophile-Jules Pelouze. *DSB* 10:499.

——— (1975). Pierre-Jean Robiquet. *DSB* 11:494–495.

Berman, M. (1978). *Social Change and Scientific Organization: The Royal Institution, 1799–1844.* Ithaca, N.Y.: Cornell University Press.

Bernal, J. D. (1932). Crystal structures of vitamin D and related compounds. *Nature* 129:277–278.

——— (1939). *The Social Function of Science.* London: Routledge.

——— (1963). William Thomas Astbury. *Biog. Mem. FRS* 9:1–35.

——— (1968). The material theory of life. *Labour Monthly,* pp. 324–326.

Bernal, J. D., and Crowfoot, D. (1934). X-ray photographs of crystalline pepsin. *Nature* 133:794.

Bernal, J. D., Fankuchen, I., and Perutz, M. F. (1938). An X-ray study of chymotrypsin and hemoglobin. *Nature* 141:523–524.

Bernard, C. (1848). De l'origine du sucre dans l'économie animale. *Archives générales de médecine* 18:303–319.

——— (1853). *Nouvelle fonction du foie considéré comme organe producteur de matière sucrée chez l'homme et les animaux.* Paris: Baillière.

——— (1855a). *Leçons de physiologie expérimentale appliquée a la médecine.* Paris: Baillière.

——— (1855b). Sur le mécanisme de la fonction du sucre dans le foie. *Compt. Rend.* 41:461–469.

——— (1857a). *Leçons sur les effets des substances toxiques et médicamenteuses.* Paris: Baillière.

——— (1857b). Sur le mécanisme physiologique de la formation du sucre dans le foie. *Compt. Rend.* 44:578–586.

——— (1857c). Remarques sur la formation de la matière glycogène du foie. *Compt. Rend.* 44:1326.

——— (1859). *Leçons sur les propriétés physiologiques et les alterations pathologiques des liquides de l'organisme.* Paris: Baillière.

——— (1865). *Introduction a l'etude de la médecine expérimentale.* Paris: Baillière.

——— (1872). *Rapport sur les progrès et la marche de la physiologie générale en France.* Paris: Imprimérie Impériale.

——— (1877a). *Leçons sur le diabète et la glycogenèse animale.* Paris: Baillière.

——— (1877b). Critique expérimentale sur le mécanisme de la formation du sucre dans le foie. *Compt. Rend.* 85:519–525.

——— (1878–1879). *Leçons sur les phénomènes de la vie communs aux animaux et aux végétaux.* Paris: Baillière.

——— (1927). *An Introduction to the Study of Experimental Medicine* (H. C. Greene, trans.). New York: Macmillan.

——— (1965). *Cahier des notes, 1850–1860* (M. D. Grmek, ed.). Paris: Gallimard.

——— (1979). *Notes pour le rapport sur les progrès de la physiologie* (M. D. Grmek, ed.). Paris: Collège de France.

Bernfield, M. R., and Nirenberg, M. W. (1965). RNA code words and protein synthesis. *Science* 147:479–484.

Bernthsen, A. (1912). Heinrich Caro. *Ber. chem. Ges.* 45:1987–2042.

Berridge, M. J. (1987). Inositol triphosphate and diacylglycerol: two interacting second messengers. *Ann. Rev. Biochem.* 56:159–193.

——— (1995). Inositol triphosphate and calcium signalling. *Ann. N.Y. Acad. Sci.* 766:31–43.

Berry, A. J. (1960). *Henry Cavendish.* London: Hutchinson.

Berry, A. J., and Moelwyn-Hughes, E. A. (1963). Chemistry at Cambridge from 1901 to 1910. *Proc. Chem. Soc.*, pp. 357–362.

Berson, J. A. (1996). Erich Hückel, pioneer of organic quantum chemistry: Reflections on theory and experiment. *Ang. Chem. (Int. Ed.)* 35:2750–2764.

Bert, P. (1870). *Leçons sur la physiologie comparée de la respiration.* Paris: Baillière.

Berthelot, M. (1860a). *Chimie organique fondée sur la synthèse.* Paris: Mallet Bachelier.

——— (1860b). Sur la fermentation glucosique du sucre de canne. *Compt. Rend.* 50:980–984.

——— (1875). Sur les principes généraux de la thermochimie. *Ann. Chim.* [5]4:5–131.

——— (1898). La vie et les travaux de Brown-Séquard. *Rev. Sci.* [4]10:801–812.

Bertrand, G. (1894). Recherches sur le latex de l'arbre à laque du Tonkin. *Bull. Soc. Chim.* [3]11:718–721.

——— (1897a). Sur l'intervention du manganèse dans les oxydations provoqués par la laccase. *Compt. Rend.* 124:1032–1035.

——— (1897b). Recherches sur la laccase, nouveau ferment soluble, à propriétés oxydantes. *Ann. Chim.* [7]12:115–140.

Berzelius, J. J. (1806, 1808). *Föreläsingar i Djurkemien.* Stockholm: Delen.

——— (1813). *A View of the Progress and Present State of Animal Chemistry* (G. Brunnmark, trans.). London: Hatchard, Johnson, and Boosey.

——— (1836). Einige Ideen über der Bildung organischer Verbindungen in der lebendigen Natur wirksame, aber nicht bemerkte Kräfte. *Jahresber.* 15:237–245.

——— (1839). Weingährung. *Jahresber.* 18:400–403.

——— (1843). Thierchemie. *Jahresber.* 22:535–538.

——— (1916). *Jac. Berzelius Bref: Correspondence Between Berzelius and Mulder, 1834–1847.* (H. G. Söderbaum, ed.). Uppsala: Almquist and Wiksells.

Beveridge, W. L. B. (1951). *The Art of Scientific Investigation.* New York: Norton.

Bevilacqua, F. (1993). Helmholtz's *Ueber die Erhaltung der Kraft.* In *Hermann Helmholtz and the Foundation of Nineteenth-Century Science* (D. Cahan, ed.), pp. 291–402. Berkeley: University of California Press.

Beyerlein, B. (1991). *Die Entwicklung der Pharmazie als Hochschuldisziplin (1750–1875).* Stuttgart: Weissenschaftliche Verlagsgesellschaft.

Beytia, E. D., and Porter, J. W. (1976). Biochemistry of polyisoprenoid biosynthesis. *Ann. Rev. Biochem.* 45:113–142.

Bezel, R. (1979). *Der Physiologe Adolf Fick.* Zurich: Juris.

Bezkorovainy, A. (1973a). Pediatric research in Russia: The classical period. *Rush-Presbyterian-St. Luke's Hospital Bulletin* 14:315–321.

——— (1973b). Contributions of some early Russian scientists to the understanding of glycolysis. *J. Hist. Med.* 28:388–392.

——— (1974). Carnosine, carnitine, and Vladimir Gulevich. *J. Chem. Ed.* 51:652–654.

——— (1980). *Science and Medicine in Imperial Russia.* Chicago.

Bhaskar, R. (1989). *Reclaiming Reality.* London: Verso.

Bichat, X. (1801). *Anatomie Générale, Appliquée à la Physiologie et à la Médecine.* Paris: Brosson, Gabon et Cie.

Bickel, M. H. (1972). *Marceli Nencki.* Berne: Huber.

Bidder, F., and Schmidt, C. (1852). *Die Verdauungssaefte und der Stoffwechsel.* Mitau: Reyher.

Billig, C. (1994). *Pharmazie und Pharmaziestudium an der Universität Giessen.* Stuttgart: Wissenschaftliche Verlagsgesellschaft.

Billingham, R. E., Brent, L., Sparrow, E. M., and Medawar, P. B. (1954). Quantitative studies on tissue transplantation immunity. *Proc. Roy. Soc.* B143:43–80.

Biltz, W. (1904). Ein Versuch zur Deutung der Agglutinierungsvorgänge. *Z. physik. Chem.* 48:615–623.

Bing, F. C. (1971). The history of the word "metabolism." *J. Hist. Med.* 26:158–180.

——— (1973). Friedrich Bidder and Carl Schmidt. *J. Nutrition* 103:637–648.

Bird, A. P. (1986). CpG-rich islands and the function of DNA methylation. *Nature* 321:209–213.

Birk, Y. (1987). Proteinase inhibitors. In *Hydrolytic Enzymes* (A. Neuberger and K. Brocklehurst, eds.), pp. 257–307. Amsterdam: Elsevier.

Birkinshaw, J. H. (1972). Harold Raistrick. *Biog. Mem. FRS* 18:489–509.

Bischoff, T. L. W. (1853). *Der Harnstoff als Maass des Stoffwechsels.* Giessen: Ricker.

Bischoff, T. L. W., and Voit, C. (1860). *Die Gesetze der Ernährung des Fleischfressers durch Neue Untersuchungen Festgestellt.* Leipzig: Winter.

Bitting, A. W. (1937). *Appertizing or the Art of Canning: Its History and Development.* San Francisco: Trade Pressroom.

Bjerrum, N. (1909). Julius Thomsen. *Ber. chem. Ges.* 42:4971–4988.

Blackburn, E. H. (1990). Telomeres: Structure and synthesis. *J. Biol. Chem.* 265:5919–5921.

Blackburn, E. H., and Greider, C. W. (eds.) (1995). *Telomeres.* Plainview, N.Y.: Cold Spring Harbor Laboratory.

Blaschko, H. (1983). A biochemist's approach to autopharmacology. In *Selected Topics in the History of Biochemistry: Personal Recollections* (G. Semenza, ed.), pp. 189–230. Amsterdam: Elsevier.

Blinks, L. R. (1974). Winthrop John van Leuven Osterhout. *Biog. Mem. NAS* 44:213–249.

Bliss, M. (1982). *The Discovery of Insulin.* Chicago: University of Chicago Press.

——— (1984). *Banting: A Biography.* Toronto: McClelland and Stewart.

Bloch, K. (1965). The biological synthesis of cholesterol. *Science* 150:19–28.

——— (1982). The structure of cholesterol and of the bile acids. *TIBS* 7:334–336.

——— (1987). Summing Up. *Ann. Rev. Biochem.* 56:1–9.

——— (1992). Sterol molecules: Structure, biosynthesis, and function. *Steroids* 57:378–383.

——— (1994). *Blondes in Venetian Paintings, the Nine-Banded Armadillo, and Other Essays in Biochemistry.* New Haven: Yale University Press.

Bloch, K., and Schoenheimer, R. (1941). The biological precursors of creatine. *J. Biol. Chem.* 138:167–194.

Bloch, M. (1954). *The Historian's Craft.* Manchester: Manchester University Press.

Bloch, W. (1991). A biochemical perspective of the polymerase reaction. *Biochemistry* 30:2735–2747.

Bloor, D. (1976). *Knowledge and Social Imagery.* London: Routledge.

Blow, D., Chayen, W. E., Lloyd, L. F., and Saridakes, E. (1994). Control of nucleation of protein crystals. *Protein Science* 3:1638–1643.

Bock, K. D. (1972). *Strukturgeschichte des Assistenten.* Düsseldorf: Bertelsmann.

Bode, W., and Huber, R. (1992). Natural proteinase inhibitors and their interaction with proteinases. *Eur. J. Biochem.* 204:433–451.

Bodländer, G. (1895). Moritz Traube. *Ber. chem. Ges.* 28(4):1085–1108.

——— (1899). Ueber langsame Verbrennung. *Sammlung chemischer und chemischtechnischer Vorträge* 3:385–488.

Bodo, G. (1955). Crystalline cytochrome c from the king penguin. *Nature* 176:829–830.

Boerhaave, H. (1735). *Elements of Chemistry* (T. Dallowe, trans.). London: Pemberton, Clarke, Miller, and Gray.

Bogert, M. T. (1931). Charles Frederick Chandler. *Biog. Mem. NAS* 14:127–181.

Böhler, C., Nielsen, P. E., and Orgel, L. E. (1995). Template switching between PNA and RNA oligonucleotides. *Nature* 376:578–581.

Bohley, P. (1987). Intracellular proteolysis. In *Hydrolytic Enzymes* (A. Neuberger and K. Brocklehurst, eds.), pp. 307–332. Amsterdam: Elsevier.

Bohr, C. (1904). Theoretische Behandlung der quantitativen Verhältnisse bei der Sauerstoffaufnahme des Hämoglobins. *Z. Physiol.* 17:682–688.

Bohr, N. (1933). Light and life. *Nature* 131:421–423, 457–459.

Boivin, A. (1947). Directed mutation in colon bacilli, by an inducing principle of desoxyribonucleic acid nature: Its meaning for the general biochemistry of heredity. *Cold Spring Harb. Symp.* 11:7–17.

Boivin, A., Vendrely, R., and Vendrely, C. (1948). L'Acide desoxyribonucleique du noyau cellulaire dépositaire des charactères héréditaires: D'Ordre analytique. *Compt. Rend.* 226:1061–1063.

Bokland, U. (1975). Carl Wilhelm Scheele. *DSB* 12:143–150.

Bokorny, T. (1922). Zum Nachweis von aktivem Eiweiss. *Z. allgem. Physiol.* 20:74–84.

Boll, F. (1876). Zur Anatomie und Physiologie der Retina. *Monatsberiche der Akademie der Wissenschaften Berlin* 23:783–787.

——— (1877). Zur Anatomie und Physiologie der Retina. *Arch. Anat. Physiol.*, pp. 4–36. English trans. by R. Hubbard (1977), *Vision Research* 17:1253–1265.

Bond, J. S., and Butler, P. E. (1987). Intracellular proteases. *Ann. Rev. Biochem.* 56:333–364.

Bondzynski, S., and Humnicki, V. (1896). Ueber das Schicksal des Cholestrins im thierischen Organismus. *Z. physiol. Chem.* 22:396–410.

Bonner, D. M. (1946). Biochemical mutations in Neurospora. *Cold Spring Harb. Symp.* 11:14–24.

——— (1956). The genetic unit. *Cold Spring Harb. Symp.* 21:163–170.

Bonner, J., and Arreguin, B. (1949). The biochemistry of rubber formation in the guayule. *Arch. Biochem.* 21:109–124.

Bonner, T. N. (1953). *American Doctors and German Universities.* Lincoln: University of Nebraska Press.

Bonnier, G. (1886). Recherches expérimentales sur la synthèse des lichens dans un milieu privé des germes. *Compt. Rend.* 103:942–944.

Booth, H., and Hey, M. J. (1996). DNA before Watson and Crick: The pioneering studies of J. M. Gulland and D. O. Jordan at Nottingham. *J. Chem. Ed.* 73:928–931.

Bopp, F. (1849). Einiges über Albumin, Casein, und Fibrin. *Ann.* 69:16–35.

Bordet, J. (1925). Le problème de l'autolyse microbienne ou du bactériophagie. *Ann. Inst. Pasteur* 39:717–763.

Borell, M. (1976a). Organotherapy, British physiology, and discovery of the internal secretions. *J. Hist. Biol.* 9:235–268.

——— (1976b). Brown-Séquard's organotherapy and its appearance in America at the end of the nineteenth century. *Bull. Hist. Med.* 50:309–320.

——— (1976c). Origins of the Hormone Concept: Internal Secretions and Physiological Research, 1889–1905. Ph.D. diss., Yale University.

——— (1978). Setting the standards for a new science: Edward Schäfer and endocrinology. *Med. Hist.* 22:282–290.

——— (1987). Marey and d'Arsonval: The exact sciences in late nineteenth-century French medicine. *Acta Hist. Sci. Nat. Med.* 39:225–237.

Born, G. V. R. (1984). The effect of the scientific environment in Britain on refugee scientists from Germany and their effects on science in Britain. *Ber. Wissen.* 7:129–143.

Bornhauer, S. (1951). *Zur Geschichte der Schilddrüsen und Kropftforschung im 19. Jahrhundert.* Aarau: Sauerländer.

Borscheid, P. (1976). *Naturwissenschaft, Staat und Industrie in Baden, 1848–1914.* Stuttgart: Ernst Klett.

Borsook, H., and Dubnoff, J. W. (1940). Creatine formation in liver and in kidney. *J. Biol. Chem.* 134:635–639.

——— (1941). The formation of glycocyamine in animal tissues. *J. Biol. Chem.* 138:389–403.

——— (1947). The role of oxidation in the methylation of guanidoacetic acid. *J. Biol. Chem.* 171:363–375.

Borsook, H., and Keighley, G. L. (1935). The "continuing" metabolism of nitrogen in animals. *Proc. Roy. Soc.* B118:488–521.

Borsook, H., and Schott, H. F. (1931). The role of the enzyme in the succinate-enzyme-fumarate equilibrium. *J. Biol. Chem.* 92:535–557.

Bos, L. (1995). The embryonic beginning of virology: Unbiased thinking and dogmatic stagnation. *Arch. Virol.* 140:613–619.

Botting, N. P. (1995). Chemistry and neurochemistry of the kynurenine pathway of tryptophan metabolism. *Chem. Soc. Revs.* 24:401–412.

Bouchardat, A., and Sandras, C. M. (1842). Recherches sur la digestion. *Ann. Sci. Nat.* 18:225–241.

Boudon, R. (1990). *L'Art de persuader des idées fausses, fragiles ou douteuses.* Paris: Fayard.

Bougault, J., and Hérissey, H. (1921). *Notice sur la vie et les travaux de Emile Bourquelot.* Paris: Société Générale d'Imprimérie et d'Edition.

Bourdieu, P. (1984). *Homo Academicus.* Paris: Editions de Minuit.

Bourne, H. R., Sanders, D. A., and McCormick, F. (1991). The GTPase superfamily: Conserved structure and molecular mechanism. *Nature* 349:117–127.

Bourquelot, E. (1896). *Les ferments solubles (diastases-enzymes).* Paris: Société Editions Scientifiques.

Boussingault, J. B. (1838). Recherches chimiques sur la végétation, entreprises dans le but d'examiner si les plantes prennent de l'azote de l'atmosphère. *Ann. Chim.* [2]67:5–54; 69:353–367.

——— (1839). Analyses comparées des aliments consommés et les produits rendus par une vache laitière: Recherches entreprises dans le but d'examiner si les animaux herbivoires empruntent de l'azote à l'atmosphère. *Ann. Chim.* [2]71:113–127.

——— (1854). Recherches sur la végétation: Dans le but d'examiner si les plantes fixent dans leur organisme l'azote qui est à l'état gazeux dans l'atmosphère. *Compt. Rend.* 38:580–606; 39:601–613.

——— (1868). Etudes sur les fonctions des feuilles. *Ann. Chim.* [4]13:282–416.

Boveri, T. (1904). *Ergebnisse über die Konstitution der chromatischen Substanz des Zellkerns.* Jena: Fischer.

Bowen, E. J. (1965). Chemistry in Oxford: The development of the university laboratories. *Chem. Brit.* 1:517–520.

Bowers, J. Z. (1972). *Western Medicine in a Chinese Palace: Peking Union Medical College, 1917–1951*. New York: Josiah Macy Jr. Foundation.

Bowie, N. E. (ed.) (1994). *University-Business Partnerships.* Lanham, Md: Rowman and Littlefield.

Bowker, G., and Latour, B. (1987). A booming discipline short of a discipline: (Social) studies of science in France. *Soc. Stud. Sci.* 17:715–748.

Bowler, P. J. (1989). *The Mendelian Revolution: The Emergence of Hereditarian Concepts in Modern Science and Society.* Baltimore: Johns Hopkins University Press.

Boyde, T. R. C. (1977). Roots of "enzyme." *Nature* 269:194.

——— (1980). *Foundation Stones of Biochemistry.* Hong Kong: Voile et Aviron.

Boyer, P. D. (1977). Conformational coupling in oxidative phosphorylation and photophosphorylation. *TIBS* 2:38–41.

——— (1981). An autobiographical sketch related to my efforts to understand oxidative phosphorylation. In *Of Oxygen, Fuels, and Living Matter,* part 1 (G. Semenza, ed.), pp. 229–264. Chichester: Wiley.

——— (1993). The binding change mechanism in ATP synthase. Some probabilities and possibilities. *Biochim. Biophys. Acta* 1140:215–250.

——— (1997). The ATP synthase: A splendid molecular machine. *Ann. Rev. Biochem.* 66:717–749.

Boyer, P. D., Chance, B., Ernster, L., Mitchell, P., Racker, E., and Slater, E. C. (1977). Oxidative phosphorylation and photophosphorylation. *Ann. Rev. Biochem.* 46:955–1026.

Boyer, P. D., Falcone, A. B., and Harrison, W. H. (1954). Reversal and mechanism of oxidative phosphorylation. *Nature* 174:401–402.

Boyland, E. (1954). Mutagens. *Pharmacol. Revs.* 6:345–364.

Boyle, R. (1680). *The Sceptical Chymist,* 2d ed. London: Hall.

Brachet, A. (1923). Notice sur Edouard van Beneden. *Ann. Acad. Roy. Belg.* 89:167–242.

Brachet, J. (1941a). La détection histochimique et la microdosage des acides pentose-nucléiques. *Enzymologia* 10:87–96.

——— (1941b). La localisation des acides pentose-nucléiques dans les tissus animaux et les oeufs d'amphibiens en voie de développement. *Arch. Biol.* 53:207–256.

——— (1942). Le role des acides nucléiques dans la cellule. *Annales de la Société Royale Zoologique de Belgique* 733:93–100.

——— (1947). The metabolism of nucleic acids. *Cold Spring Harb. Symp.* 12:18–27.

——— (1988). Albert Claude. *Ann. Acad. Roy. Belg.* 165:93–135.

Bradbury, S. (1967). *Evolution of the Microscope.* Oxford: Pergamon.

Bradley, J. (1955). On the operational interpretation of classical chemistry. *Brit. J. Phil. Sci.* 6:32–42.

——— (1992). *Before and After Cannizzaro.* Leatheronwheel: Whittles.

Bragg, W. L. (1975). *The Development of X-ray Analysis* (D. C. Phillips and H. F. Lipson, eds.). New York: Hafner.

Bragg, W. L., Kendrew, J. C., and Perutz, M. F. (1950). Polypeptide configurations in crystalline proteins. *Proc. Roy. Soc.* A203:321–357.

Brammill, D., and Kornberg, A. (1988). Duplex opening by dnaA protein at novel sequences in initiation of replication of the *E. coli* chromosome. *Cell* 52:743–755.

Brand, E., Saidel, L. J., Goldwater, W. H., Kassell, B., and Ryan, F. J. (1945). *J. Am. Chem. Soc.* 67:1524–1531.

Branden, C. I., and Jones, T. A. (1990). Between objectivity and subjectivity. *Nature* 343:687–689.

Brassley, P. (1994). Agricultural research in Britain, 1850–1914: Failure, success, and development. *Ann. Sci.* 52:465–480.

Braun, D. (1993). Biomedical research in a period of scarcity: The United States and Great Britain. *Minerva* 31:268–290.

Braunstein, A. E. (1939). Die enzymatische Umaminierung der Aminosäuren und ihre physiologische Bedeutung. *Enzymologia* 7:25–52.

——— (1982). An eventful trail or research on pathways and enzymes of intermediary nitrogen metabolism. In *Oxygen, Fuels, and Living Matter,* part 2 (G. Semenza, ed.), pp. 251–313. Chichester: Wiley.

Braunstein, A. E., and Kritzmann, M. G. (1937). Ueber den Abund Aufbau von Aminosäuren durch Umaminierung. *Enzymologia* 2:129–146.

Brawerman, G. (1969). Role of initiation factors in the translation of messenger RNA. *Cold Spring Harb. Symp.* 34:307–312.

Brazier, M. A. B. (1987). *A History of Neurophysiology in the 19th Century.* New York: Raven.

Breathnach, C. S. (1966). George Gabriel Stokes on the function of haemoglobin. *Irish J. Med. Sci.* 6:121–125.

——— (1972). The development of blood gas analysis. *Med. Hist.* 16:51–62.

Bredig, G., and Fiske, P. S. (1912). Durch Katalysatoren bewirkte asymmetrische Synthese. *Biochem. Z.* 46:7–23.

Bredig, G., and Sommer, F. (1909). Die Schardinger-Reaktion und ähnliche Enzymkatalysen. *Z. physik. Chem.* 70:34–65.

Bredt, D. S., and Snyder, S. H. (1994). Nitric oxide: A physiologic messenger molecule. *Ann. Rev. Biochem.* 63:175–195.

Breimer, L., and Sourander, P. (1981). The discovery of the parathyroid glands in 1880: Triumph and tragedy of Ivar Sandström. *Bull. Hist. Med.* 55:558–563.

Breinl, F., and Haurowitz, F. (1930). Chemische Untersuchung des Präzipitates aus Hämoglobin und Anti-Hämoglobin Serum und Bemerkungen über die Natur der Antikörper. *Z. physiol. Chem.* 192:45–57.

Bremer, J. (1977). Carnitine and its role in fatty acid metabolism. *TIBS* 2:207–209.

Brenner, M., Müller, H. R., and Pfister, R. W. (1950). Eine neue enzymatische Peptidsynthese. *Helv. Chim. Acta* 33:568–591.

Brenner, S., Jacob, F., and Meselson, M. (1961). An unstable intermediate carrying information from genes to ribosomes for protein synthesis. *Nature* 190:576–581.

Breslow, R. (1995). Biomimetic chemistry and artificial enzymes: Catalysis by design. *Acc. Chem. Res.* 28:146–153.

Breslow, R., and Chapman, W. H. (1996). On the mechanism of action of ribonuclease A: Relevance of enzymatic studies with *p*-nitrophenylphosphate ester and a thiophosphate ester. *PNAS* 93:10018–10021.

Bretscher, M. S., and Grunberg-Manago, M. (1962). Polynucleotide-directed protein synthesis using an *E. coli* cell–free system. *Nature* 195:283–284.

Brewer, F. M. (1957). The place of chemistry at Oxford. *Proc. Chem. Soc.* pp. 185–189.

Bricmont, J. (1996). Science of chaos or chaos of science? *Ann. N.Y. Acad. Sci.* 775:131–175.

Bridges, C. B. (1916). Nondisjunction as proof of the chromosome theory of heredity. *Genetics* 1:1–52, 107–163.

——— (1935). Salivary chromosome maps. *J. Heredity* 26:60–64.

——— (1936). "The Bar Gene" a duplication. *Science* 83:210–211.

Brieger, G. H. (1974). Elie Metchnikoff. *DSB* 9:331–335.

Briggs, G. E. (1948). Frederick Frost Blackman. *Obit. Not. FRS* 5:651–657.

Briggs, G. E., and Haldane, J. B. S. (1925). A note on the kinetics of enzyme action. *Biochem. J.* 19:388–389.

Brigl, P. (1931). William Küster. *Ber. chem. Ges.* 64A:15–36.

Brillouin, L. (1962). *Science and Information Theory*, 2d ed. New York: Academic.

Britton, S. W. (1930). Adrenal insufficiency and related considerations. *Physiol. Revs.* 10:617–682.

Broad, W. J. (1980). History of science losing its science. *Science* 207:389.

Brock, T. D. (1988). *Robert Koch, a Life in Medicine and Bacteriology.* Madison, Wis.: Science Tech.

——— (1990). *The Emergence of Bacterial Genetics.* Plainview, N.Y.: Cold Spring Harbor Laboratory.

Brock, W. H. (1965). The life and work of William Prout. *Med. Hist.* 9:101–126.

——— (ed.) (1967). *The Atomic Debates.* Leicester: Leicester University Press.

——— (1970). Francis William Aston. *DSB* 1:320–322.

——— (1972). Liebig's laboratory accounts. *Ambix* 19:47–58.

——— (1981). Liebigiana: Old and new perspectives. *Hist. Sci.* 19:201–218.

——— (ed.) (1984). *Justus von Liebig und August Wilhelm Hofmann in ihren Briefen.* Weinheim: Verlag Chemie.

——— (1985). *From Protyle to Proton: William Prout and the Nature of Matter.* Bristol: Hilger.

——— (1986). The British Association committee on chemical symbols, 1834: Edward Turner's letter to British chemists and a reply by William Prout. *Ambix* 33:33–42.

——— (1990). Liebig, Gregory, and the British Association, 1837–1842. *Ambix* 37:134–147.

——— (1997). *Justus von Liebig: The Chemical Gatekeeper.* Cambridge: Cambridge University Press.

Brock, W. H., et al. (eds.) (1981). *John Tyndall.* Dublin: Royal Dublin Society.

Brocke, B. vom (ed.) (1991a). *Wissenschaftsgeschichte und Wissenschaftspolitik im Industriezeitalter: Das "System Althoff" in historischer Perspektive.* Hildesheim: Lax.

——— (1991b). Friedrich Althoff: A great figure in higher education in Germany. *Minerva* 29:269–293.

Brocklehurst, K., Willenbrock, F., and Salih, E. (1987). Cysteine proteinases. *Hydrolytic Enzymes* (A. Neuberger and K. Brocklehurst, eds.), pp. 39–158. Amsterdam: Elsevier.

Brodie, B. C. (1812). Further experiments and observations on the actions of poisons on the animal system. *Phil. Trans.* 102:205–227.

Brodie, T. G. (1910). Some new forms of apparatus for the analysis of the gases of the blood by the chemical method. *J. Physiol.* 39:391–396.

Brody, S., and Yanofsky, C. (1953). Suppressor gene alteration of protein primary structure. *PNAS* 50:9–16.

Brooke, J. H. (1968). Wöhler's urea, and its vital force: A verdict from the chemists. *Ambix* 15:84–114.

——— (1971). Organic synthesis and the unification of chemistry: A reappraisal. *Brit. J. Hist. Sci.* 5:363–392.

——— (1975). Laurent, Gerhardt and the philosophy of chemistry. *Hist. Stud. Phys. Sci.* 6:405–429.

——— (1976). Charles Adolphe Wurtz. *DSB* 14:529–532.

——— (1981). Avogadro's hypothesis and its fate: A case-study in the failure of case-studies. *Hist. Sci.* 19:235–273.

——— (1987). Methods and methodology in the development of organic chemistry. *Ambix* 34:147–155.

Brookes, P. (1990). The early history of the biological alkylating agents, 1918–1968. *Mutation Research* 233:3–14.

Brooks, N. M. (1995). Nikolai Zinin at Kazan University. *Ambix* 42:129–142.

Brooks, S. C., and Brooks, M. M. (1941). *Permeability of Living Cells.* Berlin: Bornträger.

Broomhead, J. M. (1951). The structure of pyrimidines and purines, 4: The crystal structure of guanine hydrochloride and its relation to that of adenine hydrochloride. *Acta Cryst.* 4:92–100.

Brown, A. J. (1902). Enzyme action. *J. Chem. Soc. (Trans.)* 81:373–388.

Brown, C. M. (1989). *Benjamin Silliman.* Princeton: Princeton University Press.

Brown, D. M., and Todd, A. R. (1952). Nucleotides, 10: Some observations on the structure and chemical behaviour of the nucleic acids. *J. Chem. Soc.* pp. 52–58.

——— (1955). Nucleic acids. *Ann. Rev. Biochem.* 24:311–338.

Brown, J. R. (1984). *Scientific Rationality: The Sociological Turn.* Dordrecht: Reidel.

——— (1989). *The Rational and the Social.* London: Routledge.

——— (1991). Latour's prosaic prose. *Canadian Journal of Philosophy* 21:245–261.

——— (1994). *Smoke and Mirrors: How Science Reflects Reality.* London: Routledge.

Brown, R., and Danielli, J. F. (eds.) (1954). *Active Transport and Secretion.* (Symp. Soc. Exp. Biol., vol. 8). Cambridge: Cambridge University Press.

Brown, S. C. (1979). *Benjamin Thompson, Count Rumford.* Cambridge: MIT Press.

Brown, S. W. (1966). Heterochromatin. *Science* 151:417–425.

Brown-Séquard, C. E., and d'Arsonval, J. A. (1891). Additions à une note sur l'injection des extraits liquides de divers organes, comme méthode thérapeutique. *Compt. Rend. Soc. Biol.* [9]3:265–268.

Brücke, E. T. (1928). *Ernst Brücke.* Vienna: Springer.

Brücke, E. W. (1861). Beiträge zur Theorie der Verdauung. *Sitzungsberichte der Akademie der Wissenschaften Wien, Math.-Wiss. Kl.* 43(2):601–623.

Brugsch, T., and Schittenhelm, A. (1910). *Der Nukleinstoffwechsel und seine Störungen.* Jena: Fischer.

Brunori, M., and Chance, B. (eds.) (1988). *Cytochrome oxidase: structure, function, and physiopathology. Ann. N.Y. Acad. Sci.* Vol. 550.

Brunton, T. L. (1871). Experimental investigation of the action of medicines. *Brit. Med. J.* 1:413–415.

Bruppacher-Celler, M. (1971). *Rudolf Buchheim (1820–1879) und die Entwicklung einer experimentellen Pharmakologie.* Zurich: Juris.

Brush, S. G. (1965). *Kinetic Theory.* Vol. 1, *The Nature of Gases and of Heat.* Oxford: Pergamon.

——— (1966). *Kinetic Theory.* Vol. 2, *Irreversible Processes.* Oxford: Pergamon.

——— (1974). Should the history of science be rated X? *Science* 183:1164–1172.

——— (1995). Scientists as historians. *Osiris* 10:215–231.

Brutlag, D., Schekman, R., and Kornberg, A. (1971). A possible role for RNA polymerase in the initiation of M13 DNA synthesis. *PNAS* 68:2826–2829.

Buchanan, J. M. (1986). A backward glance. In *Selected Topics in the History of Biochemistry: Personal Recollections.* Vol. 2 (A. Neuberger et al., eds.), pp. 1–69. Amsterdam: Elsevier.

——— (1994). Aspects of nucleotide enzymology and biology. *Protein Science* 3:2151–2157.

Buchanan, J. M., and Hartman, S. C. (1959). Enzymic reactions in the synthesis of purines. *Adv. Enzymol.* 21:199–261.

Buchanan, J. M., and Hastings, A. B. (1946). The use of isotopically marked carbon in the study of intermediary metabolism. *Physiol. Revs.* 26:120–155

——— (1989). Eric Glendenning Ball. *Biog. Mem. NAS* 58:49–73.

Buchdahl, G. (1973). Leading principles and induction: The methodology of Matthias Schleiden. In *Foundations of Scientific Method: The Nineteenth Century* (R. N. Giere and R. S. Westfall, eds.), pp. 23–52. Bloomington: Indiana University Press.

Bücher, T. (1947). Ueber ein phosphatübertragendes Gärungsferment. *Biochim. Biophys. Acta* 1:292–314.

——— (1989). Commentary. *Biochim. Biophys. Acta* 1000:223–227.

Buchner, E. (1897). Gährung ohne Hefezellen. *Ber. chem. Ges.* 30:117–124.

——— (1966). Cell-free fermentation. In *Nobel Lectures: Chemistry, 1901–1921*, pp. 103–120. Amsterdam: Elsevier.

Buchner, E., Buchner, H., and Hahn, M. (1903). *Die Zymase Gährung.* Munich: Oldenburg.

Buchner, E., and Meisenheimer, J. (1904). Die chemische Vorgänge bei der alkoholischen Gärung. *Ber. chem. Ges.* 37:417–428.

Buchner, E., and Rapp, R. (1898). Alkoholische Gärung ohne Hefezellen. *Ber. chem. Ges.* 31:212–213.

Buchner, R. (1963). Die politische und geistige Vorstellungswelt Eduard Buchners. *Zeitschrift für bayerische Landesgeschichte* 26:631–645.

Buchwald, J. Z. (1976). William Thompson (Baron Kelvin of Largs). *DSB* 13:374–388.

Bud, R. (1992). The zymotechnic roots of biotechnology. *Brit. J. Hist. Sci.* 25:127–144.

——— (1993). *The Uses of Life: A History of Biotechnology.* Cambridge: Cambridge University Press.

Bud, R., and Roberts, G. (1984). *Science Versus Practice: Chemistry in Victorian Britain.* Manchester: Manchester University Press.

Buican, D. (1982). Mendelism in France and the work of Lucien Cuénot. *Scientia* 117:129–136.

——— (1984). *Histoire de la Génétique et de l'Evolutionnisme en France.* Paris: PUF.

Bülbring, E., and Walker, J. M. (1984). Joshua Harold Burn. *Biog. Mem. FRS* 30:45–89.

Bunge, G. (1889). *Lehrbuch der physiologischen und pathologischen Chemie,* 2d ed. Leipzig: Vogel.

——— (1902). *Textbook of Physiological and Pathological Chemistry* (F. A. Starling and E. H. Starling, trans. from 4th German ed.). Philadelphia: Blakiston.

Bunge, M. (1962). *Intuition and Science.* Englewood Cliffs: Prentice-Hall.

——— (1967). *Scientific Research.* Berlin: Springer.

——— (1991, 1992). A critical examination of the new sociology of science. *Philosophy of the Social Sciences* 21:524–560; 22:46–76.

——— (1996). In praise of intolerance to charlatanism in academia. *Ann. N.Y. Acad. Sci.* 775:96–115.

Bünning, E. (1975). *Wilhelm Pfeffer.* Stuttgart: Wissenschaftliche Verlagsgesellschaft.

Bunzl, M., et al. (1995). Truth, objectivity, and history. *J. Hist. Ideas* 56:651–680.

Burbaum, J. J., and Schimmel, P. (1991). Structural relationships and the classifications of aminoacyl-tRNA synthases. *J. Biol. Chem.* 266:16965–16968.

Burchardt, L. (1975). *Wissenschaftspolitik in Wilhelminischen Deutschland.* Göttingen: Vanderhoeck and Rupprecht.

——— (1977). Halbstaatliche Wissenschaftsforderung im Kaiserreich und in der frühen Weimarer Republik. In *Medizin, Naturwissenschaft, und Technik und das zweite Kaiserreich* (G. Mann and R. Winau, eds.), pp. 35–51. Göttingen: Vanderhoeck and Rupprecht.

——— (1978). Die Ausbildung des Chemikers im Kaiserreich. *Tradition* 25:31–50.

——— (1980). Professionalisierung oder Berufskonstruktion? Beispiel des Chemikers im wilhelminischen Deutschland. *Geschichte und Gesellschaft* 6:326–348.

Burgess, R. R. (1971). RNA polymerase. *Ann. Rev. Biochem.* 40:711–740.

Bürgin, A. (1958). *Geschichte des Geigy-Unternehmens von 1758 bis 1939*. Basel: Birkhäuser.

Burgus, R., and Guillemin, R. (1970). Hypothalamic releasing factors. *Ann. Rev. Biochem.* 39:499–526.

Burian, R. (1906). Chemie der Spermatozooen. *Erg. Physiol.* 5:768–846.

Burian, R. M. (1993). Technique, task definition, and the transition from genetics to molecular genetics: Aspects of the work on protein synthesis in the laboratories of J. Monod and P. Zamecnik. *J. Hist. Biol.* 26:387–407.

——— (1994). Jean Brachet's cytochemical embryology: Connections with the renovation of biology in France? In *Les Sciences biologiques et médicales en France, 1920–1950* (C. Debru et al., eds.), pp. 207–220. Paris: CNRS.

——— (1996). The tools of the discipline: Biochemists and molecular biologists. *J. Hist. Biol.* 29:451–462.

Burian, R. M., and Gayon, J. (1990). Genetics after World War II: The laboratories at Gif. *Cahiers pour l'histoire du CNRS, 1939–1989*. Paris: CNRS.

Burian, R. M., Gayon, J., and Zallen, D. (1988). The singular fate of genetics in the history of French biology, 1900–1940. *J. Hist. Biol.* 21:367–402.

Burian, R. M., Richard, R. C., and van der Steen, W. J. (1996). Against generality. *Stud. Hist. Phil. Sci.* 27:1–29.

Burk, D. (1939). A colloquial consideration of the Pasteur and neo-Pasteur effects. *Cold Spring Harb. Symp.* 7:420–459.

Burke, J. G. (1966). *Origins of the Science of Crystals*. Berkeley: University of California Press.

Burnet, F. M. (1929a). Method for the study of bacteriophage multiplication in broth. *Brit. J. Exp. Path.* 10:109–114.

——— (1929b). Smooth-rough variation in bacteria in its relation to bacteriophage. *J. Path. Bact.* 32:15–42.

——— (1941). *The Production of Antibodies*. Melbourne: Macmillan.

——— (1957). A modification of Jerne's theory of antibody production, using the concept of clonal selection. *Austr. J. Sci.* 20:67–69.

——— (1959). *The Clonal Selection Theory of Acquired Immunity*. Cambridge: Cambridge University Press.

Burnet, F. M., and Fenner, F. (1949). *The Production of Antibodies*, 2d ed. Melbourne: Macmillan.

Burnet, F. M., and McKie, M. (1929). Observations on a permanently lysogenic strain of *B. enterides Gaertner*. *Austr. J. Exp. Biol. Med. Sci.* 6:277–284.

Burnett, G., and Kennedy, E. P. (1954). The enzymatic phosphorylation of proteins. *J. Biol. Chem.* 211:969–980.

Burns, D. T. (1993). London chemists and chemistry, prior to the formation of the Chemical Society in 1844. *Anal. Proc.* 30:334–337.

Burns, R. C., and Hardy, R. W. F. (1975). *Nitrogen Fixation in Bacteria and Higher Plants.* New York: Springer.

Burris, R. H. (1991). Nitrogenases. *J. Biol. Chem,* 266:9339–9342.

——— (1992). Perry William Wilson. *Biog. Mem. NAS* 61:439–467.

——— (1994). Historical developments in biological nitrogen fixation. In *Historical Perspectives in Plant Science* (K. J. Frey, ed.), pp. 23–41. Ames: Iowa State University Press.

Burris, R. H., Baumann, C. A., and Potter, V. R. (1990). Conrad Arnold Elvehjem. *Biog. Mem. NAS* 59:135–167.

Burris, R. H., and Wilson, P. W. (1945). Biological nitrogen fixation. *Ann. Rev. Biochem.* 14:685–708.

Burrows, M. T. (1911). The growth of tissues of the chick-embryo outside the animal body with special reference to the nervous system. *J. Exp. Zool.* 10:63–84.

Busch, A. (1959). *Die Geschichte der Privatdozenten.* Stuttgart: Enke.

——— (1963). The vicissitudes of the *Privatdozent:* Breakdown and adaptation in the recruitment of the German university teacher. *Minerva* 1:319–341.

Buskirk, J. J., et al., (1984). Aminomalonic acid: Identification in *Escherichia coli* and atherosclerotic plaque. *PNAS* 81:722–728.

Butenandt, A. (1961). The Windaus memorial lecture. *Proc. Chem. Soc,* pp. 131–138.

——— (1973). *Ernst Klenk.* Krefeld: Scherpe.

——— (1979). The discovery of oestrone. *TIBS* 4:215–216.

Butenandt, A., Biekert, E., and Linzen, B. (1956). Ueber Ommochrome. VII. Modellversuche zur Bildung des Xanthochromatins in vivo. *Z. physiol. Chem.* 305:284–289.

Butenandt, A., Weichert, R., and Derjugin, W. V. (1943). Ueber Kynurenin. *Z. physiol. Chem.* 279:27–43.

Butler, J. A. V., and Davison, P. F. (1957). Deoxyribonucleoprotein, a genetic material. *Adv. Enzymol.* 18:161–190.

Butler, S. V. F. (1988). Centers and peripheries: The development of British physiology. *J. Hist. Biol.* 21:473–500.

Büttner, J. (ed.). *History of Clinical Chemistry.* Berlin: Walter de Gruyter.

Buzan, J. M., and Frieden, C. (1996). Yeast actin: Polymerization studies of wild type and a poorly polymerizing mutant. *PNAS* 93:91–95.

Bykov, G. V. (1962). The origin of the theory of chemical structure. *J. Chem. Ed.* 39:220–224.

Bylebyl, J. J. (1970). William Beaumont, Robley Dunglison, and the "Philadelphia physiologists." *J. Hist. Med.* 25:3–21.

Bynum, W. (1970). Chemical structure and pharmacological action. *Bull. Hist. Med.* 44:518–538.

Bywaters, H. W. (1916). Frederick William Pavy. *Biochem. J.* 19:1–4.

Cadena, D., and Gill, G. (1992). Receptor tyrosine kinases. *FASEB J.* 6:2332–2337.

Caetano-Anollés, G., and Gresshoff, P. M. (1991). Plant genetic control of nodulation. *Ann. Rev. Microbiol.* 45:345–382.

Cagniard-Latour, C. (1838). Mémoire sur la fermentation vineuse. *Ann. Chim.* [2]68:206–223.

Cahan, D. (ed.) (1993). *Hermann von Helmholtz and the Foundations of Nineteenth-Century Science.* Berkeley: University of California Press.

Cairns, J. (1963). The bacterial chromosome and its manner of replication as seen by autoradiography. *J. Mol. Biol.* 6:208–213.

Cairns, J., Overbaugh, J., and Miller, S. (1988). The origin of mutants. *Nature* 335:142–145.

Cairns, J., Stent, G. S., and Watson, J. D. (eds.) (1966). *Phage and the Origins of Molecular Biology.* Plainview, N.Y.: Cold Spring Harbor Laboratory.

Caldin, E. F. (1961). *The Structure of Chemistry.* London: Sheed and Ward.

Calvin, M. (1953). The path of carbon in photosynthesis; the quantum conversion in photosynthesis. *Chem. Eng. News* 31:1622–1625; 1735–1738.

——— (1956). The photosynthetic carbon cycle. *J. Chem. Soc.*, pp. 1895–1915.

——— (1984). Gilbert Newton Lewis: His influence on physical-organic chemists at Berkeley. *J. Chem. Ed.* 61:14–18.

Calvin, M., and Seaborg, G. T. (1984). The college of chemistry in the G. N. Lewis era, 1912–1946. *J. Chem. Ed.* 61:11–13.

Cambrosio, A., Jacobi, D., and Keating, P. (1993). Ehrlich's "beautiful pictures" and the controversial beginnings of immunological imagery. *Isis* 84:662–699.

Cambrosio, A., and Keating, P. (1995). *Exquisite Specificity: The Monoclonal Antibody Revolution.* New York: Oxford University Press.

Campbell, J. L. (1993). Yeast DNA replication. *J. Biol. Chem.* 268:25261–25264.

Campbell, M. (1980). Did de Vries discover the law of segregation independently? *Ann. Sci.* 37:639–655.

Campbell, P. N. (1960). The synthesis of proteins by the cytoplasmic components of animal cells. *Biol. Revs.* 35:413–458.

——— (1995). The importance of asking questions. In *Selected Topics in the History of Biochemistry: Personal Recollections.* Vol. 4 (E. C. Slater et al., eds.), pp. 403–437.

Campbell, P. N., and Work, T. S. (1953). Biosynthesis of proteins. *Nature* 171:997–1001.

Caneva, K. L. (1993). *Robert Mayer and the Conservation of Energy.* Princeton University Press.

——— (1997). Physics and *Naturphilosophie. Hist. Sci.* 35:35–106.

Canguilhem, G. (1970). Marie François Xavier Bichat. *DSB* 2:122–123.

——— (1988). *Ideology and Rationality in the History of the Life Sciences* (A. Goldhammer, trans.). Cambridge: MIT Press.

Cannon, W. B. (1922). Henry Pickering Bowditch. *Biog. Mem. NAS* 17:183–196.

——— (1926). Physiological regulation of normal states: Some tentative postulates concerning biological homeostasis. In *A. Richet. Ses Amis, ses Collègues, ses Elèves.* Paris: Auguste Petit.

——— (1929). Organization for physiological homeostasis. *Physiol. Revs.* 9:399–431.

——— (1932). *The Wisdom of the Body.* New York: Norton.

——— (1940). The role of chance in discovery. *Sci. Mon.* 50:204–209.

——— (1943). Lawrence Joseph Henderson. *Biog. Mem. NAS* 23:31–58.

——— (1945). *The Way of an Investigator.* New York: Norton.

Cantoni, G. L. (1953). S-adenosylmethionine: A new intermediate formed enzymatically from L-methionine and adenosine triphosphate. *J. Biol. Chem.* 204:403–416.

Cantor, G. (1991). *Michael Faraday: Sandemanian and Scientist.* London: Macmillan.

Capaldi, R. A. (1990). Structure and function of cytochrome c oxidase. *Ann. Rev. Biochem.* 59:569–596.

Capecchi, M. R. (1966). Initiation of *E. coli* proteins. *PNAS* 55:1517–1524.

Capecchi, M. R., and Klein, H. A. (1969). Characterization of three proteins involved in polypeptide chain termination. *Cold Spring Harb. Symp.* 34:469–477.

Carafoli, E. (1987). Intracellular calcium homeostasis. *Ann. Rev. Biochem.* 56:395–433.

Cardinale, G. J., and Udenfriend, S. (1974). Prolyl hydroxylase. *Adv. Enzymol.* 41:245–300.

Cardwell, D. S. L. (ed.) (1968). *John Dalton and the Progress of Science.* Manchester: Manchester University Press.

——— (1971). *From Watt to Clausius.* Manchester: Manchester University Press.

——— (1972). *The Organisation of Science in England.* 2d ed. London: Heinemann.

——— (1989). *James Joule.* Manchester: Manchester University Press.

Carlier, M. F. (1991). Actin: Protein structure and filament dynamics. *J. Biol. Chem.* 266:1–4.

Carlson, E. A. (1966). *The Gene: A Critical Study.* Philadelphia: Saunders.

——— (1971). An unacknowledged founding of molecular biology: H. J. Muller's contributions to gene theory, 1910–1936. *J. Hist. Biol.* 4:149–170.

——— (1981). *Genes, Radiation, and Society: The Life and Work of H. J. Muller.* Ithaca, N.Y.: Cornell University Press.

Carlson, E. T. (1975). Benjamin Rush. *DSB* 11:616–618.

Carnahan, J. E., Mortenson, L. E., Mower, H. F., and Castle, J. E. (1960). Nitrogen fixation in cell-free extracts of *Clostridium pasteurianum. Biochim. Biophys. Acta* 44:520–535.

Carneiro, A. (1993). Adolphe Wurtz and the atomic controversy. *Ambix* 40:75–95.

Caro, H. (1892). Ueber die Entwicklung der Teerfarben-Industrie. *Ber. chem. Ges.* 25:955–1105.

Caroe, G. M. (1978). *William Henry Bragg.* Cambridge: Cambridge University Press.

Caron. J. A. (1988). "Biology" in the life sciences: A historiographical contribution. *Hist. Sci.* 26:223–268.

Carothers, W. H. (1931). Polymerization. *Chem. Revs.* 8:353–426.

Carpenter, G., and Cohen, S. (1990). Epidermal growth factor. *J. Biol. Chem.* 265:7709–7712.

Carpenter, K. J. (1994). *Protein and Energy.* Cambridge: Cambridge University Press.

Carpenter, K. J., and Sutherland, B. (1995). Eijkman's contribution to the discovery of vitamins. *J. Nutrition* 125:155–163.

Carpino, L. A., Beyermann, M., Wenschuh, H., and Bienert, M. (1996). Peptide synthesis via amino acid halides. *Acc. Chem. Res.* 29:268–274.

Carrel, A. (1935). *Man the Unknown.* New York: Harper.

Carrington, A., Hills, G. J., and Webb, K. R. (1974). Neil Kensington Adam. *Biog. Mem. FRS* 20:1–26.

Carter, C. E. (1950). Paper chromatography of purine and pyrimidine derivatives of yeast nucleic acid. *J. Am. Chem. Soc.* 72:1466–1471.

Carter, H. E. (1979). Identification and synthesis of threonine. *Fed. Proc.* 38:2684–2686.

Caruthers, M. H. (1985). DNA synthesis machines: DNA chemistry and its uses. *Science* 230:281–285.

Casey, P. J., and Gilman, A. G. (1988). G protein involvement in receptor-effector coupling. *J. Biol. Chem.* 263:2577–2580.

Caskey, T. H. (1980). Peptide chain termination. *TIBS* 5:234–237.

Caspari, E. W., et al. (1980). Controversial geneticist and creative biologist [Richard B. Goldschmidt]. *Experientia* Suppl. 35:1–154.

Caspersson, T. (1936). Ueber den chemischen Aufbau der Strukturen des Zellkerns. *Skand. Arch. Physiol.* 23(suppl. 8):1–151.

——— (1941). Studien über den Eiweissumsatz der Zelle. *Naturwiss,* 28:33–43.

——— (1947). The relation between nucleic acid and protein synthesis. *Symp. Soc. Exp. Biol.* 1:127–151.

Caspersson, T., and Schultz, J. (1951). Cytochemical studies in the study of the gene. In *Genetics in the 20th Century* (L. C. Dunn, ed.), pp. 155–171. New York: Macmillan.

Casselton, L. A. (1971). Suppressor genes. *Sci Prog.* 59:143–160.

Castle, W. E. (1903). The laws of heredity of Galton and Mendel and some laws governing race improvement by selection. *Proc. Am. Acad. Arts Sci.* 39:223–242.

——— (1919). Is the arrangement of the genes in the chromosome linear? *PNAS* 5:25–32, 500–506.

Cat, J., Cartwright, N., and Chang, H. (1996). Otto Neurath: Politics and the unity of science. In *The Disunity of Science* (P. Galison and D. J. Stump, eds.), pp. 347–369. Stanford: Stanford University Press.

Cate, J. H., Gooding, A. R., Podell, E., Zhou, K., Golden, B. L., Kundrot, C. E., Cech, T. R., and Doudna, J. A. (1996). Crystal structure of a group I ribozyme domain: Principles of RNA packing. *Science* 273:1678–1685.

Cathcart, E. P. (1912). *The Physiology of Protein Metabolism.* London: Longmans Green.

——— (1929). Diarmid Noel Paton. *Proc. Roy. Soc.* B104:ix–xii.

Cato, A. C. B., Ponta, H., and Herrlich, P. (1992). Regulation of gene expression by steroid hormones. *Prog. Nucleic Acid Res.* 43:1–36.

Cavarelli, J., and Moras, D. (1993). Recognition of tRNAs by aminoacyl-tRNA synthetases. *FASEB J.* 7:70–86.

Cech, T. R. (1987). The chemistry of self-splitting RNA and RNA enzymes. *Science* 236:1532–1539.

——— (1990). Self-splicing of group I introns. *Ann. Rev. Biochem.* 59:543–568.

Cech, T. R., Zaug, A. J., and Grabowski, P. J. (1981). *In vitro* splicing of the ribosomal RNA precursor of Tetrahymena: Involvement of a guanosine nucleotide in the excision of the intervening sequence. *Cell* 27:487–496.

Cecil, R., and Ogston, A. G. (1948). The sedimentation of thymus nucleic acid in the ultracentrifuge. *J. Chem. Soc.*, pp. 1382–1386.

Cerutti, L. (1990). Giulio Natta. *DSB* 18:661–663.

Chadarevian, S. de (1996). Sequences, conformations, information: Biochemists and molecular biologists in the 1950s. *J. Hist. Biol.* 29:361–386.

Challenger, F. (1951). Biological methylation. *Adv. Enzymol.* 12:429–491.

Challey, J. F. (1971). Nicolas Léonard Sadi Carnot. *DSB* 3:79–84.

Chamberlin, M., and Berg, P. (1962). Deoxyribonucleic acid-directed synthesis of ribonucleic acid by an enzyme from *Escherichia coli. PNAS* 48:81–94.

——— (1964). Mechanism of RNA polymerase action: Characterization of the DNA-dependent synthesis of polyadenylic acid. *J. Mol. Biol.* 8:708–726.

Chamberlin, M., McGrath, J., and Waskell, C. (1970). New RNA polymerase from *Escherichia coli* infected with bacteriophage T4. *Nature* 228:227–231.

Chambers, D. A. (ed.) (1995). *DNA, the Double Helix: Perspective and Prospective at Forty Years. Ann. N.Y. Acad. Sci.* Vol. 758.

Champe, S. P., and Benzer, S. (1962). Reversal of mutant phenotypes by 5-fluorouracil: An approach to nucleotide sequences in messenger-RNA. *PNAS* 48:532–546.

Chan, H. S., and Dill, K. A. (1990). Origins of structure in globular proteins. *PNAS* 87:6388–6392.

Chance, B. (1951). Enzyme-substrate compounds. *Adv. Enzymol.* 12:153–190.

——— (1953). The carbon monoxide compounds of the cytochrome oxidases. *J. Biol. Chem.* 202:383–416.

——— (1956). Interaction of adenosine diphosphate with the respiratory chain. In *Enzymes: Units of Biological Structure and Function* (O. H. Gaebler, ed.), pp. 447–463. New York: Academic.

——— (1991). Optical method. *Ann. Rev. Biophys.* 20:1–28.

Chance, B., Cohen, P., Jobsis, F., and Schoener, B. (1962). Intracellular oxidation-reduction states in vivo. *Science* 137:499–505.

Chance, B., and Williams, G. R. (1956a). Respiratory enzymes in oxidative phosphorylation. III. The steady state. *J. Biol. Chem.* 217:409–427.

——— (1956b). The respiratory chain and oxidative phosphorylation. *Adv. Enzymol.* 17:65–134.

Changeux, J. P. (1961). The feedback control mechanism of biosynthetic L-threonine deaminase by L-leucine. *Cold Spring Harb. Symp.* 26:313–318.

——— (1963). Allosteric interactions in biosynthetic L-threonine deaminase from *E. coli* K-12. *Cold Spring Harb. Symp.* 28:497–504.

——— (1981). The acetylcholine receptor: An allosteric membrane protein. *Harvey Lectures* 75:85–254.

Channon, H. J. (1926). The biological significance of the unsaponifiable matter in oils. *Biochem. J.* 20:400–418.

Chantrenne, H. (1990). Jean Brachet. In *Selected Topics in the History of Biochemistry* 3 (R. Semenza and R. Jaenicke, eds.), pp. 201–213. Amsterdam: Elsevier.

Chapeville, F., and Haenni, A. L. (1980). *Concepts of Chemical Recognition in Biology.* Berlin: Springer.

Charbonnier, J. B., Carpenter, E., Gigant, B., Golinelli-Pimpaneau, B., Eschhar, Z., Green, B. S., and Knossow, M. (1995). Crystal structure of the complex of a catalytic Fab fragment with a transition-state analog: Structural similarities in esterase-like catalytic antibodies. *PNAS* 92:11721–11725.

Chargaff, E. (1950). Chemical specificity of nucleic acids and mechanism of their enzymatic degradation. *Experientia* 6:201–209.

——— (1955). Isolation and composition of the deoxypentose nucleic acids and of the corresponding nucleoproteins. In *The Nucleic Acids* (E. Chargaff and J. N. Davidson, eds.), 1:307–407. New York: Academic.

——— (1968a). What really is DNA? Remarks on the changing aspects of a scientific concept. *Prog. Nucleic Acid Res.* 8:297–333.

——— (1968b). A quick trip up Mount Olympus. *Science* 159:1448–1449.

——— (1971). Preface to a grammar of biology: A hundred years of nucleic acid research. *Science* 172:637–642.

——— (1974). Building the tower of babble. *Nature* 248:776–779.

——— (1975a). A fever of reason: The early way. *Ann. Rev. Biochem.* 44:1–18.

——— (1975b). Voices in the labyrinth: Dialogues around the study of nature. *Persp. Biol. Med.* 18:251–285.

——— (1978). *Heraclitean Fire.* New York: Rockefeller University Press.

——— (1986). *Serious Questions.* Boston: Birkhäuser.

Chargaff, E., Lipschitz, R., Green, C., and Hodes, M. E. (1951). The composition of the desoxyribonucleic acid of salmon spleen. *J. Biol. Chem.* 192:223–230.

Chargaff, E., and Vischer, E. (1948). Nucleoproteins, nucleic acids, and related substances. *Ann. Rev. Biochem.* 17:201–226.

Charle, C. (1990). Les professeurs des facultés des sciences en France: Une comparaison Paris-Province (1880–1900). *Rev. Hist. Sci.* 43:427–450.

Charpentier-Morize, M. (1989). La contribution des laboratoires propres au CNRS à la recherche chimique en France. *Cahiers pour l'Histoire du CNRS, 1939–1989* 4:79–112.

Chaussat, C. (1820). Mémoire sur l'influence du système nerveux sur la chaleur animale. *J. Phys. Chim. Hist. Nat.* 91.6–23, 92–111.

Chen, C. K., Wieland, T., and Simon, M. I. (1996). RGS-r, a retinal specific RGS protein, binds an intermediate conformation of transducin and enhances recycling. *PNAS* 93:12885–12889.

Chen, K. K. (1981). Two pharmacological traditions: Notes from experience. *Ann. Rev. Pharmacol.* 21:1–6.

Cheng, T. H., and Doi, R. H. (1968). Advances in nucleic acid research in Communist China. *Fed. Proc.* 27:1430–1454.

Cherbuliez, E. (1980). Histoire de la chimie à Genève. *Chimia* 34:25–31.

Cheung, W. Y. (1982). Calmodulin: An overview. *Fed. Proc.* 41:2253–2257.

Chevreul, M. E. (1824). *Considérations sur l'Analyse Organique et sur ses Applications.* Paris: Levrault.

——— (1847). Rapports de l'agriculture avec les autres connaissances humaines. *Journal des Savants,* pp. 577–591.

Chevreul, M. E., et al. (1855). Rapport sur le travail de M. Georges Ville, dont l'objet est de prouver que le gaz azote de l'air s'assimule aux végétaux. *Compt. Rend.* 41:757–778.

Chibnall, A. C. (1939). *Protein Metabolism in the Plant.* New Haven: Yale University Press.

——— (1942). Amino-acid analysis and the structure of proteins. *Proc. Roy. Soc.* B131:136–160.

——— (1987). *My Early Days in Biochemistry.* London: Biochemical Society.

Chibnall, A. C., Rees, M. W., Williams, E. F., and Boyland, E. (1940). The glutamic acid of normal and malignant tissue proteins. *Biochem. J.* 34:285–300.

Chick, H., Hume, M., and Macfarlane, M. (1971). *War on Disease: A History of the Lister Institute.* London: Deutsch.

Chick, H., and Martin, C. J. (1912). On the "heat coagulation" of proteins, 4: The conditions controlling the agglutination of proteins already acted on by hot water. *J. Physiol.* 45:261–295.

Childs, B. (1970). Sir Archibald Garrod's conception of chemical individuality: A modern appreciation. *New Eng. J. Med.* 282:71–78.

Chittenden R. H. (1894). Some recent chemico-physiological discoveries regarding the cell. *Am. Nat.* 28:97–117.

——— (1928). *History of the Sheffield Scientific School of Yale University.* New Haven: Yale University Press.

——— (1930). *The Development of Physiological Chemistry in the United States.* New York: Chemical Catalog.

——— (1937). Lafayette Benedict Mendel. *Biog. Mem. NAS* 18:123–155.

——— (1945). *The First Twenty-Five Years of the American Society of Biological Chemists.* New Haven: American Society of Biological Chemists.

Chodat, R., and Bach, A. (1903). Untersuchungen Uber die Rolle der Peroxyde in der Chemie der lebenden Zelle. V. Zerlegung der sogennanten Oxydasen in Oxygenasen und Peroxydasen. *Ber. chem. Ges.* 36:606–609.

Chossat, C. (1820). Mémoire sur l'influence du système nerveux sur la chaleur animale. *Ann. Phys. Chim. Sci. Nat.* 91:6–23, 92–111.

Christianson, J. A., et al. (1949). J. N. Brønsted. *Acta Chem. Scand.* 3:1187–1276.

Christie, J. R. R. (1993). William Cullen and the practice of chemistry. In *William Cullen and the Eighteenth Century Medical World* (A. Doig et al., eds.), pp. 98–109. Edinburgh: Edinburgh University Press.

Christie, M (1994). Philosophers versus chemists concerning "laws of nature." *Stud. Hist. Phil. Sci.* 25:613–629.

Chubin, D. E. (1976). The conceptualization of scientific disciplines. *Sociological Quarterly* 17:448–476.

——— (1985). Misconduct in research: An issue in science policy and practice. *Minerva* 23:175–202.

——— (1990). Scientific malpractice and the contemporary politics of knowledge. In *Theories of Science in Society* (S. E. Cozzens and T. F. Gieryn, eds.), pp. 144–167. Bloomington: Indiana University Press.

Churchill, F. B. (1968). August Weismann and a break from tradition. *J. Hist. Biol.* 1:91–112.

——— (1969). From machine-theory to entelechy: Two studies in developmental biology. *J. Hist. Biol.* 2:165–185.

——— (1970). Hertwig, Weismann, and the meaning of the reduction division circa 1890. *Isis* 61:429–457.

——— (1974). William Johannsen and the genotype concept. *J. Hist. Biol.* 7:5–30.

——— (1986). Weismann, hydromedusae, and the biological imperative: A reconsideration. In *History of Embryology* (T. L. Horder et al., eds.), pp. 7–33. Cambridge: Cambridge University Press.

Claesson, S., and Pedersen, K. O. (1972). The Svedberg. *Biog. Mem. FRS* 18:595–627.

Clapham, D. E. (1996). The G-protein nanomachine. *Nature* 379:297–299.

Clark, A. J. (1933). *The Mode of Action of Drugs on Cells.* London: Arnold.

——— (1938). *Applied Pharmacology.* 6th ed. Philadelphia: Blakiston.

Clark, A. J., and Margulies, A. D. (1965). Isolation and characterization of recombinant deficient mutants of *Escherichia coli* K-12. *PNAS* 53:451–459.

Clark, B. R. (1983). *The Higher Education System: Academic Organization in Cross-National Perspective.* Berkeley: University of California Press.

——— (ed.) (1984). *Perspectives on Higher Education: Eight Disciplinary and Comparative Views.* Berkeley: University of California Press.

——— (ed.) (1987). *The Academic Profession: National, Disciplinary, and Institutional Settings.* Berkeley: University of California Press.

Clark, E. A., Shattil, S. J., and Brugge, J. S. (1994). Regulation of protein kinase in platelets. *TIBS* 19:464–469.

Clark, G., and Kasten, F. H. (1983). *History of Staining.* 3d ed. Baltimore: Williams and Wilkins.

Clark, J., Modgil, C., and Modgil, S. (eds.) (1990). *Robert K. Merton: Consensus and Controversy.* London: Falmer.

Clark, P. (1976). Atomism versus thermodynamics. In *Method and Appraisal in the Physical Sciences* (C. Howson, ed.), pp. 41–105. Cambridge: Cambridge University Press.

Clark, R. J. H. (1991). Thomas Graham: Would his research be unfunded today? *Chem. Soc. Rev.* 20:405–424.

Clark, R. W. (1968). *J. B. S.: The Life of J. B. S. Haldane.* London: Hodder and Stoughton.

Clark. W. M. (1925). Recent studies on reversible oxidation-reduction in organic systems. *Chem. Revs.* 2:127–178.

——— (1928). *The Determination of Hydrogen Ions.* 3d ed. Baltimore; Williams and Wilkins.

——— (1938). Walter Jennings Jones. *Biog. Mem. NAS* 20:79–139.

——— (1960). *Oxidation-Reduction Potentials of Organic Systems.* Baltimore: Williams and Wilkins.

——— (1962). Prefatory chapter: Notes on a half-century of teaching, research, and administration. *Ann. Rev. Biochem.* 31:1–24.

Clarke, A., and Fujimura, J. (eds.) (1992). *The Right Tools for the Job: At Work in the Twentieth-Century Life Sciences.* Princeton: Princeton University Press.

Clarke, E., and Jacyna, L. S. (1987). *Nineteenth-Century Origin of Neuroscientific Concepts.* Berkeley: University of California Press.

Clarke, F. W. (1910). Oliver Wolcott Gibbs. *Biog. Mem. NAS* 7:3–22.

Clarke, H. T. (1952). Henry Drysdale Dakin. *J. Chem. Soc.*, pp. 3319–3324.

——— (1955). The Journal of Biological Chemistry. *J. Biol. Chem.* 216:449–454.

——— (1956). William John Gies. *Am. Phil. Soc. Year Book.* pp. 111–115.

——— (1958). Impressions of an organic chemist in biochemistry. *Ann. Rev. Biochem.* 27:1–14.

Clarke, P. H. (1986). Roger Yate Stanier. *Biog. Mem. FRS* 32:543–568.

Claude, A. (1946). Fractionation of mammalian liver cells by differential centrifugation. *J. Exp. Med.* 84:51–89.

——— (1950). Studies on cells: Morphology, chemical constitution, and distribution of biochemical function. *Harvey Lectures* 43:121–164.

Clausius, L. (1865). Ueber verschiedene für die Anwendung bequeme Formen der Hauptgleichungen der mechanischen Wärmetheorie. *Ann. Phys.* 125:353–400.

Clay, R. S., and Court, T. H. (1932). *The History of the Microscope.* London: Griffin.

Cleland, R. E. (1923). Chromosome arrangements during meiosis in certain Oenatheras. *Am. Nat.* 57:562–566.

Cleland, W. W. (1953). The kinetics of enzyme-catalyzed reactions with two or more substrates or products. *Biochim. Biophys. Acta* 67:104–137, 178–187, 189–196.

——— (1982). Use of isotope effects to elucidate enzyme mechanisms. *Crit. Revs. Biochem.* 13:385–428.

Clementi, A. (1915). Sulla diffusione nell'organismo e nel regno dei verte brati e sulla importanza fisiologica dell'arginasi. *Arch. Fisiol.* 13:189–230.

Clericuzio, A. (1990). A redefinition of Boyle's chemistry and corpuscular philosophy. *Ann. Sci.* 47:561–589.

——— (1993). From van Helmont to Boyle: A study of the transmission of Helmontian chemical and medical theories in seventeenth-century England. *Brit. J. Hist. Sci.* 26:303–334.

Cochin, D. (1880). Recherches du ferment alcoolique soluble. *Ann. Chim.* [5]21:430–432.

Cockburn, I., and Henderson, R. (1996). Public-private interaction in pharmaceutical research. *PNAS* 93:12725–12730.

Cockcroft, J. (1967). George de Hevesy. *Biog. Mem. FRS* 13:125–166.

Cohen, E. (1908). Hartog Jacob Hamburger. *Biochem. Z.* 11:i–xxxiii.

——— (1912). *Jacobus Henricus van't Hoff: Sein Leben und Wirken.* Leipzig: Akademische Verlagsgesellschaft.

Cohen, G. N. (1986). Four decades of Franco-American collaboration in biochemistry and molecular biology. *Persp. Biol. Med.* 29:S141–S148.

Cohen, G. N., and Monod, J. (1957). Bacterial permeases. *Bact. Revs.* 21:169–194.

Cohen, H. F. (1994). *The Scientific Revolution.* Chicago: University of Chicago Press.

Cohen, I. B. (1956). *Franklin and Newton.* Philadelphia: American Philosophical Society.

——— (1990). *Benjamin Franklin's Science.* Cambridge: Harvard University Press.

Cohen, J., et al. (1995). The culture of credit. *Science* 268:1706–1718.

Cohen, J. M. (1956). *The Life of Ludwig Mond.* London: Methuen.

Cohen, J. S., and Portugal, F. H. (1975). A comment on historical analysis in biochemistry. *Persp. Biol. Med.* 18:204–207.

Cohen, K. P., et al. (1983). Harold Clayton Urey. *Biog. Mem. FRS* 29:623–659.

Cohen, L. J. (1989). *An Introduction to the Philosophy of Induction and Probability.* Oxford: Oxford University Press.

Cohen, P. (1989). The structure and regulation of protein phosphatases. *Ann. Rev. Biochem.* 58:458–508.

Cohen, P., and Cohen, P. T. W. (1989). Protein phosphatases come of age. *J. Biol. Chem.* 264:21435–21438.

Cohen, P. P., and Hayano, M. (1946). Urea synthesis by liver homogenates. *J. Biol. Chem.* 166:251–259.

Cohen, S., and Carpenter, G. (1975). Human epidermal growth factor. Isolation and chemical and biological properties. *PNAS* 72:1317–1321.

Cohen, S. N., Chang, A. C. Y., Boyer, H. W., and Hollings, R. B. (1973). Construction of biologically functional bacterial plasmids *in vitro. PNAS* 70:3240–3244.

Cohen, S. S. (1947). The synthesis of bacterial viruses in infected cells. *Cold Spring Harb. Symp.* 12:35–49.

——— (1948). The synthesis of bacterial viruses, 2: The origin of the phosphorus found in the deoxyribonucleic acids of the T2 and T4 bacteriophages. *J. Biol. Chem.* 174:295–303.

——— (1956). Molecular bases of parasitism of some bacterial viruses. *Science* 123:653–656.

——— (1968). *Virus-Induced Enzymes.* New York: Columbia University Press.

——— (1982). Two refugee chemists in the United States: How we see them. *Proc. Am. Phil. Soc.* 126:301–315.

——— (1984). The biochemical origins of molecular biology. *TIBS* 9:334–336.

——— (1986). Finally, the beginnings of molecular biology. *TIBS* 11:92–93.

——— (1990a). Frederick Charles Bawden. *DSB* 17:58–61.

——— (1990b). Wendell Meredith Stanley. *DSB* 18:841–848.

——— (1994). The Northrop hypothesis and the origin of a virus-induced enzyme. *Protein Science* 3:150–153.

——— (1998). *A Guide to the Polyamines.* New York: Oxford University Press.

Cohen, S. S., and Lichtenstein, J. (1960). Polyamines and ribosome structure. *J. Biol. Chem.* 235:2112–2116.

Cohen, S. S., and Stanley, W. M. (1942). The molecular size and shape of the nucleic acid of tobacco mosaic virus. *J. Biol. Chem.* 144:589–598.

Cohn, E. J. (1925). The physical chemistry of proteins. *Physiol. Revs.* 5:349–437.

Cohn, E. J., and Edsall, J. T. (1943). *Proteins, Amino Acids, and Peptides as Ions and Dipolar Ions.* New York: Reinhold.

Cohn, F. (1901). *Blätter der Erinnerung.* 2d ed. Breslau: Kern.

Cohn, M. (1957). Contributions of studies on the β-galactosidase of *Escherichia coli* to our understanding of enzyme synthesis. *Bact. Revs.* 21:140–168.

——— (1989). The way it was: A commentary. *Biochim. Biophys. Acta* 1000:109–112.

Cohn, M. (1992). Carl Ferdinand Cori. *Biog. Mem. NAS* 61:79–109.

——— (1995). Some early tracer experiments with stable isotopes. *Protein Science* 4:2444–2447.

——— (1996). Carl and Gerty Cori: A personal recollection. In *Creative Couples in the Sciences* (H. M. Pycior, N. G. Slack, P. G. Abir-Am, eds.), pp. 72–84. New Brunswick, N. J.; Rutgers University Press.

Cohn, M., and Drysdale, G. R. (1955). A study with O^{18} of ATP formation in oxidative phosphorylation. *J. Biol. Chem.* 216:831–846.

Cohn, M., and Reed, G. H. (1982). Magnetic resonance studies of active sites of enzymic complexes. *Ann. Rev. Biochem.* 51:365–394.

Cohn, W. E. (1950). The anion-exchange separation of ribonucleotides. *J. Am. Chem. Soc.* 72:1471–1478.

Cohn, W. E., and Volkin, E. (1953). On the structure of ribonucleic acids. I. Degradation with snake venom diesterase and the isolation of pyrimidine diphosphates. *J. Biol. Chem.* 203:319–332.

Cohnheim, O. (1901). Die Umwandlung des Eiweiss durch die Darmwand. *Z. physiol. Chem.* 33:451–465.

——— (1911). *Chemie der Eiweisskörper.* 3d ed. Braunschweig: Vieweg.

Cole, S. (1992). *Making Science.* Cambridge: Harvard University Press.

——— (1996). Voodoo sociology: recent developments in the sociology of science. *Ann. N. Y. Acad. Sci.* 775:274–287.

Cole, S., and Cole, J. R. (1967). Scientific output and recognition: A study of the reward system in science. *Am. Soc. Rev.* 32:377–390.

Coleman, D. E., and Sprang, S. R. (1996). How G proteins work: A continuing story. *TIBS* 21:41–44.

Coleman, W. (1965). Cell, nucleus, and inheritance: An historical study. *Proc. Am. Phil. Soc.* 109:124–158.

——— (1970). Bateson and chromosomes: Conservative thought in science. *Centaurus* 15:228–314.

——— (1971). *Biology in the Nineteenth Century.* New York: Wiley.

Coleman, W., and Holmes, F. L. (eds.) (1988). *The Investigative Enterprise: Experimental Physiology in Nineteenth Century Medicine.* Berkeley: University of California Press.

Coley, N. G. (1967a). Alexander Marcet. *Med. Hist.* 12:394–402.

——— (1967b). The Animal Chemistry Club: Assistant society to the Royal Society. *Notes Roy. Soc.* 22:173–185.

——— (1971). Animal chemists and urinary calculi. *Ambix* 18:69–93.

——— (1973). Henry Bence Jones, M. D., F. R. S. *Notes Roy. Soc.* 28:31–56.

——— (1988). Medical chemistry at Guy's Hospital (1720–1850). *Ambix* 35:155–168.

——— (1996). Studies in the history of animal chemistry and its relation to physiology. *Ambix* 43:164–187.

Colin, J. J. (1825). Mémoire sur la formation du sucre. *Ann. Chim.* [2]28:128–142.

Collander, P. R. (1962–1963). Ernst Overton (1865–1933), a pioneer to remember. *Leopoldina* [3]8/9:242–254.

Collard, P. (1976). *The Development of Microbiology.* Cambridge: Cambridge University Press.

Collatz, K. G. (1976). Die Harnbiosynthese. Entschlüsslung eines Stoffwechselzyklus. *Naturw. Rund.* 29:224–230.

Collie, J. N. (1893). The production of naphthalene derivatives from dehydroacetic acid. *J. Chem. Soc.* 63:329–337.

——— (1907). Derivatives of the multiple keten group. *J. Chem. Soc.* 91:1806–1813.

Collins, F. D. (1954). The chemistry of vision. *Biol. Revs.* 29:453–477.

Collins, H. M., and Pinch, T. (1982). *Frames of Meaning: The Social Construction of Extraordinary Science.* London: Routledge.

——— (1993). *The Golem: What Everyone Should Know about Science.* Cambridge: Cambridge University Press.

Collman, J. P. (1996). Coupling H_2 to electron transfer. *Nature Structural Biology* 3:213–217.

Colowick, S. P., Cori, G. T., and Slein, M. W. (1947). The effect of adrenal cortex and anterior pituitary extracts and insulin on the hexokinase reaction. *J. Biol. Chem.* 168:583–596.

Colvin, J. R., Smith, D. B., and Cook, W. H. (1954). The microheterogeneity of proteins. *Chem. Revs.* 54:687–711.

Combes, R. (1946). Marin Molliard. *Rev. Gen. Bot.* 53:145–157.

Comfort, N. C. (1995). Two genes, no enzyme: A second look at Barbara McClintock and the 1951 Cold Spring Harbor Symposium. *Genetics* 140:1161–1166.

Conant, J. B. (1950). *The Overthrow of the Phlogiston Theory: The Chemical Revolution of 1775–1789.* Cambridge: Harvard University Press.

——— (1970). Theodore William Richards and the periodic table. *Science* 168:425–428.

——— (1974). Theodore William Richards. *Biog. Mem. NAS* 44:251–286.

Conant, J. B., and Fieser, L. F. (1925). Methemoglobin. *J. Biol. Chem.* 62:595–622.

Conaway, R. C., and Conaway, J. W. (1993). General initiation factors for RNA polymerase II. *Ann. Rev. Biochem.* 62:161–190.

Conklin, E. G. (1905). Organ-forming substances in the eggs of Ascidians. *Biol. Bull.* 8:205–230.

——— (1913). William Keith Brooks. *Biog. Mem. NAS* 7:23–88.

Connstein, W., and Lüdecke, K. (1919). Ueber Glyceringewinnung durch Zucker. *Ber. chem. Ges.* 52:1385–1391.

Consden, R., Gordon, A. H., and Martin, A. J. P. (1944). Quantitative analysis of proteins: A partition chromatographic method using paper. *Biochem. J.* 38:476–484.

Consden, R., Gordon, A. H., Martin, A. J. P., and Synge, R. L. M. (1947). Gramicidin S: The sequence of the amino acid residues. *Biochem. J.* 41:596–602.

Cook, A. H. (1960). Ian Morris Heilbron. *Biog. Mem. FRS* 6:65–86.

Cookson, G. H., and Rimington, C. (1954). Porphobilinogen. *Biochem. J.* 57:476–484.

Cooter, R., and Pumphrey. S. (1994). Separate spheres and public places: On the history of science popularization and science in popular culture. *Hist. Sci.* 32:237–267.

Copeman, W. S. C. (1969). The story of gout. *Medical College of Virginia Quarterly* 5:2–8.

Corey, R. B. (1948). X-ray studies on amino acids and peptides. *Adv. Protein Chem.* 4:385–406.

Corey, R. B., and Pauling, L. (1953). Fundamental dimensions of peptide chains. *Proc. Roy. Soc.* B141:10–20.

Cori, C. F. (1931). Mammalian carbohydrate metabolism. *Physiol. Revs.* 11:143–275.

——— (1946). Enzymatic reactions in carbohydrate metabolism. *Harvey Lectures* 41:253–272.

——— (1969). The call of science. *Ann. Rev. Biochem.* 38:1–20.

——— (1978). Earl Wilbur Sutherland. *Biog. Mem. NAS* 49:319–350.

——— (1983). Embden and the glycolytic pathway. *TIBS* 8:257–259.

Cori, C. F., and Cori, G. T. (1929). Glycogen formation in the liver from *d*- and *l*-lactic acid. *J. Biol. Chem.* 81:389–403.

——— (1936). Mechanism of formation of hexose-monophosphate in muscle and isolation of a new phosphate ester. *Proc. Soc. Exp. Biol. Med.* 34:702–705.

Cori, G. T. (1954). Glycogen structure and enzyme deficiencies in glycogen storage disease. *Harvey Lectures* 48:145–171.

Cori, G. T., and Cori, C. F. (1940). Kinetics of the enzymatic synthesis of glycogen from glucose-1-phosphate. *J. Biol. Chem.* 135:733–756.

——— (1952). Glucose-6-phosphatase of the liver in glycogen storage disease. *J. Biol. Chem.* 199:661–667.

Cori, G. T., Cori, C. F., and Schmidt, G. (1939). The role of glucose-1 phosphate in the formation of blood sugar and synthesis of glycogen in the liver. *J. Biol. Chem.* 129:629–630.

Cori, G. T., Slein, M. W., and Cori, C. F. (1945). Isolation and crystallization of *d*-glyceraldehyde-3-phosphate dehydrogenase from rabbit muscle. *J. Biol. Chem.* 159:565–566.

Corner, G. W. (1958). *Anatomist at Large.* New York: Basic.

——— (1965). *A History of the Rockefeller Institute, 1901–1953: Origins and Growth.* New York: Rockefeller University Press.

——— (1981). *The Seven Ages of a Medical Scientist.* Philadelphia: University of Pennsylvania Press.

Cornforth, J. W. (1976). Asymmetry and enzyme action. *Science* 193:121–125.

——— (1993). The trouble with synthesis. *Austr. J. Chem.* 46:157–170.

Corran, H. S., Green, D. E., and Straub, F. B. (1939). On the catalytic function of heart flavoprotein. *Biochem. J.* 33:793–801.

Correns, C. (1900). G. Mendel's Regel über das Verhalten der Nachkommen schaft der Rassenbastarde. *Ber. bot. Ges.* 18:158–168.

Corrigan, J. J. (1969). D-amino acids in animals. *Science* 164:142–149.

Coryell, C. D. (1940). The proposed terms "exergonic" and "endergonic" for thermodynamics. *Science* 92:380.

Coser, A. (ed.) (1975). *The Idea of Social Structure: Papers in Honor of Robert K. Merton.* New York: Harcourt, Brace and Jovanovich.

Costa, A. B. (1962). *Michel Eugène Chevreul: Pioneer of Organic Chemistry.* Madison: University of Wisconsin Press.

——— (1971a). Adolf Carl Ludwig Claus. *DSB* 3:299–301.

——— (1971b). James Dewar. *DSB* 4:78–81.

——— (1971c). Arthur Michael (1853–1942): The meeting of thermodynamics and organic chemistry. *J. Chem. Ed.* 48:243–246.

Cotton, R. G. H., and Gibson, F. (1967). The biosynthesis of tyrosine in *Aero bacter aerogenes:* Partial purification of the T protein. *Biochim. Biophys. Acta* 147:222–237.

Coulson, C. A. (1961). *Valence.* 2d ed. Oxford: Oxford University Press.

Couper, A. S. (1858). On a new chemical theory. *Phil. Mag.* [4]53:469–489.

Courrier, R. (1957a). Lucien Cuénot. *Not. Acad. Sci.* 3:332–389.

——— (1957b). Sergei Nikolaevich Winogradsky. *Not. Acad. Sci.* 3:677–713.

——— (1972). Gabriel Bertrand. *Not. Acad. Sci.* 5:111–171.

——— (1974). Georges Urbain. *Not. Acad. Sci.* 6:1–28.

Courtois, J. E., and Malengeau, P. (1952). *Histoire de la Société de Chimie Biologique.* Paris: Société de Chimie Biologique.

Courtot, C. (1936). Victor Grignard. *Bull. Soc. Chim.* [5]3:1433–1472.

Coux, O., Tanaka, K., and Goldberg, A. L. (1996). Structure and function of the 20S and 26S proteasomes. *Ann. Rev. Biochem.* 65:801–847.

Cowen, D. L. (1983). The nineteenth-century German immigrant and American pharmacy. In *Perspektiven der Pharmaziegeschichte* (P. Dilg et al., eds.), pp. 13–28. Graz: Akademische Druckund Verlaganstalt.

Cox, M. M., and Lehman, I. R. (1987). Enzymes of general recombination. *Ann. Rev. Biochem.* 56:229–262.

Crabtree, B., and Taylor, D. J. (1979). Thermodynamics and metabolism. In *Biochemical Thermodynamics* (M. N. Jones, ed.), pp. 333–378. Amsterdam: Elsevier.

Crafts, J. M. (1900). Friedel memorial lecture. *J. Chem. Soc.* 77:993–1018.

Craig, J. E. (1983). Higher education and social mobility in Germany. In *The Transformation of the Higher Learning* (K. H. Jarausch, ed.), pp. 219–244. Stuttgart: Klett-Cotta.

——— (1984). *Scholarship and Nation Building: The Universities of Strasbourg and Alsatian Society, 1870–1939.* Chicago: University of Chicago Press.

Crandall, D. I., and Gurin, S. (1949). Studies of acetoacetate formation with labeled carbon, 1: Experiments with pyruvate, acetate, and fatty acids in washed liver homogenates. *J. Biol. Chem.* 181:829–843.

Crane, F. L., Widmer, C., Lester, R. L., and Hatefi, Y. (1959). Coenzyme Q and the succinoxidase activity of the electron transfer particle. *Biochim. Biophys. Acta* 31:476–489.

Crane, R. K. (1965). Na^{+}-dependent transport in the intestine and other animal tissues. *Fed. Proc.* 24:1000–1006.

——— (1983). The road to ion-coupled membrane processes. In *Selected Topics in the History of Biochemistry: Personal Recollections* (G. Semenza, ed.), pp. 43–72. Amsterdam: Elsevier.

Cranefield, P. F. (1957). The organic physics of 1847 and the biophysics of today. *J. Hist. Med.* 12:407–423.

——— (1966). The philosophical and cultural interests of the biophysics movement of 1847. *J. Hist. Med.* 21:1–7.

——— (1988). Carl Ludwig and Emil du Bois-Reymond: A study in contrasts. *Gesnerus* 45:271–282.

Crawford, E. (1990). The universe of international science, 1880–1939. In *Solomon's House Revisited* (T. Frängsmyr, ed.), pp. 251–269. Canton, Mass.: Science History Publications.

——— (1996). *Arrhenius: From Ionic Theory to the Greenhouse Effect.* Canton, Mass.: Science History Publications.

Crawford, E., Heilbron, J. L., and Ullrich, H. (1987). *The Nobel Population.* Berkeley: Office of Science and Technology.

Creagar, A. (1995). Review of *A Skeptical Biochemist. J. Hist. Biol.* 28:174–176.

——— (1996). Wendell Stanley's dream of a free-standing biochemistry department at the University of California, Berkeley. *J. Hist. Biol.* 29:331–360.

Creager, A. N. H., and Gaudillière, J. P. (1996). Meanings in search of experiment and vice-versa: The invention of allosteric regulation in Paris and Berkeley, 1959–1968. *Stud. Hist. Biol. Sci.* 27:1–89.

Creese, M. R. S. (1991). British women of the nineteenth and early twentieth centuries who contributed to research in the chemical sciences. *Brit. J. Hist. Sci.* 24:275–305.

Creighton, H. B., and McClintock, B. (1931). A correlation of cytological and genetical crossing-over in *Zea mays. PNAS* 17:492–497.

Creighton, T. E. (ed.) (1992). *Protein Folding.* New York: Freeman.

Crick, F. H. C. (1958). On protein synthesis. *Symp. Soc. Exp. Biol.* 12:138–163.

——— (1959). The present position of the coding problem. In *Structure and Function of Genetic Elements,* pp. 35–38. Upton, N.Y.: Brookhaven National Laboratory.

——— (1962). The recent excitement in the coding problem. *Prog. Nucleic Acid Res.* 1:163–217.

——— (1966). *Of Molecules and Men.* Seattle: University of Washington Press.

——— (1970). The Central Dogma of molecular biology. *Nature* 227:561–563.

——— (1974). The double helix: A personal view. *Nature* 248:766–769.

——— (1988). *What Mad Pursuit.* New York: Basic.

——— (1993). Looking backwards: A birthday card for the double helix. *Gene* 135:15–18.

——— (1995). DNA: A cooperative discovery. *Ann. N. Y. Acad. Sci.* 758:198–199.

Crick, F. H. C., Barnett, L., Brenner, S., and Watts-Tobin, R. J. (1961). General nature of the genetic code for proteins. *Nature* 192:1227–1232.

Crick, F. H. C., Griffith, J. S., and Orgel, L. E. (1957). Codes without commas. *PNAS* 43:416–421.

Crick, F. H. C., Wang, J. C., and Bauer, W. R. (1979). Is DNA really a double helix? *J. Mol. Biol.* 129:449–461.

Crick, F. H. C., and Watson, J. D. (1954). The complementary structure of deoxyribonucleic acid. *Proc. Roy. Soc.* A223:80–96.

Crofts, A. R. (1979). Peter Mitchell and the chemiosmotic hypothesis. *Biochem. Soc. Bull.* 1:4–7.

Crombie, A. C. (ed.) (1963). *Scientific Change.* London: Heinemann.

——— (1988). Designed in the mind: Western visions of science, nature, and humankind. *Hist. Sci.* 26:1–10.

——— (1994). *Styles of Scientific Thinking in the European Tradition.* London: Duckworth.

——— (1995). Commitments and styles of European scientific thinking. *Hist. Sci.* 33:225–238.

Crombie, J. M. (1886). On the algo-lichen hypothesis. *J. Linnean Soc.* 21:259–283.

Crook, E. M. (ed.) (1959). *Glutathione.* Cambridge: Cambridge University Press.

Crosland, M. P. (1962). *Historical Studies in the Language of Chemistry.* London: Heinemann.

——— (1967). *The Society of Arcueil: A View of French Science at the Time of Napoleon I.* London: Heinemann.

——— (1970a). Amedeo Avogadro. *DSB* 1:343–350.

——— (1970b). Jean-Baptiste Biot. *DSB* 2:133–140.

——— (1973). Lavoisier's theory of acidity. *Isis* 64:306–325.

——— (1975). The development of a professional career in science in France. *Minerva* 13:38–57.

——— (1978). *Gay-Lussac, Scientist and Bourgeois.* Cambridge: Cambridge University Press.

——— (1992). *Science Under Control: The French Academy of Sciences, 1795–1914.* Cambridge: Cambridge University Press.

——— (1994). *In the Shadow of Lavoisier: The Annales de Chemie and the Establishment of a New Science.* London: British Society for the History of Science.

Cross, C. R. (1919). James Mason Crafts. *Biog. Mem. NAS* 9:159–177.

Cross, S. J., and Albury, W. R. (1987). Walter B. Cannon, L. J. Henderson, and the organic analogy. *Osiris* [2]3:165–192.

Crowther, J. G. (1974). *The Cavendish Laboratory, 1874–1974.* New York: Science History Publications.

Cruickshank, D. W. J., Juretschko, H. J., and Kato, N. (eds.) (1992). *P. P. Ewald and His Dynamical Theory of X-ray Diffraction.* Oxford: Oxford University Press.

Cruse, W. B. T., Saludjian, P., Biala, E., Strazewski, P., Prange, T., and Kennard, O. (1994). Structure of a mispaired RNA double helix at 1.6 Å resolution and implication for the prediction of RNA secondary structure. *PNAS* 91:4160–4164.

Cuénot, L. (1902). La loi de Mendel et l'hérédité de la pigmentation chez les souris. *Arch. Zool. Exp. Gen.* [3]10:xxvii–xxx.

Culotta, C. A. (1970a). Arsène d'Arsonval. *DSB* 1:302–305.

——— (1970b). Tissue oxidation and theoretical physiology: Bernard, Ludwig, and Pflüger. *Bull. Hist. Med.* 44:109–140.

——— (1970c). On the color of blood from Lavoisier to Hoppe-Seyler, 1777–1864: A theoretical dilemma. *Episteme* 4:219–233.

——— (1972). *Respiration and the Lavoisier Tradition: Theory and Modification, 1777–1850*. Philadelphia: American Philosophical Society.

Culp, S., and Kitcher, P. (1989). Theory structure and theory change in contemporary molecular biology. *Brit. J. Phil. Sci.* 40:459–483.

Cunningham, A., and Jardine, N. (eds.) (1990). *Romanticism and the Sciences.* Cambridge: Cambridge University Press.

Cunningham, A., and Williams, P. (eds.) (1992). *The Laboratory Revolution in Medicine.* Cambridge: Cambridge University Press.

Cunningham, G. J., and McGowan, G. K. (1992). John Simon and public health, 1840–1871. In *The History of British Pathology*, pp. 204–226. Bristol: White Tree.

Curtis, R. C. (1992). Popularizing science: Polanyi or Popper? *Minerva* 29:116–130.

Curtius, T. (1904). Verkettung von Amidosäuren. *J. prakt. Chem.* [2]70:57–108.

Cuthbertson, W. F. J., and Page, J. E. (1994). Ernest Lester Smith. *Biog. Mem. FRS* 40:348–365.

Cuvier, G. (1828). *Rapport historique sur les progrès des sciences naturelles depuis 1789, et sur leur Etat Actuel.* 2d ed. Paris: Verdière et Ladrange.

Czworkowski, J., and Moore, P. B. (1996). The elongation phase of protein synthesis. *Prog. Nucleic Acid Res.* 54:293–332.

Dagognet, F. (1967). *Méthodes et doctrine dans l'oeuvre de Pasteur.* Paris: Presses Universitaires.

——— (1969). *Tableaux et langages de la chimie.* Paris: Seuil.

——— (1994). *Pasteur dans la légende.* Paris: Les Empécheurs de Penser en Rond

Dahl, P. F. (1978). Ludvig August Colding. *DSB* 15:84–87.

Dahmus, M. E. (1994). The role of multisite phosphorylation in the regulation of RNA polymerase II activity. *Prog. Nucleic Acid Res.* 48:143–179.

Dakin, H. D. (1909). The mode of oxidation in the animal organism of phenyl derivatives of fatty acids, 5: Studies on the fate of phenylvaleric acid and its derivatives. *J. Biol. Chem.* 6:221–233.

——— (1912). *Oxidations and Reductions in the Animal Body.* London: Longmans, Green.

Dakin, H. D., and Dudley, H. W. (1913). Glyoxylase, 3: The distribution of the enzyme and its relation to the pancreas. *J. Biol. Chem.* 15:463–474.

Dale, H. H. (1914). The action of certain esters and ethers of choline, and their relation to muscarine. *J. Pharm. Exp. Ther.* 6:147–190.

——— (1923). The physiology of insulin. *Lancet* i, pp. 989–993.

——— (1935). Harold Ward Dudley. *Obit. Not. FRS* 1:595–606.

——— (1955). Edward Mellanby. *Biog. Mem. FRS* 1:193–222.

——— (1957). Percival Hartley. *Biog. Mem. FRS* 3:81–100.

——— (1961). Thomas Renton Elliott. *Biog. Mem. FRS* 7:53–73.

——— (1962). Otto Loewi. *Biog. Mem. FRS* 8:67–90.

Dale, H. H., and Dudley, H. W. (1929). The presence of histamine and acetyl choline in the spleen of the ox and the horse. *J. Physiol.* 68:97–123.

Dalgleish, C. E. (1953). The template theory and the role of transpeptidation in protein biosynthesis. *Nature* 171:1027–1028.

Dalziel, K. (1983). Axel Hugo Theorell. *Biog. Mem. FRS* 29:585–621.

Dane, E., et al. (1942). The Arbeiten Heinrich Wielands. *Naturwiss.* 30:333–373.

Danielli, J. F. (1975). The bilayer hypothesis of membrane structure. In *Cell Membranes* (G. Weissmann and R. Claiborne, eds.). pp. 3–11. New York: HP.

Daniels, G. H. (1971). *Science in American Society.* New York: Knopf.

D'Arcy Thompson, R. (1974). *The Remarkable Gamgees.* Edinburgh: Ramsey Head.

Darden, L. (1991). *Theory Change in Science: Strategies for Molecular Genetics.* New York: Oxford University Press.

Darlington, C. D. (1947). Nucleic acid and the chromosomes. *Symp. Soc. Exp. Biol.* 1:252–269.

——— (1980). The evolution of genetic systems: Contributions of cytology to evolutionary theory. In *The Evolutionary Synthesis* (E. Mayr and W. B. Provine, eds.), pp. 70–80. Cambridge: Harvard University Press.

Darr, S. C., Brown, J. W., and Pace, N. R. (1992). The varieties of ribonuclease P. *TIBS* 17:178–182.

Darrigol, O. (1990). Léon Nicolas Brillouin. *DSB* 17:104–109.

Das, A. (1993). Control of transcription termination by DNA-binding proteins. *Ann. Rev. Biochem.* 62:893–930.

Daston, L. (1995). The moral economy of science. *Osiris* 10:3–24.

Daub, E. E. (1976). Gibbs Phase Rule: A centenary retrospect. *J. Chem. Ed.* 53:747–751.

Daum, G., et al. (1994). The ins and outs of Raf kinase. *TIBS* 19:474–480.

Davenport, C. B. (1917). The personality, heredity, and work of Charles Otis Whitman. *Am. Nat.* 51:5–30.

Davenport, D. A. (ed.) (1991). Michael Faraday: Chemist and Popular Lecturer. *Bull. Hist. Chem.* 11:3–104.

Davenport, H. A., and Sacks, J. (1929). Muscle phosphorus, 2: The acid hydrolysis of lactacidogen. *J. Biol. Chem.* 81:469–477.

Davenport, H. W. (1982a). Physiology, 1850–1923: The view from Michigan. *The Physiologist* 24 suppl.

——— (1982b). Epinephrin(e). *The Physiologist* 25:76–82.

——— (1992). *A History of Gastric Secretion and Digestion.* New York: Oxford University Press.

David, E. E., Jr., et al. (1992, 1993). *Responsible Science.* Washington, D.C.: National Academy Press.

Davie, E. W., and Fujikawa, K. (1975). Basic mechanisms in blood coagulation. *Ann. Rev. Biochem.* 44:799–829.

Davies, D. R., and Cohen, G. H. (1996). Interactions of protein antigens with antibodies. *PNAS* 93:7–12.

Davies, D. R., Padlan, E. A., and Sheriff, S. (1990). Antibody-antigen complexes. *Ann. Rev. Biochem.* 59:439–473.

Davies, J., Gilbert, W., and Gorini, L. (1964). Streptomycin, suppression, and the code. *PNAS* 51:883–890.

Davies, J. E., and Gassen, H. G. (1983). Synthetic gene fragments in genetic engineering: The renaissance of chemistry in molecular biology. *Ang. Chem. (Int. Ed.)* 22:12–31.

Davies, M. (1970). Peter Joseph William Debye. *Biog. Mem. FRS* 16:175–232.

——— (1990). W. T. Astbury, Rosie Franklin, and DNA: A memoir. *Ann. Sci.* 47:607–618.

Davies, P. J. (1995). Alexander Porfir'yevich Borodin (1833–1887): Composer, chemist, physician, and social reformer. *J. Med. Biog.* 3:207–217.

Davis, B. D. (1948). Isolation of biochemically deficient mutants of bacteria by penicillin. *J. Am. Chem. Soc.* 70:4267.

——— (1955). Biosynthesis of the aromatic amino acids. In *Amino Acid Metabolism* (W. D. McElroy and B. Glass, eds.), pp. 799–811. Baltimore: Johns Hopkins University Press.

——— (1989). Transcriptional bias: A non-Lamarckian mechanism for substrate induced mutations. *PNAS* 86:5005–5009.

——— (1990). The human genome and other initiatives. *Science* 249:342.

Davis, H. (1970). Joseph Erlanger. *Biog. Mem. NAS* 41:111–139.

Davis, J. S. (ed.) (1948). *Carl Alsberg, Scientist at Large.* Stanford University Press.

Davis, T. L. (1929). Paul Schützenbeger. *J. Chem. Ed.* 6:1403–1414.

Davson, H., and Danielli, J. F. (1943). *The Permeability of Natural Membranes.* Cambridge: Cambridge University Press.

Davy de Virville, A. (1970). Gaston Bonnier. In *Essays in Biohistory* (P. Smit and J. C. V. ter Laage, eds.), pp. 1–13. Utrecht: International Association for Plant Taxonomy.

Dawson, C. R., and Mallette, M. F. (1945). The copper proteins. *Adv. Protein Chem.* 2:179–248.

Dawson, M. H. (1930). The transformation of pneumococcal types. *J. Exp. Med.* 51:99–147.

Dawson, M. H., and Sia, R. H. P. (1931). The transformation of pneumococcal types. *J. Exp. Med.* 54:681–710.

Dawson, P. E., Muir, T. W., Clark-Lewis, I., and Kent, S. B. H. (1994). Synthesis of proteins by native chemical ligation. *Science* 266:776–779.

Dawson, R. F. (1948). Alkaloid biogenesis. *Adv. Enzymol.* 8:203–251.

Day, C. R. (1992). Science, applied science, and higher education in France, 1870–1945: An historiographic survey since the 1950s. *J. Soc. Hist.* 26:367–384.

Day, H. G. (1974). Elmer Verner McCollum. *Biog. Mem. NAS* 45:263–335.

Day, S. B. (1976). Edward Stevens. *DSB* 13:46–47.

Dean, A. C. R., and Hinshelwood, C. (1953). Observations on bacterial adaptation. *Symp. Soc. Gen. Microbiol.* 3:21–45.

——— (1966). *Growth, Function, and Regulation in Bacterial Cells.* Oxford: Clarendon.

Debru, C. (1983). *L'Esprit des protéines.* Paris: Hermann.

——— (1987). *Philosophie moléculaire: Monod, Wyman, Changeux.* Paris: Vrin.

——— (1990). Victor Henri. *DSB* 17:412–413.

——— (ed.) (1995). *Essays in the History of the Physiological Sciences.* Amsterdam: Rodopi.

Debru, C., Gayon, J., and Picard, J. F. (eds.) (1994). *Les Sciences biologiques et médicales en France, 1920–1950.* Paris: CNRS Editions.

Debus, A. G. (1965). *The English Paracelsians.* London: Oldbourne.

De Duve, C., and Beaufay, H. (1981). A short history of tissue fractionation. *J. Cell. Biol.* 91:293s–299s.

Deichmann, U. (1992). *Biologen unter Hitler.* Frankfurt: Campus Verlag.

Deichmann, U., and Müller-Hill, B. (1994). Biological research at universities and Kaiser-Wilhelm institutes in Nazi Germany. In *Science, Technology, and National Socialism* (M. Renneberg and M. Walker, eds.), pp. 169–183. Cambridge: Cambridge University Press.

Delage, Y. (1895). *La Structure du protoplasma, et les théories de l'hérédité et les grands problèmes de la biologie.* Paris: Reinwald.

DeLange, R. J., and Smith, E. L. (1971). Histones: Structure and function. *Ann. Rev. Biochem.* 40:279–314.

Delaunay, A. (1962). *L'Institut Pasteur des origines à aujourd'hui.* Paris: France-Empire.

Delbrück, M. (1941). A theory of autocatalytic synthesis of polypeptides and its application to the problem of chromosome reproduction. *Cold Spring Harb. Symp.* 9:122–126.

——— (1949). A physicist looks at biology. *Trans. Conn. Acad. Sci.* 38:175–191.

Delbrück, M., and Schrohe, A. (1904). *Hefe, Gärung, und Fäulnis.* Berlin: Parey.

Delbrück, M., and Stent, G. S. (1957). On the mechanism of DNA replication. In *The Chemical Basis of Heredity* (W. D. McElroy and B. Glass, eds.), pp. 699–743. Baltimore: Johns Hopkins University Press.

Delépine, M. (1951). Joseph Bienaimé Caventou. *J. Chem. Ed.* 28:454–461.

Delezenne, C. (1917). Albert Dastre. *Bull. Acad. Med.* 78:477–480.

DeLorey, T. M., and Olsen, R. W. (1992). γ-Aminobutyric acid receptor: structure and function. *J. Biol. Chem.* 267:16747–16750.

Deloume, L. (1939). *De Claude Bernard a d'Arsonval.* Paris: Baillière.

Deltete, R. J., and Thorsell, D. L. (1996). Josiah Willard Gibbs and Wilhelm Ostwald: A contrast in scientific style. *J. Chem. Ed.* 73:289–295.

DeLucia, P., and Cairns, J. (1969). Isolation of an *E. Coli* strain with a mutation affecting DNA polymerase. *Nature* 224:1164–1166.

Delwiche, C. C. (1956). Denitrification. In *Nitrogen Metabolism* (W. D. McElroy and B. Glass, eds.), pp. 233–259. Baltimore: Johns Hopkins University Press.

Demerec, M. (1935). Role of genes in evolution. *Am. Nat.* 69:125–138.

Denbigh, K. G. (1989a). Note on entropy, disorder, and disorganisation. *Brit. J. Phil. Sci.* 40:323–332.

——— (1989b). The many faces of irreversibility. *Brit. J. Phil. Sci.* 49:501–518.

Denbigh, K. G., and Denbigh, J. S. (1985). *Entropy in Relation to Incomplete Knowledge.* Cambridge: Cambridge University Press.

Denis, P. S. (1856). *Nouvelles études chimiques, physiologiques, et médicales sur les substances albuminoides, etc.* Paris: Baillière.

Dennstedt, M. (1899). Die Entwicklung der organischen Elementaranalyse. *Sammlung chemischer und chemisch-technischer Vorträge.* 4:1–114.

Denton, R. M. (1986). Early events in insulin action. *Adv. Cyclic Nucl. Res.* 29:293–341.

De Pamphilis, M. L. (1993). Origins of DNA replication in metazoan chromosomes. *J. Biol. Chem.* 268:1–4.

De Romo, A. C. (1989). Tallow and the time capsule: Claude Bernard's discovery of the pancreatic digestion of fat. *Hist. Phil. Life Sci.* 11:253–274.

Desmazières, J. B. (1827). Recherches microscopiques et physiologiques sur le genre Mycoderma. *Ann. Sci. Nat.* 10:42–67.

Desmond, A. (1994). *Huxley: The Devil's Disciple.* London: Michael Joseph.

Desnuelle, P. (1958). Claude Fromageot. *Bull. Soc. Chim. Biol.* 40:1688–1709.

——— (1983). Survey of a French biochemist's life. In *Selected Topics in the History of Biochemistry: Personal Recollections* (G. Semenza, ed.), pp. 283–311. Amsterdam: Elsevier.

Despretz, C. (1824). Recherches expérimentales sur les causes de la chaleur animale. *Ann. Chim.* [2]26:337–364.

Dessoix, D., and Arber, W. (1962). Host specificity of DNA produced by *Escherichia coli,* 2: Control over acceptance of DNA from infecting phage lambda. *J. Mol. Biol.* 5:37–49.

Determann, H., Bonhard, K., and Wieland, T. (1963). Untersuchungen über die Plastein Reaktion, 6: Einfluss der Kettenlänge und der Endgruppen des Monomeren auf die Kondensierbarkeit. *Helv. Chim. Acta* 46:2489–2509.

Deuticke, H. J. (1933). Gustav Embden. *Erg. Physiol.* 35:32–49.

Deutscher, M. (1993). Ribonuclease multiplicity, diversity, and complexity. *J. Biol. Chem.* 268:13011–13014.

De Vitt, M. (1991). *Realism and Truth.* 2d ed. Oxford: Blackwell.

Diamond, L. K. (1971). Edwin J. Cohn memorial lecture: The fulfillment of his prophecy. *Vox Sanguinis* 20:433–440.

Diénert, F. V. (1900). *Sur la fermentation du galactose et sur l'accoutumance des levures à ce sucre.* Sceaux: Charaire.

Diesenhofer, J., Epp, O., Sinning, I., and Michel, H. (1995). Crystallographic refinement at 2.7 Å resolution and refined model of the photosynthetic reaction centre from *Rhodopseudomonas viridis. J. Mol. Biol.* 246:429–457.

Dietrich, M. R. (1995). Richard Goldschmidt's "heresies" and the evolutionary synthesis. *J. Hist. Biol.* 28:431–461.

——— (1996). On the mutability of genes and geneticists. *Persp. Sci.* 4:321–345.

Dilleman, G., et al. (1989). La vie de Joseph Pelletier. *Rev. Hist. Pharm.* 36:129–214.

Dimroth, K. (1986). Adolf Windaus. *Chem. Ber.* 119:xxxi–lviii.

Dintzis, H. M. (1961). Assembly of the peptide chains of hemoglobin. *PNAS* 47:247–261.

Dische, Z. (1937). Mit dem Hauptreduktionsprozess der Blutglykolyse gekoppelte Synthese von Adenosintriphosphorsäure. *Enzymologia* 1:288–310.

——— (1953). Color reactions of nucleic acid components. In *The Nucleic Acids* (E. Chargaff and J. N. Davidson, eds.), 1:285–305. New York: Academic.

Dixon, J. E. (1995). Structure and catalytic properties of protein tyrosine phosphatases. *Ann. N.Y. Acad. Sci.* 766:18–22.

Dixon, M. (1929). Oxidation mechanisms in animal tissues. *Biol. Revs.* 4:352–397.

Dixon, M., and Elliott, K. A. C. (1929). The effect of cyanide on the respiration of animal tissues. *Biochem. J.* 23:812–830.

Dixon, M., and Thurlow, S. (1925). Studies on xanthine oxidase, 6: A cell oxidation system independent of iron. *Biochem. J.* 19:672–675.

Dixon, M., and Zerfas, L. G. (1940). The role of coenzymes in dehydrogenase systems. *Biochem. J.* 34:371–391.

Djerassi, C. (1992). *The Pill, Pigmy Chimps, and Degas' Horse.* New York: Basic.

Dobbin, L. (1934). The Couper quest. *J. Chem. Ed.* 11:331–338.

Dobbs, B. J. (1975). *The Foundations of Newton's Alchemy.* Cambridge: Cambridge University Press.

——— (1991). *The Janus Faces of Genius: The Role of Alchemy in Newton's Thought.* Cambridge: Cambridge University Press.

Dobry, A., Fruton, J. S., and Sturtevant, J. M. (1952). Thermodynamics of hydrolysis of peptide bonds. *J. Biol. Chem.* 195:149–154.

Dobzhansky, T. (1936). Position effect on genes. *Biol. Revs.* 11:364–382.

Dodge, B. O. (1927). Nuclear phenomena associated with heterothallism and homothallism in the ascomycete Neurospora. *J. Agr. Res.* 35:289–305.

——— (1939). Some problems in the genetics of the fungi. *Science* 90:379–385.

Doherty, F. J., and Mayer, R. J. (1992). *Intracellular Protein Degradation.* Oxford: Oxford University Press.

Döhren, H. von, and Kleinkauf, H. (1988). Research on nonribosomal systems: Biosynthesis of peptide antibiotics. In *The Roots of Modern Biochemistry* (H. Kleinkauf et al., eds.), pp. 355–367. Berlin: de Gruyter.

Doke, T. (1969). Establishment of biochemistry in Japan. *Jap. Stud. Hist. Sci.* 8:145–153.

Dolby, R. G. A. (1976). The case of physical chemistry. In *Perspectives on the Emergence of Scientific Disciplines* (G. Lemaine et al., eds.). The Hague: Mouton.

——— (1984). Thermochemistry versus thermodynamics: The nineteenth century. *Hist. Sci.* 22:375–400.

Dolman, C. E. (1970). Julius Oscar Brefeld. *DSB* 2:436–438.

——— (1973). Heinrich Hermann Robert Koch. *DSB* 7:420–435.

——— (1975). Lazzaro Spallanzani. *DSB* 12:553–567.

Dolphin, D., et al. (eds.) (1989). *Glutathione.* New York: Wiley.

Dombrowski, A. J., Johnson, B. D., Lometto, M., and Gross, C. A. (1996). The sigma subunit of *Escherichia coli* RNA polymerase senses promoter spacing. *PNAS* 93:8858–8862.

Donnan, F. G. (1933). Ostwald memorial lecture. *J. Chem. Soc.* 136:316–332.

——— (1948). Ernst Julius Cohen. *Obit. Not. FRS* 5:667–687.

Donohue, J. (1976). Honest Jim? *Quart. Rev. Biol.* 51:285–289.

Donovan, A. (1994). *Antoine Lavoisier.* Oxford: Blackwell.

Donovan, A. L. (1975). *Philosophical Chemistry in the Scottish Enlightenment.* Edinburgh: Edinburgh University Press.

Doolittle, W. F., and Brown, J. R. (1995). Tempo, mode, the progenitor, and the universal root. In *Tempo and Mode in Evolution* (W. M. Fitch and F. J. Ayala, eds.), pp. 3–24. Washington, D.C.: National Academy Press.

Doty, P., Marmur, J., Eigner, J., and Schildkraut, C. (1960). Strand separation and specific recombination in deoxyribonucleic acids. *PNAS* 46:461–476.

Doudna, J. A., and Szostak, J. W. (1989). RNA-catalyzed synthesis of complementary-strand RNA. *Nature* 339:519–522.

Douglas, C. G. (1936). John Scott Haldane. *Biog. Mem. FRS* 2:115–139.

Douglas, K. T. (1987). Mechanism of action of glutathione-dependent enzymes. *Adv. Enzymol.* 59:103–167.

Douglass, J., Civelli, O., and Herbert, E. (1984). Polyprotein gene expression: Generation of diversity of neuroendocrine peptides. *Ann. Rev. Biochem.* 53:665–715.

Dounce, A. L. (1952). Duplicating mechanism for peptide chain and nucleic acid synthesis. *Enzymologia* 15:251–258.

Douzou, P. (1979). The study of enzyme mechanisms by a combination of cosolvent, low-temperature, and high-pressure techniques. *Quarterly Review of Biophysics* 12:521–569.

——— (1982). Developments in low-temperature biochemistry and biology. *Proc. Roy. Soc.* B217:1–28.

Dowmont, Y. P., and Fruton, J. S. (1952). Chromatography of peptides as applied to transamidation reactions. *J. Biol. Chem.* 197:271–283.

Drabkin, D. (1958). *Thudichum: Chemist of the Brain.* Philadelphia: University of Pennsylvania Press.

——— (1978). Selected landmarks in the history of porphyrins and their biologically functional derivatives. In *The Porphyrins* (D. Dolphin, ed.), 1A:29–83. New York: Academic.

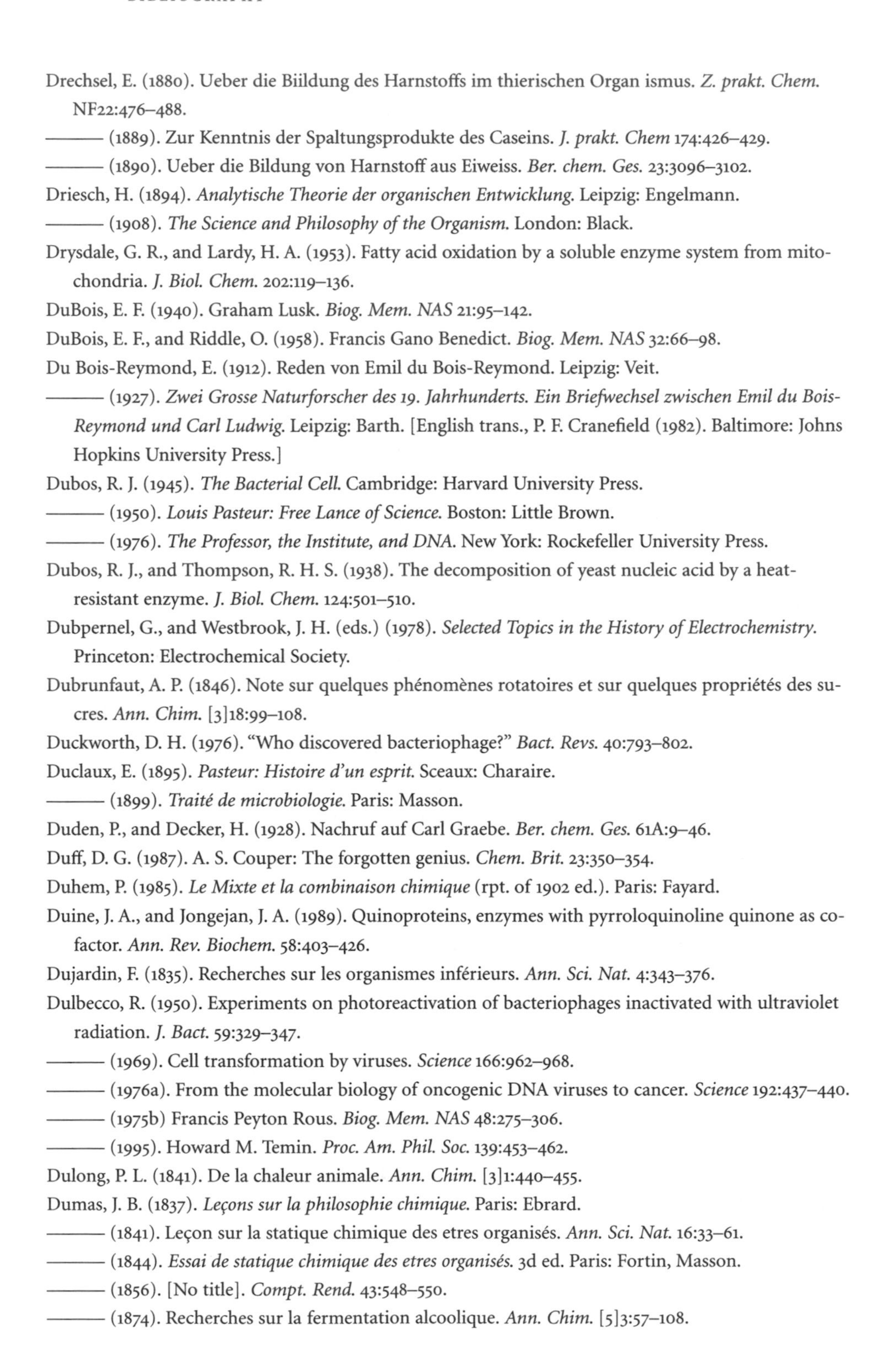

Drechsel, E. (1880). Ueber die Biildung des Harnstoffs im thierischen Organ ismus. *Z. prakt. Chem.* NF22:476–488.

——— (1889). Zur Kenntnis der Spaltungsprodukte des Caseins. *J. prakt. Chem* 174:426–429.

——— (1890). Ueber die Bildung von Harnstoff aus Eiweiss. *Ber. chem. Ges.* 23:3096–3102.

Driesch, H. (1894). *Analytische Theorie der organischen Entwicklung.* Leipzig: Engelmann.

——— (1908). *The Science and Philosophy of the Organism.* London: Black.

Drysdale, G. R., and Lardy, H. A. (1953). Fatty acid oxidation by a soluble enzyme system from mitochondria. *J. Biol. Chem.* 202:119–136.

DuBois, E. F. (1940). Graham Lusk. *Biog. Mem. NAS* 21:95–142.

DuBois, E. F., and Riddle, O. (1958). Francis Gano Benedict. *Biog. Mem. NAS* 32:66–98.

Du Bois-Reymond, E. (1912). Reden von Emil du Bois-Reymond. Leipzig: Veit.

——— (1927). *Zwei Grosse Naturforscher des 19. Jahrhunderts. Ein Briefwechsel zwischen Emil du Bois-Reymond und Carl Ludwig.* Leipzig: Barth. [English trans., P. F. Cranefield (1982). Baltimore: Johns Hopkins University Press.]

Dubos, R. J. (1945). *The Bacterial Cell.* Cambridge: Harvard University Press.

——— (1950). *Louis Pasteur: Free Lance of Science.* Boston: Little Brown.

——— (1976). *The Professor, the Institute, and DNA.* New York: Rockefeller University Press.

Dubos, R. J., and Thompson, R. H. S. (1938). The decomposition of yeast nucleic acid by a heat-resistant enzyme. *J. Biol. Chem.* 124:501–510.

Dubpernel, G., and Westbrook, J. H. (eds.) (1978). *Selected Topics in the History of Electrochemistry.* Princeton: Electrochemical Society.

Dubrunfaut, A. P. (1846). Note sur quelques phénomènes rotatoires et sur quelques propriétés des sucres. *Ann. Chim.* [3]18:99–108.

Duckworth, D. H. (1976). "Who discovered bacteriophage?" *Bact. Revs.* 40:793–802.

Duclaux, E. (1895). *Pasteur: Histoire d'un esprit.* Sceaux: Charaire.

——— (1899). *Traité de microbiologie.* Paris: Masson.

Duden, P., and Decker, H. (1928). Nachruf auf Carl Graebe. *Ber. chem. Ges.* 61A:9–46.

Duff, D. G. (1987). A. S. Couper: The forgotten genius. *Chem. Brit.* 23:350–354.

Duhem, P. (1985). *Le Mixte et la combinaison chimique* (rpt. of 1902 ed.). Paris: Fayard.

Duine, J. A., and Jongejan, J. A. (1989). Quinoproteins, enzymes with pyrroloquinoline quinone as cofactor. *Ann. Rev. Biochem.* 58:403–426.

Dujardin, F. (1835). Recherches sur les organismes inférieurs. *Ann. Sci. Nat.* 4:343–376.

Dulbecco, R. (1950). Experiments on photoreactivation of bacteriophages inactivated with ultraviolet radiation. *J. Bact.* 59:329–347.

——— (1969). Cell transformation by viruses. *Science* 166:962–968.

——— (1976a). From the molecular biology of oncogenic DNA viruses to cancer. *Science* 192:437–440.

——— (1975b) Francis Peyton Rous. *Biog. Mem. NAS* 48:275–306.

——— (1995). Howard M. Temin. *Proc. Am. Phil. Soc.* 139:453–462.

Dulong, P. L. (1841). De la chaleur animale. *Ann. Chim.* [3]1:440–455.

Dumas, J. B. (1837). *Leçons sur la philosophie chimique.* Paris: Ebrard.

——— (1841). Leçon sur la statique chimique des etres organisés. *Ann. Sci. Nat.* 16:33–61.

——— (1844). *Essai de statique chimique des etres organisés.* 3d ed. Paris: Fortin, Masson.

——— (1856). [No title]. *Compt. Rend.* 43:548–550.

——— (1874). Recherches sur la fermentation alcoolique. *Ann. Chim.* [5]3:57–108.

——— (1885). Théophile Jules Pelouze. *Discours et Eloges Académiques I*, pp. 127–198. Paris: Gauthiers-Villars.

Dunn, L. C. (1965). William Ernest Castle. *Biog. Mem. NAS* 38:31–80.

——— (1969). Genetics in historical perspective. In *Genetics Organization* (E. W. Caspari and A. W. Ravin, eds.), pp. 1–90. New York: Academic.

——— (1973). Xenia and the origin of genetics. *Proc. Am. Phil. Soc.* 117:105–111.

Dupré, J. (1993). *The Disorder of Things: Metaphysical Foundations of the Disunity of Science.* Cambridge: Harvard University Press.

Dupré, P. (1993). Thudichum and Dupré: Brothers-in-law. *J. Roy. Soc. Med.* 86:417–420.

Durell, J., and Fruton, J. S. (1954). Proteinase-catalyzed transamidation and its efficiency. *J. Biol. Chem.* 207:487–500.

Durell, J., and Garland, J. T. (1969). Acetylcholine-stimulated phosphoesteratic cleavage of phosphoinositides: Hypothetical role in membrane depolarization. *Ann. N.Y. Acad. Sci.* 165:743–754.

Du Vigneaud. V. (1952). *A Trail of Research.* Ithaca, N.Y.: Cornell University Press.

Eagle, H. (1959). Amino acid metabolism in mammalian cell culture. *Science* 130:432–437.

——— (1960). Metabolic studies with normal and malignant human cells in culture. *Harvey Lectures* 54:156–175.

Earle, W. R., Bryant, J. C., Schilling, E. L., and Evans, V. J. (1956). Growth of cell suspensions in tissue culture. *Ann. N.Y. Acad. Sci* 63:666–682.

Easson, L., and Stedman, E. (1933). Studies on the relationship between chemical constitution and physiological action, 5: Molecular dissymetry and physiological activity. *Biochem. J.* 27:1257–1266.

Eckart, W. U. (1991). Friedrich Althoff und die Medizin. In *Wissenschafts geschichte und Wissenschaftspolitik im Industriezeitalter: Das "System Althoff" in Historischer Perspektive* (B. vom Brocke, ed.), pp. 375–494. Hildesheim: Lax.

Ede, A. (1993). When is a tool not a tool? Understanding the role of laboratory equipment in the early colloidal chemistry laboratory. *Ambix* 40:11–24.

——— (1996). Colloids and quantification: The ultracentrifuge and its transformation of colloid chemistry. *Ambix* 43:32–45.

Edelman, A. M., Blumenthal. D. K., and Krebs, E. G. (1987). Protein serine/threonine kinases. *Ann. Rev. Biochem.* 56:567–613.

Edelman, G. M. (1873). Antibody structure and molecular immunology. *Science* 180:830–840.

Edelman, J. (1956). The formation of oligosaccharides by enzymic transglycosylation. *Adv. Enzymol.* 17:189–232.

Edge, D. (1994). On keeping bouncing. *Science, Technology, and Human Values* 19:366–385.

Edman, P. (1950). Method for determination of the amino acid sequence of peptides. *Acta Chim. Scand.* 4:283–293.

Edsall, J. T. (1961). Edwin Joseph Cohn. *Biog. Mem. NAS* 35:47–84.

——— (1962). Proteins as macromolecules: An essay on the development of the macromolecule concept and some of its vicissitudes. *Arch. Biochem. Biophys.* Suppl. 1:12–20.

——— (1972). Blood and hemoglobin: The evolution of knowledge of function and adaptation in a biochemical system. *J. Hist. Biol.* 5:205–257.

——— (1977). Immigrant scientists and American biochemistry. *TIBS* 2:N51–N53.

——— (1980a). The Journal of Biological Chemistry after seventy-five years. *J. Biol. Chem.* 255:8939–8951.

——— (1980b). Hemoglobin and the origins of the concept of allosterism. *Fed. Proc.* 39:226–235.

——— (1985a). Carbon dioxide in the blood: Equilibrium between red cells and plasma. The work of D. D. van Slyke and L. J. Henderson. *Hist. Phil. Life Sci.* 7:105–120,

——— (1985b). Jeffries Wyman and myself: A story of two interacting lives. In *Selected Topics in the History of Biochemistry* (G. Semenza, ed.), pp. 99–195. Amsterdam: Elsevier.

——— (1990). Jeffries Wyman: Scientist, philosopher, and adventurer. *Biophys. Chem.* 37:7–14.

——— (1995a). Hsien Wu and the first theory of protein denaturation. *Adv. Protein Chem.* 46:1–5.

——— (1995b). Shiro Akabori. *Protein, Nucleic Acid, and Enzyme.* 40:217–221.

Edsall, J. T., and Stockmayer, W. H. (1980). George Scatchard. *Biog. Mem. NAS* 52:335–377.

Edson, N. L., and Leloir, L. F. (1936). Ketogenesis-antiketogenesis, 5: Metabolism of ketone bodies. *Biochem. J.* 30:2319–2332.

Edwards, P. A. W. (1981). Some properties and applications of monoclonal antibodies. *Biochem. J.* 200:1–10.

Eggleton, P., and Eggleton, G. P. (1927). The physiological significance of phosphagen. *J. Physiol.* 63:155–161.

——— (1928). Further observations on phosphagen. *J. Physiol.* 65:15–24.

Ehrlich, F. (1904). Ueber das natürliche Isomere des Leucins. *Ber. chem. Ges.* 37:1809–1840.

Ehrlich, P. (1877). Beiträge zur Kenntnis der Anilinfärbungen ihrer Verwendung in der mikroskopischen Technik. *Arch. mikr. Anat.* 13:263–277.

——— (1885). *Das Sauerstoff-Befürfnis des Organismus. Eine Farbenanalytische Studie.* Berlin: Hirschwald.

——— (1897). Die Wertmessung des Diphtherieheilserums und deren theoretischen Grundlagen. *Klinisches Jahbuch* 6:299–326.

——— (1900). On immunity with special reference to cell life. *Proc. Roy. Soc.* 66:424–448.

——— (1910). *Studies on Immunity.* 2d ed. New York: Wiley.

Ehrlich, P., and Morgenroth, J. (1900). Ueber Haemolysine. Dritte Mitteilung. *Berlin-Klinische Wochenschrift* 37:453–458.

Einbeck, H. (1914). Ueber das Vorkommen der Fumarsäure im frischen Fleische. *Z. Physiol. Chem.* 90:301–308.

Eisenberg, D. (1994). Max Perutz's achievements: How did he do it? *Protein Science* 3:1625–1628.

Eisenberg, H. (1990). Never a dull moment: Peripatetics through the gradient of science and life. In *Selected Topics in the History of Biochemistry* (G. Semenza and R. Jaenicke, eds.), pp. 265–348. Amsterdam: Elsevier.

Eisenhart, C. (1974). Karl Pearson. *DSB* 10:447–473.

Eisner, T., and Meinwald, J. (eds.) (1995). *Chemical Ecology: The Chemistry of Biotic Interaction.* Washington, D.C.: National Academy Press.

Eley, D. D. (1976). Eric Keightley Rideal. *Biog. Mem. FRS* 22:381–413.

Elias, N., Martins, H., and Whitley, R. (1982). *Scientific Establishments and Hierarchies.* Dordrecht: Reidel.

Ellinger, A. (1904). Die Entstehung der Kynureninsäure. *Z. physiol. Chem.* 43:325–337.

——— (1913). Max Jaffé. *Ber. chem. Ges.* 46:831–847.

Ellinger, P., and Koschara, W. (1934). The lyochromes: A new group of animal pigments. *Nature* 133:553–556.

Elliott, T. R. (1905). On the action of adrenalin. *J. Physiol.* 32:401–467.

——— (1933). Walter Morley Fletcher. *Obit. Not. FRS* 1:153–163.

Ellis, E., and Delbrück, M. (1939). The growth of bacteriophage. *J. Gen. Physiol.* 22:365–384.

Elsden, S. R. (1981). Hydrogenase, 1931–1981. *TIBS* 6:251–253.

——— (1982). Roy Markham. *Biog. Mem. FRS* 28:319–345.

Elvehjem, C. A. (1935). The biological significance of copper and its relation to iron metabolism. *Physiol. Revs.* 15:471–507.

——— (1954). Edwin Bret Hart. *Biog. Mem. NAS* 28:117–161.

Elvehjem, C. A., Madden, R. J., Strong, F. M., and Woolley, D. W. (1937). Relation of nicotinic acid and nicotinic acid amide to canine black tongue. *J. Am. Chem. Soc.* 59:1767–1768.

Emala, C. W., Schwindinger, W. F., Wand, G. S., and Levine, M. A. (1994). Signal transducing G proteins: Basic and clinical implications. *Prog. Nucleic Acid Res.* 47:81–111.

Embden, G. (1924). Untersuchungen über des Verlauf der Phosphorsäuren und Milchsäurebildung bei der Muskeltätigkeit. *Klin. Wchschr.* 3:1393–1396.

Embden, G., Deuticke, H. J., and Kraft, G. (1933). Ueber die intermediären Vorgänge bei der Glykolyse in der Muskelatur. *Klin. Wchschr.* 12:213–215.

Embden, G., and Laquer. F. (1914). Ueber die Chemie des Lactacidogens, 1: Isolierungsversuche. *Z. Physiol. Chem.* 93:94–123.

——— (1921). Ueber die Chemie des Lactacidogens. *Z. physiol. Chem.* 113: 1–9.

Embden, G., and Michaud, L. (1908). Ueber den Abbau der Acetessigsäure im Tierkörper. *Beitr. chem. Physiol. Path.* 11:332–347.

Embden, G., and Oppenheimer, M. (1912). Ueber den Abbau der Benztraubensäure im Tierkörper. *Biochem. Z.* 45:186–206.

Embden, G., and Schmitz, E. (1910). Ueber synthetische Bildung von Amino säuren in der Leber. *Biochem. Z.* 29:423–428.

Emberger, L. (1946). Alexandre Guillermond. *Rev. Gen. Bot.* 53:337–361.

Emerson, R., and Arnold, W. (1933). The photochemical reaction in photosynthesis. *J. Gen. Physiol.* 16:191–205.

Emerson, R., and Lewis, C. M. (1943). The dependence of the quantum yield of Chlorella photosynthesis on the wave length of light. *Am. J. Bot.* 30:165–178.

Emerson, S. (1945). Genetics as a tool for studying gene structure. *Ann. Missouri Bot. Garden* 32:243–249.

Emery, V. C., and Akhtar, M. (1987). Pyridoxal phosphate dependent enzymes. In *Enzyme Mechanisms* (M. I. Page and A. Williams, eds.), pp. 345–389. London: Royal Society of Chemistry.

Engel, M. (1989). Aus der Frühgeschichte der Biochemie in Berlin, 1790 1850. *Mitt. Ges. Dtsch. Chem. Fachgr. Gesch. Chem.* 3:11–26.

——— (1996). Enzymologie und Gärungschemie: Alfred Wohls und Carl Neubergs Reaktionschemata der alkoholischen Gärung. *Mitt. Ges. Dtsch. Chem. Fachgr. Gesch. Chem.* 12:3–29.

Engelhardt, D. von (1976). *Hegel und die Chemie.* Wiesbaden: Pressler.

——— (1994). Paracelsus im Urteil des 18. Jahrhunderts. *Gesnerus* 51:165–182.

Engelhardt, V. A. (1932). Die Beziehungen zwischen Atmung und Pyrophosphatumsatz in Vogelerythrocyten. *Biochem. Z.* 45:186–206.

——— (1942). Enzymatic and mechanical properties of muscle proteins. *Yale J. Biol. Med.* 15:21–38.

——— (1974). On the dual role of respiration. *Mol. Cell. Biochem.* 5:25–33.

——— (1982). Life and science. *Ann. Rev. Biochem.* 51:1–19.

Engelhardt, V. A., and Lyubimova, M. N. (1939). Myosine and adenosine triphosphatase. *Nature* 144:668–669.

Engelhardt, W., and Decker-Hauff, H. (1963). *Quellen zur Gründung der Natur wissenschaftliche Fakultät in Tübingen, 1859–1863*. Tübingen: Mohr.

Engelmann, T. W. (1881). Neue Methode zur Untersuchung der Sauerstoffaus scheidung pflanzlicher und thierischer Organismen. *Bot. Z.* 39:441–448.

——— (1882). Ueber Sauerstoffausscheidung von Pflanzenzellen im Mikro spektrum. *Bot. Z.* 40:419–426.

Englesberg, E., Irr, J., Power, J., and Lee, N. (1965). Positive control of enzyme synthesis by gene C in the L-arabinose system. *J. Bact.* 90:946–957.

Enkvist, T. (1972). *The History of Chemistry in Finland, 1828–1918*. Helsinki: Finnish Academy of Sciences.

Enroth-Cugell, C. (1994). Ragnar Granit. *Proc. Am. Phil. Soc.* 138:329–332.

Ephrussi, B. (1953). *Nucleo-cytoplasmic Relations in Microorganisms*. Oxford: Oxford University Press.

Epstein, E., Nabors, M. W., and Stowe, B. B. (1967). Origin of indigo of woad. *Nature* 216:547–549.

Erdmann, A. M. (1964). Cornelis Adrianus Pekelharing. *J. Nutrition* 83:1–9.

Erlanger, J. (1951). William Henry Howell. *Biog. Mem. NAS* 26:153–180.

Erlenmeyer, E., and Schöffer, A. (1859). Ein experimentellkritischer Beitrag zur Kenntnis der Eiweiskörper. *Zeitschrift für Chemie* 2:315–343.

Ernster, L. (ed.) (1984). *Bioenergetics*. Amsterdam: Elsevier.

——— (1993). P/O ratio: The first fifty years. *FASEB J.* 7:1520–1524.

Ernster, L., and Schatz, G. (1981). Mitochondria: A historical review. *J. Cell. Biol.* 91:227s–255s.

Erxleben, C. P. F. (1818). *Ueber die Güte und Stärke des Bieres, und die Mittel, diese Eigenschaften zu würdigen*. Prague.

Eschenmoser, A., and Wintner, C. E. (1977). Natural product synthesis and vitamin B_{12}. *Science* 196:1410–1420.

Esposito, J. L. (1977). *Schelling's Idealism and Philosophy of Nature*. Cranbury, N.J.: Associated University Press.

Ettre, L. S. (1995). Early petroleum chemists and the beginning of chromatography. *Chromatography* 40:207–216.

Ettre, L. S., and Zlatkis, A. (1979). *75 Years of Chromatography*. Amsterdam: Elsevier.

Etzkowitz, H. (1996). Conflicts of interest and commitment in academic science in the United States. *Minerva* 34:259–277.

Eugster, C. H. (1983). 150 Jahre Chemie an der Universität Zurich. *Chymia* 37:194–237.

Euler, H. von. (1936). Die Cozymase. *Erg. Physiol.* 38:1–30.

——— (1938). Bedeutung der Wirkstoffe (Ergone), Enzyme und Hilfstoffe om Zellenleben. *Ergebnisse der Vitaminund Hormonforschung* 1:159–190.

——— (1952). In van't Hoff's laboratory in Berlin, 1899 and 1900. *Chemisch Weekblad* 48:644–645.

Euler, H. von, and Kullberg, S. (1911). Ueber die Wirkung der Phosphatese. *Z. Physiol Chem.* 74:15–28.

Euler, H. von, and Myrbäck, K. (1923). Gärungs-co-Enzym (Co-Zymase) der Hefe. I. *Z. Physiol. Chem.* 131:179–203.

Euler, H. von, Myrbäck, K., and Nilsson, R. (1928). Neuere Forschungen über den enzymatischen Kohlenhydratabbau (I). Die Mutation als einleitende Reaktion des Glucose-Abbaues und das daran beteiligte Enzymsystem. *Erg. Physiol.* 26:531–567.

Euler, H. S. von. (1968). Goeran Liljestrand. *Acta Physiol. Scand.* 72:1–8.

Eulner, H. H. (1970). *Die Entstehung der Medizinischen Spezialfächer an der Universitäten des Deutschen Sprachgebietes.* Stuttgart: Enke.

Evans, E. A. (1940). The metabolism of pyruvate in pigeon liver. *Biochem. J.* 34:829–837.

Evans, E. A., and Slotin, I. (1941). Carbon dioxide utilization by pigeon liver. *J. Biol. Chem.* 141:439–450.

Evans, M. A., and Evans, H. E. (1970). *William Morton Wheeler, Biologist.* Cambridge: Harvard University Press.

Eve, A. S., and Creasey, C. H. (1945). *Life and Work of John Tyndall.* London: Macmillan.

Ewald, P. P. (ed.) (1962). *Fifty Years of X-Ray Diffraction.* Utrecht: International Union of Crystallography.

Eyre, J. V. (1958). *Henry Edward Armstrong.* London: Butterworths.

Fabbroni, G. (1799). D'un mémoire du cit. Fabroni [sic] sur les fermentations etc. [by Fourcroy]. *Ann. Chim.* 31:299–327.

Fairclough, G. F., and Fruton, J. S. (1966). Peptide-protein interaction as studied by gel filtration. *Biochemistry* 5:673–683.

Falck, R. (1925). Oskar Brefeld. *Bot. Arch.* 11:1–25.

Fano, U. (1942). On the interpretation of radiation experiments in genetics. *Quart. Rev. Biol.* 17:244–252.

Fantini, B. (1984). Chemical and biological classification of proteins. *Hist. Phil. Life Sci.* 5:3–32.

——— (1988). Utilisation par la génétique moléculaire du vocabulaire de la théorie de l'information. In *Transfert de Vocabulaire dans les Sciences* (M. Groult, ed.), pp. 159–170. Paris: Editions du CNRS.

Faraday, M. (1839). *Experimental Researches in Electricity.* London: Taylor and Francis.

Farley, J. (1972). The spontaneous generation controversy (1700–1860): The origin of parasitic worms. *J. Hist. Biol.* 5:95–125.

——— (1974). *The Spontaneous Generation Controversy from Descartes to Oparin.* Baltimore: Johns Hopkins University Press.

Farley, J., and Geison, G. L. (1974). Science, politics, and spontaneous generation in nineteenth-century France: The Pasteur-Pouchet debate. *Bull. Hist. Med.* 48:161–198.

Fasman, G. (ed.) (1989). *Prediction of Protein Structure and the Principles of Protein Conformation.* New York: Plenum.

Fauré-Fremiet, E. (1948). L'Oeuvre de Félix Dujardin et la notion du protoplasme. *Protoplasma* 23:250–269.

——— (1964). Maurice Caullery. *Not. Acad. Sci.* 4:429–480.

Fayet, J. (1960). *La Révolution Française et la Science.* Paris: Rivière.

Federoff, N., and Botstein, D. (eds.) (1992). *The Dynamic Genome: Barbara McClintock's Ideas in the Century of Genetics.* Plainview, N.Y.: Cold Spring Harbor Laboratory Press.

Feldberg, W. (1970). Henry Hallett Dale. *Biog. Mem. FRS* 16:77–174.

Feldman, G. D. (1973). A German scientist between illusion and reality: Emil Fischer, 1909–1919. In *Deutschland in der Weltpolitik des 19. und 20. Jahrhundert* (I. Geiss and B. J. Wendt, eds.), pp. 341–362. Düsseldorf: Bertelsmann.

Felix, K. (1957). Robert Feulgen zum Gedächtnis. *Z. Physiol. Chem.* 307:1–13.

Fenn, W. O. (1961). John Raymond Murlin. *Am. Phil. Soc. Year Book,* pp. 145–152.

Fenton, H. J. R. (1894). Oxidation of tartaric acid in the presence of iron. *J. Chem. Soc.* 65:899–910.

Ferber, C. von (1956). *Die Entwicklung des Lehrkörpers der Deutschen Universitäten und Hochschulen, 1864–1954.* Göttingen: Vanderhoeck and Rupprecht.

Fermi, G., Perutz, M., Shannan, B., and Fourme, R. (1984). The crystal structure of human deoxyhaemoglobin at 1.74 Å resolution. *J. Mol. Biol.* 175:159–174.

Ferretti, L., Karnik, S. S., Khorana, H. G., Nassal, M., and Oprian, D. D. (1986). Total synthesis of a gene for bovine rhodopsin. *PNAS* 83:599–603.

Fersht, A. R. (1977). *Enzyme Structure and Mechanism.* San Francisco: Freeman.

——— (1988). Relationships between apparent binding energies measured in site-directed mutagenesis experiments and energetics of binding and catalysis. *Biochemistry* 27:1577–1587.

Fersht, A. R., Shi, J. P., Wilkinson, A. J., Blow, D. M., Carter, P., Waye, M. M. Y., and Winter, G. P. (1984). Analysis of enzyme structure and activity by protein engineering. *Ang. Chem. (Int. Ed.)* 23:467–473.

Feulgen, R. (1914). Ueber die "Kohlenhydratgruppe" in der echten Nukleinsäure. *Z. physiol. Chem.* 92:154–158.

——— (1923). *Chemie und Physiologie der Nukleinstoffe.* Berlin: Bornträger.

Feulgen, R., and Rossenbeck, H. (1924). Mikroskopisch-chemischer Nachweis einer Nukleinsäure vom Typus Thymusnukleinsäure und die darauf beruhende elektive Färbung von Zellkernen in mikroskopischen Präparaten. *Z. physiol. Chem.* 135:203–248.

Feyerabend, P. (1975). *Against Method.* London: NLB.

——— (1987). *Farewell to Reason.* London: Verso.

——— (1995). *Killing Time.* Chicago: University of Chicago Press.

Fichter, F. (1911). Rudolph Fittig. *Ber. chem. Ges.* 44:1339–1401.

Fick, A. (1882). *Mechanische Arbeit und Wärmeentwicklung bei der Muskeltätigkeit.* Leipzig: Brockhaus.

——— (1905). Ueber das Wesen der Muskelarbeit [1876]. In *Gesammelte Abhandlungen* 4:41–63. Würzburg: Stahel.

Fick, A., and Wislicenus, J. (1865). Ueber die Entstehung der Muskelkraft. *Vierteljahrschrift der Züricher naturforschenden Gesellschaft* 10:317–348.

Fieser, L. F. (1975). Arthur Michael. *Biog. Mem. NAS* 46:331–366.

Figurovski, N. R., and Soloviev, Y. I. (1954). Aleksei Ivanovich Khodnev. *Trudy Inst. Ist. Est.* 2:19–45.

Fildes, P. (1941). Inhibition of bacterial growth by indoleacrylic acid and its relation to tryptophan: An illustration of the inhibitory action of substances chemically related to an essential metabolite. *Brit. J. Exp. Path.* 22:293–298.

Findlay, A. (1947). James Dewar. In *British Chemists* (A. Findlay and W. H. Mills, eds.), pp. 30–57. London: Chemical Society.

Finger, S. (1994). *Origins of Neuroscience.* New York: Oxford University Press.

Fink, C. (1989). *Marc Bloch: A Life in History.* Cambridge: Cambridge University Press.

Finkelman, H. E., Rgast, G. M., Foord, O., and Fischer, E. H. (1996). Expression and characterization of glycogen synthase kinase-3 mutants and their effect on glycogen synthase activity in intact cells. *PNAS* 93:10228–10233.

Finlay, M. R. (1988). The German agricultural experiment stations and the beginnings of American agricultural research. *Agr. Hist.* 62(2):41–50.

——— (1991). The rehabilitation of an agricultural chemist: Justus von Liebig and the seventh edition. *Ambix* 38:155–167.

——— (1995). Early marketing of the theory of nutrition: The science and culture of Liebig's extract of meat. In *The Science and Culture of Nutrition, 1840–1940* (H. Kamminga and A. Cunningham, eds.), pp. 48–74. Amsterdam: Rodopi.

Fisch, M. (1991). *William Whewell, Philosopher of Science.* Oxford: Clarendon.

Fischer, A. (1946). *Biology of Tissue Cells.* Cambridge: Cambridge University Press.

Fischer, E. (1894). Einfluss der Konfiguration auf die Wirkung der Enzyme. *Ber. chem. Ges.* 27:2986–2993.

——— (1898). Bedeutung der Stereochemie für die Physiologie. *Z. physiol. Chem.* 26:60–87.

——— (1899a). Synthesen in der Puringruppe. *Ber. chem. Ges.* 32:435–504.

——— (1899b). Ueber die Spaltung einiger racemischer Amidosäuren in die optisch-aktiven Componenten. *Ber. chem. Ges.* 32:2451–2471.

——— (1901). Ueber die Hydrolyse des Caseins durch Salzsäure. *Z. Physiol. Chem,* 33:161–176.

——— (1902). Ueber die Hydrolyse der Proteinstoffe. *Chem. Z.* 26:939–940.

——— (1906a). Untersuchungen über Aminosäuren, Polypeptide und Proteine. *Ber. chem. Ges.* 39:530–610.

——— (1906b). *Untersuchungen über Aminosäuren, Polypeptide und Proteine,* 1: *1899–1906.* Berlin: Springer.

——— (1907a). Synthetical chemistry in its relation to biology. *J. Chem. Soc.* 91:1749–1765.

——— (1907b). Synthesen von Polypeptiden, 17. *Ber. chem. Ges.* 40:1754–1767.

——— (1907c). *Untersuchungen in der Puringruppe.* Berlin: Springer.

——— (1913). Synthese von Depsiden, Flechtenstoffen, und Gerbstoffe. *Ber. chem. Ges.* 46:3253–3289.

——— (1916). Isomerie der Polypeptide. *Sitzungsberichte der Preussischen Akademie der Wissenschaften zu Berlin,* pp. 990–1008.

——— (1922). *Aus meinem Leben.* Berlin: Springer.

——— (1923). *Untersuchungen über Aminosäuren, Polypeptide und Proteine,* 2: *1907–1919* (M. Bergmann, ed.). Berlin: Springer.

——— (1924). Die Kaiser-Wilhelm-Institute und der Zusammenhang von organ ischer Chemie und Biologie. In *Untersuchungen aus verschiedenen Gebieten* (M. Bergmann, ed.), pp. 796–816. Berlin: Springer.

Fischer, E. (1958). Meister, Lucius und Brüning, die Gründer der Farbwerke Höchst AG. *Tradition* 3:65–78.

Fischer, E., Bergmann, M., and Schotte, H. (1920). Ueber das Glucal und seine Umwandlung in eine neue Gruppe des Traubenzuckers. *Ber. chem. Ges.* 53:509–547.

Fischer, E., and Fourneau, E. (1901). Ueber einige Derivate des Glycocolls. *Ber. chem. Ges.* 34:2868–2877.

Fischer, E., and Freudenberg, K. (1913), Ueber das Tannin und die Synthese ähnlicher Stoffe, 3: Hochmolekulare Verbindungen. *Ber. chem. Ges.* 46:1116–1138.

Fischer, E., and Roeder, G. (1901). Synthese des Uracils, Thymins, und Phenyluracils. *Ber. chem. Ges.* 34:3751–3764.

Fischer, E., and Schmidmer, E. (1893). Ueber das Aufsteigen von Salzlösungen in Filterpapier. *Ann. Chem.* 272:156–163.

Fischer, E., and Thierfelder, H. (1894). Verhalten der verschiedenen Zucker gegen reine Hefe. *Ber. chem. Ges.* 27:2031–2037.

Fischer, E. H., Charbonneau, H., and Tonks, N. K. (1991). Protein tyrosine phosphatases: A diverse family of intracellular and transmembrane enzymes. *Science* 253:401–406.

Fischer, E. P. (1985). *Licht und Leben. Ein Bericht über Max Delbrück, den Wegbereiter der Molekularbiologie.* Constance: Universitätsverlag.

Fischer, E. P., and Lipson, C. (1988). *Thinking about Science: Max Delbrück and the Origins of Molecular Biology.* New York: Norton.

Fischer, H. (1934–1940). *Die Chemie des Pyrrols.* Leipzig: Akademische Verlagsgesellschaft.

Fischer, H. O. L. (1960). Fifty years "Synthetiker" in the service of biochemistry. *Ann. Rev. Biochem.* 29:1–14.

Fischer, H. O. L., and Baer, E. (1932). Ueber die 3-Glycerinaldehyd-phosphorsäure. *Ber. chem. Ges.* 65:337–345.

Fischer, J. L. (1979). Yves Delage. *Revue de Synthèse* [3]100:443–461.

——— (1995). Emile Guyénot: Connaissances biologique et théorie de la vie. In *Les Savants et l'epistémologie vers la fin du XIXe siècle* (M. Panza and J. C. Pont, eds.), pp. 221–231. Paris: Blanchard.

Fischer, J. L., and Smith J. (1984). French embryology and the "mechanics of development" from 1887 to 1910. *Hist. Phil. Life Sci.* 6:25–39.

Fischer, W., et al. (eds.) (1994). *Exodus von Wissenschaften aus Berlin.* Berlin: de Gruyter.

Fisher, J. M., and Scheller, R. H. (1988). Prohormone processing and the secretory pathway. *J. Biol. Chem.* 263:16515–16518.

Fisher, N. W. (1973). Organic classification before Kekulé. *Ambix* 20:106–131, 209–233.

——— (1974). Kekulé and organic classification. *Ambix* 21:29–52.

Fisher, R. A. (1936). Has Mendel's work been rediscovered? *Ann. Sci.* 1:115–137.

Fisher, S. (1996). William Odling: "Interpreter and liaison-officer," advocate of a new system of chemistry. *Ambix* 43:145–163.

Fishmann, E. (1985). A reconstruction of the first experiments in stereochemistry. *Janus* 72:131–156.

Fiske, C. H., and SubbaRow. Y. (1927). The nature of the "inorganic phosphate" in involuntary muscle. *Science* 65:401–403.

——— (1929). Phosphocreatine. *J. Biol. Chem.* 81:629–679.

Fittig, R. (1870). *Das Wesen und die Ziele der chemischen Forschung und des chemischen Studium.* Leipzig: Quandt und Handel.

Flechtner, H. J. (1981). *Carl Duisberg.* Düsseldorf: Econ Verlag.

Fleck, L. (1935). *Entstehung und Entwicklung einer wissenschaftlichen Tatsache.* Basel: Benno Schwalbe. English trans. 1979 (J. Trenn and R. K. Merton, eds.), Chicago: University of Chicago Press.

Fleisch, A. (1924). Some oxidation processes in normal and cancer tissue. *Biochem. J.* 18:294–311.

Fleming, D. (1968). Emigré physicists and the biological revolution. *Perspectives in American History* 2:152–189.

Flemming, H. W. (ed.) (1965). *Wie die ersten Heilmittel nach Höchst kamen.* Dokumente aus Höchster Archiven no. 8. Frankfurt: Höchst.

——— (1967). *Dr. Sells Teerdistillation in Offenbach.* Dokumente aus Höchster Archiven no. 26. Frankfurt: Höchst.

——— (1968). *Ludwig Knorr, Begründer Hoechster wissenschaftlicher Tradition.* Dokumente aus Höchster Archiven no. 31. Frankfurt: Höchst.

Flemming, W. (1879). Ueber das Verhalten des Kerns bei der Verteilung und über die Bedeutung mehrkerniger Zellen. *Arch. path. Anat. Physiol.* 77:1–29.

——— (1882). *Zellsubstanz, Kern und Zellteilung.* Leipzig: Vogel.

Fletcher, H. G. (1940). Augustin Pierre Dubrunfaut: An early sugar chemist. *J. Chem. Ed.* 17:153–156.

Fletcher, M. (1957). *The Bright Countenance: A Personal Biography of Walter Fletcher.* London: Hodder and Stoughton.

Fletcher, W. M., and Hopkins, F. G. (1907). Lactic acid in amphibian muscle. *J. Physiol.* 35: 247–309.

Fleury, P. (1966). Esquisse d'une histoire de la biochimie en France au XIXe siècle. *Biologie Médicale* 55:457–485.

Flood, W. E. (1963). *The Origins of Chemical Names.* London: Oldbourne.

Florkin, M. (1972). *A History of Biochemistry,* 1: *Proto-Biochemistry;* 2: *From Proto-Biochemistry to Biochemistry.* Amsterdam: Elsevier.

——— (1975). *A History of Biochemistry,*3: *History of the Identification of the Sources of Free Energy in Organisms.* Amsterdam: Elsevier.

——— (1979). *L'Ecole liègeoise de physiologie et son maitre Léon Fredericq.* Liège: Vaillant-Carmanne.

Fluck, E. (1989). Leopold Gmelin: Ein Heidelberger Chemiker und sein Werk. *Naturw. Rund.* 42:435–441.

Fodor, J. A. (1974). Special sciences (or the disunity of science as a working hypothesis). *Synthese* 28:97–115.

Foglia, V. G. (1980). The history of Bernardo A. Houssay's research laboratory, Instituto de Biologia y Medicina Experimental. *J. Hist. Med.* 35:380–396.

Fol, H. (1879). Recherches sur la fécondation et le commencement de l'hènologie chez divers animaux. *Mémoires de la Société de Physique et d'Histoire Naturelle de Genève* 26:89–397.

Folin, O. (1905). A theory of protein metabolism. *Am. J. Physiol.* 13:117–138.

Forbes, G. S. (1971). Josiah Parsons Cooke, Jr. *DSB* 3:397–399.

Forbes, R. J. (1948). *A Short History of the Art of Distillation.* Leiden: Brill.

——— (1954). Chemical, culinary and cosmetic arts. In *A History of Technology* (C. Singer, E. J. Holmyard, and A. R. Hall, eds.), 1:238–298. Oxford: Clarendon.

Foreman, D. (1992). The concept of negative feedback: Moore and Price. *Endocrinology* 131:543–545.

Forman, P. (1991). Independence, not transcendence for the history of science. *Isis* 82:71–86.

Forrest, D. W. (1974). *Francis Galton.* New York: Taplinger.

Foster, M. (1877). *A Text Book of Physiology.* London: Macmillan.

Foster, W. D. (1970). *A History of Medical Bacteriology.* London: Heinemann.

Fourcroy, A. F. (1789a). Extrait d'un mémoire ayant pour titre, Recherches pour servir a l'histoire du gaz azote ou de la mofette, comme principe des matières animales. *Ann. Chim.* 1:40–46.

——— (1789b). Mémoire sur l'existence de la matière albumineuse dans les végétaux. *Ann. Chim.* 3:252–262.

——— (1801). *Système des connaissances chimiques.* 3d ed. Paris: Baudouin.

Fournier, M. (1996). *The Fabric of Life: Microscopy in the Seventeenth Century.* Baltimore: Johns Hopkins University Press.

Fowden, L. (1958). New amino acids of plants. *Biol. Revs.* 33:393–441.

Fox, M. S., and Allen, M. K. (1964). On the mechanism of deoxyribonucleate integration in pneumococcal transformation. *PNAS* 52:412–419.

Fox, R. (1971). *The Caloric Theory from Lavoisier to Regnault.* Oxford: Oxford University Press.

——— (1973). Scientific enterprise and the patronage of research in France, 1800–1870. *Minerva* 11:440–494.

——— (1975). Henri Victor Regnault. *DSB* 11:352–354.

——— (1984). Science, industry, and the social order in Mulhouse, 1798–1871. *Brit. J. Hist. Sci.* 17:127–168.

Fox, R., and Weisz, G. (eds.) (1980). *The Organization of Science and Technology in France, 1808–1914.* Cambridge: Cambridge University Press.

Fox, S. W. (1945). Terminal amino acids in peptides and proteins. *Adv. Protein Chem.* 2:155–177.

——— (1956). Evolution of protein molecules and thermal synthesis of biochemical substances. *Am. Sci.* 44:347–359.

Fox, S. W., and Dose, K. (1972). *Molecular Evolution and Life.* San Francisco: Freeman.

Fox, T. D. (1987). Natural variation in the genetic code. *Ann. Rev. Genetics* 21:67–91.

Fraenkel, G., and Friedman, S. (1957). Carnitine. *Vit. Horm.* 15:73–118.

Fraenkel, S. (1901). *Die Arzneimittel-Synthese auf Grundlege der Beziehungen zwischen chemischen Aufbau und Wirkung.* Berlin: Springer.

Fraenkel-Conrat, H., Singer, B., and Williams, R. C. (1957). Infectivity of virus nucleic acid. *Biochim. Biophys. Acta* 25:87–96.

Frank, A. B. (1877). Ueber die biologische Verhältnisse des Thallus einiger Krustenflechten. *Beiträge zur Biologie der Pflanzen* 2:123–200.

Frank, F. C. (1936). [no title]. *Nature* 138:242.

Frank, O. (1908). Carl Voit. *Z. Biol.* 51:i–xxiv.

Frank, R. G. (1979). Thomas Willis. *DSB* 14:404–409.

——— (1987). American physiologists in German laboratories, 1865–1914. In *Physiology in the American Context, 1850–1940* (G. Geison, ed.), pp. 11–46. Bethesda, Md.: American Physiological Society.

Franke, W. W. (1988). Matthias Jacob Schleiden and the definition of the cell nucleus. *European Journal of Cell Biology* 47:145–156.

Franklin, A. (1986). *The Neglect of Experiment.* Cambridge: Cambridge University Press.

Franklin, K. J. (1953). *Joseph Barcroft.* Oxford: Blackwell.

Franklin, R. E., and Gosling, R. G. (1953). Molecular configuration of sodium thymonucleate. *Nature* 171:740–741.

Fraser, T. (1872). The connection between the chemical properties and the the physiological action of active substances. *Brit. Med. J.* 2:401–403.

Fred, E. B., Baldwin, I. L., and McCoy, E. (1932). *Root-Nodule Bacteria and Leguminous Plants* (supplemental bibliography, 1939). Madison: University of Wisconsin Press.

Fredericq, L., and Massart, Z. (1908). Léo Abram Errera. *Ann. Acad. Roy. Belg.* 74:131–277.

Freeth, F. A. (1957). Frederick George Donnan. *Biog. Mem. FRS* 3:23–39.

Fresco, J. R., Alberts, B. M., and Doty, P. (1960). Some molecular details of the secondary structure of ribonucleic acids. *Nature* 188:98–100.

Freudenthal, G. (1984). The role of shared knowledge in science: The failure of the constructivist programme in the sociology of science. *Soc. Stud. Sci.* 14:285–295.

——— (ed.) (1990). *Etudes sur Hélène Metzger.* Leiden: Brill.

——— (1995). *Aristotle's Theory of Material Substance: Heat and Pneuma, Heat and Soul.* Oxford: Clarendon.

Freudenthal, H. (1976). Norbert Wiener. *DSB* 14:344–347.

Freund, H., and Berg, A. (eds.) (1963–1966). *Geschichte der Mikroskopie.* Frankfurt: Umschau Verlag.

Freund, I. (1904). *The Study of Chemical Composition.* Cambridge: Cambridge University Press.

Freundlich, M. M. (1963). Origin of the electron microscope. *Science* 142:185–188.

Freyhofer, H. H. (1982). *The Vitalism of Hans Driesch.* Frankfurt: Lang.

Frické, M. (1976). The rejection of Avogadro's hypotheses. In *Method and Approach in the Physical Sciences* (C. Howson, ed.), pp. 177–307. Cambridge: Cambridge University Press.

Fridovich, I. (1974). Superoxide dismutase. *Adv. Enzymol.* 41:35–97.

——— (1986). Superoxide dismutase. *Adv. Enzymol.* 58:61–97.
——— (1995). Superoxide radical and superoxide dismutase. *Ann. Rev. Biochem.* 64: 97–112.
Friedberg, E. C., Walker, G. C., and Siede, W. (1995). *DNA Repair and Mutagenesis.* Washington, D.C.: ASM.
Friedel, C. (1878). La vie et les travaux de Delafosse. *Rev. Sci.* [2]8:481–484.
——— (1885). Notice sur la vie et les travaux de Charles Adolphe Wurtz. *Bull. Soc. Chim.* 43:i–lxxx.
——— (1898). Paul Schützenberger. *Bull. Soc. Chim.* [3]19:i–xliii.
Frieden, C., and Nichol, L. W. (eds.) (1981). *Protein-Protein Interactions.* New York: Wiley-Interscience.
Friedländer. P. (1915). Die Bedeutung der Baeyerschen Indigosynthesen. *Naturw.* 3:573–576.
Friedman, M. (1992). *Kant and the Exact Sciences.* Cambridge: Harvard University Press.
Friedmann, E. (1913). Zur Kenntnis der Abbaues der Karbonsäuren im Tier körper, 17: Ueber die Bildung von Acetessigsäure aus Essigsäure bei der Leberdurchblutung. *Biochem. Z.* 55:436–442.
Friedmann, H. C. (ed.) (1981). *Enzymes.* Stroudsburg, Pa.: Hutchinson Ross.
Fries, N. (1938). Ueber die Bedeutung von Wuchsstoffen für das Wachstum verschieden Pilze. *Symbolae Botanicae Upsaliensis* 3(2):iii–188.
Fromm, H. J. (1975). *Initial Rate Enzyme Kinetics.* Berlin: Springer.
Fruton, J. S. (1949). The synthesis of peptides. *Adv. Protein Chem.* 6:1–82.
——— (1950). The role of proteolytic enzymes in the synthesis of peptide bonds. *Yale J. Biol. Med.* 22:263–271.
——— (1951). The place of biochemistry in the university. *Yale J. Biol. Med.* 23:305–310.
——— (1957). Enzymic hydrolysis and synthesis of peptide bonds. *Harvey Lectures.* 51: 64–87.
——— (1963). Chemical aspects of protein synthesis. In *The Proteins,* 2d ed. (H. Neurath, ed.), 1:189–310. New York: Academic.
——— (1970). The specificity and mechanism of pepsin action. *Adv. Enzymol.* 33:401–433.
——— (1972). *Molecules and Life.* New York: Wiley.
——— (1974). The active site of pepsin. *Acc. Chem. Res.* 7:241–246.
——— (1976a). The emergence of biochemistry. *Science* 192:327–334.
——— (1976b). The mechanism of the catalytic action of pepsin and related proteinases. *Adv. Enzymol.* 44:1–36.
——— (1978). Enzymes in the Middle Ages? *TIBS* 3:N281.
——— (1979a). Early theories of protein structure. *Ann. N.Y. Acad. Sci.* 325:1–15.
——— (1979b). P. A. Levene and 2-deoxy-D-ribose. *TIBS* 3:49–50.
——— (1982a). The education of a biochemist. In *Of Oxygen, Fuels, and Living Matter.* Part 2 (G. Semenza, ed.), pp. 315–360. Chichester: Wiley.
——— (1982b). Proteinase-catalyzed synthesis of peptide bonds. *Adv. Enzymol.* 53:239–306.
——— (1986). Albert Lester Lehninger. *Am. Phil. Soc. Year Book,* pp. 141–144.
——— (1987a). From peptones to peptides. In *Peptides 1986* (D. Theodoropoulos, ed.), pp. 25–34. Berlin: Walter de Gruyter.
——— (1987b). Aspartyl proteinases. In *Hydrolytic Enzymes* (A. Neuberger and K. Brocklehurst, eds.), pp. 1–37. Amsterdam: Elsevier.
——— (1988a). The Liebig research group: A reappraisal. *Proc. Am. Phil. Soc.* 132:1–66.
——— (1988b). Energy-rich bonds and enzymatic peptide synthesis. In *The Roots of Modern Biochemistry* (A. Kleinkauf, H. von Döhren, and L. Jaenicke, eds.), pp. 165–180. Berlin: Walter de Gruyter.
——— (1990a). *Contrasts in Scientific Style.* Philadelphia: American Philosophical Society.

——— (1990b). Leonor Michaelis. *DSB* 18:620–625.

——— (1990c). Thomas Burr Osborne and protein chemistry. In *Perspectives in Biochemical and Genetic Regulation of Photosynthesis* (I. Zelitch, ed.), pp. 1–14. New York: Liss.

——— (1990d). Rudolf Schoenheimer. *DSB* 18:791–795.

——— (1992a). *A Skeptical Biochemist.* Cambridge: Harvard University Press.

——— (1992b). An episode in the history of protein chemistry: Pehr Edman's method of the sequential degradation of peptides. *Int. J. Peptide Protein Res.* 39:189–194.

——— (1994). *Eighty Years.* New Haven: Epikouros.

——— (1995a). *A Bio-bibliography for the History of the Biochemical Sciences since 1800.* 2d ed. Philadelphia: American Philosophical Society.

——— (1995b). Thomas Burr Osborne and Chemistry. *Bull. Hist. Chem.* 17/18:1–8.

Fruton, J. S., and Bergmann, M. (1939). The specificity of pepsin. *J. Biol. Chem.* 127:627–641.

Fruton, J. S., and Simmonds, S. (1958). *General Biochemistry.* 2d ed. New York: Wiley.

Fruton, J. S., Stein, W. H., and Bergmann, M. (1946). Chemical reactions of the nitrogen mustard gases, 5: The reaction of the nitrogen mustard gases with protein constituents. *J. Org. Chem.* 11:559–570.

Fry, I. (1996). On the biological significance of the properties of matter: L. J. Henderson's theory of the fitness of the environment. *J. Hist. Biol.* 29:155–196.

Fuerst, J. A. (1982). The role of reductionism in the development of molecular biology: Peripheral or central? *Soc. Stud. Sci.* 12:241–278.

——— (1984). The definition of molecular biology and the definition of policy: The role of the Rockefeller Foundation's policy for molecular biology. *Soc. Stud. Sci.* 14:225–237.

Fulton, J. F., and Thomson, E. H. (1947). *Benjamin Silliman.* New York: Schuman.

Fuoss, R. M. (1971). Charles August Kraus. *Biog. Mem. NAS* 42:119–159.

Furberg, S. (1952). On the structure of nucleic acids. *Acta Chem. Scand.* 6:634–640.

Furchgott, R. F. (1996). The discovery of endothelium derived relaxing factor and its importance in the identification of nitric oxide. *J. Am. Med. Assn.* 276:1186–1188.

Furchgott, R. F., and Zawadski, J. V. (1980). The obligatory role of endothelial cells in the relaxation of arterial smooth muscle by acetylcholine. *Nature* 280:373–376.

Furth, J. J., Hurwitz, J., and Anders, M. (1962). The role of deoxyribonucleic acid in ribonucleic acid synthesis. *J. Biol. Chem.* 237:2611–2619.

Fürth, O. von (1903). Die chemische Zustandsänderungen des Muskels. *Erg. Physiol.* 2:574–611.

——— (1912–1913). *Probleme der physiologischen und pathologischen Chemie.* Leipzig: Vogel.

——— (1919). Die Kolloidchemie des Muskels und ihre Beziehung zu den Problemen der Kontraction und der Starre. *Erg. Physiol.* 17:363–571.

Futai, M., Noumi, T., and Maeda, M. (1989). ATP synthase (H^+-ATPase): Results by combined biochemical and molecular biological approaches. *Ann. Rev. Biochem.* 58:111–136.

Fye, W. B. (1985). H. Newell Martin: A remarkable career destroyed by neurasthenia and alcoholism. *J. Hist. Med.* 40:133–136.

——— (1986). Carl Ludwig and the Leipzig physiological institute: "A factory of new knowledge." *Circulation* 74:920–928.

——— (1987). Growth of American physiology, 1850–1900. In *Physiology in the American Context* (G. Geison, ed.), pp. 92–129. Baltimore: American Physiological Society.

Gaissinovich, A. E. (1985). Contradictory appraisal by K. A. Timiriazev of Mendelian principles and its subsequent perception. *Hist. Phil. Life Sci.* 7:257–286.

Galambos, L., and Sturchio, J. L. (1996). The pharmaceutical industry in the twentieth century: A reappraisal of the sources of innovation. *History and Technology* 13:83–100.

Galaty, D. H. (1974). The philosophical basis of mid-nineteenth-century reductionism. *J. Hist. Med.* 29:295–316.

Gale, E. F. (1953). Assimilation of amino acids by Gram-positive bacteria and some actions of antibiotics thereon. *Adv. Protein Chem.* 8:285–391.

——— (1957a). Nucleic acids and protein synthesis. *Harvey Lectures* 51:25–63.

——— (1957b). Nucleic acids and the incorporation of amino acids. In *The Structure of Nucleic Acids and Their Role in Protein Synthesis* (E. M. Crook, ed.), pp. 47–59. Cambridge: Cambridge University Press.

Gale, E. F., and Fildes, P. G. (1965). Donald Devereux Woods. *Biog. Mem. FRS* 11:203–219.

Gale, E. F., and Folkes, J. P. (1953). The assimilation of amino-acids by bacteria, 14: Nucleic acid and protein synthesis in *Staphylococcus aureus. Biochem. J.* 53: 483–492.

——— (1955). The assimilation of amino-acids by bacteria, 20: The incorporation of labeled amino acids by disrupted staphylococcal cells. *Biochem. J.* 59:661–675.

Galison, P. (1987). *How Experiments End.* Chicago: University of Chicago Press.

Galison, P., and Stump, D. J. (eds.) (1996). *The Disunity of Science.* Stanford: Stanford University Press.

Gallagher, R. B., Gilder, J., Nossal, G. J. V., and Salvatore, G. (eds.) (1995). *Immunology: The Making of a Modern Science.* London: Academic Press.

Galzi, J. L., Edelstein, S. J., and Changeux, J. P. (1996). The multiple phenotypes of allosteric receptor mutants. *PNAS* 93:1853–1858.

Gamgee, A. (1877). On the photochemical processes in the retina. *Nature* 15:296, 477–478.

Gamow, G. (1954). Possible relation between deoxyribonucleic acid and protein structures. *Nature* 173:318.

——— (1955). Information transfer in the living cell. *Sci. Am.* 193(4):70–78.

Gamow, G., and Ycas, M. (1955). Statistical correlation of protein and ribonucleic acid composition. *PNAS* 41:1011–1019.

Ganss, G. A. (1937). *Geschichte der pharmazeutischen Chemie an der Universität Göttingen.* Marburg: Euker.

Garbers, D. L. (1989). Guanylate cyclase, a cell surface receptor. *J. Biol. Chem.* 264:9103–9106.

Garbers, D. L., and Lowe, D. G. (1994). Guanylyl cyclase receptors. *J. Biol. Chem.* 269:30741–30774.

Garen, A. (1968). Sense and nonsense in the genetic code. *Science* 160:149–159.

Garratt, B. (1962). Deuterium: Harold C. Urey. *J. Chem. Ed.* 39:583–584.

Garrod, A. E. (1902). The incidence of alkaptonuria: A study in chemical individuality.

——— (1909). *Inborn Errors of Metabolism.* London: Frowde, Hodder, and Stoughton. (A second edition appeared in 1929.)

Garrod, A. E., and Hele, T. S. (1905). The uniformity of the homogentisic acid excretion in alcaptonuria. *J. Physiol.* 33:198–205.

Garthwaite, J. (1991). Glutamate, nitric oxide, and cell-cell signalling in the nervous system. *Trends in Neurosciences* 14:60–67.

Gasking, E. B. (1967). *Investigations into Generation.* London: Hutchinson.

Gassen, H. G. (1982). The bacterial ribosome: A programmed enzyme. *Ang. Chem. (Int. Ed.)* 21:23–36.

Gaudillière, J. P. (1989). Chimie biologique ou biologie moléculaire? La biochimie au CNRS dans les années soixante. *Cahiers pour l'Histoire du CNRS* 7:91–147.

——— (1991). Catalyse enzymatique et oxydations cellulaires: L'oeuvre de Gabriel Bertrand et son héritage. In *Instutut Pasteur* (M. Morange, ed.), pp. 118–136. Paris: La Découverte.

——— (1992). J. Monod, S. Spiegelman et l'adaptation enzymatique. Programmes de recherches, cultures locales, et traditions disciplinaires. *Hist. Phil. Life Sci.* 14:23–71.

——— (1993). Molecular biology in the French tradition? Redefining local traditions and disciplinary patterns. *J. Hist. Biol.* 26:473–498.

——— (1996). Molecular biologists, biochemists, and messenger RNA: The birth of a scientific network. *J. Hist. Biol.* 29:417–445.

Gaunt, R., and Eversole, W. J. (1949). Notes on the history of the adrenal cortical problem. *Ann. N.Y. Acad. Sci.* 50:511–521.

Gaupp, E. (1917). *August Weismann.* Jena: Fischer.

Gause, G. F., and Brazhnikova, M. G. (1944). Gramicidin S. Origin and mode of action. *Lancet,* pp. 715–718.

Gautier, A. (1869). *Les Fermentations.* Paris: Savy.

Gautschi, F., and Bloch, K. (1957). On the structure of an intermediate in the biological demethylation of lanosterol. *J. Am. Chem. Soc.* 79:684–689.

Gay, H. (1976). Radicals and types. *Stud. Hist. Phil. Sci.* 7:1–51.

Gay-Lussac, J. L. (1809). Mémoire sur la combinaison des substances gazeuses, les uns avec les autres. *Mémoires de la Société d'Arcueil* 2:207–234.

——— (1810). Extrait d'un mémoire sur la fermentation. *Ann. Chim.* 76:245–259.

Gayon, J. (1994). Génétique de la pigmentation de l'oeil de la drosophile: la contribution specifique de Boris Ephrussi. In *Les Sciences biologiques et médicales en France, 1920–1950* (C. Debru et al., eds.), pp. 187–206. Paris: CNRS Editions.

Gayon, U., and Dupetit, G. (1886). Recherches sur la réduction des nitrates par les infiniments petits. *Mémoires de la Société des Sciences Physiques et Naturelles de Bordeaux* [3]2:201–307.

Gefter, M. L., Hirota, Y., Kornberg, T., Wechsler, J. A., and Barnoux, C. (1971). Analysis of DNA polymerases II and III in mutants of *Escherichia coli* thermosensitive for DNA synthesis. *PNAS* 68:3150–3153.

Geider, K., and Hoffmann-Berling, H. (1981). Proteins controlling the helical structure of DNA. *Ann. Rev. Biochem.* 50:233–260.

Geison, G. L. (1969). The protoplasmic theory of life and the vitalist mechanist debate. *Isis* 60:273–292.

——— (1971). Ferdinand Cohn. *DSB* 3:336–341.

——— (1972). Social and institutional factors in the stagnancy of English physiology. *Bull. Hist. Med.* 46:30–58.

——— (1974). Louis Pasteur. *DSB* 10:350–416.

——— (1978). *Michael Foster and the Cambridge School of Physiology.* Princeton: Princeton University Press.

——— (1981). Scientific change, emerging specialties, and research schools. *Hist. Sci.* 19:20–40.

——— (1987). International relations and domestic elites in American physiology, 1900–1940. In *Physiology in the American Context* (G. L. Geison, ed.), pp. 115–154. Bethesda, Md.: American Physiological Society.

——— (1995). *The Private Life of Louis Pasteur.* Princeton: Princeton University Press.

Geison, G. L., and Holmes, F. L. (eds.) (1993). Research schools: Historical reappraisals. *Osiris,* vol. 8.

Gelius, R. (1996). Carl Schorlemmer als Wissenschaftshistoriker: Zur Kenntnis seines unvollendeten Manuskript "Beiträge zur Geschichte der Chemie." *NTM* NS4:65–81.

Gellert, M. (1981). Gene expression. *Ann. Rev. Biochem.* 50:233–260.

Gellert, M., Little, J. W., Oshinsky, K., and Zimmerman, S. B. (1968). Joining of DNA strands by DNA ligase of *E. coli*. *Cold Spring Harbor Symp.* 31:21–25.

Gellner, E. (1985). *Relativism and the Social Sciences.* Cambridge: Cambridge University Press.

Genevois, L. (1956). Henri Devaux. *Rev. Gen. Bot.* 63:341–346.

George, P., and Rutman, R. J. (1980). The "high energy phosphate bond" concept. *Prog. Biophys. Biophys. Chem.* 19:1–53.

Gerabek, W. (1991). Der Leipziger Physiologe Carl Ludwig und die medizinische Instrumenten. *Sudhoffs Arch.* 75:171–179.

Gerber, G., and Sauer, G. (1985). Die Entdeckung des Adenin durch Albrecht Kossel—sein Leben und Wirken. *Charité Annalen* NF5:355–365.

Gerhardt, C. (1853–1856). *Traité de chimie organique.* Paris: Firmin Didot.

Gerhart, J. C., and Pardee, A. B. (1962). The enzymology of control by feedback inhibition. *J. Biol. Chem.* 237:891–896.

——— (1963). The effect of the feedback inhibitor, CTP, on subunit interactions in aspartate transcarbamylase. *Cold Spring Harbor Symp.* 28:491–496.

Gerhart, J. C., and Schachman, H. K. (1965). Distinct subunits for the regulation and catalytic activity of aspartate transcarbamylase. *Biochemistry* 4:1054–1062.

——— (1968). Allosteric interactions in aspartate transcarbamylase, 2: Evidence for different conformational states of the protein in the presence and absence of specific ligands. *Biochemistry* 7:538–552.

Gest, H. (1991). Sunbeams, cucumbers, and purple bacteria: The discovery of photosynthesis revisited. *Persp. Biol. Med.* 34:254–274.

Gesteland, R. F., and Atkins, J. F. (1996). Recoding: Dynamic reprogramming of translation. *Ann. Rev. Biochem.* 65:741–768.

Geyl, P. (1958). *Debates with Historians.* Cleveland: World.

Ghiretti, F. (1994). Bartolomeo Bizio and the rediscovery of Tyrian purple. *Experientia* 50:802–807.

Gholson, B., Shadish, W. R., Jr., Niemeyer, R. A., and Houts, A. C. (1989). *Psychology of Science.* Cambridge: Cambridge University Press.

Giacomoni, D. (1993). The origin of DNA:RNA hybridization. *J. Hist. Biol.* 26:89–107.

Gibbs, F. W. (1965). *Joseph Priestley.* London: Nelson.

Gibbs, J. W. (1906). *The Scientific Papers of J. Willard Gibbs* (H. A. Bumstead and R. G. van Name, eds.). London: Longmans, Green.

Gibson, F. (1995). Chorismic acid and beyond. In *Selected Topics in the History of Biochemistry: Personal Recollections.* Vol. 4 (E. C. Slater, R. Jaenicke, and G, Semenza, eds.), pp, 259–301. Amsterdam: Elsevier.

Giegé, R., Puglisi, J. D., and Florentz, C. (1993). tRNA structure and amino acylation efficiency. *Prog. Nucleic Acid Res.* 45:129–206.

Giere, R. N. (1989). Scientific rationality as instrumental rationality. *Stud. Hist. Phil. Sci.* 20:377–384.

Gierer, A., and Schramm, G. (1956). Die Infektiosität der Nukleinsäure aus Tabakmosaic virus. *Z. Naturforsch.* 11b:138–142.

Gilbert, G. N., and Mulkay, M. (1984). *Opening Pandora's Box: A Sociological Analysis of Scientists' Discourse.* Cambridge: Cambridge University Press.

Gilbert, W. (1981). DNA sequencing and gene structure. *Science* 214:1305–1312.

——— (1986). The RNA world. *Nature* 319:618.

Gilbert, W., and Maxam, A. (1973). The nucleotide sequence of the *Lac* operator. *PNAS* 70:3581–3584.

Gilbert, W., and Müller-Hill, B. (1966). Isolation of the *Lac* repressor. *PNAS* 56:1891–1898.

Gilboa, E., Mitra, S. W., Goff, E., and Baltimore, D. (1979). A detailed model of reverse transcriptase and test of crucial aspects. *Cell* 18:93–100.

Gillespie. D., and Spiegelman, S. (1965). A quantitative assay for DNA-RNA hybrids with DNA immobilized on a membrane. *J. Mol. Biol.* 12:829–842.

Gillespie, L. J., and Liu, T. H. (1931). The reputed dehydrogenation of hydroquinone by palladium black. *J. Am. Chem. Soc.* 53:3969–3972.

Gillis, J. (1966). August Kekulé et son oeuvre, réalisée a Gand de 1858 a 1867. *Mem. Acad. Roy. Belg.* 37(1):1–40.

Gillispie, C. C. (1980). History of science: Perceptions. *Science* 207:934.

——— (1994). Recent trends in the historiography of science. *Bull. Hist. Chem.* 15/16:19–26.

Gilman, A. G. (1987). G proteins: Transducers of receptor-generated signals. *Ann. Rev. Biochem.* 56:615–649.

Gilpin, R. (1968). *France in the Age of the Scientific State.* Princeton: Princeton University Press.

Gimmler, H. (1982). *Julius Sachs, Würzburger Botaniker und Pflanzenphysiologe.* Würzburg: Universitätsbibliothek.

Ginsburg, A. (1972). Glutamine synthetase of *Escherichia coli:* Some physical and chemical properties. *Adv. Protein Chem.* 26:1–79.

Ginsburg, A., and Stadtman, E. R. (1970). Multienzyme systems. *Ann. Rev. Biochem.* 39:429–472.

Giordano, A. (1958). Pietro Rondoni. *Verhandl. path. Ges.* 41:417–426.

Giri, L., Hill, W. E., Wittmann, H. G., and Wittmann-Liebold, B. (1984). Ribosomal proteins: Their structure and spatial arrangement in prokaryotic ribosomes. *Adv. Protein Chem.* 36:1–78.

Glas, E. (1975). The protein theory of G. J. Mulder. *Janus* 62:289–308.

——— (1978). Methodology and the emergence of physiological chemistry. *Stud. Hist. Phil. Sci.* 9:291–312.

Glass, B. (1986). Geneticists embattled: Their stand against rampant eugenics and racism in America during the 1920s and 1930s. *Proc. Am. Phil. Soc.* 130:130–154.

——— (1990). Nikolai Vladimirovich Timoféeff-Ressovsky. *DSB* 18:919–926.

Glasstone, G. P., Knight, B. C., and Wilson, G. (1973). Paul Gordon Fildes. *Biog. Mme. FRS* 19:317–347.

Glazer, A. N., and Melis, A. (1987). Photochemical reaction centers: Structure, organization, and function. *Ann. Rev. Plant Physiol.* 38:11–45.

Gley, E. (1929). Le thyroide, les progrès de l'endocrinologie et la biologie. *Endokrinologie* 5:73–81.

Glomset, J. A. (1959). The further purification and properties of a phosphatase from beef spleen able to hydrolyze completely the phosphorus of α-casein. *Biochim. Biophys. Acta* 32:349–357.

Glover, D. M., and Hames, B. D. (1995). *DNA Cloning.* 2d ed. Oxford: Oxford University Press.

Glover, J., Pennock, J. F., Pitt, G. A., et al. (1978). Richard Alan Morton. *Biog. Mem. FRS* 24:409–442.

Glusker, J. P. (1981). *Structural Crystallography in Chemistry and Biology.* London: Hutchinson Ross.

——— (1991). Structural aspects of metal liganding to functional groups in proteins. *Adv. Protein Chem.* 42:1–76.

——— (1994). Dorothy Crowfoot Hodgkin. *Protein Science* 3:2465–2469.
Goebel, K. von (1924). *Wilhelm Hofmeister.* Leipzig: Akademische Verlags gesellschaft.
Goebel, W. (1984). *August Kekulé.* Leipzig: Teubner.
Goebel, W. F. (1975). The golden era of immunology at the Rockefeller Institute. *Persp. Biol. Med.* 18:419–426.
Goertzel, T., and Goertzel, B. (1995). *Linus Pauling: A Life in Science and Politics.* New York: Basic.
Gohau, G. (1979). Alfred Giard. *Rev. Syn.* [3]95/96:393–406.
Gold, L., Politsky, B., Uhlenbeck, O., and Yarus, M. (1995). Diversity of oligonucleotide function. *Ann. Rev. Biochem.* 64:763–797.
Goldberg, N. D., O'Dea, R. F., and Haddox, M. K. (1973). Cyclic AMP. *Adv. Cyclic Nucl. Res.* 3:155–223.
Goldner, M. G. (1955). Adolf Magnus-Levy. *Proc. Virchow Med. Soc.* 14:29–35.
Goldschmidt, R. (1916). Genetic factors and enzyme reaction. *Science* 43:98–100.
——— (1938). The theory of the gene. *Sci. Mon.* 46:268–273.
Goldschmidt, R., and Kodani, M. (1942). The structure of the salivary gland chromosomes and its meaning. *Am. Nat.* 76:529–551.
Goldsmith, M. (1980). *Sage: A Life of J. B. S. Haldane.* London: Hutchinson.
Goldstein, J. L., and Brown, M. S. (1977). The low-density lipoprotein pathway and its relation to atherosclerosis. *Ann. Rev. Biochem.* 46:897–930.
Goldthwait, D. R., Peabody, R. A., and Greenberg, G. R. (1955). The biosynthesis of the purine ring. In *Amino Acid Metabolism* (W. D. McElroy and B. Glass, eds.), pp. 765–781. Baltimore: Johns Hopkins University Press.
Golinski, J. (1992). *Science as Public Culture.* Cambridge: Cambridge University Press.
——— (1994). Precision instruments and the demonstration of order of proof in Lavoisier's chemistry. *Osiris* [2]9:30–47.
Golovin, N. E. (1963). The creative person in science. In *Scientific Creativity* (C. W. Taylor and F. Barron, eds.), pp. 7–23. New York: Wiley.
Gomez-Sanchez, M. (1989). *Severo Ochoa.* Caja de Ahorros de Asturias.
Goodfield, J. (1975). Changing strategies: A comparison of reductionist attitudes in biological and medical research in the nineteenth and twentieth centuries. In *Studies in the Philosophy of Biology* (F. J. Ayala and T. Dobzhansky, eds.), pp. 65–86. London: Macmillan.
Gooding, D. (1990). *Experiment and the Making of Meaning.* Dordrecht: Kluwer.
Gooding, D., and James, F. A. J. L. (1985). *Faraday Rediscovered.* London: Macmillan.
Gooding, D., Pinch, T, and Schaffer, S. (1989). *The Uses of Experiment.* Cambridge: Cambridge University Press.
Goodman, D. C. (1969). Problems in crystallography in the early nineteenth century. *Ambix* 16:152–166.
——— (1972a). Chemistry and the two kingdoms of nature during the nineteenth century. *Med. Hist.* 16:113–130.
——— (1972b). The discovery of Brownian motion. *Episteme* 6:12–29.
——— (1976). William Hyde Wollaston. *DSB* 14:486–494.
Goodman, M., and Kenner, G. W. (1957). The synthesis of peptides. *Adv. Protein Chem.* 12:466–638.
Goodman, N. (1958). The test of simplicity. *Science* 128:1064–1069.
Goodwin, T. W. (1954). *Carotenoids: Their Comparative Biochemistry.* New York: Chemical.
——— (ed.) (1968). *Porphyrins and Related Compounds.* London: Academic Press.
Goppelsroeder, F. (1901). *Capillaranalyse.* Basel: Birkhäuser.

Gordon, A. H. (1996). Richard Laurence Millington Synge. *Biog. Mem. FRS* 42:453–479.

Gordon, A. H., Martin, A. J. P., and Synge, R. L. M. (1941). A study of the partial acid hydrolysis of some proteins, with special reference to the mode of linkage of the basic amino acids. *Biochem. J.* 35:1369–1387.

Gorter, E., and Grendel, F. (1925). On bimolecular layers of lipoids on the the chromocytes of blood. *J. Exp. Med.* 41:439–450.

Gortner, R. A. (1929). *Outlines of Biochemistry.* New York: Wiley.

Goss, S. J. (1993). A fresh look at the somatic cell genetics of hepatoma differentiation. *J. Cell. Science* 104: 231–235.

Gots, J. S. (1950). Accumulation of 5(4)-amino-4(5)-imidazole carboxamide in relation to sulfonamide bacteriostasis and purine metabolism in *Escherichia coli. Fed. Proc.* 9:178–179.

——— (1957). Purine metabolism in bacteria, 1: Feedback inhibition. *J. Biol. Chem.* 278:57–66.

Gotthelf, A., and Lennox, J. G. (eds.) (1987). *Philosophical Issues in Aristotle's Biology.* Cambridge: Cambridge University Press.

Gottschalk, A. (1921). *Ueber den Begriff des Stoffwechsels in der Biologie.* Berlin: Bornträger.

Gould, R. F. (1994). Francis Patrick Garvan. In *American Chemists and Chemical Engineers* (W. D. Miles and R. F. Gould, eds.). pp. 110–112. Guilford, Conn.: Gould.

Goulian, M. (1971). Biosynthesis of DNA. *Ann. Rev. Biochem.* 40:855–898.

Goulian, M., Kornberg, A., and Sinsheimer, R. L. (1967). Enzymatic synthesis of DNA, 24: Synthesis of infectious phage ϕX174 DNA. *PNAS* 58:2321–2326.

Goupil, M. (1991). *Du Flou au Clair? Histoire de l'affinité chimique de Cardan à Prigogine.* Paris: Editions du CTHS.

Graham, J. (1995). *Revolutionary in Exile: The Emigration of Joseph Priestley to America.* Philadelphia: American Philosophical Society.

Graham, T. (1861). Liquid diffusion applied to analysis. *Phil. Trans.* 151:183–224.

——— (1876). *Chemical and Physical Researches.* Edinburgh: University Press.

Granger, G. (1971). Marie-Jean-Antoine-Nicolas Caritat, Marquis de Condorcet. *DSB* 3:383–388.

Granick, S., and Beale, I. (1978). Hemes, chlorophyll, and related compounds: Biosynthesis and metabolic regulation. *Adv. Enzymol.* 46:33–203.

Granit, R. (1966). *Charles Scott Sherrington: An Appraisal.* London: Nelson.

Grassmann, W. (1969). Richard Kuhn. *Bayer. Akad. Wiss. Jahrbuch*, pp. 231–253.

Gratia, A. (1921). Studies on the d'Hérelle phenomenon. *J. Exp. Med.* 34:115–131.

——— (1931). Sur l'identité du phénomène de Twort et du phénomène d'Hérelle. *Ann. Inst. Pasteur* 46:1–16.

Graves, J. D., Campbell, J. S., and Krebs, E. G. (1995). Protein serine/threonine kinases of the MAPK cascade. *Ann. N.Y. Acad. Sci.* 766:320–343.

Gray, C. H. (1983). The bile pigments. *TIBS* 8:381–383.

Gray, H. B., and Solomon, E. I. (1981). Electronic structure of the blue copper centers in proteins. In *Copper Proteins* (T. G. Spiro, ed.), pp. 1–34. New York: Wiley.

Gray, H. B., and Winkler, J. R. (1996). Electron transfer in proteins. *Ann. Red. Biochem.* 65:537–561.

Gray, J. A. (1979). *Ivan Pavlov.* New York: Viking.

Gray, R. D. (1952). *Goethe the Alchemist.* Cambridge: Cambridge University Press.

Green, A. A., and Cori, G. T. (1943). Crystalline muscle phosphorylase, 1: Preparation, properties, and molecular weight. *J. Biol. Chem.* 151:21–29.

Green, D. E. (1945). Enzymes and trace substances. *Adv. Enzymol.* 1:177–198.

——— (1951). The cyclophorase complex of enzymes. *Biol. Revs.* 26:410–455.

——— (1954). Fatty acid oxidation in soluble systems of animal tissues. *Biol. Revs.* 29:330–366.

——— (1959). Electron transport and oxidative phosphorylation. *Adv. Enzymol.* 21:73–129.

Green, D. E., Dewan, J. G., and Leloir, L. F. (1937). The β-hydroxybutyric dehydrogenase of animal tissues. *Biochem. J.* 31:934–939.

Green, D. E., Needham, D. M., and Dewan, J. G. (1937). Dismutations and oxidoreductions. *Biochem. J.* 31:2327–2352.

Green, J. H. S. (1966). Sir Christopher Ingold and the Chemistry Department, University College, London. In *Studies on Chemical Structure and Reactivity* (J. H. Ridd, ed.), pp. 265–274. London: Methuen.

Green, J. R. (1898). The alcohol-producing enzyme of yeast. *J. Botany* 12:491–497.

——— (1899). *The Soluble Ferments and Fermentation.* Cambridge: Cambridge University Press.

Green, R., and Noller, H. F. (1997). Ribosomes and translation. *Ann. Rev. Biochem.* 66:679–716.

Greenaway, A. J., et al. (1932). *The Life and Work of William Henry Perkin Jr.* London: Chemical Society.

Greenberg, D. M. (1972). Distinguished biochemical discoveries and biochemists on the Berkeley campus. *Persp. Biol. Med.* 16:136–153.

Greengard, P. (1978). *Cyclic Nucleotides, Phosphorylated Proteins, and Neuronal Function.* New York: Raven.

Greenstein, G. (1990). The magician. *American Scholar* 59:118–125.

Greenstein, J. P. (1944). Nucleoproteins. *Adv. Protein Chem.* 1:209–287.

Greenwald, I. (1925). A new type of phosphoric acid compound isolated from blood, with some remarks on the effect of substitution on the rotation of *l*-glyceric acid. *J. Biol. Chem.* 63:339–349.

Gregory, F. (1977). *Scientific Materialism in Nineteenth-Century Germany.* Dordrecht: Reidel.

Griffith, F. (1928). The significance of pneumococcal types. *J. Hygiene* 27:113–159.

Griffiths, P. A., et al. (1995). *On Being a Scientist.* 2d ed. Washington, D.C.: National Academy Press.

Grigorian, N. A. (1974). Ivan Petrovich Pavlov. *DSB* 10:431–436.

Grimaux, E. (1882). Sur les colloides azotés. *Bull. Soc. Chim.* [2]38:64–69.

Grimaux, E., and Gerhardt, C. (1900). *Charles Gerhardt, sa Vie, son Oeuvre, sa Correspondence.* Paris: Masson.

Griminger, P. (1972). Casimir Funk. *J. Nutrition* 102:1105–1132.

Grimm, S., Stanger, B. Z., and Leder, P. (1996). RIP and FADD: Two "death domain"–containing proteins can induce apoptosis by convergent, but dissociable pathways. *PNAS* 93:10923–10927.

Grindley, N. D. F., and Reed, R. R. (1985). Transpositional recombination in prokaryotes. *Ann. Rev. Biochem.* 54:863–896.

Grmek, M. D. (1962). *L'Introduction de l'expérience quantitative dans les sciences biologiques.* Paris: Palais de la Découverte.

——— (1967a). *Catalogue des manuscripts de Claude Bernard,* Paris: Masson.

——— (1967b). Evolution des conceptions de Claude Bernard sur le milieu intérieur. In *Philosophie et méthodologie scientifiques de Claude Bernard,* pp. 117–150. Paris: Masson.

——— (1968a). First steps in Claude Bernard's discovery of the glycogenic function of the liver. *J. Hist. Biol.* 1:141–154.

——— (1968b). La glucogenèse et le diabète dans l'oeuvre de Claude Bernard. In *Commentaires sur six grands livres de la médecine française,* pp. 187–234. Paris: Cercle des Livres Précieuses.

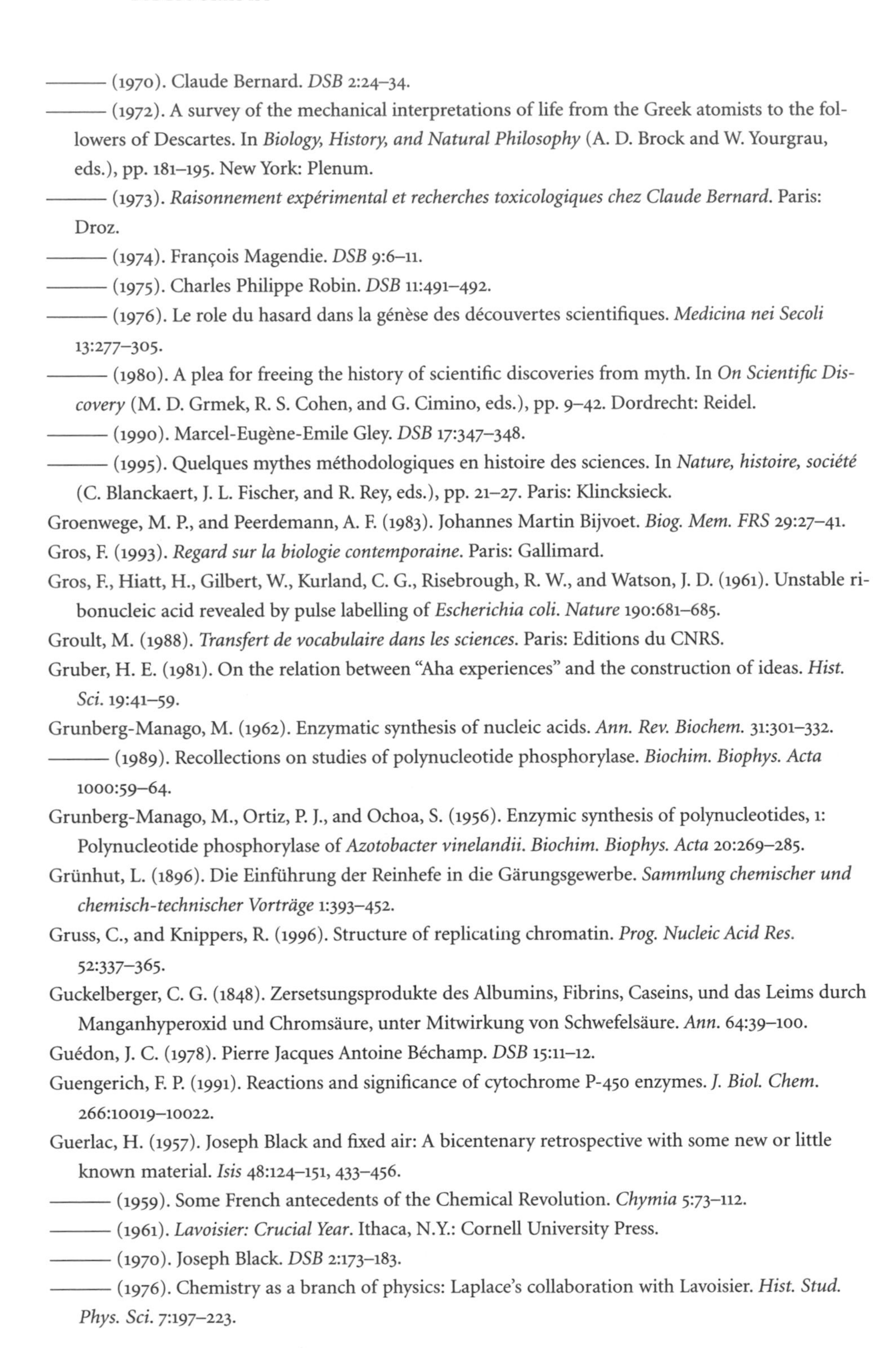

——— (1970). Claude Bernard. *DSB* 2:24–34.

——— (1972). A survey of the mechanical interpretations of life from the Greek atomists to the followers of Descartes. In *Biology, History, and Natural Philosophy* (A. D. Brock and W. Yourgrau, eds.), pp. 181–195. New York: Plenum.

——— (1973). *Raisonnement expérimental et recherches toxicologiques chez Claude Bernard*. Paris: Droz.

——— (1974). François Magendie. *DSB* 9:6–11.

——— (1975). Charles Philippe Robin. *DSB* 11:491–492.

——— (1976). Le role du hasard dans la génèse des découvertes scientifiques. *Medicina nei Secoli* 13:277–305.

——— (1980). A plea for freeing the history of scientific discoveries from myth. In *On Scientific Discovery* (M. D. Grmek, R. S. Cohen, and G. Cimino, eds.), pp. 9–42. Dordrecht: Reidel.

——— (1990). Marcel-Eugène-Emile Gley. *DSB* 17:347–348.

——— (1995). Quelques mythes méthodologiques en histoire des sciences. In *Nature, histoire, société* (C. Blanckaert, J. L. Fischer, and R. Rey, eds.), pp. 21–27. Paris: Klincksieck.

Groenwege, M. P., and Peerdemann, A. F. (1983). Johannes Martin Bijvoet. *Biog. Mem. FRS* 29:27–41.

Gros, F. (1993). *Regard sur la biologie contemporaine*. Paris: Gallimard.

Gros, F., Hiatt, H., Gilbert, W., Kurland, C. G., Risebrough, R. W., and Watson, J. D. (1961). Unstable ribonucleic acid revealed by pulse labelling of *Escherichia coli*. *Nature* 190:681–685.

Groult, M. (1988). *Transfert de vocabulaire dans les sciences*. Paris: Editions du CNRS.

Gruber, H. E. (1981). On the relation between "Aha experiences" and the construction of ideas. *Hist. Sci.* 19:41–59.

Grunberg-Manago, M. (1962). Enzymatic synthesis of nucleic acids. *Ann. Rev. Biochem.* 31:301–332.

——— (1989). Recollections on studies of polynucleotide phosphorylase. *Biochim. Biophys. Acta* 1000:59–64.

Grunberg-Manago, M., Ortiz, P. J., and Ochoa, S. (1956). Enzymic synthesis of polynucleotides, 1: Polynucleotide phosphorylase of *Azotobacter vinelandii*. *Biochim. Biophys. Acta* 20:269–285.

Grünhut, L. (1896). Die Einführung der Reinhefe in die Gärungsgewerbe. *Sammlung chemischer und chemisch-technischer Vorträge* 1:393–452.

Gruss, C., and Knippers, R. (1996). Structure of replicating chromatin. *Prog. Nucleic Acid Res.* 52:337–365.

Guckelberger, C. G. (1848). Zersetsungsprodukte des Albumins, Fibrins, Caseins, und das Leims durch Manganhyperoxid und Chromsäure, unter Mitwirkung von Schwefelsäure. *Ann.* 64:39–100.

Guédon, J. C. (1978). Pierre Jacques Antoine Béchamp. *DSB* 15:11–12.

Guengerich, F. P. (1991). Reactions and significance of cytochrome P-450 enzymes. *J. Biol. Chem.* 266:10019–10022.

Guerlac, H. (1957). Joseph Black and fixed air: A bicentenary retrospective with some new or little known material. *Isis* 48:124–151, 433–456.

——— (1959). Some French antecedents of the Chemical Revolution. *Chymia* 5:73–112.

——— (1961). *Lavoisier: Crucial Year*. Ithaca, N.Y.: Cornell University Press.

——— (1970). Joseph Black. *DSB* 2:173–183.

——— (1976). Chemistry as a branch of physics: Laplace's collaboration with Lavoisier. *Hist. Stud. Phys. Sci.* 7:197–223.

——— (1977). Some historical assumptions in the history of science. In *Essays and Papers in the History of Modern Science*, pp. 27–39. Baltimore: Johns Hopkins University Press.

Guéron, J., and Magat, M. (1971). A history of physical chemistry in France. *Ann. Rev. Phys. Chem.* 22:1–23.

Guerrier-Takada, C., and Altman, S. (1993). A physical assay for and kinetic analysis of the interactions between M1 RNA and tRNA precursor substrates. *Biochemistry* 32:7152–7161.

Guerrier-Takada, C., Gardiner, K., Marsh, T., Pace, N., and Altman, S. (1983). The RNA moiety of ribonuclease P is the catalytic subunit of the enzyme. *Cell* 35:849–857.

Guggenheim, E. A. (1960). The Niels Bjerrum memorial lecture. *Proc. Chem. Soc.*, pp. 104–114.

Guillemin, R. (1978). Pioneering in neuroendocrinology, 1952–1969. In *Pioneers in Neuroendocrinology.* Vol. 2 (J. Meites, ed.), pp. 221–239. New York: Plenum.

Gukasyan, A. G. (1948). *G. A. Sakharin.* Moscow.

Guldberg, C. M., and Waage, P. (1899). Untersuchungen über die chemischen Affinitäten (R. Abegg, trans. and ed.). *Ostwald's Klassiker der exakten Wissenschaften,* no. 104. Leipzig: Engelmann.

Gulick, A. (1938). What are the genes? *Quart. Rev. Biol.* 13:1–18, 140–168.

——— (1939). Growth and reproduction as problems in the synthesis of protein molecules. *Growth* 3:241–260.

Gulland, J. M. (1938). Nucleic acids. *J. Chem. Soc.*, pp. 1722–1734.

——— (1947a). The structures of nucleic acids. *Symp. Soc. Exp. Biol.* 1:1–14.

——— (1947b). The structures of nucleic acids. *Cold Spring Harbor Symp.* 12:95–103.

Gulland, J. M., Jordan, D. O., and Taylor, H. F. W. (1947). Deoxypentose nucleic acids, 2: Electrometric tittration of the acidic and basic groups of the deoxypentose nucleic acid of calf thymus. *J. Chem. Soc.*, pp. 1131–1141.

Gullväg, I., and Wetlesen, J. (eds.) (1982). *In Sceptical Wonder.* Oslo: Universtetsforlaget.

Gunsalus, I. C., Horecker, B. L., and Wood, W. A. (1955). Pathways of carbohydrate metabolism in microorganisms. *Bact. Revs.* 19:79–128.

Gunsalus, I. C., and Sligar, S. G. (1978). Oxygen reduction by the P450 mono-oxygenase systems. *Adv. Enzymol.* 47:1–44.

Günther, H. Ueber den Muskelfarbstoff. *Virchows Arch.* 230:146–178.

Gupta, S. P., Milford, E. L., and SubbaRow, Y. (1987). *In Search of Panacea.* Nanuet, N.Y.: Evelyn.

Gustin, B. H. (1973). Charisma, recognition, and the motivation of scientists. *Am. J. Sociol.* 78:1119–1134.

——— (1975). *The Emergence of the German Chemical Profession, 1790–1867.* Ph.D. diss., University of Chicago.

Guston, D. H., and Keniston, K. (1994). *The Fragile Contract: University Science and the Federal Government.* Cambridge: MIT Press.

Gutte, B. (ed.) (1995). *Peptides.* San Diego: Academic Press.

Güttler, H. (1972). Die Begriffe Plasma und Protoplasma: Ihre Entwicklung und Wandlung in der Biologie. *Rete* 1:365–375.

Haake, P. (1987). Thiamine-dependent enzymes. In *Enzyme Mechanisms* (M. I. Page and A. Williams, eds.), pp. 390–403. London: Royal College of Chemistry.

Haas, E. (1938). Isolierung eines neuen gelben Ferments. *Biochem. Z.* 298:378–390.

Haber, F. (1901). Bemerkungen über Elektrodenpotentiale. *Z. Elektrochem,* 7:1042–1053.

Haber, L. F. (1958). *The Chemical Industry During the Nineteenth Century.* Oxford: Clarenden.

Haberling, W. (1924). *Johannes Müller.* Leipzig: Akademische Verlags gesellschaft.

Habricht, C. (1978). Zur Geschichte der klinischen Enzymdiagnostik und ihren naturwissenschaftlichen Vorausstezungen. In *Medizinische Diagnostik in Geschichte und Gegenwart* (C. Habicht et al., eds.), pp. 549–571. Munich: Fritsch.

Hacking, I. (1982). Language, truth, and reason. In *Rationality and Relativism* (M. Hollis and S. Lukes, eds.), pp. 48–66. Oxford: Blackwell.

——— (1983). *Representing and Intervening.* Cambridge: Cambridge University Press.

——— (1985). Do we see through a microscope? In *Images of Science* (P. M. Churchland and C. A. Hooker, eds.), pp. 132–152. Chicago: University of Chicago Press.

——— (1988). The participant irrealist at large in the laboratory. *Brit J. Phil. Sci.* 39:277–294.

——— (1990). *The Taming of Chance.* Cambridge: Cambridge University Press.

——— (1996). The disunities of the sciences. In *The Disunity of Science* (P. Gallison and D. J. Stump, eds.), pp. 37–74. Stanford: Stanford University Press.

Hadden, R. (1988). Mathematics, relativism, and David Bloor. *Phil. Soc. Sci.* 18:433–445.

Hadelich, W. (1862). Ueber die Bestandtheile des Guaiakharzes. *J. prakt. Chem.* 87:321–343.

Hadley, P. (1927). Microbic dissociation. *J. Inf. Dis.* 40:1–132.

Haeckel, E. (1866). *Generelle Morphologie der Organismen.* Berlin: Reimer.

——— (1917). *Kristallseelen.* Leipzig: Kroner.

Hafner, K. (1981). 150 Jahre Annalen. *Ann. Chem.* 12:i–xii.

Hagedoorn, A. L. (1911). *Autokatalytical Substances as Determinants for the Inheritable Characters.* Leipzig: Engelmann.

Hagelhans, U. (1985). *Jacob Moleschott als Physiologe.* Frankfurt: Peter Lang.

Hagenau, F. (1958). The ergastoplasm: Its history, ultrastructure, and biochemistry. *International Review of Cytology* 7:425–483.

Hager, T. (1995). *Force of Nature: The Life of Linus Pauling.* New York: Simon and Schuster.

Hagstrom, W. O. (1965). *The Scientific Community.* New York: Basic.

Hahn, A. (1931). Zur Thermodynamik des Erholungsvorganges im Muskel. *Z. Biol.* 91:444–448.

Haigh, E. (1984). *Xavier Bichat and the Medical Theory of the Eighteenth Century.* London: Wellcome Institute for the History of Medicine.

Haiser, F. (1895). Zur Kenntnis der Inosinsäure. *Mon. Chem.* 16:190–206.

Hakfoort, C. (1992). Science deified: Wilhelm Ostwald's energeticist worldview and the history of scientism. *Ann. Sci.* 49:525–544.

Halary, C. (1994). *Exiles du Savoir. Les Migrations Scientifiques et leurs Mobiles.* Paris: Harmattan.

Haldane, J. B. S. (1930). *Enzymes.* London: Longmans, Green.

——— (1932). *The Inequality of Man.* London: Chatto and Windus.

——— (1937). The biochemistry of the individual. In *Perspectives in Biochemistry* (J. Needhan and D. E. Green, eds.), pp. 1–10. Cambridge: Cambridge University Press.

——— (1938). *The Marxist Philosophy and the Sciences.* New York: Random House.

——— (1942). *New Paths in Genetics.* New York: Harper.

——— (1945). A physicist looks at genetics. *Nature* 155:375–376.

——— (1966). An autobiography in brief. *Persp. Biol. Med.* 9:476–481.

Haldane, J. B. S., et al. (1954). The origin of life. *New Biology* 16:9–67.

Haldane, J. S. (1895). The relation of the action of carbonic oxide to oxygen tension. *J. Physiol.* 18:201–217.

——— (1908). The relation of physiology to physics and chemistry. *Nature* 78:553–556.

——— (1917). *Organism and Environment as Illustrated by the Physiology of Breathing.* New Haven: Yale University Press.

——— (1931). *The Philosophical Basis of Biology.* London: Hodder and Stoughton.

——— (1935). *Respiration.* New Haven: Yale University Press.

Haldane, J. S., and Priestley, J. G. (1905). The regulation of the lung ventilation. *J. Physiol.* 32:225–266.

Haldane, J. S., and Smith, J. L. (1896). The oxygen tension of arterial blood. *J. Physiol.* 20:497–520.

Hall, A. (1990). The cellular functions of small GTP binding proteins. *Science* 249:635–640.

Hall, A. R. (1969). Can the history of science be history? *Brit. J. Hist. Sci.* 4:207–220.

——— (1983). On Whiggism. *Hist. Sci.* 21:45–59.

Hall, B. D., and Spiegelman, S. (1961). Sequence complementarity of T2 DNA and T2 specific RNA. *PNAS* 47:137–146.

Hall, C. E. (1956). Method for the observation of macromolecules with the electron microscope, illustrated with micrographs of DNA. *J. Biophys. Biochem. Cytol.* 2:625–627.

Hall, C. R. (1934). *A Scientist in the Early Republic: Samuel Latham Mitchill.* New York: Columbia University Press.

Hall, D. L. (1971). The iatromechanical background of Lagrange's theory of animal heat. *J. Hist. Biol.* 4:245–248.

Hall, M. N., and Patrick, L. (1993). *The Early Days of Yeast Genetics.* Plainview, N.Y.: Cold Spring Harbor Laboratory Press.

Hall, T. S. (1969). *Ideas of Life and Matter.* Chicago: University of Chicago Press.

Hall, V. M. D. (1980). The role of force or power in Liebig's physiological chemistry. *Med. Hist.* 24:20–59.

Hallé, J. N. (1791). Essai de théorie sur l'animalisation et l'assimilation des alimens. *Ann. Chim.* 11:158–174.

Haltenhoff, G. (1880). Franz Boll, sa vie et ses travaux. *Annales d'Oculistique* 83:90–102.

Hamblin, A. S. (1988). *Lymphokines.* Oxford: IRL.

Hamburger, V. (1980). Ramon y Cajal, R. G. Harrison, and the beginnings of neuroembryology. *Persp. Biol. Med.* 23:600–616.

——— (1984). Hilde Mangold, co-discoverer of the organizer. *J. Hist. Biol.* 17:1–11.

——— (1988). *The Heritage of Experimental Embryology: Hans Spemann and the Organizer.* New York: Oxford University Press.

Hamilton, L. D. (1968). DNA: Models and reality. *Nature* 218:633–637.

Hamilton, W. C. (1970). The revolution in crystallography. *Science* 169:133–141.

Hamlin, C. (1990). *A Science of Impurity.* Bristol: Hilger.

Hammarsten, E. (1924). Zur Kenntnis der biologischen Bedeutung der Nucleinsäureverbindungen. *Biochem. Z.* 144:383–469.

Hammarsten, O. (1894). Zur Kenntnis der Nucleoproteide. *Z. physiol. Chem.* 19:19–37.

——— (1895). *Lehrbuch der physiologischen Chemie.* 3d ed. Wiesbaden: Bergmann.

Hammes, G. G., and Wu, C. W. (1974). Kinetics of allosteric enzymes. *Ann. Rev. Biophys.* 3:1–33.

Hammett, L. P. (1940). *Physical Organic Chemistry.* New York: McGraw-Hill.

Hamoir, G. (1994). *La Découverte de la méiose et du centrosome par Edouard van Beneden.* Brussels: Académie Royale de Belgique.

Hampton, R., Dimster-Denk, D., and Rine, J. (1996). The biology of HMG-CoA reductase: The pros of contraregulation. *TIBS* 21:140–145.

Häner, R., and Dervan, P. B. (1990). Single-strand DNA triple helix formation. *Biochemistry* 29:9761–9765.

Hanes, C. S. (1937). The action of amylases in relation to the structure of starch and its metabolism in the plant, 4: Starch degradation by the component amylases of malt. *New Phytologist* 36:189–239.

——— (1940). The breakdown and synthesis of starch by an enzyme system from pea seeds. *Proc. Roy. Soc.* B128:421–450; B129:174–208.

Hankins, T. L. (1979). In defence of biography: The use of biography in the history of science. *Hist. Sci.* 17:1–16.

Hannaway, O. (1975). *The Chemists and the Word.* Baltimore: Johns Hopkins University Press.

——— (1976). The German model of chemical education in America: Ira Remsen at Johns Hopkins (1876–1913). *Ambix* 23:145–164.

Hannun, Y. A. (1994). The sphingomyelin cycle and the second messenger function of ceramide. *J. Biol. Chem.* 269:3125–3218.

Hanson, H. (1970). Emil Abderhalden. *Nova Acta Leopoldina* NS36:257–317.

Hanson, J., and Huxley, H. E. (1955). The structural basis of contraction in striated muscle. *Symp. Soc. Exp. Biol.* 9:228–264.

Hanson, N. R. (1958). *Patterns of Discovery.* Cambridge: Cambridge University Press.

——— (1969). *Perception and Discovery.* San Francisco: Freeman, Cooper.

——— (1971a). *Observation and Explanation.* New York: Harper and Row.

——— (1971b). *What I Do Not Believe and Other Essays* (S. Toulmin and H. Woolf, eds.). Dordrecht: Reidel.

Haraway, D. (1976). *Crystals, Fabrics, and Fields.* New Haven: Yale University Press.

Harden, A. (1903). Ueber alkoholische Gärung mit Hefe-Prestoff (Buchner's Zymase) bei Gegenwart von Blutserum. *Ber. chem. Ges.* 36:715–716.

——— (1923). *Alcoholic Fermentation.* 3d ed. London: Longmans, Green.

——— (1930). Samuel Barnett Schryver. *J. Chem. Soc.* 133:901–905.

Harden A., and Maclean, H. (1911). On the alleged presence of an alcoholic enzyme in animal tissues and organs. *J. Hysiol.* 42:64–92.

Harden, A., and Young, W. J. (1906). The alcoholic ferment of yeast-juice, 2: The co-ferment of yeast juice. *Proc. Roy. Soc.* N75:369–375.

Hardie, D. W. F., and Pratt, J. D. (1966). *A History of the Modern Chemical Industry.* London: Pergamon.

Hardy, W. B. (1899). Structure of cell protoplasm. *J. Physiol.* 24:158–210.

——— (1900). A preliminary investigation of the conditions which determine the stability of irreversible hydrosols. *Proc. Roy. Soc.* 66:110–125.

——— (1903). Colloidal solution, the globulin system. *J. Physiol.* 29:xxvi–xxix.

——— (1906). The physical basis of life. *Sci. Prog.* 1:177–205.

——— (1936). To remind: A biological essay. In *Collected Scientific Papers,* pp. 896–917. Cambridge: Cambridge University Press.

Hargittai, I., and Hargittai, M. (1986). *Symmetry Through the Eyes of a Chemist.* Weinheim: VCH Verlagsgesellschaft.

Harington, C. R., and Barger, G. (1927). Chemistry of thyroxine, 3: Constitution and synthesis of thyroxine. *Biochem. J.* 21:169–183.

Harold, F. M. (1978). The 1978 Nobel Prize in Chemistry. *Science* 202:1174, 1176.

——— (1986). *The Vital Force: A Study of Bioenergetics.* New York: Freeman.

Harré, R. (1972). *The Philosophies of Science.* Oxford: Oxford University Press.

Harries, C. (1905). Zur Kenntnis der Kautschukarten: Ueber Abbau und Constitution des Parakautschuks. *Ber. chem. Ges.* 38:1195–1203.

——— (1917). Eduard Buchner. *Ber. chem. Ges.* 50:1843–1876.

Harrington, R. E., and Winicov, I. (1994). New concepts in protein-DNA recognition: Sequence-directed DNA binding and flexibility. *Prog. Nucleic Acid Res.* 47:195–270.

Harris, D. L. (1946). Phosphoprotein phosphatase, a new enzyme from the frog egg. *J. Biol. Chem.* 165:541–550.

Harris, H. (1963). *Garrod's Inborn Errors of Metabolism.* Oxford: Oxford University Press.

Harris, H. (1970). *Cell Fusion.* Oxford: Oxford University Press.

——— (1995). *The Cells of the Body: A History of Somatic Cell Genetics.* Plainview, N.Y.: Cold Spring Harbor Laboratory Press.

Harris, H. (1992). To serve mankind in peace and the fatherland in war: The case of Fritz Haber. *German History* 10:24–38.

Harris, H., and Watkins, J. H. (1965). Hybrid cells derived from mouse and man: Artificial heterokaryons of mammalian cells from different species. *Nature* 205:640–646.

Harris, J., and Brock, W. H. (1976). From Giessen to Gower Street: Toward a biography of Williamson. *Ann. Sci.* 31:95–130.

Harris, J. I., and Waters, M. (1976). Glyceraldehyde-3-phosphate dehydrogenase. In *The Enzymes.* 3d ed. (P. B. Boyer, ed.), 13:1–49. New York: Academic Press.

Harrison, E. (1987). Whigs, prigs, and historians of science. *Nature* 329:213–214.

Harrison, R. G. (1910). The outgrowth of the nerve fiber as a mode of protoplasmic movement. *J. Exp. Zool.* 9:787–946.

——— (1937). Embryology and its relations. *Science* 85:369–374.

Harrow, B. (1955). *Casimir Funk.* New York: Dodd Mead.

Harteck, P. (1960). Physical chemists in Berlin, 1919–1933. *J. Chem. Ed.* 37:462–466.

Hartl, F. U., and Neupert, W. (1990). Protein sorting to mitochondria: Evolutionary conservations of folding and assembly. *Science* 247:930–938.

Hartley, H. (1930). Theodore William Richards. *J. Chem. Soc.*, pp. 1930–1968.

——— (1965). Chemistry in Oxford: The contribution of the college laboratories. *Chem. Brit.* 1:521–524.

Hartley, P. (1952). Henry Drysdale Dakin. *Obit. Not. FRS* 8:129–148.

——— (1953). Henry Stanley Raper. *Obit. Not. FRS* 8:567–582.

Hartman, P. E. (1957). Transduction: A comparative review. In *The Chemical Basis of Heredity* (W. D. McElroy and B. Glass, eds.), pp. 408–462. Baltimore: Johns Hopkins University Press.

Hartree, E. F. (1981). Keilin, cytochrome, and the concept of the respiratory chain. In *Of Oxygen, Fuels, and Living Matter* (G. Semenza, ed.), pp. 161–227. Chichester: Wiley.

Harvey, A. M. (1976). The Department of Physiological Chemistry: Its historical evolution. *Johns Hopkins Medical Journal* 139:237–273.

Harvey, E. N. (1957). *A History of Luminescence from the Earliest Times until 1900.* Philadelphia: American Philosophical Society.

——— (1958). Edwin Grant Conklin. *Biog. Mem. NAS* 31:54–91.

Harwood, J. (1993). *Styles of Scientific Thought: The German Genetics Community, 1900–1933.* Chicago: University of Chicago Press.

Haskins, F. A., and Mitchell, H. K. (1949). Evidence for a tryptophane cycle in Neurospora. *PNAS* 35:500–506.

Hassenfratz, J. H. (1791). Mémoire sur la combinaison de l'oxigène avec le carbone et l'hydrogène du sang, sur la dissolution de l'oxigène dans le sang, et sur la manière dont le calorique se dégage. *Ann. Chim.* 9:261–274.

Hassid, W. Z. (1969). Biosynthesis of oligosaccharides and polysaccharides in plants. *Science* 165:137–144.

Hastings, A. B. (1970). A biochemist's anabasis. *Ann. Rev. Biochem.* 39:1–24.

——— (1976). Donald Dexter Van Slyke. *Biog. Mem. NAS* 48:309–360.

——— (1989). *Crossing Boundaries.* Grand Rapids: Four Centers.

Hastings, J. J. H. (1971). Development of the fermentation industries in Great Britain. *Advances in Applied Microbiology* 14:1–45.

Haug, T. (1985). *Friedrich August Flückiger.* Stuttgart: Deutscher Apotheker Verlag.

Hauptmann, H. A., and Blessing, R. H. (1987). Fünfundsiebzig Jahre Röntgenstrahlenbeugung und Kristallstrukturanalyse. *Naturw. Rund.* 40:463–470.

Haurowitz, F. (1937). Antigene, Antikörper, und Immunität. *Klin. Wchschr.* 16:257–261.

Haworth, R. D. (1948). John Masson Gulland. *Obit. Not. FRS* 6:67–82.

Haworth, W. N., Hirst, E. L., and Sherwood, F. A. (1937). Polysaccharides, 23: Determination of the chain length of glycogen. *J. Chem. Soc.* 140:577–581.

Haworth, W. N., and Percival, E. G. V. (1932). Polysaccharides, 11: Molecular structure of glycogen. *J. Chem. Soc.* 135:2277–2282.

Hawthorne, R. M. (1983). Henry Drysdale Dakin, biochemist (1880–1952): The option of obscurity. *Persp. Biol Med.* 26:553–566.

Hayaishi, O. (ed.) (1974). *Molecular Mechanisms of Oxygen Activation.* New York: Academic Press.

Hayes, W. (1953). The mechanism of genetic recombination in *Escherichia coli. Cold Spring Harbor Symp.* 18:75–93.

——— (1966). Genetic transformation: A retrospective appreciation. *J. Gen. Microbiol.* 45:386–397.

——— (1968). *The Genetics of Bacteria and Their Viruses.* 2d ed. New York: Wiley.

——— (1980). Portrait of viruses: Bacteriophage lambda. *Intervirology* 13:133–152.

Haynes, R. H. (1989). Genetics and the unity of biology. *Genome* 31:1–7.

Hecht, L. I., Stephenson, M. L., and Zamecnik, P. C. (1959). Binding of amino acids to the end group of a soluble ribonucleic acid. *PNAS* 45:505–512.

Hecht. S. (1937). Rods, cones, and the chemical basis of vision. *Physiol. Revs.* 17:239–290.

——— (1938). The photochemical basis of vision. *Journal of Applied Physics* 9:156–164.

Hecht, S., Shlaer, S., and Pirenne, M. H. (1942). Energy, quanta, and vision. *J. Gen. Physiol.* 25:819–840.

Hecht, S., and Williams, R. E. (1922). The visibility of monochromatic radiation and the absorption spectrum of visual purple. *J. Gen. Physiol.* 5:1–33.

Hechter, O. (1955). Concerning possible mechanisms of hormone action. *Vit. Horm.* 13:293–345.

Heelis, P. F., Hartman, R. F., and Rose, S. D. (1995). Photodynamic repair of UV-damaged DNA: A chemist's perspective. *Chem. Soc. Revs.*, pp. 289–297.

Heffter, A. (1908). Gibt es reduzierende Fermente im Tierkörper? *Arch. exp. Path. Pharm.* Suppl., pp. 253–260.

Heidegger, M. (1953). *Einführung in die Metaphysik.* Tübingen: Niemeyer.

Heidelberger, M. (1977). A "pure" organic chemist's downward path. *Ann. Rev. Microbiol.* 31:1–12.

——— (1979). A "pure" organic chemist's downward path—chapter 2. *Ann. Rev. Biochem.* 48:1–21.

Heidelberger, M., and Avery, O. T. (1923). The specific soluble substance of pneumococcus. *J. Exp. Med.* 38:73–79.

Heidelberger, M., and Landsteiner, K. (1923). On the antigenic properties of hemoglobin. *J. Exp. Med.* 38:561–571.

Heidenhain, M. (1894). Neue Untersuchungen über die Centralkörper und ihre Beziehung zum Kern und Zellenprotoplasma. *Arch. mikr. Anat.* 43:423–758.

Heilbron, I. M., Kamm, E. D., and Owens, W. M. (1926). The unsaponifiable matter from the oils of elasmobranch fish. *J. Chem. Soc.* 129:1630–1650.

Heilbron, J. L. (1976). Alessandro Volta. *DSB* 14:69–82.

Heilbron, J. L., and Kuhn, T. S. (1969). The genesis of the Bohr atom. *Hist. Stud. Phys. Sci.* 1:211–290.

Heim, R., et al. (1959). *A la Mémoire de quinze Savants Français Lauréats de l'Institut Assassinés par les Allemands,* pp. 133–145. Paris.

Heimann, P. M. (1976). Mayer's concept of "force": The "axis" of a new science of physics. *Hist. Stud. Phys. Sci.* 7:277–296.

Heimans, J. (1962). Hugo de Vries and the gene concept. *Am. Nat.* 96:93–104.

Hein, F. (1941). Arthur Hantzsch. *Ber. chem. Ges.* 74A:147–163.

Hein, H. (1972). The endurance of the mechanism-vitalism controversy. *J. Hist. Biol.* 5:159–188.

Heinecke, B. (1995). The mysticism and science of Johann Baptista van Helmont (1579–1644). *Ambix* 42:65–78.

Heinemann, K. (1939). Zur Geschichte der Entdeckung der roten Blutkörperchen. *Janus* 43:1–42.

Heinig, K. (1979). *Carl Schorlemmer.* 2d ed. Leipzig: Teubner.

Heitz, E., and Bauer, H. (1933). Beweise für der Chromosomennatur der Kernschleifen in der Knäuelkernen von *Bibio hortulanus. Z. wiss. Zellforsch.* 17:67–82.

Heizmann, C. W. (ed.) (1991). *Novel Calcium Binding Proteins.* New York: Springer.

Helinski, D. R., and Yanofsky, C. (1963). A genetic and biochemical analysis of second-site reversion. *J. Biol. Chem.* 238:1043–1048.

Helmholtz, H. (1845). Ueber den Stoffverbrauch bei der Muskelarbeit. *Arch. Anat. Physiol.* pp. 72–83.

——— (1847). *Ueber die Erhaltung der Kraft.* Berlin: Reimer.

——— (1848). Ueber die Wärmeentwicklung bei der Muskelaction. *Arch. Anat. Physiol.*, pp. 147–164.

——— (1881). On the modern development of Faraday's conception of electricity. *J. Chem. Soc.* 39:277–304.

——— (1882). *Wissenschaftliche Abhandlungen.* Leipzig: Barth.

Helmont, J. B. van (1662). *Oriatrike or Physik Refined* (J. Chandler, trans.). London: Lodowick Lloyd.

Helmreich, E. J. M. (1995). Recollections: Vacillations of a classical enzymologist. In *Selected Topics in the History of Biochemistry: Personal Recollections.* Vol. 4 (E. C. Slater, L. Jaenicke, and G. Semenza, eds.), pp. 163–191. Amsterdam: Elsevier.

Helvoort, T. van (1991). What is a virus? The case of tobacco mosaic disease. *Stud. Hist. Phil. Sci.* 22:557–588.

——— (1992). Bacteriological and physiological research styles in the early controversy on the nature of the bacteriophage phenomenon. *Med. Hist.* 36:243–270.

Hempel, C. G. (1965). *Aspects of Scientific Explanation.* New York: Free Press.

——— (1988). On the cognitive status and the rationale of scientific methodology. *Poetics Today* 9:1–27.

Henderson, L. J. (1909). Das Gleichgewicht zwischen Basen und Säuren im Tierischen Organismus. *Erg. Physiol.* 8:254–325.

——— (1913). *The Fitness of the Environment.* New York: Macmillan.

——— (1917). *The Order of Nature.* Cambridge: Harvard University Press.

——— (1928). *Blood: A Study in General Physiology.* New Haven: Yale University Press.

——— (1935). *Pareto's General Sociology: A Physiologist's Interpretation.* Cambridge: Harvard University Press.

Henderson, P. J. F. (1991). Studies of translocation catalysis. *Bioscience Reports* 11:477–538.

Henderson, R., Baldwin, J. M., Ceska, T., Zemlin, F., Beckmann, E., and Downing, K. (1990). Model for the structure of bacteriorhodopsin based on high resolution electron cryomicroscopy. *J. Mol. Biol.* 213:899–929.

Henderson, W. O. (1976). *The Life of Friedrich Engels.* London: Frank Cass.

Hendry, J. (1980). Weimar culture and quantum mechanics. *Hist. Sci.* 18:115–180.

Henikoff, S. (1987). Multifunctional polypeptides for purine *de novo* synthesis. *BioEssays* 6:8–13.

Henking, H. (1891). Ueber Spermatogenese bei *Pyrrhocoris apterus. Z. wiss. Zool.* 51:685–736.

Henri, V. (1903). *Lois Générales de l'Action des Diastases.* Paris: Hermann.

Henriques, V., and Hansen, C. (1905). Ueber Eiweißsynthese im Tierkörper. *Z. Physiol. Chem.* 43:417–446.

Heppel, L. A., Singer, M. F., and Hilmoe, R. J. (1959). The mechanism of the action of polynucleotide phosphorylase. *Ann. N.Y. Acad. Sci.* 81:635–644.

Herbst, R. M. (1944). The transamination reaction. *Adv. Enzymol.* 4:75–97.

Herbst, R. M., and Engel, L. L. (1934). A reaction between α-ketonic acids and α-amino acids. *J. Biol. Chem.* 107:505–512.

Hérelle, F. d'. (1926). *The Bacteriophage and Its Behavior.* Baltimore: Williams and Wilkins.

Hering, E. (1888). Zur Theorie der Vorgänge in der lebendigen Substanz. *Lotos* 37:35–70.

Hermann, L. (1867). *Untersuchungen über den Stoffwechsel der Muskeln, ausgehend vom Gaswechsel derselben.* Berlin: Hirschwald.

Herrick, J. B. (1949). *Memories of Eighty Years.* Chicago: University of Chicago Press.

Herriott, R. M. (1947). Reactions of native proteins with chemical reagents. *Adv. Protein Chem.* 3:169–225.

——— (1955). John Maurice Nelson. *J. Chem. Ed.* 32:513–517.

——— (1989). Moses Kunitz. *Biog. Mem. NAS* 58:305–317.

Herschel, J. F. W. (1822). On the rotation impressed by plates of rock crystal on the planes of polarization of the rays of light, as connected with certain peculiarities in its crystallization. *Transactions of the Cambridge Philosophical Society* 1:43–52.

Herschlag, D. (1995). RNA chaperones and the RNA folding problem. *J. Biol. Chem.* 270:20871–20874.

Hershey, A. D., and Chase, M. (1952). Independent functions of viral protein and nucleic acid in growth of bacteriophage. *J. Gen. Physiol.* 36:39–56.

Hershey, J. W. B. (1989). Protein phosphorylation controls translation rates. *J. Biol. Chem.* 264:20823–20826.

Hershko, A. (1988). Ubiquitin-mediated protein degradation. *J. Biol. Chem.* 263:15237–15240.

Hertwig, O. (1876). Beiträge zur Kenntnis der Bildung, Befruchtung, und Theilung des thierischen Eies. *Morphologisches Jahrbuch* 1:347–434.

——— (1885). Das Problem der Befruchtung und der Isotropie des Eies, eine Theorie der Vererbung. *Jenaische Zeitschrift für Medizin und Naturwissenschaft* 18:276–318.

Hess, G. H. (1840). Thermochemische Untersuchungen. *Ann. Phys.* 50:385–404.

Hesse, F. (1976). *Julius Eugen Schlossberger.* Düsseldorf: Tritsch.

Hesse, M. (1988). Socializing epistemology. In *Construction and Constraint* (E. McMullin, ed.), pp. 97–122. Notre Dame: University of Notre Dame Press.

Hevesy, G. (1923). The absorption and translocation of lead by plants. A contribution to the application of the method of radioactive indicators in the investigation of change of substance in plants. *Biochem. J.* 17:439–445.

——— (1948). Historical sketch of the biological application of tracer elements. *Cold Spring Harbor Symp.* 13:129–150.

Hibberd, M. G., and Trentham, D. R. (1986). Relationships between chemical and mechanical events during muscular contraction. *Ann. Rev. Biophys.* 15:119–151.

Hickel, E. (1975). Pepsin, Veteran der Enzymchemie. *Naturw. Rund.* 28:14–18.

——— (1979). Die organische Elementaranalyse. *Pharmazie in unserer Zeit* 8:1–10.

Hicks, D., and Potter, J. (1991). Sociology of scientific knowledge: A reflexive citation analysis. *Soc. Stud. Sci.* 21:459–501.

Hiebert, E. N. (1964). The problem of organic analysis. In *Mélanges Alexandre Koyré* (I. B. Cohen and R. Taton, eds.), pp. 303–325. Paris: Hermann.

——— (1973). Ernst Mach. *DSB* 8:595–607.

——— (1978). Hermann Walther Nernst. *DSB* 15:432–453.

——— (1987). The scientist as philosopher of science. *NTM* 24:7–17.

Hiebert, E. N., and Körber, H. G. (1978). Friedrich Wilhelm Ostwald. *DSB* 15:455–469.

Hildebrandt, G. (1993). The discovery of the diffraction of X-rays in crystals: A historical review. *Crystal Structure and Technology* 28:747–766.

Hill, A. C. (1898). Reversible zymohydrolysis. *J. Chem. Soc.* 73:634–658.

Hill, A. V. (1912). The heat-production of surviving amphibian muscles during rest, activity, and rigor. *J. Physiol.* 44:466–513.

——— (1913). The combination of haemoglobin with oxygen and with carbon monoxide, 1. *Biochem. J.* 7:471–480.

——— (1932). The revolution in muscle physiology. *Physiol. Revs.* 12:56–67.

——— (1940). George Barger. *Obit. Not. FRS* 3:63–85.

——— (1950). A challenge to biochemists. *Biochim. Biophys. Acta* 4:4–11.

——— (1956). Why biophysics? *Science* 124:1233–1237.

——— (1959). The heat production of muscle and nerve, 1848–1914. *Ann. Rec. Physiol.* 21:1–18.

——— (1969). Bayliss and Starling and the happy fellowship of physiologists. *J. Physiol.* 204:1–13.

Hill, J. W. (1977). Wallace Hume Carothers. *Welch Foundation Conferences on Chemical Research* 20:232–250.

Hill, L. (1935). Edward Albert Sharpey-Shafer. *Obit. Not. FRS* 1:401–407.

Hill, R. (1926). The chemical nature of haemochromogen and its carbon monoxide compound. *Proc. Roy. Soc.* B100:419–430.

——— (1939). Oxygen produced by isolated chloroplasts. *Proc. Roy. Soc.* B127:192–210.

——— (1965). The biochemists' green mansions: The electron transport chain in plants. *Essays in Biochemistry* 1:121–152.

——— (1970). The growth of knowledge of photosynthesis. In *The Chemistry of Life* (J. Needham, ed.), pp. 1–14. Cambridge: Cambridge University Press.

Hilt, W., and Wolf, D. H. (1996). Proteasomes: Destruction as a programme. *TIBS* 21:96–102.

Hilvert, D. (1994). Chemical synthesis of proteins. *Chem. Biol.* 1:201–203.

Himsworth, H., and Pitt-Rivers, R. V. (1972). Charles Robert Harington. *Biog. Mem. FRS* 18:267–308.

Hinshelwood, C. N. (1946). *The Chemical Kinetics of the Bacterial Cell.* Oxford: Clarendon.

Hinz, H. J. (ed.) (1986). *Thermodynamic Data for Biochemistry and Biotechnology.* Berlin: Springer.

Hirsch, R. (1950). The invention of printing and the diffusion of alchemical and chemical knowledge. *Chymia* 3:115–141.

Hirst, E. L. (1951). Walter Norman Haworth. *Obit. Not. FRS* 7:373–404.

His, W. (1874). *Unsere Körperform und das physiologische Problem ihrer Entstehung. Briefe an einen befreundenten Naturforscher.* Leizig: Vogel.

His, W., Jr. (1887). Ueber den Stoffwechselprodukt des Pyridins. *Arch. exp. Path. Pharm.* 22:253–260.

Hjelt, E. (1916). *Geschichte der organischen Chemie von ältesten Zeit bis zur Gegenwart.* Braunschweig: Vieweg.

Hlasewitz, H., and Habermann, J. (1871). Ueber die Proteinstoffe. *Ann.* 159:304–333.

——— (1873). Ueber die Proteinstoffe. *Ann.* 169:150–166.

Hoagland, M. B. (1960). The relationship of nucleic acid and protein synthesis as revealed by studies in cell-free systems. In *The Nucleic Acids* (E. Chargaff and J. N. Davidson, eds.) 3:349–408. New York: Academic Press.

——— (1990). *Toward the Habit of Truth: A Life in Science.* New York: Norton.

——— (1996). Biochemistry or molecular biology? The discovery of "soluble RNA." *TIBS* 21:77–80.

Hoagland, M. B., Keller, E. B., and Zamecnik, P. C. (1956). Enzymatic carboxyl activation of amino acids. *J. Biol. Chem.* 218:345–358.

Hoagland, M. B., Zamecnik, P. C., and Stephenson, M. L. (1957). Intermediate reactions in protein synthesis. *Biochim. Biophys. Acta* 24:215–216.

Hoch, F. L., and Lipmann, F. (1954). The uncoupling of respiration and phosphorylation by thyroid hormones. *PNAS* 40:909–921.

Hoch, P. K. (1987). Migration and the generation of new scientific ideas. *Minerva* 28:209–217.

Hodgkin, D. C. (1965). The X-ray analysis of complicated molecules. *Science* 150:979–988.

——— (1976). Kathleen Lonsdale. *Biog. Mem. FRS* 21:447–484.

——— (1979). Crystallographic measurements and the structure of protein molecules as they are. *Ann. N.Y. Acad. Sci.* 325:121–145.

——— (1980). John Desmond Bernal. *Biog. Mem. FRS* 26:17–84.

Hodgkin, D. C., and Riley, D. P. (1968). Some ancient history of protein X-ray analysis. In *Structural Chemistry and Molecular Biology* (A. Rich and N. Davidson, eds.), pp. 15–48. San Francisco: Freeman.

Hoefer, J. (1872). *Histoire de la physique et de la chimie.* Paris: Hachette.

Hoerr, N. L., et al. (1957). Robert Russell Bensley. *Anat. Rec.* 128:1–18.

Hoesch, K. (1921). *Emil Fischer.* Berlin: Springer.

Hoff, J. H. van't (1875). *La Chimie dans l'espace.* Rotterdam: Bazendijk.

——— (1884). *Etudes de dynamique chimique.* Amsterdam: Muller.

——— (1887). Die Rolle des osmotischen Druckes in der Analogie zwischen Lösungen und Gasen. *Z. physik. Chem.* 1:481–508.

——— (1898). Ueber die zunehmende Bedeutung der anorganischen Chemie. *Z. anorg. Chem.* 18:1–13.

——— (1960). The role of imagination in science. *J. Chem. Ed.* 37:467–470.

Hoffmann, R. (1988). Under the surface of the chemical article. *Ang. Chem. (Int. Ed.)* 27:1593–1602.

——— (1995). *The Same and Not the Same.* New York: Columbia University Press.

Hoffmann, R., and Laszlo, P. (1991). Representation in chemistry. *Ang. Chem. (Int. Ed.)* 30:1–16.

Hoffmann-Ostenhof, O. (1978). The origin of the word enzyme. *TIBS* 3:186–188.

Hofmann, A. W. (1870). Zur Erinnerung an Gustav Magnus. *Ber. chem. Ges.* 3:993–1101.

——— (ed.) (1888). *Aus Justus Liebig's und Friedrich Wöhler's Briefwechsel.* Braunschweig: Vieweg.

Hofmann, J. R. (1996). *André-Marie Ampère.* Cambridge: Cambridge University Press.

Hofmann, K. (1943). The chemistry and biochemistry of biotin. *Adv. Enzymol.* 3:289–313.

Hofmeister, F. (1882). Zur Lehre vom Pepton, 5: Das Verhalten des Peptons in der Magenschleimhaut. *Z. physiol. Chem.* 6:69–73.

——— (1894). Ueber Methylierung im Thierkörper. *Arch. esp. Path. Pharm.* 33:198–215.

——— (1900). Willy Kühne. *Ber. chem. Ges.* 33:3875–3880.

——— (1901). *Die Chemische Organisation der Zelle.* Braunschweig: Vieweg.

——— (1902a). Ueber den Bau der Eiweisskörper. *Naturw. Rund.* 17:529–533, 545–549.

——— (1902b). Ueber Bau und Gruppierung der Eiweisskörper. *Erg. Physiol.* 1:759–802.

——— (1908). Einiges über die Bedeutung und den Abbau der Eiweisskörper. *Arch. exp. Path. Pharm.* Suppl., pp. 273–281.

——— (1914). Vom chemisch-morphologischen Grenzgebiet. *Zeitschrift für Morphologie und Anthropologie* 18:717–724.

Hogeboom, G. H., Claude, A., and Hotchkiss, R. D. (1946). The distribution of cytochrome oxidase in the cytoplasm of the mammalian liver cell. *J. Biol. Chem.* 165:615–629.

Hogeboom, G. H., Schneider, W. C., and Palade, G. E. (1948). Cytochemical studies of mammalian tissues, 1: Isolation of intact mitochondria and submicroscopic particulate material. *J. Biol. Chem.* 172:619–635.

Hogness, D. S., Cohn, M., and Monod, J. (1955). Studies on the induced synthesis of β-galactosidase in *Escherichia coli:* The kinetics and mechanism of sulfur incorporation. *Biochem. Biophys. Acta* 16:99–116.

Hohmann, G. (1963). Zum Andenken an Professor Dr. Siegfried Thannhauser. *Münch. med. Wchschr.* 105:357–359.

Hokin, L. E. (1984). Receptors and the phosphoinositide-generated second messengers. *Ann. Rev. Biochem.* 54:205–235.

Holden, C., et al. (1993). Careers: A survival guide. *Science* 260:1765–1813.

Holiday, E. R. (1930). The characteristic absorption of ultraviolet radiation by certain purines. *Biochem. J.* 24:619–625.

Holley, R. W. (1968). Experimental approaches to the determination of the nucleotide sequences of large oligonucleotides and small nucleic acids. *Prog. Nucleic Acid Res.* 8:37–47.

Holley, R. W., Apgar, J., Everett, G. A., Madison, J. T., Marquisse, M., Merrill, S. H., Penswick, J. R., and Zamir, A. (1965). Structure of a ribonucleic acid. *Science* 147:1462–1465.

Holliday, R. (1964). A mechanism for gene conversion in fungi. *Genetics Research* 5:282–304.

——— (1990). The history of the DNA heteroduplex. *BioEssays* 12:133–142.

Hollinger, D. A. (1983). The defense of democracy and Robert K. Merton's formulation of the scientific ethos. *Knowledge and Society* 4:1–15.

Hollinger, R., and Depew, D. (eds.) (1995). *Pragmatism: From Progressivism to Postmodernism.* Westport, Conn.: Praeger.

Hollis, M., and Lukes, S. (1982). *Rationality and Relativism.* Oxford: Blackwell.

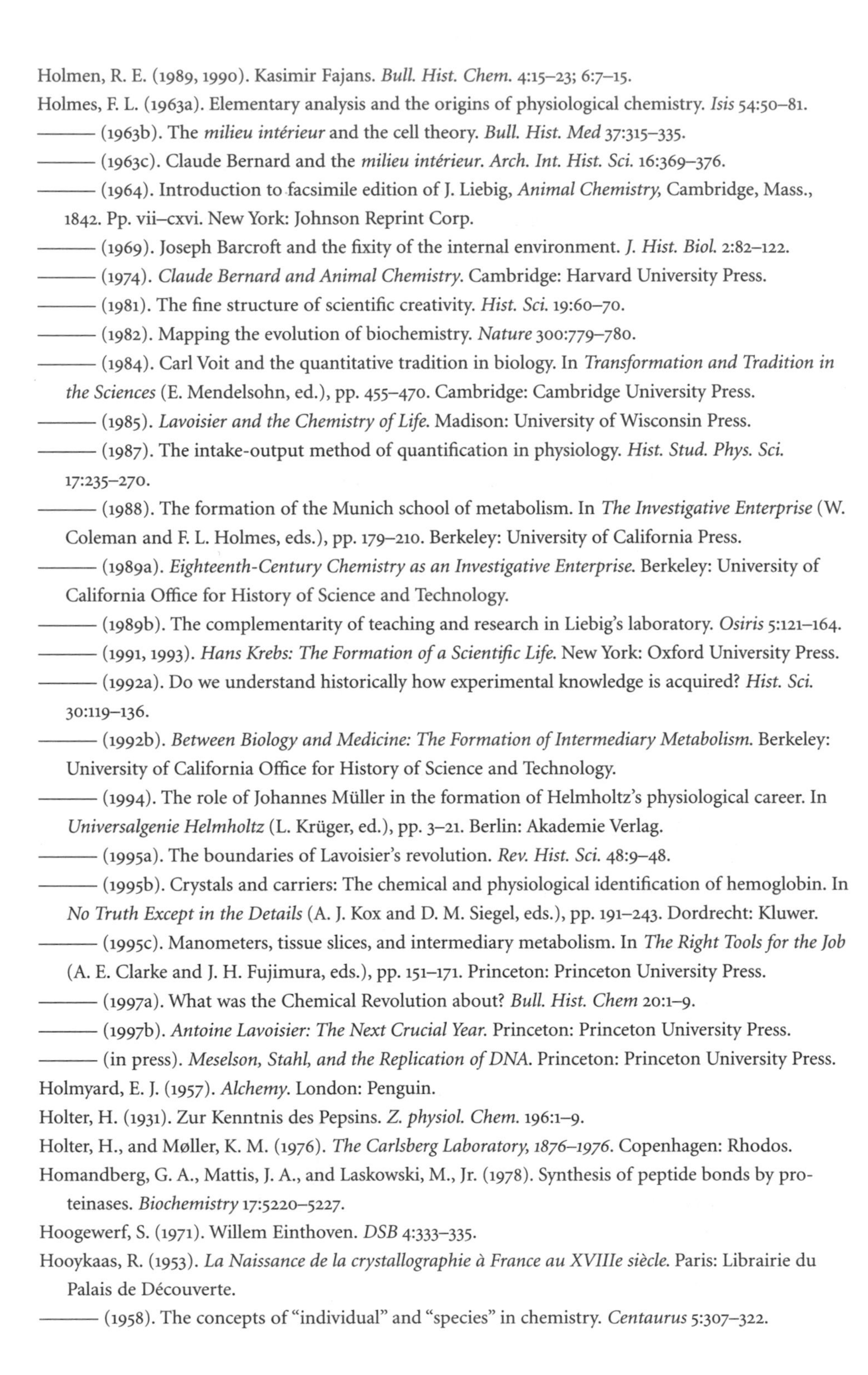

Holmen, R. E. (1989, 1990). Kasimir Fajans. *Bull. Hist. Chem.* 4:15–23; 6:7–15.

Holmes, F. L. (1963a). Elementary analysis and the origins of physiological chemistry. *Isis* 54:50–81.

——— (1963b). The *milieu intérieur* and the cell theory. *Bull. Hist. Med* 37:315–335.

——— (1963c). Claude Bernard and the *milieu intérieur. Arch. Int. Hist. Sci.* 16:369–376.

——— (1964). Introduction to facsimile edition of J. Liebig, *Animal Chemistry*, Cambridge, Mass., 1842. Pp. vii–cxvi. New York: Johnson Reprint Corp.

——— (1969). Joseph Barcroft and the fixity of the internal environment. *J. Hist. Biol.* 2:82–122.

——— (1974). *Claude Bernard and Animal Chemistry.* Cambridge: Harvard University Press.

——— (1981). The fine structure of scientific creativity. *Hist. Sci.* 19:60–70.

——— (1982). Mapping the evolution of biochemistry. *Nature* 300:779–780.

——— (1984). Carl Voit and the quantitative tradition in biology. In *Transformation and Tradition in the Sciences* (E. Mendelsohn, ed.), pp. 455–470. Cambridge: Cambridge University Press.

——— (1985). *Lavoisier and the Chemistry of Life.* Madison: University of Wisconsin Press.

——— (1987). The intake-output method of quantification in physiology. *Hist. Stud. Phys. Sci.* 17:235–270.

——— (1988). The formation of the Munich school of metabolism. In *The Investigative Enterprise* (W. Coleman and F. L. Holmes, eds.), pp. 179–210. Berkeley: University of California Press.

——— (1989a). *Eighteenth-Century Chemistry as an Investigative Enterprise.* Berkeley: University of California Office for History of Science and Technology.

——— (1989b). The complementarity of teaching and research in Liebig's laboratory. *Osiris* 5:121–164.

——— (1991, 1993). *Hans Krebs: The Formation of a Scientific Life.* New York: Oxford University Press.

——— (1992a). Do we understand historically how experimental knowledge is acquired? *Hist. Sci.* 30:119–136.

——— (1992b). *Between Biology and Medicine: The Formation of Intermediary Metabolism.* Berkeley: University of California Office for History of Science and Technology.

——— (1994). The role of Johannes Müller in the formation of Helmholtz's physiological career. In *Universalgenie Helmholtz* (L. Krüger, ed.), pp. 3–21. Berlin: Akademie Verlag.

——— (1995a). The boundaries of Lavoisier's revolution. *Rev. Hist. Sci.* 48:9–48.

——— (1995b). Crystals and carriers: The chemical and physiological identification of hemoglobin. In *No Truth Except in the Details* (A. J. Kox and D. M. Siegel, eds.), pp. 191–243. Dordrecht: Kluwer.

——— (1995c). Manometers, tissue slices, and intermediary metabolism. In *The Right Tools for the Job* (A. E. Clarke and J. H. Fujimura, eds.), pp. 151–171. Princeton: Princeton University Press.

——— (1997a). What was the Chemical Revolution about? *Bull. Hist. Chem* 20:1–9.

——— (1997b). *Antoine Lavoisier: The Next Crucial Year.* Princeton: Princeton University Press.

——— (in press). *Meselson, Stahl, and the Replication of DNA.* Princeton: Princeton University Press.

Holmyard, E. J. (1957). *Alchemy.* London: Penguin.

Holter, H. (1931). Zur Kenntnis des Pepsins. *Z. physiol. Chem.* 196:1–9.

Holter, H., and Møller, K. M. (1976). *The Carlsberg Laboratory, 1876–1976.* Copenhagen: Rhodos.

Homandberg, G. A., Mattis, J. A., and Laskowski, M., Jr. (1978). Synthesis of peptide bonds by proteinases. *Biochemistry* 17:5220–5227.

Hoogewerf, S. (1971). Willem Einthoven. *DSB* 4:333–335.

Hooykaas, R. (1953). *La Naissance de la crystallographie à France au XVIIIe siècle.* Paris: Librairie du Palais de Découverte.

——— (1958). The concepts of "individual" and "species" in chemistry. *Centaurus* 5:307–322.

——— (1972). René Just Haüy. *DSB* 6:178–183.

——— (1975). Jean-Baptiste Romé de Lisle. *DSB* 11:520–524.

Hopkins, B. S. (1944). William Albert Noyes. *J. Am. Chem. Soc.* 66:1045–1066.

Hopkins, F. G. (1906). The analyst and the medical man. *Analyst* 31:385–404.

——— (1912). Feeding experiments illustrating the importance of accessory factors in normal dietaries. *J. Physiol.* 44:425–460.

——— (1913). The dynamic side of biochemistry. *Nature* 92:213–223.

——— (1921a). On an autoxidizable constituent of the cell. *Biochem. J.* 15:286–305.

——— (1921b). The chemical dynamics of muscle. *Johns Hopkins Hospital Bulletin.* 32:359–367.

——— (1926). On current views concerning the mechanisms of biological oxidation. *Skand. Arch. Physiol.* 49:33–59.

——— (1927). Benjamin Moore. *Proc. Roy. Soc.* B101:xvii–xix.

——— (1929). On glutathione: A reinvestigation. *J. Biol. Chem.* 84:269–320.

——— (1930a). Denaturation of proteins by urea and related substances. *Nature* 120:328–330, 383–384.

——— (1930b). The earlier history of vitamin research. In *Les Prix Nobel en 1929*, pp. 1–12. Stockholm: Nobel Foundation.

——— (1934). William Bate Hardy. *Biochem. J.* 28:1149–1152.

——— (1949). A lecture on organicism. In *Hopkins and Biochemistry* (J. Needham and E. Baldwin, eds.), pp. 179–190. Cambridge: Heffer.

Hopkins, F. G., and Cole, S. W. (1902). A contribution to the chemistry of proteids, 1: A preliminary study of a hitherto undescribed product of tryptic digestion. *J. Physiol.* 27:418–428.

Hopkins, F. G., and Dixon, M. (1922). On glutathione: A thermostable oxidation-reduction system. *J. Biol. Chem.* 54:527–563.

Hoppe, B. (1983). Die Biologie der Mikroorganismen von F. J. Cohn. *Sudhoffs Arch.* 67:158–189.

——— (1987). Lichenologia Schwendeneriana. *Ber. bot. Ges.* 100:305–326.

Hoppe-Seyler, F. (1866). Beiträge zur Kenntnis der Constitution des Blutes, 1: Ueber die Oxydation im lebenden Blute. *Med. Chem. Unt.*, pp. 133–140.

——— (1870). Ueber Fäulnissprocesse und Desinfektion. *Med. Chem. Unt.*, pp. 561–581.

——— (1871). Ueber die chemische Zusammensetzung des Eiters. *Med. Chem. Unt.*, pp. 486–501.

——— (1873). Ueber den Ort der Zersetzung von Eiweissund andere Nährstoffen im thierischen Organismus. *Pflügers Arch.* 7:399–417.

——— (1876). Ueber die Processe der Gährungen und ihre Beziehung zum Leben der Organismen. *Pflügers Arch.* 12:1–17.

——— (1877a). Vorwort. *Z. physiol. Chem.* 1:I–II.

——— (1877b). Ueber die Stellung der physiologischen Chemie zur Physiologie im Allgemeinem. *Z. physiol. Chem.* 1:270–273.

——— (1877c–1878). *Physiologische Chemie.* Berlin: Hirschwald.

——— (1884). *Ueber die Entwicklung der physiologischen Chemie und ihre Beziehung für die Medicin.* Strassburg: Trübner.

Horder, T. J., Witkowski, J. A., and Wylie, C. C. (eds.) (1985). *A History of Embryology.* Cambridge: Cambridge University Press.

Horecker, B. L. (1962). Interdependent pathways of carbohydrate metabolism. *Harvey Lectures* 57:35–71.

——— (1982). Cytochrome reductase, the pentose phosphate pathway, and Schiff base mechanisms. In *Of Oxygen, Fuels, and Living Matter.* Part 2 (G. Semenza, ed.), pp. 59–172. Chichester: Wiley.

Horecker, B. L., Tsolas, O., and Lai, C. Y. (1972). Aldolases. In *The Enzymes* 3d ed. (P. D. Boyer, ed.), pp. 213–258. New York: Academic Press.

Horowitz, N. H. (1948). The one-gene one-enzyme hypothesis. *Genetics* 33:612–613.

——— (1995). George Wells Beadle. *Biog. Mem. FRS* 41:44–54.

Horowitz, N. H., and Leupold, U. (1951). Some recent studies bearing on the one gene–one enzyme hypothesis. *Cold Spring Harbor Symp.* 16:65–72.

Horvath-Peterson, S. (1984). *Victor Duruy and French Education.* Baton Rouge: Louisiana State University Press.

Hoskins, R. G. (1949). The thyroid-pituitary apparatus as a servo (feedback) mechanism. *Journal of Clinical Endocrinology* 9:1429–1451.

Hoswell, A. E. (1882). Vincenz Kletzinsky. *Ber. chem. Ges.* 15:3310–3315.

Hotchkiss, R. D. (1952). The role of deoxyribonucleotides in bacterial transformations. In *Phosphorus Metabolism* (W. D. McElroy and B. Glass, eds.), pp. 426–436. Baltimore: Johns Hopkins University Press.

——— (1966). Gene, transforming principle, and DNA. In *Phage and the Origin of Molecular Biology* (J. Cairns, G. S. Stent, and J. D. Watson, eds.), pp. 184–200. Plainview, N.Y.: Cold Spring Harbor Laboratory Press.

——— (1979). The identification of nucleic acids as genetic determinants. *Ann. N.Y. Acad. Sci.* 325:321–342.

Hotchkiss, R. D., and Dubos, R. (1941). The isolation of bacteriocidal substances from cultures of *Bacillus brevis. J. Biol. Chem.* 141:155–162.

Houssay, B. A. (1936). What we have learned from the toad concerning hypophyseal function, and the hypophysis and metabolism. *New Eng. J. Med.* 214:913–926, 961–986.

——— (1952). The discovery of pancreatic diabetes: The role of Oscar Minkowski. *Diabetes* 1:112–116.

Howard, J. B., and Rees, D. C. (1991). Perspectives on non-heme iron protein chemistry. *Adv. Protein Chem.* 42:199–280.

Höxtermann, E. (1984). *Otto Heinrich Warburg: Ein "Architekt" der Naturwissenschaften.* Berlin: Humblodt Universität.

——— (1990). Zur Geschichte der Chlorophyllisolation. *NTM* 17:80–107.

——— (1994). Zur Geschichte des Hormonbegriffs in der Botanik und zur der Entdeckungsgeschichte der "Wuchsstoffen." *Hist. Phil. Life Sci.* 16:311–314.

——— (1995). Die ersten Wirkungsspektren der Photosynthese im 19. Jahrhundert. *Sudhoffs Arch.* 79:22–53.

——— (1996). Ueber "Gährung" und "gährende Pflänzchen"; Aus den Anfängen der Gärungsforschung im 19. Jahrhundert. *NTM* 4:31–51.

Höxtermann, E., and Sucker, U. (1989). *Otto Warburg.* Leipzig: Teubner.

Hoyningen-Huene, P. (1993). *Remembering Scientific Revolutions: Thomas S. Kuhn's Philosophy of Science.* Chicago: University of Chicago Press.

Hoytinck, G. J. (1970). Physical chemistry in the Netherlands after van't Hoff. *Ann. Rev. Phys. Chem.* 21:1–16.

Huang, C. Y. F., and Ferrell, J. E. (1996). Ultrasensitivity in the mitogen-activated protein kinase cascade. *PNAS* 93:10078–10083.

Hubbard, R. (1976). 100 years of rhodopsin. *TIBS* 1:154–168.

——— (1977). Preface to English translation of F. Boll, *On the Anatomy and Physiology of the Retina,* and of W. Kühne, *Chemical Processes in the Retina. Vision Research* 17:1247–1248.

Hubbard, R., and Wald, G. (1952). Cis-trans isomers of vitamin A and retinine in the rhodopsin system. *J. Gen. Physiol.* 36:269–315.

Hubbard, S. R., Wei, L., Ellis, L., and Hendrickson, W. A. (1994). Crystal structure of the tyrosine kinase domain of the human insulin receptor. *Nature* 372:746–754.

Huber, R. (1988). Flexibility and rigidity of proteins and of protein-pigment complexes. *Ang. Chem. (Int. Ed.)* 27:79–88.

Hubicki, W. (1973). Andreas Libavius. *DSB* 8:309–312.

Hucho, F. (1975). The pyruvate dehydrogenase multienzyme complex. *Ang. Chem. (Int. Ed.)* 14:591–601.

Hückel, W. (1941). Ossian Aschan. *Ber. chem. Ges.* 74A:189–220.

——— (1961). Otto Wallach. *Chem. Ber.* 94:vii–cviii.

Hudson, C. S. (1948). Historical aspects of Fischer's fundamental conventions for writing stereoformulas in a plane. *Adv. Carb. Chem.* 3:1–22.

Hufbauer, K. (1982). *The Formation of the German Chemical Community.* Berkeley: University of California Press.

Hüfner, G. (1872). *Betrachtungen über die Wirkungsweise der ungeformten Fermente.* Leipzig: Barth.

——— (1873). *Ueber die Entwicklung des Begriffs Lebenskraft.* Tübingen: Fues.

Huggins, M. L. (1971). 50 years of hydrogen bond theory. *Ang. Chem. (Int. Ed.)* 10:147–152.

Hughes, A. (1959). *A History of Cytology.* New York: Abelard-Schuman.

Huhle-Kreutzer, G. (1989). *Die Entwicklung arzneilichen Produktionstätten aus Apothekerlaboratorien.* Stuttgart: Deutscher Apotheker Verlag.

Huijing, F. (1979). Discovering of a defect in glycogen degradation. *TIBS* 4:192.

Huisgen, R. (1986). Adolf von Baeyer's scientific achievements: A legacy. *Ang. Chem. (Int. Ed.).* 25:297–311.

Hulin, N. (1990). Les doctorats dans les disciplines au XIXe siècle. *Rev. Hist. Sci.* 43:401–426.

Hull, D. L. (1972). Reduction in genetics: Biology or philosophy. *Phil. Sci.* 39:491–499.

——— (1974). *Philosophy of Biological Science.* Englewood Cliffs, N.J.: Prentice Hall.

——— (1979). In defense of presentism. *History and Theory* 18:1–14.

——— (1988). *Science as a Process.* Chicago: University of Chicago Press.

Hulme, E., Birdsall, N., and Buckley, N. (1990). Muscarinic receptor subtypes. *Ann. Rev. Pharmacol.* 30:633–673.

Humboldt, W. von (1970). University reform in Germany. *Minerva* 8:242–297.

Hume, E. D. (1923). *Béchamp or Pasteur? A Lost Chapter in the History of Biology.* Chicago: Covici-McGee.

Hummel, J. P., and Dryer, W. J. (1962). Measurement of protein-binding phenomena by gel filtration. *Biochim. Biophys. Acta* 63:530–532.

Hund, F. (1977). Early history of the quantum mechanical treatment of the chemical bond. *Ang. Chem. (Int. Ed.)* 16:87–91.

Hünefeld, F. L. (1840). *Der Chemismus in der thierischen Organisation.* Leipzig: Brockhaus.

Hünemörder, C., and Scheele, I. (1977). Das Berufsbild des Biologen im zweiten deutschen Kaiserreich—Anspruch und Wirklichkeit. In *Medizin, Naturwissenschaft, und Technik und das zweite Kaiserreich* (G. Mann and R. Winau, eds.), pp. 119–151. Göttingen: Vandenbroeck and Rupprecht.

Hunter, A. (1912). On urocanic acid. *J. Biol. Chem.* 11:537–545.

——— (1928). *Creatine and Creatinine.* London: Longmans, Green.

Hunter, J. (1837). *The Works of John Hunter.* Vol. 1, *Lectures on the Principles of Surgery* (J. F. Palmer, ed.). London: Longmans.

Hunter, M. (ed.) (1994). *Robert Boyle Reconsidered.* Cambridge: Cambridge University Press.

Hunter, T., and Cooper, J. A. (1985). Protein-tyrosine kinases. *Ann. Rev. Biochem.* 54:897–930.

Hupke, T. (1994). *Die Zeitschrift für physikalische Chemie.* Herzberg: Verlag Traugott Bautz.

Hurd, C. D. (1961). The general philosophy of chemical nomenclature. *J. Chem. Ed.* 38:43–47.

Hurwitz, J., Bresler, A., and Diringer, R. (1960). The enzymic incorporation of ribonucleotides into polyribonucleotides and the effect of DNA. *Biochim. Biophys. Acta* 3:15–19.

Husserl, E. (1977). *Cartesian Meditations* (E. Ströker, ed.). Hamburg: Meiner.

Hutchinson, J. F. (1985). Tsarist Russia and the bacteriological revolution. *J. Hist. Med.* 40:420–439.

Huxley, A. F. (1953). X-ray analysis and the problem of muscle. *Proc. Roy. Soc.* B141:59–62.

——— (1957). Muscle structure and theories of contraction. *Prog. Biophys.* 7:255–318.

——— (1980). *Reflections on Muscle.* Liverpool: Liverpool University Press.

Huxley, H. E. (1969). The mechanism of muscular contraction. *Science* 164:1356–1366.

——— (1971). The structural basis of muscular contraction. *Proc. Roy. Soc.* B178:131–149.

——— (1990). Sliding filaments and molecular motile systems. *J. Biol. Chem.* 265:8347–8350.

Huxley, T. H. (1869). The physical basis of life. *Fortnightly Review* NS5:129–145.

Ichihara, M. (1933). Ueber die Phosphoamidase. *J. Biochem.* 18:87–106.

Ihde, A. J. (1969). Theodore William Richards and the atomic weight problem. *Science* 164:647–651.

——— (1971). Stephen Moulton Babcock: Benevolent skeptic. In *Perspectives in the History of Science and Technology* (D. H. D. Roller, ed.), pp. 271–282. Norman: University of Oklahoma Press.

——— (1990). *Chemistry as Viewed from Bascom's Hill: A History of the Chemistry Department at the University of Wisconsin.* Madison: University of Wisconsin Press.

Ihde, A. J., and Janssen, J. F. (1974). Early American studies on respiration calorimetry. *Mol. Cell. Biochem.* 5:11–16.

Iltis, H. (1932). *Life of Mendel* (E. Paul and C. Paul, trans.). New York: Norton.

Ingencamp, C. (1886). Die geschichtliche Entwicklung unserer Kenntnis vom Fäulnis und Gärung. *Z. klin. Med.* 10:59–107.

Ingen-Housz, J. (1780). *Expériences sur les végétaux.* Paris: Didot.

Ingle, D. J. (1975). Edward Calvin Kendall. *Biog. Mem. NAS* 47:249–290.

——— (1979). Anton J. Carlson; A biographical sketch. *Persp. Biol. Med.* 22:S114–S136.

Ingram, V. M. (1957). Gene mutation in human hemoglobin: The chemical difference between normal and sickle-cell hemoglobin. *Nature* 180:326–328.

Inoué, S. (1969). The physics of the structural organization in living cells. In *Biology and the Physical Sciences* (S. Devons, ed.), pp. 139–171. New York: Columbia University Press.

——— (1981). Cell division and the mitotic spindle. *J. Cell Biol.* 91:131S–147S.

Inouye, M. (1988). Antisense RNA: Its functions and application in gene regulation, a review. *Gene* 72:25–34.

Ippen, K., Miller, J. H., Scaife, J., and Beckwith, J. (1968). New controlling elements in the *lac* operon of *E. coli. Nature* 217:825–827.

Ironmonger, K. D. C. (1957). The Royal Institution of Great Britain. *J. Chem. Ed* 34:607–610.

——— (1958). The Royal Institution and the teaching of chemistry in the nineteenth century. *Nature* 182:80–83.

Isbell, H. S. (1974). The Haworth-Hudson controversy and the development of Haworth's concepts of ring conformation and neighbouring group effects. *Chem. Soc. Revs.* 3:1–16.

Ishimi, Y., Claude, A., Bullock, P., and Hurwitz, J. (1988). Complete enzymatic replication of the DNA containing the SV40 origin of replication. *J. Biol. Chem.* 163:19723–19733.

Isler, H. (1965). *Thomas Willis, Doctor and Scientist.* New York: Hafner.

Itakura, K., Rossi, J. J., and Wallace, R. B. (1984). Synthesis and use of synthetic oligonucleotides. *Ann. Rev. Biochem.* 53:323–356.

Iterson, G. van, et al. (1940). *Martinus Willem Beijerinck: His Life and Work.* The Hague: Nijhoff.

Ivanovski, D. I. (1899). Ueber die Mosaikkrankheit der Tabakpflanzen. *Z. Bakt. Parasit.* 2(5):250–254.

Iwanow, L. (1903). Ueber die fermentative Zersetzung der Thymonukleinsäure durch Schimmelpilze. *Z. Physiol. Chem,* 39:31–43.

Iwata, S., Ostermeier, C., Ludwig, B., and Michel, H. (1995). Structure at 2.8 Å resolution of cytochrome oxidase from *Paracoccus denitrificans. Nature* 376:660–669.

Izant, J. G., and Weintraub, H. (1984). Inhibition by thymidine kinase gene expression by antisense RNA: A molecular approach to genetic analysis. *Cell* 36:1007–1015.

Jackson, D. A., Symons, R. H., and Berg, P. (1972). Biochemical method for inserting new genetic information of simian virus 40: Circular SV40 molecules containing phage genes and the galactose operon of *E. coli. PNAS* 69:2904–2909.

Jacob, F. (1970). *La Loqique du Vivant.* Paris: Gallimard.

Jacob, F., and Monod, J. (1959). Gènes de structure et gènes de regulation dans la biosynthèse des protéines. *Compt. Rend.* 249:1282–1284.

——— (1961). Genetic regulatory mechanisms in the biosynthesis of proteins. *J. Mol. Biol.* 3:318–356.

Jacob, F., Perrin, D., Sanchez, C., and Monod, J. (1960). L'opéron: Groupe de gènes coordinéc par un opérateur. *Compt. Rend.* 250:1727–1729.

Jacob, F., and Wollman, E. L. (1958). Genetic and physical determination of chromosomal segments in *Escherichia coli. Symp. Soc. Exp. Biol.* 12:75–92.

Jacobson, K., Sheets, E. D., and Simson, R. (1995). Revisiting the fluid mosaic model of membranes. *Science* 268:1441–1442.

Jacquemin, C. (1988). Paul Bert ou la science positive au service d'une politique nationale. *Arch. Inter. Physiol. Biochim.* 96:A34–A42.

Jacques, J. (1950). Le vitalisme et la chimie organique pendant la première moitié du XIXe siècle. *Rev. Hist. Sci.* 3:32–66.

——— (1987). *Berthelot: Autopsie d'un mythe.* Paris: Belon.

Jadin, F. (1933). Notice historique sur la Faculté de Pharmacie de Strasbourg (1803–1933). *Strasbourg Médical* 93:509–528.

Jaffé, M. (1874). Ueber einen neuen Bestandtheil des Hundesharns. *Ber. chem. Ges.* 7:1669–1673.

——— (1877). Ueber das Verhalten der Benzoesäure im Organismus der Vogel. *Ber. chem. Ges.* 10:1925–1930.

——— (1878a). Zur Kenntnis der synthetischen Vorgänge im Thierkörper. *Z. Physiol. Chem.* 2:47–64.

——— (1878b). Weitere Mittheilung über die Ornithursäure und ihre Derivate. *Ber. chem. Ges.* 11:406–409.

Jagendorf, A. T. (1967). Acid-base transitions and phosphorylation by chloroplasts. *Fed. Proc.* 26:1361–1369.

Jagendorf, A. T., and Uribe, E. (1966). ATP formation caused by acid-base transition of spinach chloroplasts. *PNAS* 55:170–177.

Jager, L. de (1890). Erklärungsversuch ueber die Wirkungsart der ungeformte Fermente. *Virchows Arch.* 121:182–187.

Jahn, I., et al. (1985). *Geschichte der Biologie.* 2d ed. Jena: VEB Fischer Verlag.

Jaki, S. L. (1984). *Uneasy Genius: The Life and Work of Pierre Duhem.* The Hague: Martinus Nijhoff.

Jakoby, W. B. (1955). An interrelationship between tryptophan, tyrosine, and phenylalanine in Neurospora. In *Amino Acid Metabolism* (W. D. McElroy and B. Glass, eds.), pp. 909–913. Baltimore: Johns Hopkins University Press.

Jaksch, R. von (1883). Ueber das Vorkommen der Acetessigsäure im Harn. *Z. Physiol. Chem.* 7:485–490.

Jakubke, H. D., Kuhl, P., and Könnecke, A. (1985). Grundprizipien der proteasekatalysierten Knüpfung der Peptidbindung. *Ang. Chem.* 97:70–140.

James, F. A. J. L. (1983). The establishment of spectro-chemical analysis as a practical method of qualitative analysis, 1854–1861. *Ambix* 30:30–53.

——— (1985). The creation of a Victorian myth: The historiography of spectroscopy. *Hist. Sci.* 23:1–24.

James, M. N. G., Sielecki, A., Salituro, F., Rich, D. H., and Hofmann, T. (1982). Conformational flexibility in the active site of aspartyl proteinases revealed by a pepstatin fragment binding to penicillopepsin. *PNAS* 79:6137–6141.

James, R. R. (1994). *Henry Wellcome.* London: Hooder and Stoughton.

James, W. (1880). Great men, great thoughts, and the environment. *Atlantic Monthly* 46:441–459.

Janin, J. (1990). Errors in three dimensions. *Biochimie* 72:706–709.

——— (1995). Principles of protein-protein interaction from structure to thermodynamics. *Biochimie* 77:497–505.

Janin, J., and Chothia, C. (1990). The structure of protein-protein recognition sites. *J. Biol. Chem.* 265:16027–16030.

Janko, J., Tesinska, E., Deutscher, H. J., and Remane, H. (1989). Zur Entdeckungsgeschichte der flüssigen Krystalle. *NTM* 26:19–29.

Jaquet, A. (1892). Ueber die Bedingungen der Oxydationsvorgänge in den Geweben. *Arch. exp. Path. Pharm,* 29:386–396.

Jarausch, K. H. (1982). *Students, Society, and Politics in Imperial Germany.* Princeton: Princeton University Press.

Jarausch, K. H., et al. (eds.) (1983). *The Transformation of Higher Learning, 1860–1930.* Stuttgart: Klett-Cotta.

Jencks, W. P. (1969). *Catalysis in Chemistry and Enzymology.* New York: McGraw-Hill.

——— (1975). Binding energy, specificity, and enzyme catalysis: The Circe effect. *Adv. Enzymol.* 43:219–410.

——— (1980). The utilization of binding energy in coupled vectorial processes. *Adv. Enzymol.* 51:75–106.

——— (1994). Reaction mechanisms, catalysis, and movement. *Protein Science* 4:2459–2464.

Jensen, E. V. (1992). Remembrance: Gregory Pincus: Catalyst for early receptor studies. *Endocrinology* 131:1581–1582.

Jerne, N. K. (1955). The natural selection theory of antibody formation. *PNAS* 41:849–857.

——— (1985). The generative grammar of the immune system. *Ang. Chem. (Int. Ed.)* 24:810–816.

Jevons, F. R. (1962). Boerhaave's biochemistry. *Med. Hist.* 6:343–362.

Job, P. (1939). Georges Urbain. *Bull. Soc. Chim.* [5]6:745–766.

Joffe, J. S. (1948). Dmitri Nikolaevich Prianishnikov. *Soil Science* 66:165–169.

Johannsen, W. (1909). *Elemente der exakten Erblichkeitslehre.* Jena: Fischer.

Johnson, H. A. (1970), Information theory in biology after 18 years. *Science* 168:1545–1550.

Johnson, I. S. (1983). Human insulin from recombinant DNA technology. *Science* 219:632–637.

Johnson, J. A. (1985a). Academic chemistry in Imperial Germany. *Isis* 76:500–524.

——— (1985b). Academic self-regulation and the chemical profession in Imperial Germany. *Minerva* 23:241–271.

——— (1989). Hierarchy and creativity in chemistry, 1871–1914. *Osiris* 5:214–240.

——— (1990). *The Kaiser's Chemists: Science and Modernization in Imperial Germany.* Chapel Hill: University of North Carolina Press.

Johnson, L. N. (1992). Glycogen phosphorylase: Control by phosphorylation and allosteric effectors. *FASEB J.* 6:2274–2282.

Johnson, L. N., and Barford, D. (1994). Electrostatic effects in the control of glycogen phosphorylase by phosphorylation. *Protein Science* 3:1726–1730.

Johnson, P., and Perutz, M. (1981). Gilbert Smithson Adair. *Biog. Mem. FRS* 27:1–17.

Johnson, S. F. (1990). Fourier transform infrared spectroscopy. *Chem. Brit.* 26:573–577.

Johnson, W. B. (1803). *History of the Progress and Present State of Animal Chemistry.* London: J. Johnson.

Jolly, J. (1923). Louis Ranvier. *Arch. Anat. Micr.* 19:i–lxxii.

Jones, D. F. (1944). Edward Murray East. *Biog. Mem. NAS* 22:217–242.

Jones, D. S., Nishimura, S., and Khorana, H. G. (1966). Studies on polynucleotides, 56: Further syntheses, *in vitro,* of copolypeptides containing two amino acids in alternating sequence etc. *J. Mol. Biol.* 16:454–472.

Jones, E. R. H., et al. (1992). A history of steroid chemistry. *Steroids* 57:353–424, 577–664.

Jones, J. H. (1994). Henry Prentiss Armsby. In *American Chemists and Chemical Engineers* (W. D. Miles and R. F. Gould, eds.), pp. 6–7. Guilford, Conn.: Gould.

Jones, M. E. (1953). Albrecht Kossel: A biographical sketch. *Yale J. Biol. Med.* 26:80–97.

Jones, M. E., Hearn, W. R., Fried, M., and Fruton, J. S. (1952). Transamidation reactions catalyzed by cathepsin C. *J. Biol. Chem.* 195:645–656.

Jones, M. E., Spector, L., and Lipmann, F. (1955). Carbamyl phosphate, the carbamyl donor in enzymatic citrulline synthesis. *J. Am. Chem. Soc.* 77:819–820.

Jones, P. G. (1984). Crystal structure determination: A critical view. *Chem. Soc. Revs.* 13:157–172.

Jones, P. R. (1983). *Bibliographie der Dissertationen amerikanischer und britischer Chemiker an deutschen Universitäten, 1840–1914.* Munich: Forschungsinstitut des Deutsches Museums.

——— (1991). The training in Germany of English-speaking chemists in the nineteenth century and its profound influence in America and Britain. In *World Views and Discipline Formation* (W. R. Woodward and R. S. Cohen, eds.), pp. 299–308. Dordrecht: Kluwer.

Jones, S., and Thornton, J. M. (1996). Principles of protein-protein interactions. *PNAS* 93:13–20.

Jones, W. (1920a). *Nucleic Acids.* 2d ed. London: Longmans, Green.

——— (1920b). The action of boiled pancreas extract on yeast nucleic acid. *Am. J. Physiol.* 52:203–207.

Jorgensen, B. S. (1965). Berzelius und die Lebenskraft. *Centaurus* 10:258–281.

Jorgensen, C. B. (1976). August Putter, August Krogh, and modern ideas on the use of dissolved organic matter in aqueous environments. *Biol. Revs.* 51:291–328.

Jorissen, W. P., and Reicher, L. T. (1912). *J. H. van't Hoffs Amsterdamer Periode 1877–1895*. Helder: De Boer.

Jorpes, E. (1951). Alexander Schmidt. *J. Chem. Ed.* 28:578–579.

——— (1966). *Jac. Berzelius: His Life and Work.* Stockholm: Almquist and Wiksell

——— (1969). Einar Hammarsten. *Nordisk Medicin* 80:1701–1703.

Jorpes, E., and Mutt, V. (1973). Secretin and cholestokinin. *Handbook of Experimental Pharmacology* 34:1–179. Berlin: Springer.

Jost, J. P., and Rickenberg, H. V. (1971). Cyclic AMP. *Ann. Rev. Biochem.* 40:741–774.

Jost, L. (1930). Zum hundersten Geburtstag Anton de Barys. *Z. Bot.* 24:1–74.

Jost, W. (1966). The first 45 years of physical chemistry in Germany. *Ann. Rev. Phys. Chem.* 17:1–14.

Jowett, M., and Quastel, J. H. (1935). Studies on fat metabolism. *Biochem. J.* 29:2143–2180.

Judson, H. F. (1979). *The Eighth Day of Creation: The Making of the Revolution in Biology.* New York: Simon and Schuster.

Jukes, T. H. (1980a). The discovery of folic acid. *TIBS* 5:112–113.

——— (1980b). Lecithin, choline, methionine and "labile methyl" groups in biochemistry. *TIBS* 5:307–308.

——— (1980c). Vitamin K: A reminiscence. *TIBS* 5:140–141.

Julian, M. M. (1984). Dorothy Wrinch and a search for the structure of proteins. *J. Chem. Ed.* 61:890–892.

Jumelle, M. H. (1924). L'Oeuvre scientifique de Gaston Bonnier. *Rev. gen. Bot.* 26:289–307.

Jung, C. G. (1968). *Psychology and Alchemy.* 2d ed. Princeton: Princeton University Press.

Jungfleisch, E. (1913). Notice sur la vie et les travaux de Marcelin Berthelot. *Bull. Soc. Chim.* [4]13:i–cclx.

Jungner, G., Jungner, J., and Allgren, L. G. (1949). Molecular weight determination on thymonucleic acid compounds by dielectric measurements. *Nature* 163:849–850.

Jungnickel, C., and McCormmach, R. (1996). *Cavendish.* Philadelphia: American Philosophical Society.

Kabat, E. A. (1943). Immunochemistry of the proteins. *J. Immunol.* 47:513–587.

——— (1992). Michael Heidelberger. *J. Immunol.* 148:301–307.

Kabat, E. A., and Mayer, M. M. (1961). *Quantitative Immunochemistry.* 2d ed. Springfield, Ill.: Thomas.

Kadenbach, B. (1983). Structure and evolution of the "Atmungsferment" cytochrome c oxidase. *Ang. Chem. (Int. Ed.)* 22:275–283.

Kagawa, Y. (1984). Proton motive ATP synthesis. In *Bioenergetics* (L. Ernster, ed.), pp. 149–186. Amsterdam: Elsevier.

Kahane, E. (1978). *Parmentier ou la dignité de la pomme de terre.* Paris: Blanchard.

——— (1988). *Boussingault entre Lavoisier et Pasteur.* Argueil: Jonas.

Kahlbaum, G. W. A. (ed.) (1900). *Letters of Jöns Jacob Berzelius and Christian Friedrich Schönbein, 1836–1847* (F. V. Darbyshire and N. V. Sidgwick, trans.). London: Williams and Norgate.

——— (1904). *Justus von Liebig und Friedrich Mohr in ihren Briefen von 1834–1870.* Leipzig: Barth.

Kahlbaum, G. W. A., and Darbyshire, F. V. (eds.) (1899). *Letters of Faraday and Schönbein, 1836–1862.* London: Williams and Norgate.

Kahlbaum, G. W. A., and Thun, E. (eds.) (1900). *Justus von Liebig und Christian Friedrich Schönbein. Briefwechsel 1853–1868.* Leipzig: Barth.

Kahlson, G. (1953). Thorsten Thunberg. *Acta Physiol. Scand.* 30 Suppl. 111:1–24.

Kaiser, E. T., Lawrence, D. S., and Rokita, S. E. (1985). The chemical modification of enzymatic specificity. *Ann. Rev. Biochem.* 54:565–595.

Kalckar, H. M. (1937). Phosphorylation in kidney tissue. *Enzymologia* 2:47–52.

——— (1939). The nature of phosphoric esters found in kidney extracts. *Biochem. J.* 33:631–641.

——— (1941). The nature of energetic coupling in biological syntheses. *Chem. Revs.* 28:71–178.

——— (1942). The enzymatic action of myokinase. *J. Biol. Chem.* 143:299–300.

——— (1969). *Biological Phosphorylation: Development of Concepts.* Englewood Cliffs, N.J.: Prentice-Hall.

——— (1983). The isolation of Cori-ester, "the Saint Louis Gateway," to a first approach of a dynamic formulation of macromolecular biosynthesis. In *Selected Topics in the History of Biochemistry* (G. Semenza, ed.), pp. 1–24. Amsterdam: Elsevier.

——— (1990). Autobiographical notes from a nomadic scientist. In *Selected Topics in the History of Biochemistry.* Vol. 3 (G. Semenza and R. Jaenicke, eds.), pp. 101–176. Amsterdam: Elsevier.

Kalmus, H. (1983). The scholastic origins of Mendel's concepts. *Hist. Sci.* 21:61–83.

Kamen, M. D. (1963). The early history of carbon-14. *J. Chem. Ed.* 40:234–242.

——— (1986a). On creativity of eye and ear: A commentary on the career of T. W. Engelmann. *Proc. Am. Phil. Soc.* 130:232–246.

——— (1986b). A cupful of luck, a pinch of sagacity. *Ann. Rev. Biochem.* 55:1–34.

——— (1994). Reflections on the first half-century of long-lived radioactive carbon (^{14}C). *Proc. Am. Phil. Soc.* 138:48–59.

——— (1995). Liebling's Law ("ILL"). *Proc. Am. Phil. Soc.* 139:358–367.

Kamminga, H. (1993). Taking antecedents seriously: A lesson in heuristics from biology. In *Correspondence, Invariance, and Heuristics* (S. French and H. Kamminga, eds.), pp. 65–82. Dordrecht: Kluwer.

Kamminga, H., and Weatherall, M. W. (1996). The making of a scientist, 1: Frederick Gowland Hopkins' construction of dynamic biochemistry. *Med. Hist.* 40:269–292.

Kamp, A. F., et al. (1959). *Albert Jan Kluyver: His Life and Work.* Amsterdam: North-Holland.

Kandel, E., and Abel, T. (1995). Neuropeptides, adenylyl cyclase, and memory storage. *Science* 268:825–826.

Kant, I. (1902). *Kritik der Urteilskraft* (K. Vorländer, ed,). Leipzig: Dürr.

——— (1906). *Kritik der reinen Vernunft* (T. Valentiner, ed.). Leipzig: Dürr.

Kantrowitz, E. R., and Lipscomb, W. N. (1988). *Escherichia coli* aspartate transcarbamoylase: The relation between structure and function. *Science* 241:669–674.

Kanz, K. T. (ed.) (1994). *Philosophie des Organischen in der Goethezeit.* Stuttgart: Franz Steiner.

Kaplan, G. (1956). Henry Devaux, plant physiologist, pioneer of surface physics. *Science* 124:1017–1018.

Kaplan, N. O., and Kennedy, E. P. (eds.) (1966). *Current Aspects of Biochemical Energetics.* New York: Academic.

Kapoor, S. G. (1969). Dumas and organic classification. *Ambix* 16:1–65.

——— (1973). Auguste Laurent. *DSB* 8:54–61.

Kargon, R. H. (1966). *Atomism in England from Hariot to Newton.* Oxford: Clarendon.

——— (1977). *Science in Victorian Manchester.* Baltimore: Johns Hopkins University Press.

Karlin, A., Holtzman, E., Yodh, N., Lobel, P., Wall, Jt and Hainfeld, J. (1983). The arrangement of the subunits of the acetylcholine receptor of *Torpedo californica.* *J. Biol. Chem.* 258:6678–6681.

Karlson, P. (1963). New concepts on the mode of action of hormones. *Persp. Biol. Med.* 6:203–214.

——— (1977). 100 Jahre Biochemie im Spiegel von Hoppe-Seyler's Zeitschrift für Physiologische Chemie. *Z. Physiol. Chem.* 358:710–752.

——— (1978). Carotene as provitamin A. *TIBS* 3:235–236.

——— (1982). Was sind Hormone? Der Hormonbegriff in Geschichte und Gegenwart. *Naturwiss.* 69:3–14.

——— (1986). Wie und warum entstehen wissenschaftliche Irrtümer. *Naturw. Rund.* 39:380–389.

——— (1990). *Adolf Butenandt.* Stuttgart: Wissenschaftliche Verlagsgesellschaft.

——— (1996). On the hormonal control of insect metamorphosis: A historical review. *Int. J. Dev. Biol.* 40:93–96.

Karp, B., and Karp, V. (1988). *Louis Rapkine.* North Bennington, Vt.: Orpheus.

Karrer, P., and Wehrli, H. (1933). 25 Jahre Vitamin A Forschung. *Nova Acta Leopoldina* NF1:175–275.

Karström, H. (1930). Ueber die Enzymbildung in Bakterien. *Ann. Acad. Sci. Finn,* 33:1–147.

——— (1937). Enzymatische Adaption bei Mikroorganismen. *Erg. Enzymforsch.* 7:350–376.

Kase, K. (1931). Untersuchungen über die Harnstoffbildung. *Biochem. Z.* 233:258–282.

Kasich, A. M. (1946). William Prout and the discovery of hydrochloric acid in the gastric juice. *Bull. Hist. Med.* 20:340–358.

Kasten, F. H. (1964). Robert Feulgen. *Acta Histochimica* 17:88–99.

Kastle, J. H., and Loevenhart, A. S. (1900). Concerning lipase, the fat-splitting enzyme, and the reversibility of its action. *Am. Chem. J.* 24:491–525.

Kästner, I. (1996). Kein Nobelpreis für Maria Manassein. In *Dilettanten und Wissenschaft* (E. Strauss, ed.), pp. 123–134. Amsterdam: Rodopi.

Katchalski-Katzir, E. (1993). Poly-alpha-amino acids as the simplest protein models: Recollections of a retired state president. *Protein Science* 2:476–482.

Katsoyannis, P. G. (ed.) (1972). *The Chemistry of Polypeptides.* New York: Wiley.

Katz, B. (1978). Archibald Vivian Hill. *Biog. Mem. FRS* 24:71–149.

Kauffman, G. B. (1960). Sophus Mads Jørgensen and the Werner-Jørgensen controversy. *Chymia* 6:180–204.

——— (1966). *Alfred Werner, Founder of Coordination Chemistry.* Berlin: Springer.

——— (1976). Marcel Delépine. *Coordination Chemistry Reviews* 21:181–219.

——— (1980). Niels Bjerrum (1879–1958): A centennial evaluation. *J. Chem. Ed.* 57:779–782, 863–867.

——— (ed.) (1994). *Coordination Chemistry: A Century of Progress.* Washington, D.C.: American Chemical Society.

——— (1997). A stereochemical achievement of the first order: Alfred Werner's resolution of cobalt complexes, 85 years later. *Bull. Hist. Chem.* 20:50–59.

Kaufman, S. (1971). The phenylalanine hydroxylating system from mammalian liver. *Adv. Enzymol.* 35:245–319.

Kay, L. E. (1985). Conceptual models and analytical tools: The biology of the physicist Max Delbrück. *J. Hist. Biol.* 18:207–246.

——— (1986). W. M. Stanley's crystallization of the tobacco mosaic virus, 1930–1940. *Isis* 77:450–472.

——— (1988). Laboratory technology and biological knowledge: The Tiselius electrophoresis apparatus. *Hist. Phil. Life Sci.* 10:51–72.

——— (1989). Selling pure science in wartime: The biochemical genetics of G. W. Beadle. *J. Hist. Biol.* 22:73–101.

——— (1992). Quanta of life: Atomic physics and the reincarnation of phage. *Hist. Phil. Life Sci.* 14:3–21.

——— (1993). *The Molecular Vision of Life.* Oxford: Oxford University Press.

——— (1995). Who wrote the book of life? Information and the transformation of molecular biology, 1945–1955. *Science in Context* 8:609–634.

——— (1997). Cybernetics, information, life: The emergence of scriptural representations of heredity. *Configurations* 5:23–91.

Kaziro, Y., Itoh, H., Kozasa, T., Nakafuku, M., and Satoh, T. (1991). Structure and function of signal-transducing GTP-binding proteins. *Ann. Rev. Biochem.* 60:349–400.

Keating, P., and Ousman, A. (1991). The problem of natural antibodies. *J. Hist. Biol.* 24:245–263.

Keen, R. (1976). Friedrich Wöhler. *DSB* 14:474–479.

Keilin, D. (1925). On cytochrome, a respiratory pigment, common to animals, yeast, and higher plants. *Proc. Roy. Soc.* B98:312–339.

——— (1927). Influence of carbon monoxide and light on indophenol oxidase of yeast cells. *Nature* 119:670–671.

——— (1929). Cytochrome and respiratory enzymes. *Proc. Roy. Soc.* B104:206–252

——— (1930). Cytochrome and intracellular oxidase. *Proc. Roy. Soc.* B106:408–444.

——— (1950). Heavy metal catalysis in cellular metabolism. *Nature* 165:4–5.

——— (1966). *History of Cell Respiration and Cytochrome.* Cambridge: Cambridge University Press.

Keilin, D., and Hartree, E. F. (1939). Cytochrome and cytochrome oxidase. *Proc. Roy. Soc.* B127:167–191.

Kekulé, A. (1858). Ueber die Constitution und die Metamorphosen der chemischen Verbindungen. *Ann.* 106:129–159.

——— (1861). *Lehrbuch der organischen Chemie.* Erlangen: Enke.

——— (1867). On some points of chemical philosophy. *The Laboratory* 1:303–306.

——— (1878). On the scientific aims and achievements of chemistry. *Nature* 18:210–213.

Kekwick, R. O., and Pedersen, K. O. (1974). Arne Tiselius. *Biog. Mem. FRS* 20:401–428.

Kelker, H. (1986). Flüssige Kristalle und die Theorie des Lebens. *Naturw. Rund.* 39:239–247.

Kellenberger, E. (1966). Electron microscopy of developing bacteriophage. In *Phage and the Origin of Molecular Biology* (J. Cairns, G. S. Stent, and J. D. Watson, eds.), pp. 116–127. Plainview, N.Y.: Cold Spring Harbor Laboratory Press.

——— (1995). History of phage research as viewed by a European. *FEMS Microbiology Reviews* 17:7–24.

Keller, E. F. (1990). Physics and the emergence of molecular biology: A history of cognitive and political synergy. *J. Hist. Biol.* 23:389–409.

Kelley, W. N., and Weiner, I. M. (eds.) (1978). *Uric Acid.* Berlin: Springer.

Kelley, W. N., and Wyngaarden, J. B. (1974). The enzymology of gout. *Adv. Enzymol.* 41:1–33.

Kelly, T. J. (1988). SV40 DNA replication. *J. Biol. Chem.* 163:17889–17892.

Kelner, A. (1949). Effect of visible light on the recovery of *Streptomyces griseus* conidia from ultraviolet irradiation injury. *PNAS* 35:73–79.

Kemp, B. E., Parker, M. W., Hu, S., Tiganis, T, and House, C. (1994). Substrate and substrate interaction with protein kinases: Determinants of specificity. *TIBS* 19:440–444.

Kendall, E. C. (1919). Isolation of the iodine compound which occurs in the thyroid. *J. Biol. Chem.* 39:125–147.

——— (1971). *Cortisone.* New York: Scribner's.

Kendall, I. (1935). James Walker. *Obit. Not. FRS* 1:537–549.

Kendrew, J. C. (1968). Information and conformation in biology. In *Structural Chemistry and Molecular Biology* (A. Rich and N. Davidson, eds.), pp. 187–197. San Francisco: Freeman.

Kendrew, J. C., Bodo, C., Dintzis, H. M., Parrish, R. G., and Wyckoff, H. (1958). A three-dimensional model of the myoglobin molecule obtained by X-ray analysis. *Nature* 181:662–666.

Kendrew, J. C., Dickerson, R. E., Strandberg, B. E., Hart, R. G., Davies, D. R., Phillips, D. C., and Shore, V. C. (1960). Structure of myoglobin. A three-dimensional Fourier synthesis at 2 Å resolution. *Nature* 185:422–427.

Kennard, O., and Hunter, W. N. (1991). Single-crystal X-ray diffraction studies of oligonucleotides and oligonucleotide-drug complexes. *Ang. Chem. (Int. Ed.)* 30:1254–1277.

Kennard, O., and Salisbury, S. A. (1993). Oligonucleotide X-ray structures in the study of the conformation and interactions of nucleic acids. *J. Biol. Chem.* 268:10701–10704.

Kennedy, E. P., and Lehninger, A. L. (1949). Oxidation of fatty acids and tricarboxylic acid cycle intermediates in rat liver mitochondria. *J. Biol. Chem.* 179:957–972.

Kennelly. P. J., and Krebs, E. G. (1991). Consensus sequences as substrate specific determinants for protein kinases and protein phosphatases. *J. Biol. Chem.* 266:15555–15558.

Kent, A. (ed.) (1950). *An Eighteenth Century Lectureship in Chemistry.* Glasgow: Glasgow University Press.

Kent, S. B. H. (1988). Chemical synthesis of peptides and proteins. *Ann. Rev. Biochem.* 57:957–989.

Kerker, M. (1976). The Svedberg and molecular reality. *Isis* 67:190–216.

Kersaint, G. (1958). Sur une correspondance inedite de Nicolas Louis Vauquelin. *Bull. Soc. Chim.*, pp. 1603–1619.

——— (1966). *Antoine François Fourcroy: Sa vie et son oeuvre.* Paris: Editions du Musée.

Kessler, C., and Holtke, W. J. (1986). Specificity of restriction endonucleases and methylases. *Gene* 47:1–52.

Kestner, O. (1925). *Chemie der Eiweisskörper.* 4th ed. Braunschweig: Vieweg.

Kevles, D. J. (1980). Genetics in the United States and Great Britain, 1890–1930. *Isis* 71:441–455.

——— (1993). Renato Dulbecco and the new animal virology: Methods, medicine, and molecules. *J. Hist. Biol.* 26:409–442.

Kevles, D. J., and Hood, L. (1992). *The Code of Codes: Scientific and Social Issues in the Human Genome Project.* Cambridge: Harvard University Press.

Keyser, B. W. (1990). Between science and craft: The case of Berthollet and dyeing. *Ann. Sci.* 47:213–260.

Khorana, H. G. (1968). Polynucleotide synthesis and the genetic code. *Harvey Lectures* 62:79–105.

——— (1979). Total synthesis of a gene. *Science* 203:614–625.

——— (1988). Bacteriorhodopsin, a membrane protein that uses light to translocate protons. *J. Biol. Chem.* 263:7439–7442.

——— (1989). Nucleotide synthesis and the years at the Enzyme Institute. In *One Hundred Years of Agricultural Chemistry and Biochemistry at Wisconsin* (D. L. Nelson and B. C. Siltvedt, eds.), pp. 225–237. Madison, Wis.: Science Tech.

——— (1992). Rhodopsin, photoreceptor of the rod cell. *J. Biol. Chem.* 267:1–4.

Khorana, H. G., et al. (1976). Total synthesis of the structural gene for the precursor of tyrosine suppressor transfer RNA from *Escherichia coli. J. Biol. Chem.* 251:565–694.

Kikuchi, S. (1997). A history of the structural theory of benzene: The aromatic sextet rule and Hückel's rule. *J. Chem. Ed.* 74:194–201.

Kim, D. W. (1995). J. J. Thomson and the emergence of the Cavendish school, 1885–1990. *Brit. J. Hist. Sci.* 28:191–226.

Kim, J., Woo, D., and Rees, D. C. (1993). X-ray crystal structure of the nitrogenase molybdenum-iron protein from *Clostridium pasteurianum* at 3.0 Å resolution. *Biochemistry* 32:7104–7115.

Kim, J. L., Geiger, J. H., Hahn, S., and Sigler, P. B. (1993). Crystal structure of a yeast TBP/TATA box complex. *Nature* 365:512–520.

Kim, J. L., Nikolov, D. B., and Burley, S. K. (1993). Co-crystal structure of a TBP recognizing the minor groove of a TATA element. *Nature* 365:520–527.

Kim, K. M. (1994). *Explaining Scientific Consensus: The Case of Mendelian Genetics.* New York: Guilford.

Kim, M. G. (1992). The layers of chemical language. *Hist. Sci.* 30:69–76, 397–437.

King, C. G. (1979). The isolation of vitamin C from lemon juice. *Fed. Proc.* 38:2681–2683.

King, H. (1956). Sigmund Otto Rosenheim. *Biog. Mem. FRS* 2:257–267.

King, L. S. (1975). Georg Ernst Stahl. *DSB* 12:599–606.

King, T. E., Mason, H. S., and Morrison, M. (eds.) (1988). *Oxidases and Related Redox Systems.* New York: Liss.

Kingreen, H. (1972). *Theodor Wilhelm Engelmann.* Münster: Institut für Geschichte der Medizin.

Kingzett, C. T. (1878). *Animal Chemistry.* London: Longmans, Green.

Kisselev, L. L. (1990). Wladimir Engelhardt: The man and the scientist. In *Selected Topics in the History of Biochemistry: Personal Recollections.* Vol. 3 (G. Semenza and R. Jaenicke, eds.), pp. 67–99. Amsterdam: Elsevier.

Kisselev, L. L., and Wolfson, A. D. (1994). Aminoacyl-tRNA synthetases from higher eukaryotes. *Prog. Nucleic Acid Res.* 48:83–142.

Kitcher, P. (1983). "Genes." *Brit. J. Phil. Sci.* 33:337–359.

Kleckner, N. (1996). Meiosis: How could it work? *PNAS* 93:8167–8174.

Kleiber, M. (1961). *The Fire of Life.* New York: Wiley.

Klein, D. J., and Trinajstic, N. (1990). Valence-bond theory and chemical structure. *J. Chem. Ed.* 67:633–637.

Klein, H. M., and Helmreich, E. J. M. (1985). The role of pyridoxal-5′-phosphate in carbohydrate metabolism. *Curr. Top. Cell Regul.* 26:281–294.

Klein, J. R., and Harris, J. S. (1938). The acetylation of sulfanilamide in vitro. *J. Biol. Chem.* 124:613–626.

Klein, M. (1936). *Histoire des origines de la théorie cellulaire.* Paris: Hermann.

——— (1970). Arnold Adolph Berthold. *DSB* 2:72–73.

——— (1980). *Regards d'un biologiste.* Paris: Hermann.

Klein, M. J. (1972a). Mechanical explanation at the end of the nineteenth century. *Centaurus* 17:59–82.

——— (1972b). Josiah Willard Gibbs. *DSB* 5:386–393.

Klein, U. (1994a). Origin of the concept of the chemical compound. *Science in Context* 7:163–204.

——— (1994b). *Verbindung und Affinität.* Basel: Birkhäuser.

——— (1995). E. F. Geoffroy's table of different "rapports" observed between different chemical substances: A reinterpretation. *Ambix* 42:79–100.

Kleinkauf, H., Döhren, H. von, and Jaenicke, L. (eds.) (1988). *The Roots of Modern Biochemistry.* Berlin: Walter de Gruyter.

Kleinkauf, H., and Gevers, W. (1969). Nonribosomal polypeptide synthesis: The biosynthesis of a cyclic peptide antibiotic, gramicidin S. *Cold Spring Harbor Symp.* 34:805–813.

Kleinschmidt, A., Lang, D., Jacherts, D., and Zahn, R. K. (1962). Darstellung und Längenmessungen des Deoxyribonucleinsäureinhalts von T2-Bakteriophagen. *Biochim. Biophys. Acta* 61:857–864.

Kleinschmidt, A., Lang, D., and Zahn, R. K. (1961). Ueber die intrazellulare Formation von Bakterien-DNS. *Z. Naturforsch.* 16b:730–739.

Klenk, E. (1931). Hans Thierfelder. *Z. physiol. Chem.* 203:1–9.

Kletzinsky, V. (1858). *Compendium der Biochemie.* Vienna: Braumüller.

Kleywegt, G. T., and Jones, T. A. (1995). Where freedom is given, liberties are taken. *Structure* 3:535–540.

Klinkowski, M. (1941). Oscar Loew. *Ber. chem. Ges.* 74A:115–136.

Klinman, J. P. (1978). Kinetic isotopic effects in enzymology. *Adv. Enzymol.* 46:415–494.

Klinman, J. P., and Mu, D. (1994). Quinoenzymes in biology. *Ann. Rev. Biochem* 63:299–344.

Klooster, H. S. van (1973). Frans Maurits Jaeger. *DSB* 7:59.

Klosterman, L. J. (1985). A research school of chemistry in the nineteenth century: Jean-Baptiste Dumas and his research students. *Ann. Sci.* 42:1–40.

Klug, A. (1968). Rosalind Franklin and the discovery of the structure of DNA. *Nature* 219:808–810, 843–844, 879, 1192.

——— (1974). Rosalind Franklin and the double helix. *Nature* 248:787–788.

Klug, A., and Rhodes, D. (1987). "Zinc fingers": A novel protein motif for nucleic acid recognition. *TIBS* 12:464–469.

Kluyver, A. J., and Donker, H. J. L. (1925). The catalytic transference of hydrogen as the basis of the chemistry of dissimilation processes. *Proceedings of the Royal Academy of Sciences, Amsterdam* 28:605–618.

——— (1926). Die Einheit in der Biochemie. *Chemie der Zelle und Gewebe* 13:134–190.

Knieriem, W. (1874). Beiträge zur Kenntnis der Bildung des Harnstoffs im thierischen Organismus. *Z. Biol.* 10:263–294.

Knight, D. M. (1967). *Atoms and Elements.* London: Hutchinson.

——— (ed.) (1968). *Classical Scientific Papers: Chemistry.* London: Mills and Boon.

——— (1992). *Humphry Davy.* Oxford: Blackwell.

Knoop, F. (1904). *Der Abbau aromatischen Fettsäuren im Tierkörper.* Freiburg: Kuttruff.

——— (1910). Ueber den physiologischen Abbau der Säuren und die Synthese einer Aminosäure im Tierkörper. *Z. physiol. Chem.* 67:489–520.

Knoop, F., and Martius, C. (1936). Ueber die Bildung der Citronensäure. *Z. physiol. Chem.* 242:I.

Knoop, F., and Oesterlin, H. (1925). Ueber die natürliche Synthese der Aminosäuren und ihre experimentelle Reproduktion. *Z. physiol. Chem.* 148:294–315.

Knorr-Cetina, K. D. (1979). Tinkering toward success. *Theory and Society* 8:347–376.

——— (1981). *The Manufacture of Knowledge.* Oxford: Pergamon.

——— (1983). The ethnographic study of scientific work: Toward a constructivist interpretation of science. In *Science Observed* (K. Knorr-Cetina and M. Mulkay, eds.), pp. 113–140. Beverly Hills: Sage.

Knowles, J. R., and Albery, W. J. (1977). Perfection in enzyme catalysis: The energetics of triose phosphate isomerase. *Acc. Chem. Res.* 10:105–111.

Knox, W. E. (1955). The metabolism of phenylalanine and tyrosine. In *Amino Acid Metabolism* (W. D. McElroy and B. Glass, eds.). pp. 836–866. Baltimore: Johns Hopkins University Press.

——— (1958). Sir Archibald Garrod's "inborn errors of metabolism." *Am. J. Hum. Gen.* 10:9–32, 95–124, 249–267, 385–397.

——— (1960). Glutathione. In *The Enzymes* 2d ed. (P. D. Boyer et al., eds.), 2:253–294. New York: Academic.

Kober, F. (1980). Die Geschichte des ersten und zweiten Hauptsatzes der Thermodynamik. *Chem. Z.* 104:195–200.

——— (1982). Die Theorie der chemischen Bindung in "vor-quantenmechnischer" Zeit. *Chem. Z.* 106:1–11.

Koch, R. (1929). Wilhelm Roux. *Sudhoffs Arch.* 22:114–150.

Koch, R. (1995). The case of Latour. *Configuration* 3:319–347.

Koch-Weser, J., and Schechter, P. (1978). Schmiedeberg in Strassburg, 1872–1918: The making of modern pharmacology. *Life Sciences* 22:1361–1372.

Koerting, W. (1967), Die Deutsche Universität in Prag. *Bayerisches Aerzteblatt* 22:802–818.

Kögl, F., and Erxleben, H. (1939). Zur Aetiologie der malignen Tumoren, 1: Mitteilung über die Chemie der Tumoren. *Z. physiol. Chem.* 258:57–95.

Köhler, G. (1985). Derivation and diversification of monoclonal antibodies. *Ang. Chem. (Int. Ed.)* 24:827–833.

Köhler, G., and Milstein, C. (1975). Continuous cultures of fused cells secreting antibody of predefined specificity. *Nature* 256:495–497.

Kohler, R. E. (1971). The background to Eduard Buchner's discovery of cell-free fermentation. *J. Hist. Biol.* 4:35–61.

——— (1972). The reception of Eduard Buchner's discovery of cell-free fermentation. *J. Hist. Biol.* 5:327–353.

——— (1973a). The enzyme theory and the origin of biochemistry. *Isis* 64:181–196.

——— (1973b). Review of *Molecules and Life. Isis* 64:389–390.

——— (1975). The history of biochemistry: A survey. *J. Hist. Biol.* 8:275–318.

——— (1977a). Warren Weaver and the Rockefeller Foundation program in molecular biology: A case study in the management of science. In *The Scientist in the American Context: New Perspectives* (N. Reingold, ed.), pp. 249–293. Washington, D.C.: Smithsonian Institution.

——— (1977b). Rudolf Schoenheimer, isotopic tracers, and biochemistry in the 1930s. *Hist. Stud. Phys. Sci.* 8:257–298.

——— (1980). History of science: Perceptions. *Science* 207:934.

——— (1982). *From Medical Chemistry to Biochemistry: The Making of a Biomedical Discipline.* Cambridge: Harvard University Press.

——— (1985a). Bacterial physiology: The medical context. *Bull. Hist. Med.* 59:54–74.

——— (1985b). Innovation in normal science: Bacterial physiology. *Isis* 76:162–181.

——— (1991). *Partners in Science: Foundations and Natural Scientists.* Chicago: University of Chicago Press.

——— (1994). *Lords of the Fly: Drosophila and the Experimental Life.* Chicago: University of Chicago Press.

Kohler, R. E., Ron, E. Z., and Davis, B. D. (1958). Significance of the free 70S ribosomes in *Escherichia coli* extracts. *J. Mol. Biol.* 36:71–82.

Kolbe, H. (1845). Beiträge zur Kenntnis der gepaarten Verbindungen. *Ann.* 54:145–188.

——— (1858). *Ueber die chemische Constitution der organischen Verbindungen.* Marburg: Elwert.

——— (1877). Zeichen der Zeit. *J. Prakt. Chem.* 123:473–477.

Kolesko, A. J., and Young, R. A. (1995). The RNA polymerase II holoenzyme and its implications for gene regulation. *TIBS* 20:113–116.

Koller, G. (1958). *Das Leben des Biologen Johannes Müller.* Stuttgart: Wissenschaftliche Verlagsgesellschaft.

Kölliker, A. (1885). Die Bedeutung der Zellkerne für die Vorgänge der Vererbung. *Z. wiss. Zool.* 42:1–46.

Kondo, M., Gallerini, R., and Weissmann, C. (1970). Subunit structure of Qβ replicase. *Nature* 228:525–527.

Königsberger, L. (1902–1903). *Hermann von Helmholtz.* Braunscheig: Vieweg.

Kopp, H. (1843–1847). *Geschichte der Chemie.* Braunschweig: Vieweg.

Kopperl, S. J. (1976). T. W. Richards' role in American graduate education in chemistry. *Ambix* 23:165–174.

Kornberg, A. (1957). Pathways of enzymatic synthesis of nucleotides and polynucleotides. In *The Chemical Basis of Heredity* (W. D. McElroy and B. Glass, eds.), pp. 579–608. Baltimore: Johns Hopkins University Press.

——— (1960). Biologic synthesis of deoxyribonucleic acid. *Science* 131:1503–1508.

——— (1969). The active center of DNA polymerase. *Science* 163:1410–1418.

——— (1988). DNA replication. *J. Biol. Chem.* 263:1–4.

——— (1989). *For the Love of Enzymes.* Cambridge: Harvard University Press.

——— (1995). *The Golden Helix: Inside Biotech Ventures:* Sausalito, Calif.: University Science.

Kornberg, A., and Baker, T. A. (1992). *DNA Replication,* 2d ed. New York: Freeman.

Kornberg, A., Lieberman, I., and Simms, E. S. (1955). Enzymatic synthesis of purine nucleotides. *J. Biol. Chem.* 215:417–423.

Kornberg, A., and Pricer, W. E. (1950). On the structure of triphosphopyridine nucleotide. *J. Biol. Chem.* 186:557–567.

Kornberg, H. L. (1989). Travelling to, and along, the glyoxylate pathway. *Biochim. Biophys. Acta* 1000:271–278.

Kornberg, H. L., and Elsden, S. R. (1961). The metabolism of 2-carbon compounds by microorganisms. *Adv. Enzymol.* 23:401–470.

Kornberg, H. L., and Williamson, D. H. (1984). Hans Adolf Krebs. *Biog. Mem. FRS* 30:351–385.

Kornberg, R. D. (1974). Chromatic structure: A repeating unit of histones and DNA. *Science* 184:868–871.

Koshland, D. E. (1960). Active site and enzyme action. *Adv. Enzymol.* 22:45–97.

——— (1963). The role of flexibility in enzyme action. *Cold Spring Harbor Symp.* 28:473–482.

——— (1970). The molecular basis for enzyme regulation. In *The Enzymes,* 3d ed. (P. D. Boyer, ed.), 1:341–396. New York: Academic.

Koshland, D. E., and Levitski, A. (1974). CTP synthase and related enzymes. In *The Enzymes,* 3d Ed. (P. D. Boyer, ed.), 10:539–559. New York: Academic.

Koshland, D. E., and Neet, K. E. (1968). The catalytic and regulatory properties of enzymes. *Ann. Rev. Biochem.* 37:359–410.

Koshland, D. E., Nemethy, G., and Filmer, D. (1966). Comparison of binding data and theoretical models in proteins containing subunits. *Biochemistry* 5:365–385.

Kossel, A. (1879). Ueber das Nuklein der Hefe. *Z. physiol. Chem.* 3:284–291.

——— (1881). *Untersuchungen über die Nucleine und ihre Spaltprodukte.* Strassburg: Trübner.

——— (1884). Ueber ein peptonartigen Bestandteil des Zellkerns. *Z. Physiol. Chem.* 8:511–515.

——— (1908). *Die Probleme der Biochemie.* Heidelberg: Horning.

——— (1928). *The Protamines and the Histones.* London: Longmans, Green.

Kossel, A., and Dakin, H. D. (1904). Ueber die Arginase. *Z. physiol. Chem.* 41:321–331; 42:181–188.

Kossel, A., and Neumann, A. (1893). Ueber des Thymin, ein Spaltprodukt der Nukleinsäure. *Ber. chem. Ges.* 26:2753–2756.

——— (1894). Darstellung und Spaltprodukte der Nukleinsäure (Adenylsäure). *Ber. chem. Ges.* 27:2215–2222.

——— (1896). Ueber Nukeinsäure und Thyminsäure. *Z. physiol. Chem.* 22:74–82.

Kostychev, S. (1926). Ueber die Nichtexistenz einiger Fermente. *Z. physiol. Chem.* 154:262–275.

Kotake, Y. (1935). Zum intermediären Stoffwechsel des Tryptophans. *Erg. Physiol.* 37:245–263.

Kottler, D. B. (1978). Louis Pasteur and molecular dissymmetry. *Stud. Hist. Biol.* 2:57–98.

Kozak, M. (1991). Structural features in eukaryotic mRNAs that modulate the initiation of translation. *J. Biol. Chem.* 266:19867–19870.

Kragh, H. (1982). Julius Thomsen and 19th-century speculations on the complexity of atoms. *Ann. Sci.* 39:37–60.

——— (1984). Julius Thomsen and classical thermochemistry. *Brit. J. Hist. Sci.* 17:255–272.

——— (1987). *An Introduction to the Historiography of Science.* Cambridge: Cambridge University Press.

——— (1989). The Aether in late 19th-century chemistry. *Ambix* 36:49–65.

——— (1993). Between physics and chemistry: Helmholtz's route to a theory of chemical thermodynamics. In *Helmholtz and the Foundations of Nineteenth-Century Science* (D. Cahan, ed.), pp. 403–431. Berkeley: University of California Press.

——— (1996). Gamow's game: The road to the hot big bang. *Centaurus* 39:335–361.

Kragh, H., and Weininger, S. J. (1996). Sooner silence than confusion: The tortuous entry of entropy into chemistry. *Hist. Stud. Phys. Biol. Chem.* 27:91–130.

Krämer, A. (1996). The structure and function of proteins involved in mammalian pre-mRNA splicing. *Ann. Rev. Biochem.* 65:367–409.

Krayevsky, A. A., and Kukhanova, M. K. (1979). The peptidyltransferase center of ribosomes. *Prog. Nucleic Acid Res.* 23:1–51.

Krebs, A. T. (1955). Early history of the scintillation counter. *Science* 122:17–18.

Krebs, E. G. (1994). The growth of research on protein phosphorylation. *TIBS* 19:439.

Krebs, E. G., and Beavo, J. A. (1979). Phosphorylation-dephosphorylation of enzymes. *Ann. Rev. Biochem.* 48:923–959.

Krebs, E. G., and Fischer, E. H. (1962). Molecular properties and transformations of glycogen phosphorylase in animal tissues. *Adv. Enzymol.* 24:273–290.

Krebs, H. A. (1935a). Metabolism of amino-acids, 3: Deamination of amino-acids. *Biochem. J.* 29:1620–1644.

——— (1935b). Metabolism of amino-acids, 4: The synthesis of glutamine from glutamic acid and ammonia, and the enzymic hydrolysis of glutamine in animal tissues. *Biochem. J.* 29:1951–1969.

——— (1942). Urea formation in the mammalian liver. *Biochem. J.* 36:758–767.

——— (1943). The intermediary stages in the biological oxidation of carbohydrate. *Adv. Enzymol.* 3:191–252.

——— (1947). Cyclic processes in living matter. *Enzymologia* 12:88–100.

——— (1956). Die Steuerung der Stoffwechselvorgänge. *Deutsche med. Wchschr.* 81:4–8.

——— (1957). Control of metabolic processes. *Endeavour* 16:125–132.

——— (1970). The history of the tricarboxylic acid cycle. *Persp. Biol. Med.* 14:154–170.

——— (1972a). Otto Heinrich Warburg. *Biog. Mem. FRS* 18:629–699.

——— (1972b). The Pasteur effect and the relations between respiration and fermentation. *Essays in Biochemistry* 8:1–34.

——— (1974). The discovery of carbon dioxide fixation in animal tissues. *Mol. Cell. Biochem.* 5:70–62.

——— (1979). *Otto Warburg: Zellphysiologe—Biochemiker—Mediziner.* Stuttgart: Wissenschaftliche Verlagsgesellschaft. [English ed. 1981, Oxford: Clarendon.]

——— (1981). *Reminiscences and Reflections.* Oxford: Clarendon.

Krebs, H. A., and Decker, K. (1982). Feodor Lynen. *Biog. Mem. FRS* 28:261–317.

Krebs, H. A., and Eggleston, L. V. (1940). The oxidation of pyruvate in pigeon breast muscle. *Biochem. J.* 34:442–459.

Krebs, H. A., and Henseleit, K. (1932). Untersuchungen über Harnstoffbildung im Tierkörper. *Z. Physiol. Chem.* 210:33–66.

Krebs, H. A., and Johnson, W. A. (1937). The role of citric acid in the intermediate metabolism in animal tissues. *Enzymologia* 4:148–156.

Krebs, H. A., and Kornberg, H. L. (1957). A survey of energy transformations in living matter. *Erg. Physiol.* 49:212–298.

Krebs, H. A., and Lipmann, F. (1974). Dahlem in the nineteen twenties. In *Energy, Respiration, and Biosynthesis in Molecular Biology* (D. Richter, ed.), pp. 7–27. Berlin: Walter de Gruyter.

Krebs, H. A., and Shelley, J. W. (eds.) (1975). *The Creative Process in Science.* Amsterdam: Excerpta Medica.

Kreil, G. (1990). The processing of precursors by dipeptidylaminopeptidases: A case of molecular ticketing. *TIBS* 15:23–26.

——— (1994). Peptides containing a D-amino acid from frogs and mollusks. *J. Biol. Chem.* 269:10967–10970.

Kremer, R. L. (1990). *The Thermodynamics of Life and Experimental Physiology, 1770–1880.* New York: Garland.

Kretsinger, R. H., Rudnick, S. E., and Weissman, L. J. (1986). Crystal structure of calmodulin. *J. Inorg. Biochem.* 28:289–302.

Krishna, R. G., and Wold, F. (1993). Post-translational modification of proteins. *Adv. Enzymol.* 67:265–299.

Krishnamurthy, R. S. (ed.) (1996). *The Pauling Symposium.* Corvallis: Oregon State University Libraries.

Kritzmann, M. G. (1938). Ueber den Abund Aufbau von Aminosäuren durch Umaminierung. *Enzymologia* 5:44–51.

Krizenecky, J. (1965). *Fundamenta Genetica.* Brno: Moravian Museum.

Krohn, R. (1982). On Gieryn on the "relativist/constructivist" programme in the sociology of science: Naïveté and reaction. *Soc. Stud. Sci.* 12:325–328.

Kröner, H. P. (1989). Die Emigration deutschsprachiger Mediziner im Nationalsozialismus. *Ber. Wissen.* 12:1*–44*.

Krueger, A. P. (1931). The sorption of bacteriophage by living and dead susceptible bacteria, 1: Equilibrium reactions. *J. Gen. Physiol.* 14:493–516.

Krüger, L. (ed.) (1994). *Universalgenie Helmholtz: Rückblick nach 100 Jahren.* Berlin: Akademie Verlag.

Kruhoffer, P., and Crone, C. (1972). Einar Lundsgaard. *Erg. Physiol.* 65:1–14.
Kruta, V. (1969). *J. E. Purkyne: Physiologist.* Prague: Academia.
——— (1976). Friedrich Tiedemann. *DSB* 13:402–404.
Kubowitz, F. (1937). Ueber die Zusammensetzung der Kartoffeloxydase. *Biochem. Z.* 292:221–229.
Kuechler, E., and Rich, A. (1970). Position of the initiator and peptidyl sites in the *E. coli* ribosome. *Nature* 225:920–924.
Kühl, S. (1994). *The Nazi Connection: Eugenics, American Racism, and German National Socialism.* New York: Oxford University Press.
Kühling, O. (1899). Ueber die Reduktion des Tolualloxazins. *Ber. chem. Ges.* 32:1650–1653.
Kuhn, R. (1935). Sur les flavines. *Bull. Soc. Chim. Biol.* 17:905–926.
Kuhn, R., Rudy, H., and Weygand, F. (1936). Synthese der Lactoflavin-5′-phosphorsäure. *Ber. chem. Ges.* 69:1543–1547.
Kuhn, T. S. (1959). Energy conservation as an example of simultaneous discovery. In *Critical Problems in the History of Science* (M. Clagett, ed.), pp. 321–356. Madison: University of Wisconsin Press.
——— (1962). *The Structure of Scientific Revolutions.* Chicago: University of Chicago Press.
——— (1963). The function of dogma in scientific research. In *Scientific Change* (A. C. Crombie, ed.), pp. 347–369. London: Heinemann.
——— (1971). The relations between history and history of science. *Daedalus* 100:271–304.
——— (1977). *The Essential Tension.* Chicago: University of Chicago Press.
——— (1978). *Black-Body Theory and the Quantum Discontinuity, 1894–1912.* New York: Oxford University Press.
——— (1993). Afterword. In *World Changes: Thomas Kuhn and the Nature of Science* (P. Horwich, ed.). Cambridge: MIT Press.
Kuhn, W. (1962). Georg Bredig. *Chem. Ber.* 95:xlii–lxii.
Kühne, W. (1864). *Untersuchungen über das Protoplasma und die Contractilität.* Leipzig: Engelmann.
——— (1866). *Lehrbuch der physiologischen Chemie.* Leipzig: Engelmann.
——— (1876). Ueber das Verhalten verschiedener organisierte und sog. ungeformte Fermente. *Verhandl. Nat. Heidelberg* NF1:190–193.
——— (1878a). Erfahrungen und Bemerkungen über Enzyme und Fermente. *Untersuchungen aus dem physiologischen Instutut Heidelberg* 1:291–324.
——— (1878b). *On the Photochemistry of the Retina and on Visual Purple* (M. Foster, trans. and ed.). London: Macmillan.
Kullmann, W. (1987). *Enzymatic Peptide Synthesis.* Boca Raton: CRC.
Kunitz, M. (1940). Crystalline ribonuclease. *J. Gen. Physiol.* 24:15–32.
——— (1950). Crystalline desoxyribonuclease. *J. Gen. Physiol.* 33:349–377.
Kunkel, H. G. (1964). Myeloma proteins and antibodies. *Harvey Lectures* 59:219–242.
Kunkel, T. A. (1992). DNA replication fidelity. *J. Biol. Chem.* 267:18251–18254.
Kurland, C. G. (1960). Molecular characterization of ribonucleic acid from *E. coli* ribosomes, 1: Isolation and molecular weight. *J. Mol. Biol.* 2:83–91.
Kurzer, F., and Sanderson, P. M. (1956). Urea in the history of organic chemistry. *J. Chem. Ed.* 33:452–459.
Kuschinsky, G. (1968). The influence of Dorpat on the emergence of pharmacology as a distinct discipline. *J. Hist. Med.* 23:258–271.

Kuslan, L. I. (1975). François Marie Raoult. *DSB* 11:297–300.

Kützing, F. (1837). Mikroskopische Untersuchungen über Hefe und Essigmutter, nebst mehreren andered dazu gehörigen vegetabilischen Gebilden. *J. prakt. Chem.* 11:385–409.

——— (1960). *Aufzeichnungen und Erinnerungen* (H. W. Müller and R. Zaunick, eds.). Leipzig: Barth.

Kuznick, P. J. (1987). *Beyond the Laboratory: Scientists as Political Activists in 1930s America.* Chicago: University of Chicago Press.

Lahue, R. S., and Schachman, H. K. (1986). Communication between polypeptide chains in aspartate transcarbamoylase. Conformational changes at the active sites of unliganded chains resulting from ligand binding to other chains. *J. Biol. Chem.* 261:3079–3084.

Laidler, K. J. (1993). *The World of Physical Chemistry.* Oxford: Oxford University Press.

Laidler, K. J., and King, M. C. (1983). The development of transition-state theory. *J. Phys. Chem.* 87:2657–2664.

Lakatos, I. (1974). History of science and its rational reconstructions. In *The Interaction Between Science and Philosophy* (Y. Elkana, ed.), pp. 196–241. Atlantic Highlands, N. J.: Humanities.

Lakatos, I., and Musgrave, A. (eds.) (1970). *Criticism and the Growth of Knowledge.* Cambridge: Cambridge University Press.

Laland, S. G., and Zimmer, T. L. (1973). The protein thiotemplate mechanism of synthesis for peptide antibiotics produced by *Bacillus brevis. Essays in Biochemistry* 9:31–57.

Lalli, E., and Sassone-Corsi, P. (1994). Signal transduction and gene regulation: The nuclear response to cAMP. *J. Biol. Chem.* 296:17359–17362.

Lambert, R. (1963). *Sir John Simon.* London: McGibbon and Kee.

Lamond, A. I., and Gibson, T. J. (1990). Catalytic RNA and the origin of genetic systems. *Trends in Genetics* 6:145–149.

Landsteiner, K. (1945). *The Specificity of Serological Reactions.* 2d ed. Cambridge: Harvard University Press.

Landsteiner, K., and Heidelberger, M. (1923). On the antigenic properties of hemoglobin. *J. Exp. Med.* 38:561–571.

Langan, T. A. (1973). Protein kinases and protein kinase substrates. *Adv. Cyclic Nucl. Res.* 3:99–153.

Langdon, R. G., and Bloch, K. (1953). On the biosynthesis of squalene: The utilization of squalene in the biosynthesis of cholesterol. *J. Biol. Chem.* 200:129–144.

Langley, J. N. (1905). On the reactions of cells and of nerve-endings to certain poisons, chiefly as regards the reaction of striated muscle to nicotine and to curari. *J. Physiol.* 33:374–413.

——— (1906). On nerve-endings and on special excitable substances in cells. *Proc. Roy. Soc.* B78:170–194.

——— (1908). On the contraction of muscle, chiefly in relation to the presence of "receptive" substances, 3: The reaction of frog's muscle to nicotine after denervation. *J. Physiol.* 37:285–300.

——— (1917). Arthur Sheridan Lea. *Proc. Roy. Soc.* B89:xxv–xxvii.

Langley, L. L. (ed.) (1973). *Homeostasis: Origin of a Concept.* Stroudsburg: Dowden, Hutchinson, and Ross.

Langmuir, I. (1939). The structure of proteins. *Proceedings of the Physical Society* 51:592–612.

Lankester, E. R. (1871). Ueber das Vorkommen von Haemoglobin in den Muskeln der Mollusken und die Verbreitung desselben in den lebenden Organismen. *Pflügers Arch.* 4:315–320.

——— (1872). A contribution the knowledge of haemoglobin. *Proc. Roy. Soc.* 21:70–81.

Lanni, F. (1964). The biological coding problem. *Adv. Genetics* 12:1–141.

Larder, D. F. (1967). Historical aspects of the tetrahedron in chemistry. *J. Chem. Ed.* 44:661–666.

Lardy, H. A. (1978). Copper as an essential element in iron utilizatiom. *TIBS* 3:93–94.

Lardy, H. A., and Ferguson, S. M. (1969). Oxidative phosphorylation in mitochondria. *Ann. Rev. Biochem.* 38:991–1034.

Lardy, H. A., and Phillips, P. H. (1943). The effect of thyroxine and dinitrophenol on sperm metabolism. *J. Biol. Chem.* 149:177–182.

Larner, J. (1990). Insulin and the stimulation of glycogen synthesis: The road from glycogen structure to glycogen synthase to cyclic AMP-dependent kinases to insulin mediators. *Adv. Enzymol.* 63:173–231.

——— (1992). Gerty Theresa Cori. *Biog. Mem. NAS* 61:111–135.

Larson, J. L. (1994). *Interpreting Nature: The Science of Living Form from Linneaus to Kant.* Baltimore: Johns Hopkins University Press.

Larsson, S., Källebring, B., Wittung, P., and Malmström, B. (1995). The Cu_A center of cytochrome c oxidase: Electronic structure and spectra of models compared to the properties of the Cu_A demain. *PNAS* 92:7167–7171.

Laskowski, M. (1967). DNases and their use in studies of primary structure of nucleic acids. *Adv. Enzymol.* 29:165–220.

——— (1982). Nucleases: Historical perspectives. In *Nucleases* (S. M. Linn and R. J. Roberts, eds.), pp. 1–21. Plainview, N.Y.: Cold Spring Harbor Laboratory Press.

Laskowski, M., Jr., and Kato, I. (1980). Protein inhibitors of proteinases. *Ann. Rev. Biochem.* 49:593–626.

Laskowski, N. (1846). Ueber die Proteintheorie. *Ann.* 58:129–166.

Laszlo, P. (1993). *La Parole des choses ou le langage de la chimie.* Paris: Hermann.

Latchman, D. S. (1995). *Eukaryotic Transcription Factors.* 2d ed. New York: Academic.

Latham, P. W. (1887). *Lectures on some Points in the Pathology of Rheumatism, Gout, and Diabetes.* Cambridge: Deighton Bell.

Latour, B. (1984). *Guerre et paix.* Paris: Métaillié. [English translation (1988) entitled *The Pasteurization of France.* Cambridge: Harvard University Press].

——— (1987). *Science in Action.* Cambridge: Harvard University Press.

——— (1990). Drawing things together. In *Representation in Scientific Practice* (M. Lynch and S. Woolgar, eds.), pp. 19–68. Cambridge: MIT Press.

——— (1992). Pasteur on lactic acid yeast: A partial semiotic analysis. *Configurations* 1:129–145.

——— (1995). A propos de la "science privée de Louis Pasteur." *La Recherche,* November, pp. 33–34.

Latour, B., and Woolgar, S. (1979). *Laboratory Life: The Social Construction of Scientific Facts.* Beverly Hills: Sage. [2d ed. 1986, Princeton: Princeton University Press].

Laudan, L. (1971). Isidore Auguste Marie François Comte. *DSB* 3:375–380.

——— (1981). The pseudo-science of science? *Phil. Soc. Sci.* 11:173–198.

Laupheimer, P. (1992). *Phlogiston oder Sauerstoff.* Stuttgart: Wissenschaftliche Verlagsgesellschaft.

Laurent, A. (1837). Suites de recherches diverses de chimie organique. *Ann. Chim.* [2]66:314–335.

Lavieties, P. H. (1956). John Punnett Peters: An appreciation. *Yale J. Biol. Med.* 29:175–190.

Lavoisier, A. (1799). *Elements of Chemistry.* 4th ed. (R. Kerr, trans.). Edinburgh: Creech.

——— (1862). *Oeuvres de Lavoisier.* Paris: Imprimerie Impériale.

Law, J. (1973). The development of specialties in science: The case of X-ray protein crystallography. *Science Studies* 3:275–303.

Lawes, J. B., and Gilbert, J. W. (1863). On the amounts of, and methods of estimating ammonia and nitric acid in rain water. *J. Chem. Soc.* 26:100–186.

——— (1866). On the source of fat of the animal body. *Phil. Mag.* [4]32:439–451.

Lawes, J. B., Gilbert, J. W., and Pugh, E. (1861). On the sources of nitrogen of vegetation, with special reference to the question whether plants assimilate free or uncombined nitrogen. *Phil. Trans.* 151:431–577.

Layton, D. (1965). The original observations of Brownian motion. *J. Chem. Ed.* 42:367–368.

Leatherbarrow, R. J., and Fersht, A. (1987). Use of protein engineering to study enzyme mechanisms. In *Enzyme Mechanisms* (M. I. Page and A. Williams, eds.), pp. 78–96. London: Royal Society of Chemistry.

Leathes, J. B. (1906). *Problems in Animal Metabolism.* Philadelphia: Blakiston.

Lebedev, A. (1912). Ueber den Mechanismus der alkoholischen Gärung. *Biochem. Z.* 46:483–489.

LeBel, J. A. (1874). Sur les relations entre les formules atomiques des corps organiques et le pouvoir rotatoire de leurs dissolutions. *Bull. Soc. Chim.* 22:337–347.

Leberman, K. R., and Dahlberg, A. E. (1995). Ribosome-catalyzed peptide-bond formation. *Prog. Nucleic Acid Res.* 50:1–23.

Lecanu, L. R. (1838). Etudes chimiques sur le sang humain. *Ann. Chim.* [2]67:54–70.

Le Châtelier, H. (1888). Recherches expérimentales et théoriques sur les equilibres chimiques. *Annales des Mines et des Carburants* [8]13:157–380.

Lechevalier, H. (1972). Dmitri Iosifovich Ivanovski. *Bact. Revs.* 36:135–145.

Lechner, T. (1965). Prof. Siegfried Thannhauser zum Gedächnis. *Med. Welt,* pp. 226–229.

Lecomte, J. (1985). Zenon Marcel Bacq. *Ann. Acad. Roy. Belg.* 151:53–99.

Lecomte du Nouy, P. (1964). *Entre savoir et croire.* Paris: Gonthier.

Lecourt, D. (1974). *L'Epistemologie historique de Gaston Bachelard.* Paris: Vrin.

Lederberg, E. M. (1951). Lysogenicity in *E. coli* k-12. *Genetics* 36:560.

Lederberg, E. M., and Lederberg, J. (1953). Genetic studies of lysogenicity in *Escherichia coli. Genetics* 38:51–64.

Lederberg, J. (1947). Gene recombination and linked segregations in *Escherichia coli. Genetics* 32:505–525.

——— (1950). Isolation and characterization of biochemical mutants of bacteria. In *Methods in Medical Research* (J. H. Comroe Jr., ed.), 3:5–22. Chicago: Year Book Publishers.

——— (1951). Genetic studies with bacteria. In *Genetics in the Twentieth Century* (L. C. Dunn, ed.), pp. 263–289. New York: Macmillan.

——— (1956a). Bacterial protoplasts induced by penicillin. *PNAS* 42:574–577.

——— (1956b). Genetic transduction. *Am. Sci.* 44:264–280.

——— (1959). Genes and antibodies. *Science* 129:1649–1653.

——— (1987). Genetic recombination of bacteria: A discovery account. *Ann. Rev. Genetics* 21:23–46.

——— (1988). Ontogeny of the clonal selection theory of antibody production. *Ann. N.Y. Acad. Sci.* 546:175–182.

——— (1989). Replica plating and individual selection of bacterial mutants: Isolation of preadaptive mutants in bacteria by sib selection. *Genetics* 121:395–399.

——— (1990). Edward Lawrie Tatum. *Biog. Mem. NAS* 59:357–386.

——— (1993). Genetic maps: Fruit flies, people, bacteria, and molecules. A tribute to Morgan and Sturtevant. In *The Biological Century* (R. Barlow, ed.), pp. 26–49. Woods Hole, Mass.: Marine Biological Laboratory.

Lederberg, J., and Lederberg, E. M. (1952). Replica plating and indirect selection of bacterial mutants. *J. Bact.* 63:399–406.

Lederberg, J., and Tatum, E. L. (1946). Gene recombination in *Escherichia coli. Nature* 158:558.

Leegwater, A. (1986). The development of Wilhelm Ostwald's chemical energetics. *Centaurus* 29:314–337.

Leete, E. (1965). Biosynthesis of alkaloids. *Science* 147:1000–1006.

——— (1969). Alkaloid biosynthesis. *Adv. Enzymol.* 32:373–422.

Legallois, J. J. C. (1817). Deuxième mémoire sur la chaleur animale. *Ann. Chim.* [2]4:5–23, 113–127.

Legée, G. (1987). La physiologie allemande à Strasbourg de 1872 à 1914. *Histoire et Nature* 28/29:69–88.

——— (1991). La physiologie dans l'oeuvre de Jean Senebier. *Gesnerus* 49:307–322.

Lehman. I. R. (1974). DNA ligase: Structure, mechanism, and function. *Science* 186:790–797.

Lehman, I. R., Bessman, M. J., Simms, E. S., and Kornberg, A. (1958). Enzymatic synthesis of deoxyribonucleic acid. *J. Biol. Chem.* 233:163–177.

Lehman, I. R., and Kaguni, L. S. (1989). DNA polymerase. *J. Biol. Chem.* 264:4265–4268.

Lehmann, C. G. (1855). *Physiological Chemistry* [trans. of 2d ed. (1852) by G. E. Day; R. E. Rogers, ed.]. Philadelphia: Blanchard and Lea.

Lehmann, J. (1930). Zur Kenntnis Biologischer Oxydations- Reduktionspontentiale. Messungen im System: Succinat-Fumarat- Succindehydrogenase. *Skand. Arch. Physiol.* 58:173–312.

Lehmann, O. (1907). *Die scheinbar lebende Kristalle.* Munich: Schreiber.

——— (1911). *Die neue Welt der flüssigen Kristalle und deren Bedeutung fUr Physik, Chemie, Technik und Biologie.* Leipzig: Akademische Verlagsgesellschaft.

Lehn, J. M. (1988). Supramolecular chemistry: Scope and perspectives. Molecules, supermolecules, and molecular devices. *Ang. Chem. (Int. Ed.)* 27:90–112.

Lehnartz, E. (1933). Die chemische Vorgänge bei der Muskelkontraktion. *Erg. Physiol.* 35:874–966.

Lehninger, A. L. (1945). On the activation of fatty acid oxidation. *J. Biol. Chem.* 161:437–451.

——— (1951). Phosphorylation coupled to the oxidation of dihydrophosphopyridine nucleotide. *J. Biol. Chem.* 190:345–359.

——— (1954). *The Mitochondrion.* New York: Benjamin.

——— (1955). Oxidative phosphorylation. *Harvey Lectures* 49:176–215.

Leibowitz, H. (1989). *Fabricating Lives.* New York: Knopf.

Lejeune, P. (1989). *On Autobiography.* Minneapolis: University of Minnesota Press.

Leloir, L. F. (1964). Nucleoside diphosphate sugars and saccharide synthesis. *Biochem. J.* 91:1–8.

——— (1971). Two decades of research on the biosynthesis of saccharides. *Science* 172:1299–1302.

Leloir, L. F., and Cardini, C. E. (1957). Biosynthesis of glycogen from uridine diphosphate glucose. *J. Am. Chem. Soc.* 79:6340–6341.

Leloir, L. F., and Munoz, J. M. (1939). Fatty acid oxidation in liver. *Biochem. J.* 33:734–746.

——— (1944). Butyrate oxidation by liver enzymes. *J. Biol. Chem.* 153:53–60.

Leloir, L. F., Olavarria, J. M., Soldemberg, S. H., and Carminati, H. (1959). Biosynthesis of glycogen from uridine diphosphate glucose. *Arch. Biochem. Biophys.* 81:508–520.

Leloir, L. F., and Paladini, A. C. (1983). The discovery of sugar nucleotides. In *Selected Topics in the History of Biochemistry* (G. Semenza, ed,), pp. 25–42. Amsterdam: Elsevier.

Lemaine, G. R., et al. (eds.) (1976). *Perspectives on the Emergence of Scientific Disciplines.* The Hague: Mouton.

Lemay, P. (1949). Desormes et Clement decouvrent et expliquent la catalyse. *Chymia* 2:45–49.

Lembeck, F., and Giere, W. (1968). *Otto Loewi, ein Lebensbild in Dokumenten.* Berlin: Springer.

Lemberg, R. (1961). Cytochromes of group A and their prosthetic groups. *Adv. Enzymol.* 23:265–321.

Lenay, C. (1995). Préhistoire de la génétique: Hugo de Vries et l'idée d'indépendance des caractères. In *Nature, Histoire, Société* (C. Blanckaert, J. L. Fischer, and R. Rey, eds.), pp. 133–145. Paris: Klincksieck.

Lendner, A. (1934). Prof. Dr. R. Chodat. *Verhandl. Schw. Nat. Ges.* 115:529–550.

Lengyel, P., Speyer, J. F., and Ochoa, S. (1961). Synthetic polypolynucleotides and the amino acid code. *PNAS* 47:1936–1942.

Lenoir, T. (1980). Kant, Blumenbach, and vital materialism in German biology. *Isis* 71:77–108.

——— (1982). *The Strategy of Life: Teleology and Mechanism in Nineteenth Century German Biology.* Dordrecht: Reidel.

——— (1988). Science for the clinic: Science policy and the formation of Carl Ludwig's institute in Leipzig. In *The Investigative Enterprise* (W. Coleman and F. L. Holmes, eds.), pp. 139–178. Berkeley: University of California Press.

Lenz, M. (1910). *Geschichte der Königlichen Friedrich-Wilhelm-Universität zu Berlin.* Vol. 3. Halle: Buchhandlung des Waisenhauses.

Leonetti, J. P., Degols, G., Clarence, J. P., Mechti, N., and Lebleu, B. (1993). Cell delivery and action of antisense oligonucleotides. *Prog. Nucleic Acid Res.* 44:143–166.

Leplin, J. (ed.) (1984). *Scientific Realism.* Berkeley: University of California Press.

Leprieur, F., and Papon, P. (1979). Synthetic dyestuffs: The relations between academic chemistry and chemical industry in nineteenth-century France. *Minerva* 17:197–224.

Lepsius, B. (1918). August Wilhelm von Hofmann. *Ber. chem. Ges.* Suppl. 51:154.

Lerner, R. A., and Benkovic, S. J. (1988). Principles of antibody catalysis. *BioEssays* 9:107–112.

Lerner, R. A., Benkovic, S. J., and Schultz, P. G. (1991). At the crossroads of chemistry and immunology: Catalytic antibodies. *Science* 252:659–667.

Lesch, J. E. (1984). *Science and Medicine in France: The Emergence of Experimental Physiology, 1790–1855.* Cambridge: Harvard University Press.

——— (1990). Hans Thierfelder. *DSB* 18:904–906.

Lesky, E. (1970). Ernst Wilhelm von Brücke. *DSB* 2:530–532.

——— (1976). *The Vienna Medical School of the 19th Century.* Baltimore: Johns Hopkins University Press.

Lester, J. (1995). *E. Ray Lankester and the Making of Modern British Biology.* Oxford: British Society for the History of Science.

Lestradet, H. (1993). Historique de la découverte de l'insuline. *Hist. Sci. Med.* 27:61–68.

Leuhrsen, K. R., Taha, S., and Walbot, V. (1994). Nuclear pre-mRNA processing in higher plants. *Prog. Nucleic Acid Res.* 47:149–193.

Levene, P. A. (1903). On the chemistry of the nerve cell. *Journal of Medical Research* 10:204–211.

——— (1909). Ueber die Hefenucleinsäure. *Biochem. Z.* 17:120–131.

——— (1921). On the structure of thymus nucleic acid and on its possible bearing on the structure of plant nucleic acid. *J. Biol. Chem.* 48:119–125.

Levene, P. A., and Bass, L. W. (1931). *Nucleic Acids.* New York: Chemical Catalogue.

Levene, P. A., and Jacobs, W. A. (1908, 1909). Ueber die Inosinsäure. *Ber. chem. Ges.* 41:2703–2797; 42:335–338, 1196–1203.

——— (1909). Ueber Guanylsäure. *Ber. chem. Ges.* 42:2469–2473.

——— (1912). On the structure of thymus nucleic acid. *J. Biol. Chem.* 12:411–420.

Levene, P. A., Mikeska, L. A., and Mori, T. (1930). The carbohydrate of thymonucleic acid. *J. Biol. Chem.* 85:785–787.

Levene, P. A., and Mori, T. (1929). Ribodesose and xylodesose and their bearing on the structure of thyminose. *J. Biol. Chem.* 83:803–816.

Levene, P. A., and Raymond, A. L. (1928). Hexose diphosphate. *J. Biol. Chem.* 80:633–638.

Levene, P. A., and Tipson, R. S. (1935). The ring structure of thymidine. *J. Biol. Chem.* 109:623–630.

Levere, T. H. (1971). *Affinity and Matter.* Oxford: Oxford University Press.

——— (1996). Romanticism, natural philosophy, and the sciences. *Persp. Sci.* 4:463–488.

Levi-Montalcini, R. (1987). The nerve growth factor: Twenty-five years later. *Bioscience Reports* 7:681–699.

——— (1988). *In Praise of Imperfection: My Life and Work.* New York: Basic.

Levine, L. (ed.) (1995). *The Continuing Importance of Theodosius Dobzhansky.* New York: Columbia University Press.

Levine, R., and Goldstein, R. S. (1955). On the mechanism of the action of insulin. *Hormone Research* 11:343–380.

Levy, L. (1889). Ueber Farbstoffe in den Muskeln. *Z. Physiol. Chem.* 13:309–325.

Lewis, D. (1983). Cyril Dean Darlington. *Biog. Mem. FRS* 29:113–157.

Lewis, D. E. (1994). The university of Kazan: Provincial cradle of Russian organic chemistry. *J. Chem. Ed.* 71:29–42, 93–97.

——— (1995). Aleksandr Mikhailovich Zaitsev. *Bull. Hist. Chem.* 17/18:21–30.

Lewis, E. B. (1951). Pseudoallelism and gene evolution. *Cold Spring Harbor Symp.* 16:159–174.

Lewis, G. N., and Randall, M. (1923). *Thermodynamics and the Free Energy of Chemical Substances.* New York: McGraw-Hill.

Lewis, I. M. (1934). Bacterial variation with special reference to behavior of some mutabile strains of colon bacteria in synthetic media. *J. Bact.* 26:619–639.

Lewitsky, G. A. (1931). Sergei Gavrilovich Navashin. *Ber. bot. Ges.* 40:149–163.

Li, A. (1992). J. B. Collip, A. M. Hanson, and the isolation of the parathyroid hormone, or endocrines and enterprise. *J. Hist. Med.* 47:405–438.

Li, C. H. (1956, 1957). Hormones of the anterior pituitary gland. *Adv. Protein Chem.* 11:101–190; 12:269–317.

Libavius, A. (1964). *Die Alchemie des Andreas Libavius, ein Lehrbuch der Chemie aus dem Jahre 1597* (E. Pietsch and A. Kotowski, trans.). Weinheim: Verlag Chemie.

Licht, S., Gurfen, G. J., and Stubbe, J. (1996). Thiyl radicals in ribonucleotide reductases. *Science* 271:477–481.

Lichtenthaler, F. W. (1992). Emil Fischer's proof of the configuration of sugars. *Ang. Chem. (Int. Ed.)* 31:1541–1556.

Liébecq, C. (1977). Bref historique de la Société Belge de biochimie. *Arch. Int. Physiol. Biochim.* 85:381–386.

——— (1992). A brief history of the *European Journal of Biochemistry* on the occasion of its 25th anniversary. *Eur. J. Biochem.* 204:421–432.

Lieben, F. (1935). *Geschichte der physiologischen Chemie.* Leipzig: Deuticke.

Liebenau, J. (1987). *Medical Science and Medical Industry: The Formation of the American Pharmaceutical Industry.* Baltimore: Johns Hopkins University Press.

——— (1989). Paul Ehrlich as commercial scientist and research administrator. *Med. Hist.* 34:65–78.

——— (1990). The rise of the British pharmaceutical industry. *Brit. Med. J.*, pp. 724–728, 733.

Liebenau, J., Higby, G. J., and Stroud, E. C. (eds.) (1990). *Pill Peddlers: Essays on the History of the Pharmaceutical Industry.* Madison, Wis.: American Institute for the History of Pharmacy.

Lieberman, I., Kornberg, A., and Simms, E. S. (1955). Enzymatic synthesis of pyrimidine nucleotides. *J. Biol. Chem.* 215:403–415.

Lieberman, K. R., and Dahlberg, A. E. (1995). Ribosome-catalyzed peptide-bond formation. *Prog. Nucleic Acid Res.* 50:1–23.

Liebig, G. (1850). Ueber die Respiration der Muskeln. *Arch. anat. Physiol.* pp. 393–416.

Liebig, J. (1834). Ueber einige Stickstoffverbindungen. *Ann.* 10:1–47.

——— (1839). Ueber die Erscheinung der Gährung, Fäulnis, und Verwesung und ihre Ursachen. *Ann.* 30:250–287.

——— (1840). *Organic Chemistry in Its Applications to Agriculture and Physiology* (trans. of 1840 ed. by L. Playfair). London: Taylor and Walter.

——— (1841). Ueber die stickstoffhaltigen Nahrungsmittel des Pflanzenreiches. *Ann.* 39:129–169.

——— (1842). *Die Organische Chemie in ihrer Anwendung auf Physiologie und Pathologie.* Braunschweig: Vieweg.

——— (1843, 1846a). *Die Thier-Chemie oder the organische Chemie in ihrer Anwendung auf Physiologie und Pathologie.* Braunschweig: Vieweg.

——— (1846b). Ueber den Schwefelgehalt des stickstoffhaltigen Bestandtheils der Erbsen. *Ann.* 57:131–133.

——— (1846c). *Chemistry and Physics in Relation to Physiology and Pathology.* London: Baillière.

——— (1847a). *Researches on the Chemistry of Food* (W. Gregory, trans.). London: Taylor and Walton.

——— (1847b). Ueber die Bestandteile der Flüssigkeiten des Fleisches. *Ann.* 62:257–369.

——— (1853). Ueber Kynurensäure. *Ann.* 86:125–126.

——— (1870). Ueber die Gährung und die Quelle der Muskelkraft. *Ann.* 153:1–47, 137–228.

[Liebig, J., and Wöhler, F.] (1839). Das enträthselte Geheimnis der geistigen Gährung. *Ann.* 29:100–104.

Liebmann, A. J. (1956). History of distillation. *J. Chem. Ed.* 33:166–173.

Lillie, F. R. (1944). *The Woods Hole Marine Biological Laboratory.* Chicago: University of Chicago Press.

Lillie, R. S. (1924). Reactivity of the cell. In *General Cytology* (E. V. Cowdry, ed.), pp. 167–233. Chicago: University of Chicago Press.

Lindauer, M. W. (1962). The evolution of the concept of chemical equilibrium from 1775 to 1923. *J. Chem. Ed.* 39:384–390.

Lindeboom, G. A. (1968). *Herman Boerhaave: The Man and His Work.* London: Methuen.

——— (1971). Christiaan Eijkman. *DSB* 4:310–312.

Lindegren, C. C. (1932). The genetics of Neurospora II: Segregation of the sex factors in asci of N. crassa, N. sitophila, and N. tetrasperma. *Bull. Torrey Bot. Club* 59:119–138.

——— (1936). A six-point map of the sex-chromosome of *Neurospora crassa. J. Genetics* 32:243–256.

Linderstrøm-Lang, K. (1939). S. P. L. Sørensen. *Compt. Rend. Carlsberg* 23:I–XXI.

Linderstrøm-Lang, K., Hotchkiss, R. D., and Johansen, G. (1938). Peptide bonds in globular proteins. *Nature* 142:996.

Lindroth, S. (1973). Carl Linneaus. *DSB* 8:374–381.

Lingrel, J. B., and Kuntzweiler, T. (1994). Na^+,K^+-ATPase. *J. Biol. Chem.* 269:19659–19662.

Linser, H. (1966). The hormonal system of plants. *Ang. Chem. (Int. Ed.)* 5:776–784.

Lipkin, D., Cook, W. H., and Markham, R. (1959). Adenosine-3′,5′-phosphoric acid: a proof of structure. *J. Am. Chem. Soc.* 81:6198–6203.

Lipman, T. O. (1964). Wöhler's preparation of urea and the fate of vitalism. *J. Chem. Ed.* 41:452–458; 42:394–397.

——— (1966). The response to Liebig's vitalism. *Bull. Hist. Med.* 40:511–524.

——— (1967). Vitalism and reductionism in Liebig's physiological thought. *Isis* 58:167–185.

Lipmann, F. (1933). Ueber die oxydative Hemmbarkeit der Glykolyse und den Mechanismus der Pasteurschen Reaktion. *Biochem. Z.* 265:133–140.

——— (1939a). Coupling between pyruvic acid dehydrogenation and adenylic acid phosphorylation. *Nature* 143:281.

——— (1939b). An analysis of the pyruvic acid oxidation system. *Cold Spring Harbor Symp.* 7:248–259.

——— (1941). Metabolic generation and utilization of phosphate bond energy. *Adv. Enzymol.* 1:99–162.

——— (1945). Acetylation of sulfanilamide by liver homogenates and extracts. *J. Biol. Chem.* 160:173–190.

——— (1950). Biosynthetic mechanisms. *Harvey Lectures* 44:99–123.

——— (1954). On the mechanism of some ATP-linked reactions and certain aspects of protein synthesis. In *The Mechanism of Enzyme Action* (W. D. McElroy and B. Glass, eds.), pp. 599–604. Baltimore: Johns Hopkins University Press.

——— (1969). Peptide chain elongation in protein biosynthesis. *Science* 164:1024–1031.

——— (1971). *Wanderings of a Biochemist.* New York: Wiley-Interscience.

——— (1973). Nonribosomal polypeptide synthesis on polyenzyme templates. *Acc. Chem. Res.* 6:361–367.

——— (1977). Discovery of creatine phosphate in muscle. *TIBS* 2:21–22.

——— (1984). A long life in times of great upheaval. *Ann. Rev. Biochem.* 53:1–33.

Lippmann, E. O. (1919, 1931, 1954). *Entstehung und Ausbreitung der Alchemie.* Berlin: Springer; Weinheim: Verlag Chemie.

——— (1934). Alter und Herkunft des Namens "Organische Chemie." *Chem. Z.* 100:1009–1016.

Lipson, H. (1990). The introduction of Fourier methods into crystal-structure determination. *Notes Roy. Soc.* 44:257–264.

Lithgow, T., Glick, B. S., and Schatz, G. (1995). The protein import receptors of mitochondria. *TIBS* 20:98–101.

Liu, X., Garriga, P., and Khorana, H. G. (1996). Structure and function of rhodopsin: Correct folding and misfolding in two point mutants in the intradiscal domain of rhodopsin identified in retinitis pigmentosa. *PNAS* 93:4554–4564.

Lloyd, D. J., and Shore, A. (1938). *Chemistry of the Proteins.* 2d ed. Philadelphia: Blakiston.

Lloyd, G. E. R. (1968). *Aristotle: The Growth and Structure of His Thought.* Cambridge: Cambridge University Press.

——— (1987). Empirical research in Aristotle's biology. In *Philosophical Issues in Aristotle's Biology* (A. Gotthelf and J. G. Lennox, eds.), pp. 53–63. Cambridge: Cambridge University Press.

——— (1996). *Aristotelian Explorations.* Cambridge: Cambridge University Press.

Lobban, P., and Kaiser, A. D. (1973). Enzymatic end-to-end joining of DNA molecules. *J. Mol. Biol.* 79:453–471.

Locke, D. (1992). *Science as Writing.* New Haven: Yale University Press.

Lockemann, G. (1949). *Robert Wilhelm Bunsen.* Stuttgart: Wissenschaftliche Verlagsgesellschaft.

Lodish, H. F. (1970). Specificity in bacterial protein synthesis: Role of initiation factors and ribosomal subunits. *Nature* 226:705–707.

——— (1988). Transport of secretory and membrane glycoproteins from the rough endoplasmic reticulum to the Golgi. *J. Biol. Chem.* 263:2107–2110.

Loeb, J. (1906). *The Dynamics of Living Matter.* New York: Columbia University Press.

——— (1912). *The Mechanistic Conception of Life.* Chicago: University of Chicago Press.

——— (1916). *The Organism as a Whole.* New York: Putnam.

——— (1922). *Proteins and the Theory of Colloidal Behavior.* New York: McGraw-Hill.

Loeb, L. A., and Kunkel, T. A. (1982). Fidelity and DNA synthesis. *Ann. Rev. Biochem.* 52:429–457.

Loew, O. (1880). Eine Hypothese über die Bildung des Albumins. *Pflügers Arch.* 22:503–512.

——— (1883). Ein weiterer Beweis dass Eiweiss des lebenden Protoplasma eine andere chemische Constitution besitzt, als das abgestorbenen. *Pflügers Arch.* 30:348–368.

——— (1896). *The Energy of Living Protoplasm.* London: Kegan Paul, Trench, Trübner.

Loew, O., and Bokorny, T. (1881). Ein chemischer Unterscheid zwischen lebendigen und totem Protoplasma. *Pflügers Arch.* 25:150–164.

——— (1882). Einige Bemerkungen über Protoplasma. *Pflügers Arch.* 28:94–98.

Loewi, O. (1898). Ueber das "harnstoffbildende" Ferment der Leber. *Z. physiol. Chem.* 25:511–522.

——— (1902). Ueber Eiweißsynthese im Tierkörper. *Arch. exp. Path. Pharm.* 48:303–330.

——— (1921). Ueber humorale Uebertragbarkeit der Herznervenwirkung. *Pflügers Arch.* 189:239–242.

Löffler, W. (1917). Desaminierung und Harnstoffbildung im Tierkörper. *Biochem. Z.* 85:230–294.

——— (1920). Zur Kenntnis der Leberfunktion unter experimentell pathologischen Bedingungen. *Biochem. Z.* 112:164–187.

Loftfield, R. B. (1972). The mechanism of the aminoacylation of transfer RNA. *Prog. Nucleic Acid Res.* 12:87–128.

Lohman, T. M. (1993). Helicase-catalyzed DNA unwinding. *J. Biol. Chem.* 268:2269–2272.

Lohman, T. M., and Bjornson, K. P. (1996). Mechanisms of helicase-catalyzed DNA unwinding. *Ann. Rev. Biochem.* 65:169–214.

Lohmann, K. (1928). Ueber das Vorkommen und Umsatz von Pyrophosphat in der Zelle. *Biochem. Z.* 202:466–493; 203:164–207.

——— (1934). Ueber die enzymatische Spaltung der Kreatinphosphorsäure; zugleich ein Beitrag sum Chemismus der Muskelkontraktion. *Biochem. Z* 264–277.

——— (1935). Konstitution der Adenylpyrophosphorsäure und Adenosindiphosphorsäure. *Biochem. Z.* 282:120–123.

Lohmann, K., and Schuster, P. (1937). Untersuchungen über Cocarboxylase. *Biochem. Z.* 294:188–214.

Lolis, E., and Petsko, G. A. (1990). Transition-state analogues in protein crystallography: Probes of the structural source of enzyme catalysis. *Ann. Rev. Biochem.* 59:597–630.

Long, E. R. (1949). Harry Gideon Wells. *Biog. Mem. NAS* 26:233–261.

Longsworth, L. G., and Shedlovsky, T. (1970). Duncan Arthur MacInnes. *Biog. Mem. NAS* 41:295–317.

Longuet-Higgins, H. C., and Fischer, M. E. (1978). Lars Onsager. *Biog. Mem. FRS* 24:443–471.

Loomis, W. F., and Lipmann, F. (1948). Reversible inhibition of the coupling between phosphorylation and oxidation. *J. Biol. Chem.* 173:807–808.

Lorch, J. (1974). The charisma of crystals in biology. In *The Interaction Between Science and Philosophy* (Y. Elkana, ed.), pp. 445–461. Atlantic Highlands, N.J.: Humanities.

Lorenz, I. (1984). Die Entwicklung eines eigenständigen Faches und Institutes für Physiologische Chemie an der Universität Leipzig. *Z. ges. inn. Med.* 39:585–590.

Lorsch, J. R., and Szostak, J. W. (1996). Chance and necessity in the selection of nucleic acid catalysts. *Acc. Chem. Res.* 29:103–110.

Lotze, S. (1986). Die Chemie in Kurhessen von 150 Jahren. *Zeitschrift des Vereins für Hessische Geschichte und Landeskunde* 91:105–121.

Low, S. C., and Berry, M. J. (1996). Knowing when to stop: Selenocysteine incorporation in eukaryotes. *TIBS* 21:203–207.

Lowndes, J. (1956). Robert Henry Aders Plimmer. *Biochem. J.* 62:353–357.

Löwy. I. (1992). The strength of loose concepts—boundary concepts, federative experimental strategies, and disciplinary growth: The case of immunology. *Hist. Sci.* 39:371–396.

——— (1994). On hybridizations, networks, and new disciplines: The Pasteur Institute and the development of microbiology in France. *Stud. Hist. Phil. Sci.* 25:655–688.

Lübke, K., Schillinger, E., and Töpert, M. (1976). Hormone receptors. *Ang. Chem. (Int. Ed.)* 15:741–748.

Lucas-Lenard, J., and Lipmann, F. (1971). Protein biosynthesis. *Ann. Rev. Biochem.* 40:409–448.

Ludmerer, K. M. (1972). *Genetics and American Society.* Baltimore: Johns Hopkins University Press.

Ludwig, K. (1852). *Lehrbuch der Physiologie des Menschen.* Heidelberg: Winter.

Luisi, P. L. and Thomas, R. M. (1990). The pictographic molecular paradigm. *Natrurwiss.* 77:67–74.

Lumb, E. S. (1950). Cytochemical reactions of nucleic acids. *Quart. Rev. Biol.* 25:278–291.

Lund, E. W. (1965). Guldberg and Waage and the law of mass action. *J. Chem. Ed.* 42:548–550.

Lundblad, R. L., and Noyes, C. M. (1984). *Chemical Reagents for Protein Modification.* Boca Raton, Fla.: CRC.

Lundgren, A. (1975). The changing role of numbers in 18th-century chemistry. In *The Quantifying Spirit in the 18th Century* (T. Frängsmyr et al., eds.), pp. 245–266. Berkeley: University of California Press.

Lundsgaard, E. (1930). Untersuchungen über Muskelkontraktion ohne Milchsäurebildung. *Biochem. Z.* 217:162–177.

Lunin, N. (1881). Ueber die Bedeutung der anorganische Salze für die Ernährung des Thieres. *Z. physiol. Chem.* 5:31–39.

Lurçat, F. (1989). Timidité de l'histoire des sciences. *Mesure* 1:125–136.

Luria, S. E. (1939). Actions des radiations sur le *Bacterium coli. Compt. Rend.* 209:604–606.

——— (1970). Molecular biology: Past, present, and future. *BioScience* 20:1289–1293, 1296.

——— (1984). *A Slot Machine, a Broken Test Tube.* New York: Harper and Row.

Luria, S. E., and Anderson, T. F. (1942). The identification and characterization of bacteriophages with the electron microscope. *PNAS* 28:127–130.

Luria, S. E., and Delbrück, M. (1943). Mutations of bacteria from virus sensitivity to virus resistance. *Genetics* 28:491–511.

Lurie, E. (1960). *Louis Agassiz: A Life in Science.* Chicago: University of Chicago Press.

Lusk, G. (1910). The fate of amino acids in the organism. *J. Am. Chem. Soc.* 32:671–680.

——— (1928). *Elements of the Science of Nutrition.* 4th ed. Philadelphia: Saunders.

Lüthje, H. (1904). Die Zuckerbildung aus Eiweiss. *Deutsches Archiv für klinische Medizin* 79:498–513.

Luttenberger, F. (1992). Arrhenius vs. Ehrlich on immunochemistry: Decisions about scientific progress in the context of the Nobel Prize. *Theoretical Medicine* 13:137–173.

Lutwak-Mann, C., and Mann, T. (1981). The Parnas School. *TIBS* 6:309–310.

Lwoff, A. (1953). Lysogeny. *Bact. Revs.* 17:269–337.

——— (1962). *Biological Order.* Cambridge: MIT Press.

——— (1966). The prophage and I. In *Phage and the Origins of Molecular Biology* (J. Cairns, G. S. Stent, and J. D. Watson, eds.), pp. 88–99. Plainview, N.Y.: Cold Spring Harbor Laboratory Press.

——— (1968). Review of *The Double Helix. Sci. Am.* 219:133–138.

——— (1971). From protozoa to bacteria and viruses: Fifty years with microbes. *Ann. Rev. Microbiol.* 25:1–26.

——— (1977). Jacques Lucien Monod. *Biog. Mem. FRS* 23:385–412.

Lwoff, A., and Gutmann, A. (1950). Recherches sur un *Bacillus mégatherium* lysogène. *Ann. Inst. Pasteur* 78:711–739.

Lwoff, A., and Ullmann, A. (eds.) (1978). *Selected Papers in Molecular Biology by Jacques Monod.* New York: Academic.

——— (1979). *Origins of Molecular Biology: A Tribute to Jacques Monod.* New York: Academic.

Lyle, R. E., and Lyle, G. G. (1964). A brief history of polarimetry. *J. Chem. Ed.* 41:308–313.

Lynch, M. (1985). *Art and Artifact in Laboratory Science.* London: Routledge.

Lynen, F. (1954). Acetyl coenzyme A and the fatty acid cycle. *Harvey Lectures* 48:210–244.

——— (1955). Lipide metabolism. *Ann. Rev. Biochem.* 24:653–688.

——— (1959). Fritz Kögl. *Bayer. Akad. Wiss. Jahrbuch,* pp. 187–188.

——— (1965). Hans von Euler-Chelpin. *Bayer. Akad. Wiss. Jahrbuch,* pp. 206–212.

Lynen, F., Knappe, J., Lorch, E., Jütting, G., and Ringelmann, E. (1959). Die biochemische Funktion des Biotins. *Ang. Chem.* 71:481–486.

Lynen, F., Reichert, E., and Rueff, L. (1951). Zum biologischen Abbau der Essigsäure. VI. Aktivierte Essigsäure, ihre Isolierung aus Hefe und ihre chemische Natur. *Ann.* 574:1–32.

Lyons, P. (1981). Lars Onsager. *Am. Phil. Soc. Year Book,* pp. 485–495.

Lythgoe, B., and Todd, A. R. (1945). Structure of adenosine di- and triphosphate. *Nature* 155:695–696.

McAllister, J. W. (1989). Truth and beauty in scientific reason. *Synthese* 78:25–51.

——— (1991). The simplicity of theory: Its degree and form. *Journal for General Philosophy of Science* 22:1–14.

Macalpine, I., Hunter, R., and Rimington, C. (1968). Porphyria in the royal houses of Stuart, Hanover, and Prussia. A follow-up study of George III's illness. *Brit. Med. J.* (1):7–18.

McBride, J. M. (1974). The hexaphenylethane riddle. *Tetrahedron* 30:2009–2022.

McCann, S. M. (ed.) (1988). *Endocrinology: People and Ideas.* Bethesda, Md.: American Physiological Society.

McCarthy, R. J., and Church, R. B. (1970). The specificity of molecular hybridization reactions. *Ann. Rev. Biochem.* 39:131–150.

McCarty, M. (1946). Purification and properties of desoxyribonuclease isolated from beef pancreas. *J. Gen. Physiol.* 29:123–139.

——— (1985). *The Transforming Principle.* New York: Norton.

Mach, B., Reich, E., and Tatum, E. L. (1963). Separation of the biosynthesis of the antibiotic tyrocidine from protein synthesis. *PNAS* 50:175–181.

Mach, E. (1896). On the part played by accident in invention and discovery. *Monist* 6:161–175.

Machin, K. E. (1964). Feedback theory and its application to biological systems. *Symp. Soc. Exp. Biol.* 18:421–446.

McClain, W. H., Guerrier-Takada, C., and Altman, S. (1987). Model substrates for an RNA enzyme. *Science* 238:527–530.

McClelland, C. E. (1980). *State, Society, and University in Germany, 1700–1914*. Cambridge: Cambridge University Press.

——— (1983). Professionalization and higher education in Germany. In *The Transformation of Higher Learning* (K. H. Jarausch, ed.), pp. 306–320. Stuttgart: Klett-Cotta.

McClintock, B. (1941). Spontaneous alterations in chromosome size and form in *Zea mays*. *Cold Spring Harbor Symp.* 9:72–81.

——— (1950). The origin and behavior of mutable loci in maize. *PNAS* 36:344–355.

McClung, C. E. (1902). The accessory chromosome: Sex determinant? *Biol. Bull.* 3:43–84.

McClure, W. R. (1985). Mechanism and control of transcription in prokaryotes. *Ann. Rev. Biochem.* 54:171–204.

McCollum, E. V. (1956). C. J. M. Méhu, a forgotten man of science. *J. Chem. Ed.* 33:507.

——— (1957). *A History of Nutrition*. Boston: Houghton Mifflin.

McCosh, T. W. J. (1975). Boussingault versus Ville: The social, political, and scientific aspects of their dispute. *Ann. Sci.* 32:475–490.

——— (1984). *Boussingault, Chemist and Agriculturist*. Dordrecht: Reidel.

McCoy, R. H., Meyer, C. E., and Rose, W. C. (1935). Feeding experiments with mixtures of highly purified amino acids, 8: Isolation and identification of a new essential amino acid. *J. Biol. Chem.* 112:283–302.

McCullough, D. (1969). W. K. Brooks' role in the history of American biology. *J. Hist. Biol.* 2:411–438.

McElheny, V. K. (1966). Pasteur Institute scientists demand sweeping reforms: Pasteur Institute rebels lose a round. *Science* 151:809; 153:1226–1228.

McElroy, W. D., and Glass, B. (eds.) (1957). *A Symposium on the Chemical Basis of Heredity*. Baltimore: Johns Hopkins University Press.

McEvoy, J. G. (1975). A "revolutionary" philosophy of science: Feyerabend and the degeneration of critical rationalism into sceptical fallibilism. *Phil. Sci.* 42:49–66.

——— (1978). Joseph Priestley, "aerial philosopher": Metaphysics and methodology in Priestley's chemical thought from 1772 to 1781. *Ambix* 25:1–55, 93–116, 153–175.

——— (1997). Positivism, whiggism, and the chemical revolution: A study in the historiography of chemistry. *Hist. Sci.* 35:1–33.

MacFadyen, A., Morris, G. H., and Rowland, S. (1900). On expressed yeast-cell plasma (Buchner's "zymase"). *Proc. Roy. Soc.* 67:250–266.

MacGillivray, H. (1968). A personal biography of Arthur Robertson Cushny. *Ann. Rev. Pharmacol.* 8:1–24.

McGlashan, M. L. (1966). The use and misuse of the laws of thermodynamics. *J. Chem. Ed.* 43:226–232.

McGovern, P. E., and Michel, R. H. (1990). Royal purple dye: The chemical reconstruction of the ancient Mediterranean industry. *Acc. Chem. Res.* 23:152–158.

McGrath, M. E., Gillmer, S. A., and Fletterick, R. J. (1995). Ecotin: Lessons on survival in a protein-filled world. *Protein Science* 4:141–148.

McGucken, W. (1969). *Nineteenth-Century Spectroscopy*. Baltimore: Johns Hopkins University Press.

——— (1984). *Scientists, Society, and State: The Social Relations of Science Movement in Great Britain, 1931–1947*. Columbus: Ohio State University Press.

McGuiness, B. (ed.) (1987). *Unified Science*. Dordrecht: Reidel.

McHenry, C. S. (1991). DNA polymerase III holoenzyme. *J. Biol. Chem.* 266:19127–19130.

McIlwain, H. (1990). Biochemistry and neurochemistry in the 1800s: Their origins in comparative animal chemistry. *Essays in Biochemistry* 25:197–224.

MacInnes, D. A., and Granick, S. (1958). Leonor Michaelis. *Biog. Mem. NAS* 31:282–321.

MacKay, E. M., Barnes, R. H., Carne, H. O., and Wick, A. N. (1940). Ketogenic activity of acetic acid. *J. Biol. Chem.* 135:157–163.

McKenzie, H. A. (1994). The Kjeldahl determination of nitrogen: retrospect and prospect. *Trends in Analytical Chemistry* 13:138–144.

Mackenzie, N. A., Malthouse, J. P. G., and Scott, I. A. (1984). Studying enzyme mechanisms by ^{13}C nuclear magnetic resonance. *Science* 225:883–889.

McKie, D. (1944). Wöhler's "synthetic" urea and the rejection of vitalism: A chemical legend. *Nature* 153:608–610.

——— (1952). *Antoine Lavoisier.* London: Constable.

McKie, D., and Heathcote, N. H. deV. (1935). *The Discovery of Specific and Latent Heats.* London: Arnold.

McKusick, V. A. (1960). Walter S. Sutton and the physical basis of Mendelism. *Bull. Hist. Med.* 34:487–497.

McLaughlin, P. (1982). Blumenbach and der Bildungstrieb. *Med. Hist. J.* 17:357–372.

——— (1989). *Kant's Kritik der teleologischen Urteilskraft.* Bonn: Bouvier.

——— (1994). Kant's Organismusbegriff in der Kritik der Urteilskraft, In *Philosophie des Organischen in der Goethezeit* (K. T. Kanz, ed.), pp. 100–110. Stuttgart: Steiner.

MacLeod, R. M. (1971). The support of Victorian science: The endowment of the research movement in Great Britain, 1868–1900. *Minerva* 9:197–230.

——— (1994). "Instructed Men" and mining engineers: The associates of the Royal School of Mines and British imperial science. *Minerva* 32:422–439.

MacMunn, C. A. (1886). Researches on myohaematin and the histohaematins. *Phil. Trans.* 177:267–298.

——— (1887). Further observations on myohaematin and the histohaematins. *J. Physiol.* 8:51–65.

——— (1890). Ueber das Myohämatin. *Z. physiol. Chem.* 14:328–329.

——— (1914). *Spectrum Analysis Applied to Biology and Medicine.* London: Longmans, Green.

MacNider, W. deB. (1946). John Jacob Abel. *Biog. Mem. NAS* 24:231–257.

McPherson, A. (1991). A brief history of protein crystal growth. *Journal of Crystal Growth* 110:1–10.

Macquer, P. J. (1777). *A Dictionary of Chemistry.* 2d English ed. (J. Keir, trans.). London: Cadell and Elmsly.

McRae, S. F. (1987). A. B. Macallum and physiology at the University of Toronto. In *Physiology in the American Context* (G. L. Geison, ed.), pp. 97–114. Bethesda, Md.: American Physiological Society.

Macrakis, K. (1993a). *Surviving the Swastika: Scientific Research in Nazi Germany.* New York: Oxford University Press.

——— (1993b). The survival of basic biological research in National Socialist Germany. *J. Hist. Biol.* 26:519–543.

McRee, D. E. (1993). *Practical Protein Crystallography.* San Diego: Academic.

Madden, S. C., and Whipple, G. H. (1940). Plasma proteins: Their sources, production, and utilization. *Physiol. Revs.* 20:194–217.

Maehle, A. H., Glase, M., and Tröhler, U. (1990). Der Göttinger Weg von der medizinischen zur physiologischen Chemie 1840–1940. *Z. Physiol. Chem.* 371:447–454.

Magendie, F. (1816–1817). *Précis Elémentaire de Physiologie.* Paris: Méquignon-Marvis.

——— (1841). Rapport fait a l'Académie des Sciences au nom de la Commission dite de la gélatine. *Compt. Rend.* 13:237–295.

Magnus, G. (1837). Ueber die im Blute enthaltenen Gase, Sauerstoff, Stickstoff, und Kohlensäure. *Ann. Phys.* 40:583–606.

Magnus-Levy, A. (1899). *Die Oxybuttersäure und ihre Beziehungen zum Coma diabeticus.* Leipzig: Vogel.

Maienschein, J. (1983). Experimental biology in transition: Harrison's embryology, 1895–1910. *Stud. Hist. Biol.* 6:107–127.

——— (1991). *Transforming Traditions in American Biology, 1880–1915.* Baltimore: Johns Hopkins University Press.

——— (1996). From presentation to representation in E. B. Wilson's *The Cell. Biology and Philosophy* 6:227–254.

Mains, R. E., Epper, B. A., and Ling, N. (1977). Common precursor to corticotropins and endorphins. *PNAS* 74:3014–3018.

Maitland, P. (1950). Biogenetic origin of the pyrrole pigments. *Quart. Revs. Chem. Soc.* 4:45–68.

Majerus, P. W., Connolly, T. M., Bansal, V. S., Inhorn, R. C., Ross, T. S., and Lips, D. L. (1988). Inositol phosphates: Synthesis and degradation. *J. Biol. Chem.* 263:3051–3054.

Makino, K. (1931a). Ueber die Konstitution der Adenosintriphosphorsäure. *Biochem. Z.* 278:161–163.

——— (1931b). Ueber den Nucleinstoffwechsel, 5: Ueber die Konstitution der Nucleinsäure. *Z. physiol. Chem.* 232:229–235.

Malfatti, H. (1891). Beiträge zur Kenntnis der Nucleine. *Z. physiol. Chem.* 16:68–86.

Malmström, B. G. (1982). Enzymology of oxygen. *Ann. Rev. Biochem.* 51:21–59.

Malone, D. (1926). *The Public Life of Thomas Cooper.* New Haven: Yale University Press.

Maly, R. (1873). *Jahresberichte über die Fortschritte der Thierchemie.* 1:14.

Manassein, M. (1897). Zur Frage von der alkoholischen Gärung. *Ber. chem. Ges.* 30:3061–3062.

Manchester, K. L. (1995). Did a tragic accident delay the discovery of the double helical structure of DNA? *TIBS* 20:126–128.

Manchot, W. (1902). Ueber Peroxydbildung beim Eisen. *Ann.* 325:10–124.

Mandeles, S. (1972). *Nucleic Acid Sequence Analysis.* New York: Columbia University Press.

Mandelstam, J. (1958). Turnover of protein in growing and non-growing populations of *Escherichia coli. Biochem. J.* 69:110–119.

Mani, N. (1956). Das Werk von Friedrich Tiedemann und Leopold Gmelin; "Die Ernährung nach Versuchen," und seine Bedeutung für die Entwicklung der Ernährungslehre in ersten Hälfte des 19. Jahrhunderts. *Gesnerus* 13:190–214.

——— (1964). Die Entdeckung des Glykogens durch Claude Bernard. *Z. klin. Chem.* 2:97–104.

——— (1966). Paul Bert als Politiker, Pädagog, und Begründer der Höhenphysiologie. *Gesnerus* 23:109–116.

——— (1967). *Die Historischen Grundlagen der Leberforschung, 2: Die Geschichte der Leberforschung von Galen bis Claude Bernard.* Basel: Schwalbe.

——— (1976). Die wissenschaftliche Ernährungslehre im 19. Jahrhundert. In *Ernährung und Ernährungslehre im 19. Jahrhundert* (E. Heischkelt-Artelt, ed.), pp. 22–75. Göttingen: Vanderhoeck and Rupprecht.

Manley, J. L. (1995). Messenger RNA polyadenylation: A universal phenomenon. *PNAS* 92:1800–1801.

Mann, F. G. (1957). The place of chemistry at Cambridge. *Proc. Chem. Soc.,* pp. 190–193.

Mann, G. (1906). *Chemistry of the Proteids.* London: Macmillan.

Mann, T. (1964). David Keilin. *Biog. Mem. FRS.* 10:183–205.

Mann, T., and Keilin, D. (1938). Haemocuprein and hepatocuprein, copper-protein compounds of blood and liver in mammals. *Proc. Roy. Soc.* B126:303–315.

Mannheim, K. (1952). *Essays on the Sociology of Knowledge* (O. Peeskemeti, ed.). New York: Oxford University Press.

Mannherz, H. G. (1992). Crystallization of actin in complex with actin-binding proteins. *J. Biol. Chem.* 267:11661–11664.

Mannitzer, H. G., and Goody, R. S. (1976). Proteins of contractile systems. *Ann. Rev. Biochem.* 45:427–465.

Marchand, R. F. (1838). Fortgesetzte Versuche über die Bildung des Harnstoffes im thierischen Körper. *J. prakt. Chem.* 14:490–497.

Marcus, A. I. (1988). The wisdom of the body politic: The changing nature of publicly sponsored American agricultural research since the 1830s. *Agricultural History* 62(2):4–26.

——— (1990). From Ehrlich to Waksman: Chemotherapy and the seamed web of the past. In *Beyond History of Science* (E. Garber, ed.), pp. 266–283. Bethlehem, Pa.: Lehigh University Press.

Margoliash, E., and Schejter, A. (1966). Cytochrome c. *Ann. Rev. Biochem.* 21:113–286.

——— (1984). " . . ., and 70 years ago": Myohematins and histohematins (cytochromes). *TIBS* 9:364–367.

Margolis, H. (1993). *Paradigms and Barriers: How Habits of Mind Govern Scientific Belief.* Chicago: University of Chicago Press.

Mark, H. (1980). Aus den frühen Tagen der makromolecularen Chemie. *Naturwiss.* 67:477–483.

Markert, C. L., and Whitt, G. S. (1968). Molecular varieties of enzymes. *Experientia* 24:977–991.

Markert, C. L., Shaklee, J. B., and Whitt, G. S. (1976). The evolution of a gene. *Science* 189:102–114.

Markham, R. (1977). Landmarks in plant virology: Genesis of concepts. *Ann. Rev. Phytopathol.* 15:17–39.

Markham, R., and Smith, J. D. (1951). Structure of ribonucleic acid. *Nature* 168:406–408.

——— (1954). Nucleoproteins and viruses. In *The Proteins* (H. Neurath and K. Bailey, eds.), 2A:1–122. New York: Academic.

Marletta, M. A. (1993). Nitric oxide synthase structure and mechanism. *J. Biol. Chem.* 268:12231–12234.

Marrack, J. R. (1938). *The Chemistry of Antigens and Antibodies.* London: HMSO.

Marseille, J. (1968). *Das physiologische Lebenswerk von Emil du Bois-Reymond mit besonderer Berücksichtigung seiner Schüler.* Münster: Dr. med. diss.

Marsh, J. E. (1921). William Odling. *J. Chem. Soc.* 119:553–564.

Marshall, L. H. (1983). The fecundity of aggregates: The axonologists at Washington University, 1922–1942. *Persp. Biol. Med.* 26:613–636.

Marsonnet, M. (1995). *Science, Reality, and Language.* Albany: State University of New York Press.

Martin, A. J. P. (1948). Partition chromatography. *Ann. N.Y. Acad. Sci.* 49:249–264.

Martin, A. J. P., and Synge, R. L. M. (1941). A new form of chromatogram employing two liquid phases. *Biochem. J.* 35:1358–1368.

——— (1945). Analytical chemistry of the proteins. *Adv. Protein Chem.* 2:1–83.

Martini, G. A., and Schmidt, G. (1955). Professor S. Thannhauser zum 70. Geburtstag. *Deutsche med. Wchschr.* 80:987–989.

Martius, C. (1937). Ueber den Abbau der Citronensäure. *Z. Physiol. Chem.* 247:104–110.

——— (1982). How I became a biochemist. In *Of Oxygen, Fuels, and Living Matter.* Part 2 (G. Semenza, ed.), pp. 1–57. Chichester: Wiley.

Martius, C., and Knoop, F. (1937). Der physiologische Abbau der Citronensäre. *Z. physiol. Chem.* 246: I–II.

Martius, C., and Lynen, F. (1950). Probleme des Citronensäurecyclus. *Adv. Enzymol.* 10:167–222.

Marton, L. (1968). *Early History of the Electron Microscope.* San Francisco: San Francisco Press.

Maruyama, K. (1991). The discovery of adenosine triphosphate and the establishment of its structure. *J. Hist. Biol.* 24:145–154.

Marwick, A. (1989). *The Nature of History.* 3d ed. London: Macmillan.

Marx, J. (1974). L'art d'observer au XVIIIe siècle: Jean Senebier et Charles Bonnet. *Janus* 61:201–220.

Mason, G. G. F., and Rivett, A. J. (1994). Proteasomes: The changing face of proteolysis. *Chem. Biol.* 1:197–199.

Mason, J. (1992). The admission of the first women to the Royal Society of London. *Notes. Roy. Soc.* 46:279–300.

Mason, S. F. (1976). The foundations of classical stereochemistry. *Topics in Stereochemistry* 9:1–34.

——— (1987). From molecular morphology to universal dissymetry. In *Essays on the History of Organic Chemistry* (J. G. Traynham, ed.), pp. 35–53. Baton Rouge: Louisiana State University Press.

——— (1989). The development of concepts of chiral discrimination. *Chirality* 1:183–191.

——— (1991). From Pasteur to parity violation: Cosmic dissymmetry and the origins of biomolecular handedness. *Ambix* 38:85–99.

Massey, V. (1994). Activation of molecular oxygen by flavins and flavoproteins. *J. Biol. Chem.* 269:22359–22462.

Massini, R. (1907). Ueber einen in biologischer Beziehung interessanten Kolistamm *(Bacillus coli mutabile):* Ein Beitrag zur Variation bei Bakterien. *Arch. Hyg.* 61:250–290.

Mathews, A. P. (1924). Some general aspects of the chemistry of cells. In *General Cytology* (E. V. Cowdry, ed.), pp. 13–95. Chicago: University of Chicago Press.

Mathews, A. P., and Walker, S. (1909). The action of cyanide and nitriles on the spontaneous oxidation of cystein. *J. Biol. Chem.* 6:29–37.

Mathias, P. (1959). *The Brewing Industry in England, 1700–1830.* Cambridge: Cambridge University Press.

Matteucci, C. (1856). Recherches sur les phénomènes de la contraction musculaire. *Ann. Chim.* [3]47:129–153.

Matthaei, J. H., Jones, O. W., Martin, R. G., and Nirenberg, M. (1962). Characteristics and composition of RNA coding units. *PNAS* 48:666–677.

Matthews, D. M. (1975). Intestinal absorption of peptides. *Physiol. Revs.* 55:537–608.

——— (1978). Otto Cohnheim, the forgotten physiologist. *Brit. Med. J.* 2:618–619.

Mattis, J. A., Henes, J. B., and Fruton, J. S. (1977). Interaction of papain with derivatives of phenylalanylglycinal. *J. Biol. Chem.* 252:6776–6782.

Maulitz, R. C. (1971). Schwann's way: Cells and crystals. *J. Hist. Med.* 26:422–436.

Mauskopf, S. (1976). Crystals and compounds: Molecular structure and composition in nineteenth-century France. *Trans. Am. Phil. Soc.* 66:1–82.

——— (1988). Molecular geometry in 19th-century France: Shifts in guiding assumptions. In *Scrutinizing Science* (A. Donovan et al., eds.), pp. 125–144. Dordrecht: Kluwer.

Maxam, A. M., and Gilbert, W. (1977). A new method for sequencing DNA. *PNAS* 74:560–564.

May, G. (1979). *L'Autobiographie.* Paris: PUF.

Mayall, B. H., et al. (1984). Torbjörn Oskar Caspersson. *Cytometry* 5:314–318.

Mayer, A. (1882). *Die Lehre von den Chemischen Fermenten oder Enzymologie.* Heidelberg: Winter.

Mayer, R. (1893). *Die Mechanik der Wärme.* 3d ed. (J. J. Weyrauch, ed.). Stuttgart: Cotta.

Mayer, R. J., and Doherty, F. J. (1992). Ubiquitin. *Essays in Biochemistry* 27:37–48.

Maynard, L. A. (1958). James Batcheller Sumner. *Biog. Mem. NAS* 31:376–396.

——— (1962). Wilbur O. Atwater: A biographical sketch. *J. Nutrition* 78:3–9.

Mayr, E. (1988). *Toward a New Philosophy of Biology.* Cambridge: Harvard University Press.

——— (1992). The autonomy of biology: The position of biology among the sciences. *Quart. Rev. Biol.* 71:97–106.

Mayr, E., and Provine, W. B. (eds.) (1980). *The Evolutionary Synthesis.* Cambridge: Harvard University Press.

Mazia, D. (1956a). The life history of the cell. *Am. Sci.* 44:1–32.

——— (1956b). Nuclear products and nuclear replication. In *Enzymes: Units of Biological Structure and Function* (O. H. Gaebler, ed.), pp. 261–278. New York: Academic.

——— (1987). The chromosome cycle and the centrosome cycle. *Int. Rev. Cytol.* 100:49–72.

Mazia, D., and Dan, K. (1952). The isolation and biochemical characterization of the mitotic apparatus. *PNAS* 38:826–838.

Mazumdar, P. M. H. (1972). Immunity in 1890. *J. Hist. Med.* 27:312–324.

——— (1974). The antigen-antibody reaction and physics and chemistry of life. *Bull. Hist. Med.* 48:1–21.

——— (1976). *Karl Landsteiner and the Problem of Species.* Baltimore: Johns Hopkins University Press.

——— (ed.) (1989). *Immunology, 1930–1980.* Toronto: Wall and Thompson.

——— (1995). *Species and Specificity.* Cambridge: Cambridge University Press.

Médard, L., and Tachoire, H. (1994). *Histoire de la thermochimie: Prelude à la thermodynamique chimique.* Aix-en-Provence: Université de Provence.

Medawar, J. S. (1990). *A Very Decided Preference.* Oxford: Oxford University Press.

Medawar, P. B. (1958). The immunology of tissue transplantation. *Harvey Lectures* 52:144–176.

——— (1963). Is the scientific paper a fraud? *The Listener,* 12 September, pp. 377–378.

——— (1967). *The Art of the Soluble.* London: Methuen.

——— (1968). Lucky Jim. *The New York Review of Books,* 28 March, pp. 3–5.

——— (1969). *Induction and Intuition in Scientific Thought.* Philadelphia: American Philosophical Society.

——— (1979). *Advice to a Young Scientist.* New York: Harper and Row.

——— (1986). *Memoir of a Thinking Radish.* Oxford: Oxford University Press.

Medawar, P. B., and Medawar, J. S. (1983). *Aristotle to Zoos.* London: Weidenfeld and Nicholson.

Medvei, V. C. (1982). *A History of Endocrinology.* Lancaster, Pa.: MTP Press.

Medzihradszky, K. (1981). Victor Bruckner. *Acta Chim. Acad. Sci. Hung.* 107:287–314.

Meerwein, H. (1939). Karl Friedrich von Auwers. *Ber. chem. Ges.* 72A:111–121.

Meier, P. (1993). *Paracelsus, Arzt und Prophet.* Zurich: Amman Verlag.

Meijer, O. G. (1986). Hugo de Vries und Johann Gregor Mendel: Die Geschichte einer Vereinigung. *Folia Mendeliana* 21:69–90.

Meinel, C. (1978). *Die Chemie an der Universität Marburg seit Beginn des 19. Jahrhundert.* Marburg: Elwert.

——— (1988). Early seventeenth-century atomism: Theory, epistemology, and the insufficiency of experiment. *Isis* 79:68–103.

——— (1992). August Wilhelm Hofmann: "Reigning chemist-in-chief." *Ang. Chem. (Int. Ed.)* 31:1254–1282.

Meinel, C., and Scholz, H. (1992). *Die Allianz von Wissenschaft und Industrie. August Wilhelm Hofmann.* Weinheim; VCH Verlagsgesellschaft.

Meinel, C., and Voswinckel, P. (eds.) (1994). *Medizin, Naturwissenschaft, Technik, und National Sozialismus. Kontinuitäten und Diskontinuitäten.* Stuttgart: Verlag für Geschichte der Naturwissenschaft und Technik.

Meister, A. (1955). Transamination. *Adv. Enzymol.* 16:185–246.

——— (1956). Metabolism of glutamine. *Physiol. Revs.* 36:103–127.

Meister, A., and Tate, S. S. (1976). Glutathione and related γ-glutamyl compounds: Biosynthesis and utilization. *Ann. Rev. Biochem.* 45:559–604.

Meister, E., and Thompson, N. R. (1976). Physical-chemical methods for the recovery of protein from waste effluents of potato chip processing. *J. Agr. Food Chem.* 24:919–923.

Meites, J., Donovan, B. T., and McCann, S. M. (eds.) (1975, 1978). *Pioneers in Neuroendocrinology.* New York: Plenum.

Meites, S. (1989). *Otto Folin: America's First Clinical Chemist.* Washington, D.C.: American Association for Clinical Chemistry.

Melchers, G. (1987). Fritz von Wettstein. *Ber. bot. Ges.* 100:373–405.

Meldola, R. (1908). William Henry Perkin. *J. Chem. Soc.* 93:2214–2257.

Melhado, E. M. (1980). Mitscherlich's discovery of isomorphism. *Hist. Stud. Phys. Sci.* 11:87–123.

——— (1981). *Jacob Berzelius.* Stockholm: Almquist and Wiksell.

——— (1989). Toward an understanding of the chemical revolution. *Knowledge and Society* 8:123–137.

Melhado, E. M., and Frängsmyr, T. (eds.) (1992). *Enlightenment Science in the Romantic Era: The Chemistry of Berzelius and Its Cultural Setting.* Cambridge: Cambridge University Press.

Mendel, G. (1865). Versuche über Pflanzen-hybriden. *Verhandlungen des Naturforschenden Vereines in Brünn* 4:3–47. [Facsimile reprint, *J. Heredity* (1951) 42:3–47.]

Mendel, L. B., and Jackson, H. C. (1898). On the excretion of kynurenic acid. *Am. J. Physiol.* 2:1–28.

Mendelsohn, E. (1963). Cell theory and the development of general physiology. *Arch. Int. Hist. Sci.* 6:419–429.

——— (1964). *Heat and Life.* Cambridge: Harvard University Press.

——— (1989). Robert K. Merton: The celebration and defense of science. *Science in Context* 3:269–289.

Mendelsohn, K. (1973). *The World of Walther Nernst.* London: Macmillan.

Mering, J. van, and Minkowski, O. (1890). Diabetes mellitus nach Pankreasextirpation. *Arch. exp. Path. Pharm.* 26:371–387.

Merrick, L. (1996). *The World Made New: Frederick Soddy. Science, Politics, Environment.* Oxford: Oxford University Press.

Merrifield, R. B. (1993). *Life During a Golden Age of Peptide Chemistry.* Washington, D.C.: American Chemical Society.

——— (1995). Solid-phase peptide synthesis. In *Peptides* (B. Gutte, ed.), pp. 93–169. San Diego: Academic.

Merton, R. K. (1957). *Social Theory and Social Structure.* 2d ed. Glencoe, Ill.: Free Press. [3d ed., 1968.]

——— (1961). Singletons and multiples in scientific discovery: A chapter in the sociology of science. *Proc. Am. Phil. Soc.* 105:470–486.

——— (1963). The ambivalence of scientists. *Bull. Johns Hopkins Hosp.* 112:77–97.

——— (1965). *On the Shoulders of Giants: A Shandean Postscript.* New York: Free Press. [2d ed., 1985.]

——— (1968a). The Matthew effect in science. *Science* 159:56–63.

——— (1968b). Making it scientifically. *New York Times Book Review,* 25 February, pp. 41–43, 45.

——— (1973). *The Sociology of Science: Theoretical and Empirical Investigations.* Chicago: University of Chicago Press.

——— (1979). *The Sociology of Science: An Episodic Memoir.* Carbondale: Southern Illinois University Press.

——— (1987). Three fragments from a sociologist's notebooks: Establishing the phenomena, specified ignorance, and strategic research materials. *Am. J. Sociol.* 13:1–28.

——— (1993). *On the Shoulders of Giants: The Post-Italiante Edition.* Chicago: University of Chicago Press.

Mertz, J. E., and Davis, R. W. (1972). Cleavage of DNA by R_I restriction endonuclease generates cohesive ends. *PNAS* 69:3370–3374.

Meselson, M., and Stahl, F. W. (1958). The replication of DNA in *Escherichia coli. PNAS* 44:671–682.

Meselson, M., Stahl, F. W., and Vinograd, J. (1957). Equilibrium sedimentation of macromolecules in density gradients. *PNAS* 43:581–588.

Metchnikoff, I. (1893). *Lectures on the Comparative Pathology of Inflammation.* London: Kegan Paul, Trench Trübner.

——— (1901). *L'Immunité dans les maladies infectieuses.* Paris: Masson. [English trans. by F. G. Binnie, Cambridge: Cambridge University Press, 1905.]

Metzger, H. (1918). *La Génèse des sciences des cristaux.* Paris: Alcan.

——— (1930). *Newton, Stahl, Boerhaave et la Doctrine Chimique.* Paris: Alcan.

Meuron-Landot, M. de (1965). Friedrich Miescher, l'homme qui a découvert les acides nucléiques. *Histoire de la Médecine* 15:2–25.

Meyer, E. von (1884). Zur Erinnerung an Hermann Kolbe. *J. prakt. Chem.* NF30:417–466.

——— (1889). *Geschichte der Chemie von den ältesten Zeit bis zur Gegenwart.* Leipzig: Veit.

Meyer, H. H. (1922). Schmiedebergs Werk. *Arch. exp. Path. Pharm.* 92:i–xxvii.

Meyer, J., and Weinmann, J. P. (1957). Distribution of phosphoamidase activity in the male albino rat. *J. Histochem. Cytochem.* 5:354–397.

Meyer, K. H. (1943). The chemistry of glycogen. *Adv. Enzymol.* 3:100–135.

Meyer, R. (1917). *Victor Meyer: Leben und Wirken eines deutschen Chemiker und Naturforscher.* Leipzig: Akademische Verlagsgesellschaft.

Meyerhof, O. (1918). Ueber das Vorkommen des Coferments der alkoholischen Hefegärung im Muskelgewebe und seine mutmassliche Bedeutung im Atmungsmechanismus. *Z. Physiol. Chem.* 101:165–175.

——— (1925). Ueber den Zusammenhang der Spaltungsvorgängen mit der Atmung in der Zelle. *Ber. chem. Ges.* 58:991–1001.

——— (1926a). Ueber die enzymatische Milchsäurebildung im Muskelextrakt. *Biochem. Z.* 178:395–418, 462–490.

——— (1926b). Thermodynamik des Lebensprozesses. In *Handbuch der Physik* (H. Geiger and K. Scheel, eds.), 2:238–271. Berlin: Springer.

——— (1930). *Die Chemische Vorgänge im Muskel.* Berlin: Springer.

——— (1944). Energy relationships in glycolysis and phosphorylation. *Ann. N.Y. Acad. Sci.* 45:377–393.

——— (1945). The origin of the reaction of Harden and Young in cell-free alcoholic fermentation. *J. Biol. Chem.* 157:105–117.

Meyerhof, O., and Kiessling, W. (1935). Ueber den Hauptweg der Milchsäurebildung in der Muskulatur. *Biochem. Z.* 283:83–113.

Meyerhof, O., and Lohmann, K. (1931). Ueber die Energetik der anaeroben Phosphagensynthese ("Kreatinphosphorsäure") im Muskelextrakt. *Naturwiss.* 19:575–576.

——— (1934). Ueber die enzymatische Gleichgewichtsreaktion zwischen Hexosediphosphorsäure und Dioxyacetonphosphorsäure. *Biochem. Z.* 271:89–110.

Meyerhof, O., Ohlmeyer, P., and Möhle, W. (1938). Ueber die Koppelung zwischen Oxydoreduktion und Phosphorylierung bei der anaeroben Kohlenhydrat spaltung. *Biochem. Z.* 297:90–133.

Meyerson, E. (1921). *De l'Explication dans les sciences.* Paris: Payot. [English trans. by M. A. Sipfle and D. A. Sipfle, Dordrecht: Kluwer, 1991.]

——— (1926). *Identité et réalité.* 3d ed. Paris: Alcan.

Mialhe, L. (1856). *Chimie appliquée a la physiologie et a la thérapeutique.* Paris: Masson.

Michaelis, L. (1900). Die vitale Färbung, eine Darstellungsmethode der Zellgranula. *Arch. mikr. Anat.* 55:558–575.

——— (1914). *Die Wasserstoffionenkonzentration: Ihre Bedeutung für die Biologie und die Methode ihrer Messung.* Berlin: Springer.

——— (1929). *Oxidation-Reduktions-Potentiale, mit besondere Berücksichtigung ihrer physiologischen Bedeutung.* Berlin: Springer. [2d ed., 1933.]

——— (1935). Semiquinones, the intermediate steps of reversible oxidationreduction. *Chem. Revs.* 16:243–286.

Michaelis, L., and Menten, M. L. (1913). Die Kinetik der Invertasewirkung. *Biochem. Z.* 49:333–369.

Michel, J., et al. (1991). *La Nécessité de Claude Bernard.* Paris: Méridien.

Miescher, F. (1871a). Ueber die chemische Zusammensetzung des Eiters. *Med. chem. Unt.*, pp. 441–460.

——— (1871b). Die Kerngebilde im Dotter des Hühnereies. *Med. Chem. Unt.* pp. 502–509.

——— (1874). Das Protamin, eine neue organische Basis aus der Samenfaden des Rheinlachses. *Ber. chem. Ges.* 7:376–379.

——— (1896). Physiologisch-chemische Untersuchungen über das Lachsmilch. *Arch. exp. Path. Pharm.* 37:100–155.

——— (1897). *Die Histochemischen und Physiologischen Arbeiten.* Leipzig: Vogel.

Mihalyi, E. (1978). *Application of Proteolytic Enzymes to Protein Structure Studies.* West Palm Beach, Fla.: CRC.

Milas, N. A. (1932). Auto-oxidation. *Chem. Revs.* 10:295–364.

Miles, E. W. (1979). Tryptophan synthase: Structure, function, and subunit interaction. *Adv. Enzymol.* 49:127–186.

Miles, W. (1953). Benjamin Rush, chemist. *Chymia* 4:37–77.

——— (1994). John Maclean. In *American Chemists and Chemical Engineers* (W. D. Miles and R. F. Gould, eds.), 2:174–175. Guilford, Conn.: Gould.

Millar, M., and Millar, I. T. (1983, 1985, 1988). Chemists as autobiographers. *J. Chem. Ed.* 60:365–370; 62:275–281; 65:847–853.

Miller, A., and Waelsch, H. (1957). The formation of N^{10}-formyl folic acid from formamidinoglutaric acid and folic acid: Formimino transfer from formamidinoglutaric acid to tetrahydrofolic acid. *J. Biol. Chem.* 228:383–417.

Miller, D. G. (1971). Pierre Duhem. *DSB* 4:225–233.

Miller, S. G., and Orgel, L. E. (1974). *The Origins of Life on Earth.* Engelwood Cliffs, N.J.: Prentice-Hall.

Mills, H. W. (1939). Arthur Hutchinson. *J. Chem. Soc.*, pp. 210–213.

Milne Edwards, H. (1862). *Leçons sur la physiologie et l'anatomie comparée de l'homme et des animaux.* Vol. 7. Paris: Masson.

Milstein, C. (1985). From the structure of antibodies to the diversification of the immune response. *Ang. Chem. (Int. Ed.)* 24:816–826.

Minding, J. (1844). Johann Franz Simon. *Beitr. physiol. path. Chemie* 1:547–552.

Minkowski, O. (1884). Ueber das Vorkommen von Oxybuttersäure im Harn bei Diabetes mellitus. *Arch. exp. Path. Pharm.* 18:35–48, 147–150.

Mirsky, A. E. (1951). Some chemical aspects of the cell nucleus. In *Genetics in the 20th Century* (L. C. Dunn, ed.), pp. 127–153. New York: Macmillan.

Mirsky, A. E., and Pauling, L. (1936). On the structure of native, denatured, and coagulated proteins. *PNAS* 22:439–447.

Mitchell, J. S. (1956). Prof. E. J. Friedmann. *Nature* 178:397.

Mitchell, P. (1961). Coupling of phosphorylation to electron and hydrogen transfer by a chemiosmotic type of mechanism. *Nature* 191:144–149.

——— (1979a). Compartmentation and communication in living systems. Ligand conduction: A general catalytic principle in chemical, osmotic and chemiosmotic reaction systems. *Eur. J. Biochem.* 95:1–20.

——— (1979b). Keilin's respiratory chain concept and its chemi-osmotic consequences. *Science* 206:1148–1159.

——— (1981). Bioenergetic aspects of unity in biochemistry: evolution of the concept of ligand conduction in chemical, osmotic, and chemiosmotic reaction mechanisms. In *Oxygen, Fuels, and Living Matter.* Part 1 (G. Semenza, ed.), pp. 1–160. Chichester: Wiley.

——— (1991). Foundations of vectorial metabolism and osmochemistry. *Bioscience Reports* 11:297–346.

Mitchison, N. A. (1990). Peter Brian Medawar. *Biog. Mem. FRS* 35:283–301.

Mitscherlich, E. (1834). Ueber die Aetherbildung. *Ann. Phys.* 31:273–282.

Mizrahi, V., and Benkovic, S. J. (1988). The dynamics of DNA polymerasecatalyzed reaction. *Adv. Enzymol.* 61:437–457.

Mizushima, S. (1972). A history of physical chemistry in Japan. *Ann. Rev. Phys. Chem.* 23:1–14.

Mizuuchi, K. (1992). Polynucleotidyl transfer reactions in tranpositional DNA recombination. *J. Biol. Chem.* 267:21273–21276.

Möbius, M. (1904). *Matthias Jacob Schleiden.* Leipzig: Engelmann.

Mocek, R. (1974). *Wilhelm Roux, Hans Driesch.* Jena: Fischer.

Mochnacka, I. (1956). Prace Jakuba Karola Parnasa. *Acta Biochim. Polon.* 3:3–39.

Modrich, P. (1987). DNA mismatch correction. *Ann. Rev. Biochem.* 56:435–466.

Moffatt, J. G., and Khorana, H. G. (1961). Nucleoside polyphosphates, 12: The total synthesis of coenzyme A. *J. Am. Chem. Soc.* 83:663–675.

Mokrushin, S. G. (1962). Thomas Graham and the definition of colloids. *Nature* 195:861.

Moldave, K. (1985). Eukaryotic protein synthesis. *Ann. Rev. Biochem.* 54:1109–1149.

Möller, H. (1984). Wissenschaft in der Emigration—Quantitative und geographische Aspekte. *Ber. Wiss.* 7:1–9.

Monaghan, D., Bridges, R., and Cotman, C. (1989). The excitatory amino acid receptors. *Ann. Rev. Pharmacol.* 29:365–402.

Moncada, S., Palmer, R. M. J., and Higgs, E. A. (1991). Nitric oxide: Physiology, pathophysiology, and pharmacology. *Pharmacol. Rev.* 43:109–142.

Monnier, M. (1942). Federico Battelli. *Rivista di biologia* 33:267–280.

Monod, J. (1942). *La Croissance des cultures bactériennes.* Paris: Hermann.

——— (1947). The phenomenon of enzymatic adaptation and its bearings on problems of genetics and cellular differentiation. *Growth Symposium* 11:223–289.

——— (1950). Adaptation, mutation, and segregation in the formation of bacterial enzymes. In *Biochemical Aspects of Genetics* (R. T. Williams, ed.), pp. 51–58. Cambridge: Cambridge University Press.

——— (1956). Remarks on the mechanism of enzyme induction. In *Enzymes: Units of Biological Structure and Function* (O. H. Gaebler, ed.), pp. 7–28. New York: Academic.

——— (1966). From enzymatic adaptation to allosteric transitions. *Science* 154:475–483.

——— (1970). *Le Hasard et la nécessité.* Paris: Seuil.

——— (1971). Du microbe à l'homme. In *Of Microbes and Life* (J. Monod and E, Borek, eds.), pp. 1–9. New York: Columbia University Press,

Monod, J., Changeux, J. P., and Jacob, F. (1963). Allosteric proteins and cellular control systems. *J. Mol. Biol.* 6:306–329.

Monod, J., and Cohn, M. (1952). La biosynthèse induite des enzymes (adaptation enzymatique). *Adv. Enzymol.* 13:67–119.

Monod, J., and Jacob, F. (1961). General conclusions: Teleonomic mechanisms in cellular metabolism, growth, and differentiation. *Cold Spring Harbor Symp.* 26:389–401.

Monod, J., Jacob, F., and Gros, F. (1962). Structural and rate-determining factors in the biosynthesis of adaptive enzymes. In *The Structure and Biosynthesis of Macromolecules* (D. J. Bell and J. K. Grant, eds.), pp. 104–132. Cambridge: Cambridge University Press.

Monod, J., Wyman, J., and Changeux, J. P. (1965). On the nature of allosteric transitions: A plausible model. *J. Mol. Biol.* 12:88–118.

Moolenaar, W. H. (1995). Lysophosphatidic acid, a multifunctional phospholipid messenger. *J. Biol. Chem.* 270:12949–12952.

Moore, A. R. (1910). A biochemical conception of dominance. *University of California Publications in Physiology* 4:9–15.

——— (1912). On Mendelian dominance. *Wilhelm Roux Archiv für Entwicklungsmechanik* 34:168–175.

Moore, B. (1898). Chemistry of the digestive process. In *Textbook of Physiology* (E. A. Schäfer, ed.), 1:312–474. Edinburgh: Pentland.

Moore, B., and Whitley, E. (1909). The properties and classification of the oxidizing enzymes, and analogies between enzymic activity and the effects of immune bodies and complement. *Biochem. J.* 4:136–167.

Moore, C. R., and Price, D. (1932). Gonad hormone functions and the reciprocal influence between gonads and hypophysis with its bearing on the problem of sex hormone antagonism. *Am. J. Anat.* 50:12–67.

Moore, P. B. (1995). Molecular mimicry in protein synthesis? *Science* 270:1453–1454.

Moore, S. (1978). On the artificial formation of urea. *TIBS* 3:17–18.

——— (1981). Pancreatic DNase. In *The Enzymes* (P. D. Boyer, ed.), 14A:281–296. New York: Academic.

Moore, S., and Stein, W. H. (1951). Chromatography of amino acids on sulfonated polystyrene resins. *J. Biol. Chem.* 192:663–681.

——— (1973). Chemical structures of pancreatic ribonuclease and deoxyribonuclease. *Science* 180:458–464.

Moore, W. (1989). *Schrödinger: Life and Thought.* Cambridge: Cambridge University Press.

Morales, M. F. (1995). A brief and subjective history of contractility. *Protein Science* 4:130–132.

Moran, S., Ren, R. X., and Kool, E. T. (1997). A thymidine triphosphate shape analog lacking Watson-Crick pairing ability is replaced with high sequence selectivity. *PNAS* 94:10506–10511.

Morange, M. (1982). La révolution silencieuse de la biologie moléculaire. *Le Débat* no. 18, pp. 62–75.

——— (1991). *L'Institut Pasteur.* Paris: La Découverte.

——— (1994). *Histoire de la biologie moléculaire.* Paris: La Découverte.

Morawetz, H. (1985). *Polymers: The Origins and Growth of a Science.* New York: Wiley.

Morgan, G., and Fruton, J. S. (1978). Kinetics of the action of thermolysin on peptide substrates. *Biochemistry* 17:3562–3568.

Morgan, G. T. (1935). Charles Thomas Kingzett. *J. Chem. Soc.*, pp. 1899–1902.

Morgan. N. (1980). The development of biochemistry in England through botany and the brewing industry, 1870–1890. *Hist. Phil. Life Sci.* 2:141–166.

——— (1983). William Dobinson Halliburton. *Notes Roy. Soc.* 38:129–145.

——— (1990). The strategy of biological research programmes: Reassessing the "dark age" of biochemistry, 1910–1930. *Ann. Sci.* 47:139–150.

Morgan, T. H. (1910). Sex limited inheritance in Drosophila. *Science* 32:120–122.

——— (1919). *The Physical Basis of Heredity.* Philadelphia: Lippincott.

——— (1926a). *The Theory of the Gene.* New Haven: Yale University Press. [2d ed., 1928.]

——— (1926b). Genetics and the physiology of development. *Am. Nat.* 60:489–515.

——— (1940). Edmund Beecher Wilson. *Biog. Mem. NAS* 21:315–342.

Morgan, T. H., Sturtevant, A. H., Muller, H. J., and Bridges, C. B. (1915). *The Mechanism of Mendelian Heredity.* New York: Holt.

Morgan, W. T., and Francis, G. E. (1966). Arthur Wormall. *Biog. Mem. FRS* 12:543–564.

Moritz, K. B. (1993). *Theodor Boveri.* Stuttgart: Gustav Fischer.

Morrell, J. B. (1969). Thomas Thomson: Professor of Chemistry and university reformer. *Brit. J. Hist. Sci.* 4:245–265.

——— (1972). The chemist breeders: The research schools of Liebig and Thomas Thomson. *Ambix* 19:1–46.

——— (1993). W. H. Perkin, Jr., at Manchester and Oxford: From Irwell to Isis. *Osiris* 8:104–126.

Morris, R. J. (1972). Lavoisier and the caloric theory. *Brit. J. Hist. Sci.* 6:1–38.

Morris-Suzuki, T. (1994). *The Technological Transformation of Japan.* Cambridge: Cambridge University Press.

Morrow, J. F., Cohen, S. N., Chang, A. C. Y., Boyer, H. W., Goodman, H. M., and Helling, R. B. (1974). Replication and transcription of eukaryotic DNA in *Escherichia coli. PNAS* 71:1743–1747.

Morse, D. E., and Horecker, B. L. (1958). The mechanism of action of aldolases. *Adv. Enzymol.* 31:125–181.

Morse, M. L., Lederberg, E. M., and Lederberg, J. (1956). Transduction in *Escherichia coli* K-12. *Genetics* 41:142–156.

Morsier, G. de, and Monnier, M. (1977). *La Vie et l'oeuvre de Frédéric Battelli.* Basel: Schwabe.

Mortenson, L. E., Morris, J. A., and Jeng, D. Y. (1967). Purification, metal composition, and the properties of molybdoferrodoxin and also ferrodoxin, two of the components of the nitrogen-fixing system of *Clostridium pausteurianum*. *Biochim. Biophys. Acta* 141:516–522.

Morton, R. A. (1969). *The Biochemical Society: Its History and Activities, 1911–1969*. London: Biochemical Society.

——— (1972). Biochemistry at Liverpool, 1902–1971. *Med. Hist.* 16:321–353.

Morton, R. A., and Pitt, G. A. J. (1955). Studies on rhodopsin, 9: pH and the hydrolysis of indicator yellow. *Biochem. J.* 59:128–134.

Moser, L. (1842). Ueber das Process des Sehens und die Wirkung des Lichts auf alle Körper. *Ann. Phys.* 56:177–234.

Moser, W. (1967). *Der Physiologe Jacob Moleschott und seine Philosophie*. Zurich: Juris.

Moss, J., and Lane, M. D. (1971). The biotin-dependent enzymes. *Adv. Enzymol.* 35:321–442.

Moss, R. W. (1988). *Free Radical: Albert Szent-Györgyi and the Battle over Vitamin C*. New York: Paragon.

Moulin, A. M. (1991). *Le Dernier langage de la médecine: Histoire de l'immunologie de Pasteur au SIDA*. Paris: PUF.

Mounce, H. O. (1985). Review of D. Bloor, *Wittgenstein: A Social Theory of Knowledge*. *Brit. J. Phil. Sci.* 36:344–346.

Moutier, J. (1885). *La Thermodynamique et ses principales applications*. Paris: Gauthier-Villars.

Moy, T. D. (1989). Emil Fischer as "chemical mediator": Science, industry, and government in World War One. *Ambix* 36:109–120.

Mudd, S, (1932). A hypothetical mechanism of antibody formation. *J. Immunol.* 23:423–427.

Mueller, J. H. (1923). A new sulfur-containing amino acid isolated from the hydrolytic products of a protein. *J. Biol. Chem.* 56:157–169.

Muir, H. M., and Neuberger, A. (1950). The biogenesis of porphyrins, 2: The origin of the methylene carbon atoms. *Biochem. J.* 47:97–104.

Mulder, G. J. (1838). Ueber die Zusammensetzung von Fibrin, Albumin, Leimzucker, Leucin u. s. w. *Ann.* 28:73–82.

——— (1839). Ueber die Zusammensetzung einiger thierischen Substanzen. *J. prakt. Chem.* 16:129–152.

——— (1844–1851). *Versuch einer allgemeinen physiologische Chemie*. Braunschweig: Vieweg.

——— (1846). *Liebig's Question to Mulder, Tested by Morality and Science*. London: Blackwood. (P. F. H. Fromberg, trans. from the original Dutch edition [1846]. A German translation also appeared in 1846.)

Mulkay, M. (1976). Norms and ideology in science. *Social Science Information* 15:637–656.

——— (1991). *Sociology of Science: A Sociological Pilgrimage*. Bloomington: Indiana University Press.

Mulkay, M., and Gilbert, G. N. (1984). *Opening Pandora's Box*. Cambridge: Cambridge University Press.

Müller, F. C. (1902). *Geschichte der organischen Naturwissenschaften*. Berlin: Bondi.

Müller, H. (1856). *Anatomisch-physiologische Untersuchungen über die Retina des Menschen und der Wirbelthiere*. Leipzig: Engelmann.

Muller, H. J. (1922). Variation due to change in the individual gene. *Am. Nat.* 56:32–50.

——— (1927). Artificial transmutation of the gene. *Science* 66:84–87.

——— (1935). On the dimensions of chromosomes and genes in dipteran salivary glands. *Am. Nat.* 69:405–411.

——— (1936). Physics in the attack on the fundamental problems of genetics. *Sci. Mon.* 44:210–214.

——— (1943). Edmund B. Wilson: An appreciation. *Am. Nat.* 77:5–37, 142–172.

——— (1946). A physicist stands amazed at genetics. *J. Heredity* 37:90–92.

——— (1947). The gene. *Proc. Roy. Soc.* B134:1–37.

Müller, I. (1996). The impact of the Zoological Station in Naples on developmental physiology. *Int. J. Dev. Biol.* 40:103–111.

Müller, J. (1843). *Elements of Physiology* (W. Baly, trans.; arranged from 2d London ed. by J. Bell). Philadelphia: Lea and Blanchard.

Müller, K. (1970). *Moritz Traube und seine Lehre der Fermente.* Zurich: Juris.

Mullins, N. C. (1972). The development of a scientific specialty: the phage group and the origins of molecular biology. *Minerva* 10:51–82.

Mullis, K. B. (1990). The unusual origin of the polymerase chain reaction. *Sci. Am.* 262(4):56–65.

Mullis, K. B., Ferré, F., and Gibbs, R. A. (eds.) (1994). *The Polymerase Chain Reaction.* Boston: Birkhäuser.

Multhauf, R. P. (1955). J. B. van Helmont's reformation of the Galenic doctrine of digestion. *Bull. Hist. Med.* 154–163.

——— (1966). *The Origins of Chemistry.* London: Oldbourne.

——— (1976). Crystals and compounds. *Trans. Am. Phil. Soc.* NS66, part 3.

Munro, H. N. (1963). The origin and growth of our present concepts of protein metabolism. In *Mammalian Protein Metabolism* (H. N. Munro and J. B. Allison, eds.), pp. 1–34. New York: Academic.

Munro, R. E., and Marcker, K. E. (1967). Ribosome-catalyzed reaction of puromycin with a formylmethionine-containing oligonucleotide. *J. Mol. Biol.* 25:347–350.

Muralt, A. von (1952). Otto Meyerhof. *Erg. Physiol.* 47:i–xx.

Muralt, A. von, and Edsall, J. T. (1930). Studies on the physical chemistry of muscle globulin, 3, 4. *J. Biol. Chem.* 89:315–386.

Murnaghan, J. H., and Talalay, P. (1967). John Jacob Abel and the crystallization of insulin. *Persp. Biol. Med.* 10:334–330.

Murray, I. (1971). Paulesco and the isolation of insulin. *J. Hist. Med.* 26:150–157.

Musgrave, A. (1976). Why did oxygen supplant phlogiston? Research programmes in the Chemical Revolution. In *Method and Appraisal in the Physical Sciences* (C. Howson, ed,), pp. 181–209. Cambridge: Cambridge University Press.

——— (1993). *Common Sense, Science, and Scepticism.* Cambridge: Cambridge University Press.

Musser, S. M., Stowell, M. H. B., and Chan, S. I. (1995). Cytochrome c oxidase: Chemistry of a molecular machine. *Adv. Enzymol.* 71:79–208.

Mutter, M. (1985). The construction of new proteins and enzymes: A prospect for the future? *Ang. Chem. (Int. Ed.)* 24:639–653.

Myers, J. (1974). Conceptual developments in photosynthesis, 1924–1974. *Plant Physiology* 54:420–426.

Nachmansohn, D., and Berman, M. (1946). Studies on choline acetylase, 3: The preparation of the coenzyme and its effect on the enzyme. *J. Biol. Chem.* 165:551–563.

Naess, A. (1972). *The Pluralist and Possibilist Aspect of the Scientific Enterprise.* Oslo: Universitetsforlaget.

Nagata, S. (1994). Fas and Fas ligand: A death factor and its acceptor. *Adv. Immunol.* 57:129–144.

Nagel, E. (1961). *Structure of Science: Problems in the Logic of Scientific Explanation.* New York: Harcourt, Brace and World.

Nägeli, C. (1879). Theorie der Gärung. *Abhandlungen der königlichen Akademie der Wissenschaften München* 13(2):77–205.

——— (1884). *Mechanisch-physiologische Theorie der Abstammungslehre.* Leipzig: Oldenbourg.

Nägeli, C., and Loew, O. (1878). Ueber die chemische Zusammensetzung der Hefe. Ann. 193:322–328.

Najjar, V., and Chung, C. W. (1956). Enzymatic steps in denitrification. In *Inorganic Nitrogen Metabolism* (W. D. McElroy and B. Glass, eds.), pp. 260–291. Baltimore: Johns Hopkins University Press.

Naquet, A. (1867). *Principes de chimie fondées sur les théories modernes.* 2d ed. Paris: Pavy.

Nash, K. (1950). *The Atomic-Molecular Theory.* Cambridge: Harvard University Press.

——— (1963). *The Nature of the Natural Sciences.* Boston: Little, Brown.

Nasini, R. (1926). Giacomo Luigi Ciamician. *J. Chem. Soc.* 129:996–1004.

Nason, A. (1956). Enzymatic steps in the assimilation of nitrate and nitrite in fungi and green plants. In *Inorganic Nitrogen Metabolism* (W. D. McElroy and B. Glass, eds.), pp. 109–136. Baltimore: Johns Hopkins University Press.

Nasse, O. (1869). Beiträge zur Physiologie der contractilen Substanz. *Pflügers Arch.* 2:97–121.

Nathan, C., and Xie, Q. (1994). Regulation of the biosynthesis of nitric oxide. *J. Biol. Chem.* 269:13725–13728.

Nathans, D. (1979). Restriction endonucleases, simian virus 40, and the new genetics. *Science* 206:903–909.

Nathans, D., and Smith, H. O. (1975). Restriction endonucleases in the analysis and restructuring of DNA molecules. *Ann. Rev. Biochem.* 44:273–293.

Nathans, J. (1987). Molecular biology of visual pigments. *Ann. Rev. Neurosci.* 10:163–194.

Natta, G. (1965). Macromolecular chemistry. *Science* 147:261–272.

Naunyn, B. (1921). Oswald Schmiedeberg. *Arch. esp. Path. Pharm.* 90:i–viii.

Needham, D. M. (1930). A quantitative study of succinic acid in muscle, 3: Glutamic and aspartic acids as precursors. *Biochem. J.* 24:208–227.

——— (1971). *Machina Carnis: The Biochemistry of Muscular Contraction in Its Historical Development.* Cambridge: Cambridge University Press.

Needham, D. M., and Pillai, R. K. (1937). The coupling of oxido-reductions and dismutations with esterification of phosphate in muscle. *Biochem. J.* 31:1837–1851.

Needham, J. (1936). *Order and Life.* New Haven: Yale University Press.

——— (1954–1974). *Science and Civilisation in China.* Cambridge: Cambridge University Press.

——— (1962). Frederick Gowland Hopkins. *Persp. Biol. Med.* 6:2–46.

——— (1978). Science reborn in China. *Nature* 274:832–834.

Needham, J., and Baldwin, E. (eds.) (1949). *Hopkins and Biochemistry.* Cambridge: Cambridge University Press.

Needham, P. (1996). Substitution: Duhem's explication of a chemical paradigm. *Persp. Sci.* 4:408–433.

Negelein, E., and Brömel, H. (1939a). Protein der *d*-Aminosäureoxydase. *Biochem. Z.* 300:225–239.

——— (1939b). R-Diphosphoglycerinsäure, ihre Isolierung und Eigenschaften. *Biochem. Z.* 303:132–144.

Negelein, E., and Haas, E. (1935). Ueber die Wirkungsweise des Zwischenferments. *Biochem. Z.* 282:206–220.

Negelein, E., and Wulff, H. (1937). Diphosphopyridinproteid: Alkohol, Acetaldehyd. *Biochem. Z.* 293:351–389.

Nelson, D. L., and Soltvedt, B. C. (eds.) (1989). *One Hundred Years of Agricultural Chemistry and Biochemistry at Wisconsin.* Madison, Wis.: Science Tech Publishers.

Nencki, M. (1872). Die Wasserentziehung im Thierkörper. *Ber. chem. Ges.* 5:890–893.

——— (1878). Ueber den chemischen Mechanismus des Fäulniss. *J. Prakt. Chem.* 17:105–124.

——— (1885). Ueber das Parahämoglobin. *Arch. exp. Path. Pharm.* 20:332–343.

——— (1904). *Marceli Nencki Opera Omnia.* Braunschweig: Vieweg.

Nencki, M., and Sieber, N. (1882). Untersuchungen über die biologische Oxydation. *J. prakt. Chem.* 26:1–41.

Nernst, W. (1904). Ueber die Anwendbarkeit der Gesetze des chemischen Gleichgewichts auf Gemische von Toxin und Antitoxin. *Z. Elektrochem.* 10:377–380.

——— (1906). Ueber die Berechnung chemischer Gleichgewichte aus thermischen Messungen. *Nachrichten von der Gesellschaft der Wissenschaften zu Göttingen,* pp. 1–40.

Nersessian, N. J. (ed.) (1987). *The Process of Science.* Dordrecht: Nijhoff.

Neubauer, O. (1909). Ueber den Abbau der Aminosäuren im gesunden und kranken Organismus. *Deutsches Arch. klin. Med.* 95:211–256.

Neubauer, O., and Falta, W. (1904). Ueber den Schicksal einiger aromatischen Säuren bei der Alkaptonurie. *Z. physiol. Chem.* 42:81–101.

Neubauer, O., and Fromherz, K. (1911). Ueber den Abbau der Aminosäuren bei der Hefegärung. *Z. physiol. Chem.* 70:326–350.

Neuberg, C., and Brahm, B. (1907). Ueber die Inosinsäure. *Biochem. Z.* 5:438–450.

Neuberg, C., and Kerb, J. (1913). Ueber zuckerfreie Gärungen, 12: Ueber die Vorgänge bei der Hefegärung. *Biochem. Z.* 53:406–419.

Neuberg, C., and Reinfurth, E. (1919). Weitere Untersuchungen über die korrelative Bildung von Acetaldehyd und Glycerin bei der Zuckerspaltung und neue Beiträge zur Theorie der alkoholischen Gärung. *Ber. chem. Ges.* 52:1677–1703.

Neuberger, A. (1973). James Norman Davidson. *Biog. Mem. FRS* 19:281–303.

——— (1985). The Lister Institute of Preventive Medicine. *BioEssays* 3:234–235.

——— (1990). An octogenarian looks back. In *Selected Topics in the History of Biochemistry: Personal Recollections.* Vol. 3 (G. Semenza and L. Jaenicke, eds.), pp. 21–65. Amsterdam: Elsevier.

——— (1996). Claude Rimington. *Biog. Mem. FRS* 42:363–378.

Neumann, A. (1898). Zur Kenntnis der Nukleinsubstanzen. *Arch. Anat. Physiol.,* pp. 374–378.

Neumann, J. (1989). Der historisch-soziale Ansatz medizinischer Wissenschafts-theorie von Ludwik Fleck. *Sudhoffs Arch.* 73:12–25.

Neumeister, R. (1897a). *Lehrbuch der Physiologischen Chemie.* Jena: Fischer.

——— (1897b). Bemerkungen zu Eduard Buchner's Mitteilungen über "Zymase." *Ber. Chem. Ges.* 30:2963–2966.

——— (1903). *Betrachtungen über das Wesen der Lebenserscheinungen. Beitrag zum Begriff des Protoplasma.* Jena: Fischer.

Neurath, H. (1957). The activation of zymogens. *Adv. Protein Chem.* 12:320–386.

——— (1975). Limited proteolysis and zymogen activation. In *Proteases and Biological Control* (E. Reich, D. B. Rifkin, and E. Shaw, eds.), pp. 51–64. Plainview, N.Y.: Cold Spring Harbor Laboratory Press.

Neurath, O. (1944). *Foundations of the Social Sciences.* Chicago: University of Chicago Press.

Newell, L. C. (1976). Chemical education in America from the earliest days to 1820. *J. Chem. Ed.* 53:402–404.

Newton, A. C. (1995). Protein kinase C. *J. Biol. Chem.* 270:28495–28498.

Nicklès, J. (1856). Henri Braconnot. *Mémoires de l'Académie Stanislas* [3]22:xxii–cxlix.

Nickles, T. (1995). Philosophy of science and history of science. *Osiris* 10:139–163.

Nielsen, P. E., Egholm, M., Berg, R. H., and Buchardt, O. (1991). Sequence-selective recognition of DNA by strand displacement with a thymine-substituted polyamide. *Science* 254:1497–1500.

Nier, A. O. (1955). Determination of isotopic masses and abundances by mass spectrometry. *Science* 121:737–744.

Nierenstein, M. (1934). A missing chapter in the history of organic chemistry: The link between elementary analysis by dry-distillation and combustion. *Isis* 21:123–130.

Nierhaus, K. H., Jünemann, R., and Spahn, C. M. T. (1997). Are the current three-site models valid descriptions of the ribosomal elongation cycle? *PNAS* 94:10499–10500.

Nipperdey, T., and Schmugge, L. (1970). *50 Jahre Forschungsförderung in Deutschland.* Berlin: Deutsche Forschungsgemeinschaft.

Nirenberg, M., and Leder, P. (1964). RNA codewords and protein synthesis. *Science* 145:1399–1407.

Nirenberg, M., and Matthaei, J. H. (1961). The dependence of cell-free protein synthesis in *E. coli* upon naturally occurring or synthetic polyribonucleotides. *PNAS* 47:1588–1602.

Nirenberg, M., Matthaei, J. H., and Jones. O. W. (1962). An intermediate in the biosynthesis of polyphenylalanine directed by synthetic template RNA. *PNAS* 48:104–109.

Nissani, M. (1994). Psychological, historical, and ethical reflections on the Mendelian paradox. *Persp. Biol. Med.* 37:182–196.

Nola, R. (1995). There are more things in heaven and earth, Horatio, than are dreamt of in your philosophy: A dialogue on realism and constructivism. *Stud. Hist. Phil. Sci.* 25:689–727.

Noller, H. F. (1984). Structure of ribosomal RNA. *Ann. Rev. Biochem.* 53:119–162.

——— (1991). Ribosomal RNA and translation. *Ann. Rev. Biochem.* 60:191–227.

——— (1993). tRNA-rRNA interactions and peptidyl transfer. *FASEB J.* 7:87–89.

Noller, H. F., Hoffarth, V., and Zimniak, L. (1992). Unusual resistance of peptidyl transferase to protein extraction procedures. *Science* 256:1416–1419.

Nomura, M. (1973). Assembly of bacterial ribosomes. *Science* 179:864–873.

——— (1988). Initiation of protein synthesis: Early participation and recent revisit. In *The Roots of Modern Biochemistry* (H. Kleinkauf, H. van Döhren, and L. Jaenicke, eds.), pp. 493–504. Berlin: de Gruyter.

Nomura, M., Gourse, R., and Baughman, G. (1984). Regulation of the synthesis of ribosomes and ribosomal components. *Ann. Rev. Biochem.* 53:75–117.

Nomura, M., Hall, B. D., and Spiegelman, S. (1960). Characterization of the RNA synthesized after bacteriophage T2 infection. *J. Mol. Biol.* 2:306–326.

Nomura, M., and Lowry, C. V. (1967). Phage F2 RNA-directed binding of formylmethionyl-RNA to ribosomes and the role of the 30S ribosomal subunits in the initiation of protein synthesis. *PNAS* 58:946–953.

Nomura, M., Mizushima, S., Ozaki, M., Traub, P., and Lowry, C. V. (1969). Structure and function of ribosomes and their molecular components. *Cold Spring Harbor Symp.* 34:49–61.

Norberg, B. (1950). Phosphoprotein phosphatase in the rat. *Acta Chem. Scand.* 4:1206–1215.

Nord, F. F. (1940). Facts and interpretations in the mechanism of alcoholic fermentation. *Chem. Revs.* 26:423–472.

Nordenskiold, E. (1928). *History of Biology.* New York: Tudor.

Norrish, R. G. W. (1969). Fifty years of physical chemistry in Great Britain. *Ann. Rev. Phys. Chem.* 20:1–24.

North, J. D. (1996). Alistair Cameron Crombie. *Hist. Sci.* 34:245–248.

Northrop, D. B. (1981). The expression of isotope effects on enzyme-catalyzed reactions. *Ann. Rev. Biochem.* 50:103–131.

Northrop, J. H. (1924). The kinetics of trypsin digestion, 5: Schütz's rule. *J. Gen. Physiol.* 6:723–729.

——— (1930). Crystalline pepsin, 1: Isolation and tests of purity. *J. Gen. Physiol.* 13:739–766.

——— (1935). The chemistry of pepsin and trypsin. *Biol. Revs.* 10:263–282.

——— (1951). Growth and phage production of lysogenic B. megatherium. *J. Gen. Physiol.* 34:715–735.

——— (1961). Biochemists, biologists, and William of Occam. *Ann. Rev. Biochem.* 20:1–10.

Northrop, J. H., Kunitz, M., and Herriott, R. M. (1948). *Crystalline Enzymes.* 2d ed. New York: Columbia University Press.

Nossal, G. J. V. (1986). Turning points in cellular immunology: The skein untangled through a global invisible college. *Persp. Biol. Med.* 29:S167–S177.

——— (1995). One cell—one antibody. In *Immunology: The Making of a Modern Science* (R. B. Gallagher et al., eds.), pp. 39–47. London: Academic.

Nossal, G. J. V., and Lederberg, J. (1958). Antibody production by single cells. *Nature* 181:1419–1420.

Novikoff, A. B. (1945). The concept of integrative levels and biology. *Science* 101:209–215.

Novitski, M. (1992). *Auguste Laurent and the Prehistory of Valence.* Chur, Switzerland: Harwood Academic Publishers.

Noyes, W. A. (1939). Julius Oscar Stieglitz. *Biog. Mem. NAS* 21:275–314.

Noyes, W. A., and Norris, J. F. (1932). Ira Remsen. *Biog. Mem. NAS* 14:207–257.

Nuland, S. B. (1980). Doctors and historians. *J. Hist. Med.* 43:137–140.

Nutman, P. S. (ed.) (1976). *Symbiotic Nitrogen Fixation in Plants.* Cambridge: Cambridge University Press.

Nuttall, R. H. (1977). Chemical microscopy, 1800–1840. *Chem. Brit.* 13:289–295.

Nye, M. J. (1972). *Molecular Reality.* London: Macdonald.

——— (1977). Nonconformity and creativity: A study of Paul Sabatier, chemical theory, and the French scientific community. *Isis* 68:375–391.

——— (1986). *Science in the Provinces.* Berkeley: University of California Press.

——— (1993a). *From Chemical Philosophy to Theoretical Chemistry.* Berkeley: University of California Press.

——— (1993b). Philosophies of chemistry since the eighteenth century. In *Chemical Sciences in the Modern World* (S. H. Mauskopf, ed.), pp. 3–24. Philadelphia: University of Pennsylvania Press.

——— (1993c). National styles? French and English chemistry in the nineteenth and early twentieth centuries. *Osiris* 8:30–49.

Nygaard, A. P., and Hall, B. D. (1964). Formation and properties of RNA:DNA complexes. *J. Mol. Biol.* 9:125–142.

Obermayer, F., and Pick, E. P. (1906). Ueber die chemischen Grundlagen der Arteigenschaften der Eiweisskörper. Bildung von Immunpräzipitinen durch chemisch veränderte Eiweisskörper. *Wiener med. Wchschr.* 19:327–333.

Obermayer, M., and Lynen, F. (1976). Structure of biotin enzymes. *TIBS* 1:169–171.

Ochoa, S. (1941). "Coupling" of phosphorylation with oxidation in brain. *J. Biol. Chem.* 138:751–773.

——— (1943). Efficiency of aerobic phosphorylation in cell-free heart extracts. *J. Biol. Chem.* 151:493–505.

——— (1954). Enzymic mechanisms in the citric acid cycle. *Adv. Enzymol.* 15:183–270.

——— (1985). Carl Ferdinand Cori. *TIBS* 10:147–150.

——— (1990). Luis Federico Leloir. *Biog. Mem. FRS* 35:203–208.

O'Dell, T. J., Hawkins, R. D., Kandel, E. R., and Arancio, O. (1991). Tests of the roles of two diffusible substances in long-term potentiation: Evidence for nitric oxide as a possible early retrograde messenger. *PNAS* 88:11285–11289.

Oelsner, R. (1989). *August Wilhelm Hofmann*. Berlin: David.

Oertmann, E. (1877). Ueber den Stoffwechsel entbluteter Frösche. *Pflügers Arch.* 15:381–398.

Oesper, P. (1964). The history of the Warburg apparatus: Some reminiscences on its use. *J. Chem. Ed.* 41:294–296.

Oesper, R., and Lemay, P. (1950). Henri Sainte-Claire Deville. *Chymia* 3:205–221.

Ogata, K., Nohara, H., and Morita, T. (1957). The effect of ribonuclease on the amino acid-dependent exchange between labeled inorganic pyrophosphate and ATP by the pH5 enzyme. *Biochim. Biophys. Acta* 26:656–657.

Ogilvie, J. F. (1990). The nature of the chemical bond. *J. Chem. Ed.* 67:280–289.

Ogston, A. G. (1948). Interpretation of experiments on metabolic processes, using isotopic tracer elements. *Nature* 162:963.

Ogston, A. G., and Smithies, O. (1948). Some thermodynamic and kinetic aspects of metabolic phosphorylation. *Physiol. Revs.* 28:283–303.

Ohle, H. (1931). Die Chemie der Monosaccharide und die Glykolyse. *Erg. Physiol.* 33:558–701.

Ohlmeyer, P. (1948). Probleme des Zwischenstoffwechsels. Gedenkblatt für Franz Knoop. *Ang. Chem.* 60:29–35.

Okazaki, R., Okazaki, T., Sakabe, K., Sugimoto, K., and Sugino, A. (1968). Mechanism of DNA chain growth, 1: Possible discontinuity and unusual secondary structure of newly synthesized chains. *PNAS* 59:598–605.

Okazaki, R., Sugimoto, K., Okazaki, T., Imae, Y., and Sugino, A. (1970). DNA chain growth: *In vivo* and *in vitro* synthesis in a DNA polymerasenegative mutant of *E. coli*. *Nature* 228:223–226.

Olby, R. C. (1966). *The Origins of Mendelism*. London: Constable.

——— (1970). Francis Crick, DNA, and the Central Dogma. *Daedalus* 99:938–987.

——— (1971). Schrödinger's problem: What is life? *J. Hist. Biol.* 4:119–148.

——— (1972). Avery in retrospect. *Nature* 239:295–296.

——— (1974). *The Path to the Double Helix*. London: Macmillan.

——— (1979). Mendel no Mendelian? *Hist. Sci.* 17:53–72.

——— (1985). The "mad pursuit": X-ray crystallographers' search for the structure of haemoglobin. *Hist. Phil. Life Sci.* 7:171–193.

——— (1990). Cyril Dean Darlington. *DSB* 17:203–209.

Olby, R., and Posner, E. (1967). An early reference to genetic coding. *Nature* 215:556.

O'Leary, M. H. (1989). Multiple isotope effects on enzyme-catalyzed reactions. *Ann. Rev. Biochem.* 58:377–401.

Olesko, K. M. (1988). On institutes, investigations, and scientific training. In *The Investigative Enterprise in Experimental Physiology* (W. Coleman and F. L. Holmes, eds.), pp. 295–332. Berkeley: University of California Press.

Olesko, K. M., and Holmes, F. L. (1993). Experiment, quantification, and discovery: Helmholtz's early physiological researches, 1843–1850. In *Hermann Helmholtz and the Foundations of Nineteenth-Century Science* (D. Cahan, ed.), pp. 54–74. Berkeley: University of California Press.

Oleson, A., and Voss, J. (eds.) (1979). *The Organization of Knowledge in Modern America, 1860–1920*. Baltimore: Johns Hopkins University Press.

Oliver, G., and Schäfer, E. B. (1895). The physiological action of extracts of the suprarenal capsules. *J. Physiol.* 18:230–275.

Olmsted, J. M. D. (1944). *François Magendie.* New York: Schuman.

——— (1946). *Charles-Edouard Brown-Séquard: A Nineteenth-Century Neurologist and Endocrinologist,* Baltimore: Johns Hopkins University Press.

Olsen, S. (1962). Otto Diels. *Chem. Ber.* 95:v–xlvi.

Olson, R. E. (1979). Discovery of vitamin K. *TIBS* 4:118–120.

Olwell, R. (1996). "Condemned to footnotes": Marxist scholarship in the history of science. *Science and Society* 60:7–26.

Omura, T., and Sato, R. (1964). The carbon monoxide-binding pigment of liver microsomes. *J. Biol. Chem.* 239:2370–2385.

O'Neil, K. T., and DeGrado, W. F. (1990). How calmodulin binds its targets: Sequence dependent recognition of amphiphilic α-helices. *TIBS* 15:59–64.

Onsager, L. (1969). The motion of ions: Principles and concepts. *Science* 166:1359–1364.

Onslow, H. (1915). A contribution to our knowledge of the chemistry of coat colour in animals and dominant and recessive whiteness. *Proc. Roy. Soc.* B89:36–58.

Oparin, A. I. (1938). *The Origin of Life* (S. Morgulis, trans.). New York: Macmillan.

——— (1968). *Genesis and Evolutionary Development of Life* (E. Maass, trans.). New York: Academic.

Oppenheimer, C. (1900). *Die Fermente und ihre Wirkungen.* 5th ed. Leipzig: Vogel.

Oppenheimer, J. M. (1966). Ross Harrison's contributions to experimental embryology. *Bull. Hist. Med.* 40:525–543.

——— (1972). Karl Wilhelm Richard von Hertwig. *DSB* 6:336–337.

Orel, V. (1968). Will the story on "too good" results of Mendel's data continue? *Bioscience* 18:776–778.

——— (1995). *Gregor Mendel: The First Geneticist.* Oxford: Oxford University Press.

Orel, V., and Hartl, D. L. (1994). Controversies in the interpretations of Mendel's discovery. *Hist. Phil. Life Sci.* 16:423–464.

Orgel, L. E., and Crick, F. H. C. (1993). Anticipating an RNA world. Some past speculations on the origin of life: Where are they today? *FASEB J.* 7:238–239.

Orten, J. M. (1956). Eugen Baumann. *J. Nutrition* 58:3–10.

——— (1976). George Raymond Cowgill. *J. Nutrition* 106:1227–1234.

Ortony, A. (ed.) (1970). *Metaphor and Thought.* Cambridge: Cambridge University Press.

Osann, G. (1853). Ueber eine Modification des Wasserstoffs. *J. prakt. Chem.* 48:385–391.

Osborn, H. F. (1931). *Cope: Master Naturalist.* Princeton: Princeton University Press.

Osborne, T. B. (1892). Crystallized vegetable proteins. *Am. Chem. J.* 14:662–689.

——— (1908). Our present knowledge of the plant proteins. *Science* 28:417–427.

——— (1910). Die Pflanzenproteine. *Erg. Physiol.* 10:47–215.

——— (1911). Samuel William Johnson. *Biog. Mem. NAS* 7:204–222.

——— (1913). Heinrich Ritthausen. *Biol. Bull.* 2:335–346.

——— (1924). *The Vegetable Proteins.* 2d ed. London: Longmans, Green.

Osborne, T. B., and Harris, I. F. (1902). Die Nucleinsäure des Weizenembryos. *Z. physiol. Chem.* 36:85–133.

Osborne, T. B., and Jones, D. B. (1910). A consideration of the sources of loss in analyzing the products of protein hydrolysis. *Am. J. Physiol.* 26:305–328.

Osborne, T. B., and Mendel, L. B. (1914). Amino-acids in nutrition and growth. *J. Biol. Chem.* 17:325–349.

Osler, M. J. (ed.) (1991). *Atoms, Pneuma, and Tranquility: Epicurean and Stoic Themes in European Thought.* Cambridge: Cambridge University Press.

Osterhout, W. J. V. (1930). Jacques Loeb. *Biog. Mem. NAS* 13:318–401.

Ostrowski, P. F. (1986). Who discovered vitamins? *Polish Review* 31:171–183.

Ostrowski, W. (1966). Leon Marchlewski. *Folia Biologica* 14:341–356.

Ostwald, W. (1900). Ueber Oxydationen mittels freien Sauerstoffs. *Z. physik. Chem.* 34:248–252.

O'Sullivan, C., and Tompson, F. W. (1890). Invertase: A contribution to the history of an enzyme or unorganized ferment. *J. Chem. Soc.* 57:834–931.

Oswald, A. (1899). Die Eiweisskörper der Schilddrüse. *Z. Physiol. Chem.* 27:14–49.

Oudshoorn, N. (1990). On the making of sex hormones: Research materials and the production of knowledge. *Soc. Stud. Sci.* 20:5–33.

Ovchinnikov, Y. A. (1979). Physico-chemical basis of ion transport through biological membranes: Ionophores and ion channels. *Rur. J. Biochem.* 94:321–336.

Overbeek, J. T. G. (1960). Hugo Rudolph Kruyt. *Proc. Chem. Soc.*, pp. 437–440.

Overton, E. (1899). Ueber die allgemeinen osmotischen Eigenschaften der Zelle, ihre vermuthlichen Ursachen und ihre Bedeutung für die Physiologie. *Viert. Nat. Ges. Zurich* 44:88–125.

——— (1901). *Studien über die Narkose.* Jena: Fischer.

Ozeki, H. (1956). Abortive transduction in purine-requiring mutants of *Salmonella typhimurium.* Washington, D.C.: *Carnegie Institution of Washington Publications*

Pabo, C. O., and Sauer, R. T. (1984). Protein-DNA recognition. *Ann. Rev. Biochem.* 53:293–321.

——— (1992), Transcription factors: Structural families and principles of DNA recognition. *Ann. Rev. Biochem.* 61:1053–1095.

Pace, N. R., and Smith, D. (1990). Ribonuclease P: Function and variation. *J. Biol. Chem.* 265:3587–3590.

Padgett, R. A., Grabowski, P. J., Konarska, M. M., Seiler, S., and Sharp, P. A. (1986). Splicing of messenger RNA precursors. *Ann. Rev. Biochem.* 55:1119–1150.

Page, I. H. (1937). *Chemistry of the Brain.* London: Baillière, Tindall, and Cox.

Page, M. I. (1979). Entropy, binding energy, and enzymic catalysis. *Ang. Chem. Int. Ed.* 16:449–459.

Pagel, W. (1944). The religious and philosophical aspects of van Helmont's science and medicine. *Bull. Hist. Med.* Suppl. 2.

——— (1955). J. B. van Helmont's reformation of the Galenic theory of digestion—and Paracelsus. *Bull. Hist. Med.* 29:563–568.

——— (1956). Van Helmont's ideas on gastric digestion and the gastric acid. *Bull. Hist. Med.* 30:524–536.

——— (1962). The wild spirit (Gas) of van Helmont and Paracelsus. *Ambix* 10:1–13.

——— (1972). Johann Baptista van Helmont. *DSB* 6:253–259.

——— (1982a). *Paracelsus.* 2d ed. Basel: Karger.

——— (1982b). *Johann Baptista van Helmont.* Cambridge: Cambridge University Press.

——— (1984). *The Smiling Spleen: Paracelsianism in Storm and Stress.* New York: Karger.

Pagliotti, A. L., and Santi, D. V. (1977). The catalytic mechanism of thymidylate synthetase. *Bioorganic Chemistry* 1:277–311.

Painter, T. S. (1933). A new method for the study of chromosome rearrangements and the plotting of chromosome maps. *Science* 78:585–586.

Painter, T. S., and Muller, H. J. (1929). Parallel cytology and genetics of induced translocations and deletions in Drosophila. *J. Heredity* 20:287–298.

Pais, A. *Niels Bohr's Times in Physics, Philosophy, and Polity.* Oxford: Clarendon.

Palade, G. E. (1952a). The fine structure of the mitochondrion. *Anat. Rec.* 114:427–451.

——— (1952b). A study of fixation in electron microscopy. *J. Exp. Med.* 95:285–297.

——— (1975). Intracellular aspects of the process of protein synthesis. *Science* 189:347–358.

Palladin, N. (1909). Ueber das Wesen der Pflanzenatmung. *Biochem. Z.* 18:151–206.

Palladino, P. (1994). Wizards and devotees: On the Mendelian theory and the professionalization of agricultural science in Great Britain and the United States, 1880–1930. *Hist. Sci.* 32:409–444.

Palmer, R. M. J., Ferrige, M., and Moncada, S. (1987). Nitric oxide release accounts for the biological activity of endothelium-derived relaxing factor. *Nature* 327:524–526.

Palmer, W. G. (1965). *A History of the Concept of Valency to 1930.* Cambridge: Cambridge University Press.

Paneth, F. A. (1962). The epistemological status of the chemical concept of the element. *Brit. J. Phil. Sci* 13:1–14, 144–169.

Pang, A. S. K. (1989). Oral history and the history of science: A review essay with speculation. *Journal of Oral History* 10:270–285.

Paolini, C. (1968). *Justus Liebig, eine Bibliographie sämtlicher Veröffentlichtungen.* Heidelberg: Carl Winter.

Parascandola, J. (1971a). Organismic and holistic concepts in the thought of L. J. Henderson. *J. Hist. Biol.* 4:63–113.

——— (1971b). L. J. Henderson and the theory of buffer action. *Med. Hist. J.* 6:297–309.

——— (1974). Dinitrophenol and bioenergetics: A historical perspective. *Mol. Cell. Biochem.* 5:69–77.

——— (1975). The evolution of stereochemical concepts in pharmacology. In *Van't Hoff–LeBel Centennial* (O. B. Ramsay, ed.), pp. 143–158. Washington, D.C.: American Chemical Society.

——— (ed.) (1980). *The History of Antibiotics.* Madison, Wis.: American Institute for the History of Pharmacy.

——— (1981). The theoretical basis of Paul Ehrlich's chemotherapy. *J. Hist. Med.* 36:19–43.

——— (1982). John J. Abel and the early development of pharmacology at Johns Hopkins University. *Bull. Hist. Med.* 56:512–527.

——— (1986). The development of receptor theory. In *Discoveries in Pharmacology* (M. J. Parnham and J. Bruinvels, eds.), 3:129–156. Amsterdam: Elsevier.

——— (1992). *The Development of American Pharmacology: John J. Abel and the Shaping of a Discipline.* Baltimore: Johns Hopkins University Press.

Pardee, A. B. (1959). Mechanisms for control of enzyme synthesis and enzyme activity in bacteria. In *Regulation of Cell Metabolism* (G. E. W. Wolstenholme and C. M. O'Connor, eds.), pp. 295–304. Boston: Little, Brown.

Pardee, A. B., Jacob, F., and Monod, J. (1959). The genetic control and cytoplasmic expression of "inducibility" in the synthesis of β-galactosidase of *E. coli. J. Mol. Biol.* 1:165–178.

Parkinson, E. M. (1976). George Gabriel Stokes. *DSB* 13:74–79.

Parnas, J. (1910). Ueber fermentative Beschleunigung der Cannizzaroschen Aldehydumlagerung durch Gewebesaft. *Biochem. Z.* 28:274–294.

——— (1937). Der Mechanismus der Glycogenolyse im Muskel. *Erg. Enzymforsch.* 6:57–110.

——— (1943). Coenzymatic reactions. *Nature* 151:577–580.

Parnas, J. (1981). Thorvald Madsen, leader in international public health. *Danish Medical Bulletin* 28:82–86.

Parnas, J., and Baranowski, T. (1935). Sur les phosphorylations initiales du glycogène. *Compt. Rend. Soc. Biol.* 120:307–310.

Parnas, J., Ostern, P., and Mann, T. (1934). Ueber die Verkettung der chemischen Vorgänge im Muskel. *Biochem. Z.* 272:64–70.

Partington, J. R. (1956). The life and work of John Mayow. *Isis* 47:217–230.

——— (1960). Joseph Black's "Lectures on the Elements of Chemistry." *Chymia* 6:27–67.

——— (1961–1970). *A History of Chemistry.* London: Macmillan.

Partington, J. R., and McKee, D. (1937–1939). Historical studies on the phlogiston theory. *Ann. Sci.* 2:361–404; 3:1–58, 337–371; 4:113–149.

Partridge, S. M., and Blombäck, B. (1979). Pehr Edman. *Biog. Mem. FRS* 25:241–265.

Pascal, P. (1937). Notice sur la vie et les travaux de Henry Le Châtelier. *Bull. Soc. Chim.* [5]4:1557–1611.

Pascal, R. (1960). *Design and Truth in Autobiography.* Cambridge: Harvard University Press.

Pasteur, L. (1857). Mémoire sur la fermentation appelée lactique. *Compt. Rend.* 45:913–916.

——— (1858a). Mémoire sur la fermentation appelée lactique. *Ann. Chim.* [3]52:404–418.

——— (1858b). Nouveaux faits concernant l'histoire de la fermentation alcoolique. *Compt. Rend.* 47:1011–1013.

——— (1860a). Mémoire sur la fermentation alcoolique. *Ann. Chim.* [3]58:323–426.

——— (1860b). Recherches sur la dissymétrie moléculaire des produits organiques naturels. In *Leçons de Chimie professés au 1860 . . .*, pp. 1–48. Paris: Hachette.

——— (1860c). Note sur la fermentation alcoolique. *Compt. Rend.* 50:1083–1084.

——— (1861). Expériences et vues nouvelles sur la nature des fermentations. *Compt. Rend.* 52:1260–1264.

——— (1862). Mémoire sur les corpuscules organisées qui existent dans l'atmosphère. Examin de la doctrine des generations spontanées. *Ann. Chem.* [3]64:5–110.

——— (1871). Note sur un mémoire de M. Liebig relatif aux fermentations. *Compt. Rend.* 73:1419–1424.

——— (1876a). Sur la fermentation de l'urine. *Compt. Rend.* 83:5–8.

——— (1876b). Réponse a M. Berthelot. *Compt. Rend.* 83:10.

——— (1876c). *Etudes sur la Bière.* Paris: Gauthier-Villars.

——— (1878). Première réponse a M. Berthelot. *Compt. Rend.* 87:1053–1058.

——— (1879). *Examen critique d'un écrit postume de Claude Bernard sur la fermentation.* Paris: Gauthier-Villars.

——— (1886). La dissymétrie moléculaire. In *Conférences faites a la Société Chimique*, pp. 24–37. Paris: Hachette.

——— (1922–1939). *Oeuvres de Pasteur.* 7 vols. Paris: Masson.

Patai, R. (1994). *The Jewish Alchemists.* Princeton: Princeton University Press.

Patin, G. (1846). *Lettres* (J. H. Reveille-Parisse, ed.). Paris: Baillière.

Patterson, T. S. (1937). Jean-Beguin and his "Tyrocinium Chymicum." *Ann. Sci.* 2:243–298.

Paul, E. R. (1978). Alexander W. Williamson on the atomic theory: A study of nineteenth-century atomism. *Ann. Sci.* 35:17–31.

Paul, H. W. (1972a). *The Sorcerer's Apprentice: The French Scientist's Image of German Science, 1840–1919.* Gainesville: University of Florida Press.

——— (1972b). The issue of decline in nineteenth-century French science. *French Historical Studies* 7:416–450.

——— (1976). Scholarship and ideology: The chair of the general history of science at the Collège de France, 1892–1913. *Isis* 67:376–397.

——— (1978). Emile Meyerson. *DSB* 15:422–425.

——— (1985). *From Knowledge to Power: The Rise of the Science Empire in France, 1860–1939*. Cambridge: Cambridge University Press.

——— (1990). Die Entwicklung der Forschungsförderung im modernen Frankreich. In *Forschung im Spannungsfeld von Politik und Gesellschaft* (R. Vierhaus and B. von Brocke, eds.), pp. 695–725. Stuttgart: Deutsche Verlagsanstalt.

——— (1996). *Science, Vine, and Wine in Modern France.* Cambridge: Cambridge University Press.

Paul, J. R., and Long, C. N. H. (1958). John Punnett Peters. *Biog. Mem. NAS* 31:347–375.

Pauli, W. (1907). *Physical Chemistry at the Service of Medicine.* New York: Wiley.

——— (1934). The chemistry of amino acids and proteins. *Ann. Rev. Biochem.* 3:111–132.

Pauling, L. (1939). *The Nature of the Chemical Bond.* Ithaca, N.Y.: Cornell University Press.

——— (1940). A theory of the structure and process of formation of antibodies. *J. Am. Chem. Soc.* 62:2643–2657.

——— (1946). Molecular architecture and biological reactions. *Chem. Eng. News* 24:1375–1377.

——— (1948). Nature of the forces between large molecules of biological interest. *Nature* 161:707–709.

——— (1955). The stochastic method and the structure of proteins. *Am. Sci.* 43:285–297.

——— (1958). Arthur Amos Noyes. *Biog. Mem. NAS* 31:322–346.

——— (1970). Fifty years of progress in structural chemistry and molecular biology. *Daedalus* 99:988–1014.

——— (1971). Roscoe Gilkey Dickinson. *DSB* 4:82.

——— (1974). Molecular basis of biological specificity. *Nature* 248:769–771.

——— (1996). The discovery of the alpha-helix. *Chem. Intell.* 2:32–38.

Pauling, L., and Campbell. D. H. (1942). The manufacture of antibodies in vitro. *J. Exp. Med.* 76:211–230; *Science* 95:440–441.

Pauling, L., and Corey, R. B. (1950). Two hydrogen-bonded spiral configurations of the polypeptide chain. *J. Am. Chem. Soc.* 72:5349.

——— (1951). The structure of proteins. *PNAS* 37:205–211, 275–285.

——— (1953). A proposed structure for the nucleic acids. *PNAS* 39:84–97.

Pauling, L., and Delbrück, M. (1940). The nature of the intermolecular forces operative in biological processes. *Science* 92:77–79.

Pauling, L., and Niemann, C. (1939). The structure of proteins. *J. Am. Chem. Soc.* 61:1860–1867.

Pauly, P. J. (1984). The appearance of academic biology in later nineteenth-century America. *J. Hist. Biol.* 17:369–397.

——— (1987a). *Controlling Life: Jacques Loeb and the Engineering Ideal in Biology.* New York: Oxford University Press.

——— (1987b). General physiology and the discipline of physiology. In *Physiology in the American Context* (G. L. Geison, ed.), pp. 195–207. Bethesda, Md.: American Physiological Society.

Pavlov, I. P. (1902). *The Work of the Digestive Glands* (W. H. Thompson, trans.). London: Griffin.

Payen, A., and Persoz, J. F. (1833). Mémoire sur la diastase, les principaux produits de ses reactions, et leurs applications aux arts industriels. *Ann. Chim.* [2]53:73–91.

Peacocke, A. R. (1983). *An Introduction to the Physical Chemistry of Biological Organization.* Oxford: Oxford University Press.

Pearson, K. (1906). Walter Frank Raphael Weldon. *Biometrika* 5:1–50.

Pedersen, J. (1941). La Fondation Carlsberg. *Le Nord* 2/3:176–204.

Pedersen, K. O. (1983). The Svedberg and Arne Tiselius: The early development of modern protein chemistry at Uppsala. In *Selected Topics in the History of Biochemistry: Personal Recollections* (G. Semenza, ed.), pp. 233–281. Amsterdam: Elsevier.

Pedersen, P. L., and Amzel, L. M. (1993). ATP synthases. *J. Biol. Chem.* 268:9937–9940.

Pekelharing, C. A. (1902). Mittheilungen über Pepsin. *Z. physiol. Chem.* 35:8–30.

Pelikan, J. (1984). Scholarship: A sacred vocation. *Scholarly Publishing* 16:3–22.

Pellegrin, P. (1987). Logical difference and biological difference: The unity of Aristotle's thought. In *Philosophical Issues in Aristotle's Biology* (A. Gotthelf and J. E. Lennox, eds.), pp. 313–338. Cambridge: Cambridge University Press.

Pelletier, J., and Caventou, J. B. (1817). Sur la matière verte des feuilles. *Journal de Pharmacie* 3:486–491.

Pelling, M. (1978). *Cholera, Fever, and English Medicine, 1825–1865.* Oxford: Oxford University Press.

Pelz, W. (1987). *Zellenlehre: Der Einfluss Hugo von Mohls auf die Entwicklung der Zellenlehre.* Frankfurt: Peter Lang.

Penefsky, H. S., and Cross, R. L. (1991). Structure and mechanism of F_0F_1-type ATP synthases and ATPases. *Adv. Enzymol.* 64:173–214.

Peng Loh, Y. (ed.) (1992). *Mechanisms of Intracellular Trafficking and Processing of Proproteins.* Boca Raton, Fla.: CRC.

Penman, S. (1995). Rethinking cell structure. *PNAS* 92:5251–5257.

Pérez-Ramos, A. (1988). *Francis Bacon's Idea of Science and the Maker's Knowledge Tradition.* Oxford: Oxford University Press.

Perkin, W. H. (1896). The origin of the coal-tar colour industry, and the contributions of Hofmann and his pupils. *J. Chem. Soc.* 69:596–637.

——— (1905). Johannes Wislicenus. *J. Chem. Soc.* 87:501–534.

——— (1923). Baeyer memorial lecture. *J. Chem. Soc.* 123:1520–1546.

Perkins, D. D. (1992). Neurospora: The organism behind the molecular revolution. *Genetics* 130:687–701.

Perkins, J. P. (1973). Adenyl cyclase. *Prog. Nucleic Acid Res.* 3:1–64.

Perlman, R. L. (1977). Homeostasis. *TIBS* 2:259.

Perlmann, G. (1955). The nature of phosphorus linkages in proteins. *Adv. Protein Chem.* 10:1–30.

Perone, N. (1993). The history of steroidal contraceptive development: The progestins. *Persp. Biol. Med.* 36:347–368.

Perrin, C. E. (1987). Revolution or reform: The Chemical Revolution and eighteenth-century concepts of scientific change. *Hist. Sci.* 25:395–423.

——— (1989). Document, text, and myth: Lavoisier's crucial year revisited. *Brit. J. Hist. Sci.* 22:3–25.

Perry, S. V. (1989). The interaction of actin with myosin 40 years ago. *Biochim. Biophys. Acta* 1000:159–162.

Perutz, M. (1970). Stereochemistry of cooperative effects in haemoglobin: The Bohr effect and combination with organic phosphates. *Nature* 228:728–739.

——— (1980). The origins of molecular biology. *New Scientist* 85:326–329.

——— (1985). Early days of protein crystallography. *Methods in Enzymology* 114:3–18.

——— (1986). Keilin and the Molteno. *Cambridge Review* 107:152–156.

——— (1989). *Is Science Necessary?* New York: Dutton.

——— (1990a). Physics and the riddle of life. In *Selected Topics in the History of Biochemistry: Personal Recollections* (G. Semenza and L. Jaenicke, eds.), pp. 1–20. Amsterdam: Elsevier.

——— (1990b). *Mechanisms of Cooperativity and Allosteric Regulation in Proteins.* Cambridge: Cambridge University Press.

——— (1994). Polar zippers: Their role in human disease. *Protein Science* 3:1629–1637.

——— (1995). Review of *The Private Science of Louis Pasteur. New York Review of Books,* 21 December, pp. 54, 56–58.

Perutz, M. F., Wilkins, M. H. F., and Watson, J. D. (1969). DNA helix. *Science* 164:1537–1539.

Peterkofsky, A., Reizer, A., Reizer, J., Gollop, N., Peng-Ping, Z., and Amin, N. (1993). Bacterial adenyl cyclases. *Prog. Nucleic Acid Res.* 44:32–65.

Petermann, M. L. (1964). *Physical and Chemical Properties of Ribosomes.* Amsterdam: Elsevier.

Peters, G. (1968). *Walther Flemming, sein Leben und Werk.* Neumünster: Wachholz.

Peters, R. (1898). Ueber Oxydations- und Reduktionsketten und den Einfluss komplexer Ionen auf ihre elektromotorische Kraft. *Z. physik. Chem.* 26:193–236.

Peters, R. A. (1930). Surface structure in the integration of cell activity. *Trans. Faraday Soc.* 26: 797–807.

——— (1950). *Biochemical Lesions and Lethal Synthesis.* Oxford: Pergamon.

——— (1958). John Beresford Leathes. *Biog. Mem. FRS* 4:185–191.

Petit, C. (1990). Georges Teissier. *DSB* 18:901–904.

Petrunkevitch, A. (1920). Russia's contribution to science. *Trans. Conn. Acad. Sci.* 23:211–241.

Pettenkofer, M., and Voit, C. (1866). Untersuchungen über den Stoffverbrauch des normalen Menschen. *Z. Biol.* 2:459–573.

Pettijohn, D. E. (1988). Histone-like proteins and bacterial chromosome structure. *J. Biol. Chem.* 263:12793–12796.

Peumery, J. J. (1986). Charles-Louis-Arthur Barreswil (1817–1870) et les sciences médicales. *Hist. Sci. Med.* 20:243–248.

Peyer, U. (1972). *Rudolf Schönheimer (1898–1941) und das Beginn der Tracer-Technik bei Stoffwechseluntersuchungen.* Zurich: Juris.

Pfeffer, W. (1872). Untersuchungen über die Proteinkörper und die Bedeutung des Asparagins beim Keimen der Samen. *Jahrb. wiss. Bot.* 8:429–574.

——— (1877). *Osmotische Untersuchungen.* Leipzig: Engelmann.

——— (1881). *Pflanzenphysiologie.* Leipzig: Engelmann.

Pfetsch, F. (1970). Scientific organisation and science policy in Imperial Germany. *Minerva* 8:557–580.

——— (1974). *Zur Entstehung der Wissenschaftspolitik in Deutschland, 1750–1914.* Berlin: Duncker and Humblot.

Pflüger, E. (1872). Ueber die Diffusion des Sauerstoffs, den Ort und die Gesetze der Oxydationsprocesse im thierischen Organismus. *Pflügers Arch.* 6:43–64.

——— (1875). Beiträge zur Lehre von der Respiration, 1: Ueber die physiologische Verbrennung in dem lebenden Organismen. *Pflügers Arch.* 10:251–369, 641–644.

——— (1877). Die Physiologie und ihre Zukunft. *Pflügers Arch.* 15:361–365.

——— (1878). Ueber Wärme und Oxydation der lebenden Materie. *Pflügers Arch.* 18:247–380.

——— (1893). Ueber einige Gesetze des Eiweisstoffwechsels. *Pflügers Arch.* 54:333–419.

——— (1903). Glykogen. *Pflügers Arch.* 96:1–398.

——— (1910). Nachschrift. *Pflügers Arch.* 131:302–305.

Philippi, E. (1962). Fritz Pregl. *Microchem. J.* 6:5–16.

Phillips, D., et al. (1989). *The Evaluation of Scientific Research.* Chichester: Wiley.

Phillips, D. C. (1972). On the stereochemical basis of enzyme action: Lessons from lysozyme. *Harvey Lectures* 66:135–160.

——— (1986). Development of concepts of protein structure. *Persp. Biol. Med.* 29:124–130.

Piccard, J. (1874). Ueber Protamin, Guanin und Sarkin als Bestandteile des Lachsspermas. *Ber. chem. Ges.* 7:1714–1719.

Picco, G. (1981). *Das Biochemische Institut der Universität Zürich.* Aarau: Sauerländer.

Pickart, C. M., and Jencks, W. P. (1984). Energetics of the calcium-transporting ATPase. *J. Biol. Chem.* 259:1629–1643.

Pickels, E. G. (1942). The ultracentrifuge. *Chem. Revs.* 30:341–355.

Pickering, A. (1984). *Constructing Quarks.* Chicago: University of Chicago Press.

——— (1995). *The Mangle of Practice.* Chicago: University of Chicago Press.

Pickles, S. S. (1910). The constitution and synthesis of caoutchouc. *J. Chem. Soc.* 97:1085–1090.

Pickstone, J. V. (1973). Globules and coagula: Concepts of tissue formation in the early nineteenth century. *J. Hist. Med.* 28:336–356.

——— (1976). Vital actions and organic physics: Henri Dutrochet and French physiology during the 1820s. *Bull. Hist. Med.* 50:191–212.

——— (1995). Past and present knowledges in the practice of the history of science. *Hist. Sci.* 33:203–224.

Picton, H., and Linder, S. E. (1892). Solution and pseudo-solution. *J. Chem. Soc.* 61:148–172.

Pierson, G. W. (1952, 1955). *Yale College: An Educational History.* New Haven: Yale University Press.

Pietsch, E. (1937). Sinn und Aufgaben der Geschichte der Chemie. *Ang. Chem.* 50:939–948.

Pietsch, E., and Beyer, E. (1939). Leopold Gmelin. *Ber. chem. Ges.* 72:5–33.

Pigman, W. W. (1944). Specificity, classification, and mechanism of the glycosidases. *Adv. Enzymol.* 4:41–74.

Pilet, P. E. (1962). Jean Senebier, un des précurseurs de Claude Bernard. *Arch. Int. Hist. Sci.* 15:303–313.

——— (1975). Nicolas-Théodore de Saussure. *DSB* 12:123–124.

Pilkington, A. E. (1976). *Bergson and His Influence.* Cambridge: Cambridge University Press.

Pinch, T. (1986). *Confronting Nature.* Dordrecht: Reidel.

Pincus, G., and Thimann, K. V. (eds) (1948). *The Hormones.* New York: Academic.

Piria, R. (1844). Note sur l'asparagine. *Compt. Rend.* 19:575–577.

Pirie, N. W. (1937). The meaninglessness of the terms life and living. In *Perspectives in Biochemistry* (J. Needham and D. E. Green, eds.) pp. 11–22. Cambridge: Cambridge University Press.

——— (1940). The criteria of purity used in the study of large molecules of biological origin. *Biol. Revs.* 15:377–404.

——— (1962). Patterns of assumptions about large molecules. *Arch. Biochem. Biophys.* Suppl. 1:21–29.

——— (1966). John Burdon Sanderson Haldane. *Biog. Mem. FRS* 12:219–249.

——— (1970). Retrospect on the biochemistry of viruses. *Biochem. Soc. Symp.* 30:43–56.

——— (1983). Sir Frederick Gowland Hopkins. In *Selected Topics in the History of Biochemistry* (G. Semenza, ed.), pp. 103–128. Amsterdam: Elsevier.

——— (1984). The false dawn of nucleic acids. *TIBS* 9:35–36.

——— (1986). Recurrent luck in research. In *Selected Topics in the History of Biochemistry* (G. Semenza, ed.), pp. 491–522. Amsterdam: Elsevier.

——— (1990). Jean-Louis Brachet. *Biog. Mem. FRS* 36:85–99.

Pittendrigh, C. S. (1958). Adaptation, natural selection, and behavior. In *Behavior and Evolution* (A. Roe and G. G. Simpson, eds.), pp. 390–417. New Haven: Yale University Press.

Pitt-Rivers, R. (1978). The thyroid hormones: Historical aspects. *Hormonal Peptides and Hormones* 6:391–422.

Plantefol, L. (1968). Le genre du mot enzyme. *Compt. Rend.* 266C:41–46.

Platt, B. S. (1956). Sir Edward Mellanby. *Ann. Rev. Biochem.* 25:1–28.

Plimmer, R. H. A. (1908). *The Constitution of the Proteins.* Part 2. London: Longmans, Green.

Podolsky, S. (1996). The role of the virus in origin-of-life theorizing. *J. Hist. Biol.* 29:79–126.

Poggi, S., and Bossi, M. (1994). *Romanticism in Science.* Dordrecht: Kluwer.

Pohl, J. (1893). Ueber die Oxydation des Methyl- und Aethylalcohols im Thierkörper. *Arch. exp. Path. Pharm.* 31:281–302.

——— (1896). Ueber den oxydative Abbau der Fettsäuren im thierischem Organismus. *Arch. exp. Path. Pharm.* 37:413–425.

Pohl, J., and Spiro, K. (1923). Franz Hofmeister, sein Leben und Wirken. *Erg. Physiol.* 22:1–50.

Poirier, J. P. (1993). *Antoine Laurent de Lavoisier.* Paris: Pygmalion.

Polanyi, M. (1921). Die chemische Konstitution der Zellulose. *Naturwiss.* 9:288.

Polgar, L. (1987). Structure and function of serine proteases. In *Hydrolytic Enzymes* (A. Neuberger and K. Brocklehurst, eds.), pp. 159–200. Amsterdam: Elsevier.

Pollack, S. J., Jacobs, J. W., and Schultz, P. G. (1986). Selective chemical catalysis by an antibody. *Science* 234:1570–1573.

Pollard, T. D., and Cooper, J. A. (1986). Actin and actin-binding proteins: A critical evaluation of mechanisms and functions. *Ann. Rev. Biochem.* 55:987–1035.

Pollister, A. W., and Leuchtenberger, C. (1949). The nucleoprotein content of whole nuclei. *PNAS* 35:66–71.

Pollock, M. R. (1950). Penicillinase adaptation in *B. cereus:* Adaptive enzyme formation in the absence of free substrate. *Brit. J. Exp. Pathol.* 31:739–753.

——— (1951). The relation between fixation of penicillin and penicillinase adaptation in *B. cereus. Brit. J. Exp. Pathol.* 32:387–396.

——— (1953). Stages in enzyme adaptation. *Symp. Gen. Microbiol. Soc.* 3:150–183.

——— (1970). The discovery of DNA: An ironic tale of chance, prejudice, and insight. *J. Gen. Microbiol.* 63:1–20.

——— (1976). From pangens to polynucleotides: The evolution of ideas on the mechanism of biological replication. *Persp. Biol. Med.* 19:455–472.

Polonovski, J. (1993). Eloge de Jean Roche. *Bull. Acad. Med.* 177:355–361.

Pontecorvo, G. (1952). Genetic formulation of gene structure and gene action. *Adv. Enzymol.* 13:121–149.

Pontremoli, S., and Melloni, E. (1986). Extralysosomal protein degradation. *Ann. Rev. Biochem.* 455–481.

Popielski, L. (1901). Ueber das peripherisches reflectorische Nervencentrum des Pankreas. *Pflügers Arch.* 86:215–246.

Popjak, G., and Cornforth, J. W. (1960). The biosynthesis of cholesterol. *Adv. Enzymol.* 22:281–335.

Popkin, R. H. (1970). Scepticism and the study of history. In *Physics, Logic, and History* (W. Yourgrau and A. D. Breck, eds.), pp. 209–230. New York: Plenum.

Popper, K. R. (1959). *The Logic of Scientific Discovery.* London: Hutchinson.

——— (1962). *Conjectures and Refutations.* London: Routledge.
——— (1972). *Objective Knowledge.* Oxford: Oxford University Press.
Porra, R. J., and Meisch, H. U. (1984). The biosynthesis of chlorophyll. *TIBS* 9:99–104.
Porse, A. T., and Garrett, R. A. (1995). Mapping important nucleotides in the peptidyl transferase centre of 23S rRNA using a random mutagenesis approach. *J. Mol. Biol.* 249:1–10.
Porter, R. R. (1973). Structural studies of immunoglobulins. *Science* 180:713–716.
Porter, T. D., and Coon, M. J. (1991). Cytochrome P-450. *J. Biol. Chem.* 266:13469–13472.
Portin, P. (1993). The concept of the gene: Short history and present status. *Quart. Rev. Biol.* 68:173–223.
Portmann, M. L. (1974). Neue Aspekte zur Biographie des Basler Biochemikers Gustav von Bunge (1844–1920) aus seinem handschriflichen Nachlass. *Gesnerus* 31:39–46.
Potter, V. H., and Heidelberger, C. (1949). Biosynthesis of "asymmetric" citric acid: A substantiation of the Ogston concept. *Nature* 164:180–181.
Pouchet, G. (1886). Robin, sa vie, son oeuvre. *J. Anat. Physiol.* 22:i–clxv.
Prabhakar, N. R., Dinerman, J. L., Agassi, F. H., and Snyder, S. H. (1995). Carbon monoxide: A role in carotid body chemoreception. *PNAS* 92:1994–1997.
Prandtl, W. (1949). Das chemische Laboratorium der Bayerischen Akademie der Wissenschaften in München. *Chymia* 2:81–97.
——— (1952). *Die Geschichte des chemischen Laboratorium der Bayerischen Akademie der Wissenschaften in München.* Weinheim: Verlag Chemie.
Pratt, J. H. (1954). A reappraisal of researches leading to the discovery of insulin. *J. Hist. Med.* 9:281–289.
Prelog, V., and Jeger, O. (1980). Leopold Ruzicka. *Biog. Mem. FRS* 26:411–501.
Pressman, B. C. (1965). Induced active transport of ions in mitochondria. *PNAS* 53:1076–1083.
Prévost, J. L., and Dumas, J. B. (1823). Examen du sang et de son action dans les divers phénomènes de la vie. *Ann. Chim.* [2]23:90–104.
Preyer, W. (1871). *Die Blutkrystalle.* Jena: Fischer.
Prianishnikov, D. (1904). Zur Frage der Asparaginbildung. *Ber. bot. Ges.* 22:35–43.
——— (1924). Asparagin und Harnstoff. *Biochem. Z.* 150:407–423.
Price, D. (1974). Carl Richard Moore. *Biog. Mem. NAS* 45:385–412.
Price, D. J. deS. (1963). *Little Science, Big Science.* New York: Columbia University Press.
Price, W. H., Cori, C. F., and Colowick, S. P. (1945). The effect of anterior pituitary extract and of insulin on the hexokinase reaction. *J. Biol. Chem.* 160:633–634.
Priestley, J. (1774). *Experiments and Observations on Different Kinds of Air.* London: Johnson.
Prigogine, I., and Stengers, I. (1979). *La Nouvelle Alliance: Métamorphose de la science.* Paris: Gallimard.
Primas, H. (1983). *Chemistry, Quantum Mechanics, and Reductionism.* 2d ed. Berlin: Springer.
Prince, R. C. (1996). Photosynthesis: The Z-scheme revisited. *TIBS* 21:121–122.
Pringsheim, E. G. (1932). *Julius Sachs, der Begründer der neuen Pflanzenphysiologie.* Jena: Fischer.
Prins, J. A. (1976). Johannes Diderik van der Waals. *DSB* 14:109–111.
Proctor, D. F. (ed.) (1995). *A History of Breathing Physiology.* New York: Marcel Dekker.
Pross, H. (1955). *Die Deutsche Akademische Emigration nach den Vereinigten Staaten 1933–1941.* Berlin: Duncker und Humblot.
Prusiner, S. B. (1982). Novel proteinaceous infectious particles cause scrapie. *Science* 216:136–144.
——— (1991). Molecular biology of prion diseases. *Science* 252:1515–1522.
Przibram, H. (1926). *Die anorganische Grenzgebiete der Biologie.* Berlin: Bornträger.

Ptitsyn, O. B. (1995). Molten globule and protein folding. *Adv. Protein Chem.* 47:83–229.

Puck, T. T., and Marcus, P. I. (1955). A rapid method for viable cell titration and clone production with Hela cells in tissue culture. *PNAS* 41:432–437.

Pullman, M. E., Penefsky, H. S., Datta, A., and Racker, E. (1960). Partial resolution of enzymes catalyzing oxidative phosphorylation, 1: Purification and properties of soluble dinitrophenol-stimulated adenosine triphosphatase. *J. Biol. Chem.* 190:685–696.

Pumphrey, S. (1995). Who did the work? Experimental philosophers and public demonstrators in Augustan England. *Brit. J. Hist. Sci.* 28:131–156.

Purich, D. L, Fromm, H. J., and Rudolph, F. B. (1973). The hexokinases: Kinetic, physical, and regulatory properties. *Adv. Enzymol.* 39:249–326.

Putnam, F. W. (1969). Immunoglobulin structure: Variability and homology. *Science* 163:633–644.

——— (1993). Henry Bence Jones: The best chemical doctor in London. *Persp. Biol. Med.* 36:565–579.

——— (1994). Felix Haurowitz. *Biog. Mem. NAS* 64:135–163.

Pyenson, L. (1977). "Who the guys were": Prosopography in the history of science. *Hist. Sci.* 15:155–188.

Quagliariello, G. (1944). Filippo Bottazzi. *Erg. Physiol.* 45:16–33.

Quane, D. (1990). The reception of hydrogen bonding by the chemical community, 1920–1937. *Bull. Hist. Chem.* 7:3–13.

Quastel, J. H. (1926). Dehydrogenations produced by bacteria, 4: A theory of the mechanisms of oxidation and reduction *in vivo*. *Biochem. J.* 20:166–193.

——— (1974). Fifty years of biochemistry: A personal account. *Can. J. Biochem.* 52:71–82, 433.

——— (1977). The constitution and synthesis of thyroxine. *TIBS* 2:69.

Quastel, J. H., and Whetham, M. D. (1924). The equilibria existing between succinic, fumaric, and malic acids in the presence of resting bacteria. *Biochem. J.* 18:519–534.

Quastel, J. H., and Wooldridge, W. R. (1927). Experiments on bacteria in relation to the mechanism of enzyme action. *Biochem. J.* 21:1224–1251.

Quastler, H. (ed.) (1953). *Essays on the Use of Information Theory in Biology.* Urbana: University of Illinois Press.

Querner, H. (1972a). Probleme der Biologie um 1900 auf der Versammlung der Deutschen Naturforscher und Aerzte. In *Wege der Naturforschung, 1822–1972* (H. Querner and H. Schipperges, eds.), pp. 186–202. Berlin: Springer.

——— (1972b). Wilhelm His. *DSB* 6:434–436.

Queruel, A. (1994). *Vauquelin et son temps.* Paris: L'Harmattan.

Quine, W. V. O. (1960). *Word and Object.* Cambridge: MIT Press.

——— (1969). *Ontological Relativity and Other Essays.* New York: Columbia University Press.

——— (1976). *The Ways of Paradox.* Cambridge: Harvard University Press.

Raacke, I. D. (1983). Herbert McLean Evans. *J. Nutrition* 113:927–943.

Rabinow, P. (1996). *Making PCR.* Chicago: University of Chicago Press.

Rabinowitz, M., and Lipmann, F. (1960). Reversible phosphate transfer between yolk phosphoprotein and adenosine triphosphate. *J. Biol. Chem.* 235:1043–1050.

Racker, E. (1954). Formation of acyl and carbonyl complexes associated with electron-transport and group-transfer reactions. In *Mechanisms of Enzyme Action* (W. D. McElroy and B. Glass, eds.), pp. 464–470. Baltimore: Johns Hopkins University Press.

——— (1955). Mechanism of action and properties of pyridine nucleotide-linked enzymes. *Physiol. Revs.* 35:1–56.

——— (1965). *Mechanisms in Bioenergetics.* New York: Academic.

——— (1967). Resolution and reconstitution of the inner mitochondrial membrane. *Fed. Proc.* 26:1335–1340.

——— (1972). Reconstitution of a calcium pump with phospholipids and a purified adenosine triphosphatase from sarcoplasmic reticulum. *J. Biol. Chem.* 247:8198–8200.

——— (1974). History of the Pasteur effect and its pathobiology. *Mol. Cell. Biochem.* 5:17–23.

——— (1976). *A New Look at Mechanisms in Bioenergetics.* New York: Academic.

——— (1980). From Pasteur to Mitchell: A hundred years of bioenergetics. *Fed. Proc.* 39:210–215.

——— (1983). Resolution and reconstitution of biological pathways from 1919 to 1984. *Fed. Proc.* 42:2899–2909.

——— (1985). *Reconstitution of Transporters, Receptors, and Pathological States.* Orlando, Fla.: Academic.

——— (1988). Regulation of function of membrane proteins by phosphorylation and dephosphorylation. In *The Roots of Modern Biochemistry* (H. Kleinkauf, H. van Döhren, and L. Jaenicke, eds.), pp. 295–304. Berlin: Walter de Gruyter.

Racker, E., and Horstman, L. L. (1967). Partial resolution of the enzymes catalyzing oxidative phosphorylation, 13: Structure and function of submitochondrial particles completely resolved with respect to coupling factor 1. *J. Biol. Chem.* 242:2547–2551.

Racker, E., and Krimsky, I. (1952). The mechanism of oxidation of aldehydes by glyceraldehyde-3-phosphate dehydrogenase. *J. Biol. Chem.* 248:731–743.

Racker, E., and Racker, F. W. (1981). Resolution and reconstitution: A dual autobiographical sketch. In *Of Oxygen, Fuels, and Living Matter* (G. Semenza, ed.), pp. 265–323. Chichester: Wiley.

Racker, E., and Stoeckenius, W. (1974). Reconstitution of the purple membrane vesicles catalyzing light-driven proton uptake and adenosine triphosphate function. *J. Biol. Chem.* 249:662–663.

Radding, C. M. (1982). Homologous pairing and strand exchange in genetic recombination. *Ann. Rev. Genetics* 16:405–437.

——— (1991). Helical interactions in homologous pairing and strand exchange driven by recA protein. *J. Biol. Chem.* 266:5355–5358.

Radman, M., and Wagner, R. (1988). The high fidelity of DNA replication. *Sci. Am.* 259(2):40–46.

Rafaelson, O. J. (1982). Thudichum: The founder of neurochemistry. In *Historical Aspects of the Neurosciences* (F. C. Rose and W. F. Bynum, eds.), pp. 293–305. New York: Raven.

Rainger, R., Benson, K. R., and Maienschein, J. (eds.) (1988). *The American Development of Biology.* Philadelphia: University of Pennsylvania Press.

Rajagopalan, K. V., and Johnson, J. L. (1992). The pterin molybdenum cofactors. *J. Biol. Chem.* 267:10199–10202.

RajBhandary, U. L., and Stuart, A. (1966). Nucleic acids: Sequence analysis. *Ann. Rev. Biochem.* 35:759–788.

Rall, J. E. (1990). Solomon Aaron Berson. *Biog. Mem. NAS* 59:55–70.

Ramart, P. (1926). Albin Haller. *Bull. Soc. Chim.* [4]39:1037–1072.

Ramberg, P. J. (1994). Johannes Wislicenus, atomism, and the philosophy of chemistry. *Bull. Hist. Chem.* 15/16:45–54.

——— (1995). Arthur Michael's critique of stereochemistry, 1887–1899. *Hist. Stud. Phys. Sci.* 26:89–139.

Ramirez, B. E., Malmström, B. G., Winkler, J. R., and Gray, H. B. (1995). The currents of life: The terminal electron-transfer complex of respiration. *PNAS* 92:11949–11951.

Ramsay, W. *The Life and Letters of Joseph Black.* London: Constable.

Ramsey, E. M. (1994). George Washington Corner. *Biog. Mem. NAS* 65:57–93.

Randall, J. T. (1951). An experiment in biophysics. *Proc. Roy. Soc.* A208:1–24.

——— (1975). Emmeline Jean Hanson. *Biog. Mem. FRS* 21:313–344.

Randle, P. J. (1986). Carl Ferdinand Cori. *Biog. Mem. FRS* 32:67–95.

——— (1990). Frank George Young. *Biog. Mem. FRS* 36:583–599.

Raper, H. S. (1907). The condensation of acetaldehyde and its relation to the biochemical synthesis of fatty acids. *J. Chem. Soc.* 91:1831–1838.

——— (1935). Julius Berend Cohen. *J. Chem. Soc.*, pp. 1331–1335.

Rapkine, L. (1938). Role des groupements sulfhydrilés dans l'activité de l'oxyréduction du triosephosphate. *Compt. Rend.* 207:301–304.

Rappaport, R. (1960). G. F. Rouelle: An eighteenth-century chemist and teacher. *Chymia* 6:68–101.

——— (1961). Rouelle and Stahl: The phlogistic revolution in France. *Chymia* 7:73–102.

Rasmussen, N. (1993). Facts, artifacts, and mesosomes. *Stud. Hist. Phil. Sci.* 24:227–265.

——— (1995). Mitochondrial structure and the practice of cell biology in the 1950s. *J. Hist. Biol.* 28:381–429.

——— (1997a). The mid-century biophysics bubble: Hiroshima and the biological revolution in America revisited. *Hist. Sci.* 35:245–293.

——— (1997b). *Picture Control: The Electron Microscope and the Transformation of Biology in America, 1940–1960*. Stanford: Stanford University Press.

Raspail, F. V. (1830). *Essai de chimie microscopique appliqué à la physiologie.* Paris: Meilhac.

——— (1833). *Nouveau Système de chimie organique.* Paris: Baillière.

Ratliff, R. L. (1981). Terminal deoxynucleotidyl transferase. In *The Enzymes.* 3d ed. (P. D. Boyer, ed.), 14A:105–118. New York: Academic.

Ratner, S. (1954). Urea synthesis and metabolism of arginine and citrulline. *Adv. Enzymol.* 15:319–387.

——— (1977). A long view of nitrogen metabolism. *Ann. Rev. Biochem.* 46:1–24.

Rattansi, P. M. (1970). Jean Beguin. *DSB* 1:571–572.

Raven, C. E. (1950). *John Ray, His Life and Works.* 2d ed. Cambridge: Cambridge University Press.

Ravetz, J. R. (1971). *Scientific Knowledge and Its Social Problems.* Oxford: Clarendon.

Ravin, A. W. (1977). The gene as catalyst, the gene as organism. *Stud. Hist. Biol.* 1:1–45.

——— (1978). Genetics in America: A historical overview. *Persp. Biol. Med.* 21:214–231.

Ray, W. J., and Peck, E. J. (1972). Phosphoglyceromutase. In *The Enzymes.* 3d ed. (P. D. Boyer, ed.), 6:459–475. New York: Academic.

Razin, A., and Kafri, T. (1994). DNA methylation from embryo to adult. *Prog. Nucleic Acid Res.* 48:53–81.

Rechsteiner, M., Hoffman, L., and Dubiel, W. (1993). The multicatalytic and 26S proteases. *J. Biol. Chem.* 268:6065–6068.

Record, M. T., Mazur, S. J., Melançon, P., Roe, H. H., Shaner, S. L, and Unger, L. (1981). Double helical DNA: Conformations, physical properties, and interaction with ligands. *Ann. Rev. Biochem.* 50:997–1024.

Redfield, A. C. (1934). The hemocyanins. *Biol. Revs.* 9:175–212.

Redhead, M. (1990). Quantum theory. In *Companion to the History of Modern Science* (R. C. Olby et al., eds.), pp. 458–478. London: Routledge.

Reed, H. S. (1942). *A Short History of the Plant Sciences.* Waltham: Chronica Botanica.

——— (1949). Jan Ingenhousz, plant physiologist, with a history of the discovery of photosynthesis. *Chronica Botanica* 11:285–393.

Reed, L. J. (1957). The chemistry and function of lipoic acid. *Adv. Enzymol.* 18:319–347.

——— (1974). Multienzyme systems. *Acc. Chem. Res.* 7:40–46.

Reed, L. J., DeBurk, B. G., Gunsalus, I. C., and Schnakenberg, G. H. F. (1951). Chemical nature of α-lipoic acid. *Science* 114:93–94.

Rees, M. (1987). Warren Weaver. *Biog. Mem. NAS* 57:493–530.

Regan, L. (1995). Protein design: Novel metal-binding sites. *TIBS* 20:280–285.

Regnault, V., and Reiset, J. (1849). Recherches chimiques sur la respiration des animaux des diverses classes. *Ann. Chim.* [3]26:299–519.

Rehberg, P. B. (1951). August Krogh. *Yale J. Biol. Med.* 24:83–102.

Rehbock, P. F. (1987). Prosopography? *Ideas and Production* 6:116–121.

Reichard, P. (1959). The enzymic synthesis of pyrimidines. *Adv. Enzymol.* 21:263–294.

——— (1993). From RNA to DNA, why so many ribonucleotide reductases? *Science* 260:1773–1777.

Reichert, E. T., and Brown, A. P. (1909). *The Differentiation and Specificity of Corresponding Proteins and Other Vital Substances in Relation to Biological Classification and Organic Evolution: The Crystallography of Hemoglobins.* Washington, D.C.: Carnegie Institution.

Reichlin, S. (1986). Somatostatin: Historical aspects. *J. Gestroent.* 21 (Suppl. 119):1–10.

Reichvarg, D., and Jacques, J. (1991). *Savants et ignorants: Une Histoire de la vulgarisation des sciences.* Paris: Seuil.

Reil, J. C. (1799). Veränderte Mischung unf Form der thierischen Materie, als Krankheit oder nächste Ursache der Krankheitsfälle Betrachtet. *Reil's Archiv für Physiologie* 3:424–461.

Reingold, N. (1972). American indifference to basic research: A reappraisal. In *Nineteenth Century American Science: A Reappraisal* (G. H. Daniels, ed.), pp. 38–62. Evanston: Northwestern University Press.

——— (1980). Through paradigm-land to a normal history of science. *Soc. Stud. Sci.* 10:475–496.

——— (1981). Science, scientists, and historians of science. *Hist. Sci.* 19:274–283.

——— (ed.) (1990). Symposium on documents, interpretations, and the history of the sciences. *Proc. Am. Phil. Soc.* 134:338–441.

——— (1991). *Science, American Style.* New Brunswick, N.J.: Rutgers University Press.

Reinke, J. (1873). Zur Kenntnis des Rhizoma von *Corallorhiza* und *Epipogon. Flora* 31:145–208.

——— (1883). Die Autoxydation in der lebenden Pflanzenzelle. *Bot. Z.* 41:65–76, 89–103.

——— (1925). *Mein Tagewerk.* Freiburg: Herder.

Remane, H. (1984). *Emil Fischer.* Leipzig: Teubner.

Remane, H., and Wiese, F. (1992). Zur Fusionierung des "Journals für praktische Chemie" mit der "Chemiker-Zeitung." *J. prakt. Chem.* 334:5–13.

Renneberg, M., and Walker, M. (1994). *Science, Technology, and National Socialism.* Cambridge: Cambridge University Press.

Renner, O. (1919). Zur Biologie und Morphologie der männlichen Haplonten einiger Oenotheren. *Z. Bot.* 11:305–308.

Resneck, S. (1970). The European education of an American scientist and its influence in 19th-century America: Eben Norton Horsford. *Tech. Cult.* 11:366–388.

Rey-Pailhade, J. de (1888). Nouvelles recherches physiologiques sur la substance organique hydrogénant le soufre a froid. *Compt. Rend.* 107:43–44.

Rheinberger, H. J. (1993). Experiment and orientation: Early systems of in vitro protein synthesis. *J. Hist. Biol.* 26:443–471.

——— (1995a). From microsomes to ribosomes: "Strategies" of "representation." *J. Hist. Biol.* 28:49–89.

——— (1995b). When did Carl Correns read Gregor Mendel's paper? *Isis* 86:612–616.

——— (1996). Comparing experimental systems: Protein synthesis in microbes and in animal tissues at Cambridge (Ernest F. Gale) and at the Massachusetts General Hospital (Paul C. Zamecnik), 1945–1960. *J. Hist. Biol.* 29:387–416.

Rhoades, M. M. (1954). Chromosomes, mutations, and cytoplasm in maize. *Science* 120:115–120.

Rhoads, R. E. (1993). Regulation of eukaryotic protein synthesis by initiation factors. *J. Biol. Chem.* 268:3017–3020.

Rhodes, G. (1993). *Crystallograohy Made Crystal Clear.* San Diego: Academic.

Rich. A. (1960). A hybrid helix containing both deoxyribose and ribose polynucleotides and its relation to the transfer of information between the nucleic acids. *PNAS* 46:1044–1053.

——— (1989). Looking back at a *Biochimica et Biophysics Acta* paper published 32 years ago. *Biochim. Biophys. Acta* 1000:83–86.

Rich, A., Nordheim, A., and Wang, A. K. J. (1984). Chemistry and biology of left-handed Z-DNA. *Ann. Rev. Biochem.* 53:791–846.

Richardson, J. S. (1981). Anatomy and physiology of protein structures. *Ann. Rev. Biochem.* 34:167–270.

Riddle, O. (1909). Our knowledge of melanin color formation and its bearing on the Mendelian description of heredity. *Biol. Bull.* 16:316–350.

Riedl, R. (1978). *Order in Living Organisms.* New York: Wiley.

Riesenfeld, E. H. (1931). *Svante Arrhenius.* Leipzig: Akademische Verlagsgesellschaft.

Riggs, A. R., Reiness, G., and Zubay, G. (1971). Purification and DNA-binding properties of catabolite gene activator protein. *PNAS* 68:1222–1225.

Rimington, C. (1995). A biochemical autobiography. In *Selected Topics in the History of Biochemistry: Personal Recollections.* Vol. 3 (E. C. Slater et al., eds.), pp. 349–401. Amsterdam: Elsevier.

Ris, H., and Mirsky, A. (1949). Quantitative cytochemical determination of desoxyribonucleic acid with the Feulgen nucleal reaction. *J. Gen. Physiol.* 33:125–146.

Rittenberg, D., and Schoenheimer, R. (1937). Deuterium as an indicator in the study of intermediary metabolism, 11: Further studies on the biological uptake of deuterium into organic substances, with special reference to fat and cholesterol formation. *J. Biol. Chem.* 121:235–253.

Ritter, H., and Zerweck, W. (1956). Arthur von Weinberg. *Chem. Ber.* 89:xix–xli.

Ritthausen, H. (1866). Ueber die Glutaminsäure. *J. prakt. Chem.* 99:454–462.

——— (1872). *Die Eiweisskörper der Getreiden, Hülsenfrüchte, und Oelsamen.* Bonn: Cohen.

Robbins, E. A., and Boyer, P. D. (1957). Determination of the equilibrium of the hexokinase reaction and the free energy of hydrolysis of adenosine triphosphate. *J. Biol. Chem.* 224:121–135.

Roberts, G. C. K. (1993). *NMR of Macromolecules.* New York: Oxford University Press.

Roberts, G. K. (1966). C. K. Ingold at University College London: Educator and department head. *Brit. J. Hist. Sci.* 29:65–82.

——— (1976). The establishment of the Royal College of Chemistry: An investigation of the social context of early-Victorian chemistry. *Hist. Stud. Phys. Sci.* 7:437–485.

——— (1980). The liberally-educated chemist: Chemistry in the Cambridge Natural Science Tripos, 1857–1914. *Hist. Stud. Phys. Sci.* 11:157–183.

Roberts, H. F. (1929). *Plant Hybridization Before Mendel.* Princeton: Princeton University Press.
Roberts, J. D. (1996). The beginnings of physical organic chemistry in the United States. *Bull. Hist. Chem.* 19:48–56.
Roberts, J. W. (1969). Termination factor for RNA synthesis. *Nature* 224:1168–1174.
Roberts, K., and Hyams, J. S. (eds.) (1979). *Microtubules.* New York: Academic.
Roberts, R. J. (1976). Restriction endonucleases. *CRC Crit. Revs. Biochem.* 4:123–164.
Robertson, J. M. (1972). Molecules and crystals, 1926–1970. *Helv. Chim. Acta* 55:119–127.
Robertson, J. M., and Woodward, I. (1937). An X-ray study of phthalocyanins, 3: Quantitative structure determination of metal phthalocyanine. *J. Chem. Soc.*, pp. 219–230.
Robertson, M. (1949). Marjory Stephenson. *Obit. Not. FRS* 6:563–575.
Robertson, R. N. (1995). Charge separation: A personal involvement in a fundamental biological process. In *Selected Topics in the History of Biochemistry: Personal Recollections.* Vol. 3 (E. C. Slater et al., eds.), pp. 303–348. Amsterdam: Elsevier.
Robin, C., and Verdeil, F. (1853). *Traité de chimie anatomique et physiologie normale et pathologique.* Paris: Baillière.
Robin, E. D. (ed.) (1979). *Claude Bernard and the Internal Environment.* New York: Dekker.
Robinow, C. F. (1956). The chromatin bodies of bacteria. *Bact. Revs.* 29:207–242.
Robinow, C. F., and Kellenberger, E. (1994). The bacterial nucleoid revisited. *Microbiol. Revs.* 38:211–232.
Robinson, G. (1971). Anton de Bary. *DSB* 3:611–614.
——— (1976). Eduard Adolf Strasburger. *DSB* 13:87–90.
——— (1979). *A Prelude to Genetics.* Lawrence, Kan.: Coronado.
Robinson, J. D. (1984). The chemiosmotic hypothesis of energy coupling and the path of scientific opportunity. *Persp. Biol. Med.* 27:367–383.
——— (1997). *Moving Questions: A History of Membrane Transport and Bioenergetics.* Oxford: Oxford University Press.
Robinson, R. (1917). A theory of the mechanism of the phytochemical synthesis of certain alkaloids. *J. Chem. Soc.* 111:876–899.
——— (1927). An aspect of the biochemistry of sugars. *Nature* 120:44, 656.
——— (1931). Constitution of cholesterol. *Nature* 130:540–541.
——— (1934). Structure of cholesterol. *Chem. Ind.* 53:1062–1063.
——— (1947). Arthur Lapworth. In *British Chemists* (A. Findlay and W. H. Mills, eds.), pp. 353–368. London: Chemical Society.
——— (1955). *The Structural Relations of Natural Products.* Oxford: Clarendon.
——— (1976). *Memoirs of a Minor Prophet: 70 Years of Organic Chemistry.* Amsterdam: Elsevier.
Robinson, T. (1960). Michael Tswett. *Chymia* 6:146–151.
Robison, G. R., Butcher, R. W., and Sutherland, E. W. (eds.) (1971). *Cyclic AMP.* New York: Academic.
Robison, R. (1922). A new phosphoric acid ester produced by the action of yeast juice on hexoses. *Biochem. J.* 16:809–824.
Robison, R., and King, E. J. (1931). Hexosemonophosphate esters. *Biochem. J.* 25:323–338.
Roche, A. (1935). Le pois moléculaire des protéines. *Bull. Soc. Chim. Biol.* 17:704–744.
Rocke, A. J. (1981). Kekulé, Butlerov, and the historiography of the theory of chemical structure. *Brit. J. Hist. Sci.* 14:27–57.
——— (1984). *Chemical Atomism in the Nineteenth Century: From Dalton to Cannizzaro.* Columbus: Ohio University Press.

——— (1985). Hypothesis and experiment in the early development of Kekulé's benzene theory. *Ann. Sci.* 42:355–381.

——— (1990). "Between two stools": Kopp, Kolbe, and the history of chemistry. *Bull. Hist. Chem.* 12:19–24.

——— (1993a). Group research in German chemistry: Kolbe's Marburg and Leipzig institutes. *Osiris* 8:53–79.

——— (1993b). *The Quiet Revolution: Hermann Kolbe and the Science of Organic Chemistry.* Berkeley: University of California Press.

——— (1994). History and science, history of science: Adolphe Wurtz and the renovation of the academic profession in France. *Ambix* 41:20–32.

——— (1995). Pride and prejudice in chemistry: Kolbe, Hofmann, and German antisemitism. In *The Interaction of Scientific and Jewish Cultures in Modern Times* (Y. Rabkin and I. Robinson, eds.), pp. 127–159. Lewiston, Mass.: Mellen.

Rocke, A. J., and Heuser, E. (eds.) (1994). *Justus von Liebig und Hermann Kolbe in ihren Briefen, 1846–1873.* Mannheim: Bionomica Verlag.

Rodbell, M., Birnbaumer, L., Pohl, S., and Krans, H. M. G. (1971). The glucagon-sensitive adenyl cyclase system in plasma membranes of rat liver, 5: An obligatory role of guanyl nucleotides in glucagon action. *J. Biol. Chem.* 246:1877–1882.

Rodd, E. H. (1947). Henry Edward Armstrong. In *British Chemists* (A. Findley and W. H. Mills, eds.), pp. 58–95. London: Chemical Society.

Roden, L., and Feingold, D. S. (1985). A vintage year for Jorpes, Crafoord, and heparin. *TIBS* 10:407–409.

Roe, D. A. (1981). William Cumming Rose. *J. Nutrition* 111:1313–1320.

Roesler, W. J., Vandenmark, G. R., and Hanson, R. W. (1988). Cyclic AMP and the induction of eukaryotic gene transcription. *J. Biol. Chem.* 263:9063–9066.

Roger, J. (1963). *Les Sciences de la vie dans la pensée française du XVIIIe siècle.* Paris: Armand Colin.

——— (1979). Chimie et biologie des "molécules organiques" de Buffon à la "physico-chimie" de Lamarck. *Hist. Phil. Life Sci.* 1:43–64.

Rolleston, H. D. (1936). *The Endocrine Organs in Health and Disease.* Oxford: Oxford University Press.

Roll-Hansen, N. (1976). Critical teleology: Immanuel Kant and Claude Bernard on the limitations of experimental biology. *J. Hist. Biol.* 9:59–91.

——— (1978a). The genotype theory of Wilhelm Johannsen and its relation to plant breeding and the study of evolution. *Centaurus* 22:201–235.

——— (1978b). Drosophila genetics: A reductionist research program. *J. Hist. Biol.* 11:159–210.

——— (1979). Reductionism in biological research: Reflections on some historical case studies in experimental biology. In *Perspectives in Metascience* (J. Bärmark, ed.), pp. 157–172. Göteborg: Kungl. Vetenskapsoch Vitterhets Samhället.

——— (1998). Studying natural science without nature? Reflections on the realism of so-called laboratory studies. *Hist. Stud. Phys. Biol. Sci.* 29:165–187.

Roscoe, H. E. (1906). *The Life and Experiences of Henry Enfield Roscoe . . . Written by Himself.* London: Macmillan.

Rose, I. A. (1995). Isotope strategies for the study of enzymes. *Protein Science* 4:1430–1433.

Rose, W. C. (1938). The nutritive significance of amino acids. *Physiol. Revs.* 18:109–136.

——— (1969). Recollections of personalities involved in the early history of American biochemistry. *J. Chem. Ed.* 46:759–763.

Rose, W. C., and Coon, M. J. (1974). Howard Bishop Lewis. *Biog. Mem. NAS* 44:139–173.

Rosen, G. (1936). Carl Ludwig and his American students. *Bull. Hist. Med.* 4:609–649.

——— (1959). The conservation of energy and the study of metabolism. In *The Historical Development of Physiological Thought* (C. M. Brooks and P. F. Cranefield, eds.), pp. 243–263. New York: Hafner.

——— (1970). William Beaumont. *DSB* 1:542–545.

Rosenberg, A. (1985). *The Structure of Biological Science.* Cambridge: Cambridge University Press.

——— (1989). From reductionism to instrumentalism? In *What the Philosophy of Biology Is* (M. Ruse, ed.), pp. 245–262. Dordrecht: Kluwer.

——— (1994). *Instrumental Biology or the Disunity of Science.* Chicago: University of Chicago Press.

Rosenberg, C. E. (1970). Wilbur Olin Atwater. *DSB* 1:325–326.

Rosenberg, T. (1954). The concept and definition of active transport. *Symp. Soc. Exp. Biol.* 8:27–41.

Rosenfeld, A. (1962). *Langmuir: The Man and the Scientist.* Oxford: Pergamon.

——— (1966). *The Quintessence of Irving Langmuir.* New York: Oxford University Press.

Rosenfeld, G. (1895). Die Grundgesetze der Acetonurie und ihre Behandlung. *Zent. inn. Med.*, pp. 1233–1244.

Rosenfeld, L. (1973). James Prescott Joule. *DSB* 7:180–182.

Rosenheim, O., and King, H. (1932). Ring system of sterols and bile acids. *J. Soc. Chem. Ind.* 51:464–466.

——— (1934). The chemistry of the sterols, bile acids, and other cyclic constituents of natural fats and oils. *Ann. Rev. Biochem.* 3:87–110.

Rosenmund, K. W., and Zetzsche, F. (1921). Ueber die Beeinflussung der Wirksamkeit der Katalysatoren. *Ber. chem. Ges.* 54:2038–2942.

Ross, S. (1975). Walthère Victor Spring. *DSB* 12:592–594.

Rossiter, M. W. (1971). *Liebig and the Americans: A Study in the Transit of Science, 1840–1880.* Ph.D. diss., Yale University.

——— (1975). *The Emergence of Agricultural Science: Liebig and the Americans, 1840–1880.* New Haven: Yale University Press.

——— (1982). *Women Scientists in America: Struggles and Strategies to 1940.* Baltimore: Johns Hopkins University Press.

——— (1993). The Matthew Matilda effect in science. *Soc. Stud. Sci.* 23:325–341.

——— (1994). Mendel the mentor: Yale women doctorates in biochemistry, 1898–1937. *J. Chem. Ed.* 71:215–219.

——— (1995). *Women Scientists in America: Before Affirmative Action, 1940–1972.* Baltimore: Johns Hopkins University Press.

Rossmann, M. G. (1994). The beginnings of structural biology. *Protein Science* 3:1731–1733.

Roth, J. (1973). Peptide hormone binding to receptors: A review of direct studies in vitro. *Metabolism* 33:1059–1073.

——— (1988). Receptors: Birth, eclipse, and rediscovery. In *Endocrinology: People and Ideas* (M. S. McCann, ed.), pp. 369–396. Bethesda, Md.: American Physiological Society.

Roth, P. A. (1987). *Meaning and Method in the Social Sciences: A Case for Methodological Pluralism.* Ithaca, N.Y.: Cornell University Press.

——— (1996). Will the real scientists stand up? Dead ends and live issues in the explanation of scientific knowledge. *Stud. Hist. Phil. Sci.* 27:43–68.

Rothberg, M. (1965). *Jacques Loeb and His Research Activities.* Zurich: Juris.

Rothen, A. (1948). Long range enzymatic action on films of antigen. *J. Am. Chem. Soc.* 70:2732–2740.

Rothschuh, K. E. (1963). Carl August Sigismund Schultze in Freiburg i. Br. und sein Brief aus Greifswald von 12.3.1833. *Sudhoffs Arch.* 47:334–359.

——— (1971). Emil Heinrich du Bois-Reymond. *DSB* 4:200–205.

——— (1976). Max Verworn. *DSB* 14:2–3.

——— (1978). *Konzepte der Medizin in Vergangenheit und Gegenwart.* Stuttgart: Hippocrates Verlag.

Rothstein, A. (1984). Membrane mythology: Technical versus conceptual developments in the progress of research. *Can. J. Biochem.* 62:1111–1120.

Roughton, F. J. W. (1970). Some recent work on the interactions of oxygen, carbon dioxide, and haemoglobin. *Biochem. J.* 117:801–812.

Roughton, F. J. W., and Kendrew, J. C. (eds.) (1949). *Haemoglobin.* London: Butterworths.

Rouvray, D. H. (1975). The pioneers of isomer enumeration. *Endeavour* 35:28–33.

——— (1977). The changing role of the symbol in the evolution of chemical notation. *Endeavour* NS1:23–31.

——— (1992). Some key historical highlights in the evolution of the modern concept of valence. *Journal of Molecular Structure* 259:1–28.

Roux, E. (1898). La fermentation alcoolique et l'evolution de la microbie. *Rev. Sci.* [4]10:833–840.

——— (1904). Notice sur la vie et les travaux d'Emile Duclaux. *Ann. Inst. Pasteur* 18:337–362.

Roux, W. (1883). *Ueber die Bedeutung der Kerntheilungsfiguren.* Leipzig: Engelmann.

Ruben, S. (1943). Photosynthesis and phosphorylation. *J. Am. Chem. Soc.* 65:279–282.

Ruben, S., Kamen, M. D., and Hassid, W. Z. (1940). Photosynthesis with radioactive carbon, 2: Chemical properties of the intermediates. *J. Am. Chem. Soc.* 62:3443–3450.

Ruben, S., Randall, M., Kamen, M. D., and Hyde, J. L. (1941). Heavy oxygen (O^{18}) as a tracer in the study of photosynthesis. *J. Am. Chem. Soc.* 63:877.

Rubin, L. P. (1980). Styles in scientific explanation: Paul Ehrlich and Svante Arrhenius on immunochemistry. *J. Hist. Med.* 35:397–426.

Rubner, M. (1894). Die Quelle der thierischen Wärme. *Z. Biol.* 30:73–142.

——— (1913). Die Ernährungsphysiologie der Hefezelle bei der alkoholischer Gärung. *Archiv für Physiologie* Suppl., pp. 1–392.

Rudé, G. (1966). *The Crowd in History.* New York: Wiley.

Rudolph, G. (1972). Die Physiologie auf den Versammlungen der Deutschen Naturforscher und Aerzte von der Gründung bis zum Jahre 1890 etc. In *Wege der Naturforschung, 1822–1972* (H. Querner and H. Schipperges, eds.), pp. 147–170. Berlin: Springer.

——— (1983). Der Mechanismusproblem in der Physiologie des 19. Jahrhunderts. *Ber. Wissen.* 6:7–28.

Ruestow, E. G. (1996). *The Microscope in the Dutch Republic.* Cambridge: Cambridge University Press.

Rüger, W. (1972). Transcription of genetic information and its regulation by protein factors. *Ang. Chem. (Int. Ed)* 11:883–893.

Rundle, R. E., and French, D. (1943). The configuration of starch in the starch-iodine complex, 3: X-ray diffraction studies of the starch-iodine complex. *J. Am. Chem. Soc.* 65:1707–1710.

Runnström, J., and Michaelis, L. (1935). Correlation of oxidation and phosphorylation in hemolyzed blood in the presence of methylene blue and pyocyanine. *J. Gen. Physiol.* 18:717–727.

Rupe, H. (1932). *Adolf von Baeyer als Lehrer und Forscher.* Stuttgart: Enke.

Ruska, H. (1941). Ueber ein neues bei der Bakteriophagen Lyse auftretender Formelement. *Naturwiss.* 19:367–368.

Russell, C. A. (1971). *The History of Valency.* Leicester: Leicester University Press.

——— (1984). Whigs and professionals. *Nature* 308:777–778.

——— (1986). *Lancastrian Chemist: The Early Years of Edward Frankland.* Milton Keynes, England: Open University Press.

——— (1987). The changing role of synthesis in organic chemistry. *Ambix* 34:169–180.

——— (1996). *Edward Frankland.* Cambridge: Cambridge University Press.

Russell, E. J. (1921). *Soil Conditions and Plant Growth.* 4th ed. London: Longmans.

——— (1966). *A History of Agricultural Science in Great Britain, 1620–1954.* London: Allen and Unwin.

Russell, E. S. (1985). A history of mouse genetics. *Ann. Rev. Genetics* 19:1–28.

——— (1989). Sewall Wright's contributions to physiological genetics and to inbreeding theory and practice. *Ann. Rev. Genetics* 23:1–18.

Russell, N. (1988a). Toward a history of biology in the twentieth century: Directed autobiographies as historical sources. *Brit. J. Hist. Sci.* 21:77–89.

——— (1988b). Oswald Avery and the origin of molecular biology. *Brit. J. Hist. Biol.* 21:393–400.

Russett, C. E. (1966). *The Concept of Equilibrium in American Social Thought.* New Haven: Yale University Press.

Russo, F. D., and Silhavy. T. J. (1992). Alpha: The Cinderella subunit of RNA polymerase. *J. Biol. Chem.* 267:14515–14518.

Ruzicka, L. (1932). The life and work of Otto Wallach. *J. Chem. Soc.* 135:1582–1597.

——— (1959). History of the isoprene rule. *Proc. Chem. Soc.*, pp. 341–360.

——— (1973). In the borderland between bioorganic chemistry and biochemistry. *Ann. Rev. Biochem.* 42:1–20.

Ryan, F. J. (1952). Adaptation to use lactose by *Escherichia coli. J. Gen. Microbiol.* 7:69–88.

Ryle, A. P., Sanger, F., Smith, L. F., and Kitai, R. (1955). The disulfide bonds of insulin. *Biochem. J.* 60:541–556.

Ryman, B. E., and Whelan, W. J. (1971). New aspects of glycogen metabolism. *Adv. Enzymol.* 34:285–443.

Sabatini, D. D., and Blobel, G. (1970). Controlled proteolysis of nascent polypeptides in rat liver cell fractions, 2: Location of polypeptides in rough microsomes. *J. Cell Biol.* 45:146–153.

Sabatini, D. D., Kreibich, G., Morimoto, T., and Adesnik, M. (1982). Mechanisms for the incorporation of proteins in membranes and organelles. *J. Cell Biol.* 92:1–22.

Sachel, A., and Wahle, E. (1993). Poly(A) tail metabolism and function in eukaryotes. *J. Biol. Chem.* 268:22955–22958.

Sachs, F. (1905). Ueber die Nuklease. *Z. physiol. Chem.* 46:337–353.

Sachs, J. (1862). Ueber den Einfluss des Lichtes auf die Bildung des Amylum in den Chlorophyllknöten. *Bot. Z.* 44:365–373.

——— (1890). *History of Botany* (H. E. F. Gurney, trans.). Oxford: Clarendon.

Sachse, A. (1928). *Friedrich Althoff und sein Werk.* Berlin: Mittler.

Sachse, H. (1890). Ueber die geometrische Isomerien der Hexamethylenderivate. *Ber. chem. Ges.* 23:1365–1366.

Sakaguchi, K., and Ikeda, Y. (1970). Applied microbiology in Japan. In *Profiles of Japanese Science and Industry* (H. Yukawa, ed.), pp. 90–105. Tokyo: Kodansha.

Sakodynski, K. (1972). The life and scientific works of Michael Tswett. *J. Chromat.* 73:303–360.

Salaman, R. N. (1949). *The History and Social Influence of the Potato.* Cambridge: Cambridge University Press.

Saldanha, R., Mohr, G., Belfort, M., and Lambowitz, A. M. (1993). Group I and group II introns. *FEBS J.* 7:15–24.

Salkowski, E. (1877). Ueber den Vorgang der Harnstoffbildung im Thierkörper und der Einfluss der Ammoniaksalzen auf denselben. *Z. Physiol. Chem.* 1:1–59.

Saltzman, M. D. (1974). Benzene and the triumph of the octet theory. *J. Chem. Ed.* 51:498–502.

——— (1986). The development of physical organic chemistry in the United States and the United Kingdom, 1919–1939: Parallels and contrasts. *J. Chem. Ed.* 63:588–593.

——— (1997). Thomas Martin Lowry and the mixed multiple bond. *Bull. Hist. Chem.* 20:10–17.

Sambursky, S. (1959). *Physics of the Stoics.* London: Routledge.

Samuel, C. E. (1993). The eIF-2 protein kinases, regulators of translation in eukaryotes from yeasts to humans. *J. Biol. Chem.* 268:7603–7606.

Sancar, A., and Sancar, G. B. (1988). DNA repair enzymes. *Ann. Rev. Biochem.* 57:29–67.

Sander, K. (ed.) (1985). *August Weismann (1834–1914) und die theoretische Biologie des 19. Jahrhunders.* Freiburg: Verlag Rombach.

Sandler, I. (1979). Some reflections on the protean nature of the scientific precursor. *Hist. Sci.* 17:170–190.

Sandler, I., and Sandler, L. (1985). A conceptual ambiguity that contributed to the neglect of Mendel's paper. *Hist. Phil. Life Sci.* 7:3–70.

Sandritter, W., and Kasten, F. H. (1964). *100 Years of Histochemistry in Germany.* Stuttgart: Schattauer.

Sanfelice, F. (1928). Ueber die Natur des Virus der Taubenpocke. *Z. Immun.* 54:487–495.

Sanger, F. (1945). The free amino acids of insulin. *Biochem. J.* 39:507–515.

——— (1952). The arrangement of amino acids in proteins. *Adv. Protein Chem.* 7:1–67.

——— (1981). Determination of nucleotide sequences in DNA. *Science* 214:1205–1210.

——— (1988). Sequences, sequences, sequences. *Ann. Rev. Biochem.* 57:1–28.

Sanger, F., Air, G. M., Barrell, B. G., Brown, N. L., Coulson, A. R., Fiddes, J. C., Hutchinson, C. A., Slocombe, P. M., and Smith, M. (1977). Nucleotide sequence of bacteriophage ϕX174 DNA. *Nature* 265:687–695.

Sanger, F., Donelson, J. E., Coulson, A. R., Kössel, H., and Fischer, D. (1973). Use of DNA polymerase primed by a synthetic oligonucleotide to determine a nucleotide sequence in phage f1 DNA. *PNAS* 70:1209–1213.

Sanger, F., Nicklen, S., and Coulson, A. R. (1977). DNA sequencing with chain-terminating inhibitors. *PNAS* 74:5463–5467.

Sanson, A. (1857). Note sur la formation physiologique du sucre dans l'économie animale. *Compt. Rend.* 44:1323–1325.

Sapp, J. (1983). The struggle for authority in the field of heredity, 1900–1932. *J. Hist. Biol.* 16:311–342.

——— (1987). *Beyond the Gene: Cytoplasmic Inheritance and the Struggle for Authority in Genetics.* New York: Oxford University Press.

——— (1990). The nine lives of Gregor Mendel. In *Experimental Inquiries* (E. LeGrand, ed.), pp. 137–166. Dordrecht: Kluwer.

——— (1994). *Evolution by Association: A History of Symbiosis.* New York: Oxford University Press.

Sarabhai, A. S., Stretton, A. O. W., Brenner, S., and Bolle, A. (1964). Colinearity of the gene with the polypeptide chain. *Nature* 201:13–17.

Sarges, R., and Witkop, B. (1964). Gramicidin A, 4: Primary structure of valine and isoleucine gramicidin A. *J. Am. Chem. Soc.* 86:1862–1863.

Sarkar, S. (1992). Haldane as biochemist: The Cambridge decade, 1923–1933. In *Founders of Evolutionary Genetics* (S. Sarkar, ed,), pp. 53–81. Dordrecht: Kluwer.

——— (1996). Biological information: A skeptical look at some central dogmas of molecular biology. In *The Philosophy and History of Molecular Biology: New Perspectives* (S. Sarkar, ed.), pp. 187–231. Dordrecht: Kluwer.

Sarton, G. (1938). The scientific basis of the history of science. In *Cooperation in Research* (publication no. 601), pp. 465–481. Washington, D.C.: Carnegie Institution.

Sasso, J. (1980). The stages of the creative process. *Proc. Am. Phil. Soc.* 124:119–132.

Sauer, G., Rapoport, S., and Rost, G. (1961). Zur Geschichte der Biochemie in Berlin. *NTM* 1:119–147.

Saussure, T. de (1804). *Recherches chimiques sur la végétation.* Paris: Nyon.

——— (1819). Sur la décomposition de l'amidon à la temperature atmosphérique, par l'action de l'air et de l'eau. *Ann. Chim.* [2]11:379–408.

Saville, J. (1993). Edward Palmer Thompson. *International Review of Social History* 38:i–vii.

Sawin, C. T. (1988). Defining thyroid hormone: Its nature and control. In *Endocrinology: People and Ideas* (S. M. McCann, ed.), pp. 149–199. Bethesda, Md.: American Physiological Society.

Sawyalow, W. W. (1901). Zur Theorie der Eiweissverdauung. *Pflügers Arch.* 88:171–225.

Sayre, A. (1975). *Rosalind Franklin and DNA.* New York: Norton.

Scatchard, G. (1949). The attractions of proteins for small molecules and ions. *Ann. N.Y. Acad. Sci.* 51:660–672.

——— (1960). John Gamble Kirkwood. *Journal of Chemical Physics* 33:1279–1281.

——— (1969). Edwin J. Cohn and protein chemistry. *Vox Sanguinis* 17:37–44.

Scerri, E. R. (1993). Correspondence and reduction in chemistry. In *Correspondence, Invariance, and Heuristics* (S. French and H. Kamminga, eds.), pp. 45–64. Dordrecht: Kluwer.

Schachman, H. K. (1988). Can a simple model account for the allosteric transition of aspartate transcarbamoylase? *J. Biol. Chem.* 263:18583–18586.

Schadewaldt, H. (1972). Hermann Hellriegel. *DSB* 6:237–238.

Schaffner, K. F. (1967). Antireductionism and molecular biology. *Science* 157:644–647, 857–862.

——— (1974). The peripherality of reductionism in the development of modern biology. *J. Hist. Biol.* 7:111–139.

Schalck, A. (1940). *Das Leben und Wirken des Heidelberger Physiologen Willy Kühne.* Düsseldorf: Nolte.

Schaller, M. D., Bouton, A. H., Flynn, D. C., and Parsons, J. T. (1993). Identification and characterization of novel substrates for protein tyrosine kinases. *Prog. Nucleic Acid Res.* 44:205–227.

Schardinger, F. (1902). Ueber das Verhalten der Kuhmilch gegen Methylenblau und seine Verwendung zur Unterscheidung von ungekochter und gekochter Milch. *Z. Unt. Nahrung* 5:1113–1121.

Scharrer, B. (1975). Neurosecretion and its role in neuroendocrine regulation. In *Pioneers in Neuroendocrinology* (J. Meites, B. T. Donovan, and S. M. McCann, eds.), pp. 257–265. New York: Plenum.

Scharrer, E., and Scharrer, B. (1937). Ueber Drüsen-Nervenzellen und neurosekretorische Organe bei Wirbellosen und Wirbeltieren. *Biol. Revs.* 12:185–216.

Schatz, G. (1996). Efraim Racker. *Biog. Mem. NAS* 70:321–346.

Scheer, B. T. (1939). The development of the concept of tissue respiration. *Ann. Sci.* 4:295–305.

Scheer, H. (1991). *Chlorophylls.* Boca Raton, Fla.: CRC.

Schekman, R., Weiner, A., and Kornberg, A. (1974). Multienzyme systems of DNA replication. *Science* 186:987–993.

Schekman, R., Weiner, J. H., Weiner, A., and Kornberg, A. (1975). Ten proteins required for conversion of ϕX174 single-stranded DNA to duplex form *in vitro*. *J. Biol. Chem.* 250:5859–5865.

Schelar, V. M. (1966). Thermochemistry and the third law of thermodynamics. *Chymia* 11:99–124.

Schenck. F. (1902). Adolf Fick. *Pflügers Arch.* 90:313–361.

Schiemann, E. (1934). Erwin Baur. *Ber. bot. Ges.* 51–114.

Schiller, J. (1967a). Wöhler, l'urée et le vitalisme. *Sudhoffs Arch.* 51:229–243.

——— (1967b). *Claude Bernard et les problèmes scientifiques de son temps.* Paris: Cèdre.

——— (1971). Les frustrés de l'histoire—Henri Dutrochet. *Episteme* 5:188–200.

——— (1980). *Physiology and Classification: Historical Relations.* Paris: Maloine.

Schiller, J., and Schiller, T. (1975). *Henri Dutrochet: Le Matérialisme mécaniste et la physiologie générale.* Paris: Blanchard.

Schimmel, P. (1987). Aminoacyl-tRNA synthetases: General scheme of structure-function relationships in the polypeptides and recognition of tRNA. *Ann. Rev. Biochem.* 56:125–158.

——— (1990). Hazards and their exploitation in the applications of molecular biology to structure-function relationships. *Biochemistry* 29:9495–9502.

Schimmel, P., and Söll, D. (1979). Aminoacyl-tRNA synthetases. *Ann. Rev. Biochem.* 48:601–648.

Schimper, A. F. W. (1878). *Untersuchungen über die Proteinkrystalloide der Pflanzen.* Strassburg: Trübner.

Schleiden, M. (1842). *Grundzüge der wissenschaftlichen Botanik.* Vol. 1. Leipzig: Engelmann.

Schlenk, F. (1942). Nicotinamide nucleotide enzymes. In *A Symposium on Respiratory Enzymes,* pp. 104–133. Madison: University of Wisconsin Press.

——— (1984). The discovery of enzymatic transmethylation. *TIBS* 9:34–35.

——— (1987). The ancestry, birth, and adolescence of adenosine triphosphate. *TIBS* 12:367–368.

Schlenk, F., and Snell, E. E. (1945). Vitamin B_6 and transamination. *J. Biol. Chem.* 157:425–426.

Schlesinger, M. (1932). Ueber die Bindung der Bacteriophagen an homologe Bakterien. *Z. Hyg.* 114:136–160.

——— (1934a). Zur Frage der chemischen Zusammensetzung der Bacteriophagen. *Biochem. Z.* 273:306–311.

——— (1934b). The Feulgen reaction of the bacteriophage substance. *Nature* 138:508–509.

Schlesinger, M. J. (1981). Proteolipids. *Ann. Rev. Biochem.* 50:193–206.

——— (1990). Heat shock proteins. *J. Biol. Chem.* 265:12111–12114.

Schloesing, A. T., and Laurent. E. (1890). Sur la fixation de l'azote gazeux par les legumineuses. *Compt. Rend.* 111:750–753.

Schloesing, J. J. T., and Müntz, A. (1877). Sur la nitrification par les ferments organiques. *Compt. Rend.* 84:301–303.

Schmauderer, E. (1969). Die Stellung des Wissenschaftlers zwischen chemischer Forschung und chemischer Industrie im 19. Jahrhundert. *Technikgeschichte in Einzeldarstellungen* 11:37–93.

——— (1971). Der Einfluss der Chemie auf die Entwicklung des Patentwesen in der zweiten Hälfte des 19. Jahrhunderts. *Tradition* 16:144–176.

——— (ed.) (1973). *Der Chemiker im Wandel der Zeiten.* Weinheim: Verlag Chemie.

Schmauss, W. (1994). *Durkheim's Philosophy of Science and the Sociology of Knowledge.* Chicago: University of Chicago Press.

Schmidt, G. (1928). Ueber fermentative Desaminierung im Muskel. *Z. physiol. Chem.* 179:243–249, 264–269.

Schmidt, G. (1974). *Das geistige Vermächtnis von Gustav von Bunge.* Zurich: Juris.

——— (1991). Zur Entwicklung der Fächer Klinische Chemie und Laboratoriumsdiagnostik in der Wiener Schule. *Ber. Wissen.* 14:231–239.

Schmidt, G., and Levene, P. A. (1938). Ribonucleodepolymerase (the Jones-Dubos enzyme). *J. Biol. Chem.* 126:423–434.

Schmidt, O. T. (1959). Géza Zemplén. *Chem. Ber.* 92:i–xix.

Schmidt-Nielsen, B. (1995). *August and Marie Krogh: Lives in Science.* New York: American Physiological Society.

Schmidt-Nielsen, K. (1967). The unusual animal, or to expect the unexpected. *Fed. Proc.* 26:981–986.

Schmiedeberg, O. (1877). Ueber das Verhältnis des Ammoniaks und der primären Aminosäuren zur Harnstoffbildung im Thierkörper. *Arch. exp. Path. Pharm.* 8:1–14.

——— (1881). Ueber Oxydationen und Synthesen im Thierkörper. *Arch. exp. Path. Pharm.* 14:288–312.

——— (1912). Rudolf Buchheim. *Arch. exp. Path. Pharm.* 67:1–54.

Schmitt-Fiebig, J. (1988). *Einflusse und Leistungen Deutscher Pharmazeuten, Naturwissenschaftler und Aerzte seit dem 18. Jahrundert in Chile.* Stuttgart: Deutscher Apotheker Verlag.

Schmorl, K. (1952). *Adolf von Baeyer.* Stuttgart: Wissenschaftliche Verlagsgesellschaft.

Schneider, H. A. (1973). Harry Steenbock. *J. Nutrition* 103:1233–1247.

Schoen, H. M., Grove, C. S., and Palermo, J. A. (1956). The early history of crystallization. *J. Chem. Ed.* 33:373–375.

Schoen, M. (1928). *The Problem of Fermentation* (H. L. Hind, trans.). London: Chapman and Hall

Schoenheimer, R. (1926). Ein Beitrag zur Bereitung der Peptide. *Z. physiol. Chem.* 154:203–224.

——— (1929). Ueber eine eigenartige Störung des Kohlenhydrat-Stoffwechsel. *Z. physiol. Chem.* 182:148–150.

——— (1937). The investigation of intermediary metabolism with the aid of heavy hydrogen. *Harvey Lectures* 32:122–144.

——— (1941). *The Dynamic State of Body Constituents.* Cambridge: Harvard University Press.

Schoenheimer, R., and Breusch, F. (1933). Synthesis and destruction of cholesterol in the organism. *J. Biol. Chem.* 103:430–444.

Schoenheimer, R., Ratner, S., and Rittenberg, D. (1939). Studies in protein metabolism, 10: The metabolic activity of body proteins investigated with *l*(-)-leucine containing two isotopes. *J. Biol. Chem.* 130:703–732.

Schoenheimer, R., and Rittenberg, D. (1935). Deuterium as an indicator in the study of intermediary metabolism. *Science* 82:156–157.

——— (1938). The application of isotopes to the study of intermediary metabolism. *Science* 87:221–226.

Schoepfer, G. J., Jr. (1981, 1982). Sterol biosynthesis. *Ann. Rev. Biochem.* 50:585–621; 51:555–585.

Schofield, K. (1994). The development of Ingold's system of organic chemistry. *Ambix* 41:87–107.

——— (1995). Some aspects of the work of Arthur Lapworth. *Ambix* 42:160–186.

Schofield, R. E. (1967). Joseph Priestley: Natural philosopher. *Ambix* 14:1–15.

Scholz, H. (1991). Friedrich Althoffs Einfluss auf die Entwicklung der Chemie. In *Wissenschaftsgeschichte und Wissenschsftspolitik im Industriezeitalter: Das System Althoff in historischer Perspektive* (B. von Brocke, ed.), pp. 337–354. Hildesheim: Lax.

Schönbein, C. F. (1848). Ueber einige chemische Wirkungen der Kartoffel. *Ann. Phys.* 75:357–361.

——— (1860). Fortsetzung der Beiträge zur näheren Kenntnis der Sauerstoffes. *J. prakt. Chem.* 81:1–20.

——— (1861). Ueber einige durch das Haarchenanziehung des Papieres hervorgebrachte Trennungswirkungen. *Ann. Phys.* 114:275–280.

——— (1863). Ueber die katalytische Wirksamkeit organischer Materien und deren Verbreitung in der Pflanzenund Thierwelt. *Z. prakt. Chem.* 89:323–344.

Schöpf, C. (1937). Die Synthese von Naturstoffen, insbesondere die Alkaloide, unter physiologischen Bedingungen und ihre Bedutung für die Frage der der Entstehung einiger Naturstoffe in der Zelle. *Ang. Chem.* 50:779–787, 797–805.

Schorlemmer, C. (1894). *The Rise and Development of Organic Chemistry.* Rev. ed., A. Smithells. London: Macmillan.

Schott, T. The movement of science and of scientific knowledge: Joseph Ben-David's contribution to its understanding. *Minerva* 31:455–477.

Schrader, F. Three quarter-centuries of cytology. *Science* 107:155–159.

Schrödinger, E. (1944). *What is Life? The Physical Aspect of the Living Cell.* Cambridge: Cambridge University Press.

Schroeder, W. von (1882). Ueber die Bildungstätte des Harnstoffs. *Arch. exp. Path. Pharm.* 15:364–402.

Schröer, H. (1967). *Carl Ludwig.* Stuttgart: Wissenschaftliche Verlagsgesellschaft.

Schryver, S. B. (1909). *The General Characters of the Proteins.* London: Longmans, Green.

Schuchert, C., and LeVene, C. M. (1940). *O. C. Marsh: Pioneer in Paleontology.* New Haven: Yale University Press.

Schuett, H. W. (1984). *Die Entstehung des Isomorphismus.* Hildesheim: Gerstenberg.

——— (1988). Pohl für Windaus: Zum optischen Nachweis eines Vitamins. *Sudhoffs Arch.* 72:98–105.

——— (1992). *Eilhard Mitscherlich.* Munich: Oldenburg.

Schulte, B. (1984). Johann Ludwig Wilhelm Thudichum (1829–1901), protagonist and antiprotagonist of brain chemistry. In *De Novo Inventis* (A. H. M. Kerkhoff et al., eds.), pp. 389–400. Amsterdam: North-Holland University Press.

Schultz, J. (1941). The evidence of the nucleoprotein nature of the gene. *Cold Spring Harbor Symp.* 9:55–65.

——— (1947). The nature of heterochromatin. *Cold Spring Harbor Symp.* 12:179–191.

——— (1967). Innovators and controversies. *Science* 157:296–301.

Schultzen, O., and Nencki, M. (1869). Die Vorstufen des Harnstoffs im Organismus. *Ber. chem. Ges.* 2:566–571.

Schulz, F. N. (1903). *Die Grösse des Eiweissmoleküls.* Jena: Fischer.

Schulze, E., and Barbieri, J. (1879). Ueber die Eiweisszersetzung in Kürbiskeimlinge. *J. prakt. Chem.* NF20:385–418.

Schulze, E., and Bosshard, E. (1883). Ueber das Glutamin. *Landwirtschaftliche Versuchstationen* 29:295–307.

Schulze, E., and Steiger, E. (1887). Ueber das Arginin. *Z. physiol. Chem.* 11:43–65.

Schulze, E., and Winterstein, E. (1897). Ueber die Spaltungsproducte des Arginins. *Ber. chem. Ges.* 30:2879–2882.

Schütte, E. (1976). Free from physiology. *TIBS* 1:N205-N206.

Schütz, E. (1885). Eine Methode zur Bestimmung der relativen Pepsinmenge. *Z. Physiol. Chem.* 9:577–590.

Schütz, J. (1900). Zur Kenntnis der quantitativaten Pepsinwirkung. *Z. physiol. Chem.* 30:1–14.

Schwabe, K. (1988). *Deutsche Hochschullehrer als Elite, 1815–1945.* Boppard am Rhein: Harald Boldt Verlag.

Schwann, T. (1836). Ueber das Wesen des Verdauungsprocesses. *Arch. Anat. Physiol.*, pp. 90–138.

——— (1837). Vorläufige Mitteilung, betreffend Versuche über die Weingährung. *Ann. Phys.* 41:184–193.

——— (1839). *Mikroskopische Untersuchungen über die Uebereinstimmung in der Struktur und dem Wachstum der Tiere und Pflanzen.* Berlin: Sander.

Schwarzenbach, G. (1966). Die Entstehung der Valenzlehre und Alfred Werner. *Experientia* 22:633–646.

——— (1974). Chemische Systematik. *Chimia* 28:101–108.

Schweiger, H. G. (1986). Otto Meyerhof. In *Semper Apertus* (W. Doerr, ed.), 2:359–375. Berlin: Springer.

Schwendener, S. (1868). Untersuchungen über den Flechtenthallus. *Beiträge zur wissenschaftlichen Botanik* 6:195–207.

Schwerin, E. A. (1867). *Zur Kenntnis von der Verdauung der Eiweisskörper.* Berlin: Lange.

Schwyzer, R. (1975). Studies on polypeptide hormone receptors. In *Molecular Aspects of Membrane Phenomena* (H. R. Kaback et al., eds.), pp. 123–131. Berlin: Springer.

——— (1995). 100 years lock-and-key concept: Are peptide keys shaped and guided to their receptors by the target cell membrane? *Protein Science* 3:5–16.

Scott, A. I. (1994). Towards a total, genetically engineered synthesis of vitamin B_{12}. *Synlett.* 11:871–883.

Scott, M. P. (1987). Complex loci of Drosophila. *Ann. Rev. Biochem.* 56:195–227.

Scott-Moncrieff, R. (1936). A biochemical survey of some Mendelian factors for flower color. *J. Genetics* 32:117–170.

Scriven, E. F. V., and Turnbull, K. (1988). Azides: Their preparation and synthetic uses. *Chem. Revs.* 88:297–368.

Scully, M. F. (1992). The biochemistry of blood clotting: The digestion of a liquid to form a solid. *Essays in Biochemistry* 27:17–36.

Sechenov, I. M. (1965). *Autobiographical Notes.* (D. B. Lindsay, ed.; K. Hanes, trans.). Washington. D.C.: American Institute of Biological Sciences.

Seeder, H. L. (1954). Prof. Dr. Evert Gorter. *Koll. Z.* 136:99–102.

Segal, H. L. (1959). The development of enzyme kinetics. In *The Enzymes* (P. D. Boyer et al., eds,), 1:1–48. New York: Academic.

Segel, I. H. (1975). *Enzyme Kinetics.* New York: Wiley-Interscience.

Segerstrale, U. (1993). Bringing the scientist back in: The need for an alternative sociology of knowledge. In *Controversial Science* (T. Brante et al., eds.), pp. 57–82. Albany: University of the State of New York Press.

Séguin, A. (1791). Mémoire sur l'eudométrie. *Ann. Chim.* 9:293–303.

Seidler, E. (1991). *Die Medizinische Fakultät der Albert-Ludwigs-Universität Freiburg im Breisgau.* Berlin: Springer.

Sela, M. (1987). A peripatetic and personal view of molecular immunology for one third of a century. *Ann. Rev. Immunol.* 5:1–19.

Semenza, G., et al. (1984). Biochemistry of the Na^+,D-glucose cotransporter of the small-intestinal brush-border membrane. *Biochim. Biophys. Acta* 779:343–349.

Senchenkova, E. M. (1960). Andrei Sergeevich Famintzin. *Bot. Zhur.* 45:309–317.

——— (1974). Vladimir Ivanovich Palladin. *DSB* 10:281–283.

——— (1975). Dmitri Nikolaevich Prianishnikov. *DSB* 11:179–180.

——— (1976). Mikhail Semenovich Tsvet. *DSB* 13:486–488.

Senebier, J. (1788). *Expériences sur l'action de la lumière solaire dans la végétation.* Geneva: Barde, Manget et Cie.

——— (1804). *Essai sur l'art d'observer et de faire des expériences.* Geneva: Paschoud.

Senior, A. E. (1988). ATP synthesis by oxidative phosphorylation. *Physiol. Revs.* 58:177–231.

Serafini, A. (1989). *Linus Pauling: A Man and His Science.* New York: Paragon.

Sernka, T. J. (1979). Claude Bernard and the nature of gastric acid. *Persp. Biol. Med.* 2:523–530.

Servos, J. W. (1976). The knowledge corporation: A. A. Noyes and chemistry at Cal-Tech, 1915–1930. *Ambix* 23:175–186.

——— (1980). The industrial relations of science: Chemical engineering at MIT, 1900–1939. *Isis* 71:531–549.

——— (1982). A disciplinary program that failed: Wilder D. Bancroft and the *Journal of Physical Chemistry,* 1896–1933. *Isis* 73:207–232.

——— (1983). Review of R. E. Kohler, *From Medical Chemistry to Biochemistry. Isis* 74:273–275.

——— (1984). G. N. Lewis: The disciplinary setting. *J. Chem. Ed.* 61:5–10.

——— (1990). *Physical Chemistry from Ostwald to Pauling.* Princeton: Princeton University Press.

——— (1994). Wilder Dwight Bancroft. *Biog. Mem. NAS* 65:3–39.

Setlow, R. B. (1968). The photochemistry, photobiology, and repair of polynucleotides. *Prog. Nucleic Acid Res.* 8:257–293.

Severin, S. E. (1983). Sergei E. Severin: Life and scientific activity. In *Selected Topics in the History of Biochemistry* (G. Semenza, ed.), pp. 365–390. Amsterdam: Elsevier.

Shaffer, P. A. (1923). Intermediary metabolism of carbohydrates. *Physiol. Revs.* 3:394–437.

Shaffer, P. A., and Ronzoni, E. (1932). Carbohydrate metabolism. *Ann. Rev. Biochem.* 1:247–266.

Shamin, A. N. (1970). Aleksei Nikolaevich Bach. *DSB* 1:360–363.

——— (1990a). Andrei Nikolaevich Belozerski. *DSB* 17:68–69.

——— (1990b). Mikhail Mikhailovich Shemiakin. *DSB* 18:813–814.

Shanahan, T. (1989). Kant, *Naturphilosophie,* and Oersted's discovery of electromagnetism: A reassessment. *Stud. Hist. Phil. Sci.* 20:287–305.

Shannon, C. E., and Weaver, W. (1949). *The Mathematical Theory of Communication.* Urbana: University of Illinois Press.

Shapere, D. (1969). Biology and the unity of science. *J. Hist. Biol.* 2:3–18.

——— (1984). *Reason and the Search for Knowledge.* Dordrecht: Reidel.

——— (1989). Evolution and continuity in scientific change. *Phil. Sci.* 56:419–437.

Shapin, S. (1989). The invisible technician. *Am. Sci.* 77:554–563.

Shapin, S., and Thackray, A. (1974). Prosopography as a research tool in history of science: The British scientific community, 1700–1900. *Hist. Sci.* 12:1–28.

Sharp, P. A. (1987). Splicing of messenger RNA precursors. *Science* 235:766–771.

Sharpey-Shafer, E. (1927). *History of the Physiological Society During Its First Fifty Years.* Cambridge: Cambridge University Press.

Shedlovsky, T. (1943). Criteria of purity of proteins. *Ann. N.Y. Acad. Sci.* 43:259–272.

Sheehan, H. (1985). *Marxism and the Philosophy of Science: The First Hundred Years.* Atlantic Highlands: Humanities.

Shemin, D. (1956). The biosynthesis of porphyrins. *Harvey Lectures* 50:258–284.

——— (1974). Hans Thacher Clarke. *Am. Phil. Soc. Year Book,* pp. 134–137.

Shemin, D., and Rittenberg, D. (1945). The utilization of glycine for the synthesis of a porphyrin. *J. Biol. Chem.* 159:567–568.

Shemin, D., and Wittenberg, J. (1950). The location in protoporphyrin of the carbon atoms derived from the α-carbon of glycine. *J. Biol. Chem.* 185:103–116.

——— (1951). The mechanism of porphyrin formation: The role of the tricarboxylic acid cycle. *J. Biol. Chem.* 192:315–334.

Sherrington, C. (1906). *The Integrative Action of the Nervous System.* New York: Scribner's.

Shimazono, N. (1991). The history of the Japanese Biochemical Society from 1925 to 1990. *J. Biochem.* 112:S1–S18.

Shine, J., and Delgarno, L. (1974). The 3′-terminal sequence of *Escherichia coli* 16S ribosomal RNA: Complementarity to nonsense triplets and ribosome binding sites. *PNAS* 71:1342–1346.

Shinn, T. (1979). The issue of decline in nineteenth century French science. *Hist. Stud. Phys. Sci.* 10:271–332.

Shippen-Lentz, D., and Blackburn, E. H. (1990). Functional evidence for an RNA template in telomerase. *Science* 247:546–552.

Shoppee, C. W. (1972). Christopher Ke Ingold. *Biog. Mem. FRS* 18:349–411.

Shorter, J. (1978). The conversion of ammonium cyanate into urea: A saga in reaction mechanisms. *Chem. Soc. Revs.* 7:1–14.

——— (1990). Hammett memorial lecture. *Prog. Phys. Org. Chem.* 17:1–29.

——— (1996). The sigma culture. *Chem. Intell.* 2:39–47.

Shortland, M., and Yeo, R. (eds.) (1996). *Telling Lives in Science.* Cambridge: Cambridge University Press.

Shortle, D., and Botstein, D. (1985). Strategies and applications of in vitro mutagenesis. *Science* 229:1193–1201.

Shortle, D., et al. (1996). Protein folding for realists: A timeless phenomenon. *Protein Science* 5:991–1000.

Shryock, R. H. (1948). American indifference to basic science during the nineteenth century. *Arch. Int. Hist. Sci.* 5:50–65.

——— (1971). The medical reputation of Benjamin Rush: Contrasts over two centuries. *Bull. Hist. Med.* 45:507–552.

Sibatani, A. (1981). Two faces of molecular biology: Revolution and normal science. *Rivista di Biologia* 74:279–296.

Siegfried, M. (1916). *Ueber partielle Eiweisshydrolyse.* Berlin: Bornträger.

Siegfried, R., and Dobbs, B. J. (1968). Composition, a neglected aspect of the chemical revolution. *Ann. Sci.* 24:275–293.

Siekevitz, P., et al. (1972). The social responsibility of scientists. *Ann. N.Y. Acad. Sci.* 196:197–291.

Siekevitz, P., and Zamecnik, P. C. (1981). Ribosomes and protein synthesis. *J. Cell. Biol.* 91:53s–65s.

Signer, R., Caspersson, T., and Hammarsten, E. (1938). Molecular shape and size of thymonucleic acid. *Nature* 141:122.

Siilivask, K. (ed.) (1985). *History of Tartu University, 1632–1982.* Tallinn: Periodika.

Silliman, B. (1886). John Lawrence Smith. *Biog. Mem. NAS* 2:217–248.

Silverstein, A. M. (1982). Development of the concept of immunological specificity. *Cellular Immunology* 67:396–409; 71:183–195.

——— (1989). *A History of Immunology.* San Diego: Academic.

Siminovich, L. (1976). On the nature of the heritable variation in cultured somatic cells. *Cell* 7:1–11.

Simmer, H. H. (1955). Aus den Anfängen der physiologichen Chemie in Deutschland. *Sudhoffs Arch.* 39:216–236.

——— (1958). Zur Entwicklung der physiologischen Chemie. *Ciba Zeitschrift* 8:3013–3044.

——— (1978). Die Entdeckung und Entdecker des Sekretins. *Med. Welt* 29:1991–1996.

Simmonds, S. (1950). The metabolism of phenylalanine and tyrosine in mutant strains of *Escherichia coli. J. Biol. Chem.* 185:755–762.

——— (1972). Peptidase activity and peptide metabolism in *Escherichia coli.* In *Peptide Transport in Bacteria and Mammalian Gut* (K. Elliott and M. O'Connor, eds.), pp. 43–57. Amsterdam: Elsevier.

Simmonds, S., Cohn, M., Chandler, J. P., and du Vigneaud, V. (1943). The utilization of the methyl group of choline in the biological synthesis of methionine. *J. Biol. Chem.* 149:519–525.

Simmonds, S., and du Vigneaud, V. (1945). A further study of the lability of the methyl group of creatine. *Proc. Soc. Exp. Biol. Med.* 59:293–294.

Simmonds, S., and Fruton, J. S. (1951). The action of penicillin on the growth response of a Gram-negative bacillus to amino acids and peptides. *Yale J. Biol. Med.* 23:407–426.

Simms, G. R. (1963). *The Scientific Work of Karl Landsteiner.* Biberstein: Freundes-Dienst.

Simonds, A. P. (1978). *Karl Mannheim's Sociology of Knowledge.* Oxford: Oxford University Press.

Simpson, A. D. C. (ed.) (1980). *Joseph Black: A Commemorative Symposium.* Edinburgh: Royal Scottish Museum.

Simpson, G. G. (1964). *This View of Life.* New York: Harcourt, Brace and World.

Simpson, R. T. (1973). Structure and function of chromatin. *Adv. Enzymol.* 38:41–107.

Sinding, C. (1996). Literary genres and the construction of knowledge in biology: Semantic shifts and scientific change. *Soc. Stud. Sci.* 26:43–70.

Singer, M. F. (1958). Phosphorolysis of oligoribonucleotides by polynucleotide phosphorylase. *J. Biol. Chem.* 232:211–228.

——— (1995). Unusual reverse transcription. *J. Biol. Chem.* 270:24623–24626.

Singer, M. F., and Berg, P. (1991). *Genes and Genomes.* Mill Valley, Calif.: University Science Books.

——— (1997). Biosynthesis of intercellular messenger peptides. In *Exploring Genetic Mechanisms* (M. Singer and P. Berg, eds.), pp. 405–458. Sausalito, Calif.: University Science Books.

Singer, M. F., and Fruton, J. S. (1957). Some properties of beef spleen phosphoamidase. *J. Biol. Chem.* 229:111–119.

Singer, S. J., and Nicolson, G. L. (1972). The fluid mosaic model of the cell membrane. *Science* 175:720–731.

Singleton, R. (1997a). Heterotropic CO_2-fixation, mentors, and students: The Wood-Werkman reactions. *J. Hist. Biol.* 30:91–120.

——— (1997b). Harland Goff Wood: An American biochemist. In *Selected Topics in the History of Biochemistry.* Vol. 4 (G. Semenza and R. Jaenicke, eds.), pp. 1–50. Amsterdam: Elsevier.

Sinsheimer, R. L. (1957). First steps toward a genetic chemistry. *Science* 125:1123–1128.

——— (1960). The nucleic acids of the bacterial viruses. In *The Nucleic Acids* (E. Chargaff and J. N. Davidson, eds.), 3:187–244. New York: Academic.

——— (1968). Bacteriophage ϕX174 and related viruses. *Prog. Nucleic Acid Res.* 8:115–169.

Sirks, M. J. (1952). The earliest illustrations of chromosomes. *Genetica* 26:65–76.

Sismondo, S. (1992). Some social constructions. *Soc. Stud. Sci.* 23:515–553.

Sjöstrand, F. (1943). Fixation and preparation of tissues for electron microscopic examination of tissues. *Nature* 151:725–726.

——— (1953). Electron microscopy of mitochondria and cytoplasmic double membranes. *Nature* 171:30–32.

Skou, J. C. (1965). Enzymatic basis for active transport of Na^+ and K^+ across the cell membrane. *Physiol. Revs.* 45:596–618.

——— (1989). The identification of the sodium pump as the membrane Na^+/K^+ATPase: A commentary. *Biochim. Biophys. Acta* 1000:435–438.

Skulachev, V. P. (1991). Chemiosmotic systems in bioenergetics: H^+-cycles and Na^+ cycles. *Bioscience Reports* 11:387–444.

Slater, E. C. (1953). Mechanism of phosphorylation in the respiratory chain. *Nature* 172:975–978.

——— (1977). Cytochrome. *TIBS* 2:138–139.

——— (1983). The Q cycle, an ubiquitous mechanism of electron transfer. *TIBS* 8:239–242.

——— (1986). The BAL-labile factor in the respiratory chain. In *Selected Topics in the History of Biochemistry: Personal Recollections* (A. Neuberger et al., eds.), pp. 197–253. Amsterdam: Elsevier.

——— (1994). Peter Dennis Mitchell. *Biog. Mem. FRS* 40:283–303.

Slayman, C. L. (1993). Channels, pumps, and osmotic machines in plants: A tribute to Osterhout. In *The Biological Century* (R. B. Brown et al., eds.), pp. 120–149. Woods Hole, Mass.: Marine Biological Laboratory.

Sloan, P. R. (1977). Descartes, the sceptics, and the rejection of vitalism in seventeenth-century physiology. *Stud. Hist. Phil. Sci.* 8:1–28.

Smeaton, W. A. (1962). *Fourcroy, Chemist and Revolutionary.* Cambridge: Heffer.

——— (1963). Guyton de Morveau and chemical affinity. *Ambix* 11:55–64.

——— (1970). Torbern Olof Bergman. *DSB* 2:4–8.

Smedley-Maclean, I. (1941). Arthur Harden. *Biochem. J.* 35:1071–1081.

Smith, C. U. M. (1983). The earliest biochemistry. *TIBS* 8:193–195.

Smith, C. W., and Wise, M. N. (1989). *Energy and Empire: A Biographical Study of Lord Kelvin.* Cambridge: Cambridge University Press.

Smith, D. (1995). Henry Dale, scepticism, theory, luck, strategy, and tactics. In *Essays in the History of the Physiological Sciences* (C. Debru, ed.), pp. 195–198. Amsterdam: Rodopi.

Smith, D., and Nicolson, M. (1989). The "Glasgow School" of Paton, Findlay, and Cathcart: Conservative thought in chemical physiology, nutrition, and public health. *Soc. Stud. Sci.* 19:195–218.

Smith, E. F. (1917). *The Life of Robert Hare, an American Chemist.* Philadelphia: Lippincott.

——— (1918). *James Woodhouse: A Pioneer in Chemistry.* Philadelphia: Winston.

Smith, E. L. (1955). Isolation and chemistry of vitamin B_{12}. In *The Biochemistry of Vitamin B_{12}* (R. T. Williams, ed.), pp. 2–16. Cambridge: Cambridge University Press.

Smith, E. L. (1982). The evolution of a biochemist. In *Of Oxygen, Fuels, and Living Matter* (G. Semenza, ed.), pp. 361–445. Chichester: Wiley.

Smith, E. L., DeLange, R. J., and Bonner, J. (1970). Chemistry and biology of the histones. *Physiol. Revs.* 50:159–170.

Smith, H. O., Danner, D. B., and Deich, R. A. (1981). Genetic transformation. *Ann. Rev. Biochem.* 50:41–68.

Smith, H. O., and Wilcox, K. W. (1970). A restriction enzyme from *Haemophilus influenzae. J. Mol. Biol.* 51:379–392.

Smith, L. M., et al. (1986). Fluorescence detection in automated DNA sequence analysis. *Nature* 321:674–679.

Smith, M. (1985). In vitro mutagenesis. *Ann. Rev. Genetics* 19:423–462.

Smith, O. L. K., and Hardy, J. D. (1975). Cyril Norman Hugh Long. *Biog. Mem. NAS* 46:265–309.

Smith, P. H. (1994). *The Business of Alchemy: Science and Culture in the Holy Roman Empire.* Princeton: Princeton University Press.

Smith, S. (1994). The animal fatty acid synthase: One gene, one polypeptide, seven enzymes. *FASEB J.* 8:1248–1259.

Smith, W. C. (1939). Arthur Hutchinson. *Obit. Not. FRS* 2:483–491.

Smocovitis, V. B. (1996). *Unifying Biology.* Princeton: Princeton University Press.

Smythe, C. V., and Gerischer, W. (1933). Ueber die Vergärung der Hexosemonohexosephosphorsäure. *Biochem. Z.* 260:414–416.

Snelders, H. A. M. (1975). Christian Friedrich Schönbein. *DSB* 12:196–198.

——— (1982). The Mulder-Liebig controversy elucidated by their correspondence. *Janus* 69:199–221.

——— (1984). J. H. van't Hoff's research school in Amsterdam, 1877–1895. *Janus* 71:1–30.

——— (1986). *The Letters from Gerrit Jan Mulder to Justus Liebig, 1838–1846.* Amsterdam: Rodopi.

——— (1994). Analogieschlüsse in der chemischen Vergangenheit: Irrwege und Wegweiser. *NTM* NS2:65–75.

Snell, E. E. (1943). Growth promotion on tryptophan-deficient media by *o*-aminobenzoic acid and its attempted reversal with orthanilamide. *Arch. Biochem.* 2:389–394.

——— (1945). The microbiological assay of amino acids. *Adv. Protein Chem.* 2:85–118.

——— (1958). Chemical structure in relation to biological activities of vitamin B_6. *Vit. Horm.* 16:77–125.

——— (1975). Tryptophanase: Structure, catalytic activities, and mechanism of action. *Adv. Enzymol.* 42:287–333.

——— (1979). Lactic acid bacteria and identification of B-vitamins: Some historical notes. *Fed. Proc.* 38:2690–2693.

——— (1993). From bacterial nutrition to enzyme structure: A personal odyssey. *Ann. Rev. Biochem.* 62:1–27.

Snellen, H. A. (1995). *Willem Einthoven (1860–1927): Father of Electrocardiography.* Dordrecht: Kluwer.

Snow, C. P. (1959). *The Two Cultures and the Scientific Revolution.* Cambridge: Cambridge University Press.

Snyder, S. H., Sklar, P. B., and Pevsner, J. (1988). Molecular mechanisms of olfaction. *J. Biol. Chem.* 263:13971–13974.

Sober, E. (1975). *Simplicity.* Oxford: Clarendon.

Sobotta, J. (1923). Wilhelm von Waldeyer-Hartz. *Anat. Anz.* 56:1–53.

Société Industrielle de Mulhouse (1902). *Histoire documentale de l'industrie de Mulhouse et ses environs aux XIXe siècle.* Mulhouse: Bader.

——— (1994). Chemie et l'Alsace. *Bull. Soc. Ind. Mulhouse* No. 833.

Soda, K., and Tanizawa, K. (1979). Kynureninases: Enzymological properties and regulation mechanism. *Adv. Enzymol.* 49:1–40.

Soddy, F. (1913). Intra-atomic charge. *Nature* 92:399–400.

Söderquist, T. (1994). Darwinian overtones: Niels K. Jerne and the selection theory of antibody formation. *J. Hist. Biol.* 27:481–529.

Solomon, M. (1995). The pragmatic turn in the naturalistic philosophy of science. *Persp. Sci.* 3:206–230.

Soloviev, Y. I. (1970). Aleksandr Porfirevich Borodin. *DSB* 2:316–317.

Sols, A. (1981). Multimodulation of enzyme activity. *Curr. Top. Cell. Regul.* 19:77–101.

——— (1990). Enzyme regulation: From allosteric sites to intracellular behavior. In *Selected Topics in the History of Biochemistry: Personal Recollections.* Vol. 3 (G. Semenza and R. Jaenicke, eds.), pp. 177–199. Amsterdam: Elsevier.

Sonderhoff, R., and Thomas, H. (1937). Die enzymatische Dehydrierung der Trideutero-essigsäure. *Ann.* 530:195–213.

Sonneborn, T. M. (1938). Mating types in *Paramecium aurelia. Proc. Am. Phil. Soc.* 79:411–434.

——— (1975). Herbert Spencer Jennings. *Biog. Mem. NAS* 47:143–223.

Sorby, H. C. (1867). On a definite method of qualitative analysis of animal and vegetable colouring-matters by means of the spectrum microscope. *Proc. Roy. Soc.* 15:433–455.

——— (1873). On comparative vegetable chromatology. *Proc. Roy. Soc.* 21:442–483.

Sørensen, M., and Sørensen, S. P. L. (1925). On the coagulation of proteins by heating. *Compt. Rend. Carlberg* 15:1–26.

Sørensen, S. P. L. (1909). Enzymstudien, 2: Ueber die Messung und Bedeutung der Wasserionenkonzentration bei enzymatischen Prozessen. *Biochem. Z.* 21:131–200.

——— (1930). Die Konstitution der löslichen Proteinstoffe als reversibel dissoziable Komponentensysteme. *Koll. Z.* 53:102–124, 170–199, 306–318.

Sourkes, T. L. (1955). Moritz Traube, 1826–1894: His contribution to biochemistry. *J. Hist. Med.* 10:379–391.

——— (1993a). How Thudichum came to study the brain. *J. Hist. Neurosci.* 2:107–119.

——— (1993b). John Simon, Robert Lowe, and the origin of state-supported biomedical research in nineteenth-century England. *J. Hist. Med.* 48:436–453.

——— (1995). The protagon paradox. *J. Hist. Neurosci.* 4:37–62.

Southern, E. N. (1975). Detection of specific sequences among DNA fragments separated by gel electrophoresis. *J. Mol. Biol.* 98:503–518.

Sowden, J. C. (1962). Hermann Otto Laurenz Fischer. *Adv. Carb. Chem.* 17:1–14.

Sowers, J. R. (ed.) (1980). *Hypothalamic Hormones.* Stroudsburg: Dowden, Hutchington and Ross.

Soyfer, V. N. (1994). *Lysenko and the Tragedy of Soviet Science.* New Brunswick, N.J.: Rutgers University Press.

Spallanzani, L. (1789). *Dissertations Relative to the Natural History of Animals and Vegetables.* 2d ed. London: Murray.

——— (1807). *Rapports de l'air avec les etres organisées* (J. Senebier, ed.). Geneva: Paschoud.

Spaude, M. (1973). *Eugen Albert Baumann.* Zurich: Juris.

Spee, W. von (1906). Walther Flemming. *Anat. Anz.* 28:41–59.

Speiser, P. (1961). *Karl Landsteiner.* Vienna: Hollinek. [English trans., 1975.]

Speyer, J. F., Lengyel, P., Basilio, C., and Ochoa, S. (1962). Synthetic polynucleotides and the amino acid code. *PNAS* 48:441–448.

Spiegel, A. (1892). Carl Schorlemmer. *Ber. chem. Ges.* 25:1107–1123.

Spiegelman, S. (1945). The physiology and genetic significance of enzymatic adaptation. *Ann. Missouri Bot. Gar.* 32:139–163.

Spitzer, W. (1897). Die Bedeutung gewisser Nucleoproteide für die oxydative Leistung der Zelle. *Pflügers Arch.* 67:615–656.

Sprinson, D. B. (1960). The biosynthesis of aromatic compounds from D-glucose. *Adv. Carb. Chem.* 15:235–270.

Srb, A., and Horowitz, N. H. (1944). The ornithine cycle in Neurospora and its genetic control. *J. Biol. Chem.* 154:129–139.

Srere, P. (1987). Complexes of sequential metabolic enzymes. *Ann. Rev. Biochem.* 56:89–124.

Stacey, M. (1973). The consequences of some projects initiated by Sir Norman Haworth. *Chem. Soc. Revs.* 2:145–161.

——— (1994). Michael Heidelberger. *Biog. Mem. FRS* 39:178–197.

Stadie, W. C. (1945). The intermediary metabolism of fatty acids. *Physiol. Revs.* 25:395–441.

——— (1954). Current concepts of the action of insulin. *Physiol. Revs.* 34:52–100.

Stadler, L. J. (1928). Mutations in barley induced by X-rays or radium. *Science* 68:186–187.

——— (1954). The gene. *Science* 120:811–819.

Stadtman, E. R. (1966). Allosteric regulation of enzyme activity. *Adv. Enzymol.* 28:41–154.

Stadtman, E. R., Doudoroff, M., and Lipmann, F. (1951). The mechanism of acetoacetate synthesis. *J. Biol. Chem.* 191:377–382.

Stadtman, T. C. (1979). Some selenium-dependent biochemical processes. *Adv. Enzymol.* 48:1–28.

——— (1991). Biosynthesis and function of selenocysteine-containing enzymes. *J. Biol. Chem.* 266:16257–16260.

Stahl, E. (1877). *Beiträge zur Entwicklungsgeschichte von Flechten.* Leipzig: Felix.

Stahnke, J. (1979). Blutfarbstoff-Kristalle von Reichert bis Hoppe-Seyler. *Sudhoffs Arch.* 63:154–189.

——— (1981). Häminkristalle—Ludwik Teichmanns Blutnachweis. *Med. Hist. J.* 16:391–413.

Stanbury, J. B., Wyngaarden, J. B., Frederickson, D. S., Goldstein, J. L., and Brown, M. S. (1983). *The Metabolic Basis of Inherited Disease.* 5th ed. New York: McGraw-Hill.

Stanier, R. Y. (1980). The journey, not the arrival matters. *Ann. Rev. Microbiol.* 34:1–48.

Stanley, W. M. (1935). Isolation of a crystalline protein possessing the properties of tobacco-mosaic virus. *Science* 81:644–645.

——— (1936). Chemical studies on the virus of tobacco mosaic, 6: The isolation from diseased Turkish tobacco plants of a crystalline protein possessing the properties of tobacco-mosaic virus. *Phytopathology* 26:305–320.

——— (1937). Chemical studies on the virus of tobacco mosaic, 8: The isolation of a crystalline protein possessing the properties of Acuba mosaic virus. *J. Biol. Chem.* 117:325–340.

Stanley, W. M., and Hassid, W. Z. (1969). Hermann Otto Laurenz Fischer. *Biog. Mem. NAS* 40:91–112.

Stare, F. J., and Baumann, C. A. (1936). The effect of fumarate on respiration. *Proc. Roy. Soc.* B121:338–357.

Starling, E. H. (1905). The Croonian lectures on the chemical correlations of the functions of the body. *Lancet* 2:339–341, 423–425, 501–503, 579–583.

Staudinger, H. (1920). Ueber Polymerisation. *Ber. chem. Ges.* 53:1073–1085.

——— (1929). Die Chemie der hochmolekularen organische Stoffe im Sinne der Kekuléschen Strukturlehre. *Z. ang. Chem.* 42:37–40, 67–73.

——— (1961). *Arbeitserinnerungen.* Heidelberg: Hüthig.

Staudinger, H. J. (1978). Ascorbic acid. *TIBS* 3:211–212.

Stebbins, G. L. (1972). The evolutionary significance of biological templates. In *Biology, History, and Natural Philosophy* (A. D. Breck and W. Yourgrau, eds.), pp. 79–102. New York: Plenum.

Stedman, E., and Stedman, E. (1950). Cell specificity of histones. *Nature* 166:780–781.

Stein, W. D. (1986). James Frederic Danielli. *Biog. Mem. FRS* 32:117–135.

Steinberg, D., and Anfinsen, C. B. (1952). Evidence for intermediates in ovalbumin synthesis. *J. Biol. Chem.* 199:25–42.

Steinberg, D., Vaughan, M., and Anfinsen, C. B. (1956). Kinetic aspects of assembly and degradation of proteins. *Science* 124:389–395.

Steiner, D. F., et al. (1975). Proteolytic mechanisms in the biosynthesis of polypeptide hormones. In *Proteases and Biological Control* (E. Reich et al., eds.), pp.531–549. Plainview, N.Y.: Cold Spring Harbor Laboratory Press.

——— (1992). The new enzymology of precursor processing endoproteases. *J. Biol. Chem.* 267:23435–23438.

Steitz, J. A. (1986). Shaping research in gene expression: Role of the Cambridge Medical Research Council Laboratory of Molecular Biology. *Persp. Biol. Med.* 29:S90–S95.

Stenflo, J., Ferlund, P., Egan, W., and Roepstorff, P. (1974). Vitamin K dependent modifications of glutamic acid residues in prothrombin. *PNAS* 71:2730–2733.

Stent, G. S. (1968a). That was the molecular biology that was. *Science* 160:390–395.

——— (1968b). What they are saying about Honest Jim. *Quart. Rev. Biol.* 43:179–184.

——— (1969). *The Coming of the Golden Age.* Garden City, N.Y.: Natural History Press.

——— (1972). Prematurity and uniqueness of scientific discovery. *Sci. Am.* 228(12):84–93.

——— (1979). A thought-collective. *Quart. Rev. Biol.* 54:421–427.

——— (1982). Max Delbrück. *Genetics* 101:1–16.

——— (1989). Light and life: Niels Bohr's legacy to contemporary biology. *Genome* 31:11–15.

Stephen, W. I. (1984). Determination of nitrogen in organic compounds in the years before Kjeldahl's method. *Anal. Proc.* 21:215–220.

Stephenson, M. (1949). *Bacterial Metabolism.* 3d ed. London: Longmans, Green.

Stephenson, M., and Strickland, L. H. (1931). Hydrogenase, 1: A bacterial enzyme activating molecular hydrogen. *Biochem. J.* 25:205–214.

Stern, C. (1931). Zytologisch-genetische Untersuchungen als Beweise für die Morgansche Theorie des Faktorenaustausches. *Biol. Z.* 51:547–587.

——— (1967). Richard Benedict Goldschmidt. *Biog. Mem. NAS* 39:141–192.

Stern, C., and Sherwood, E. R. (1966). *The Origins of Genetics: A Mendel Source Book.* San Francisco: Freeman.

Stern, F. (1987). *Dreams and Delusions.* New York: Knopf.

Stern, J. R., Ochoa, S., and Lynen, F. (1952). Enzymatic synthesis of citric acid, 5: Reaction of acetyl-coenzyme A. *J. Biol. Chem,* 198:313–321.

Stern, K. G. (1939). Respiratory catalysis in heart muscle. *Cold Spring Harbor Symp.* 7:312–322.

Stern, K. G., and Holiday, E. R. (1934). Zur Konstitution des Photo-flavins; Versuche in der Alloxazinreihe. *Ber. chem. Ges.* 67:1104–1106.

Stern, K. G., and Melnick, J. L. (1941). The photochemical spectrum of the Pasteur enzyme in retina. *J. Biol. Chem.* 139:301–323.

Sternglanz, R. (1996). Histone acylation: A gateway to transcriptional activation. *TIBS* 21:357–358.

Stetten, D. (1982). Rudi. *Persp. Biol. Med.* 25:354–368.

Steudel, H. (1901). Die Konstitution des Thymins. *Z. physiol. Chem.* 32:241–244.

——— (1906). Die Zusammensetzung der Nukleinsäuren aus Thymus und aus Heringmilch. *Z. physiol. Chem.* 49:406–409.

——— (1907). Ueber die Guanylsäure aus der Pankreasdrüse. *Z. Physiol. Chem.* 53:539–544.

Stevens, N. M. (1905). Studies in spermatogenesis with special reference to the "accessory chromosome." *Carnegie Institution of Washington,* publ. 36.

Stock, A. (1935). Carl Duisberg. *Ber. chem. Ges.* 68A:111–148.

Stock, J. T. (1989). Einar Biilmann. *J. Chem. Ed.* 66:910–912.

Stock. J. T., and Orna, M. V. (1989). *Electrochemistry, Past and Present.* Washington, D.C.: American Chemical Society.

Stoddart, J. F. (1979). From carbohydrates to enzyme analogues. *Chem. Soc. Revs.* 8:85–142.

Stoeckenius, W. (1994). From membrane structure to bacteriorhodopsin. *J. Membrane Biol.* 139:139–148.

Stokes, G. G. (1864). On the reduction and oxidation of the colouring matter of the blood. *Proc. Roy. Soc.* 13:355–364.

Stokes, T. D. (1982). The double helix and the warped zipper: An exemplary tale. *Soc. Stud. Sci.* 12:207–240.

Stokstad, E. L. R. (1979). Early work with folic acid. *Fed. Proc.* 38:2696–2698.

Stolz, F. (1904). Ueber Adrenalin und Alkylaminoacetobrenzcatechin. *Ber. chem. Ges.* 37:4149–4154.

Stolz, R. (1980, 1981). Zur Geschichte synthetischer organischer Farbstoffe in der zweiten Hälfte des 19. Jahrhunderts. *NTM* 17(2):84–101; 18(1):52–83; (2):29–48.

Stolzenberg, D. (1994). *Fritz Haber.* Weinheim: VCR.

Stone, J. R., and Marletta, M. A. (1994). Soluble guanylate cyclase from bovine lung: Activation with nitric oxide and carbon monoxide and spectral characterization of the ferrous and ferric states. *Biochemistry* 33:5636–5640.

Stone, L. (1971). Prosopography. *Daedalus* 100:46–79.

Stone, R. (1995). Rockefeller strikes fast deal with Amgen. *Science* 268:631.

Stone, R. L., and Dixon, J. E. (1994). Protein-tyrosine phosphatases. *J. Biol. Chem.* 269:31323–31326.

Stoney, G. J. (1881). On the physical units of nature. *Phil. Mag.* [5]11:353–358.

Storer, N. W. (1966). *The Social System of Science.* New York: Holt, Rinehart.

Stranges, A. N. (1984). Reflections on the electron theory of the chemical bond, 1900–1925. *J. Chem. Ed.* 61:185–190.

Strasburger, E. (1880). *Zellbildung und Zelltheilung.* 3d ed. Jena: Fischer.

——— (1884). Die Controversen der indirekten Zelltheilung. *Arch. mikr. Anat.* 23:246–304.

——— (1901). Ueber Befruchtung. *Bot. Z.* 59:353–358.

——— (1909). The minute structure of cells in relation to heredity. In *Darwin and Modern Science* (A. C. Steward, ed.), pp. 102–111. Cambridge: Cambridge University Press.

Straub, F. B. (1981). From respiration of muscle to muscle actin (1934–1943). In *Of Oxygen, Fuels, and Living Matter* (G. Semenza, ed.), pp. 325–344. Chichester: Wiley.

Straub, F. B., and Cohen, S. S. (1987). The charismatic teacher at Szeged: Albert Szent-Györgyi. *Acta Biochim. Biophys. Hung.* 22:135–148.

Straub, W. (1912). Die Bedeutung des Zellmembrans für die Wirkung chemischer Stoffe auf den Organismus. *Verh. Ges. Deutsch. Naturforsch. Aerzte,* pp. 1–25.

Straus, F. (1927). Johannes Thiele. *Ber. chem. Ges.* 60A:75–132.

Strauss, H. A., Fischer, K., Hoffmann, C., and Söllner, A. (1991). *Die Emigration der Wissenschaften nach 1933.* Munich: Saur.

Strecker, A. (1850). Ueber die künstliche Bildung der Milchsäure aus einen neuen, dem Glycocoll ähnlichen Körper. *Ann.* 75:27–45.

Strickland, S. P. (1972). *Politics, Science, and Dread Disease.* Cambridge: Harvard University Press.

Ströker, E. (1968). Element and compound: On the scientific history of two fundamental chemical concepts. *Ang. Chem. (Int. Ed.)* 7:718–724.

Strominger, J. L. (1970). Penicillin-sensitive enzymatic reactions in bacterial cell wall synthesis. *Harvey Lectures* 64:179–213.

Stryer, L. (1986). Cyclic cGMP cascades of vision. *Ann. Rev. Neurosci.* 9:87–119.

——— (1988). *Biochemistry.* 3d ed. New York: Freeman.

——— (1991). Visual excitation and recovery. *J. Biol. Chem.* 266:10711–10714.

Stryer, L., et al. (1994). Vision: From photon to perception. *PNAS* 93:557–595.

Stubbe, H. (1963). *Kurze Geschichte der Genetik bis zur Wiederentdeckung der Vererbungsregeln Gregor Mendels.* Jena: Fischer.

Stubbe, J. (1990). Ribonucleotide reductases. *Adv. Enzymol.* 63:349–417.

——— (1994). Binding site revealed of Nature's most beautiful cofactor. *Science* 166:1663–1664.

Studnicka, F. L. (1933). Matthias Jacob Schleiden und die Zelltheorie von Theodor Schwann. *Anat. Anz.* 76:80–95.

Stuehr, D. J., and Griffith, O. W. (1992). Mammalian nitric oxide synthetases. *Adv. Enzymol.* 65:287–346.

Sturtevant, A. H. (1913). The linear association of six sex-linked factors in Drosophila, as shown by their mode of association. *J. exp. Zool.* 14:43–59.

——— (1925). The effects of unequal crossing over at the Bar locus of Drosophila. *Genetics* 10:117–147.

——— (1948). The evolution and function of genes. *Am. Sci.* 36:225–236.

——— (1959). Thomas Hunt Morgan. *Biog. Mem. NAS* 33:283–325.

——— (1965). *A History of Genetics.* New York: Harper and Row.

Sturtevant, J. M. (1980). Herbert Spencer Harned. *Biog. Mem. NAS* 51:215–244.

Sucker, U. (1988). Wilhelm Pfeffer (1845–1920) und die Physiologie seiner Zeit. *NTM* 25(2):43–57.

Suckling, C. J., et al. (1978). *Chemistry Through Models.* Cambridge: Cambridge University Press.

Sudhoff, K. (1932). Ludwig Thudichum. *Nachrichten der Giessener Hochschulgesellschaft.* 9:33–45.

Summers, W. C. (1991). From culture as organism to organism as cell: Historical origins of bacterial genetics. *J. Hist. Biol.* 24:171–190.

——— (1993). How bacteriophage came to be used by the phage group. *J. Hist. Biol.* 26:255–267.

Sumner, J. B. (1926). The isolation and crystallization of the enzyme urease. *J. Biol. Chem.* 69:435–441.

Suppes, P. (1993). *Models and Methods in the Philosophy of Science.* Dordrecht: Kluwer.

Surrey, A. R. (1954). *Name Reactions in Organic Chemistry.* New York: Academic.

Süsskind, C. (1976). Irving Langmuir. *DSB* 8:22–25.

Sutherland, E. W. (1972). Studies on the mechanism of hormone action. *Science* 177:401–408.

Sutherland, E. W., Oye, I., and Butcher, R. W. (1965). The action of epinephrine and the role of the adenylcyclase system in hormone action. *Rec. Prog. Hormine Res.* 21:623–642.

Sutherland, E. W., Posternak, T., and Cori, C. F. (1949). Mechanism of the phosphoglyceric acid mutase reaction. *J. Biol. Chem.* 181:153–159.

Sutherland, E. W., and Rall, T. W. (1958). Fractionation and characterization of a cyclic adenine ribonucleotide formed by tissue particles. *J. Biol. Chem.* 232:1077–1091.

——— (1960). The relation of adenosine-3′,5′-phosphate and phosphorylase to the action of catecholamines and other hormones. *Pharmacol. Revs.* 12:265–299.

Sutherland, G., et al. (1963). *Emigration of Scientists from the United Kingdom.* London: Royal Society.

Sutton, B. (ed.) (1996). *A Century of Mycology.* Cambridge: Cambridge University Press.

Sutton, G. (1984). The physical and chemical path to vitalism: Xavier Bichat's Physiological Researches on Life and Death. *Bull. Hist. Med.* 58:53–71.

Sutton, W. S. (1903). The chromosomes in heredity. *Biol. Bull.* 4:231–251.

Svedberg, T. (1937). The ultra-centrifuge and the study of high-molecular compounds. *Nature* 139:1051–1062.

Svedberg, T., and Pedersen, K. O. (1940). *The Ultracentrifuge.* Oxford: Clarendon.

Swann, J. P. (1988). *American Scientists and the Pharmaceutical Industry.* Baltimore: Johns Hopkins University Press.

Swanson, J. C., and Suttie, J. W. (1985). Prothrombin: Characterization of processing events in rat liver microsomes. *Biochemistry* 24:3890–3897.

Swazey, J. P. (1975). Charles Scott Sherrington. *DSB* 12:395–403.

Swazey, J. P., Anderson, M. S., and Lewis, K. S. (1993). Ethical problems in academic research. *Am. Sci.* 81:542–553.

Symonds, N. (1988). Schrödinger and Delbrück: Their status in biology. *TIBS* 13:232–234.

Synge, R. L. M. (1943). Partial hydrolysis products derived from proteins and their significance for protein structure. *Chem. Revs.* 32:135–172.

——— (1962). Tsvet, Willstätter, and the use of adsorption for the purification of proteins. *Arch. Biochem. Biophys.* Suppl. 1, pp.1–6.

Synge, R. L. M., and Williams, E. F. (1990). Albert Charles Chibnall. *Biog. Mem. FRS* 35:57–96.

Szabadvary, F. (1964). Development of the pH concept: A historical survey. *J. Chem. Ed.* 41:105–107.

Szent-Györgyi, A. (1924). Ueber den Mechanismus der Succin- und Paraphenylendiaminoxydation. Ein Beitrag zur Theorie der Zellatmung. *Biochem. Z.* 150:195–210.

——— (1928). Observations on the function of peroxidase systems and the chemistry of the adrenal cortex: Description of a new carbohydrate derivative. *Biochem. J.* 22:1387–1409.

——— (1937). *Studies in Biological Oxidation and Some of Its Catalysts.* Budapest: Eggenberg and Barth.

——— (1951). *Chemistry of Muscular Contraction.* 2d ed. New York: Academic.

Szent-Györgyi, A. G. (1953). Meromyosins, the subunits of myosin. *Arch. Biochem. Biophys. Acta.* 42:305–320.

Szöllosi-Janze, M. (1996). Berater, Agent, Interessent? Fritz Haber, die BASF und die Stickstoffpolitik im ersten Weltkrieg. *Ber. Wissen.* 19:105–117.

Tabor, S. W., and Tabor, H. (1984). Methionine adenosyltransferase (S-adenosylmethionine synthetase) and S-adenosylmethionine decarboxylase. *Adv. Enzymol.* 56:251–282.

Taborsky, G. (1974). Phosphoproteins. *Adv. Protein Chem.* 28:1–210.

Tait, J. F., and Tait, S. A. S. (1988). A decade (or more) of electrocortin (aldosterone). *Steroids* 51:213–250.

Takahashi, K. (ed.) (1995). *Aspartic Proteinases.* New York: Plenum.

Takayama, K. M., and Inouye, M. (1990). Antisense RNA. *Crit. Rev. Biochem.* 25:155–184.

Talmage, D. W. (1957). Allergy and immunology. *Ann. Rev. Med.* 8:239–256.

——— (1959). Immunological specificity. *Science* 129:1643–1648.

Tamiya, H. (1973). Early days (1883–1930) of the study of oxidoreductive enzymes in Japan. In *Molecular Mechanisms of Enzyme Action* (Y. Ogura et al., eds.), pp. 1–13. Baltimore: University Park.

Tammann, G. (1895). Zur Wirkung ungeformten Fermente. *Z. Physik. Chem.* 18:426–442.

Tammann, G. A. (1976). Gustav Heinrich Johann Apollon Tammann. *DSB* 13:242–248.

Tanford, C. (1980). *The Hydrophobic Effect.* 2d ed. New York: Wiley.

——— (1989). *Ben Franklin Stilled the Waves.* Durham, N.C.: Duke University Press.

Tang, H., Sun, X., Reinberg, D., and Enright, R. H. (1996). Protein-protein interactions in eukaryotic transcription initiation: Structure of the pre-initiation complex. *PNAS* 93:1119–1124.

Tang, J. (ed.) (1977). *Acid Proteases.* New York: Plenum.

Tang, J., and Lee, Y. C. (1980). Biochemistry in the People's Republic of China. *TIBS* 5:vi–xi.

Tansey, E. M. (1989). The Wellcome Physiological Research Laboratories, 1894–1904. *Med. Hist.* 33:1–41.

Tansey, T. (1995). Sir Henry Dale and autopharmacology: The role of acetylcholine as a neurotransmitter. In *Essays in the History of the Physiological Sciences* (C. Debru, ed.), pp. 179–193. Amsterdam: Rodopi.

Tarbell, D. S., and Tarbell, A. T. (1981). *Roger Adams, Scientist and Statesman.* Washington, D.C.: American Chemical Society.

Tata, J. R. (1990). Rosalind Pitt-Rivers and the discovery of T_3. *TIBS* 15:282–284.

——— (1994). Rosalind Venetia Pitt-Rivers. *Biog. Mem. FRS* 39:326–348.

Tate, P. (1965). David Keilin. *Parasitology* 55:1–28.

Tattersall, R. (1995). Pancreatic organotherapy for diabetes, 1889–1921. *Med. Hist.* 39:288–316.

——— (1996). Frederick Pavy (1829–1911): The last of the physician chemists. *J. Roy. Coll. Phys. London* 30:238–245.

Tatum, E. L. (1945). X-ray induced mutant strains of *Escherichia coli. PNAS* 31:215–219.

Tatum, E. L., Bonner, D., and Beadle, G. W. (1944). Anthranilic acid and the biosynthesis of indole and tryptophan in Neurspora. *Arch. Biochem.* 3:477–478.

Tatum, E. L., and Lederberg, J. (1947). Gene recombination in the bacterium *Escherichia coli. J. Bact.* 53:673–684.

Taube, H. (1965). *Oxygen: Chemistry, Structure, and Excited States.* Boston: Little, Brown.

Tauber, A. I., and Chernyak, L. (1991). *Metchnikoff and the Origins of Immunology.* Oxford: Oxford University Press.

Tauber, A. I., and Podolsky, S. H. (1993). Frank Macfarlane Burnet and the immune self. *J. Hist. Biol.* 27:531–573.

Taussig, R., and Gilman, A. G. (1995). Mammalian membrane-bound adenylyl cyclases. *J. Biol. Chem.* 270:1–4.

Taylor, A. L., and Thoman, M. S. (1964). The genetic map of *Escherichia coli* K-12. *Genetics* 50:659–677.

Taylor, C. W., and Barron, F. (1963). *Scientific Creativity: Its Recognition and Development.* New York: Wiley.

Taylor, D. W. (1971). The life and teaching of William Sharpey. *Med. Hist.* 15:126–153, 241–259.

Taylor, F. S. (1949). *The Alchemists.* New York: Schuman.

——— (1953). The idea of the quintessence. In *Science, Medicine, and History* (E. A. Underwood, ed.), 1:247–265. Oxford: Oxford University Press.

Taylor, K. L. (1978). Gabriel Delafosse. *DSB* 15:114–115.

Taylor, S. S. (1989). cAMP-dependent protein kinase: Model for an enzyme family. *J. Biol. Chem.* 264:8443–8446.

Taylor, W. P., and Widlanski, T. S. (1995). Phosphoprotein phosphatases. *Chem. Biol.* 2:713–718.

Teich, M. (1965). On the historical foundations of modern biochemistry. *Clio Medica* 1:41–57.

——— (1973). From "enchyme" to "cytoskeleton." In *Changing Perspectives in the History of Science* (M. Teich and R. Young, eds.), pp. 439–471. London: Heinemann.

——— (1975). A single path to the double helix? *Hist. Sci.* 13:264–283.

——— (1980). A history of biochemistry. *Hist. Sci.* 18:46–67.

——— (1981). Ferment or enzyme: What's in a name? *Hist. Phil. Life Sci.* 3:193–215.

——— (1983). Fermentation theory and practice: The beginnings of pure yeast cultivation and English brewing, 1883–1913. In *History of Technology* (N. Smith, ed.), 8:117–133. London: Mansell.

——— (1984). Review of R. E. Kohler, *From Medical Chemistry to Biochemistry. Brit. J. Hist. Sci.* 17:239–241.

Teich, M., and Needham, D. M. (1992). *A Documentary History of Biochemistry 1770–1940*. Leicester: Leicester University Press.

Teichman, S. L., and Aldea, P. A. (1985). Nicolas Constantin Paulescu (1869–1931) discovered insulin before Banting and Best and inspired Harvey Cushing's work on the pituitary gland. *Physiologie* 22:121–134.

Teilhard de Chardin, P. (1955). *Le Phénomène humain.* Paris: Seuil.

Telkes, E. (1990). Présentation de la faculté des sciences et de son personnel à Paris (1901–1939). *Rev. Hist. Sci.* 43:452–476.

——— (1993). *Maurice Caullery, 1868–1958, un biologiste au quotidien.* Lyon: Presses Universitaires.

Temin, H. M. (1976). The DNA provirus hypothesis. *Science* 192:1075–1080.

Temin, H. M., and Baltimore, D. (1972). RNA directed DNA synthesis and RNA tumor viruses. *Adv. Virus Res.* 17:129–186.

Temin, H. M., and Mizutani, S. (1970). RNA-dependent DNA polymerase in virions of Rous sarcoma virus. *Nature* 226:1211–1213.

Temkin, O. (1946). Materialism in French and German physiology in the early nineteenth century. *Bull. Hist. Med.* 20:322–327.

Templeton, W. (1982). Two centuries of chemistry at St. Bartholomew's. *Chem. Brit.* 18:263–267.

Templeton, W. S. (1992). The polymer chain reaction. *Diagnostic Molecular Pathology* 1:58–72.

Ten Hoor, M. J. (1996). The formation of urea: Controversies and confusion. *J. Chem. Ed.* 73:42–45.

Tepperman, J. (1988). A view of the history of biology from an islet of Langerhans. In *Endocrinology: People and Ideas* (S. M. McCann, ed.), pp. 285–333. Bethesda, Md.: American Physiological Society.

Teske, A. (1969). Einstein and Smoluchowski. *Sudhoffs Arch.* 53:292–305.

Thach, R. E., et al. (1969). Purification and properties of initiation factors F_1 and F_2. *Cold Spring Harbor Symp.* 34:277–284.

Thackray, A. W. (1966). The chemistry of history. *Hist. Sci.* 5:124–134.

——— (1970). *Atoms and Powers.* Cambridge: Harvard University Press.

——— (1980). The pre-history of an academic discipline: The study of the history of science in the United States, 1891–1941. *Minerva* 18:448–473.

Thackray, A. W., and Merton, R. K. (1972). On discipline building: The paradoxes of George Sarton. *Isis* 63:473–495.

——— (1975). George Alfred Leon Sarton. *DSB* 12:107–114.

Thackray, A. W., Sturchio, J. L., Carroll, P. T., and Bud, R. (1985). *Chemistry in America, 1876–1976.* Dordrecht: Reidel.

Thannhauser, S. J., and Dorfmüller, G. (1917). Experimentelle Studien über den Nukleinstoffwechsel, 4: Iber den Aufbau der Hefenukleinsäure . . . *Z. physiol. Chem.* 100:121–132.

Thannhauser, S. J., and Ottenstein, B. (1921). Experimentelle Studien über den Nukleinstoffwechsel, 12: Die Hydrolyse der Thymusnukleinsäure mit Pikrinsäure. Ueber die Zusammensetzung der Thymusnukleinsäure. *Z. physiol. Chem.* 114:39–50.

Thelander, L., and Reichard, P. (1979). *Ann. Rev. Biochem.* 48:133–158.

Thenard, L. J. (1803). Mémoire sur la fermentation vineuse. *Ann. Chim.* 46:294–320.

Thenard, P. (1950). *Un grand Français: Le Chimiste Thenard.* Dijon: Jobard.

Theobald, D. W. (1976). Some considerations on the philosophy of chemistry. *Chem. Soc. Revs.* 5:203–213.

Théodoridès, J., et al. (1983). François Magendie. *Hist. Sci. Med.* 17:321–380.

Theorell, H. (1935). Das gelbe Oxydationsferment. *Biochem. Z.* 278:263–290.

——— (1947). Heme-linked groups and mode of action of some hemoproteins. *Adv. Enzymol.* 7:265–303.

Theunissen, B. (1996). The beginnings of the "Delft tradition" revisited: Martinus W. Beijerinck and the genetics of microorganisms. *J. Hist. Biol.* 29:197–228.

Thieffry, D., and Burian, R. M. (1996). Jean Brachet's alternative scheme for protein synthesis. *TIBS* 21:114–117.

Thomas, J. M. (1981). *Michael Faraday and the Royal Institution.* Letchworth: Hilger.

——— (1994). Turning points in catalysis. *Ang. Chem. (Int. Ed.)* 33:913–937.

Thomas, K. (1948). Franz Knoop zum Gedächtnis. *Z. physiol. Chem.* 283:1–8.

——— (1954). Fifty years of biochemistry in Germany. *Ann. Rev. Biochem.* 23:1–16.

Thompson, E. P. (1968). *The Making of the English Working Class.* London: Penguin.

——— (1991). *Customs in Common.* London: Merlin.

Thompson, H. (1973). Cyril Norman Hinshelwood. *Biog. Mem. FRS* 19:563–582.

Thompson, R. H. S., and Ogston, A. G. (1983). Rudolph Albert Peters. *Biog. Mem. FRS* 29:495–523.

Thomson, E. H. (1975). Benjamin Silliman, Jr. *DSB* 12:434–437.

Thomson, T. (1843). *Chemistry of Animal Bodies.* Edinburgh: Black.

Thomson, W. (1852). On a universal tendency in Nature to the dissipation of mechanical energy. *Proc. Roy. Soc. Edin.* 3:139–142.

Thorn, G. W. (1968). The adrenal cortex, 1: Historical aspects. *Johns Hopkins Med. J.* 123:49–77.

Thornton, H. G. (1953). Sergei Nikolaevich Winogradsky. *Obit. Not. FRS* 8:635–644.

Thornton, J. L., and Wiles, A. (1956). William Odling. *Ann. Sci.* 12:288–295.

Thorpe, T. E. (1900). The Victor Meyer memorial lecture. *J. Chem. Soc.* 77:169–206.

——— (1916). *The Right Honourable Sir Henry Enfield Roscoe.* London: Longmans, Green.

Thudichum, J. L. W. (1872). *A Manual of Chemical Physiology.* London: Longmans, Green, Reader, and Dyer.

——— (1881a). On the colouring matters of the rods and the cells of the choroid coat of the retina. *Annals of Chemical Medicine* 2:64–76.

——— (1881b). On modern text-books as impediments to the progress of animal chemistry. *Annals of Chemical Medicine* 2:183–189.

Thunberg, T. (1911). Untersuchungen über autoxydable Substanzen und autoxydable Systeme von physiologischem Interesse. *Skand. Arch. Physiol.* 24:90–96.

——— (1920). Zur Kenntnis des intermediären Stoffwechsel und dabei wirksame Enzyme. *Skand. Arch. Physiol.* 40:1–91.

——— (1926). Das Reduktions-Oxydationspotential eines Gemisch von Succinat-Fumarat. *Skand. Arch. Physiol.* 46:339–340.

——— (1933). Olof Hammarsten. *Erg. Physiol.* 35:13–31.

——— (1937). Biologische Aktivierung, Ubertragung und endgültige Oxydation des Wasserstoffs. *Erg. Physiol.* 39:76–116.

Tiedemann, F., and Gmelin, L. (1826–1827). *Die Verdauung nach Versuchen.* Heidelberg: Groos.

Tiemann, F. (1896). Eugen Baumann. *Ber. chem. Ges.* 29:2575–2580.

Tiffeneau, M. (1916). Conférence sur l'oeuvre de Charles Gerhardt. *Bull. Soc. Chim.* [4]19 Suppl.:13–103.

——— (1918). *Correspondance de Charles Gerhardt.* Paris: Masson.

Tigerstedt, R. (1904). Magnus Blix. *Skand. Arch. Physiol.* 16:334–347.

——— (1911). Christian Bohr. *Skand. Arch. Physiol.* 25:v–xvii.

Tilden, W. A. (1884). On the decomposition of terpenes by heat. *J. Chem. Soc.* 45:410–415.

——— (1912). Cannizzaro memorial lecture. *J. Chem. Soc.* 101:1677–1693.

Tiles, M. (1984). *Science and Objectivity.* Cambridge: Cambridge University Press.

Tilley, N. (1981). The logic of laboratory life. *Sociology* 15:117–126.

Timoféeff-Ressovsky, N. W., Zimmer, K. G., and Delbrück, M. (1935). Ueber die Natur der Genmutation und der Genstruktur. *Nachrichten von der Gesellschaft für Wissenschaften zu Göttingen. Math.-Phys. Klasse, Fachgruppe VI* NF1:189–245.

Tipson, R. S. (1945). The chemistry of the nucleic acids. *Adv. Carb. Chem.* 1:193–245.

——— (1957). Phoebus Aaron Theodore Levene. *Adv. Carb. Chem.* 12:1–12.

Tischler, G. (1913). Eduard Strasburger. *Archiv für Zellforschung* 9:1–40.

Tiselius, A. (1937). A new apparatus for electrophoretic analysis of colloidal mixtures. *Trans. Faraday Soc.* 33:524–531.

Tiselius, A., and Kabat, E. A. (1939). An electrophoretic study of immune sera and purified antibodies preparations. *J. Exp. Med.* 69:119–131.

Tissières, A., Schlesinger, D., and Gros, F. (1960). Amino acid incorporation into proteins by *Escherichia coli* ribosomes. *PNAS* 46:1450–1463.

Tissières, A., Watson, J. D., Schlesinger, D., and Hollingworth, B. R. (1959). Ribonucleoprotein particles from *Escherichia coli. J. Mol. Biol.* 1:221–233.

Todd, A. R. (1983). *A Time to Remember.* Cambridge: Cambridge University Press.

Todd, A. R., and Cornforth, J. W. (1976). Robert Robinson. *Biog. Mem. FRS* 22:415–527.

——— (1981). Robert Burns Woodward. *Biog. Mem. FRS* 27:629–695.

Todd, A. R., et al. (1980). *The Social Responsibility of Scientists.* London: Royal Society.

Toennies, G. (1937). The sulfur containing amino acid methionine. *Growth* 1:337–370.

Toenniessen, E., and Brinkmann, E. (1930). Ueber den oxydativen Abbau der Kohlenhydrate im Säugetiermuskel, insbesondere über die Bildung von Bernsteinsäure und Brenztraubensäure. *Z. physiol. Chem.* 187:137–159.

Tolman, R. C. (1938). *Principles of Statistical Mechanics.* Oxford: Oxford University Press.

Tondl, L. (1979). Science as a vocation. In *Perspectives in Metascience* (J. Bärmark, ed.), pp. 173–184. Göteborg: Kungl. Vetenskapsoch Vitterhets Samhället.

Tooze, J. (1978). Bacterial transformation. *TIBS* 3:261–262.

Torchinsky, Y. I. (1987). Enzymic transamination (50 years after its discovery): History and milestones in the study of its mechanism and biological role. *Molekularnaya Biologiya* 21:581–587.

Torlais, J. (1954). *L'Abbé Nollet, physicien au siècle des lumières.* Paris: Sipuco.

Tosh, J. (1984). *The Pursuit of History.* London: Longmans.

Toulmin, S. (1972). *Human Understanding.* Princeton: Princeton University Press.

Tower, D. B. (1994). *Brain Chemistry and the French Connection, 1791–1841.* New York: Raven.

Tramontano, A., Janda, K. D., and Lerner, R. A. (1986). Catalytic antibodies. *Science* 234:1566–1570.

Traube, M. (1858a). Zur Theorie der Gährungen und Verwesungserscheinungen, wie Fermentwirkungen überhaupt. *Ann. Phys.* 103:331–344.

——— (1858b). *Theorie der Fermentwirkungen.* Berlin: Dümmler.

——— (1861). Ueber die Beziehung der Respiration zur Muskelthätigkeit und die Bedeutung der Respiration überhaupt. *Virchowa Arch.* 21:386–414.

——— (1867). Experimente zur Theorie der Zellenbildung und Endosmose. *Arch. Anat. Physiol.*, pp. 87–165.

——— (1874). Ueber das Verhalten der Alkoholhefe in sauerstoffgasfreien Medien. *Ber. chem. Ges.* 7:872–887.

——— (1878). Die chemische Theorie der Fermentwirkungen und der Chemismus der Respiration. *Ber. chem. Ges.* 11:1984–1992.

——— (1899). *Gesammelte Abhandlungen.* Berlin: Mayer and Müller.

Traut, R. R., et al. (1969). Ribosomal proteins of *E. coli:* Stoichiometry and implications for ribosome structure. *Cold Spring Harbor Symp.* 34:25–38.

Trautz, M. (1930). August Horstmann. *Ber. chem. Ges.* 63A:61–86.

Travis, A. S. (1989). Science as receptor of technology: Paul Ehrlich and the synthetic dyestuffs industry. *Science in Context* 3:383–408.

——— (1993). *The Rainbow Makers: The Origins of the Synthetic Dyestuffs Industry in Western Europe.* Bethlehem: Lehigh University Press.

Traweck, S. (1988). *Beamtimes and Lifetimes.* Cambridge: Harvard University Press.

Treibs, A. (1971). *Das Leben und Wirken von Hans Fischer.* Munich: Technische Universität.

Treneer, A. (1963). *The Mercurial Chemist: A Life of Sir Humphry Davy.* London: Methuen.

Trigg, R. (1978). The sociology of knowledge. *Phil. Soc. Sci.* 8:289–298.

——— (1989). *Reality at Risk: A Defense of Realism in Philosophy and the Sciences.* 2d ed. London: Harvester.

——— (1993). *Rationality and Science.* Oxford: Blackwell.

Trikojus, V. M. (1978). Biochemistry comes to Australia. *TIBS* 3:N174–N178.

Tristram, G. R. (1949). Amino acid composition of purified proteins. *Adv. Protein Chem.* 5:83–153.

Troland, L. T. (1914). The chemical origin and regulation of life. *Monist* 24:92–113.

——— (1917). Biological enigmas and the theory of enzyme action. *Am. Nat.* 51:321–350.

Trommsdorff, J. B. (1800–1807). *Systematisches Handbuch der gesamten Chemie.* Erfurt: Hemming.

Trommsdorff, R. (1909). Zur Frage der reduzierende Eigenschaften der Milch und die Schardinger Reaktion. *Z. Bakt.* 49:291–301.

Trott zu Sulz, A. (1912). Zur Errichtung biologischer Forschungsinstitute durch die Kaiser-Wilhelm-Gesellschaft zur Förderung der Wissenschaften. Berlin: Stenographic report of meeting on 3 January 1912.

Trout, J. D. (1994). A realist looks backward. *Stud. Hist. Phil. Sci.* 25:37–64.

Trumpower, B. I. (1990). The protonmotive Q cycle. *J. Biol Chem.* 265:11409–11412.

Tschermak, E. (1900). Ueber die künstliche Kreuzung bei Pisum sativum. *Ber. bot. Ges.* 18:232–239.

Tsou, C. L. (1990). "The Highest Grade of this Clarifying Activity has no Limit"—Confucius. In *Selected Topics in the History of Biochemistry.* Vol. 3. (G. Semenza and R. Jaenicke, eds.), pp. 349–386. Amsterdam: Elsevier.

——— (1995). Chemical synthesis of crystalline insulin: A reminiscence. *TIBS* 20:289–292.

Tsuchiya, T., et al. (1990). Hamao Umezawa. *Adv. Carb. Chem.* 48:1–20.

Tsuzuki, Y., and Yamashita. A. (1968). History of the chemistry of taste in Japan. *Jap. Stud. Hist. Sci.* 7:1–26.

Tswett, M. (1906). Physikalisch-chemische Studien über Chlorophyll. *Ber. bot. Ges.* 24:316–323.

Tuchman, A. M. (1993). *Science, Medicine, and the State in Germany: The Case of Baden, 1815–1871.* Oxford: Oxford University Press.

Turner, G. L. (1980). *Essays on the History of the Microscope.* Oxford: Senecio.

Turner, R. S. (1982). Liebig versus Prussian chemistry: Reflections on early institute-building in Germany. *Hist. Stud. Phys. Sci.* 13:129–162.

Turner, S., Kerwin, E., and Woolwine, D. (1984). Careers and creativity in nineteenth-century physiology: Zloczower redux. *Isis* 75:523–529.

Turner, S. P. (1981). Interpretative charity, Durkheim, and the "strong programme" in the sociology of science. *Phil. Soc. Sci.* 11:231–243.

Turner, W. B. (1971). *Fungal Metabolites.* London: Academic.

Turpin, P. J. F. (1838). Mémoire sur la cause et les effets de la fermentation alcoolique et aceteuse. *Compt. Rend.* 7:369–402.

Tutton, A. E. H. (1924). *The Natural History of Crystals.* London: Kegan Paul, Trench, Trübner.

Tyrrell, H. J. V. (1964). The origin and present status of Fick's diffusion law. *J. Chem. Ed.* 41:397–400.

Ullmann, A., and Monod, J. (1963). Cyclic AMP as an antagonist of catabolite repression in *Escherichia coli. FEBS Lett.* 2:57.

Umbarger, H. E. (1956). Evidence for a negative-feedback mechanism in the biosynthesis of isoleucine. *Science* 123:848.

——— (1961). Feedback control by endproduct inhibition. *Cold Spring Harbor Symp.* 26:301–312.

——— (1978). Amino acid biosynthesis and its regulation. *Ann. Rev. Biochem.* 47:533–606.

——— (1992). The origin of a useful concept: Feedback inhibition. *Protein Science* 1:1392–1395.

Umezawa, H. (1972). *Enzyme Inhibitors of Microbial Origin.* Tokyo: University of Tokyo Press.

Underwood, E. A. (1977). *Boerhaave's Men at Leyden and After.* Edinburgh: Edinburgh University Press.

Unna, P. G. (1911). Die Reduktionsorte und Sauerstofforte des tierischen Gewebes. *Arch. mikr. Anat.* 78:1–21.

Urey, H. C. (1934). Deuterium and its compounds in relation to biology. *Cold Spring Harbor Symp.* 2:47–56.

Urey, H. C., Brickwedde, F. C., and Murphy, G. M. (1932). A hydrogen isotope of mass 2. *Physical Review* 39:164–165.

Uschmann, G. (1959). *Geschichte der Zoologie und der zoologischen Anstalten in Jena, 1779–1919.* Jena: Fischer.

——— (1961). *Ernst Haeckel.* 3d ed. Leipzig: Urania.

——— (1979). Haeckel's biological materialism. *Hist. Phil. Life Sci.* 1:101–118.

Utter, M. F., Barden, R. E., and Taylor, B. L. (1975). Pyruvate carboxylase: An evaluation of the relationship between structure and mechanism and between structure and catalytic activity. *Adv. Enzymol.* 42:1–72.

Utter, M. F., and Keech, D. B. (1960). Formation of oxaloacetate from pyruvate and CO_2. *J. Biol. Chem.* 235:PC17–PC18.

Utter, M. F., and Wood, H. G. (1945). Fixation of carbon dioxide in oxaloacetate by pigeon liver. *J. Biol. Chem.* 160:375–376.

——— (1951). Mechanisms of fixation of carbon dioxide by heterotrophs and auxotrophs. *Adv. Enzymol.* 12:41–151.

Vagelos, P. R. (1964). Lipid metabolism. *Ann. Rev. Biochem.* 33:139–172.

Valentin, G. G. (1855). Ueber die Wechselwirkung der Muskeln und die sie umgebenden Atmosphäre. *Archiv für physiologische Heilkunde* 14:431–478.

Valentin, J. (1949). *Friedrich Wöhler.* Stuttgart: Wissenschaftliche Verlagsgesellschaft.

Vallee, B. L., and Geddes, A. (1984). The metallochemistry of zinc enzymes. *Adv. Enzymol.* 56:284–430.

Van der Pas, P. W. (1976). Hugo de Vries. *DSB* 14:95–105.

Van Dyke, G. V. (1993). *John Lawes of Rothamsted.* Harpenden: Hoos.

Vane, J., and Cuatrecasas, P. (1984). Genetic engineering and pharmaceuticals. *Nature* 312:303–305.

Vane, J. R. (1990). History of aspirin and its mechanism of action. *Stroke* 21:IV, 12–23.

Van Heyningen, W. E. (1939). Stable isotopes as indicators in biology. *Biol. Revs.* 14:420–450.

Van Holde, K. E. (1988). *Chromatin.* New York: Springer.

Van Holde, K. E., and Miller, K. I. (1995). Hemocyanins. *Adv. Protein Chem.* 47:1–81.

Van Holde, K. E., and Zlatanova, J. (1995). Chromatin higher structure: Chasing a mirage? *J. Biol. Chem.* 270:8373–8376.

Van Klooster, H. S. (1944). Friedrich Wöhler and his American students. *J. Chem. Ed.* 21:158–170.

Van Name, R. G. (1931). Frank Austin Gooch. *Biog. Mem. NAS* 25:105–135.

Van Niel, C. B. (1931). On the morphology and physiology of the purple and green sulphur bacteria. *Archiv für Mikrobiologie* 3:1–112.

——— (1935). Photosynthesis of bacteria. *Cold Spring Harbor Symp.* 3:138–150.

——— (1941). The bacterial photosyntheses and their importance for the general problem of photosynthesis. *Adv. Enzymol.* 1:263–328.

——— (1949). The "Delft school" and the rise of general microbiology. *Bact. Revs.* 13:161–174.

——— (1957). Albert Jan Kluyver. *J. Gen. Microbiol.* 16:499–521.

Van Slyke, D. D. (1929). Determination of urea by gasometric measurement of the carbon dioxide formed by the action of urease. *J. Biol. Chem.* 73:695–723.

Van Slyke, D. D., and Jacobs, W. A. (1944). Phoebus Aaron Theodore Levene. *Biog. Mem. NAS* 23:75–126.

Van Slyke, D. D., and Meyer, G. M. (1913). The fate of protein digestion products in the body, 3: The absorption of amino acids from the blood by the tissues. *J. Biol. Chem.* 16:197–212.

Vaughan, M., and Steinberg, D. (1959). The specificity of protein biosynthesis. *Adv. Protein Chem.* 14:115–173.

Vauquelin, N. L. (1792). Observations chimiques et physiologiques sur la respiration des insectes et des vers. *Ann. Chim.* 12:273–291.

Vaux, D. L., and Strasser, A. (1996). The molecular biology of apoptosis. *PNAS* 93:2239–2244.

Veibel, S. (1939, 1943, 1968). *Kemien i Danmark.* Copenhagen: NYT Nordisk Forlog.

Venanzi, L. M. (1994). Coordination chemistry in Europe since Alfred Werner. *Chimia* 48:16–22.

Vennesland, B. (1981). Recollections and small confessions. *Ann. Rev. Plant Physiol.* 32:1–20.

——— (1991). Isotope time: 50 years ago. *FASEB J.* 5:2868–2869.

Vennesland, B., and Westheimer, F. H. (1954). Hydrogen transport and steric specificity in reactions catalyzed by pyridine nucleotide dehydrogenases. In *Mechanism of Enzyme Action* (W. D. McElroy and B. Glass, eds.), pp. 357–379. Baltimore: Johns Hopkins University Press.

Verkade, P. E. (1985). *A History of the Nomenclature of Organic Chemistry.* Dordrecht: Reidel.

Vermenouze, P. (1992). Emile Duclaux. *Revue de la Haute Auvergne.* 54:111–128.

Verney, E. B. (1956). Some aspects of the work of Ernest Henry Starling. *Ann. Sci.* 12:30–47.

Vernon, H. M. (1911). The indophenol oxidase of mammalian and avian tissues. *J. Physiol.* 43:96–108.

——— (1912). Die Abhängigkeit der Oxydasewirkung von Lipoiden. *Biochem. Z.* 47:374–395.

Vernon, K. (1990). Pus, sewage, beer, and milk: Microbiology in Britain 1870–1940. *Hist. Sci.* 28:289–325.

——— (1994). Microbes at work: Micro-organisms, the D.S.I.R., and industry in Britain, 1900–1936. *Ann. Sci.* 51:593–613.

Vernon, K. D. C. (1957). The Royal Institution of Great Britain. *J. Chem. Ed.* 34:607–610.

Verworn, M. (1897). *Allgemeine Physiologie.* 2d ed. Jena: Fischer.

——— (1901). *Die Aufgaben des physiologischen Unterrichts.* Jena: Fischer.

——— (1903). *Die Biogenhypothese.* Jena: Fischer.

Veselkin, N. P. (1950). *Viktor Vasilievich Pashutin.* Moscow.

Vesterberg, O. (1989). History of electrophoretic methods. *J. Chromat.* 480:3–19.

Vicedo, M. (1991). Realism and simplicity in the Castle-East debate on the stability of the hereditary units. *Stud. Hist. Phil. Sci.* 22:201–221.

——— (1995). Scientific styles: Toward some common ground in the history, philosophy, and sociology of science. *Persp. Sci.* 3:231–254.

Vickery, H. B. (1931). Thomas Burr Osborne. *Biog. Mem. NAS* 14:261–304.

——— (1945). Russell Henry Chittenden. *Biog. Mem. NAS* 24:59–104.

——— (1946). The early years of the Kjeldahl method to determine nitrogen. *Yale J. Biol. Med.* 18:473–516.

——— (1952). Treat Baldwin Johnson. *Biog. Mem. NAS* 23:83–119.

——— (1950). The origin of the word protein. *Yale J. Biol. Med.* 22:387–393.

——— (1962). Rudolph John Anderson. *Biog. Mem. NAS* 36:19–50.

——— (1967). William Mansfield Clark. *Biog. Mem. NAS* 39:1–26.

——— (1972). The history of the discovery of the amino acids, 2: A review of amino acids described since 1931 as components of native proteins. *Adv. Protein Chem.* 26:81–171.

——— (1975). Hans Thacher Clarke. *Biog. Mem. NAS* 46:3–20.

Vickery, H. B., and Osborne, T. B. (1928). A review of hypotheses of the structure of proteins. *Physiol. Revs.* 8:393–446.

Vickery, H. B., and Schmidt, C. L. A. (1931). The history of the discovery of the amino acids. *Chem. Revs.* 9:169–318.

Viel, C. (1996). Le salon et le laboratoire de Lavoisier à l'Arsenal . . . *Hist. Sci. Med.* 30:32–34.

Vierhaus, R., and vom Brocke, B. (1990). *Forschunng im Spannungsfeld von Politik und Gesellschaft: Geschichte und Struktur der Kaiser-Wilhelm/ Max-Planck Gesellschaft.* Stuttgart: Deutsche Verlags-Anstalt.

Villar-Palasi, C., and Larner, J. (1970). Glycogen metabolism and glycolytic enzymes. *Ann. Rev. Biochem.* 39:639–672.

Ville, G. (1854). Absorption de l'azote de l'air par les plantes. *Compt. Rend.* 38:705–709, 723–727.

Vinogradov, S. N. (1965). Chemistry at Kazan University in the nineteenth century: A case history of intellectual lineage. *Isis* 56:168–173.

Virchow, R. (1855). Die Cellularpathologie. *Virchows Arch.* 8:1–39.

Virtanen, A. I. (1938). *Cattle Fodder and Human Nutrition with Special Reference to Biological Nitrogen Fixation.* Cambridge: Cambridge University Press.

Voegtlin, C. (1939). John Jacob Abel. *J. Pharm. Exp. Ther.* 67:373–406.

Vogel, H. (1957). Repression and induction as control mechanisms of enzyme biogenesis: The "adaptive" formation of acetylornithinase. In *Chemical Basis of Heredity* (W. D. McElroy and B. Glass, eds.), pp. 276–289. Baltimore: Johns Hopkins University Press.

Voit, C. (1870). Ueber die Entwicklung der Lehre von der Quelle der Muskelkraft und einiger Theile der Ernährung seit 25 Jahren. *Z. Biol.* 6:303–401.

——— (1891). Ueber die Glykogenbildung nach Aufname verschiedener Zuckerarten *Z. Biol.* 28:245–292.

Volhard, J. (1909). *Justus von Liebig.* Leipzig: Barth.

Volkin, E., and Astrachan, L. (1957). RNA metabolism in T2-infected *Escherichia coli.* In *The Chemical Basis of Heredity* (W. D. McElroy and B. Glass, eds,), pp. 686–694. Baltimore: Johns Hopkins University Press.

Von Hippel, P. H., Bear, D. G., Morgan, W. D., and McSwiggen, J. A. (1984). Protein-nucleic acid interactions in transcription: A molecular analysis. *Ann. Rev. Biochem.* 53:389–446.

Voswinckel, P. (1993). *Der schwarze Urin.* Berlin: Blackwell.

Vries, H. de (1889). *Intracelluläre Pangenesis.* Jena: Fischer. [English trans., C. S. Gager (1910). Chicago: Open Court.]

——— (1900). Das Spaltungsgesetz der Bastarde. Vorläufige Mittheilung. *Ber. bot. Ges.* 18:83–90.

——— (1901–1902). *Die Mutationstheorie.* Leipzig: Veit.

Wada, M. (1930). Ueber Citrullin, eine neue Aminosäure in Preásaft der Wassermelone, Citrullus vulgaris. *Biochem. Z.* 224:420–429.

Waddington, C. H. (1939). The physicochemical nature of the gene. *Am. Nat.* 73:300–314.

——— (1969). Some European contributions to the prehistory of molecular biology. *Nature* 221:318–321.

Wade, N. (1981). *The Nobel Duel.* New York: Doubleday.

——— (1994). The erosion of the academic ethos: The case of biology. In *University-Business Partnerships* (N. E. Bowie, ed.), pp. 143–158.

Wagner, A. F., and Folkers, K. (1961). Discovery and chemistry of mevalonic acid. *Adv. Enzymol.* 23:471–483.

Wagner, I., and Musso, H. (1983). New naturally occurring amino acids. *Ang. Chem. (Int. Ed.)* 22:816–828.

Wagner, R. P., and Mitchell, H. K. (1964). *Genetics and Metabolism.* New York: Wiley.

Wagner-Jauregg, T., and Rauen, H. (1935). Die Dehydrierung der Citronensäure und der Iso-citronensäure durch Gürkensamen Dehydrase. *Z. physiol. Chem.* 237:227–232.

Waite, A. E. (1888). *Lives of Alchemystical Philosophers.* London: Redway.

Wakil, S. J. (1962). Lipid metabolism. *Ann. Rev. Biochem.* 31:369–406.

Wakil, S. J., Stoops, J. K., and Joshi, V. C. (1983). Fatty acid synthesis and its regulation. *Ann. Rev. Biochem.* 52:537–579.

Waksman, S. A. (1953). *Sergei N. Winogradsky, His Life and Work.* New Brunswick, N.J.: Rutgers University Press.

Wald, G. (1936). Carotenoids and the visual cycle. *J. Gen. Physiol.* 19:351–371.

——— (1948). Selig Hecht. *J. Gen. Physiol.* 32:1–16.

——— (1964). The receptors of human color vision. *Science* 145:1007–1016.

——— (1968). Molecular basis of visual excitation. *Science* 162:230–239.

Walden, P. (1928). Die Bedeutung der Wöhlerschen Harnstoffsynthese. *Naturwiss.* 16:835–849.

——— (1941). *Geschichte der organischen Chemie seit 1880.* Berlin: Springer.

Waldenström, J. (1937). Studien über Porphyrie. *Acta med. Scand.* Suppl. 82.

Waldeyer, W. (1888). Ueber Karyokinese und ihre Beziehungen zur den Befruchtungsvorgängen. *Arch. mikr. Anat.* 32:1–122.

Waldrop, M. M. (1992). *Complexity: Emerging Science at the Edge of Order and Chaos.* New York: Simon and Schuster.

Walker, J. B. (1979). Creatine: Biosynthesis, regulation, and function. *Adv. Enzymol.* 50:177–242.

Walker, J. E., Fearnley, I. M., Lutter, R., Todd, R. J., and Runswick, M. J. (1990). Structural aspects of proton-pumping ATPases. *Phil Trans.* B326:467–478.

Walker, R. A., and Sheetz, M. P. (1993). Cytoplasmic microtubule-associated motors. *Ann. Rev. Biochem.* 62:429–451.

Wallace, D. B., and Gruber, H. E. (1989). *Creative People at Work.* Oxford: Oxford University Press.

Wallach, O. (ed.) (1901). *Briefwechsel zwischen J. Berzelius und F. Wöhler.* Leipzig: Engelmann.

Wallach, O., and P. Jacobson (1918). Carl Liebermann. *Ber. chem. Ges.* 51:1135–1204.

Walling, C. (1977). Moses Gomberg. *Proc. Welch Conf.* 20:72–84.

Walsh, C. (1979). *Enzymatic Reaction Mechanisms.* San Francisco: Freeman.

——— (1980). Flavin coenzymes: At the crossroads of biological redox chemistry. *Acc. Chem. Res.* 13:148–155.

Walsh, D. A., Perkins, J. P., and Krebs, E. G. (1968). An adenosine 3′,5′-monophosphate-dependent protein kinase from rabbit skeletal muscle. *J. Biol. Chem.* 243:3763–3765.

Walter, P., Gilmore, R., and Blobel, G. (1984). Protein translocation across the endoplasmic reticulum. *Cell* 38:5–8.

Wang, J. C. (1991). DNA topoisomerases: Why so many? *J. Biol. Chem.* 266:6659–6662.

——— (1996). DNA topoisomerases. *Ann. Rev. Biochem.* 65:635–692.

Wankmüller, A. (1980). Professoren und Dozenten der physiologischen Chemie in Tübingen. In *Physik, Physiologische Chemie und Pharmazie an der Universität Tübingen* (A. Hermann and A. Wankmüller, eds.), pp. 41–77. Tübingen: Mohr.

——— (1985). German-American links in 19th-century pharmacy. *Pharm. Hist.* 27:162–165.

Warburg, O. (1908). Beobachtungen über die Oxydationsprozesse im Seeigelei. *Z. physiol. Chem.* 57:1–16.

——— (1911). Ueber die Beeinflussung der Sauerstoffatmung. *Z. physiol. Chem.* 70:413–432.

——— (1913). Ueber Sauerstoffatmende Körnchen aus Leberzellen und über Sauerstoffatmung in Berkfeld-Filtraten wässeriger Leberextrakte. *Pflügers Arch.* 154:599–617.

——— (1914a). Beiträge zur Physiologie der Zelle, insbesondere über die Oxydationsgeschwingigkeit in Zellen. *Erg. Physiol.* 14:253–337.

——— (1914b). Ueber die Rolle des Eisens in der Atmung des Seeigeleies nebst Bemerkungen über durch Eisen geschleunigte Oxydation. *Z. physiol. Chem.* 92:231–256.

——— (1919, 1920). Ueber die Geschwindigkeit der photochemischen Kohlensäure-zerstezung in lebenden Zellen. *Biochem. Z.* 100:230–270; 103:188–217.

——— (1923). Ueber die Grundlagen der Wielandschen Atmungstheorie. *Biochem. Z.* 142:518–523.

——— (1924). Ueber Eisen, den sauerstoffübertragenden Bestandteil des Atmungsferment. *Biochem. Z.* 152:479–494.

——— (1925a). Ueber Eisen, den sauerstoffübertragenden Bestandteil des Atmungsferment. *Ber. chem. Ges.* 58:1001–1011.

——— (1925b). Bemerkung zu einer Arbeit von M. Dixon und S. Thurlow sowie einer Arbeit von G. Ahlgren. *Biochem. Z.* 163:252.

——— (1926). *Ueber den Stoffwechsel der Tumoren.* Berlin: Springer. [English trans. (1930), F. Dickens. London: Constable.]

——— (1928). *Ueber die katalytische Wirkung der lebendigen Substanz.* Berlin: Springer.

——— (1929). Atmungsferment und Oxydasen. *Biochem. Z.* 214:1–3.

——— (1932). Das sauerstoffübertragende Ferment der Atmung. *Ang. Chem.* 45:1–6.

——— (1938). Chemische Konstitution von Fermenten. *Erg. Enzymforsch.* 7:210–245.

——— (1946). *Schwermetalle als Wirkungsgruppen von Fermenten.* Berlin: Saenger. [English trans. (1949), A. Lawson. Oxford: Clarendon.]

——— (1949). *Wasserstoffübertragende Fermente.* Freiburg: Cantor.

——— (1956). On the origin of cancer cells. *Science* 123:309–314; 124:269–270.

——— (1958). Photosynthesis. *Science* 128:68–73.

——— (1962). *New Methods of Cell Physiology, Applied to Cancer, Photosynthesis, and Mechanism of X-Ray Action.* New York: Interscience.

Warburg, O., and Christian, W. (1931). Ueber Aktivierung der Robisonschen Hexomonophosphorsäure in roten Blutzellen und die Gewinnung aktivierender Fermentlösungen. *Biochem. Z.* 242:206–227.

——— (1932). Ueber ein neues Oxydationsferment und sein Absorptionsspektrum. *Biochem. Z.* 254:438–450.

——— (1933). Ueber des gelbe Ferment und seine Wirkungen. *Biochem. Z.* 266:377–411.

——— (1935). Co-Fermentproblem. *Biochem. Z.* 287:464.

——— (1936). Pyridin, der wasserstoffübertragende Bestandteil der Gärungsfermenten (Pyridin-Nucleotide). *Biochem. Z.* 287:291–328.

——— (1938). Bemerkung über gelbe Fermente. *Bichem. Z.* 298:368–377.

——— (1939). Isolierung und Kristallisation des Proteins des oxydierenden Gärungsferments. *Biochem. Z.* 303:49–68.

——— (1942). Isolierung and Kristallisation des Gärungsferment Enolase. *Biochem. Z.* 310:384–421.

Warburg, O., Christian, W., and Griese, A. (1935). Wasserstoffübertragendes Co-Ferment, seine Zusammensetzung und Wirkungsweise. *Biochem. Z.* 282:157–205.

Warburg, O., Klotzsch, H., and Gawehn, K. (1954). Ueber die Oxydationsreaktion der Gärung. *Z. Naturforsch.* 9b:391–393.

Warburg, O., Kubowitz, F., and Christian, W. (1930a). Kohlenhydratverbrennung durch Methämoglobin. *Biochem. Z.* 221:494–497.

——— (1930b). Ueber die katalytische Wirkung von Methylenblau in lebenden Zellen. *Biochem. Z.* 227:245–271.

Warburg, O., and Negelein, E. (1922). Ueber den Energieumstaz bei der Kohlensäureassimilation. *Z. physik. Chem.* 102:235–266.

——— (1923). Ueber den Einfluss der Wellenlänge auf den Energieumsatz bei der Kohlensäureasimilation. *Z. physik. Chem.* 106:191–218.

——— (1929). Ueber das Absorptionsspektrum des Atmungsferments. *Biochem. Z.* 214:64–100.

Warnecke, J. M., Fürste, J. P., Hardt, W. D., and Erdmann, V. A. (1996). Ribonuclease P (RNase P) is converted to a Cd^{2+}-ribozyme by a single Rp-phosphothioate modification in the precursor tRNA cleavage site. *PNAS* 93:8924–8928.

Warren, M. J., Jay, M., Hunt, D. M., Elder, G. H., and Röhl, J. C. G. (1996). The maddening business of King George III and porphyria. *TIBS* 21:229–234.

Waser, P. G. (1986). The cholinergic receptor. In *Discoveries in Pharmacology* (M. J. Parnham and J. Bruinvels, eds.), 3:157–202. Amsterdam: Elsevier.

Wassermann, A. (1910). Ueber den Einfluss des Spezificitätsbegriff auf die moderne Medizin. *Deutsche med. Wchschr.* 36:1860–1863.

Wasteneys, H., and Borsook, H. (1930). The enzymatic synthesis of protein. *Physiol. Revs.* 10:110–145.

Watermann, R. (1960). *Theodor Schwann: Leben und Werk.* Düsseldorf: Schwann.

Waterson, A. P., and Wilkinson, L. (1978). *An Introduction to the History of Virology.* Cambridge: Cambridge University Press.

Watson, C. J. (1965). Reminiscences of Hans Fischer and his laboratory. *Persp. Biol. Med.* 8:419–435.

Watson, J. D. (1968). *The Double Helix.* New York: Atheneum.

——— (1970). *Molecular Biology of the Gene.* 2d ed. New York: Benjamin.

——— (1980). *The Double Helix* (G. S. Stent, ed.). New York: Norton.

Watson, J. D., and Crick, F. H. C. (1953a). Molecular structure of nucleic acids: A structure for deoxyribonucleic acid. *Nature* 171:737–738.

——— (1953b). Genetical implications of the structure of deoxyribonucleic acid. *Nature* 171:964–967.

Waxman, D. J., and Strominger, J. L. (1983). Penicillin-binding proteins and the mechanism of the action of β-lactam antibiotics. *Ann. Rev. Biochem.* 52:825–869.

Way, J. T. (1856). On the quantity of nitric acid and ammonia in rain water. *J. Royal Agr. Soc.* 17:618–621.

Weatherall, M., and Kamminga, H. (1992). *Dynamic Science: Biochemistry in Cambridge, 1898–1949.* Cambridge: Wellcome Unit for the History of Medicine.

——— (1996). The making of a biochemist, 2: The construction of Frederick Hopkins' reputation. *Med. Hist.* 40:415–436.

Weaver, W. (1930a). Molecular biology: Origin of the term. *Science* 152:581–582.

——— (1970b). *Scene of Change.* New York: Scribner's.

Weber, B. H. (1991). Glynn and the conceptual development of the chemiosmotic theory. *Bioscience Reports* 11:577–617.

Weber, G. (1975). Energetics of ligand binding to proteins. *Adv. Protein Chem.* 29:1–83.

——— (1990). Whither biophysics? *Ann. Rev. Biophys.* 19:1–6.

Weber, H. H. (1955). Adenosine triphosphate and motility of living systems. *Harvey Lectures* 49:37–56.

Weber, H. H., and Portzehl, J. (1952). Muscle contraction and fibrous muscle proteins. *Adv. Protein Chem.* 7:161–252.

Weber, M. (1973). The power of the state and the dignity of the academic calling in Imperial Germany. *Minerva* 11:571–632.

Weber, P. C. (1991). Physical principles of protein crystallization. *Adv. Protein Chem.* 41:1–36.

Wedegaertner, P. B., Wilson, P. T., and Bourne, H. R. (1995). Lipid modifications of the trimeric G proteins. *J. Biol. Chem.* 270:503–506.

Weindling, P. J. (1991). *Darwinism and Social Darwinism in Imperial Germany: The Contributions of the Cell Biologist Oscar Hertwig.* Stuttgart: Fischer.

——— (1992). Scientific elites and laboratory organisation in fin de siècle Paris and Berlin. In *The Laboratory Revolution in Medicine* (A. Cunningham and P. Williams, eds.), pp. 170–188. Cambridge: Cambridge University Press.

——— (1996). The impact of German medical scientists on British medicine: A case history of Oxford, 1933–1945. In *Forced Migration and Scientific Change* (M. G. Ash and A. Söllner, eds.), pp. 86–114. Cambridge: Cambridge University Press.

Weiner, C. (1988). Oral history of science: A mushroom cloud? *J. Amer. Hist.* 75:548–559.

Weiner, D. B. (1968). *Raspail, Scientist and Reformer.* New York: Columbia University Press.

Weinhouse, S., Burk, D., and Schade, A. L. (1956). On respiratory impairment of cancer cells. *Science* 124:267–268, 270–272.

Weinhouse, S., Medes, G., and Floyd, N. F. (1944). Fatty acid metabolism: The mechanism of ketone body synthesis from fatty acids, with isotopic carbon as a tracer. *J. Biol. Chem.* 155:143–151.

Weininger, S. J. (1984). The molecular structure conundrum: Can classical chemistry be reduced to quantum chemistry? *J. Chem. Ed.* 61:939–944.

Weinstein, A. (1977). How unknown was Mendel's paper? *J. Hist. Biol.* 10:341–364.

——— (1980). Morgan and the theory of natural selection. In *The Evolutionary Synthesis* (E. Mayr and W. B. Provine, eds.), pp. 432–445. Cambridge: Harvard University Press.

Weismann, A. (1883). *Ueber die Vererbung.* Jena: Fischer.

——— (1891). *Essays upon Heredity and Kindred Biological Problems.* 2d ed. (F. R. Poulton et al., trans.). Oxford: Oxford University Press.

——— (1893). *The Germ-Plasm: A Theory of Heredity* (W. N. Parker and H. Rönnfeldt, trans.). New York: Scribner's.

Weiss, J. (1937). Reaction mechanism of some proteolytic enzymes. *Chem. Ind.*, pp. 685–686.

Weiss, P. A. (1947). The problem of specificity in growth and development. *Yale J. Biol. Med.* 19:235–278.

——— (1971). A cell is not an island entire unto itself. *Persp. Biol. Med.* 14:182–205.

Weiss, S. B. (1960). Enzymatic incorporation of ribonucleotide triphosphates into the interpolynucleotide linkages of ribonucleic acid. *PNAS* 46:1020–1030.

Weissbach, A. (1977). Eukaryotic DNA polymerases. *Ann. Rev. Biochem.* 46:25–47.

Weissberg, R. W. (1993). *Creativity: Beyond the Myth of Genius.* New York: Freeman.

Weissenberg, R. (1959). *Oscar Hertwig.* Leipzig: Barth.

Weisser, U. (1984). Das erste Hormon aus der Retorte: Arbeiten am synthetisch-en Adrenalin (Suprarenin) bei Höchst. *Dokumente aus Höchster Archiven* 62:1–147.

Weisz, G. (1983). *The Emergence of Modern Universities in France, 1863–1914.* Princeton: Princeton University Press.

Wek, R. C. (1994). eIF-2 kinases: Regulators of general and gene-specific translation initiation. *TIBS* 19:491–496.

Welch, G. R. (ed.) (1985). *Organized Multienzyme Systems: Catalytic Properties.* New York: Academic.

——— (1995). T. H. Huxley and the "protoplasmic theory of life": 100 years later. *TIBS* 20:481–485.

Wellbourn, R. B. (1990). *The History of Endocrine Surgery.* New York: Praeger.

Wellner, D., and Meister, A. (1981). A survey of inborn errors of amino acid metabolism and transport in man. *Ann. Rev. Biochem.* 50:911–968.

Wells, H. G. (1928). Immunity: The chemical warfare of existence. In *Chemistry in Medicine* (J. Stieglitz, ed.), pp. 559–577. New York: Chemical Foundation.

Wells, H. G., and Osborne, T. B. (1911). The biological reactions of the vegetable proteins. *Journal of Infectious Diseases* 8:66–124.

Wells, R. D. (1988). Unusual DNA structures. *J. Biol. Chem.* 263:1095–1098.

Welsch, M. (1967). André Gratia. In *L'Université de Liège, 1936–1966* (R. Dumoulin, ed.), pp. 613–671. Liège: University of Liège.

Wendel, G. (1975). *Die Kaiser-Wilhelm Gesellschaft, 1911–1914*. Berlin: Akademie Verlag.

Wengenmeyer, F. (1983). Synthesis of peptide hormones using recombinant DNA techniques. *Ang. Chem. (Int. Ed.)* 22:842–858.

Wenzl, A. (ed.) (1951). *Hans Driesch.* Basle: Reinhardt.

Werkman, C. H., and Wood, H. G. (1942). Heterotrophic assimilation of carbon dioxide. *Adv. Enzymol.* 2:139–182.

Werner, E. A. (1923). *The Chemistry of Urea.* London: Longmans, Green.

Werner, P., and Renneberg, R. (eds.) (1991). *Ein Genie ist seltener . . . Otto Heinrich Warburg: Ein Lebensbild in Dokumenten.* Berlin: Akademie Verlag.

Werskey, G. (1978). *The Visible College: The Collective Biography of British Scientific Socialists.* New York: Holt, Rinehart and Winston.

Wertheimer, E., and Lepage, J. L. (1901). Sur les réflexes des ganglions abdominaux du sympathetique dans l'innervation du pancreas. *J. Physiol. Path. Gen.* 3:335–348, 363–374.

West, J. B. (ed.) (1996). *Respiratory Physiology: People and Ideas.* New York: Oxford University Press.

Westheimer, F. H. (1985). The discovery of the mechanisms of enzyme action, 1947–1963. *Advances in Physical Organic Chemistry* 21:1–34.

——— (1986). Polynucleotides as enzymes. *Nature* 319:534–536.

——— (1995). Myron Lee Bender. *Biog. Mem. NAS* 66:3–19.

Westphal, O. (1968). Richard Kuhn in Memoriam. *Ang. Chem. (Int. Ed.)* 7:489–506.

Wettstein, F. von (1938). Carl Erich Correns. *Ber. bot. Ges.* 56:140–160.

Wetzel, W. (1991). *Naturwissenschaften und Chemische Industrie in Deutschland.* Stuttgart: Steiner.

Weyer, J. (1971). Neue Konzeptionen der Chemiegeschichtsbeschreibung im 19. Jahrundert: Trommsdorff, Hoefer und Kopp. *Rete* 1:17–50.

——— (1973). Chemiegeschichtsbeschreibung im 19. und 20. Jahrhundert. *Sudhoffs Arch.* 57:171–194.

——— (1974a). *Chemiegeschichtsbeschreibung von Wiegleb (1790) bis Partington (1970).* Hildesheim: Gerstenberg.

——— (1974b). A hundred years of stereochemistry: The principal development phases in retrospect. *Ang. Chem. (Int. Ed.)* 13:591–698.

——— (1977). Die Aufname der van't Hoffschen Hypothese vom asymmetrischen Kohlenstoffatom (1874) in Deutschland. In *Medizin, Naturwissenschaft, Technik und das Zwite Kaiserreich* (G. Mann and R. Winau, eds.), pp. 311–320. Göttingen: Vanderhoeck and Rupprecht.

Weyl, T. (1877). Beiträge zur Kenntnis thierischer und pflanzlicher Eiweisskörper, *Z. physiol. Chem.* 1:72–100.

Wheatley, D. N., and Agutter, P. S. (1996). Historical aspects of diffusion theory in 19th-century mechanistic materialism. *Persp. Biol. Med.* 40:139–156.

Wheeler, H. L., and Johnson, T. B. (1903). Synthesis of aminooxypyrimidines having the composition of cytosine. *Am. Chem. J.* 29:492–504.

Wheeler, L. P. (1952). *Josiah Willard Gibbs.* Rev. ed. New Haven: Yale University Press.

Wheldale, M. (1911). On the direct guaiacum reaction given by plant extracts. *Proc. Roy. Soc.* B84:121–124.

Whitaker, M. A. (1984). Science, scientists, and history of science. *Hist. Sci.* 22:421–424.

White, J. H. (1932). *The History of the Phlogiston Theory.* London: Arnold.

White, M. F., and Kahn, C. R. (1994). The insulin signaling system. *J. Biol. Chem.* 269:1–4.

White, R. E., and Coon, M. J. (1980). Oxygen activation by cytochrome P-450. *Ann. Rev. Biochem.* 49:315–356.

Whitehead, A. N. (1925). *Science and the Modern World.* New York: Macmillan.

Whitesides, G. M., and Wong, C. H. (1985). Enzymes as catalysts in organic chemistry. *Ang. Chem. (Int. Ed.)* 24:617–638.

Whitt, L. A. (1990). Atoms or affinities? The ambivalent reception of the Daltonian theory. *Stud. Hist. Phil. Sci.* 21:47–89.

Wibaut, J. P. (1955). Arnold Frederik Holleman. *Rec. Trav. Chim. Pays-Bas* 74:1271–1375.

Wichmann, A. (1899). Ueber die Kristallform der Albumine. *Z. physiol. Chem.* 27:575–593.

Wickner, W., and Kornberg, A. (1974). A holoenzyme form of deoxyribonucleic acid polymerase III. *J. Biol. Chem.* 249:6244–6249.

Wickner, W., and Leonard, M. R. (1996). *Escherichia coli* preprotein translocase. *J. Biol. Chem.* 271:29514–29516.

Wickner, W., and Lodish, H. F. (1985). Multiple mechanisms of protein insertion into and across membranes. *Science* 230:400–405.

Widmann, H. (1973). *Exil und Bildungs Hilfe: Die deutschsprachige akademische Emigration in die Turkei nach 1933.* Berne: Lang.

Wieland, H. (1912a). Uber Hydrierung und Dehydrierung. *Ber. chem. Ges.* 45:484–499.

——— (1912b). Studien über den Mechanismus der Oxydationsvorgänge. *Ber. chem. Ges.* 45:2606–2615.

——— (1913). Ueber den Mechanismus der Oxydationsvorgänge. *Ber. chem. Ges.* 46:3327–3342.

——— (1922a). Ueber den Verlauf der Oxydationsvorgänge. *Ber. chem. Ges.* 55:3639–3648.

——— (1922b). Ueber den Mechanismus der Oxydationsvorgänge. *Erg. Physiol.* 20:477–578.

Wieland, H., and Boersch, E. (1919). Untersuchungen über Gallensäuren, 5: Die Reduktion der Dehydrocholsäure und der Dehydrodeoxycholsäure. *Z. physiol. Chem.* 196:190–200.

Wieland, H., Dane, E., and Scholz, E. (1932). Untersuchungen über die Konstitution der Gallensäuren. *Z. physiol. Chem.* 210:261–274.

Wieland, T. (1995a). The history of peptide chemistry. In *Peptides* (B. Gutte, ed.), pp. 1–28. San Diego: Academic.

——— (1995b). Memories of Heidelberg—and other places. In *Selected Topics in the History of Biochemistry.* Vol. 4 (E. C. Slater et al., eds.), pp. 21–108. Amsterdam: Elsevier.

Wieland, T., and Bodanszky, M. (1991). *The World of Peptides.* Berlin: Springer.

Wieland, T., Determann, H., and Albrecht, E. (1960). Untersuchungen über die Plastein-Reaktion. *Ann.* 633:186–197.

Wiener, N. (1948). *Cybernetics, or the Control and Communication in the Animal and the Machine.* New York: Wiley.

——— (1953). *Ex-Prodigy: My Childhood and Youth.* New York: Simon and Schuster.

——— (1955). *I Am a Mathematician: The Later Life of an Ex-Prodigy.* Garden City, N.Y.: Doubleday.

Wikström, M., Krab, K., and Saraste, M. (1981). *Cytochrome Oxidase: A Synthesis.* New York: Academic.

Wilhelmi, A. E. (1985). The Endocrine Society: Origin, organization, and institutions. *Endocrinology* 123: 41–43.

Wilkie, J. S. (1960, 1961). Nägeli's work on the fine structure of living matter. *Ann. Sci.* 16:11–42, 171–207, 209–230; 17:37–62.

Wilkins, M. H. F., Stokes, A. R., and Wilson, H. R. (1953). Molecular structure of deoxypentose nucleic acids. *Nature* 171:738–740.

Wilkins, M. H. F. (1987). John Turton Randall. *Biog. Mem. FRS* 33:493–535.

Wilkinson, L. (1976). The development of the virus concept as reflected in corpora of studies on individual pathogens, 3: Lessons of the plant viruses—tobacco mosaic virus. *Med. Hist.* 20:111–134.

Willcock, E. G., and Hopkins, F. G. (1907). The importance of individual amino acids in metabolism: Observations on the effect of adding tryptophane to a dietary in which zein is the sole nitrogenous constituent. *J. Physiol.* 38:88–102.

Williams, G. (1980). *Western Reserve's Experiment in Medical Education and Its Outcome.* Oxford: Oxford University Press.

Williams, J. W. (1975). Peter Joseph William Debye. *Biog. Mem. NAS* 46:23–68.

——— (1979). The development of the ultracentrifuge and its contributions. *Ann. N.Y. Acad. Sci.* 325:77–91.

Williams, L. P. (1965). *Michael Faraday: A Biography.* London: Chapman and Hall.

——— (1966). The historiography of Victorian science. *Victorian Studies* 9:197–204.

——— (1973). Kant, *Naturphilosophie,* and scientific method. In *Foundations of Scientific Method* (R. N. Giere and R. S. Westfall, eds.), pp. 3–22. Bloomington: Indiana University Press.

——— (1991). The life of science and scientific lives. *Physis* 28:199–213.

Williams, R. (1985). *Keywords.* 2d ed. New York: Oxford University Press.

Williams, R. J. (1943). The chemistry and biochemistry of pantothenic acid. *Adv. Enzymol.* 3:253–287.

Williams, R. J. P. (1961, 1962). Possible functions of chains of catalysts. *J. Theor. Biol.* 1:1–17; 3:209–229.

——— (1979). Some unrealistic assumptions in the theory of chemi-osmosis and their consequences. *FEBS Lett.* 102:126–132.

——— (1987). The functions of structure and dynamics in proteins, peptides, and metal ion complexes and their relationship to biological recognition and the handling of information. *Carlsberg Lab. Comm.* 52:1–30.

Williams, R. R., and Spies, T. D. (1938). *Vitamin B_1 (Thiamin) and Its Use in Medicine.* New York: Macmillan.

Williams, R. T. (ed.) (1948). *The Biochemical Reactions of Chemical Warfare Agents.* Cambridge: Cambridge University Press.

——— (1950). *Biochemical Aspects of Genetics.* Cambridge: Cambridge University Press.

Williams, T. I. (1990). *Robert Robinson, Chemist Extraordinary.* Oxford: Clarendon.

Williams, T. I., and Weil, H. (1953). The phases of chromatography. *Arkiv för Kemi* 5:283–299.

Williamson, A. (1885). Charles Adolphe Wurtz. *Proc. Chem. Soc.* 38:xxiii–xxxiv.

Willink, B. (1991). Origins of the second Golden age of Dutch science after 1860: Intended and unintended consequences of educational reform. *Soc. Stud. Sci.* 21:503–526.

Willis, T. (1684). *Dr. Willis' Practice of Physik, . . .* (translated by S. Pordage). London: Dring, Harper, and Leigh.

Willstätter, R. (1922). Ueber Isolierung von Enzymen. *Ber. chem. Ges.* 55:3601–3623.

——— (1926). Ueber Sauerstoff-übertragung in der lebenden Zelle. *Ber. chem. Ges.* 59:1871–1876.

——— (1927). *Problems and Methods in Enzyme Research.* Ithaca, N.Y.: Cornell University Press.

——— (1949). *Aus Meinem Leben* (A. Stoll, ed.). Weinheim: Verlag Chemie.

Willstätter, R., and Rohdewald. M. (1940). Die enzymatische Systeme der Zuckerumwandlung im Muskel. *Enzymologia* 6:1–63.

Wilson, C. (1995). *The Invisible World: Early Modern Philosophy and the Invention of the Microscope.* Princeton: Princeton University Press.

Wilson, D. B., and Dintzis, H. M. (1970). Protein chain initiation in rabbit reticulocytes. *PNAS* 66:1282–1289.

Wilson, E. B. (1895). *An Atlas of the Fertilization and Karyokinesis of the Ovum.* New York: Macmillan.

——— (1896). *The Cell in Development and Inheritance.* New York: Macmillan.

——— (1905). Studies on chromosomes, 2. *J. Exp. Zool.* 2:507–545.

——— (1923). The physical basis of life. *Science* 57:277–286.

——— (1925). *The Cell in Development and Heredity.* 3d ed. New York: Macmillan.

Wilson, E. B., and Morgan, T. H. (1920). Chiasmatype and crossing over. *Am. Nat.* 54:193–219.

Wilson, E. B., and Ross, J. (1973). Physical chemistry in Cambridge, Massachusetts. *Ann. Rev. Phys. Chem.* 24:1–27.

Wilson, H. R. (1988). The double helix and all that. *TIBS* 13:275–278.

Wilson, J. W. (1944). Cellular tissue and the dawn of the cell theory. *Isis* 35:168–173.

——— (1947). Dutrochet and the cell theory. *Isis* 37:14–21.

Wilson, L. G. (1968). Starling's discovery of osmotic equilibrium in the capillaries. *Episteme* 2:3–25.

——— (1980). Medical history without medicine. *J. Hist. Med.* 35:5–7.

——— (1983). Review of R. E. Kohler, *From Medical Chemistry to Biochemistry. J. Hist. Med.* 38:462–464.

——— (1984). Internal secretions in disease: The historical relations of clinical medicine and scientific physiology. *J. Hist. Med.* 39:263–302.

Wilson, P. W. (1957). On the sources of nitrogen of vegetation. *Bact. Revs.* 21:215–226.

Wilson, P. W., and Burris, R. H. (1947). The mechanism of biological nitrogen fixation. *Bact. Revs.* 11:41–73.

Wimmer, M. J., and Rose, A. I. (1978). Mechanisms of enzyme-catalyzed group transfer reactions. *Ann. Rev. Biochem.* 47:1031–1078.

Windaus, A., and Neukirchen, K. (1919). Umwandlung des Cholesterins in Cholansäure. *Ber. chem. Ges.* 52:19–24.

Winge, Ø. (1935). On haplophase and diplophase in some saccharomycetes. *Compt. Rend. Carlsberg* 21:71–112.

Winiwarter, H. de (1933). Albert Brachet. *Ann. Acad. Roy. Belg.* 99:143–192.

Winogradsky, S. Ueber Schwefelbacterien. *Bot. Z.* 45:606–610.

Winsor, M. P. (1979). Louis Agassiz and the species question. *Stud. Hist. Biol.* 3:89–117.

Winter, G., and Milstein, C. (1991). Man-made antibodies. *Nature* 349:293–299.

Winterstein, E. (1914). Ernst Schulze. *Ber. chem. Ges.* 47:429–449.

Wishart, G. M. (1954). Edward Provan Cathcart. *Obit. Not. FRS* 9:35–53.

Witebsky, E. (1954). Ehrlich's side-chain theory in the light of present immunology. *Ann. N.Y. Acad. Sci.* 59:168–181.

Witkop. B. (1981). Paul Ehrlichs Leitgedanken und lebendiges Werk. *Naturwiss. Rund.* 34:361–379.

——— (1992). Remembering Heinrich Wieland (1877–1957): Portrait of an organic chemist and founder of modern biochemistry. *Medicinal Research Reviews* 12:195–274.

Witkowski, J. A. (1979). Alexis Carrel and the mysticism of tissue culture. *Med. Hist.* 23:276–296.

——— (1980a). W. T. Astbury and Ross G. Harrison: The search for the molecular determination of form in the developing embryo. *Notes Roy. Soc.* 35:195–219.

——— (1980b). Dr. Carrel's immortal cells. *Med. Hist.* 24:129–142.

——— (1985). The magic of numbers. *TIBS* 10:141.

——— (1987). Optimistic analysis: Chemical embryology in Cambridge, 1920–1942. *Med. Hist.* 31:247–268.

——— (1990). Carrel's cultures. *Science* 247:1385–1386.

Witt, T. K. (1980). A short history of porphyrins and the porphyrias. *Int. J. Biochem.* 11:189–200.

Wittgenstein, L. (1993). A lecture on ethics (Heretics' Society, Cambridge, 17 November 1929). In *Philosophical Occasions* (J. C. Klagge and A. Nordmann, eds.), pp. 37–44. Indianapolis: Hackett.

Wittmann, H. G. (1982). Structure and evolution of ribosomes. *Proc. Roy. Soc.* B216:117–135.

Wittmann, H. G., and Yonath, A. (1988). Architecture of ribosomal particles as investigated by image reconstruction and X-ray crystallographic studies. In *The Roots of Modern Biochemistry* (H. Kleinkauf, H. von Döhren, and L. Jaenicke, eds.), pp. 481–492. Berlin: de Gruyter.

Woese, C. R. (1967). *The Genetic Code.* New York: Harper and Row.

Wohl, A. (1907). Die neuere Ansichten über den chemischen Verlauf der Gärung. *Biochem. Z.* 5:45–64.

Wöhler, F. (1842). Ueber die im lebenden Organismus von sich gehende Umwandlung der Benzoesäure in Hippursäure. *Ann. Phys.* 56:638–641.

Wöhler, F., and Frerichs, F. T. (1848). Ueber die Veränderungen, welche namentlich organische Stoffe bei ihrem Uebergang in den Harn erlangen. *Ann.* 65:335–349.

Wojtkowiak, A. (1989). *Paul Sabatier, un Chimiste Indépendant.* Argueil: Jonas.

Wold, F. (1971). Enolase. In *The Enzymes* 3d ed. (P. D. Boyer, ed.), 5:499–538. New York: Academic.

——— (1981). In vivo chemical modification of proteins (post-translational modification). *Ann. Rev. Biochem.* 50:783–814.

Wolf, G. (1996). Emil Abderhalden: His contribution to the nutritional biochemistry of proteins. *J. Nutrition* 126:794–799.

Wolfenden, R. (1972). Analog approaches to the structure of the transition state in enzyme reactions. *Acc. Chem. Res.* 5:10–18.

——— (1976). Transition-state analog inhibitors and enzyme catalysis. *Ann. Rev. Biophys.* 5:271–306.

Wolff, E., et al. (1967). *Philosophie et méthodologie scientifiques de Claude Bernard.* Paris: Masson.

Wolff, S. L. (1995). Clausius' Weg zur kinetischen Gastheorie. *Sudhoffs Arch.* 79:54–72.

Wolffe, A. (1995). *Chromatin: Structure and Function.* London: Academic.

Wolfrom, M. (1960). John Ulric Nef. *Biog. Mem. NAS* 34:204–227.

Wolkow, M., and Baumann, E. (1891). Ueber das Wesen der Alkaptonurie. *Z. physiol. Chem.* 15:228–286.

Wollman, E., and Wollman, E. (1936, 1938). Recherches sur le phénomène Twort-d'Hérelle (bactériophagie ou autolyse hérédito-contagieuse). *Ann. Inst. Pasteur* 56:137–170; 60:13–57.

Wollman, E. L., Jacob, F., and Hayes, W. (1962). Conjugation and genetic recombination in *Escherichia coli* K-12. *Cold Spring Harbor Symp.* 21:141–162.

Wolstenholme, G. E. W., and O'Connor, C. M. (eds.) (1957). *The Chemistry and Biology of the Purines.* London: Churchill.

——— (1959). *Regulation of Cell Metabolism.* London: Churchill.

Womack, M., and Rose, W. C. (1935). Feeding experiments with mixtures of highly purified amino acids, 7: The dual nature of the "unknown growth essential." *J. Biol. Chem.* 112:275–282.

Wood, H. G. (1972). My life and carbon dioxide fixation. In *The Molecular Basis of Biological Transport* (J. F. Woessner and F. Huijing, eds.), pp. 1–53. New York: Academic.

——— (1979). Feodor (Fitzi) Lynen. *TIBS* 4:N300-N302.

——— (1982). The discovery of the fixation of CO_2 by heterotrophic organisms and metabolism of the propionic acid bacteria. In *Of Oxygen, Fuels, and Living Matter.* Part 2 (G. Semenza, ed.), pp. 173–250. Chichester: Wiley.

——— (1985). Then and now. *Ann. Rev. Biochem.* 54:1–41.

Wood, H. G., and Barden, R. E. (1977). Biotin enzymes. *Ann. Rev. Biochem.* 46:385–413.

Wood, H. G., and Werkman, C. H. (1936). The utilization of CO_2 in the dissimilation of glycerol by the propionic acid bacteria. *Biochem. J.* 30:48–53.

Woods, D. D. (1940). The relation of *p*-aminobenzoic acid to the mechanism of the action of sulfanilamide. *Brit. J. Exp. Path.* 21:74–89.

——— (1953). The integration of research on the nutrition and metabolism of microorganisms. *J. Gen. Microbiol.* 9:151–173.

——— (1957). Albert Jan Kluyver. *Biog. Mem. FRS* 3:109–128.

Woodward, C. E. (1989). Art and elegance in the synthesis of organic compounds: Robert Burns Woodward. In *Creative People at Work* (D. B. Wallace and H. E. Gruber, eds.), pp. 227–253. New York: Oxford University Press.

Woodward, R. (1948). Biogenesis of the strychnos alkaloids. *Nature* 162:155–156.

——— (1956). Synthesis. In *Perspectives in Organic Chemistry* (A. R. Todd, ed.), pp. 155–184. New York: Interscience.

Woodward, R. B., and Bloch, K. E. (1953). The cyclization of squalene in cholesterol synthesis. *J. Am. Chem. Soc.* 75:2023–2024.

Woolf, P. K. (1975). The second messenger: Informal communication in cyclic AMP research. *Minerva* 3:349–373.

Woolgar, S. (1983). Review of K. D. Knorr-Cetina, *The Manufacture of Knowledge. Canadian Journal of Sociology* 8:466–468.

Woolley, D. W. (1952). *Antimetabolites.* New York: Wiley.

Woolley, R. G. (1978). Must a molecule have a shape? *J. Am. Chem. Soc.* 100:1073–1078.

Worm-Müller, J. (1874). Zur Kentniss der Nucleine. *Pflügers Arch.* 8:224–235.

Wortmann, J. (1880). Ueber die Beziehung der intramolekularen zur normalen Athmung der Pflanzen. *Bot. Z.* 38:25–27.

Wotiz, J. H. (ed.) (1993). *The Kekulé Riddle.* Clearwater, Fla.: Cache River.

Wright, J. K., Seckler, R., and Overath, P. (1986). Molecular aspects of sugar:ion cotransport. *Ann. Rev. Biochem.* 55:225–248.

Wright, R. D. (1978). The origin of the word "hormone." *TIBS* 3:275.

Wright, S. (1917). Color inheritance in animals. *J. Heredity* 8:224–235.

——— (1941). The physiology of the gene. *Physiol. Revs.* 21:487–527.

——— (1945). Physiological aspects of genetics. *Ann. Rev. Physiol.* 7:75–106.

——— (1953). Genes and organisms. *Am. Nat.* 87:5–18.

Wright, S. (1986). Recombinant DNA technology and its social transformation, 1972–1982. *Osiris* 2:303–360.

——— (1994). *Molecular Politics.* Chicago: University of Chicago Press.

Wrinch, D. (1938). Is there a protein fabric? *Cold Spring Harbor Symp.* 6:122–139.

Wróblewski, A. (1901). Ueber den Buchner'schen Hefepreßsaft. *J. prakt. Chem.* 172:1–70.

Wu, D. (1959). *Hsien Wu, 1893–1959: In Loving Memory.* Boston: private printing.

Wu, H. (1931). Studies on the denaturation of proteins, 13: A theory of denaturation. *Chinese Journal of Physiology* 5:321–344. [Rpt. *Adv. Protein Chem.* 46:6–26 (1996).]

Wu, R. (1993). Development of enzyme-based methods for DNA sequence analysis and their applications in the genome project. *Adv. Enzymol.* 67:431–468.

Wu, R., and Grossman, L. (1987). Site-specific mutagenesis and protein engineering. *Methods in Enzymology* 154:329–533.

Wurmser, R. (1925). Le rendement énérgetique de la photosynthèse chlorophylliènne. *Ann. Physiol. Physicochim. Biol.* 1:47–67.

Wurtz, A. (1857). Sur la formation artificielle de la glycérine. *Ann. Chim.* [3]51:94–101.

——— (1869). *Histoire des doctrines chimiques depuis Lavoisier jusqu'à nos jours.* Paris: Hachette.

——— (1880). *Traité de chimie biologique.* Paris: Masson.

Wüthrich, K. (1990). Protein structure determination in solution by NMR spectroscopy. *J. Biol. Chem.* 265:22059–22062.

Wyatt, G. R. (1950). Occurrence of 5-methylcytosine in nucleic acids. *Nature* 166:237.

Wyatt, G. R., and Cohen, S. S. (1953). The bases of the nucleic acids of some bacterial and animal viruses: The occurrence of 5-hydroxymethylcytosine. *Biochem. J.* 55:774–782.

Wyatt, H. V. (1972). When does information become knowledge? *Nature* 235:86–89.

——— (1974). How history has blended. *Nature* 249:803–805.

——— (1975). Knowledge and prematurity: The journey from transformation to DNA. *Persp. Biol. Med.* 18:149–156.

Wyman, J. (1948). Heme proteins. *Adv. Protein Chem.* 4:407–631.

——— (1949). Some physico-chemical evidence regarding the structure of hemoglobin. In *Haemoglobin* (F. J. W. Roughton and J. C. Kendrew, eds.), pp. 96–106. London: Butterworths.

——— (1963). Allosteric effects in hemoglobin. *Cold Spring Harbor Symp.* 28:483–489.

——— (1964). Linked functions and reciprocal effects in hemoglobin: A second look. *Adv. Protein Chem.* 19:223–286.

——— (1972). On allosteric models. *Curr. Top. Cell. Regul.* 6:209–226.

——— (1979). Recollections of Jacques Monod. In *Origins of Molecular Biology: A Tribute to Jacques Monod* (A. Lwoff and A. Ullmann, eds.), pp. 221–224. New York: Academic.

Wyman, J., and Allen, D. W. (1951). Heme interactions in hemoglobin and the basis of the Bohr effect. *Journal of Polymer Science* 7:499–518.

Wyman, J., and Gill, S. J. (1990). *Binding and Linkage.* Mill Valley, Calif.: University Science.

Yagi, K. (1971). Reaction mechanism of D-amino acid oxidase. *Adv. Enzymol.* 34:41–78.

Yalow, R. S. (1978). Radioimmunoassay: A probe for the fine structure of biological systems. *Science* 200:1236–1245.

Yan, S. C. B., Grinnell, B. W., and Wold, F. (1989). Post-translational modification of proteins: Some problems left to solve. *TIBS* 14:264–268.

Yanofsky, C. (1963). Amino acid replacements associated with mutation and recombination in the A gene and their relationship to in vitro coding data. *Cold Spring Harbor Symp.* 28:581–588.

——— (1987). Tryptophan synthetase: Its charmed history. *BioEssays* 6:134–137.

——— (1989). Tryptophan synthetase of *E. coli:* A multifunctional, multicomponent enzyme. *Biochim. Biophys. Acta* 1000:133–137.

Yanofsky, C., Carlton, B. C., Guest, J. R., Helinski, D. R., and Henning, U. (1964). On the colinearity of the gene structure and protein structure. *PNAS* 51:266–272.

Yanofsky, C., and Crawford, I. P. (1872). Tryptophan synthetase. In *The Enzymes.* 3d ed. (P. D. Boyer, ed.), 7:1–31. New York: Academic.

Yanofsky, C., and Kolter, R. (1982). Attenuation in amino acid biosynthesis operons. *Ann. Rev. Genetics* 16:113–134.

Yarmolinsky, M. B., and de la Haba, G. L. (1959). Inhibition by puromycin of amino acid incorporation into proteins. *PNAS* 45:1721–1729.

Yarus, M. (1993). How many catalytic RNAs? Ions and the Cheshire Cat conjecture. *FEBS J.* 7:31–39.

Yates, F. A. (1964). *Giordano Bruno and the Hermetic Tradition.* Chicago: University of Chicago Press.

Yates, R. A., and Pardee, A. B. (1956). Control of pyrimidine biosynthesis in *Escherichia coli* by a feedback mechanism. *J. Biol. Chem.* 221:757–770.

——— (1957). Control by uracil of formation of enzymes required for orotate synthesis. *J. Biol. Chem.* 227:677–692.

Young, E. G. (1976). *The Development of Biochemistry in Canada.* Toronto: University of Toronto Press.

Young, F. G. (1937). Claude Bernard and the theory of the glycogenic function of the liver. *Ann. Sci.* 2:37–83.

——— (1954). Jack Cecil Drummond. *Obit. Not. FRS* 9:99–129.

Young, F. G., and Foglia, V. G. (1974). Bernardo Alberto Houssay. *Biog. Mem. FRS* 20: 247–270.

Yoxen. E. J. (1979). Where does Schroedinger's "What Is Life?" belong in the history of molecular biology? *Hist. Sci.* 17:17–52.

——— (1982). Giving life a new meaning: The rise of the molecular biological establishment. In *Scientific Establishments and Hierarchies* (N. Elias, H. Martins, and R. Whitley, eds.), pp. 123–143. Dordrecht: Reidel.

Yudkin, J. (1938). Enzyme variation in microorganisms. *Bact. Revs.* 13:93–106.

Zacharias, E. (1881). Ueber die chemische Beschaffenheit des Zellkerns. *Bot. Z.* 39:169–176.

Zachau, H. G., Dütting, D., and Feldmann, H. (1966). The structures of two serine transfer nucleic acids. *Z. physiol. Chem.* 347:212–235.

Zakian, V. P. (1995). Telomeres: Beginning to understand the end. *Science* 270:1601–1607.

Zaleski, S. S. (1894). Carl Schmidt. *Ber. chem. Ges.* 27:963–978.

Zalkin, H. (1993). The amidotransferases. *Adv. Enzymol.* 66:203–309.

Zallen, D. T. (1993a). Redrawing the boundaries of molecular biology: The case of photosynthesis. *J. Hist. Biol.* 26:65–87.

——— (1993b). The "light" organism for the job: Green algae and photosynthesis research. *J. Hist. Biol.* 26:269–279.

Zamecnik, P. C. (1950). The use of labeled amino acids in the study of protein metabolism of normal and malignant tissues. *Cancer Research* 10:659–667.

——— (1979). Historical aspects of protein synthesis. *Ann. N.Y. Acad. Sci.* 325:269–301.

Zamenhof, S. (1957). Properties of the transforming principles. In *Chemical Basis of Heredity* (W. D. McElroy and B. Glass, eds.), pp. 351–377. Baltimore: Johns Hopkins University Press.

Zandvoort, H. (1988). Macromolecules, dogmatism, and scientific change: The prehistory of polymer chemistry as testing ground for philosophy of science. *Stud. Hist. Phil. Sci.* 19:489–515.

Zanobio, B. (1971). Micrographie illusoire et théories sur la structure de la matière vivante. *Clio Medica* 6:25–40.

Zaug, A. J., and Cech, T. R. (1986). The intervening sequence RNA of *Tetrahymena* is an enzyme. *Science* 231:470–475.

Zaug, A. J., Grabowsky, P. J., and Cech, T. R. (1983). Autocatalytic cyclization of an excised intervening sequence RNA is a cleavage-ligation reaction. *Nature* 301:578–583.

Zaunick, R. (1942). Die Entdeckung des Nachweises der Stärke, Eiweiástoffen, und Kork durch Jod im Jahre 1814. *Sudhoffs Arch.* 35:243–254.

Zekert, O. *Carl Wilhelm Scheele.* Stuttgart: Wissenschaftliche Verlagsgesellschaft.

Zelitch, I. (1989). One hundred years of research on photorespiration and biological nitrogen fixation at Wisconsin. In *One Hundred Years of Agricultural Chemistry and Biochemistry at Wisconsin* (D. L. Nelson and B. C. Soltvedt, eds.), pp. 155–176. Madison, Wis.: Science Tech.

Zeller, S., and Bliss, M. (1990). Oskar Minkowski. *DSB* 18:626–633.

Zensen, M., and Restivo, S. (1982). The mysterious morphology of immiscible liquids: A study of scientific practice. *Social Science Information* 21:447–473.

Zerner, B. (1991). Recent advances in the chemistry of an old enzyme, urease. *Bioorganic Chemistry* 19:116–131.

——— (1992). Frank Henry Westheimer: The celebration of a lifetime in chemistry. *Bioorganic Chemistry* 20:269–284.

Zeynek, R. von (1908). Gustav Hüfner. *Z. physiol. Chem.* 58:1–38.

Zhao, Y., Muir, T. W., Kent, S. B. H., Tischer, E., Scardina, J. M., and Chait, B. T. (1996). Mapping protein-protein interactions by affinity-directed mass spectrometry. *PNAS* 93:4020–4024.

Zielinska, Z. (1987). Jakub Karol Parnas. *Acta Physiologica Polonica* 38:91–99.

Zierold, D. K. (1968). *Forschungsförderung in drei Epochen.* Wiesbaden: Steiner.

Zilva, S. S. (1928). The antiscorbutic fraction of lemon juice. *Biochem. J.* 779–785.

Ziman, J. (1978). *Reliable Knowledge.* Cambridge: Cambridge University Press.

——— (1981). What are the options? Social determinants of personal research plans. *Minerva* 19:1–42.

——— (1984). *An Introduction to Science Studies.* Cambridge: Cambridge University Press.

——— (1987). *Knowing Everything About Nothing.* Cambridge: Cambridge University Press.

——— (1994). *Prometheus Bound: Science in a Dynamic Steady State.* Cambridge: Cambridge University Press.

Zimmer, K. G. (1966). The target theory. In *Phage and the Origins of Molecular Biology* (J. Cairns, G. S. Stent, and J. D. Watson, eds.), pp. 33–42. Plainview, N.Y.: Cold Spring Harbor Laboratory Press.

Zinder, N. D., and Lederberg, J. (1952). Genetic exchange in Salmonella. *J. Bact.* 64:679–699.

Zinoffsky, O. (1886). Ueber die Grösse des Hämoglobinmoleküls. *Z. physiol. Chem.* 10:16–34.

Zirnstein, G. (1991). Friedrich Althoffs Wirken für die Biologie in der Zeit des Umbruchs der biologischen Disziplinen in Deutschland, der Erneuerung ihrer Forschung und Lehre an der Universitäten und das Rufes nach ausseruniversitäts Forschungsstätte 1882 bis 1908. In *Wissenschaftsgeschichte und Wissenschaftspolitik im Industriealter* (B. von Brocke, ed.), pp. 355–373. Hildesheim: Lax.

Zlatanova, J., and van Holde, K. (1996). The linker histones and chromatin structure: A new twist. *Prog. Nucleic Acid Res.* 52:217–259.

Zöllner, N. (1969). Siegfried J. Thannhauser. *Internist* (Boston) 10:106–109.

Zsigmondy, R. (1909). *Colloids and the Ultramicroscope.* New York: Wiley.

Zubay, G., and Doty, P. (1959). The isolation and properties of deoxyribonucleoprotein particles containing single nucleic acid molecules. *J. Mol. Biol.* 1:1–20.

Zubay, G., Schwartz, D., and Beckwith, J. (1970). Mechanism of activation of catabolite-sensitive genes: A positive control system. *PNAS* 66:104–110.

Zucker, E. A. (ed.) (1950). *The Forty-Eighters: Political Refugees of the German Revolution of 1848*. New York: Columbia University Press.

Zuckerman, H. (1967). Nobel laureates in science: Patterns of productivity, collaboration, and authorship. *Am. Soc. Rev.* 32:391–403.

——— (1977a). *Scientific Elite.* New York: Free Press.

——— (1977b). Deviant behavior and social control in science. In *Deviance and Social Change* (E. Sagarin, ed.), pp. 87–138. Beverley Hills, Calif.: Sage.

——— (1989). The other Merton thesis. *Science in Context* 1:239–267.

Zuckerman, H., and Lederberg, J. (1986). Postmature scientific discovery? *Nature* 324:629–631.

Zuckerman, H., and Merton, R. K. (1971). Patterns of evaluation in science: Institutionalisation, structure, and functions of the referee system. *Minerva* 9:66–100.

Zumbach, C. (1984). *The Transcendental Science: Kant's conception of Biological Methodoolgy.* The Hague: Nijhoff.

Zupan, P. (1987). *Der Physiologe Carl Ludwig in Zurich, 1849–1855.* Zurich: Juris.

INDEX OF PERSONAL NAMES